DATE DUE

VOLUME FIVE HUNDRED AND FIFTY EIGHT

METHODS IN ENZYMOLOGY

Structures of Large RNA Molecules and Their Complexes

METHODS IN ENZYMOLOGY

Editors-in-Chief

JOHN N. ABELSON and MELVIN I. SIMON
Division of Biology
California Institute of Technology
Pasadena, California

ANNA MARIE PYLE
Departments of Molecular, Cellular and Developmental
Biology and Department of Chemistry Investigator
Howard Hughes Medical Institute
Yale University

DAVID W. CHRISTIANSON
Roy and Diana Vagelos Laboratories
Department of Chemistry
University of Pennsylvania
Philadelphia, PA

Founding Editors

SIDNEY P. COLOWICK and NATHAN O. KAPLAN

VOLUME FIVE HUNDRED AND FIFTY EIGHT

Methods in
ENZYMOLOGY

Structures of Large RNA Molecules and Their Complexes

Edited by

SARAH A. WOODSON
*Department of Biophysics,
Johns Hopkins University,
USA*

FRÉDÉRIC H.T. ALLAIN
*Institute of Molecular Biology and Biophysics,
ETH Zürich, Zürich
Switzerland*

AMSTERDAM • BOSTON • HEIDELBERG • LONDON
NEW YORK • OXFORD • PARIS • SAN DIEGO
SAN FRANCISCO • SINGAPORE • SYDNEY • TOKYO
Academic Press is an imprint of Elsevier

Academic Press is an imprint of Elsevier
225 Wyman Street, Waltham, MA 02451, USA
525 B Street, Suite 1800, San Diego, CA 92101-4495, USA
125 London Wall, London, EC2Y 5AS, UK
The Boulevard, Langford Lane, Kidlington, Oxford OX5 1GB, UK

First edition 2015

Copyright © 2015 Elsevier Inc. All rights reserved.

No part of this publication may be reproduced or transmitted in any form or by any means, electronic or mechanical, including photocopying, recording, or any information storage and retrieval system, without permission in writing from the publisher. Details on how to seek permission, further information about the Publisher's permissions policies and our arrangements with organizations such as the Copyright Clearance Center and the Copyright Licensing Agency, can be found at our website: www.elsevier.com/permissions.

This book and the individual contributions contained in it are protected under copyright by the Publisher (other than as may be noted herein).

Notices

Knowledge and best practice in this field are constantly changing. As new research and experience broaden our understanding, changes in research methods, professional practices, or medical treatment may become necessary.

Practitioners and researchers must always rely on their own experience and knowledge in evaluating and using any information, methods, compounds, or experiments described herein. In using such information or methods they should be mindful of their own safety and the safety of others, including parties for whom they have a professional responsibility.

To the fullest extent of the law, neither the Publisher nor the authors, contributors, or editors, assume any liability for any injury and/or damage to persons or property as a matter of products liability, negligence or otherwise, or from any use or operation of any methods, products, instructions, or ideas contained in the material herein.

ISBN: 978-0-12-801934-4
ISSN: 0076-6879

For information on all Academic Press publications
visit our website at store.elsevier.com

CONTENTS

Contributors xiii
Preface xix

Section I
RNA Structure and Dynamics

1. Native Purification and Analysis of Long RNAs 3
Isabel Chillón, Marco Marcia, Michal Legiewicz, Fei Liu, Srinivas Somarowthu, and Anna Marie Pyle

1. Introduction 4
2. Native Purification of Long Noncoding RNAs 6
3. Study of the RNA Tertiary Folding by Sedimentation Velocity Analytical Ultracentrifugation 13
4. Analysis of the RNA Tertiary Folding by Analytical Size-Exclusion Chromatography 20
5. Determination of the Secondary Structure of LncRNAs by Chemical Probing 22
Acknowledgments 35
References 35

2. Characterizing RNA Excited States Using NMR Relaxation Dispersion 39
Yi Xue, Dawn Kellogg, Isaac J. Kimsey, Bharathwaj Sathyamoorthy, Zachary W. Stein, Mitchell McBrairty, and Hashim M. Al-Hashimi

1. Introduction 40
2. NMR Relaxation Dispersion 43
3. General Protocol for Characterizing RNA ESs Using Low-to-High SL Field ^{13}C and ^{15}N Off-Resonance NMR $R_{1\rho}$ and Uniformly Labeled Nucleic Acid Samples 50
Acknowledgments 68
References 69

3. Quantifying Nucleic Acid Ensembles with X-ray Scattering Interferometry 75
Xuesong Shi, Steve Bonilla, Daniel Herschlag, and Pehr Harbury

1. Ensembles and Energy Surfaces 76

2. The Experimental Challenge	77
3. Conceptual Framework for XSI	80
4. A General Protocol for XSI Measurement of Nucleic Acid Constructs	84
Acknowledgments	96
References	96

4. 2-Aminopurine Fluorescence as a Probe of Local RNA Structure and Dynamics and Global Folding — 99

Michael J. Rau and Kathleen B. Hall

1. Introduction	100
2. Conclusions: GAC Folding—Local and Global Events	121
Acknowledgments	122
References	122

5. Mod-seq: A High-Throughput Method for Probing RNA Secondary Structure — 125

Yizhu Lin, Gemma E. May, and C. Joel McManus

1. Theory	127
2. Equipment	128
3. Materials	129
4. Protocol	133
5. Step 1: RNA Fragmentation	135
6. Step 2: Phosphatase and Kinase the RNA	135
7. Step 3: Ligate the 3' Linker	136
8. Step 4: Ligate the 5' Linker	137
9. Step 5: 5' Linker Selection	138
10. Step 6: Reverse Transcription	140
11. Step 7: PAGE Purification	141
12. Step 8: Ethanol Precipitation	143
13. Step 9: Circularization	143
14. Step 10: Subtractive Hybridization	144
15. Step 11: PCR Amplification	146
16. Step 12: Purification and Analysis	147
17. Step 13: Data Analysis Using Mod-Seeker	148
References	151

6. Reproducible Analysis of Sequencing-Based RNA Structure Probing Data with User-Friendly Tools — 153

Lukasz Jan Kielpinski, Nikolaos Sidiropoulos, and Jeppe Vinther

1. Introduction	154

2. Data Analysis	159
3. Concluding Remarks	177
Acknowledgments	177
References	177

7. Computational Methods for RNA Structure Validation and Improvement   181
Swati Jain, David C. Richardson, and Jane S. Richardson

1. Introduction	182
2. General Validation Criteria	186
3. Criteria Specific to RNA Conformation	189
4. How to Interpret Validation Results	195
5. Correcting the Problems	201
6. What's Coming Next	207
Acknowledgments	208
References	209

8. Structures of Large RNAs and RNA–Protein Complexes: Toward Structure Determination of Riboswitches   213
Jason C. Grigg and Ailong Ke

1. Introduction	214
2. Rational Construct Design for Structure Determination	215
3. Crystallization and Structure Determination	223
4. Structure–Function Validation	225
5. Conclusions	227
Acknowledgments	227
References	227

Section II
RNA-Protein Interactions

9. One, Two, Three, Four! How Multiple RRMs Read the Genome Sequence   235
Tariq Afroz, Zuzana Cienikova, Antoine Cléry, and Frédéric H.T. Allain

1. Introduction	236
2. Structural Investigations of RRMs: How Many?	236
3. The RRM Fold and Its Numerous Variations	237
4. Single RRMs Recognize a Large Repertoire of RNA Sequences: Is There a Recognition Code?	239

5. Increasing Specificity and Affinity by Recognition of Longer RNA
   Sequences   251
6. RNA Recognition by Tandem RRMs   254
7. Assembly of Higher Order Ribonucleoprotein Complexes by
   Multi RRM–RNA and Multi Protein–RNA Interactions   258
8. Structural and Genome-Wide Approaches: How do They Match?   260
9. Conclusions and Perspectives   267
Acknowledgments   268
References   268

10. Combining NMR and EPR to Determine Structures of Large RNAs and Protein–RNA Complexes in Solution   279
Olivier Duss, Maxim Yulikov, Frédéric H.T. Allain, and Gunnar Jeschke

1. Introduction   280
2. Limitations of Current Techniques to Solve Structures of Large
   RNAs and Protein–RNA Complexes in Solution   282
3. EPR Theory   285
4. EPR-Aided Approach for Structure Determination of Large RNAs
   and Protein–RNA Complexes in Solution   292
5. Future Challenges and Implications   322
Acknowledgments   324
References   324

11. Structural Analysis of Protein–RNA Complexes in Solution Using NMR Paramagnetic Relaxation Enhancements   333
Janosch Hennig, Lisa R. Warner, Bernd Simon, Arie Geerlof,
Cameron D. Mackereth, and Michael Sattler

1. Introduction   334
2. Spin Labeling of Protein–RNA Complexes   336
3. Choice of Spin Label and Attachment Sites   337
4. Spin Labeling Protocol and Sample Handling   340
5. Measurement and Analysis of PRE Data   342
6. Structure Calculation Using PRE-Derived Distance Restraints   345
7. Case Study I: Structure of the U2AF65 RRM1,2-U9 RNA Complex   347
8. Case Study II: Conformational Dynamics of RRM Domains
   in Apo U2AF65-RRM1,2   351
9. Case Study III: Use of NMR and PRE Data for Validation of the
   Sxl–Unr mRNA Complex   353
10. Outlook   355

Acknowledgments	355
References	356

12. Resolving Individual Components in Protein–RNA Complexes Using Small-Angle X-ray Scattering Experiments — 363
Robert P. Rambo

1. Introduction	364
2. Technical Requirements	365
3. Theoretical Considerations	365
4. Assessing SAXS Sample Quality	374
5. Case Studies	375
6. Considerations	387
Acknowledgments	388
References	388

13. Small-Angle Neutron Scattering for Structural Biology of Protein–RNA Complexes — 391
Frank Gabel

1. Small-Angle Neutron Scattering: A Powerful Tool for Structural Biology of Protein–RNA Complexes	392
2. Principles of SANS	393
3. Practical Aspects of SANS and Requirements on the Sample State	396
4. Model-Free Parameters: Molecular Mass and Radius of Gyration	397
5. Neutron Scattering Length Densities of Proteins and RNAs and Implications on Model Building	399
6. Possibilities and Limits of *Ab Initio* Models (Shape Envelopes)	402
7. Present State-of-the-Art Examples	405
8. Conclusions and Future Perspectives	409
Acknowledgments	411
References	411

14. Studying RNA–Protein Interactions of Pre-mRNA Complexes by Mass Spectrometry — 417
Saadia Qamar, Katharina Kramer, and Henning Urlaub

1. Background	419
2. Equipment	421
3. Materials	423
4. Protocol	433
5. Step 1—*In Vitro* Transcription	434

6.	Step 2—*HeLa* Nuclear Extract Preparation	438
7.	Step 3—RNA–Protein Complex Assembly and Purification	441
8.	Step 4—UV Cross-Linking	443
9.	Step 5—Dissociation and Hydrolysis of Proteins	445
10.	Step 6—Isolation of Cross-Links	446
11.	Step 7—Hydrolysis of RNA and Proteins	450
12.	Step 8—Desalting and Removal of RNA Fragments	451
13.	Step 9—Enrichment of RNA–Protein Cross-Links	453
14.	Step 10—Analysis by LC-Coupled ESI MS	455
15.	Step 11—MS Data Analysis	457
16.	Pros and Cons of the Presented Protocol	459
	Acknowledgments	460
	References	460

15. RNA Bind-n-Seq: Measuring the Binding Affinity Landscape of RNA-Binding Proteins — 465

Nicole J. Lambert, Alex D. Robertson, and Christopher B. Burge

1.	Introduction	466
2.	Experimental Method	468
3.	Preparation of RBNS Reagents	469
4.	RBNS Assay	472
5.	Computational Analysis	477
6.	Pipeline Inputs	482
7.	*k*-mer Analysis	482
8.	Streaming *k*-mer Assignment	484
9.	Presence Fractions	487
10.	Relative Binding Affinity Determination	487
11.	RBNS Quality Control	489
12.	Conclusions	491
	References	492

Section III
Large RNA-Protein Assemblies

16. Using Molecular Simulation to Model High-Resolution Cryo-EM Reconstructions — 497

Serdal Kirmizialtin, Justus Loerke, Elmar Behrmann, Christian M.T. Spahn, and Karissa Y. Sanbonmatsu

1.	Introduction	498
2.	Theory	501

3. Methods	503
4. Results	505
5. Summary	508
Acknowledgments	510
References	510

17. *In Vitro* Reconstitution and Crystallization of Cas9 Endonuclease Bound to a Guide RNA and a DNA Target   515

Carolin Anders, Ole Niewoehner, and Martin Jinek

1. Introduction	516
2. Electrophoretic Mobility Shift Assay	518
3. Fluorescence-Detection Size Exclusion Chromatography Assay	523
4. Crystallization of Cas9–RNA–DNA Complexes	530
5. Concluding Remarks	534
Acknowledgments	535
References	535

18. Single-Molecule Pull-Down FRET to Dissect the Mechanisms of Biomolecular Machines   539

Matthew L. Kahlscheuer, Julia Widom, and Nils G. Walter

1. Introduction	540
2. Experimental Methods	543
3. Data Analysis	559
4. The Spliceosome as a Biased Brownian Ratchet Machine	562
5. Conclusions and Outlook	566
Acknowledgment	567
References	567

19. Single-Molecule Imaging of RNA Splicing in Live Cells   571

José Rino, Robert M. Martin, Célia Carvalho, Ana C. de Jesus, and Maria Carmo-Fonseca

1. Introduction	572
2. Overview of the Method	574
3. Detailed Protocol	575
4. Concluding Remarks	583
Acknowledgments	584
References	584

Author Index	*587*
Subject Index	*619*

CONTRIBUTORS

Tariq Afroz
Institute of Molecular Biology and Biophysics, ETH Zurich, Zürich, Switzerland

Hashim M. Al-Hashimi
Department of Biochemistry, Duke University School of Medicine, and Department of Chemistry, Duke University, Durham, North Carolina, USA

Frédéric H.T. Allain
Institute of Molecular Biology and Biophysics, ETH Zürich, CH-8093 Zürich, Switzerland

Carolin Anders
Department of Biochemistry, University of Zurich, Zurich, Switzerland

Elmar Behrmann
Structural Dynamics of Proteins, Center of Advanced European Studies and Research (CAESAR), Bonn, Germany

Steve Bonilla
Department of Chemical Engineering, Stanford University, Stanford, California, USA

Christopher B. Burge
Department of Biology, and Program in Computational and Systems Biology, MIT, Cambridge, Massachusetts, USA

Maria Carmo-Fonseca
Instituto de Medicina Molecular, Faculdade de Medicina, Universidade de Lisboa, Lisboa, Portugal

Célia Carvalho
Instituto de Medicina Molecular, Faculdade de Medicina, Universidade de Lisboa, Lisboa, Portugal

Isabel Chillón
Department of Molecular, Cellular and Developmental Biology, Yale University, New Haven, Connecticut, and Howard Hughes Medical Institute, Chevy Chase, Maryland, USA

Zuzana Cienikova
Institute of Molecular Biology and Biophysics, ETH Zurich, Zürich, Switzerland

Antoine Cléry
Institute of Molecular Biology and Biophysics, ETH Zurich, Zürich, Switzerland

Ana C. de Jesus
Instituto de Medicina Molecular, Faculdade de Medicina, Universidade de Lisboa, Lisboa, Portugal

Olivier Duss
Institute of Molecular Biology and Biophysics, ETH Zürich, CH-8093 Zürich, Switzerland

Frank Gabel
Université Grenoble Alpes; Commisariat à l'Energie Atomique et aux Energies Alternatives, Direction des Sciences du Vivant, Institut de Biologie Structurale, Grenoble Cedex, France; Centre National de la Recherche Scientifique, Institut de Biologie Structurale, Grenoble Cedex 9, France, and Large Scale Structures Group, Institut Laue Langevin, Grenoble Cedex, France

Arie Geerlof
Institute of Structural Biology, Helmholtz Zentrum München, Oberschleißheim, and Center for Integrated Protein Science Munich at Biomolecular NMR Spectroscopy, Department Chemie, Technische Universität München, Garching, Germany

Jason C. Grigg
The Department of Molecular Biology and Genetics, Cornell University, Ithaca, New York, USA

Kathleen B. Hall
Department of Biochemistry and Molecular Biophysics, Washington University Medical School, St. Louis, Missouri, USA

Pehr Harbury
Department of Biochemistry, Stanford University, Stanford, California, USA

Janosch Hennig
Institute of Structural Biology, Helmholtz Zentrum München, Oberschleißheim, and Center for Integrated Protein Science Munich at Biomolecular NMR Spectroscopy, Department Chemie, Technische Universität München, Garching, Germany

Daniel Herschlag
Department of Biochemistry, Stanford University, Stanford, California, USA

Swati Jain
Program in Computational Biology and Bioinformatics, Duke University; Department of Biochemistry, Duke University Medical Center, and Department of Computer Science, Duke University, Durham, North Carolina, USA

Gunnar Jeschke
Laboratory of Physical Chemistry, ETH Zürich, CH-8093 Zürich, Switzerland

Martin Jinek
Department of Biochemistry, University of Zurich, Zurich, Switzerland

C. Joel McManus
Department of Biology, Carnegie Mellon University, Pittsburgh, Pennsylvania, USA

Matthew L. Kahlscheuer
Single Molecule Analysis Group, Department of Chemistry, University of Michigan, Ann Arbor, Michigan, USA

Ailong Ke
The Department of Molecular Biology and Genetics, Cornell University, Ithaca, New York, USA

Dawn Kellogg
Department of Chemistry, Duke University, Durham, North Carolina, USA

Lukasz Jan Kielpinski
Section for RNA and Computational Biology, Department of Biology, University of Copenhagen, Copenhagen, Denmark

Isaac J. Kimsey
Department of Biochemistry, Duke University School of Medicine, Durham, North Carolina, USA

Serdal Kirmizialtin
Department of Chemistry, New York University, Abu Dhabi, United Arab Emirates; New Mexico Consortium, and Theoretical Biology and Biophysics, Theoretical Division, Los Alamos National Laboratory, Los Alamos, New Mexico, USA

Katharina Kramer
Bioanalytical Mass Spectrometry Group, Department of Cellular Biochemistry, Max Planck Institute for Biophysical Chemistry, and Bioanalytics Research Group, Institute for Clinical Chemistry, University Medical Center, Göttingen, Germany

Nicole J. Lambert
Department of Biology, MIT, Cambridge Massachusetts, USA

Michal Legiewicz
Department of Molecular, Cellular and Developmental Biology, Yale University, New Haven, Connecticut, USA

Yizhu Lin
Department of Biology, Carnegie Mellon University, Pittsburgh, Pennsylvania, USA

Fei Liu
Department of Molecular, Cellular and Developmental Biology, Yale University, New Haven, Connecticut, USA

Justus Loerke
Institut für Medizinische Physik und Biophysik, Charité—Universitätsmedizin Berlin, Berlin, Germany

Cameron D. Mackereth
Institut Européen de Chimie et Biologie, IECB, Univ. Bordeaux, Pessac, and Inserm, U869, ARNA Laboratory, Bordeaux, France

Marco Marcia
Department of Molecular, Cellular and Developmental Biology, Yale University, New Haven, Connecticut, USA

Robert M. Martin
Instituto de Medicina Molecular, Faculdade de Medicina, Universidade de Lisboa, Lisboa, Portugal

Gemma E. May
Department of Biology, Carnegie Mellon University, Pittsburgh, Pennsylvania, USA

Mitchell McBrairty
Biophysics Enhanced Program, University of Michigan Ann Arbor, Michigan, USA

Ole Niewoehner
Department of Biochemistry, University of Zurich, Zurich, Switzerland

Anna Marie Pyle
Department of Molecular, Cellular and Developmental Biology; Howard Hughes Medical Institute, Chevy Chase, Maryland, and Department of Chemistry, Yale University, New Haven, Connecticut, USA

Saadia Qamar
Bioanalytical Mass Spectrometry Group, Department of Cellular Biochemistry, Max Planck Institute for Biophysical Chemistry, and Bioanalytics Research Group, Institute for Clinical Chemistry, University Medical Center, Göttingen, Germany

Robert P. Rambo
Diamond Light Source Ltd., Harwell Science & Innovation Campus, Didcot, United Kingdom

Michael J. Rau
Department of Biochemistry and Molecular Biophysics, Washington University Medical School, St. Louis, Missouri, USA

David C. Richardson
Department of Biochemistry, Duke University Medical Center, Durham, North Carolina, USA

Jane S. Richardson
Department of Biochemistry, Duke University Medical Center, Durham, North Carolina, USA

José Rino
Instituto de Medicina Molecular, Faculdade de Medicina, Universidade de Lisboa, Lisboa, Portugal

Alex D. Robertson
Department of Biology, and Program in Computational and Systems Biology, MIT, Cambridge, Massachusetts, USA

Karissa Y. Sanbonmatsu
New Mexico Consortium, and Theoretical Biology and Biophysics, Theoretical Division, Los Alamos National Laboratory, Los Alamos, New Mexico, USA

Bharathwaj Sathyamoorthy
Department of Biochemistry, Duke University School of Medicine, Durham, North Carolina, USA

Michael Sattler
Institute of Structural Biology, Helmholtz Zentrum München, Oberschleißheim, and Center for Integrated Protein Science Munich at Biomolecular NMR Spectroscopy, Department Chemie, Technische Universität München, Garching, Germany

Xuesong Shi
Department of Biochemistry, Stanford University, Stanford, California, USA

Nikolaos Sidiropoulos
Section for RNA and Computational Biology, Department of Biology, University of Copenhagen, Copenhagen, Denmark

Bernd Simon
European Molecular Biology Laboratory, Heidelberg, Germany

Srinivas Somarowthu
Department of Molecular, Cellular and Developmental Biology, Yale University, New Haven, Connecticut, USA

Christian M.T. Spahn
Institut für Medizinische Physik und Biophysik, Charité—Universitätsmedizin Berlin, Berlin, Germany

Zachary W. Stein
Biophysics Enhanced Program, University of Michigan Ann Arbor, Michigan, USA

Henning Urlaub
Bioanalytical Mass Spectrometry Group, Department of Cellular Biochemistry, Max Planck Institute for Biophysical Chemistry, and Bioanalytics Research Group, Institute for Clinical Chemistry, University Medical Center, Göttingen, Germany

Jeppe Vinther
Section for RNA and Computational Biology, Department of Biology, University of Copenhagen, Copenhagen, Denmark

Nils G. Walter
Single Molecule Analysis Group, Department of Chemistry, University of Michigan, Ann Arbor, Michigan, USA

Lisa R. Warner
Institute of Structural Biology, Helmholtz Zentrum München, Oberschleißheim, and Center for Integrated Protein Science Munich at Biomolecular NMR Spectroscopy, Department Chemie, Technische Universität München, Garching, Germany

Julia Widom
Single Molecule Analysis Group, Department of Chemistry, University of Michigan, Ann Arbor, Michigan, USA

Yi Xue
Department of Biochemistry, Duke University School of Medicine, Durham, North Carolina, USA

Maxim Yulikov
Laboratory of Physical Chemistry, ETH Zürich, CH-8093 Zürich, Switzerland

PREFACE

RNA has emerged in the last decade as the central molecule responsible for regulating gene expression, either by itself or in complex with RNA-binding proteins. The unexpected discovery that a high proportion of the genome is transcribed into RNA has meant that a large number of noncoding RNAs are expressed. The discovery of so many different types of noncoding RNAs has created a strong demand for tools to study the structure and dynamics of RNA, protein–RNA complexes, and their assembly.

In this volume of *Methods in Enzymology*, we brought together experts to describe key methodologies for investigating RNA structure and dynamics (Part 1), RNA–protein interactions (Part 2), and large RNA–protein assemblies (Part 3).

A theme highlighted by the chapters in this volume is the increasing importance of solution methods for probing the structures of noncoding RNAs and their complexes. Although X-ray crystallography continues to be a valuable tool for determining RNA structures, many RNA complexes are intrinsically flexible or simply difficult to crystallize. Methods such as NMR, solution X-ray scattering, fluorescence spectroscopy, and cryo-electron microscopy not only avoid the need for crystallization but also provide information about molecular motions over a range of time scales that are intrinsic to the biological function of the RNA. Sequence-based methods such as footprinting and mass spectrometry can be applied to whole transcriptomes or applied *in situ*, expanding the types of questions that can be addressed.

A second theme is the growing importance of computational modeling and data analysis. First, as the volume and complexity of the data increase, there is a commensurate need for automated data processing, noise filtering, and error analysis. This is particularly evident for new techniques such as high-throughput footprinting. Nevertheless, it also applies to established methods such as X-ray crystallography. Second, the solution data report on ensembles of conformations and are often low resolution. Increasingly sophisticated modeling algorithms are able to extract reliable structural models from noisy data sets, estimate the range of likely conformations, or combine constraints from different types of experiments.

In the first part, in Chapter 1, Pyle and coworkers present new biochemical tools to purify and study large RNA molecules. The following three chapters then focus on the dynamics of RNA using NMR (Chapter 2 by Al-Hashimi and coworkers), SAXS interferometry (Chapter 3 by Herschlag and coworkers), and fluorescence (Chapter 4 by Hall and coworkers). chapters 5 and 6 describe methods for probing RNA secondary structures genome wide, either by base modification (McManus and coworkers) or general sequencing analysis tools (Vinther and coworkers). The last two chapters of this part focus on the three-dimensional (3D) structure of RNA, with Richardson and coworkers presenting new tools to validate RNA structures (Chapter 7) and Ke and coworkers describing tips and tricks to obtain 3D structures of regulatory RNAs like riboswitches (Chapter 8).

Part 2 of the volume (chapters 9–15) describes methods for detecting protein–RNA interactions. Chapter 9 opens this part with a detailed review on how the most abundant RNA recognition motif interacts with its RNA targets. This is followed by four chapters describing new approaches to solve structures of protein–RNA complexes in solution. In Chapter 10, Jeschke and coworkers describe how EPR and NMR distance constraints can be combined to solve structures of RNA and protein–RNA complexes with the example of a 70-kDa complex. Chapter 11 by Sattler and coworkers shows how PRE (Paramagnetic Relaxation Enhancement) can be used to study RNA-binding proteins and their complexes and in particular to observe conformational changes upon RNA binding. Chapters 12 and 13 describe how SAXS (Rambo) and SANS (Gabel) can be used to solve large protein–RNA complexes, respectively. Finally, this part of the volume ends by describing how two technologies often used to generate genome-wide data sets, namely mass spectrometry and next-generation sequencing, can be used to identify the sites of protein–RNA cross-links (Urlaub and coworkers) and the affinity of protein–RNA interactions (Burge and coworkers), respectively.

Finally, Part 3 contains four chapters on very large protein–RNA assemblies. Sanbonmatsu and coworkers explore how molecular dynamics calculations improve the refinements of electron microscopy maps (Chapter 16). Jinek and coworkers describe the methods they used to make the endonuclease Cas9 amenable to crystallization in complex with RNA and DNA (Chapter 17). The last two chapters describe new approaches using single-molecule spectroscopy to study molecular machines (Walter and coworkers) and the splicing reaction (Carmo-Fonseca and coworkers).

Most chapters contain detailed protocols allowing the presented method to be used by a wide range of scientists, and not only specialists. We hope that the readers will find this volume as stimulating for future investigations as it has been for both of us.

FRÉDÉRIC H.T. ALLAIN AND SARAH A. WOODSON

SECTION I

RNA Structure and Dynamics

CHAPTER ONE

Native Purification and Analysis of Long RNAs

Isabel Chillón*,†, Marco Marcia*,1, Michal Legiewicz*,2, Fei Liu*, Srinivas Somarowthu*, Anna Marie Pyle*,†,‡,3

*Department of Molecular, Cellular and Developmental Biology, Yale University, New Haven, Connecticut, USA
†Howard Hughes Medical Institute, Chevy Chase, Maryland, USA
‡Department of Chemistry, Yale University, New Haven, Connecticut, USA
3Corresponding author: e-mail address: anna.pyle@yale.edu

Contents

1. Introduction	4
2. Native Purification of Long Noncoding RNAs	6
2.1 Construct design	6
2.2 DNA plasmid linearization	6
2.3 *In vitro* transcription of RNA	7
2.4 DNase digestion	9
2.5 Proteinase K treatment	9
2.6 EDTA chelation of divalent ions (optional)	9
2.7 Buffer exchange and purification	10
2.8 Size-exclusion chromatography	11
3. Study of the RNA Tertiary Folding by Sedimentation Velocity Analytical Ultracentrifugation	13
3.1 Preparation of samples for a study of RNA folding	13
3.2 Assembly of the optical cells, sample loading, and instrument setup	14
3.3 Setting up a sedimentation velocity experiment	15
3.4 Data analysis	17
4. Analysis of the RNA Tertiary Folding by Analytical Size-Exclusion Chromatography	20
5. Determination of the Secondary Structure of LncRNAs by Chemical Probing	22
5.1 Designing and coupling of primers	22
5.2 Generation of sequencing ladders	23
5.3 SHAPE reaction	25
5.4 DMS reaction	27
5.5 Primer extension reaction	28
5.6 Reactions for mobility shift correction	30

---

[1] Present address: European Molecular Biology Laboratory, Grenoble Outstation, 71 Avenue des Martyrs, 38042 Grenoble Cedex 9, France
[2] Present adress: School of Life Sciences, University of Warwick, Coventry, United Kingdom

5.7 Spectral calibration of the instrument 30
5.8 Preparation of samples for capillary electrophoresis 32
5.9 Data analysis 33
Acknowledgments 35
References 35

Abstract

The purification and analysis of long noncoding RNAs (lncRNAs) *in vitro* is a challenge, particularly if one wants to preserve elements of functional structure. Here, we describe a method for purifying lncRNAs that preserves the cotranscriptionally derived structure. The protocol avoids the misfolding that can occur during denaturation–renaturation protocols, thus facilitating the folding of long RNAs to a native-like state. This method is simple and does not require addition of tags to the RNA or the use of affinity columns. LncRNAs purified using this type of native purification protocol are amenable to biochemical and biophysical analysis. Here, we describe how to study lncRNA global compaction in the presence of divalent ions at equilibrium using sedimentation velocity analytical ultracentrifugation and analytical size-exclusion chromatography as well as how to use these uniform RNA species to determine robust lncRNA secondary structure maps by chemical probing techniques like selective 2′-hydroxyl acylation analyzed by primer extension and dimethyl sulfate probing.

1. INTRODUCTION

Biochemical and biophysical studies on RNA molecules *in vitro* require that target molecules are synthesized and purified as homogeneous, functional species. For catalytic RNAs, enzymatic assays can confirm that the target molecules are correctly folded (Russell et al., 2002; Wan, Mitchell, & Russell, 2009). For noncatalytic RNAs, folding homogeneity needs to be assessed by nonenzymatic, biochemical, or biophysical assays (Woodson, 2011).

Traditionally, two folding procedures have been successfully used in biochemical and biophysical studies of large RNAs, such as group I and II introns, other ribozymes, riboswitches, signal recognition particle RNA, and tRNAs. The first approach involves heat denaturation and refolding of the RNA after *in vitro* transcription (Fedorova, Waldsich, & Pyle, 2007; Lomant & Fresco, 1975; Takamoto, He, Morris, Chance, & Brenowitz, 2002; Walstrum & Uhlenbeck, 1990; Zhang & Ferre-D'Amare, 2013), whereas the second approach consists of cotranscriptional folding without any denaturation step (Batey, 2014; Toor, Keating, Taylor, & Pyle, 2008). The latter method preserves the secondary and/or tertiary structure adopted by the RNA during transcription, which can be important considering that functional RNA structures are formed cotranscriptionally *in vivo* (Frieda & Block, 2012; Heilman-Miller & Woodson, 2003; Lai, Proctor, & Meyer, 2013).

Long noncoding RNAs (lncRNAs) are involved in a staggering diversity of fundamental cellular functions and they represent important subjects for ongoing research (Gutschner & Diederichs, 2012). Most lncRNAs do not have ribozyme activity, but they are essential in development, transcription, and epigenetic processes (Necsulea et al., 2014). These RNAs, which can reach tens of kilobases in length, appear to have encountered weaker evolutionary selection constraints than protein-coding genes, thus accumulating repetitive sequences (Derrien et al., 2012). LncRNAs are therefore challenging molecules for biophysical analysis because they can form alternative conformations in the absence of any structural constraint (Huthoff & Berkhout, 2002). Given these properties, cotranscriptional native purification may be particularly useful for lncRNA targets.

Here, we describe a native purification protocol that results in pure, homogeneous lncRNA preparations amenable for biochemical and biophysical studies. While other purification methods make use of affinity tags suitable to extract RNA from *in vivo* sources (Batey & Kieft, 2007; Said et al., 2009), our protocol does not necessarily require tags. The use of tags, generally added to the 3′ end of target RNAs, ensures capturing a homogeneous population of full-length molecules. In our method, we achieve a similar goal by using a T7 polymerase construct that rarely produces short abortive transcripts (Tang et al., 2014), and centrifugal filtration and size-exclusion chromatography (SEC) as final polishing steps in purification. Not using tags may simplify cloning design and avoid inclusion of nonnative sequences that may interfere with structure formation of the target RNA. However, if tags are useful for downstream applications, their inclusion is compatible with our protocol.

We additionally describe methods to study lncRNA folding based on sedimentation velocity analytical ultracentrifugation (SV-AUC), analytical SEC, and chemical probing. These analytical techniques are provided as examples, as many other techniques (such circular dichroism, small angle X-ray scattering, etc.) can be used to monitor homogeneity and oligomeric state of long RNAs (Behrouzi, Roh, Kilburn, Briber, & Woodson, 2012; Pan & Sosnick, 1997; Rambo & Tainer, 2010). SV-AUC and analytical SEC allow one to monitor global compaction of RNA preparations in the presence of divalent ions, under equilibrium conditions (Cole, Lary, Moody, & Laue, 2008; Mitra, 2014), and the equipment required is commonly available. Chemical probing facilitates the determination of lncRNA secondary structure (Athavale et al., 2013; Novikova, Hennelly, & Sanbonmatsu, 2012), and it also utilizes reagents that are available to most investigators. In this review, we describe protocols for selective 2′-hydroxyl acylation analyzed by primer extension (SHAPE) and dimethyl sulfate (DMS) chemical probing, as they have been applied for studying lncRNAs (Novikova et al., 2012; Watts et al., 2009). Again, these represent only a subset of available techniques for mapping RNA structure in solution. The SHAPE and DMS methods employ an electrophilic reagent that reacts selectively with flexible, accessible sites on ribonucleotides, facilitating detection of loops and other single-stranded regions within lncRNA molecules.

2. NATIVE PURIFICATION OF LONG NONCODING RNAs

In this section, we outline the procedures for transcription and purification of lncRNAs following nondenaturing methods. These protocols are based on those developed in our laboratory in recent years (Fedorova, Su, & Pyle, 2002; Fedorova et al., 2007; Marcia & Pyle, 2012; Toor et al., 2008), which have been modified and updated to fit the idiosyncrasies of large noncoding RNAs that range from one to several kilobases in length (up to 4 kb in our hands). The pipeline of the procedure includes the design of the construct, linearization of the DNA template and transcription of the RNA, digestion of DNA, enzyme proteolysis, buffer exchange, and finally, filtration and purification using SEC (Fig. 1A).

2.1 Construct design

The target lncRNA should be inserted into a suitable vector (e.g., pBluescript, Life Technologies) such that it is immediately flanked at the 5′ end by T7 promoter sequence and at the 3′ end by a unique and efficient restriction site (e.g., *Bam*HI). In addition, we recommend digesting the template DNA with a restriction enzyme that produces blunt or 5′ overhanging ends to limit the production of heterogeneous RNA transcripts that vary in length (Schenborn & Mierendorf, 1985).

2.2 DNA plasmid linearization

1. Mix 50 μg of DNA plasmid with 80 U of the appropriate restriction enzyme in the presence of 1 × restriction enzyme buffer in a final volume of 100 μl.

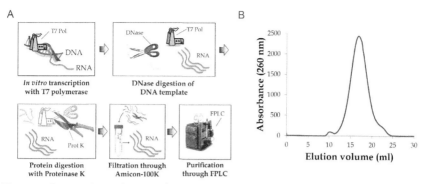

Figure 1 Enzymatic synthesis and purification of long noncoding RNAs. (A) The native purification pipeline of lncRNAs comprises the following steps (from left to right and top to bottom): the RNA is synthesized with the T7 RNA polymerase system; when the transcription reaction is finished, the DNA template is digested with DNase enzyme; enzymes in the reaction are then proteolyzed with the action of the proteinase K enzyme; RNA is captured from reaction components and buffer exchanged by ultrafiltration with Amicon centrifugal devices (Millipore); finally, the RNA is subjected to size-exclusion chromatography. (B) The resulting chromatogram represents the absorbance of the filtered RNA as a function of the elution volume. Material in the void volume of the column (10 ml elution) and shorter RNA molecules (22 ml elution) are excluded. The fraction(s) corresponding to the main peak are then selected for downstream analysis.

2. Incubate in a thermo block or mixer incubator at 37 °C overnight.

    Note: It is recommended to check that the plasmid is completely digested by running an agarose gel before starting the transcription reaction.

3. Purify the plasmid through phenol extraction followed by ethanol precipitation or through a spin column (Illustra MicroSpin G-25 column, GE Healthcare).

4. Quantify the concentration of the DNA template in a Nanodrop spectrophotometer (Thermo Scientific).

2.3 *In vitro* transcription of RNA

1. Prepare a transcription reaction in a final volume of 200 µl (Tables 1 and 2).
    a. We express our own batches of recombinant T7 RNA polymerase, which has a concentration of 2 mg/ml, approximately. We use the T7 RNA polymerase mutant P266L, which is known to produce short abortive transcripts less frequently than wild-type T7 polymerase (Tang et al., 2014). If using a commercial T7 RNA polymerase, we suggest following the manufacturer instructions.

Table 1 *In vitro* transcription reaction

Component	Final concentration	Stock	Amount
Transcription buffer (Table 2)	1×	10×	20 μl
DTT	10 mM	1 M	2 μl
ATP	3.6 mM	30 mM	24 μl
CTP	3.6 mM	30 mM	24 μl
GTP	3.6 mM	30 mM	24 μl
UTP	3.6 mM	30 mM	24 μl
RNaseOUT™	80 U	40 U/μl	2 μl
Pyrophosphatase	4 U	2 U/μl	2 μl
T7 polymerase	0.1 mg/ml	2 mg/ml	10 μl
Linearized plasmid	6.25 μg		31.25 ng/μl

Add deionized H$_2$O to 200 μl.

Table 2 Transcription buffer (10×)

Component	Final concentration	Stock (M)	Amount (μl)
MgCl$_2$	120 mM	1	120
Tris–HCl (pH 8.0)	400 mM	1	400
Spermidine	20 mM	2	10
NaCl	100 mM	5	20
Triton X-100	0.1%		1

Add deionized H$_2$O to 1 ml.

    b. Scale up the reaction above according to the amount of RNA needed in subsequent steps. We have observed that the yield of the reaction varies among transcripts and factors affecting yield can include sequence (particularly in the first several nucleotides adjacent to the promoter), the length of the transcript, and the complexity of its secondary structure.

    c. It is also advisable to optimize the concentration of magnesium for each particular transcript. The concentration indicated in the table above corresponds to that optimized for lncRNA HOTAIR.

2. Incubate the reaction for 2 h at 37 °C in a thermo block or mixer incubator.

a. Increasing the incubation time can result in increased RNA yields, although we do not recommend incubating longer than 4 h.

2.4 DNase digestion

1. After transcription, add 0.5 U of TURBO™ DNase (2 U/µl) (Life Technologies) per microgram of DNA included in the transcription reaction, as recommended by the manufacturer.
   a. For the transcription reaction described above, we would typically include 1.5 µl of DNase; however, we often add more (2.5 µl).
   b. Note that DNase enzyme is sensitive to physical denaturation. Therefore, do not vortex the reaction mixture after addition of DNase.
2. Incubate at 37 °C for 30 min.

2.5 Proteinase K treatment

1. After the DNase treatment, add 2 µl of a 30 mg/ml stock of proteinase K powder (Life Technologies) dissolved in proteinase K storage buffer (Table 3) to proteolyze the enzymes present in the reaction mixture.
   a. According to manufacturer, a final concentration of 0.25 mg/ml (1.6 µl) of proteinase K is enough for the degradation of enzymes present in the reaction mixture. However, if the presence of proteinase K is not problematic for subsequent applications, we usually add an excess of enzyme.
2. Incubate at 37 °C for 30 min.

2.6 EDTA chelation of divalent ions (optional)

In some cases, it is desirable to preserve all secondary and tertiary structure that is formed during transcription. In these cases, it is important to skip this step. However, if one wants to preserve secondary structural elements and

Table 3 Proteinase K storage buffer (1 ×)

Component	Final concentration	Stock	Amount
Tris–HCl (pH 7.5)	10 mM	1 M	100 µl
CaCl$_2$ (1 M)	1 mM	1 M	10 µl
Glycerol	40%	100%	4 ml

Add deionized H$_2$O to 10 ml.

eliminate tertiary structural features (e.g., for studies of RNA folding mechanism), then perform the chelation protocol described below.

After proteolytic digestion, divalent ions are chelated through the addition of EDTA to the solution. This treatment disrupts the RNA tertiary structure, but preserves stable secondary structure elements formed during the transcription reaction.

1. Add an amount of EDTA to the reaction that is equivalent to the number of divalent cations present in the transcription buffer.
    a. The reaction mixture described above contains 2400 nmol of $MgCl_2$. Therefore, add:
    $$2400\,nmol : 500\,nmol/\mu l\, of\, EDTA = 4.8\,\mu l\, of\, EDTA\, 0.5\,M$$
2. Vortex briefly and spin down the tubes.

2.7 Buffer exchange and purification

Buffer exchange and separation of the RNA from enzymatic products is performed using commercially available Amicon® Ultra Centrifugal Filters (Millipore) (http://www.emdmillipore.com/US/en/life-science-research/protein-sample-preparation/protein-concentration/amicon-ultra-centrifugal-filters/YKSb.qB.GXgAAAFBTrllvyxi,nav). This treatment allows the RNA to be equilibrated in filtration buffer (Table 4), a buffer solution containing physiological amounts of monovalent ions that preserve the secondary structure, while washing out unreacted nucleotides, digested DNA, protein fragments, and other components of the previous reactions.

Short macromolecules, such as abortive transcription products and the proteinase K enzyme (18.5 kDa) are also generally filtered out, although the extent of the filtration depends on the cut-off of the Amicon filter selected. All the centrifugation steps should be carried out at room temperature. It is important to keep the RNA at room temperature at all times, since changes in temperature can trap RNA molecules in alternative conformations, decreasing the homogeneity of the sample. The size of the

Table 4 Filtration buffer (1 ×)

Component	Final concentration (mM)	Stock (M)	Amount
MOPS buffer (pH 6.5)	8	1	4 ml
KCl	100	2	25 ml
EDTA (pH 8.5)	0.1	0.5	100 µl

Add deionized H_2O to 500 ml.

Table 5 Amicon-device centrifugation guide

Amicon filter	Txn volume (ml)	Rotor	Speed (rcf)	Time (min)
Amicon Ultra 0.5 ml	0.2–1	Fixed-angle rotor suitable for 2-ml tubes	4000	5 × 5 times
Amicon Ultra 15 ml	≥1	Swinging bucket rotor suitable for 50-ml tubes	3000	10–15

concentration device and the centrifugation conditions depend upon the final volume of the transcription reaction (Table 5).

The Amicon® Ultra Filters are manufactured in a range of molecular weight cut-offs (MWCO), which have been calibrated for globular proteins. We use those with a MWCO of 100 kDa for our lncRNAs, which range from 2 to 4 kb long, although the selected MWCO will vary according to the target of the study. In general, it is important to consider that selectivity of Amicon® Ultra Filters is different for proteins and for RNAs. For example, a filter that retains globular proteins of a given molecular weight will retain RNA molecules that are six times smaller (Kim, McKenna, Viani Puglisi, & Puglisi, 2007).

2.8 Size-exclusion chromatography

SEC or gel filtration chromatography is used for native purification of the RNA in a homogenous form. Alternative conformations of the RNA that are induced by temperature fluctuations, multimeric forms of the RNA formed by self-complementarity and aggregation and prematurely terminated transcripts can be resolved by SEC without altering the native secondary structure of the lncRNA (Fig. 1B).

We use different bead size resins that we pack into a Tricorn 10/300 Empty High Performance Column (GE Healthcare) (CV of approximately 25 ml). The media are selected according to the size of the RNA, following GE Healthcare recommendations (http://www.gelifesciences.co.jp/catalog/pdf/18112419.pdf) (Table 6).

All the procedures are carried out at room temperature, and empty columns should be packed at room temperature, following manufacturer recommendations (http://ge-catalog.campaignhosting.se/Documents/Tricorn/TricornInstructionsAC.pdf). We keep packed columns equilibrated with ddH$_2$O or 20% ethanol at 4 °C when they are not being used to avoid

Table 6 Selection of chromatography media and running conditions

RNA size (nt)	Media	Max. operating flow rate approx. (ml/min)	Max. operating back pressure (MPa)
1–200	Superdex 75	1	0.2
200–600	Superdex 200	1	0.2
600–2000	Sephacryl 400	1	0.2
1000–2000	Sephacryl 500	0.83	0.2
>2000	Sephacryl 1000	0.67	Not determined

bacterial contamination. In addition, all filtration solutions and packed columns should be warmed up to room temperature before the run.

Use an appropriate fast protein liquid chromatography (FPLC) system (i.e., Akta FPLC, Akta Purifier, Akta Explorer, Akta Pure) equipped with wavelength detection at 260 nm. The system should be washed with an RNase decontamination solution like RNaseZap (Life Technologies) and the column washed with at least 3 column volumes (CV) of ddH$_2$O and equilibrated with 3 CV of filtration buffer.

The following steps should be performed during SEC of lncRNA:

1. Collect fractions of 0.5 ml volume. The homogeneity of the preparation will usually be apparent from the elution chromatogram (Fig. 1B). If the RNA of interest is homogenous and elutes as a single peak, this fraction should be collected.

   Note: Self-packed columns and RNA preparations usually lead to broader peaks than those generated by protein preparations that are run in commercially packed and calibrated columns.

2. Measure the concentration of each fraction using a Nanodrop spectrophotometer (Thermo Scientific) and record absorption value at 260 nm.

3. The fraction with the highest absorption should be used for further studies. If more material is necessary, neighboring fractions should be combined and the absorption value of the mixture recorded again.

At this point, the lncRNA is used for subsequent applications, like folding studies or secondary structure probing. If long-term storage is necessary, the preservation process should be validated to be proven that it does not affect RNA biochemical and biophysical properties (e.g., formation of nonspecific interactions, multimerization, structural rearrangements, precipitation, or degradation). Based on our experience, storage of long RNAs in cold impaired the reproducibility of our assays. For that reason, we work exclusively on freshly purified RNA.

3. STUDY OF THE RNA TERTIARY FOLDING BY SEDIMENTATION VELOCITY ANALYTICAL ULTRACENTRIFUGATION

In this section, we describe the preparation of the sample for an equilibrium SV-AUC assay, and we provide a brief description of the assembly of the analytical cells and the instrument, the ProteomeLab™ XL-I analytical ultracentrifuge (Beckman Instruments). We also explain how to process the raw data in order to obtain the hydrodynamic parameters for a given RNA species. There are two basic parameters to consider when preparing an SV-AUC experiment:

The first parameter is the concentration of the sample. We typically determine the concentration of the RNA in terms of absorbance at 260 nm in the Nanodrop (Thermo Scientific). The range for maintaining a linear response is from 0.2 to 1.2 absorbance units. High sample concentrations are more prone to aggregation and can lead to artifacts; in addition, different preparations of RNA samples can vary in yield and using high concentrations of sample in every experiment reduces the number of assays that can be performed with the same RNA preparation. For these reasons, we recommend using the lowest concentration of RNA that provides a good signal. Generally, an absorbance of 0.2 at 260 nm is sufficient.

The second parameter is the rotor speed. This parameter needs to be adjusted based on the molecular weight and the expected hydrodynamic radius of the target molecule. We typically apply a speed ranging between 20,000 and 25,000 rpm for our lncRNAs. For further reference, Cole et al. (2008) provides a quick guide for selecting the rotor speed based on the molecular weight of the molecule (although the values he provides refer to globular proteins), and the estimated sedimentation coefficient of the particles, which can be assessed in a preliminary test.

3.1 Preparation of samples for a study of RNA folding

1. Take the fraction with the highest absorbance at 260 nm after the SEC step and write down the absorbance value.
2. Determine the number of samples that you can run by SV-AUC by dividing the absorbance value of the sample by the final absorbance value that you will use in the experiment, for example, 0.2. Following these recommendation, we usually have material to perform at least 12 titration measurements. If the concentration of the RNA sample is not enough for making the desired number of measurements, then mix several

fractions. If there is insufficient material in the main fraction for your experiment, we recommend taking only the two fractions eluting immediately before and immediately after the peak.

3. Set up the folding reaction in a volume of 420–500 μl. Dilute the RNA up to the desired concentration (i.e., 0.2) and add 2× AUC buffer (Table 7).
4. For each titration point, you should prepare a blank control, which does not contain RNA. The blank control contains the same volume of filtration buffer than RNA sample and the same concentration of the other components (Table 8).
5. Incubate the folding reaction at 37 °C on a heat block for a period of time that allows the RNA to reach its equilibrium structure. We usually let the RNA equilibrate for 30 min to 1 h. At this point, the samples are ready to be loaded in the optical cells of the analytical ultracentrifuge.

3.2 Assembly of the optical cells, sample loading, and instrument setup

Each analytical cell consists of a centerpiece, two window assemblies, and a cell housing. We use the charcoal-filled Epon double-sector centerpiece by

Table 7 AUC buffer (2×)

Component	Final concentration (mM)	Stock (M)	Amount
HEPES buffer (pH 7.4)	100	1	15 ml
KCl	400	2	30 ml
EDTA (pH 8.5)	0.2	0.5	60 μl

Add deionized H_2O to 150 ml.

Table 8 Preparation of an SV-AUC experiment

B1	x μl filtration buffer 250 μl of 2× AU buffer 100 μl of 5× $MgCl_2$ 50 mM x μl water
S1	x μl RNA fraction num. X 250 μl of 2× AU buffer 100 μl of 5× $MgCl_2$ 50 mM x μl water

Example: 10 mM $MgCl_2$.
$Vol_F = 500$ μl.

Beckman and sapphire windows because they are harder than quartz windows and so they have longer life. The cells are loaded into an An-60 Ti analytical rotor (Beckman Instruments).

1. Clean all components with RNaseZAP and distilled water, especially the centerpieces and the windows, which are in direct contact with the RNA sample.
2. Assemble the cells and the counterbalance following the manufacturer instructions (https://www.beckmancoulter.com/wsrportal/techdocs?docname=LXLA-TB-003).
3. Place the cell horizontally, facing the screw ring, and load 400 μl of blank sample in the left cavity and 380 μl of RNA-containing sample in the right cavity with a 500 μl Hamilton syringe. Repeat this procedure for the other two cells.

   Note: It is important that the meniscus of the blank sample is higher than that of the RNA sample so that it does not interfere with the data analysis (see Section 3.4).
4. Weigh the filled cells and make sure that opposing cells weigh within 0.5 g of each other.
5. Seal the filling holes with the red plug gaskets and screw the housing plugs in with a screwdriver.
6. Switch on the AUC instrument and place the rotor inside. Place the monochromator inside the ultracentrifuge as described by the manufacturer (https://www.beckmancoulter.com/wsrportal/techdocs?docname=LXLAI-IM-10). Activate the chamber vacuum by pressing the vacuum key on the instrument control panel.

   Note: The monochromator contains a blocking filter that can be used to eliminate wavelengths below 400 nm. For absorbance at 260 nm, position the monochromator blocking filter lever perpendicular to the stem.

3.3 Setting up a sedimentation velocity experiment

Here, we describe the parameters for a sedimentation velocity run of an lncRNA in the range of 2–3 kb in length. These parameters should be adapted for different targets under study, following sample concentration and rotor speed recommendations above to ensure complete sedimentation of the RNA molecules.

1. Open the ProteomeLab XL-A/XL-I User Interface Software (Beckman Instruments).
2. Click File > New File.

3. In the setup area, select the following values and options:
   a. Rotor: 4 hole
   b. Speed: 25,000 rpm
   c. Time: hold
   d. Temperature: 20 °C
   e. Type of scan: Velocity
   f. Optical system: Absorbance
   g. Use default values of R_{min} and R_{max}, and select 260 nm in the wavelength field.
   h. If you want to use the same scan settings for each cell, enter settings for Cell 1 and then click the All Settings Identical to Cell 1 box.
4. Click the Detail button and select the following options in the Velocity Detail dialog box:
   a. Replicates: 2
   b. Mode: Continuous
   c. Centerpiece: 2
   d. Accept the default values for the rest of the fields
5. In the setup area, click on Scan Options and select the following parameters:
   a. Leave all boxes unselected except for Stop XL after last scan.
   b. Select Overlay last 1 scan(s).
6. Click the Method button. Accept the default values in each field and select 100 scans in the number of scans field.
7. We usually perform a short- and low-speed method for the radial calibration of the instrument before performing the actual experiment. The following modifications to the setup parameters described above should be selected for the radial calibration run:
   a. In the setup area, change the following options:
      i. Speed: 3000 rpm
   b. Under Details, select the following options:
      i. Replicates: 1
   c. In Scan options, select the following box in addition to the previous ones:
      i. Radial calibration before first scan.
   d. In Method, select 1 scan.
8. Once the radial calibration is finished (around 7–10 min), let the sample stand inside the instrument for around 30 min to 1 h, to allow for temperature equilibration. After that, start the data acquisition with the Start Method Scan button.

9. When the run is complete, turn off the vacuum and take the monochromator and the rotor out. Remove the sample from the cells with a pipette and gel-loading tips and disassemble the optical cells. Clean them well with Beckman soap and distilled water, and place them back.

3.4 Data analysis

We use the software SEDFIT for data analysis (http://www.analyticalultracentrifugation.com/download.htm) (Schuck, 2000). SEDFIT is a method for size-distribution analysis that uses numerical solutions to the Lamm equations (Lebowitz, Lewis, & Schuck, 2002; Fig. 2A and B). SEDFIT also calculates hydrodynamic parameters like the Stokes radius (hydration radius) of the species sedimented in the presence of different concentrations of magnesium. The values of Stokes radius are then fit to the Hill equation and the value of the K_d or midpoint of the equilibrium folding transition is determined (Fig. 2C; Mitra, 2009, 2014).

1. Click on Data and Load new files. Select all the data corresponding to one optical cell (RA1, RA2, or RA3). In the pop-up dialog, enter the loading interval, preferably 1, if the number of scans is not very high (around 100 is good).
2. In the main software interface, drag and drop the red line in the absorbance display to indicate tentatively the position of the sample meniscus, generally located at about 6 cm and the blue line to indicate the bottom of the cell, which is usually located at around 7.2 cm.
3. In the same view, drag and drop the green lines, which indicate the limits of the analysis. We recommend placing the left green line as close as possible to the red line, whereas we usually place the right green line at around 7 cm, to avoid including convection artifacts in the analysis.
4. In the Model dialog box, select Continuous c(s) with other prior knowledge and in the pull down dialog select Continuous c(s) Conformational Change Model. This model is useful when the molar mass of the main species is known and when this species undergoes a conformational change under the different ionic conditions. In this case, the main species will sediment with different s-values under the different magnesium concentrations.
5. Click on Parameters and enter the following data:
   a. The molar mass of the RNA under study.
   b. A tentative range of s-min and s-max to display the results.

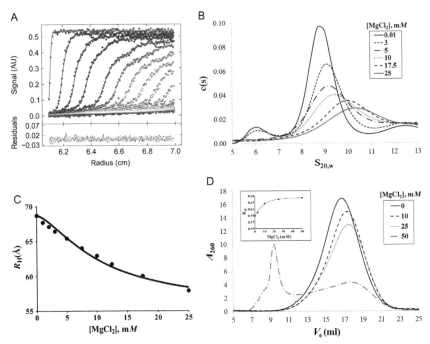

Figure 2 Study of the RNA tertiary folding by sedimentation velocity analytical ultracentrifugation (SV-AUC) and analytical size-exclusion chromatography (SEC). (A) Raw data from SV-AUC are analyzed using Sedfit program and the resulting fit is visualized using the GUSSI applet. (B) The sedimentation coefficient distribution is obtained for each magnesium titration point. The sedimentation coefficient of the main species increases as RNA is compacting in the presence of magnesium. The observed decrease in the absorbance at increasing magnesium concentrations is due to the hypochromicity effect, which occurs when RNA bases stack in the tertiary structure, thus diminishing its capacity to absorb ultraviolet light. (C) Stokes radii are calculated, represented as a function of magnesium concentration and fitted to the Hill equation. (D) Elution volumes in an analytical SEC experiment are plotted as a function of magnesium concentration. The hypochromicity effect is also observed as a decrease in the absorbance of the eluted RNA. At 50 mM magnesium chloride, the sample is mainly eluting in the void volume, which indicates that a nonspecific aggregation is occurring.

    c. The resolution or number of data points of the analysis, which is generally left as default (100).
    d. The partial specific volume, which is approximately 0.53 ml/g for RNA molecules (Mitra, 2009).
    e. The buffer density and the buffer viscosity for the experimental conditions. We use the program SEDNTERP (http://sednterp.unh.edu/) to determine the density and viscosity of each buffer

used in the magnesium titration. Select 20 °C under the Experiment tab and introduce the components of the buffer in the Solvent tab, under the Compute button (Table 9). Wherever needed, the values of density and viscosity for pure water are 0.99823 g/l and 0.01002 P, respectively.

   f. An initial value of 2 for the frictional coefficient is recommended.
   g. Start by selecting a confidence level of 0.7.
   h. Leave all boxes unchecked.
6. Click the Run command to get an initial assessment of the fit.
7. Some initial adjustment may be needed: modify the range of s-min and x-max that better fits the main peak observed. Take note of the sedimentation coefficient of the main species and use it to get a better estimation of the frictional coefficient of your RNA molecule in the Calculator included in SEDFIT software. Go to Options > Calculator > Calculate axial and frictional coefficient ratios. In the pop-up window, introduce the values required. Assume the partial specific volume and the hydration of the RNA molecule to be 0.53 ml/g and 0.59 g/g, respectively (Mitra, 2009).
8. In the Parameters tab, check the box meniscus to allow fitting of the estimated value. Click first on the Run command and then on the Fit command. Repeat these two commands assuming a confidence level of 0.95.
9. Go back to the Parameters tab and uncheck the meniscus box. Check the boxes Baseline, Fit Time Independent Noise, and Frictional Ratio. Indicate a confidence level of 0.7. Click on the Fit command to refine the model. Repeat the operation assuming a confidence level of 0.95.

Table 9 Density and viscosity values of selected SV-AUC buffers as calculated by SEDNTERP

$MgCl_2$ (mM)	Density (g/l)	Viscosity (P)
0	1.01157	0.01031
0.01	1.01167	0.01030
1	1.01168	0.01031
5	1.0119	0.01033
10	1.01223	0.01035
25	1.0133	0.01042

10. Assess the quality of the fit by considering the rms deviation, which should be well below 0.01 and the Run test Z, which should have an upper limit of 30. A visual determination of the goodness of the fit can be observed by clicking Display > Subtract Calculated Ti noise From Raw Data or by running the plug-in GUSSI (http://biophysics.swmed.edu/MBR/software.html) and selecting Plot > GUSSI data-fit-residuals plot (Fig. 2A).
11. Document the fit and the parameters displayed in the SEDFIT window by storing a screenshot of the SEDFIT window. Copy the distribution table by selecting this option under the tab Copy (Fig. 2B).
12. Calculate the Stokes radius (R_H) of the RNA at each magnesium concentration. In SEDFIT go to Options > Calculator > Calculate axial and frictional coefficient ratios, and introduce the values corresponding to the RNA and the buffer at each titration condition. In the results window, take note of the value of the Stokes radius for each condition.
13. Represent the Stokes radii of the magnesium titration as a function of the magnesium concentration and fit the curve to a Hill equation using the software GraphPad Prism 6 (GraphPad Software):

$$Y(R_H) = Y(H_{H,0}) + (Y(R_{H,f}) - Y(R_{H,0})) \times \left(\frac{[Mg]^n}{K_{Mg}^n + [Mg]^n} \right),$$

where R_H is the Stokes radius (in Å), $R_{H,f}$ and $R_{H,0}$ are the Stokes radii for unfolded and folded RNA, which correspond to the titration point with the lower and the highest concentration of magnesium, respectively. K_{Mg} value is the concentration of magnesium at which 50% of the RNA is folded and n is the Hill coefficient, which indicates the cooperativity of the folding transition (Mitra, 2009; Su, Brenowitz, & Pyle, 2003; Fig. 2C).

4. ANALYSIS OF THE RNA TERTIARY FOLDING BY ANALYTICAL SIZE-EXCLUSION CHROMATOGRAPHY

SEC can also be used as an analytical tool to observe global compaction of RNA molecules in the presence of divalent ions. The advantage of this approach lies in its ease of use and in the higher availability of necessary equipment. Specifically, we use the same chromatography columns and media as that employed during the purification step and the same

chromatography system (Akta FPLC) (see Section 2.8.). In analytical SEC there is no need to collect fractions, although it is mandatory that the FPLC system is equipped with a UV cell to monitor absorbance of the RNA sample. A description of this technique is summarized below:
1. Purify the RNA following the procedures described in Section 2.
2. Set up a folding reaction as explained in Section 3.1. Test a broad range of magnesium concentrations to determine the upper limit at which the RNA is no longer compacting, but rather aggregating. This phenomenon is especially easy to detect in analytical SEC, since large aggregates of molecules elute in the void volume of the column (Fig. 2D).
3. For every magnesium concentration tested, the FPLC system must be equilibrated in the same buffer used for the folding of RNA. The equilibration is carried out with 1 CV of buffer, provided that the experiment is starting from the lower magnesium concentration, so that no residual magnesium interferes with the assay. In other cases, equilibration should be performed with 3 CV of buffer.
4. Inject the sample and run the same protocol as that used for the purification of the RNA (Section 2.8). Note that the UV cell of the Akta FPLC system is very sensitive and very low sample volumes can be used for an analytical SEC experiment. In particular, we load 3 μg of RNA per titration condition.
5. A typical elution profile that is representative of a titration experiment is shown in Fig. 2D. A shift in the elution volume at each titration point is indicative of the compaction of the RNA at increasing magnesium concentrations. Compared to an SV-AUC profile (Fig. 2B), the resolution of the shift and that of the decrease in absorbance due to the hypochromicity effect are lower. Nevertheless, analytical SEC offers a reliable method to evaluate homogeneity of the RNA at different magnesium concentrations and to detect aggregation and degradation events.
6. To compare elution profiles between different columns, we calculate the chromatographic partition coefficient, K_{av}, or distribution constant (Fig. 2D inset; Dai, Dubin, & Andersson, 1998):

$$K_{av} = \frac{(V_e - V_0)}{(V_t - V_0)}$$

where V_e is the RNA elution volume, V_0 is the void volume of the column, and V_t is the total volume of the column. In addition, the partition coefficient also provides a quantitative determination of the highest concentration of magnesium at which the RNA ceases to continue compacting. This value also represents the concentration of magnesium that is optimal to include in the folding step of subsequent biophysical and biochemical applications. In the case of HOTAIR lncRNA, we use 25 mM of magnesium (Fig. 2D inset).

5. DETERMINATION OF THE SECONDARY STRUCTURE OF LNCRNAS BY CHEMICAL PROBING

In this section, we will provide an approach for determining the secondary structure of an lncRNA by using two chemical probing techniques: SHAPE and DMS. It is important to note that chemical probing should be performed immediately after the purification step and the samples have to be kept at room temperature at all times, avoiding any cooling down or freezing step, as this promotes the rearrangement of RNA structures. We make use of automated, high-throughput capillary electrophoresis for the analysis of fragments generated after the primer extension reactions, which are carried out with fluorescent-coupled primers. The runs are performed in a campus-located facility (http://dna-analysis.research.yale.edu/), and the raw data are analyzed by either one the following software programs: ShapeFinder (Vasa, Guex, Wilkinson, Weeks, & Giddings, 2008) and QuShape (Karabiber, McGinnis, Favorov, & Weeks, 2013).

5.1 Designing and coupling of primers

Fluorescent-coupled primers are used for primer extension detection of modifications in the RNA. Primers should be designed following standard guidelines as for DNA sequencing (approximate length of 20 nucleotides, starting and ending with purine residues, and with an approximate GC content of 50%). In addition, G should be avoided at the 5′ end if possible, as it can quench fluorescence (Behlke, Huang, Bogh, Rose, & Devor, 2005). The primers should cover the full sequence of the target gene, spaced evenly every 200–220 nucleotides, approximately. Oligonucleotides are synthesized in 50 nmol scale and modified at the 5′ end with an amine group.

We use four spectrally different fluorescent dyes in the form of succinimidyl esters: 5-carboxy-X-rhodamine (5-ROX), 5-carboxyfluorescein (5-FAM), 6-carboxy-4′,5′-dichloro-2′,7′-dimethoxyfluorescein (6-FAM), and 6-carboxytetramethylrhodamine (6-TAMRA) (Anaspec). Succinimidyl esters (SE) are recommended because they form a very stable amide bond between the dye and the amine-modified oligonucleotide primer. SE dyes are

Table 10 Primer labeling reaction

Component	Final concentration	Stock	Amount (µl)
Oligonucleotide	440–660 ng/µl	4–6 mg/ml	11
SE-dye	2.52 µg/µl	18 mg/ml	14
Sodium tetraborate buffer (pH 8.5)[a]	75 mM	0.1 M	75

[a]Prepare freshly before the coupling reaction or keep frozen aliquots at −20 °C.

hydrophobic and they are freshly dissolved in DMSO prior to the labeling reaction. We have adapted the protocol for labeling amine-modified oligonucleotides from Life Technologies (http://tools.lifetechnologies.com/content/sfs/manuals/mp00143.pdf). A summary of the dye conjugation step is provided below:

1. Prepare a labeling reaction by mixing the amine-modified oligonucleotide with the dye (Table 10).
2. Incubate the reaction in the dark for 6 h to overnight at room temperature.
3. Ethanol precipitate the oligonucleotide to ensure removal of the free dye.
4. Purify the oligonucleotide in a denaturing 20% polyacrylamide gel.
5. Locate the labeled and unlabeled oligonucleotides by UV shadowing and cut out the labeled oligonucleotide.
6. Purify the oligonucleotide by the "crush-and-soak" method using 10 mM Tris (pH 8.0) and 300 mM NaCl and precipitate with ethanol. Oligonucleotides are finally dissolved in ddH$_2$O at a concentration of 20 µM (stock) and 2 µM (working concentration).

5.2 Generation of sequencing ladders

Two sequencing ladder reactions are generated for assays to be analyzed by ShapeFinder and one or two for those to be analyzed by QuShape. In the first case, primers labeled with ROX and TAMRA are reserved for the sequencing ladders. In the second case, we usually prepare a single sequencing ladder with primers labeled with FAM. The nucleotide selected for the ladder varies with every target composition, and it is recommended to make ladder(s) complementary to the most frequent nucleotide(s). Here, we present an example for a ladder performed to detect the adenosines in the target sequence.

1. Combine the following amounts in an amber PCR tube:
   28.8 μl Master mix (Table 11)
   4 μl Labeled primer 2 μM
2. Place each PCR tube, corresponding to each sequencing ladder, into a thermocycler and run a sequencing-ladder PCR program (Table 13).
3. Stop the reaction by adding 6.4 μl of stop mixture (Table 14).
4. At this point, mix both sequencing ladders if using two for the analysis. Add 196 μl of 100% ethanol and incubate at 4 °C for 5 min.

Table 11 Sequencing-ladder master mix

Component	Final concentration	Stock	Amount (μl)
Plasmid template	1 ng/μl		
Thermo sequenase buffer			1.84
ddTTP mixture (Table 12)			16
Thermo sequenase	0.25 U/μl	4 U/μl	1.84

Add deionized H_2O to 28.8 μl.

Table 12 ddTTP mixture

Component	Final concentration (μM)	Stock (mM)	Amount (μl)
dATP	150	100	1.5
dCTP	150	100	1.5
dGTP	150	100	1.5
dTTP	150	100	1.5
ddTTP	1.5	1	1.5

Add deionized H_2O to 1 ml.

Table 13 Sequencing-ladder PCR program

Number of cycles	Temperature (°C)	Time
1	96	1 min
25	96	20 s
	55	20 s
	72	1 min
1	72	5 min

5. Spin at maximum speed for 30 min and wash two times with 70% ethanol.
6. Resuspend the pellet in 2 µl of water and 38 µl of formamide.
7. Store at −20 °C.

5.3 SHAPE reaction

The RNA purified following the procedure described in Section 2 is subjected to chemical modification using 1-methyl-7-nitroisatoic anhydride (1M7). We synthesize 1M7 in our laboratory following the protocol by Kevin Weeks and coworkers (Mortimer & Weeks, 2007), but there are other approaches and SHAPE reagents that can be employed, like NMIA and 1M6 (Rice, Leonard, & Weeks, 2014). The optimal amount of 1M7 should be titrated for every RNA target under study, since factors like the length of the molecule and the purity of each batch of reagent will influence the extent of the modification. As a starting point, use a final concentration of 1 mM 1M7 dissolved in DMSO.

1. Prepare the folding reaction (Table 15). Reactions are carried out in triplicate. For each replicate, prepare two tubes that will contain the folded RNA sample and one of the following: the reagent (1M7) for the (+) reagent sample or the solvent of the reagent (DMSO) for the (−) reagent sample. The latter sample will be used as a reference for the data analysis

Table 14 Stop mixture

Component	Final concentration	Stock	Amount (µl)
Sodium acetate (pH 5.2)	1.5 M	3 M	200
Na-EDTA (pH 8.5)	0.05 M	0.1 M	200
Glycogen	0.8 mg/ml	20 mg/ml	16

Add deionized H$_2$O to 416 µl.

Table 15 Folding reaction

Component	Final amount/conc.	(+) Reagent sample (µl)	(−) Reagent sample (µl)
RNA	20 pmol		
5 × monovalent mixture (Table 16)	1 ×	98	98
MgCl$_2$ 250 mM	25 mM	49	49

Add deionized H$_2$O to 490 µl.

because it indicates reverse transcriptase (RT) stops that are not attributable to specific modifications of the 1M7 reagent in the molecule, but rather to RNA degradation or to premature termination of reverse transcription.

2. Fold the RNA by incubating at 37 °C for 45 min. The incubation time will vary depending on the target RNA.
3. Add 1M7 or DMSO to the tubes containing the folded RNA (Table 17).
4. Incubate the reaction at 37 °C for 5 min.
5. Add 320 μl of quench mixture (Table 18) and 1 volume of isopropanol.
6. After quenching the reaction mixture, samples are incubated on ice for an hour or overnight.
7. Spin the samples at maximum speed for 30 min and wash two times with 70% ethanol.
8. Dissolve the pellet in 50 μl of RNA storage buffer (Table 19) and proceed immediately with the primer extension analysis (Section 5.5) without freezing the samples.

Table 16 Monovalent mixture (5 ×)

Component	Final concentration	Stock	Amount (ml)
KCl	$1\ M$	$2\ M$	50
HEPES (pH 7.4)	$0.25\ M$	$1\ M$	25
Na–EDTA (pH 8.5)	$0.5\ mM$	$100\ mM$	0.5

Add deionized H$_2$O to 100 ml.

Table 17 SHAPE reaction

Component	Final amount/conc.		(+) Reagent sample	(−) Reagent sample
1M7 10 mM	1 mM		54.4 μl	–
DMSO			–	54.4 μl

Table 18 Quench mixture

Component	Final concentration	Stock	Amount (μl)
Sodium acetate (pH 5.2)	$1.5\ M$	$3\ M$	200
Na–EDTA (pH 8.5)	$0.05\ M$	$0.1\ M$	200
Glycogen	0.8 mg/ml	20 mg/ml	16

Add deionized H$_2$O to 100 ml.

5.4 DMS reaction

The second chemical modifier that we use in our laboratory for probing is DMS (Sigma-Aldrich, ≥99.8% purity). The stock of DMS is prepared by dissolving 5 μl of DMS in 645 μl of cold 100% ethanol (81.3 mM). The final concentration of DMS in reaction has to be titrated for each target RNA molecule, usually ranging from 1 to 15 mM. Here, we describe a DMS probing reaction optimized for HOTAIR lncRNA.

1. The folding reaction takes place using the same volumes as the SHAPE reaction. However, HEPES buffer reacts with DMS and so we use cacodylate buffer in the 5 × monovalent ion mixture (Table 20).
2. Prepare a DMS reaction (Table 21).
3. Incubate the reaction at room temperature for 10 min.
4. Stop the reaction by adding 54.4 μl of quench mixture (Table 22).

Table 19 RNA storage buffer

Component	Final concentration (mM)	Stock (M)	Amount (μl)
MOPS (pH 6.5)	10	1	100
Na-EDTA (pH 8.5)	0.1	0.5	2

Add deionized H$_2$O to 10 ml.

Table 20 Monovalent ion mixture (5 ×)

Component	Final concentration	Stock	Amount (ml)
KCl	1 M	2 M	50
Cacodylic buffer (pH 7.0)	0.125 M	0.25 M	25
Na-EDTA (pH 8.5)	0.5 mM	100 mM	0.5

Add deionized H$_2$O to 100 ml.

Table 21 DMS reaction

Component	Final amount/conc.		(+) Reagent sample	(−) Reagent sample
DMS 81.3 mM	8.13 mM		54.4 μl	−
Ethanol			−	54.4 μl

Table 22 Quench mixture

Component	Final concentration	Stock	Amount
2-Mercaptoethanol	715 mM	14.3 M	50 μl

Add deionized H$_2$O to 1 ml.

Table 23 Precipitation reaction

Component	Final concentration	Stock	Amount (µl)
Sodium acetate (pH 5.2)	50.6 mM	3 M	20
Na-EDTA (pH 8.5)	8.4 mM	0.5 M	20
Glycogen	8.4 µg/µl	20 mg/ml	0.5
Isopropanol			600

5. Precipitate the RNA (Table 23).
6. Spin at maximum speed for 30 min and wash two times with 70% ethanol.
7. Resuspend the pellet in 50 µl of RNA storage buffer (Table 19).
8. Store at −80 °C or use in the primer extension reaction.

5.5 Primer extension reaction

The primer extension reaction is performed using the Superscript III RT (Life Technologies) to detect the specific stops of the enzyme due to the modifications in the RNA. The primers used in the reaction are labeled with fluorescent dyes. For experiments that will be analyzed with ShapeFinder, two different dyes, JOE and FAM, are used for the (+) and (−) reagent samples, respectively. For experiments that will be analyzed with QuShape, we use the same labeled primer for the reagent and the blank sample, in this case JOE because the samples will be loaded in different wells (see Section 5.8.). Here, we show an example where the (+) and the (−) reagent samples will be reverse transcribed using primers labeled with JOE and FAM.

1. Prepare an annealing reaction in amber Eppendorf tubes, protected from light (Table 24).
2. Heat the mixture at 95 °C for 2 min and then place the tubes on ice for 5 min. Incubate at 48 °C for 2 min.
3. Add 8 µl of RT mixture (Table 25) and incubate at 48 °C for 45 min.
4. Stop the reaction by adding 30 µl of the stop mixture (Table 26).
5. Precipitate by adding 1 volume (50 µl) of isopropanol.
6. At this moment, the reagent and the blank samples that have been reverse transcribed with different fluorophore-labeled primers are combined together (only for the data analysis in ShapeFinder). If the reagent and

Table 24 Annealing reaction

Component	Final amount/conc.	(+) Reagent sample (μl)	(−) Reagent sample (μl)
RNA	1 pmol	x	x
2 μM labeled primer (JOE)	0.1 μM	1	−
2 μM labeled primer (FAM)	0.1 μM	−	1
2 mM Na-EDTA (pH 8.5)	0.1 mM	1	1
5 M betaine	1 M	4	4

Add deionized H$_2$O to 12 μl.

Table 25 Reverse transcriptase (RT) mixture

Component	Final concentration	Stock	Amount (μl)
RT buffer	2.5×	5×	96
DTT	12.6 mM	100 mM	24
Betaine	944 mM	5 M	36
dNTP mix	1.26 mM	10 mM	24
Superscript III RT	10 U/μl	200 U/μl	9.6
Deionized H2O			1

Table 26 Stop mixture

Component	Final concentration	Stock	Amount (μl)
Sodium acetate (pH 5.2)	1.3 M	3 M	800
Na-EDTA (pH 8.5)	11 mM	100 mM	200
Glycogen	110 μg/μl	20 mg/ml	10
Deionized H$_2$O			800

the blank samples are reverse transcribed with the same fluorophore-labeled primer (for the data analysis in QuShape), they are not combined.
7. Spin at maximum speed for 30 min and wash two times with ethanol 70%.
8. Resuspend the pellet in 2 μl of deionized water and 48 μl of deionized formamide. Dissolve pellet by heating at 65 °C for 2 min.

5.6 Reactions for mobility shift correction

A mobility shift correction must be performed manually during data analysis with ShapeFinder for each primer set because the dyes alter the electrophoretic migration rates, and therefore cDNAs of the same length may have slightly different elution times. To generate a file for mobility shift correction, we prepare sequencing ladders as described in Section 5.2. The sequencing ladder reactions for mobility shift correction are performed with all the different labeled primers on the same DNA template and with the same type of ddNTP. Once primer extension reactions are complete, all the reactions with the different fluorophores are mixed and precipitated together, following the protocol described in Section 5.2. The pellet obtained is resuspended in 2 μl water and 38 μl formamide and dissolved at 65 °C in a heat block for 2 min. Note that if you change capillary electrophoresis conditions, a new mobility shift correction will be necessary. To avoid it, we recommend keeping the settings of the capillary instruments fixed.

5.7 Spectral calibration of the instrument

Before the first run on the instrument, a spectral calibration should be carried out with the dye set. A spectral calibration creates a matrix that corrects for the overlapping fluorescence emission spectra of the dyes used for coupling the primers. We calibrate an Applied Biosystems 3730xl DNA Genetic Analyzer for using it with the Applied Biosystems four-dye chemistry DS-20 Dye Set (5-FAM, JOE, TAMRA, ROX) (A filter set). The calibration standard is prepared as previously described (Watts et al., 2009). Briefly, four primer extension reactions are performed from primers labeled with the chosen dye set and located in different regions of a linearized control plasmid, so that four fluorescent cDNA fragments of different length are generated. The resultant primer extensions are then mixed together to obtain similar fluorescence intensity for all four dyes and loaded in the plate:

1. Linearize a pUC18 plasmid with restriction enzyme *Hind*III.
2. Label four primers complementary to pUC18 plasmid (Table 27) with the set of dyes used to calibrate the instrument, following the instructions in Section 5.1.
3. Prepare a PCR reaction for each labeled primer (Table 28).
4. Place the tubes in a thermocycler and run the spectral-calibration PCR program (Table 29).
5. Stop the primer extension reaction by adding 8 μl of stop mixture per reaction (Table 30).

Table 27 Primers used for spectral calibration

Primer[a]	Dye	Fragment (nt)
CAGAGCAGATTGTACTGAGAG	5-FAM	242
GTGTGAAATACCGCACAGAT	6-JOE	206
GCGTAAGGAGAAAATACCGCATC	6-TAMRA	188
CGCCATTCGCCATTCAGGCTGCGCAACTG	5-ROX	155

[a]Watts et al. (2009).

Table 28 Spectral-calibration PCR reaction

Component	Final concentration	Stock	Amount (μl)
Linearized pUC18 plasmid	0.1 ng/μl	0.2 ng/μl	0.5
Thermo sequenase buffer	$1\times$	$10\times$	4
dNTP mixture	75 μM	0.75 mM	4
Labeled primer	0.0125 pmol/μl	2 μM	5
Thermo sequenase	0.16 U/μl	4 U/μl	1.6

Add deionized H$_2$O to 40 μl.

Table 29 Spectral-calibration PCR program

Number of cycles	Temperature (°C)	Time
1	96	2 min
25	96 55 72	20 s 20 s 1 min
1	72	10 min

Table 30 Stop mixture

Component	Final concentration	Stock	Amount (μl)
Sodium acetate (pH 5.2)	1.5 M	3 M	100
Na–EDTA (pH 8.5)	0.25 mM	0.5 M	100
Glycogen	10 mg/ml	20 mg/ml	10

6. Add 120 μl of 100% ethanol per primer extension reaction.
7. Spin at maximum speed for 30 min and wash with 70% ethanol.
8. Dissolve each pellet in 20 μl of deionized formamide and mix the four reactions together.

5.8 Preparation of samples for capillary electrophoresis

Samples are mixed with deionized formamide and loaded in a 96-well plate according to the following guidelines, which can be modified to adjust the output signal:

1. Plates prepared for data analysis with ShapeFinder contain the (+) and the (−) reagent sample mix, together with the sequencing ladder mix, in a single well for every primer extension reaction that is carried out (Table 31).
2. Plates prepared for data analysis on QuShape contain the (+) reagent sample and a single sequencing ladder in one well whereas the (−) reagent sample and the same sequencing ladder are loaded in a different well. The advantage of this setup is that it allows the use of only two dyes (Table 32).
3. Mobility shift reactions are loaded in the same plate (Table 33).
4. Capillary electrophoresis is performed using an Applied Biosystems 3730xl DNA Genetic Analyzer with 96 capillaries equipped with a 50-cm capillary array (DNA Analysis Facility at Science Hill, Yale University). Runs were performed while maintaining an oven

Table 31 Capillary electrophoresis sample preparation (ShapeFinder)

Component	Amount (μl)
Reagent/blank mix	7
Sequencing ladder mix	2
Deionized formamide	12

Table 32 Capillary electrophoresis sample preparation (QuShape)

Component	Amount (μl)
Reagent or blank sample	7
Sequencing ladder	2
Deionized formamide	12

Table 33 Mobility shift sample preparation

Component	Amount (μl)
Mobility shift reaction	5
Deionized formamide	15

temperature of 63 °C and a run voltage of 15 kV. The instrument was prerun for 180 s before injecting the sample at 1.2 kV for 16 s. Note that very high signal should be avoided as signal cross-talk can occur despite proper spectral precalibration of the instrument.

5.9 Data analysis

The analysis of chemical probing data consists of the following steps: (i) processing of the capillary electrophoresis electropherograms to determine the absolute reactivities for each reaction, (ii) normalization of the reactivities, and (iii) use of the reactivity profiles obtained as constraints in a software program to deduce a secondary structure of the molecule (Fig. 3).

5.9.1 Determination of chemical probing reactivity profiles

The determination of chemical probing reactivity profiles can be carried out using either ShapeFinder (Vasa et al., 2008) or QuShape (Karabiber et al., 2013) software. We have used both of them successfully in our laboratory. The first one has been more extensively used, although it is no longer available online for download. The second one has been released more recently; it offers enhanced flexibility for utilizing multiple different fluorophores and it does not require import of mobility shift correction files. Regardless of the software used, the output containing the absolute nucleotide reactivity of the RNA sequence is presented as a file containing the calculated peak areas as a function of nucleotide position for the entire analyzed trace.

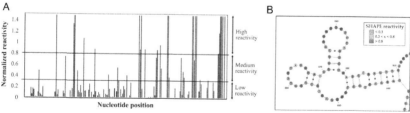

Figure 3 Determination of the secondary structure of lncRNAs by chemical probing. (A) After quantifying the absolute reactivity at each nucleotide position, SHAPE and DMS data is normalized on a scale spanning 0 to 2, approximately. Values lower than 0.3 represent constrained positions, which are part of double-stranded regions of tertiary interactions; values higher than 0.8 represent single-stranded regions; and values between 0.3 and 0.8 are likely single stranded. (B) Normalized reactivity values are used as constraints for RNA secondary structure prediction in RNAStructure software. In the figure, resulting secondary structure values are represented using VARNA software.

5.9.2 Normalization of SHAPE and DMS reactivity profiles

We normalize SHAPE and DMS data to a scale in which values ranging from 0 to 0.3 are considered not reactive; from 0.3 to 0.8, moderately reactive; and higher than 0.8, very reactive (Fig. 3A). It is important to note that the DMS reagent has different affinities for adenosine and cytosine residues and so the normalization of their reactivity profiles has to be done separately. The method described below has been modified from that previously reported by McGinnis, Duncan, and Weeks (2009):

1. Open the tab-delimited text file generated after alignment and integration in Microsoft Excel.
2. Calculate the third quartile value for the RX.area-BG.area column using the Excel function "=QUARTILE(array,quart)," where *array* is the cell range of numeric values for which you want the quartile value and *quart* is a number (3 represents the third quartile).
3. Copy the RX.area-BG.area column to a new spreadsheet and order the values in increasing order. Copy to a new column all values that fall in the range of ±10% of the third quartile value.
4. Calculate the average "=AVERAGE(number1, [number2]...)" of this new range of values.
5. Calculate and store the effective maximum reactivity value using the formula "=1.5*(AVERAGE(number1, [number2]...))," where the range is represented by the values calculated in the point 4.
6. Divide the unsorted (RX.area-BG.area) of each nucleotide by the effective maximum reactivity value to obtain the normalized reactivity values. Values greater than this value are considered outliers and are excluded.
7. Create a new text file comprising two columns: the nucleotide numbers at the left and the normalized reactivity values at the right.
8. Combine all the text files with the normalized reactivity values from all the fragments corresponding to the same RNA.

In our laboratory, we have developed a modification of this method when performing side-by-side comparisons of a wild-type lncRNA and several mutants derived of the same lncRNA. In those cases, we have observed several consistent RT stops in all molecules, independently of the mutant, in both the (+) and (−) reagent samples. This observation is likely due to the inherent flexibility of some positions for each particular type of target RNA. In those cases, our normalization protocol excludes the (−) reagent sample and compares exclusively the (+) reagent samples, thus significantly reducing the observed error in the replicates.

5.9.3 RNA secondary structure prediction and analysis

The normalized reactivity values obtained are used as constraints for the secondary structure prediction of the RNA target using the RNAStructure software (Mathews, 2014; Reuter & Mathews, 2010). The software is available as a Web server or as a program with a graphical user interface, although we prefer the second option to avoid the sequence length restraints of using the Web server. Subsequently, RNA secondary structure coordinates can be visualized and modified using the Java applet VARNA (Darty, Denise, & Ponty, 2009). Additional editing can be also performed with Adobe Illustrator (Adobe Systems) or similar editor for vector graphics.

ACKNOWLEDGMENTS

Projects that led to develop the methods described in this chapter were supported by the National Institute of Health (RO1GM50313). A. M. P. is an Investigator and I. C. is a Postdoctoral Fellow of the Howard Hughes Medical Institute. We are thankful to all members of the Pyle lab for valuable suggestions.

REFERENCES

Athavale, S. S., Gossett, J. J., Bowman, J. C., Hud, N. V., Williams, L. D., & Harvey, S. C. (2013). In vitro secondary structure of the genomic RNA of satellite tobacco mosaic virus. *PLoS One, 8*, e54384.

Batey, R. T. (2014). Advances in methods for native expression and purification of RNA for structural studies. *Current Opinion in Structural Biology, 26*, 1–8.

Batey, R. T., & Kieft, J. S. (2007). Improved native affinity purification of RNA. *RNA, 13*, 1384–1389.

Behlke, M. A., Huang, L., Bogh, L., Rose, S., & Devor, E. J. (2005). *Fluorescence quenching by proximal G-bases*. Integrated DNA Technologies.

Behrouzi, R., Roh, J. H., Kilburn, D., Briber, R. M., & Woodson, S. A. (2012). Cooperative tertiary interaction network guides RNA folding. *Cell, 149*, 348–357.

Cole, J. L., Lary, J. W., Moody, T. P., & Laue, T. M. (2008). Analytical ultracentrifugation: Sedimentation velocity and sedimentation equilibrium. *Methods in Cell Biology, 84*, 143–179.

Dai, H., Dubin, P., & Andersson, T. (1998). Permeation of small molecules in aqueous size-exclusion chromatography vis-a-vis models for separation. *Analytical Chemistry, 70*, 1576–1580.

Darty, K., Denise, A., & Ponty, Y. (2009). VARNA: Interactive drawing and editing of the RNA secondary structure. *Bioinformatics, 25*, 1974–1975.

Derrien, T., Johnson, R., Bussotti, G., Tanzer, A., Djebali, S., Tilgner, H., et al. (2012). The GENCODE v7 catalog of human long noncoding RNAs: Analysis of their gene structure, evolution, and expression. *Genome Research, 22*, 1775–1789.

Fedorova, O., Su, L. J., & Pyle, A. M. (2002). Group II introns: Highly specific endonucleases with modular structures and diverse catalytic functions. *Methods, 28*, 323–335.

Fedorova, O., Waldsich, C., & Pyle, A. M. (2007). Group II intron folding under near-physiological conditions: Collapsing to the near-native state. *Journal of Molecular Biology, 366*, 1099–1114.

Frieda, K. L., & Block, S. M. (2012). Direct observation of cotranscriptional folding in an adenine riboswitch. *Science, 338*, 397–400.

Gutschner, T., & Diederichs, S. (2012). The hallmarks of cancer: A long non-coding RNA point of view. *RNA Biology, 9*, 703–719.

Heilman-Miller, S. L., & Woodson, S. A. (2003). Effect of transcription on folding of the tetrahymena ribozyme. *RNA, 9*, 722–733.

Huthoff, H., & Berkhout, B. (2002). Multiple secondary structure rearrangements during HIV-1 RNA dimerization. *Biochemistry, 41*, 10439–10445.

Karabiber, F., McGinnis, J. L., Favorov, O. V., & Weeks, K. M. (2013). QuShape: Rapid, accurate, and best-practices quantification of nucleic acid probing information, resolved by capillary electrophoresis. *RNA, 19*, 63–73.

Kim, I., McKenna, S. A., Viani Puglisi, E., & Puglisi, J. D. (2007). Rapid purification of RNAs using fast performance liquid chromatography (FPLC). *RNA, 13*, 289–294.

Lai, D., Proctor, J. R., & Meyer, I. M. (2013). On the importance of cotranscriptional RNA structure formation. *RNA, 19*, 1461–1473.

Lebowitz, J., Lewis, M. S., & Schuck, P. (2002). Modern analytical ultracentrifugation in protein science: A tutorial review. *Protein Science, 11*, 2067–2079.

Lomant, A. J., & Fresco, J. R. (1975). Structural and energetic consequences of non-complementary base oppositions in nucleic acid helices. *Progress in Nucleic Acid Research and Molecular Biology, 15*, 185–218.

Marcia, M., & Pyle, A. M. (2012). Visualizing group II intron catalysis through the stages of splicing. *Cell, 151*, 497–507.

Mathews, D. H. (2014). RNA secondary structure analysis using RNAstructure. *Current Protocols in Bioinformatics, 46*, 12.16.1–12.16.25.

McGinnis, J. L., Duncan, C. D., & Weeks, K. M. (2009). High-throughput SHAPE and hydroxyl radical analysis of RNA structure and ribonucleoprotein assembly. *Methods in Enzymology, 468*, 67–89.

Mitra, S. (2009). Using analytical ultracentrifugation (AUC) to measure global conformational changes accompanying equilibrium tertiary folding of RNA molecules. *Methods in Enzymology, 469*, 209–236.

Mitra, S. (2014). Detecting RNA tertiary folding by sedimentation velocity analytical ultracentrifugation. *Methods in Molecular Biology, 1086*, 265–288.

Mortimer, S. A., & Weeks, K. M. (2007). A fast-acting reagent for accurate analysis of RNA secondary and tertiary structure by SHAPE chemistry. *Journal of the American Chemical Society, 129*, 4144–4145.

Necsulea, A., Soumillon, M., Warnefors, M., Liechti, A., Daish, T., Zeller, U., et al. (2014). The evolution of lncRNA repertoires and expression patterns in tetrapods. *Nature, 505*, 635–640.

Novikova, I. V., Hennelly, S. P., & Sanbonmatsu, K. Y. (2012). Structural architecture of the human long non-coding RNA, steroid receptor RNA activator. *Nucleic Acids Research, 40*, 5034–5051.

Pan, T., & Sosnick, T. R. (1997). Intermediates and kinetic traps in the folding of a large ribozyme revealed by circular dichroism and UV absorbance spectroscopies and catalytic activity. *Nature Structural Biology, 4*, 931–938.

Rambo, R. P., & Tainer, J. A. (2010). Bridging the solution divide: Comprehensive structural analyses of dynamic RNA, DNA, and protein assemblies by small-angle X-ray scattering. *Current Opinion in Structural Biology, 20*, 128–137.

Reuter, J. S., & Mathews, D. H. (2010). RNAstructure: Software for RNA secondary structure prediction and analysis. *BMC Bioinformatics, 11*, 129.

Rice, G. M., Leonard, C. W., & Weeks, K. M. (2014). RNA secondary structure modeling at consistent high accuracy using differential SHAPE. *RNA, 20*, 846–854.

Russell, R., Zhuang, X., Babcock, H. P., Millett, I. S., Doniach, S., Chu, S., et al. (2002). Exploring the folding landscape of a structured RNA. *Proceedings of the National Academy of Sciences of the United States of America, 99*, 155–160.

Said, N., Rieder, R., Hurwitz, R., Deckert, J., Urlaub, H., & Vogel, J. (2009). In vivo expression and purification of aptamer-tagged small RNA regulators. *Nucleic Acids Research, 37*, e133.

Schenborn, E. T., & Mierendorf, R. C., Jr. (1985). A novel transcription property of SP6 and T7 RNA polymerases: Dependence on template structure. *Nucleic Acids Research, 13*, 6223–6236.

Schuck, P. (2000). Size-distribution analysis of macromolecules by sedimentation velocity ultracentrifugation and lamm equation modeling. *Biophysical Journal, 78*, 1606–1619.

Su, L. J., Brenowitz, M., & Pyle, A. M. (2003). An alternative route for the folding of large RNAs: Apparent two-state folding by a group II intron ribozyme. *Journal of Molecular Biology, 334*, 639–652.

Takamoto, K., He, Q., Morris, S., Chance, M. R., & Brenowitz, M. (2002). Monovalent cations mediate formation of native tertiary structure of the Tetrahymena thermophila ribozyme. *Nature Structural Biology, 9*, 928–933.

Tang, G. Q., Nandakumar, D., Bandwar, R. P., Lee, K. S., Roy, R., Ha, T., et al. (2014). Relaxed rotational and scrunching changes in P266L mutant of T7 RNA polymerase reduce short abortive RNAs while delaying transition into elongation. *PLoS One, 9*, e91859.

Toor, N., Keating, K. S., Taylor, S. D., & Pyle, A. M. (2008). Crystal structure of a self-spliced group II intron. *Science, 320*, 77–82.

Vasa, S. M., Guex, N., Wilkinson, K. A., Weeks, K. M., & Giddings, M. C. (2008). ShapeFinder: A software system for high-throughput quantitative analysis of nucleic acid reactivity information resolved by capillary electrophoresis. *RNA, 14*, 1979–1990.

Walstrum, S. A., & Uhlenbeck, O. C. (1990). The self-splicing RNA of Tetrahymena is trapped in a less active conformation by gel purification. *Biochemistry, 29*, 10573–10576.

Wan, Y., Mitchell, D., 3rd., & Russell, R. (2009). Catalytic activity as a probe of native RNA folding. *Methods in Enzymology, 468*, 195–218.

Watts, J. M., Dang, K. K., Gorelick, R. J., Leonard, C. W., Bess, J. W., Jr., Swanstrom, R., et al. (2009). Architecture and secondary structure of an entire HIV-1 RNA genome. *Nature, 460*, 711–716.

Woodson, S. A. (2011). RNA folding pathways and the self-assembly of ribosomes. *Accounts of Chemical Research, 44*, 1312–1319.

Zhang, J., & Ferre-D'Amare, A. R. (2013). Co-crystal structure of a T-box riboswitch stem I domain in complex with its cognate tRNA. *Nature, 500*, 363–366.

CHAPTER TWO

Characterizing RNA Excited States Using NMR Relaxation Dispersion

Yi Xue*, Dawn Kellogg[†], Isaac J. Kimsey*, Bharathwaj Sathyamoorthy*, Zachary W. Stein[‡], Mitchell McBrairty[‡,1], Hashim M. Al-Hashimi*,[†,2]

*Department of Biochemistry, Duke University School of Medicine, Durham, North Carolina, USA
[†]Department of Chemistry, Duke University, Durham, North Carolina, USA
[‡]Biophysics Enhanced Program, University of Michigan Ann Arbor, Michigan, USA
[2]Corresponding author: e-mail address: hashim.al.hashimi@duke.edu

Contents

1. Introduction 40
2. NMR Relaxation Dispersion 43
   2.1 Chemical Exchange 43
   2.2 RD Experiments 46
   2.3 $R_{1\rho}$ with Low-to-High SL Fields 48
3. General Protocol for Characterizing RNA ESs Using Low-to-High SL Field ^{13}C and ^{15}N Off-Resonance NMR $R_{1\rho}$ and Uniformly Labeled Nucleic Acid Samples 50
   3.1 Construct Design 50
   3.2 Sample Preparation and Purification 51
   3.3 Protocol for Measuring $R_{1\rho}$ RD in Uniformly Labeled Nucleic Acids 51
   3.4 Data Analysis 56
   3.5 Inferring Structures of RNA ESs Using NMR Chemical Shifts and Secondary Structure Prediction 62
   3.6 Testing Model RNA ESs 65
Acknowledgments 68
References 69

Abstract

Changes in RNA secondary structure play fundamental roles in the cellular functions of a growing number of noncoding RNAs. This chapter describes NMR-based approaches for characterizing microsecond-to-millisecond changes in RNA secondary structure that are directed toward short-lived and low-populated species often referred to as "excited states." Compared to larger scale changes in RNA secondary structure, transitions toward excited states do not require assistance from chaperones, are often orders of magnitude faster, and are localized to a small number of nearby base pairs in and

---

[1] Present address: Nymirum, 3510 West Liberty Road, Ann Arbor, Michigan 48103, USA.

around noncanonical motifs. Here, we describe a procedure for characterizing RNA excited states using off-resonance $R_{1\rho}$ NMR relaxation dispersion utilizing low-to-high spin-lock fields (25–3000 Hz). $R_{1\rho}$ NMR relaxation dispersion experiments are used to measure carbon and nitrogen chemical shifts in base and sugar moieties of the excited state. The chemical shift data are then interpreted with the aid of secondary structure prediction to infer potential excited states that feature alternative secondary structures. Candidate structures are then tested by using mutations, single-atom substitutions, or by changing physiochemical conditions, such as pH and temperature, to either stabilize or destabilize the candidate excited state. The resulting chemical shifts of the mutants or under different physiochemical conditions are then compared to those of the ground and excited states. Application is illustrated with a focus on the transactivation response element from the human immune deficiency virus type 1, which exists in dynamic equilibrium with at least two distinct excited states.

1. INTRODUCTION

Many regulatory RNAs (Cech & Steitz, 2014; Serganov & Patel, 2007; Storz, 2002) undergo conformational changes that play essential roles in their biological functions (Cruz & Westhof, 2009; Dethoff, Chugh, Mustoe, & Al-Hashimi, 2012; Mandal & Breaker, 2004; Rinnenthal et al., 2011; Schwalbe, Buck, Furtig, Noeske, & Wohnert, 2007; Tucker & Breaker, 2005). Among many functionally important motional modes, changes in RNA secondary structure can expose or sequester key regulatory elements, and thereby provide the basis for molecular switches that regulate and control a wide range of biochemical processes (Breaker, 2011; Dethoff, Chugh, et al., 2012; Serganov & Patel, 2007).

For example, riboswitches (Grundy, Winkler, & Henkin, 2002; Winkler, Nahvi, & Breaker, 2002) are RNA-based genetic elements typically embedded in the 5′ untranslated region (5′ UTR) of bacterial genes that employ changes in secondary structure in order to regulate expression of metabolic genes in response to changes in cellular metabolite concentration (Schwalbe et al., 2007; Serganov & Patel, 2007). In a prototypical metabolite riboswitch (Tucker & Breaker, 2005), a ligand metabolite, such as adenine, binds the aptamer domain and induces a conformational change, which typically sequesters an RNA element into a helix (Fig. 1A). The unavailability of this element in turn changes the folding pathway of a downstream decision-making expression platform, directing it toward structures that turn off (and in some cases, on) gene expression, either by forming a transcription terminating helix or by sequestering the Shine-Dalgarno sequence where

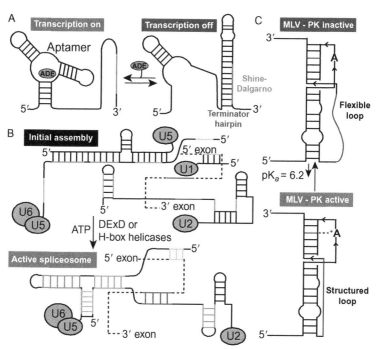

Figure 1 Secondary structural changes in RNA. (A) Adenine riboswitch secondary structure with and without the adenine metabolite, which binds to the aptamer domain stabilizing a specific secondary structure turning on transcription. In the absence of metabolite, the Shine-Dalgarno sequence (blue, gray in the print version) is sequestered turning off transcription. (B) Secondary structural transitions of mRNA (dashed line) and a spliceosome catalyzed by chaperones (DExD or H-box helicases) that are keys for splicing. (C) pH-dependent secondary structural equilibrium in the murine leukaemia virus pseudoknot (MLV-PK) mRNA that regulates translational miscoding.

the ribosome binds, thereby inhibiting translation (Fig. 1A) (Haller, Souliere, & Micura, 2011). Changes in RNA secondary structure can also affect access to RNA sites and thereby allow regulation of post-transcriptional processing, including splicing (Fig. 1B) (Cheah, Wachter, Sudarsan, & Breaker, 2007), gene silencing by microRNA (miRNA) (Kedde et al., 2010), and RNA editing (Polson, Bass, & Casey, 1996). RNA secondary structural switches are also widely used by the RNA genomes of retroviruses to transition between different roles required by various steps of the viral lifecycle (D'Souza & Summers, 2004; Huthoff & Berkhout, 2001). For example, a pH-dependent secondary structural RNA equilibrium was recently shown to regulate translational recoding in the murine leukemia virus (Fig. 1C) (Houck-Loomis et al., 2011).

Most RNA secondary structural switches require the melting of several base pairs and result in the creation and disruption of entire helices and hairpins. They can therefore be energetically costly and often require assistance from protein chaperones (Herschlag, 1995; Rajkowitsch et al., 2007) or otherwise must take place during co-transcriptional folding (Garst & Batey, 2009; Lai, Proctor, & Meyer, 2013). Recently, a new mode of RNA secondary structural transitions has been uncovered through application of solution state NMR spectroscopy (Dethoff, Petzold, Chugh, Casiano-Negroni, & Al-Hashimi, 2012; Hoogstraten, Wank, & Pardi, 2000; Lee, Dethoff, & Al-Hashimi, 2014; Venditti, Clos, Niccolai, & Butcher, 2009). These transitions are directed toward low-populated (typically <5%) and short-lived (lifetime typically <2 ms) alternative secondary structures that feature local reshuffling of base pairs (bps) in and around noncanonical motifs (Fig. 2) (Dethoff, Petzold, et al., 2012; Lee et al., 2014). These transient species are often referred to as "excited states" (ESs) (Sekhar & Kay, 2013), because they represent a higher free energy state (typically, destabilized by 1–3 kcal/mol) as compared to the energetically more favorable ground state (GS).

An example of such RNA transitions is shown in Fig. 2 for the transactivation response (TAR) element RNA from the human immune deficiency virus type 1 (HIV-1) (Aboul-ela, Karn, & Varani, 1996; Puglisi, Tan, Calnan, Frankel, & Williamson, 1992; Weeks, Ampe, Schultz,

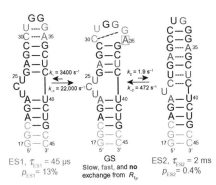

Figure 2 HIV-1 TAR exists in dynamic equilibrium with two excited states (ES1 and ES2). Shown are the ES secondary structure, population (p_{ES}), lifetime (τ), as well as the forward (k_1/k_2) and reverse (k_{-1}/k_{-2}) rate constants. The GS is labeled to show sites with slow (red, light gray in the print version), fast (green, gray in the print version), and no chemical exchange (black) measured by $R_{1\rho}$ relaxation dispersion. Note that A35 shows a combination of fast (C1′) and slow (C8) exchange. Five sites in gray which are located at the bottom were not measured due to spectral overlap.

Steitz, & Crothers, 1990). TAR has been shown to exist in dynamic equilibrium with two distinct excited states (ES1 and ES2) (Dethoff, Petzold, et al., 2012; Lee et al., 2014). ES1 is 13% populated, has a short lifetime of ~45 μs, and features localized changes in base pairing within an apical loop (Fig. 2). ES1 sequesters apical loop residues by base pairing interactions that would otherwise be available to interact with proteins (Fig. 2) (Dethoff, Petzold, et al., 2012). ES2 is only ~0.4% populated, but has a longer lifetime of ~2 ms (Fig. 2) (Lee et al., 2014). It features longer range reshuffling of base pairs that span the bulge and apical loop that are separated by a four bp helix, remodeling all noncanonical features of the TAR structure (Fig. 2) (Lee et al., 2014).

Recent studies suggest that transitions toward ESs are common in many regulatory RNAs (Dethoff, Petzold, et al., 2012; Hoogstraten et al., 2000; Lee et al., 2014; Tian, Cordero, Kladwang, & Das, 2014; Venditti et al., 2009; Zhang, Kang, Peterson, & Feigon, 2011; Zhao, Hansen, & Zhang, 2014). Compared to conventional secondary structural rearrangements, transitions toward RNA ESs are orders of magnitude faster, occur without assistance from external cofactors, and result in smaller, yet significant, changes in RNA secondary structure, which can include changes in base protonation state (Lee et al., 2014). They can therefore meet unique demands for RNA biological functions (Lee et al., 2014). In addition to playing new roles in RNA-based regulation, the unique features of RNA ESs make them potentially attractive targets for the development of RNA-targeting therapeutics (Lee et al., 2014). Here, we review NMR methods that can be used to characterize structural and energetic properties (population and lifetime) of RNA ESs.

2. NMR RELAXATION DISPERSION
2.1 Chemical Exchange

NMR relaxation dispersion (RD) experiments, which probe line-broadening contributions to NMR resonances arising from chemical exchange, can be used to characterize ESs of biomolecules. Several excellent reviews (Bothe, Nikolova, et al., 2011; Korzhnev & Kay, 2008; Palmer, 2014; Palmer, Kroenke, & Loria, 2001; Palmer & Massi, 2006; Sekhar & Kay, 2013) describe the theoretical underpinnings of RD NMR techniques. They are briefly reviewed here to make the basic concepts accessible to the general reader. Readers are referred to the above reviews for a more rigorous treatment of these experiments.

The key NMR interaction underlying the RD experiments is the chemical shift. Nuclei behave as tiny magnets and due to quantization of the nuclear spin angular momentum, they align parallel (α state) or antiparallel (β state) relative to the static NMR magnetic field (B_0). Since the parallel alignment is more energetically favorable (for nuclei with a positive gyromagnetic ratio), a net bulk magnetization over an ensemble of spins build up parallel to the magnetic field. In a basic 1D NMR experiment, radiofrequency (RF) pulses are used to realign this bulk magnetization along a direction perpendicular to the B_0 field. The bulk magnetization then precesses about the B_0 field at a characteristic resonance Larmor frequency and gives rise to a detectable oscillating magnetic field. This time-domain signal is then Fourier transformed to yield the standard frequency-domain NMR spectrum, in which unique signals at characteristic frequencies are observed for various spins for a given nuclei. These frequencies are referenced against a standard frequency (e.g., tetramethylsilane for ^{1}H) at that given B_0 field and this field-independent parameter is called the "chemical shift."

The chemical shift is directly proportional to the energy gap between the α and β states, which in turn is proportional to the magnetic field strength experienced by the nucleus. Because electronic clouds surrounding nuclei "shield" or "deshield" the nucleus from the external magnetic field by variable amounts that are highly dependent on the specific electronic environment, a wide range of chemical shifts are typically observed for different sites in a molecule. This makes the NMR chemical shifts exquisitely sensitive to structure (e.g., torsion angles, sugar pucker, etc.), protonation state, and interactions (hydrogen bonding, electrostatic interactions, etc.). For nucleic acid applications, one is typically interested in the NMR active nuclei ^{1}H, ^{13}C, ^{15}N, ^{2}H, and ^{31}P, with ^{13}C, ^{15}N, and ^{2}H introduced during synthesis, typically by using labeled nucleotide triphosphates (NTPs) in *in vitro* transcription reactions.

To understand how exchange between a GS and ES gives rise to line-broadening of the NMR resonance, consider an RNA molecule exchanging between two states: a major GS, in which a base is flipped in, and a minor ES, in which the base is flipped out (Fig. 3A). Nuclei belonging to this base (along with its immediate neighbors) will experience different electronic environments before and after the flip, therefore will be associated with distinct NMR chemical shifts, ω_{GS} and ω_{ES} (Fig. 3B). In the absence of exchange between the GS and ES, two NMR peaks are observed centered around ω_{GS} and ω_{ES} with integrated volumes reflecting the relative populations of the GS and ES (Fig. 3B). However, when the GS and ES

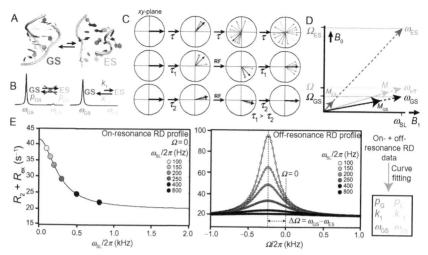

Figure 3 Characterizing chemical exchange using NMR relaxation dispersion. (A) Example of an equilibrium between a GS and ES. (B) Chemical exchange between GS and ES leads to broadening of resonances and disappearance of minor ES signal. (C) Bulk magnetization aligned along the Y-axis of the lab frame, followed by magnetization dephasing due to transverse relaxation. The middle and lower panel show the dephasing of bulk magnetization due to chemical exchange suppressed by application of RF fields for long (τ_1) and short (τ_2) delays in the CPMG experiment. (D) Resonance offset (Ω), effective field (ω), and magnetization (M) vectors for the GS and ES in $R_{1\rho}$ experiment (denoted by the subscripts GS and ES, respectively). ω_{eff}, Ω, and M represent the effective field, average offset and magnetization, respectively. (E) Simulated examples of on-resonance (left) and off-resonance (right) RD profiles showing the dependence $R_{1\rho}$ on spin-lock power ω_{SL} and offset Ω. (See the color plate.)

exchange at rates ($k_{ex} = k_1 + k_{-1}$) comparable to their NMR frequency difference ($\Delta\omega = \omega_{ES} - \omega_{GS}$), the chemical shift of a given nucleus fluctuates back and forth between ω_{GS} and ω_{ES}. Because the fluctuation is stochastic, nuclei in different molecules spend varying amounts of time in the GS and ES, and therefore spend differential amounts of times with ω_{GS} and ω_{ES} frequencies. This results in the broadening of the NMR signal, reflecting a wide range of apparent frequencies due to variable admixing of ω_{GS} and ω_{ES} (Fig. 3B). Because of its lower starting population and signal intensity, this broadening often puts the minor ES outside NMR detection limits (Fig. 3B).

It is instructive to conceptualize chemical exchange in terms of magnetization precession (Fig. 3C). Here, nuclei in the sample are associated with a bulk "magnetization" vector that is initially aligned along an axis in the laboratory frame by a "preparation" pulse. The individual magnetization vectors from nuclei in different molecules are initially coaligned along the Y- (or X-) axis via phase coherence of the nuclear spins. The magnetization vector then precesses about the B_0 field (Z-axis) with the frequency ω (Fig. 3C). In the absence of chemical exchange, two species precess at their respective chemical shift frequencies, ω_{GS} and ω_{ES}. However, in the presence of exchange, the precessing frequency of a given magnetization vector varies stochastically between ω_{GS} and ω_{ES}. Since nuclei from different molecules spend varying amounts of time precessing with frequencies ω_{GS} and ω_{ES}, they no longer precess in synchrony. Rather, they "fan out" and cause phase decoherence or "dephasing" (Fig. 3C). This dephasing leads to an additional exchange contribution referred to as R_{ex} to the observed transverse relaxation rate $R_{2,obs} = R_2 + R_{ex}$ describing the rate at which the Y (or X) component of the magnetization vector decays over time. The NMR line-width is directly proportional to $R_2 + R_{ex}$, and exchange manifests as a line-broadening contribution.

2.2 RD Experiments

RD NMR experiments can be used to characterize properties of the often invisible ES by measuring the exchange broadening contribution to the visible GS resonance (reviewed in Bothe, Nikolova, et al., 2011; Korzhnev & Kay, 2008; Palmer, 2014; Palmer et al., 2001; Palmer & Massi, 2006). In particular, experiments are used to measure the chemical exchange contribution to line-broadening (R_{ex}) during a relaxation period (T_{relax}) during which the sample is subjected to RF irradiation. Various types of RD experiments differ in the initial state of the magnetization and the type of RF irradiation that is used.

In the Carr–Purcell–Meiboom–Gill (CPMG) experiment (Carr & Purcell, 1954; Loria, Rance, & Palmer, 1999a; Meiboom & Gill, 1958; Mulder, Skrynnikov, Hon, Dahlquist, & Kay, 2001), the initial state generally consists of transverse magnetization (e.g., aligned along the Y-axis) and the RF irradiation consists of a series of equally spaced high-power 180° refocusing pulses. In the case of the spin relaxation in the rotating frame experiment, or $R_{1\rho}$ (Akke & Palmer, 1996; Deverell, Morgan, & Strange, 1970; Korzhnev, Skrynnikov, Millet, Torchia, & Kay, 2002; Mulder, de Graaf, Kaptein, & Boelens, 1998), the initial magnetization is typically aligned along an effective field direction (which is defined by both the RF irradiation power and offset) and the irradiation consists of a weaker, but continuous RF with a specified power level (ω_{SL}) and frequency offset (Ω) (Fig. 3D).

The RF field employed in this case is called a "spin-lock" (SL), as it locks the magnetization along its effective field. In both the CPMG and $R_{1\rho}$ experiments, the RF irradiation perturbs the precession of magnetization so as to diminish the efficiency with which chemical exchange results in dephasing of the magnetization, and therefore exchange broadening. For example, in the CPMG experiment, the series of 180° pulses effectively "invert" the precession of magnetization at a constant time interval (τ_{CPMG}); in this manner, some of the dephasing occurring prior to the 180° pulse is refocused in the period following the pulse, with the degree of refocusing increasing with shorter τ_{CPMG} delays (Fig. 3C). In the case of the $R_{1\rho}$ experiment, the two effective field directions associated with the GS and ES are brought into closer alignment by application of a continuous RF field (Fig. 3D), thereby decreasing the extent of dephasing arising due to precession around the GS and ES effective fields. The dependence of the exchange broadening contribution (R_{ex}) on τ_{CPMG} in the case of CPMG and ω_{SL}, as well as Ω in the case of $R_{1\rho}$ (Fig. 3E), can be used to extract exchange parameters of interest. For slow ($k_{ex} << |\Delta\omega|$) to intermediate ($k_{ex} \sim |\Delta\omega|$) exchange, these experiments can be used to determine the population (p_{ES}), lifetime ($\tau_{ES} = 1/[(1-p_{ES})k_{ex}]$), and chemical shift of the ES (ω_{ES}). For fast exchange ($k_{ex} >> |\Delta\omega|$), it becomes more difficult to reliably determine exchange parameters though one can typically accurately measure the factor $\Phi = p_{GS}p_{ES}\Delta\omega^2$, where additional experiments are needed to resolve the ES chemical shift and population (Bothe, Stein, & Al-Hashimi, 2014). It is important to note that the chemical shift of the ES carries the desired structural information.

CPMG RD experiments can be used to characterize processes with exchange rates ($k_{ex} = k_1 + k_{-1}$) in the range of $\sim 10\ s^{-1} < k_{ex} < \sim 6000\ s^{-1}$ (Korzhnev & Kay, 2008; Palmer et al., 2001). CPMG data are typically measured at multiple magnetic field strengths and combined with additional experiments (e.g., HSQC/HMQC) in order to determine the sign of the ES chemical shift (Skrynnikov, Dahlquist, & Kay, 2002). Although widely used in studies of proteins, the CPMG experiment proves difficult to apply to nucleic acids. This is due to a paucity of ideal imino nitrogens, and because ^{13}C experiments are complicated by extensive C–C interactions that are difficult to suppress due to challenges in achieving selective carbon excitation with hard pulses (Johnson & Hoogstraten, 2008; Yamazaki, Muhandiram, & Kay, 1994). The exchange timescale accessible to $R_{1\rho}$ is broader than CPMG ($\sim 60\ s^{-1} < k_{ex} < \sim 100,000\ s^{-1}$) (Palmer & Massi, 2006) and for slow-intermediate exchange, the sign of ES chemical shift sign can be deduced at a single magnetic field strength (Trott & Palmer, 2002). For processes occurring at even slower timescales (~ 20 s^{-1} < k_{ex} < ~ 300 s^{-1}) chemical-exchange saturation transfer experiments employing weak RF SL fields have recently been shown to be a robust approach to characterize lowly populated conformational states in both proteins and nucleic acids (Fawzi, Ying, Ghirlando, Torchia, & Clore, 2011; Long, Bouvignies, & Kay, 2014; Vallurupalli, Bouvignies, & Kay, 2012; Zhao et al., 2014).

2.3 $R_{1\rho}$ with Low-to-High SL Fields

The application of low ω_{SL} is required in order to characterize ESs with exchange rates slower than k_{ex} <~ 2000 s^{-1}. However, the use of lower ω_{SL} (<1000 Hz) in the conventional 2D $R_{1\rho}$ experiment is complicated due to the fact that one has many spins with a broad range of chemical shift frequencies, each with a distinct effective field (Fig. 3D) that have to be aligned along their individual effective fields during the preparation phase of the experiment. For relatively high ω_{SL} (>1000 Hz), there is a limited range of effective field orientations for the various spins, and it is possible to align the magnetization efficiently using adiabatic ramps, which are generally 3–4 ms long (Hansen & Al-Hashimi, 2007; Igumenova & Palmer, 2006; Kim & Baum, 2004; Massi, Johnson, Wang, Rance, & Palmer, 2004; Mulder et al., 1998; Zinn-Justin, Berthault, Guenneugues, & Desvaux, 1997). However, for low SL fields (<1000 Hz), one has a broader range of orientations, and proper alignment of the magnetization requires longer delays of the adiabatic ramps, resulting in severe sensitivity loss due to relaxation (Kim & Baum, 2004; Massi et al., 2004; Palmer & Massi, 2006). For a SL field of 275 Hz, the alignment of initial magnetization is efficient only for offset values >325 Hz (Kim & Baum, 2004).

Recent advances in the ^{15}N $R_{1\rho}$ experiment introduced by the groups of Palmer and Kay address these limitations and allow use of much lower SL fields (Korzhnev, Orekhov, & Kay, 2005; Massi et al., 2004), on the order of 25–150 Hz, thus extending sensitivity to exchange timescales on the order of tens of milliseconds comparable to those accessible by CPMG. These advances have been integrated into an NMR experiment for measuring ^{13}C (Hansen, Nikolova, Casiano-Negroni, & Al-Hashimi, 2009) and ^{15}N (Nikolova, Gottardo, & Al-Hashimi, 2012) $R_{1\rho}$ in nucleic acids[1] (Fig. 4). These experiments have so far been applied in the characterization of systems in fast to intermediate exchange in both proteins and

---

[1] The pulse programs are available upon request from Dr. Al-Hashimi.

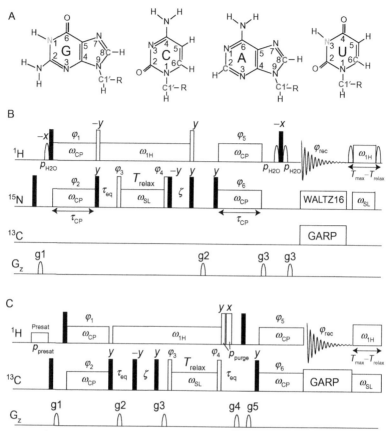

Figure 4 Pulse sequences for the measurement of ^{15}N and ^{13}C $R_{1\rho}$ in uniformly ^{13}C/^{15}N labeled nucleic acids. (A) The carbon and nitrogen nuclei that are targeted in RNA for $R_{1\rho}$ measurements are highlighted. Pulse sequences for 1D (B) ^{15}N (Nikolova et al., 2012) and (C) ^{13}C (Hansen et al., 2009) $R_{1\rho}$ experiments are as shown.

nucleic acids ($k_{ex} \sim 300 - 30,000\,\mathrm{Hz}$, $\Delta\omega \sim 20 - 800\,\mathrm{Hz}$) and will be discussed in this chapter in the context of characterizing RNA ESs.

The basic experiment uses selective Hartmann–Hahn polarization transfers developed by Bodenhausen (Ferrage, Eykyn, & Bodenhausen, 2004; Pelupessy, Chiarparin, & Bodenhausen, 1999) to excite specific spins of interest and collect data in a 1D manner. Both the GS and ES magnetization is aligned along the average effective field (ω_{eff}, Fig. 3D) through application of a hard 90° pulse (Fig. 4B and C) (Bothe et al., 2014). Upon application of the SL, magnetization in states GS and ES will evolve about their respective effective fields ω_{GS} and ω_{ES} (Fig. 3D). In cases of highly asymmetric exchange ($p_{GS} \gg p_{ES}$), the effective fields of the GS and ω_{eff} are essentially the same and it has been shown that the effective field can be considered to be on the GS (Bothe et al., 2014; Korzhnev et al., 2005).

Because only one spin is aligned along its effective field, it is possible to use significantly weaker RF fields, for a full range of 25–3500 Hz (Korzhnev et al., 2005; Massi et al., 2004). The lowest SL power attainable is predominately limited by ~ 3 times the largest homonuclear scalar coupling (J_{CC}/J_{NN}) (Zhao et al., 2014): G–N1 and U–N3 ~ 50 Hz, A/G–C8 ~ 45 Hz, C1′ ~ 150 Hz, and A–C2 ~ 40 Hz. By using a continuous low SL RF field rather than hard 180° pulses, it is also possible to suppress or eliminate unwanted C–C and N–C interactions in uniformly ^{13}C/^{15}N-labeled samples (Hansen et al., 2009). The C–H (or N–H) scalar coupling evolution and cross-correlated relaxation between C–H (or N–H) dipole–dipole and carbon chemical shift anisotropy (CSA) are efficiently suppressed by using a strong ^{1}H continuous-wave field applied on the resonance of interest (Fig. 4B and C). Consistent heating throughout a given experiment is maintained by a heat-compensation element (Hansen et al., 2009; Korzhnev et al., 2005) that applies far off-resonance ^{1}H and ^{13}C/^{15}N SLs for a given amount of time ($T_{max} - T_{relax}$), where T_{max} is the longest time delay and T_{relax} is the time delay in a given scan. Water is efficiently suppressed using either low power presaturation during interscan delay or with a WATERGATE element (Piotto, Saudek, & Sklenar, 1992) before acquisition. The same ^{15}N $R_{1\rho}$ experiment introduced by Kay and coworkers (Korzhnev et al., 2005) can be used to measure RD in imino nitrogens in DNA (Nikolova et al., 2012) and RNA (Lee et al., 2014). Unlike the ^{13}C resonances, it is generally possible to assign the imino N1/3 resonances in much larger RNA systems. This, combined with the fact that the ^{15}N chemical shift is extremely sensitive to RNA secondary structure makes imino ^{15}N $R_{1\rho}$ an ideal approach for characterizing the secondary structure of RNA ESs. These experiments have so far been applied to measure $R_{1\rho}$ data for base (N1, N3, C2, C6, and C8) and sugar (C1′) nuclei in RNA and DNA (see Fig. 4A).

3. GENERAL PROTOCOL FOR CHARACTERIZING RNA ESs USING LOW-TO-HIGH SL FIELD ^{13}C AND ^{15}N OFF-RESONANCE NMR $R_{1\rho}$ AND UNIFORMLY LABELED NUCLEIC ACID SAMPLES

3.1 Construct Design

It is common practice for transcription reactions using T7 RNA polymerase (T7 RNAP) to add two G residues at the 5′ end of the RNA in order to maximize transcription yields. RNA secondary structure prediction is used to make sure that such sequence modifications do not interfere with the RNA GS or ES.

3.2 Sample Preparation and Purification

Because the $R_{1\rho}$ experiment is sensitivity demanding, we typically require NMR samples with ≥ 1 mM concentration of $^{13}C/^{15}N$ uniformly labeled RNA. This typically yields spectra with ≥ 40 signal-to-noise ratios (S/N) at 16 scans (the minimum phase-cycling steps involved in the experiment) and with an interscan delay (d1) of ~1.4–2.0 s, allowing acquisition of ~140 $R_{1\rho}$ data points at various ω_{SL} and Ω within ~12 h. This requirement for high sensitivity can also make it challenging to record $R_{1\rho}$ data on slowly tumbling systems such as large RNAs (>70 nt) or RNA–protein complexes. TROSY-based $R_{1\rho}$ experiments can be used to improve sensitivity in such large systems (Igumenova & Palmer, 2006; Loria, Rance, & Palmer, 1999b). Samples can be prepared by *in vitro* transcription employing commercially available $^{13}C/^{15}N$-labeled NTPs, followed by polyacrylamide gel electrophoresis (PAGE) purification or other purification methods. It is important to ensure that all running buffers, reagents, glassware, and equipment used are RNAse-free. Protocols for preparation of uniformly $^{13}C/^{15}N$-labeled RNA samples are described elsewhere (Alvarado et al., 2014; Easton, Shibata, & Lukavsky, 2010; Petrov, Wu, Puglisi, & Puglisi, 2013).

3.3 Protocol for Measuring $R_{1\rho}$ RD in Uniformly Labeled Nucleic Acids

The RNA nuclei that are typically targeted for $R_{1\rho}$ measurements by the methods described here are depicted in Fig. 4A. Well-established NMR approaches (reviewed in Furtig, Richter, Wohnert, & Schwalbe, 2003) are initially used to assign resonances in the RNA.

3.3.1 Calibrating SL Power

When setting up the $R_{1\rho}$ experiment on a new spectrometer or newly installed probe/console, it is critical to calibrate ^{13}C and ^{15}N SL powers (Palmer et al., 2001). In general, SL power calibration is not necessary when switching samples or changing temperature.

Prepare a modified version of $R_{1\rho}$ pulse sequence by removing the pulses with phase φ_3 and φ_4 flanking the SL element (Fig. 4B and C). This can be done by setting their duration to zero or by omitting the pulses from the pulse program. The following protocol can then be used to calibrate ^{15}N or ^{13}C SL powers.

1. Select a sharp and well-resolved resonance to perform calibration. This resonance should exhibit minimal to no chemical exchange to avoid complication from exchange processes.
2. Compile a table containing SL powers ω_{SL} (in Hz) to be calibrated and associated power levels P (in dB). The power levels can be estimated based on the calibrated ^{15}N or ^{13}C rectangular 90° pulse values.
3. Record 1D experiments for an array of SL durations, T_{relax} (Fig. 4B and C). Typically, 33 durations are used in increments of $1/(4\omega_{SL})$ from 0 to $8/\omega_{SL}$ (Hansen et al., 2009). For example, a 100 Hz SL power would be incremented 33 times from 0 to 80 ms in increments of 2.5 ms.
4. Measure the resultant peak intensity. It should be modulated by $\exp(i\omega_{SL,real}T_{relax})$, where $\omega_{SL,real}$ is the real ω_{SL}.
5. Calculate the real ω_{SL} by fitting the intensities to the equation $I(\tau) = I_0\exp(-R\cdot T_{relax})\cos(\omega_{SL,real}T_{relax} + \varphi)$, where T_{relax} is the duration of SL, R is the relaxation rate, and I is the peak intensity.
6. Repeat Steps 4–5 for all ω_{SL} in the table and obtain $\omega_{SL,real}$ for all preset P values.
7. Fit P and $\omega_{SL,real}$ to a logarithmic equation $P = a\log_{10}(\omega_{SL,real}) + b$. Ideal curve would give an $R^2 > 0.99$. If the data points deviate from linear curve, a polynomial fitting can be adopted instead. An example is shown in Fig. 5.
8. Use the calibration equation given by the fit in 7 to back-calculate P for a desired SL power within the range of powers tested.

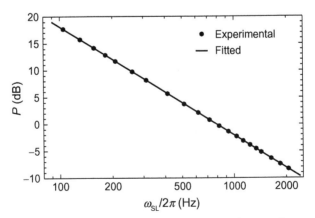

Figure 5 Example of spin-lock power calibration curve. Calibration shown was carried out for ^{15}N on Bruker 600 MHz, fitted with a linear regression.

3.3.2 Measurement of ^{15}N $R_{1\rho}$ Data

Figure 4B shows the pulse sequence employed to measure RD for imino ^{15}N nuclei. The RF pulses along with their characteristic power and duration are defined as they appear in the pulse sequence. The key parameters to be optimized are listed below:

- p_{H_2O}: Sinc-shaped pulse for manipulation of water magnetization (for flip-back and WATERGATE). p_{H_2O} should be optimized to achieve optimal peak intensity and efficient water suppression. It can be initially set to a value calculated from ^1H 90° pulse, and then optimized by arraying 5–7 values of the pulse power centered around the initial guess.
- ω_{CP}: strength of $^1H/^{15}N$ RF field for selective Hartman–Hahn polarization transfer pulses. Optimal transfer is achieved when $\omega_{CP} \approx 90$ Hz for both ^1H and ^{15}N pulses. Estimate the power levels for ^1H and ^{15}N ω_{CP} pulses corresponding to 90 Hz from the SL power calibration curve of ^{15}N (Fig. 5) and from ^1H 90° pulse. Fine tune the power levels for both the ^1H and ^{15}N pulses around that value to maximize the intensity of the desired peak with T_{relax} set to 0 ms. Efficient heteronuclear cross polarization can be achieved when ω_{CP} (in Hz) is set to values larger than $|^1J_{NH}|\sqrt{3}/4$. (Korzhnev et al., 2005).
- τ_{CP}: duration of polarization transfer delay is set to $\sim 1/|^1J_{NH}|$ and further optimized to obtain the maximum peak intensity of the desired resonance.
- ω_{1H}: ^1H SL field used to suppress scalar coupling ($^1J_{HN}$), and N–H/N DD/CSA cross-correlated relaxation during the ^{15}N SL period (T_{relax}). It is set to $\sim$5000 Hz for ^{15}N $R_{1\rho}$ experiment. ω_{1H} should also be chosen to be larger than the effective SL field to avoid polarization transfer during the SL period (Korzhnev et al., 2005). In practice, when setting up ^{15}N $R_{1\rho}$ experiment for the first time, several different ω_{1H} values are tested until dispersion data with minimum artifacts are obtained.
- ζ delay (optional calibration): nearby ^{15}N resonances that have overlapping ^1H chemical shifts can be suppressed by using the ζ delay (Fig. 4B and C), wherein $\zeta = \pi/(2\delta)$ and $\delta/(2\pi)$ are the ^{15}N offset (Hz) of the resonance to be suppressed (Hansen et al., 2009; Korzhnev et al., 2005).
- τ_{eq}: time delay for equilibration of GS and ES and is set to $\sim 3/k_{ex}$.
- T_{max}: the maximum duration of the relaxation time used to measure the $R_{1\rho}$ monoexponential decay. It varies from resonance to resonance and is typically set to a value that results in decay in the peak intensity to $\sim$30–40% of the intensity measured at $T_{min} = 0$ ms. Care must be taken to ensure the combination of ω_{SL} and T_{relax} does not exceed the tolerance of the probe.

The following protocol is employed to collect ^{15}N $R_{1\rho}$ data:
1. Optimize pulse sequence parameters as described above.
2. Record on-resonance data by setting $\Omega = 0$ Hz and measure $R_{1\rho}$ decay curves at ~20 ω_{SL} powers ranging from 50 to 2000 Hz. To do this, measure $R_{1\rho}$ monoexponential decay curves at individual power levels and zero offset by acquiring the peak intensity as a function of ~8–10 SL durations (T_{relax}). Perform duplicate measurements at T_{min} and T_{max} to help estimate $R_{1\rho}$ uncertainty. Note that the absence of RD or the absence of a dependence for R_{ex} on SL power in the on-resonance profile does not imply absence of chemical exchange. It is possible that the exchange involves a very large $\Delta\omega$ such that the carrier offset is not optimally positioned near ω_{ES}, where the R_{ex} contribution is maximum (Trott & Palmer, 2002). Off-resonance data (see Fig. 6) must be collected to verify absence of chemical exchange at a given site.
3. Measure off-resonance $R_{1\rho}$ data by varying both ω_{SL} (typically, three or more values spanning the limits described above) and Ω (≥20 values per

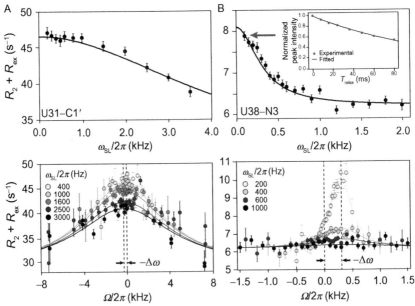

Figure 6 Representative on- (top) and off-resonance (bottom) 1D $R_{1\rho}$ RD profiles for uridine residues from HIV-1 TAR. (A) U31-C1' exhibiting fast exchange between GS and ES1 measured using 1D ^{13}C $R_{1\rho}$ experiment. (B) U38-N3 exhibiting slow exchange between GS and ES2 measured using 1D ^{15}N $R_{1\rho}$ experiment. The RD data are fitted to Laguerre equation. The inset shows the magnetization decay, as a function of relaxation delay (T_{relax}), of the first point in the on-resonance profile, which is fitted to a monoexponential to obtain the $R_{1\rho}$ value.

ω_{SL} probed). The maximum value for Ω (hereinafter referred to as Ω_{max}) is generally ~3–4 times the ω_{SL}. Beyond this offset range, $R_{1\rho}$ will be dominated by R_1 relaxation contributions, resulting in significant uncertainty when extracting R_{ex}. In general, ≥20 values of Ω are chosen with equal spacing to span ±Ω_{max} symmetrically about the zero offset.
4. It may be important to calibrate sample temperature (see note in Section 3.4.7) when carrying out the $R_{1\rho}$ measurements since small differences (~5 °C) can have significant effect on the observed RD profiles.

3.3.3 Measurement of ^{13}C $R_{1\rho}$ Data

The ^{13}C $R_{1\rho}$ pulse sequence (Fig. 4C) can be set up in a similar manner as described for ^{15}N $R_{1\rho}$ (Section 3.3.2) to target the carbon nuclei shown in Fig. 4A. A presaturation approach (PRESAT) can be used for water suppression of these nonexchangeable resonances (Hansen et al., 2009). ^{13}C nuclei typically require much shorter relaxation delays ($T_{relax} < 50\,\text{ms}$), and a broader range of ω_{SL} can be used (typically, 150–3500 Hz). Care has to be taken to limit Hartmann–Hahn matching condition with other neighboring carbons that have sizable scalar couplings. See Hansen et al. for an involved discussion (Hansen et al., 2009). Below, we describe the basic elements that differ from those of the ^{15}N $R_{1\rho}$ experiment (Section 3.3.2).

- p_{presat}: presaturation power level. Optimize power level to achieve good water suppression in combination with p_{purge} (see below). Typically, lower power (~100–150 Hz) is used to achieve selective suppression. It is critical to avoid using high-power levels for probe safety.
- ω_{CP}: strength of ^{1}H/^{13}C RF field of polarization transfer. It is set to ~100 Hz in ^{13}C $R_{1\rho}$ experiment. See Section 3.3.2.
- τ_{CP}: duration of polarization transfer. The optimal value is $\sim 1/|^1J_{CH}|$. See Section 3.3.2.
- ω_{1H}: ^{1}H SL field. See Section 3.3.2. For ^{13}C, ~8–10 kHz ω_{1H} is typically needed to completely suppress unwanted $^1J_{CH}$ and DD/CSA effects (Hansen et al., 2009).
- ζ delay (optional calibration): see Section 3.3.2. It is noted that the effectiveness of the ζ delay is significantly compromised for resonances with strong homonuclear scalar couplings (primarily applicable for ^{13}C resonances such as C6) (Hansen et al., 2009).
- p_{purge}: water purge element. p_{purge} need to be optimized by varying their power and duration for every experiment to maximize water suppression.

The protocol for ^{13}C $R_{1\rho}$ measurements is the same as the case of ^{15}N, except for the following points.

1. On-resonance data are collected at ω_{SL} ranging from 45 to 3500 or 150 to 3500 Hz for C2/C8 spins and C1'/C6 spins, respectively.
2. C–C Hartmann–Hahn transfers need to be computed as described previously (Hansen et al., 2009) in order to avoid SL powers and offset combinations that can result in transfers to neighboring or distant C resonances with significant J_{CC} couplings. For example, lower ω_{SL} are not feasible for C1' because of sizeable $^1J_{C1'C1'}$ and need to avoid Hartmann–Hahn transfers (Hansen et al., 2009).

3.3.4 Trouble Shooting $R_{1\rho}$

- *Expected peak is absent.* Either the Hartmann–Hahn polarization transfer failed or resulted in negligible signal. Make sure that ^1H and $^{13}C/^{15}N$ carrier positions have been set properly, and that a well-estimated τ_{CP} has been given based on $^1J_{CH}$ or $^1J_{NH}$. Try increasing the number of scans and array both ω_{CP} pulses around power levels estimated for a 90–100 Hz transfer.
- *Oscillations are seen in the monoexponential decay of peak intensity.* Check that the lowest $^{13}C/^{15}N$ SL power used is at least three times the largest J_{CC}/J_{NN}. For slow exchanging systems with large ES populations ($k_{ex} < 250 s^{-1}$ and $p_{ES} > 10\%$), oscillations can arise due to precession of the ES magnetization. It may be possible to suppress such oscillations by reducing the length of the τ_{eq} (see Fig. 4) to ~5 ms (Q. Zhang, University of North Carolina, personal communication).

3.4 Data Analysis

3.4.1 Fitting Monoexponentials to Obtain $R_{1\rho}$ Values

The $R_{1\rho}$ data are collected as a series of 1D spectra for each combination of ω_{SL} and Ω. The procedure below is used to extract the $R_{1\rho}$ data from each set of 1D spectra recorded using different SL durations, T_{relax}.

1. Spectra are processed by NMRPipe (Delaglio et al., 1995) and then are assembled into a single pseudo-2D spectrum using an NMRPipe script.
2. The pseudo-2D spectrum is fitted to Gaussian or Lorentzian line-shape using NMRPipe autoFit script. This fitting scheme assumes that all 1D peaks share the same line-width and resonance frequency. We find that this global approach to curve fitting improves accuracy and robustness.
3. Peak intensities are extracted from the Step 2, and are subsequently fitted against T_{relax} according to equation $I = I_0 \exp(-R_{1\rho} \cdot T_{relax})$ using a

nonlinear fitting algorithm. The nonlinear fitting could be sensitive to the initial values of I_0 and $R_{1\rho}$, especially when the signal-to-noise is poor. To address this potential problem, carry out Steps 4 and 5.

4. Normalize the peak intensities in the 1D arrays by scaling down these intensities so that the intensity at the shortest delay (T_{min}) equal to 1.0. The initial estimate for I_0 is thus set to 1.0.
5. The initial estimation for $R_{1\rho}$ is obtained by linearly fitting these intensities to the equation $\ln I = \ln I_0 - R_{1\rho} T_{relax}$. The nonlinear fitting is then performed under these initial guess, and $R_{1\rho}$ is extracted. An example of a fit is shown in Fig. 6B (inset).
6. Monte-Carlo method or Jackknife algorithms can be used to estimate the uncertainty of the fitted $R_{1\rho}$ value. These errors will be used in the fitting of $R_{1\rho}$ data described in Sections 3.4.2 and 3.4.3.

3.4.2 Fitting Off-Resonance $R_{1\rho}$ Data Using Algebraic Equations

Several algebraic expressions, derived using various approximations, can be used to fit the dependence of $R_{1\rho}$ on ω_{SL} and Ω to extract the desired exchange parameters (reviewed in Palmer & Massi, 2006). These equations can be used to extract exchange parameters from the $R_{1\rho}$ data and can also provide a powerful tool for exploring the experiment. The most versatile expression uses the Laguerre approximation developed by Palmer and coworkers (Miloushev & Palmer, 2005):

$$R_{1\rho} = R_1 \cos^2\theta + R_2 \sin^2\theta \\ + \frac{\sin^2\theta p_{GS} p_{ES} \Delta\omega_{ES}^2 k_{ex}}{\left\{ \frac{\omega_{GS}^2 \omega_{ES}^2}{\omega_{eff}^2} + k_{ex}^2 - \sin^2\theta p_{GS} p_{ES} \Delta\omega_{ES}^2 \left(1 + \frac{2k_{ex}^2(p_{GS}\omega_{GS}^2 + p_{ES}\omega_{ES}^2)}{\omega_{GS}^2 \omega_{ES}^2 + \omega_{eff}^2 k_{ex}^2}\right) \right\}} \quad (1)$$

where $\omega_{eff}^2 = \Delta\Omega^2 + \omega_{SL}^2$, $\omega_{GS}^2 = (\Omega_{GS} - \omega_{rf})^2 + \omega_{SL}^2$, $\omega_{ES}^2 = (\Omega_{ES} - \omega_{rf})^2 + \omega_{SL}^2$, and $\Delta\omega_{ES} = \Omega_{ES} - \Omega_{GS}$.

$\theta = \tan^{-1}(\omega_{SL}/\Delta\Omega)$, $\Delta\Omega = \overline{\Omega} - \omega_{rf}$, where R_1 and R_2 are the longitudinal and transverse relaxation rates, respectively, Ω_{GS} and Ω_{ES} are the resonance offsets from the SL carrier for the respective states, ω_{rf} and ω_{SL} are reference frequency and the strength of the SL carrier, $\overline{\Omega} = p_{GS}\Omega_{GS} + p_{ES}\Omega_{ES}$, and k_{ex} is the exchange rate, all in units of s^{-1}. The Laguerre equation is valid under conditions of fast exchange, and deviations can arise in slow to intermediate exchange with a large ES population

($p_{ES} > 10\%$). These deviations generally increase with slower exchange rates and increasing populations, particularly for low SL fields $\omega_{SL} \leq 400\,\text{Hz}$ (Miloushev & Palmer, 2005).

For >2-state exchange, there can be a range of exchange topologies (Palmer & Massi, 2006). Typically, the $R_{1\rho}$ data are analyzed assuming ES1 ⇌ GS ⇌ ES2 with or without minor exchange between ES1 and ES2. It should be noted that under certain exchange regimes, the $R_{1\rho}$ data will be sensitive to the presence of minor exchange between ES1 and ES2, and therefore, under favorable exchange conditions can in principle be used to define the topology of the exchange process (Palmer & Massi, 2006). For three-state without minor exchange, one can use the following equation (Palmer & Massi, 2006; Trott & Palmer, 2004):

$$R_{1\rho} = R_1 \cos^2\theta + R_2 \sin^2\theta + \sin^2\theta \left(\frac{k_1 \Delta\omega_{ES1}^2}{\Omega_{ES1}^2 + \omega_{SL}^2 + k_{-1}^2} + \frac{k_2 \Delta\omega_{ES2}^2}{\Omega_{ES2}^2 + \omega_{SL}^2 + k_{-2}^2} \right) \tag{2}$$

where $\Delta\omega_{ES1} = \Omega_{ES1} - \Omega_{GS}$, $\Delta\omega_{ES2} = \Omega_{ES2} - \Omega_{GS}$, k_1, and k_{-1} are the GS–ES1 forward and reverse rate constants, respectively, and k_2 and k_{-2} are the GS–ES2 forward and reverse rate constants, respectively.

A Laguerre approximation has also been used to analyze three-state exchange (Dethoff, Chugh, et al., 2012) and has shown excellent agreement with numerical solutions, though its range of validity has not been fully examined:

$$R_{1\rho} = R_1 \cos^2\theta + R_2 \sin^2\theta +$$

$$\sin^2\theta \left(\frac{p_{GS} p_{ES1} \Delta\omega_{ES1}^2 k_{ex1}}{\left\{ \omega_{GS}^2 \omega_{ES1}^2 / \omega_{eff}^2 + k_{ex1}^2 - \sin^2\theta p_{GS} p_{ES1} \Delta\omega_{ES1}^2 \left(1 + \frac{2k_{ex1}^2 \left(p_{GS}\omega_{GS}^2 + p_{ES1}\omega_{ES1}^2 \right)}{\omega_{GS}^2 \omega_{ES1}^2 + \omega_{eff}^2 k_{ex1}^2} \right) \right\}} \right.$$

$$\left. + \frac{p_{GS} p_{ES2} \Delta\omega_{ES2}^2 k_{ex2}}{\left\{ \omega_{GS}^2 \omega_{ES2}^2 / \omega_{eff}^2 + k_{ex2}^2 - \sin^2\theta p_{GS} p_{ES2} \Delta\omega_{ES2}^2 \left(1 + \frac{2k_{ex2}^2 \left(p_{GS}\omega_{GS}^2 + p_{ES2}\omega_{ES2}^2 \right)}{\omega_{GS}^2 \omega_{ES2}^2 + \omega_{eff}^2 k_{ex2}^2} \right) \right\}} \right) \tag{3}$$

Other expressions for three-state exchange with minor exchange have been reported (Palmer & Massi, 2006; Trott & Palmer, 2004).

When carrying out a fit to the $R_{1\rho}$ data, one starts out with broad estimates of initial exchange parameters and performs a fit using the equations described above. However, starting estimates combined with conventional local-minimum optimization methods (such as the Levenberg–Marquadt algorithm) can lead to fits being trapped in local minima. In such cases, the use of simulated annealing and basin-hopping (Wales & Doye, 1997) schemes can prove very beneficial for finding global minima.

The quality of the fit is assessed using a χ^2 analysis, for which reasonable fits typically yield values below 1.5, though care should be taken if a fit gives a χ^2 of <1 as this may indicate overfitting. A poor fit may suggest deviations from a two-state model or from the range of validity of the Laguerre equation or >2-state exchange. Statistical tests (e.g., F-test and Akaike information criterion) (Akaike, 1974) can provide insights in to the likelihood of one model over the other.

3.4.3 Fitting Off-Resonance $R_{1\rho}$ Data Using Bloch–McConnell Equations

Alternatively, one can fit the $R_{1\rho}$ data directly by numerical integration of the Bloch–McConnell (B–M) equations (McConnell, 1958). The B–M equation does not rely on any algebraic approximations and makes it possible to compute $R_{1\rho}$ data for arbitrarily complex exchange models consisting of many ESs with distinct exchange topologies. The disadvantage of this method is the high computational cost. Typically, we find that using parameters from the Laguerre fit simplifies the search for the global minimum. More typically, we use the B–M equations to establish the validity of the Laguerre approximation through comparison of computed $R_{1\rho}$ given a set of ω_{SL} and Ω around a given set of speculated exchange parameters (for example, see Bothe, Lowenhaupt, & Al-Hashimi, 2011). B–M simulations can also be used to examine effects arising from >2-state exchange.

3.4.4 Determining the Chemical Shifts of the ES

Fitting of the $R_{1\rho}$ data yields $\Delta\omega$ and p_{ES}. This together with the observed chemical shift, $\overline{\Omega}$, can be used to determine the ES chemical shift, Ω_{ES}:

$$\Omega_{ES} = \overline{\Omega} + (1 - p_{ES})\Delta\omega \qquad (4)$$

It worth noting that the same unit should be used for Ω_{ES}, $\overline{\Omega}$, and $\Delta\omega$, either in Hz or in ppm. To convert $\Delta\omega$ from Hz to ppm, one can use equation $\Delta\omega$ (ppm) = $\Delta\omega$ (Hz) × $10^6/\omega_{Larmor}$, where ω_{Larmor} (in Hz) is Larmor frequency of the spin of interest. For example, if Larmor frequency of a ^{15}N spin is 60.772 MHz, then a $\Delta\omega = -257\,\text{Hz}$ will be translated to -4.23 ppm. The GS chemical shift can be calculated by $\Omega_{GS} = \overline{\Omega} - p_{ES}\Delta\omega$.

3.4.5 Plotting RD Profiles

When displaying RD profiles, we plot $R_2 + R_{ex}$ versus SL offset frequency (Ω) for various (typically color coded) SL powers. This omits the R_1 contribution to $R_{1\rho}$, which does not provide any information regarding exchange parameters. Note that Ω is the ^{13}C or ^{15}N resonance offset frequency from the carrier position, and $\Delta\Omega = -\Delta\omega$. An example plot is shown in Fig. 6 for fast (ES1, Fig. 6A) and slow (ES2, Fig. 6B) exchange processes in HIV-1 TAR. Note that in general, the maximum R_{ex} occurs when the carrier is on-resonance with the ES chemical shift, thus simple inspection of the profile can be used to estimate the chemical shift of the ES. For fast exchange, one generally sees a broader line and it becomes more difficult to pinpoint the position of peak maximum and therefore the chemical shift of the ES (Bothe et al., 2014).

3.4.6 Estimating Uncertainties in Exchange Parameters

Bootstrap- and Monte-Carlo-based approaches can be used to estimate uncertainties in the measured exchange parameters (Bothe et al., 2014). A parent data set of $R_{1\rho}$ data points (Ω, ω_{SL}, $R_{1\rho}$, $R_{1\rho}$ error) is used for both methods.

In the Bootstrap approach:
1. 1000 child data sets of the same size as the parent data set are generated by randomly choosing data points from the parent data set. Here, each data point can be either excluded or chosen multiple times.
2. Each child data set and the parent data set are fit to the Laguerre equation to obtain the five exchange parameters: R_1, R_2, p_{ES}, k_{ex}, and $\Delta\omega$.
3. The uncertainty in the exchange parameters is then determined by calculating the standard deviation of the individual parameters of the child data sets relative to parameters of the parent data set.

In the Monte-Carlo approach:
1. The parent data set is fit to the Laguerre equation to obtain the five exchange parameters.
2. 1000 child data sets are generated using the exchange parameters, the Laguerre equation, and the Ω and ω_{SL} values of the parent data set.
3. Each child data set is noise corrupted according to the $R_{1\rho}$ error from the parent data set.
4. The child data sets are fit to the Laguerre equation.

5. The uncertainty in the exchange parameters is then determined by calculating the standard deviation of the individual parameters of the child data sets relative to parameters of the parent data set.

3.4.7 Kinetic-Thermodynamic Analysis

Valuable kinetic-thermodynamic information can be obtained from the exchange parameters (Korzhnev, Religa, Lundstrom, Fersht, & Kay, 2007; Nikolova et al., 2011). This can also provide important insights into the nature of the ES and transition state (TS). For a two-state system, the free energy difference between the GS and ES at a given temperature can readily be obtained using the following equation:

$$\Delta G_{ES}(T) = -RT \ln\left(\frac{p_{ES}}{p_{GS}}\right) \quad (5)$$

where ΔG_{ES} is the relative free energy difference between the GS (arbitrarily referenced to 0) and the ES, T is the experimental temperature (in Kelvin), and k_B is Boltzmann's constant.

The free energy difference relative to the TS can be obtained by:

$$\ln\left(\frac{k_i(T)}{T}\right) = \ln\left(\frac{k_B \kappa}{h}\right) - \frac{\Delta G_i^T(T)}{RT} \quad (6)$$

where k_i ($i=1$ or -1) are the forward and reverse rate constants, respectively. T is the experimental temperature (in K), k_B is Boltzmann's constant, κ is the transmission coefficient (assumed to be 1), h is Planck's constant, ΔG^T is the forward or reverse activation barrier, and R is the gas constant.

The enthalpies and entropies can also be obtained through temperature dependent $R_{1\rho}$ measurements.

$$\ln\left(\frac{k_i(T)}{T}\right) = \ln\left(\frac{k_B \kappa}{h}\right) - \frac{\Delta G_i^T(T_{hm})}{RT_{hm}} - \frac{\Delta H_i^T}{R}\left(\frac{1}{T} - \frac{1}{T_{hm}}\right) \quad (7)$$

where k_i ($i=1$ or -1) are the forward and reverse rate constants, respectively. T is the experimental temperature (in K), k_B is Boltzmann's constant, κ is the transmission coefficient (assumed to be 1), h is Planck's constant, R is the gas constant, ΔG^T is the forward or reverse activation barrier, ΔH^T is the forward or reverse enthalpy of activation, and T_{hm} is the harmonic mean of the experimental temperatures measured:

$$T_{\text{hm}} = n\left(\sum_{i=1}^{n}\left(\frac{1}{T_i}\right)\right)^{-1} \qquad (8)$$

Finally, $T\Delta S_i^T = \Delta H_i^T - \Delta G_i^T(T_{\text{hm}})$, where ΔS^T is the forward or reverse entropy of activation.

Note: Here, it is critical to ensure the spectrometer temperature is calibrated to match the sample temperature. This can be achieved by recording spectra of 99.8% methanol-d$_4$ (Cambridge Isotope Laboratories, Inc.) at varying spectrometer temperatures. Here, the actual sample temperature is given by $T = -16.7467\Delta\delta^2 - 52.5130\Delta\delta + 419.1381$, where $\Delta\delta$ is the difference in chemical shift (ppm) between the hydroxyl and methyl proton (Findeisen, Brand, & Berger, 2007).

3.5 Inferring Structures of RNA ESs Using NMR Chemical Shifts and Secondary Structure Prediction

3.5.1 ^{13}C and ^{15}N Chemical Shift–Structure Relationships in RNA

The ^{13}C and ^{15}N chemical shifts of the ES carry rich structural information. Base carbon chemical shifts (C2/C6/C8) are sensitive to changes in the glycosidic angle χ (*anti* vs. *syn*) as well as stacking (Dejaegere & Case, 1998; Ebrahimi, Rossi, Rogers, & Harbison, 2001; Fares, Amata, & Carlomagno, 2007; Ohlenschlager, Haumann, Ramachandran, & Gorlach, 2008; Xu & Au-Yeung, 2000). The adenine C2 is sensitive to protonation of N1. Sugar C1' chemical shifts are sensitive to both sugar pucker and the glycosidic bond angle (Dejaegere & Case, 1998; Ebrahimi et al., 2001; Fares et al., 2007; Ohlenschlager et al., 2008; Xu & Au-Yeung, 2000). The ^{15}N chemical shifts are sensitive to the properties of H-bonding (see below) (Goswami, Gaffney, & Jones, 1993) and also directly report on changes in ^{15}N protonation states (Büchner, Blomberg, F., & Rüterjans, 1978).

In Fig. 7, we show 2D HSQC-style representation that capture key ^{13}C (Fig. 7A) and ^{15}N (Fig. 7B) chemical shift–structure relationships in RNA (McBrairty, Xue, Petzold, & Al-Hashimi, 2015). As can be seen, resonances belonging to Watson–Crick (WC) bps that are surrounded by other WC bps (A-form residues) fall within a narrow region of ^{13}C chemical shifts for both base (C2, C6, C8) and sugar (C1') resonances (Dejaegere & Case, 1998; Ebrahimi et al., 2001; Fares et al., 2007; Ohlenschlager et al., 2008; Xu & Au-Yeung, 2000). Noncanonical residues have ^{13}C and ^{15}N chemical shifts that deviate from the narrow range defined by the WC A-form

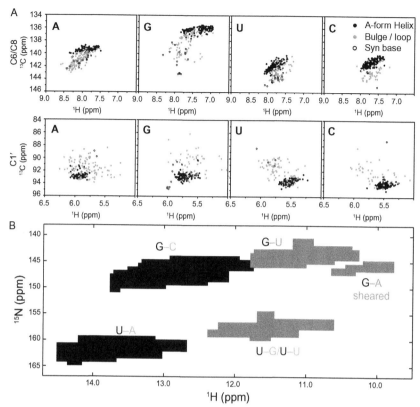

Figure 7 Chemical shift–structure relationships in RNA. (A) Distribution of C–H chemical shifts derived from BMRB database (McBrairty et al., 2015) for different regions of RNAs: A-form helical Watson–Crick base pairs (black); bulge, internal loop or apical loop (gold); residues with a *syn* base conformation (blue, black gray in the print version). (B) Distribution of N–H chemical shifts of RNA imino groups for guanine and uracil nucleotides in different base pair contexts: G–C Watson–Crick base pairs (red, dark gray in the print version), G–U wobbles (purple, gray in the print version), sheared G–A pairs (gold, gray in the print version), U–A Watson–Crick base pairs (blue, black in the print version), and U–G wobbles or U–U mispairs (green, gray in the print version).

conformation. For example, residues in bulges and apical loops have C6/C8 resonances that are downfield shifted in both carbon and proton dimensions. The C1′ chemical shifts in non-A-form regions are also upfield shifted, consistent with changes in sugar pucker from C3′-endo toward C2′-endo. In general, noncanonical bps that resemble the A-form helix geometry will show carbon chemical shifts that fall within the A-form helical distributions, whereas more flexible base-pairing and differences in backbone conformation can lead to differences in both C6/C8 (downfield) and C1′ (upfield) chemical shifts.

In addition, the imino (G–N1/U–N3) and amino (A–N6/C–N4/G–N2) ^{15}N chemical shifts are sensitive to changes in base pairing, including changes in hydrogen bonding partner, as well as the hydrogen bonding distance, which give rise to predictable changes in chemical shift from the standard WC bps to other types of base pairing (Fig. 7B). Guanine N1 in G–C bps resonates in a region downfield ($\Delta\omega_{N1} \sim 2-5$ ppm) from the chemical shifts of the same spin in G–U wobble mispairs. Likewise, uracil N3 chemical shifts of A–U pairs resonate downfield ($\Delta\omega_{N3} \sim 2-5$ ppm) from that of U–N3 in G–U mispairs. In general, G–N1 and U–N3 display upfield shift from the canonical G–C/A–U region upon partial or complete loss of stable nitrogen H-bonding. Therefore, if a base pair involving G/U changes from canonical pair to noncanonical pair or an unpaired base, or vice versa, it is likely to produce a detectable RD profile, providing a clue about the change of base-pairing in the ES.

It is also possible to capture motif-specific chemical shift signatures. For example, the UUCG apical loop features a *syn* G that has an unusual C8 resonance that is ∼7 ppm downfield shifted from the reference A-form helical chemical shifts (Dethoff, Petzold, et al., 2012). Other common motifs such as GNRA tetraloops can also be readily identified through a combination of the C8, C2, C1′ chemical shifts that remain consistent throughout these motifs.

3.5.2 Secondary Structure Prediction

The information regarding which residues undergo an exchange process and direction and magnitude of the ES change in chemical shift form a "chemical shift fingerprint," which can be used to infer potential models for the ES structure. The chemical shift fingerprints can be used to infer which residues become more/less helical upon formation of the ES. For example, in the HIV-1 TAR ES1, the G–C8 in residue G34 shows clear signs of a *syn* G analogous to that seen in the UUCG apical loop, while A35 and many other apical loop residues show signs of moving away from nonhelical sugar pucker in the GS (i.e. C2′-endo) toward more helical C3′-endo sugar pucker in ES1. In HIV-1 TAR ES2, the U38–N3 shows clear signs of a weakened hydrogen bond, as the A27-U38 moves toward a noncanonical U25–U38 bp.

RNA ESs often arise due to localized changes in secondary structure (Fig. 2). For this reason, it is instructive to test the possibility that the ES represents an alternative secondary structure. A powerful approach for obtaining a list of candidate ES structures is to use commonly available secondary structure prediction programs such as Mfold (Zuker, 2003) or MC-Fold (Parisien & Major, 2008) to examine higher energy secondary structures that can be adopted by an RNA (Fig. 8A). These alternative secondary structures are examined to see if they can qualitatively explain the observed ES chemical shift fingerprints. For example, in the case of HIV-1 TAR, one finds that the second energetically most favorable structure features a more compact loop that can explain the chemical shift fingerprints measured for ES1 (Fig. 8A) (Dethoff, Petzold, et al., 2012). A higher energy secondary structure that features remodeling of bulge, upper helix, and apical loop is also observed that can explain the more broad distribution exchange observed for ES2, and the unique chemical shift fingerprints (Lee et al., 2014). In general, a given secondary structure is interrogated to examine whether it addresses the following questions:

1. Does the alternative secondary structure result in changes in the environment of all residues showing signs of exchange broadening?
2. Does the alternative secondary structure preserve the environment of all residues showing no signs of exchange broadening?
3. Do the specific changes in structure and, specifically, transitions from helical to nonhelical residues or inter-conversion of G–C to G–U bps agree with the specific ES chemical shifts?

In this manner, a candidate ES model is generated and tested as described below.

3.6 Testing Model RNA ESs
3.6.1 Stabilizing GS and ES Using Mutations

A candidate ES model with alternative secondary structure is tested using mutations and single-atom substitutions that are designed to stabilize the ES and in some cases the GS. Generally, it is possible to design substitution mutations that, for example, convert a noncanonical bp in the ES, such as an A–C bp, into a more stable canonical bp, such as G–C or A–U. This is shown in Fig. 8B, where the HIV-1 TAR ES1 is stabilized using two mutations that convert A–C into WC bps. Likewise, mutating the A-G mismatch to a WC A–U can stabilize the TAR ES2. ESs with low energetic stability that have very low populations (<1%) can require more severe mutations in order to completely stabilize the ES. Indeed, complete stabilization of the low-populated TAR ES2 (~0.4%) required a triple mutant (Fig. 8B) (Lee et al., 2014). An ES or GS can also be stabilized through deletion of a residue.

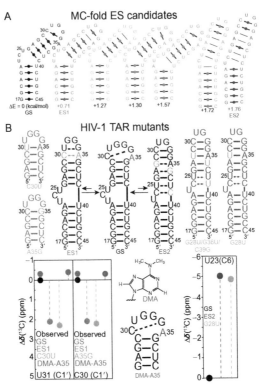

Figure 8 (A) Using secondary structure prediction to generate a list of candidate excited states. Shown are the seven lowest energy secondary structures with estimated energies referenced to the GS of HIV-1 TAR predicted from MC-Fold (Parisien & Major, 2008). The second and the seventh predicted secondary structures are ES1 (green, gray in the print version) and ES2 (red, gray in the print version), respectively as determined by $R_{1\rho}$ and chemical shift fingerprints. (B) Mutants and chemical shift fingerprint analysis of RNA ESs. Shown are TAR mutants designed to stabilize ES1 (green, gray in the print version), ES2 (red, gray in the print version), and GS (black) with residue color indicating substitution (orange, light gray in the print version), chemically modified substitution (purple, light gray in the print version). Chemical shift fingerprint is shown as a comparison of chemical shifts determined by HSQC spectra of TAR mutants and $R_{1\rho}$ relative to the chemical shift of HIV-1 TAR.

For example, in HIV-1 TAR deletion of the three-nucleotide bulge stabilizes the GS of HIV-1 TAR so there is no exchange with ES2, but still allows exchange with ES1. Replacement of the apical loop to a stable UUCG loop stabilizes the GS completely so there is no exchange between either ES1 or ES2 (Fig. 8B). All designed mutants are subjected to secondary structure prediction. Ideally, two or more ES-stabilizing mutations should be used independently, each targeting a different region of the RNA (see below). Each mutant is prepared and analyzed by NMR as described below. Because only 2D HSQC spectra are needed for chemical shift fingerprints, unlabeled RNA samples can be used and obtained by *in vitro* transcription or by solid-phase chemical synthesis. 2D HSQC NMR measurements at natural abundance can be performed with optimal sensitivity using SOFAST-HMQC (Farjon et al., 2009; Sathyamoorthy, Lee, Kimsey, Ganser, & Al-Hashimi, 2014) and BEST-TROSY (Ying et al., 2011) that yield higher sensitivity per unit time.

All mutants are subjected to the following analysis:

1. The secondary structure is determined using 2D NOESY NMR experiments. The goal is to determine whether the mutation stabilizes the proposed ES secondary structure. If the secondary structure does not match that predicted for the ES model, the mutant is not studied further, and the ES model is refined or a new mutant is selected to stabilize it.
2. The ^{13}C and ^{15}N chemical shifts are assigned and compared (see below for factors that influence this agreement) to the chemical shifts measured for the ES (or GS) in the wild-type construct. Examples of such chemical shift fingerprint comparisons are shown in Fig. 8B.
3. For a concerted exchange directed to a single ES, any ES-stabilizing mutation should affect all of the resonances experiencing exchange, including those that may be distant from the site of mutation (Fig. 8B). Moreover, these perturbations should be specifically directed toward the ES chemical shift fingerprints determined by $R_{1\rho}$ (Fig. 8B).
4. All resonances showing no signs of chemical exchange should experience little chemical shift perturbation upon mutation (typically <1 ppm). At this point, it is possible to examine if a given resonance does not sense a given ES because of too small $\Delta\omega$. For example, the HIV-1 TAR ES2 stabilizing mutant (G28U) shows small $\Delta\omega$ values for a number of resonances (U42–N3 and A22–C8) explaining why these residues show little to no sign of chemical exchange in the $R_{1\rho}$ measurements (Lee et al., 2014).

ES-stabilized mutants are not expected to have chemical shifts that perfectly match those of the ES, since the mutation itself is expected to cause changes on the order of 1 ppm (Dethoff, Petzold, et al., 2012; Lee et al., 2014). The ES-stabilized mutant could also be in fast exchange with other higher energy conformations, and this could also impact the observed chemical shifts relative to the ES.

3.6.2 Stabilizing GS and ES Using Single-Atom Substitutions

The GS or ES can be stabilized by chemical modifications, including those that occur naturally. Many chemically modified RNA samples can be obtained commercially (e.g., ChemGenes, Glen Research, Berry and Associates). Typically the sample is lyophilized, to get rid of impurities, and buffer exchanged. Chemical modifications can be used to disrupt base pairing and favor "bulged out" or "flipped out" conformations. For example, N6-N6-dimethyl substituted adenine (DMA) was used to disrupt A–C-type H-bonding in the TAR ES1 and to favor the GS in which the adenine is bulged out (Fig. 8B) (Dethoff, Chugh, et al., 2012). Indeed, the modification leads to perturbations in the chemical shifts directed toward the GS (Fig. 8B). In studies of DNA, ES that feature Hoogsteen (HG) base pairs have been stabilized in G–C and A–T base pairs through methylation of guanine and adenine N1 (Nikolova et al., 2011). Conversely, the single-atom substitution converting purine N7 to a carbon (deaza-purine) was used to destabilize the HG bps and trap the WC GS (Nikolova et al., 2012).

3.6.3 Stabilizing GS and ES by Changing pH

In some cases, an ES may be associated with a protonation event. For example, many noncanonical A–C bps are protonated, and therefore their formation can be favored by lowering pH. The HIV-1 TAR ES1 features formation of an A^+–C bp, which is absent in the GS, where it is possible lowering the pH will favor the ES over the GS. Indeed, one finds decreasing the pH results in chemical shift perturbations directed toward the ES1, while increasing the pH results in perturbations directed toward the GS. In addition, decreasing/increasing the pH results in an increase/decrease in the HIV-1 TAR ES1 population. In general, changing the pH is a powerful approach to obtain insights into any ES that may involve protonation or deprotonation.

ACKNOWLEDGMENTS

Disclosure: H.M.A. is an advisor to and holds an ownership interest in Nymirum Inc., which is an RNA-based drug discovery company. The research reported in this article was performed by the University of Michigan faculty and students and was funded by an NIH contract to H.M.A.

This work was supported by the US National Institutes of Health (R01 AI066975, NIAID R21Ai096985, and P01 GM0066275) We thank Prof Katja Petzold (Karolinska Institutet Tombtebodavägen) for input on the structure chemical shift relationships.

REFERENCES

Aboul-ela, F., Karn, J., & Varani, G. (1996). Structure of HIV-1 TAR RNA in the absence of ligands reveals a novel conformation of the trinucleotide bulge. *Nucleic Acids Research, 24*(20), 3974–3981.

Akaike, H. (1974). A new look at the statistical model identification. *IEEE Transactions on Automatic Control, 19*(6), 716–723.

Akke, M., & Palmer, A. G. (1996). Monitoring macromolecular motions on microsecond to millisecond time scales by R(1)rho-R(1) constant relaxation time NMR spectroscopy. *Journal of the American Chemical Society, 118*, 911–912.

Alvarado, L. J., Longhini, A. P., LeBlanc, R. M., Chen, B., Kreutz, C., & Dayie, T. K. (2014). Chemo-enzymatic synthesis of selectively (13)C/(15)N-labeled RNA for NMR structural and dynamics studies. *Methods in Enzymology, 549*, 133–162.

Bothe, J. R., Lowenhaupt, K., & Al-Hashimi, H. M. (2011). Sequence-specific B-DNA flexibility modulates Z-DNA formation. *Journal of the American Chemical Society, 133*(7), 2016–2018.

Bothe, J. R., Nikolova, E. N., Eichhorn, C. D., Chugh, J., Hansen, A. L., & Al-Hashimi, H. M. (2011). Characterizing RNA dynamics at atomic resolution using solution-state NMR spectroscopy. *Nature Methods, 8*(11), 919–931.

Bothe, J. R., Stein, Z. W., & Al-Hashimi, H. M. (2014). Evaluating the uncertainty in exchange parameters determined from off-resonance R1rho relaxation dispersion for systems in fast exchange. *Journal of Magnetic Resonance, 244*, 18–29.

Breaker, R. R. (2011). Prospects for riboswitch discovery and analysis. *Molecular Cell, 43*(6), 867–879.

Büchner, P., Blomberg, F., & Rüterjans, H. (1978). Nitrogen-15 nuclear magnetic resonance spectroscopy of 15N-labeled nucleotides. In B. Pullman & B. Pullman (Eds.), *Vol. 11. Nuclear magnetic resonance spectroscopy in molecular biology* (pp. 53–70). Netherlands: Springer.

Carr, H. Y., & Purcell, E. M. (1954). Effects of diffusion on free precession in nuclear magnetic resonance experiments. *Physical Review, 94*(3), 630–638.

Cech, T. R., & Steitz, J. A. (2014). The noncoding RNA revolution-trashing old rules to forge new ones. *Cell, 157*(1), 77–94.

Cheah, M. T., Wachter, A., Sudarsan, N., & Breaker, R. R. (2007). Control of alternative RNA splicing and gene expression by eukaryotic riboswitches. *Nature, 447*(7143), 497–500.

Cruz, J. A., & Westhof, E. (2009). The dynamic landscapes of RNA architecture. *Cell, 136*(4), 604–609.

Dejaegere, A. P., & Case, D. A. (1998). Density functional study of ribose and deoxyribose chemical shifts. *Journal of Physical Chemistry A, 102*(27), 5280–5289.

Delaglio, F., Grzesiek, S., Vuister, G. W., Zhu, G., Pfeifer, J., & Bax, A. (1995). NMRPipe—A multidimensional spectral processing system based on unix pipes. *Journal of Biomolecular NMR, 6*(3), 277–293.

Dethoff, E. A., Chugh, J., Mustoe, A. M., & Al-Hashimi, H. M. (2012). Functional complexity and regulation through RNA dynamics. *Nature, 482*(7385), 322–330.

Dethoff, E. A., Petzold, K., Chugh, J., Casiano-Negroni, A., & Al-Hashimi, H. M. (2012). Visualizing transient low-populated structures of RNA. *Nature, 491*(7426), 724–728.

Deverell, C., Morgan, R. E., & Strange, J. H. (1970). Studies of chemical exchange by nuclear magnetic relaxation in rotating frame. *Molecular Physics, 18*(4), 553–559.

D'Souza, V., & Summers, M. F. (2004). Structural basis for packaging the dimeric genome of Moloney murine leukaemia virus. *Nature, 431*(7008), 586–590.

Easton, L. E., Shibata, Y., & Lukavsky, P. J. (2010). Rapid, nondenaturing RNA purification using weak anion-exchange fast performance liquid chromatography. *RNA, 16*(3), 647–653.

Ebrahimi, M., Rossi, P., Rogers, C., & Harbison, G. S. (2001). Dependence of 13C NMR chemical shifts on conformations of RNA nucleosides and nucleotides. *Journal of Magnetic Resonance, 150*(1), 1–9.

Fares, C., Amata, I., & Carlomagno, T. (2007). 13C-detection in RNA bases: Revealing structure-chemical shift relationships. *Journal of the American Chemical Society, 129*(51), 15814–15823.

Farjon, J., Boisbouvier, J., Schanda, P., Pardi, A., Simorre, J. P., & Brutscher, B. (2009). Longitudinal-relaxation-enhanced NMR experiments for the study of nucleic acids in solution. *Journal of the American Chemical Society, 131*(24), 8571–8577.

Fawzi, N. L., Ying, J., Ghirlando, R., Torchia, D. A., & Clore, G. M. (2011). Atomic-resolution dynamics on the surface of amyloid-beta protofibrils probed by solution NMR. *Nature, 480*(7376), 268–272.

Ferrage, F., Eykyn, T. R., & Bodenhausen, G. (2004). Frequency-switched single-transition cross-polarization: A tool for selective experiments in biomolecular NMR. *Chemphyschem, 5*(1), 76–84.

Findeisen, M., Brand, T., & Berger, S. (2007). A 1H-NMR thermometer suitable for cryoprobes. *Magnetic Resonance in Chemistry, 45*(2), 175–178.

Furtig, B., Richter, C., Wohnert, J., & Schwalbe, H. (2003). NMR spectroscopy of RNA. *Chembiochem, 4*(10), 936–962.

Garst, A. D., & Batey, R. T. (2009). A switch in time: Detailing the life of a riboswitch. *Biochimica et Biophysica Acta, 1789*(9–10), 584–591.

Goswami, B., Gaffney, B. L., & Jones, R. A. (1993). Nitrogen-15-labeled oligodeoxynucleotides. 5. Use of 15N NMR to probe H-bonding in an O6MeG.cntdot.T base pair. *Journal of the American Chemical Society, 115*(9), 3832–3833.

Grundy, F. J., Winkler, W. C., & Henkin, T. M. (2002). tRNA-mediated transcription antitermination in vitro: Codon-anticodon pairing independent of the ribosome. *Proceedings of the National Academy of Sciences of the United States of America, 99*(17), 11121–11126.

Haller, A., Souliere, M. F., & Micura, R. (2011). The dynamic nature of RNA as key to understanding riboswitch mechanisms. *Accounts of Chemical Research, 44*(12), 1339–1348.

Hansen, A. L., & Al-Hashimi, H. M. (2007). Dynamics of large elongated RNA by NMR carbon relaxation. *Journal of the American Chemical Society, 129*(51), 16072–16082.

Hansen, A. L., Nikolova, E. N., Casiano-Negroni, A., & Al-Hashimi, H. M. (2009). Extending the range of microsecond-to-millisecond chemical exchange detected in labeled and unlabeled nucleic acids by selective carbon R(1rho) NMR spectroscopy. *Journal of the American Chemical Society, 131*(11), 3818–3819.

Herschlag, D. (1995). RNA chaperones and the RNA folding problem. *The Journal of Biological Chemistry, 270*(36), 20871–20874.

Hoogstraten, C. G., Wank, J. R., & Pardi, A. (2000). Active site dynamics in the lead-dependent ribozyme. *Biochemistry, 39*(32), 9951–9958.

Houck-Loomis, B., Durney, M. A., Salguero, C., Shankar, N., Nagle, J. M., Goff, S. P., et al. (2011). An equilibrium-dependent retroviral mRNA switch regulates translational recoding. *Nature, 480*(7378), 561–564.

Huthoff, H., & Berkhout, B. (2001). Two alternating structures of the HIV-1 leader RNA. *RNA, 7*(1), 143–157.

Igumenova, T. I., & Palmer, A. G., 3rd. (2006). Off-resonance TROSY-selected R 1rho experiment with improved sensitivity for medium- and high-molecular-weight proteins. *Journal of the American Chemical Society, 128*(25), 8110–8111.

Johnson, J. E., Jr., & Hoogstraten, C. G. (2008). Extensive backbone dynamics in the GCAA RNA tetraloop analyzed using 13C NMR spin relaxation and specific isotope labeling. *Journal of the American Chemical Society, 130*(49), 16757–16769.

Kedde, M., van Kouwenhove, M., Zwart, W., Oude Vrielink, J. A., Elkon, R., & Agami, R. (2010). A Pumilio-induced RNA structure switch in p27-3' UTR controls miR-221 and miR-222 accessibility. *Nature Cell Biology, 12*(10), 1014–1020.

Kim, S., & Baum, J. (2004). An on/off resonance rotating frame relaxation experiment to monitor millisecond to microsecond timescale dynamics. *Journal of Biomolecular NMR, 30*(2), 195–204.

Korzhnev, D. M., & Kay, L. E. (2008). Probing invisible, low-populated states of protein molecules by relaxation dispersion NMR spectroscopy: An application to protein folding. *Accounts of Chemical Research, 41*(3), 442–451.

Korzhnev, D. M., Orekhov, V. Y., & Kay, L. E. (2005). Off-resonance R(1rho) NMR studies of exchange dynamics in proteins with low spin-lock fields: An application to a Fyn SH3 domain. *Journal of the American Chemical Society, 127*(2), 713–721.

Korzhnev, D. M., Religa, T. L., Lundstrom, P., Fersht, A. R., & Kay, L. E. (2007). The folding pathway of an FF domain: Characterization of an on-pathway intermediate state under folding conditions by (15)N, (13)C(alpha) and (13)C-methyl relaxation dispersion and (1)H/(2)H-exchange NMR spectroscopy. *Journal of Molecular Biology, 372*(2), 497–512.

Korzhnev, D. M., Skrynnikov, N. R., Millet, O., Torchia, D. A., & Kay, L. E. (2002). An NMR experiment for the accurate measurement of heteronuclear spin-lock relaxation rates. *Journal of the American Chemical Society, 124*(36), 10743–10753.

Lai, D., Proctor, J. R., & Meyer, I. M. (2013). On the importance of cotranscriptional RNA structure formation. *RNA, 19*(11), 1461–1473.

Lee, J., Dethoff, E. A., & Al-Hashimi, H. M. (2014). Invisible RNA state dynamically couples distant motifs. *Proceedings of the National Academy of Sciences of the United States of America, 111*(26), 9485–9490.

Long, D., Bouvignies, G., & Kay, L. E. (2014). Measuring hydrogen exchange rates in invisible protein excited states. *Proceedings of the National Academy of Sciences of the United States of America, 111*(24), 8820–8825.

Loria, J. P., Rance, M., & Palmer, A. G. (1999a). A relaxation-compensated Carr-Purcell-Meiboom-Gill sequence for characterizing chemical exchange by NMR spectroscopy. *Journal of the American Chemical Society, 121*(10), 2331–2332.

Loria, J. P., Rance, M., & Palmer, A. G. (1999b). A TROSY CPMG sequence for characterizing chemical exchange in large proteins. *Journal of Biomolecular NMR, 15*(2), 151–155.

Mandal, M., & Breaker, R. R. (2004). Gene regulation by riboswitches. *Nature Reviews. Molecular Cell Biology, 5*(6), 451–463.

Massi, F., Johnson, E., Wang, C., Rance, M., & Palmer, A. G., 3rd. (2004). NMR R1 rho rotating-frame relaxation with weak radio frequency fields. *Journal of the American Chemical Society, 126*(7), 2247–2256.

McBrairty, M., Xue, Y., Petzold, K., & Al-Hashimi, H. M. (2015). The H-factor: A parameter that relates RNA chemical shifts and secondary structure. In preparation.

McConnell, H. M. (1958). Reaction rates by nuclear magnetic resonance. *Journal of Chemical Physics, 28*(3), 430–431.

Meiboom, S., & Gill, D. (1958). Modified spin-echo method for measuring nuclear relaxation times. *Review of Scientific Instruments, 29*(8), 688–691.

Miloushev, V. Z., & Palmer, A. G., 3rd. (2005). R(1rho) relaxation for two-site chemical exchange: General approximations and some exact solutions. *Journal of Magnetic Resonance, 177*(2), 221–227.

Mulder, F. A. A., de Graaf, R. A., Kaptein, R., & Boelens, R. (1998). An off-resonance rotating frame relaxation experiment for the investigation of macromolecular dynamics using adiabatic rotations. *Journal of Magnetic Resonance, 131*(2), 351–357.

Mulder, F. A., Skrynnikov, N. R., Hon, B., Dahlquist, F. W., & Kay, L. E. (2001). Measurement of slow (micros-ms) time scale dynamics in protein side chains by (15)N relaxation dispersion NMR spectroscopy: Application to Asn and Gln residues in a cavity mutant of T4 lysozyme. *Journal of the American Chemical Society, 123*(5), 967–975.

Nikolova, E. N., Gottardo, F. L., & Al-Hashimi, H. M. (2012). Probing transient Hoogsteen hydrogen bonds in canonical duplex DNA using NMR relaxation dispersion and single-atom substitution [Research Support, N.I.H., Extramural]. *Journal of the American Chemical Society, 134*(8), 3667–3670.

Nikolova, E. N., Kim, E., Wise, A. A., O'Brien, P. J., Andricioaei, I., & Al-Hashimi, H. M. (2011). Transient Hoogsteen base pairs in canonical duplex DNA. *Nature, 470*(7335), 498–502.

Ohlenschlager, O., Haumann, S., Ramachandran, R., & Gorlach, M. (2008). Conformational signatures of 13C chemical shifts in RNA ribose. *Journal of Biomolecular NMR, 42*(2), 139–142.

Palmer, A. G., 3rd. (2014). Chemical exchange in biomacromolecules: Past, present, and future. *Journal of Magnetic Resonance, 241*, 3–17.

Palmer, A. G., 3rd., Kroenke, C. D., & Loria, J. P. (2001). Nuclear magnetic resonance methods for quantifying microsecond-to-millisecond motions in biological macromolecules. *Methods in Enzymology, 339*, 204–238.

Palmer, A. G., 3rd., & Massi, F. (2006). Characterization of the dynamics of biomacromolecules using rotating-frame spin relaxation NMR spectroscopy. *Chemical Reviews, 106*(5), 1700–1719.

Parisien, M., & Major, F. (2008). The MC-Fold and MC-Sym pipeline infers RNA structure from sequence data. *Nature, 452*(7183), 51–55.

Pelupessy, P., Chiarparin, E., & Bodenhausen, G. (1999). Excitation of selected proton signals in NMR of isotopically labeled macromolecules. *Journal of Magnetic Resonance, 138*(1), 178–181.

Petrov, A., Wu, T., Puglisi, E. V., & Puglisi, J. D. (2013). RNA purification by preparative polyacrylamide gel electrophoresis. *Methods in Enzymology, 530*, 315–330.

Piotto, M., Saudek, V., & Sklenar, V. (1992). Gradient-tailored excitation for single-quantum NMR spectroscopy of aqueous solutions. *Journal of Biomolecular NMR, 2*(6), 661–665.

Polson, A. G., Bass, B. L., & Casey, J. L. (1996). RNA editing of hepatitis delta virus antigenome by dsRNA-adenosine deaminase. *Nature, 380*(6573), 454–456.

Puglisi, J. D., Tan, R., Calnan, B. J., Frankel, A. D., & Williamson, J. R. (1992). Conformation of the TAR RNA-arginine complex by NMR spectroscopy. *Science, 257*, 76–80.

Rajkowitsch, L., Chen, D., Stampfl, S., Semrad, K., Waldsich, C., Mayer, O., et al. (2007). RNA chaperones, RNA annealers and RNA helicases. *RNA Biology, 4*(3), 118–130.

Rinnenthal, J., Buck, J., Ferner, J., Wacker, A., Furtig, B., & Schwalbe, H. (2011). Mapping the landscape of RNA dynamics with NMR spectroscopy. *Accounts of Chemical Research, 44*(12), 1292–1301.

Sathyamoorthy, B., Lee, J., Kimsey, I., Ganser, L. R., & Al-Hashimi, H. (2014). Development and application of aromatic [C, H] SOFAST-HMQC NMR experiment for nucleic acids. *Journal of Biomolecular NMR, 60*, 77–83.

Schwalbe, H., Buck, J., Furtig, B., Noeske, J., & Wohnert, J. (2007). Structures of RNA switches: Insight into molecular recognition and tertiary structure. *Angewandte Chemie (International Ed. in English), 46*(8), 1212–1219.

Sekhar, A., & Kay, L. E. (2013). NMR paves the way for atomic level descriptions of sparsely populated, transiently formed biomolecular conformers. *Proceedings of the National Academy of Sciences of the United States of America, 110*(32), 12867–12874.

Serganov, A., & Patel, D. J. (2007). Ribozymes, riboswitches and beyond: Regulation of gene expression without proteins. *Nature Reviews. Genetics, 8*(10), 776–790.

Skrynnikov, N. R., Dahlquist, F. W., & Kay, L. E. (2002). Reconstructing NMR spectra of "invisible" excited protein states using HSQC and HMQC experiments. *Journal of the American Chemical Society, 124*(41), 12352–12360.

Storz, G. (2002). An expanding universe of noncoding RNAs. *Science, 296*(5571), 1260–1263.

Tian, S., Cordero, P., Kladwang, W., & Das, R. (2014). High-throughput mutate-map-rescue evaluates SHAPE-directed RNA structure and uncovers excited states. *RNA, 20*, 1815–1826.

Trott, O., & Palmer, A. G., 3rd. (2002). R1rho relaxation outside of the fast-exchange limit. *Journal of Magnetic Resonance, 154*(1), 157–160.

Trott, O., & Palmer, A. G., 3rd. (2004). Theoretical study of R(1rho) rotating-frame and R2 free-precession relaxation in the presence of n-site chemical exchange. *Journal of Magnetic Resonance, 170*(1), 104–112.

Tucker, B. J., & Breaker, R. R. (2005). Riboswitches as versatile gene control elements. *Current Opinion in Structural Biology, 15*(3), 342–348.

Vallurupalli, P., Bouvignies, G., & Kay, L. E. (2012). Studying "invisible" excited protein states in slow exchange with a major state conformation. *Journal of the American Chemical Society, 134*(19), 8148–8161.

Venditti, V., Clos, L., 2nd., Niccolai, N., & Butcher, S. E. (2009). Minimum-energy path for a u6 RNA conformational change involving protonation, base-pair rearrangement and base flipping. *Journal of Molecular Biology, 391*(5), 894–905.

Wales, D. J., & Doye, J. P. K. (1997). Global optimization by Basin-Hopping and the lowest energy structures of Lennard-Jones clusters containing up to 110 Atoms. *Journal of Physical Chemistry A, 101*(28), 5111–5116.

Weeks, K. M., Ampe, C., Schultz, S. C., Steitz, T. A., & Crothers, D. M. (1990). Fragments of the HIV-1 Tat protein specifically bind TAR RNA. *Science, 249*(4974), 1281–1285.

Winkler, W., Nahvi, A., & Breaker, R. R. (2002). Thiamine derivatives bind messenger RNAs directly to regulate bacterial gene expression. *Nature, 419*(6910), 952–956.

Xu, X. P., & Au-Yeung, S. C. F. (2000). Investigation of chemical shift and structure relationships in nucleic acids using NMR and density functional theory methods. *Journal of Physical Chemistry B, 104*(23), 5641–5650.

Yamazaki, T., Muhandiram, R., & Kay, L. E. (1994). NMR experiments for the measurement of carbon relaxation properties in highly enriched, uniformly 13C, 15N-labeled proteins: Application to 13Ca carbons. *Journal of the American Chemical Society, 116*, 8266–8278.

Ying, J., Wang, J., Grishaev, A., Yu, P., Wang, Y. X., & Bax, A. (2011). Measurement of (1)H-(15)N and (1)H-(13)C residual dipolar couplings in nucleic acids from TROSY intensities. *Journal of Biomolecular NMR, 51*(1–2), 89–103.

Zhang, Q., Kang, M., Peterson, R. D., & Feigon, J. (2011). Comparison of solution and crystal structures of preQ1 riboswitch reveals calcium-induced changes in conformation and dynamics. *Journal of the American Chemical Society, 133*(14), 5190–5193.

Zhao, B., Hansen, A. L., & Zhang, Q. (2014). Characterizing slow chemical exchange in nucleic acids by carbon CEST and low spin-lock field R(1rho) NMR spectroscopy. *Journal of the American Chemical Society, 136*(1), 20–23.

Zinn-Justin, S., Berthault, P., Guenneugues, M., & Desvaux, H. (1997). Off-resonance rf fields in heteronuclear NMR: Application to the study of slow motions. *Journal of Biomolecular NMR, 10*(4), 363–372.

Zuker, M. (2003). Mfold web server for nucleic acid folding and hybridization prediction. *Nucleic Acids Research, 31*(13), 3406–3415.

CHAPTER THREE

Quantifying Nucleic Acid Ensembles with X-ray Scattering Interferometry

Xuesong Shi*, Steve Bonilla[†], Daniel Herschlag*,[1], Pehr Harbury*,[1]

*Department of Biochemistry, Stanford University, Stanford, California, USA
[†]Department of Chemical Engineering, Stanford University, Stanford, California, USA
[1]Corresponding authors: e-mail address: herschla@stanford.edu; harbury@stanford.edu

Contents

1. Ensembles and Energy Surfaces	76
2. The Experimental Challenge	77
3. Conceptual Framework for XSI	80
4. A General Protocol for XSI Measurement of Nucleic Acid Constructs	84
4.1 Getting Started: The Choice of Nucleic Acid System and Construct Design	84
4.2 Stage 1: Synthesis and Purification of Au Nanocrystals	85
4.3 Stage 2: Generating Au Nanocrystal Labeled Constructs	88
4.4 Stage 3: Collecting the X-ray Scattering Data	92
4.5 Stage 4: Analyzing the Data	93
Acknowledgments	96
References	96

Abstract

The conformational ensemble of a macromolecule is the complete description of the macromolecule's solution structures and can reveal important aspects of macromolecular folding, recognition, and function. However, most experimental approaches determine an average or predominant structure, or follow transitions between states that each can only be described by an average structure. Ensembles have been extremely difficult to experimentally characterize. We present the unique advantages and capabilities of a new biophysical technique, X-ray scattering interferometry (XSI), for probing and quantifying structural ensembles. XSI measures the interference of scattered waves from two heavy metal probes attached site specifically to a macromolecule. A Fourier transform of the interference pattern gives the fractional abundance of different probe separations directly representing the multiple conformation states populated by the macromolecule. These probe–probe distance distributions can then be used to define the structural ensemble of the macromolecule. XSI provides accurate, calibrated distance in a model-independent fashion with angstrom scale sensitivity in distances. XSI data can be compared in a straightforward manner to atomic coordinates determined experimentally or predicted by molecular dynamics simulations. We describe the conceptual framework for XSI and provide a detailed protocol for carrying out an XSI experiment.

1. ENSEMBLES AND ENERGY SURFACES

The first high-resolution DNA and protein structures reported in the 1950s suggested that biological macromolecules fold into static, well-defined conformations. Fifty years later, it is clear that a continuum of order exists, ranging from rigid structures to intrinsically disordered ones. Moreover, even the most rigid macromolecules populate a dynamic ensemble of conformational states. Measuring and quantitatively modeling these ensembles remains an open challenge in biophysics.

Why care about structural ensembles? At the most basic level, they are what macromolecules look like at atomic resolution. This information is then essential for developing an accurate mental picture of macromolecules, and an intuition for how they behave. Indeed, most structured nucleic acids and proteins must adopt a multiplicity of conformations to carry out their biological functions. The ribosome is a prime example (Fischer, Konevega, Wintermeyer, Rodnina, & Stark, 2010; Frank & Agrawal, 2000). Beyond that, ensembles play a central role in biological interactions, particularly in phenomena such as allostery and conformational capture. They are linked to macromolecular folding, embodied in concepts such as folding funnels and folding intermediates (Fenwick, Esteban-Martin, & Salvatella, 2011; Salmon, Yang, & Al-Hashimi, 2014). Ensembles determine the mechanical properties of macromolecules, such as their elastic response to stretching, bending, and twisting (Mathew-Fenn, Das, & Harbury, 2008; Olson, Gorin, Lu, Hock, & Zhurkin, 1998; Shi, Herschlag, & Harbury, 2013). In this capacity, understanding ensembles will also be important in the area of engineered nanostructures (Feldkamp & Niemeyer, 2006). Perhaps most importantly, experimental characterization of ensembles will lead to conceptual advances in the physical potentials that we use for molecular modeling. Computed conformational ensembles are the simplest and most meaty predictions of molecular dynamics simulations, and experimental measurements of ensemble structure are the corresponding ground truth.

Nucleic acid conformational ensembles are particularly rich and interesting. In structured RNA, the stable secondary structures (helices) are connected through a variety of types of linkers. Each linker type imposes a different tertiary structure, or ensemble, and the resulting ensembles can be diffused, with highly populated conformations differing dramatically from one another. In addition to their intrinsic structural heterogeneity, cellular metabolism imposes diverse conformations of nucleic acids, such as D-loops, mismatched bulges, kinks, hairpins, and tight spirals (Bacolla & Wells, 2004; Palecek, 1991). The energy landscape over this complex conformational space impacts basic biology. Sequence preferences for deformed DNA states, for example, influence the patterning of nucleosomes on the genome and the binding specificity of transcription factors (Kaplan et al., 2009; Rohs et al., 2010). Despite the importance of nucleic acid ensembles, our understanding of them is quite limited.

2. THE EXPERIMENTAL CHALLENGE

Ensembles are extremely hard to study experimentally. In part, this is due to technical challenges. First, the direct study of ensembles requires a sample in aqueous solution. This is incompatible with high-resolution, solid-state approaches such as X-ray crystallography. Second, most biophysical tools provide ensemble-averaged measurements of an observable. Thus, the measurement does not provide direct information about the individual substructures that make up the ensemble. Third, for many types of measurements, extracting structural information from the raw data is indirect and model dependent. This leads to uncertainty in the conclusions, and means that many different models of the ensemble are compatible with the data. Finally, it is sometimes difficult to compare a measured observable to results from a molecular simulation, because computing the predicted value for the observable is impractical or impossible. Despite these challenges, a variety of physical techniques have been applied to the ensemble problem. The different techniques yield complementary types of information. Here, we highlight two particular approaches that have been productive in recent years: nuclear magnetic resonance (NMR) and molecular rulers. We then present methodological details for a recently developed approach.

Several different NMR experiments are useful for studying ensembles. First, NMR relaxation measurements provide incisive information about the existence, extent, and timescale of motions within an ensemble (Palmer, 2004). Specifically, in simple two-state systems, relaxation dispersion measurements can detect a well-defined alternate conformation that is infrequently populated (down to the 0.5% level) (Dethoff, Petzold, Chugh, Casiano-Negroni, & Al-Hashimi, 2012; Mittermaier & Kay, 2006). In the general multistate case, however, relaxation experiments are blind to the specific subconformations that make up the ensemble. A second type of NMR experiment measures residual dipolar couplings (RDCs) between nuclear spins (Lange et al., 2008; Zhang, Stelzer, Fisher, & Al-Hashimi, 2007). RDCs provide ensemble-averaged values for the angles between common magnetic field and interspin vectors. RDCs also provide order parameters for these polar angles. Thus, it is possible to infer averaged angular relationships between structural elements of a macromolecule and the extent to which the angles vary within an ensemble. Finally, NMR is used to measure the rate of hydrogen-deuterium exchange of labile protons in a macromolecule (Woodward, Simon, & Tuchsen, 1982). These experiments can detect high-energy states, from which exchange occurs, as well as the protons that become solvent accessible in those states. Collectively, the great strengths of the NMR approaches are their dense information content, the ability to determine timescales for motions, and the ability to determine calibrated structural parameters (i.e., angles) in a model-free way. A limitation is that the observables are often averaged over heterogeneous conformations.

Molecular rulers represent a different attack on the ensemble problem. They are used to determine the distance between probes that are attached to a macromolecule. The use of probes provides high signal-to-noise for the experimental measurements, but it reduces the information density of the data relative to NMR approaches. Intramolecular distances are complementary to the orientation angles derived from RDCs. For macromolecules with a single dominant conformation, the distance data alone can specify a unique spatial arrangement of the probes relative to each other (allowing for global rotation, translation, or reflection).

The most broadly used molecular ruler is based on fluorescence resonance energy transfer (FRET) (Jares-Erijman & Jovin, 2003). Energy transfer results from dipolar coupling between two fluorophores that are attached site specifically to a macromolecule. The efficiency of energy transfer depends strongly on the distance between the fluorophores. However, FRET data only provide an ordinal estimate of intramolecular distances. This is because energy transfer depends on fluorophore orientation and dynamics in addition to distance, and because the mapping of the FRET signal-to-distance is highly nonlinear. The FRET ruler possesses two powerful attributes. First, it can be used in complex biological environments such as whole cells. This distinguishes it from most high-resolution structural techniques. Second, it can be applied to single molecules. Different values of the FRET observable indicate the existence of different conformations of the single molecule. The approach is capable of detecting long-lived conformations that are infrequently populated. Rapidly interconverting conformations within an ensemble, however, are generally invisible to single-molecule FRET techniques.

Below, we present methods for a new class of molecular ruler based on X-ray scattering interferometry (XSI) (Mathew-Fenn, Das, Silverman, Walker, & Harbury, 2008; Shi, Beauchamp, Harbury, & Herschlag, 2014). XSI measures the interference of scattered waves from atoms in a macromolecule, which is the same physical principle that underlies X-ray crystallography. In XSI, the scattering is dominated by two heavy metal probes attached site specifically to the macromolecule. Probe scattering interference causes the intensity of the interference pattern to oscillate sinusoidally with increasing scattering angle. The frequency of these oscillations is proportional to the distance between the probes. XSI data include signals from multiple different interprobe distances, because the macromolecule populates multiple conformations with different probe separations. A Fourier transform of the interference pattern directly gives the fractional abundance of each one. These data are referred to as a distance distribution. Because scattering occurs faster than any atomic motion, the distance distribution is an unaveraged snapshot of the intramolecular probe separations that coexist in solution. Distance distributions between many different pairs of probe attachment sites can help to define the ensemble of shapes that the macromolecule adopts.

The strength of XSI is that it provides accurate, calibrated distances in a model-independent fashion. The data resolve distinct substates of an ensemble, even when the substates are rapidly interconverting. The technique is sensitive to distance differences on the angstrom scale. Calibrated distance is a simple physical observable that is directly related to three-dimensional structure. Thus, XSI data can be compared straightforwardly to atomic coordinates that have been determined experimentally or have been predicted by molecular dynamics simulations. XSI does not provide timescale information, it cannot be applied at the single-molecule level, and it is not compatible with complex environments. Its strengths and limitations are complementary to those of the FRET ruler.

Below, we describe the conceptual framework for XSI, followed by a detailed protocol for running an experiment.

3. CONCEPTUAL FRAMEWORK FOR XSI

X-ray interferometry, short for XSI, is a variant of solution X-ray scattering (SAXS). In a conventional SAXS experiment, the summed scattering inference from all atoms in a macromolecule is measured. Standard SAXS can be used to determine the overall size and rough shape of a macromolecule but provides no site-specific distance information. Point-to-point distance measurements can be obtained if a macromolecule is labeled at two locations (denoted A and B) with strongly scattering spherical probes (Fig. 1). The scattering intensity profile of the double-labeled macromolecule, $I_{AB}(S)$, includes the probe-to-probe scattering interference pattern, $I_\Delta(S)$. Here, I denotes the intensity of X-ray scattering, and S is the momentum transfer, which is proportional to the scattering angle. $I_\Delta(S)$ is a weighted sum of interference patterns corresponding to different discrete interprobe distances D:

$$I_\Delta(S) = \sum_{D=1-1000\text{Å}} P(D) * f^2_{\text{probe}}(S) * \frac{\sin(2\pi SD)}{2\pi SD}.$$

The distance distribution function, $P(D)$, is the fractional abundance of probe pairs separated by a center-to-center distance D. $P(D)$ can be determined by factoring $I_\Delta(S)$ into a sum of cardinal sine functions. The amplitude of the function with frequency D is the fractional abundance of the interprobe distance D. The $f(S)$ term is the scattering form factor of the spherical probe, which is determined experimentally; this term describes

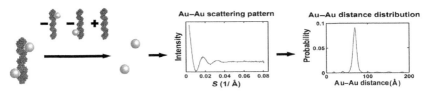

Figure 1 Obtaining a probe-probe distance distribution from X-ray scattering interferometry. (Left) A DNA duplex is labeled with gold nanocrystal probes. Subtracting the scattering profiles of the two singly labeled helices from the profiles of the doubly labeled and unlabeled helices gives the pattern of scattering interference between the two isolated gold probes (middle graph). This interference pattern is Fourier transformed into the probability distribution for the center-to-center distance between the two probes (right graph). The data shown are for two gold probes separated by 15 base steps within a 26 base-pair duplex. *Figure 1 is modified from figure 1 of Shi et al. (2013).*

the scattering properties of the probe due to its size and shape. Given an appropriate choice of distance bin resolution, there is a unique solution for P(D). Reference Mathew-Fenn, Das, Silverman, et al. (2008) presents a more detailed discussion.

One complication is that the measured scattering intensity profile, $I_{AB}(S)$, has components other than the interprobe scattering interference pattern, specifically the intramacromolecule and probe-macromolecule interference patterns. But these additional components can be independently determined from the scattering profiles of the isolated macromolecule, $I_{macro}(S)$, and of macromolecules that are singly labeled at site A and site B, $I_A(S)$ and $I_B(S)$, and the extra components can then be subtracted off:

$$I_\Delta(S) = I_{AB}(S) - c_A I_A(S) - c_B I_B(S) + c_{macro} I_{macro}(S).$$

Due to uncertainty in sample concentration, the scaling factors c_A, c_B, and c_{macro}, can deviate from unity and need to be determined by an optimization procedure (Mathew-Fenn, Das, Silverman, et al., 2008). The optimal coefficients are chosen so that the oscillations in an S-weighted $I_\Delta(S)$ profile sum to zero and so that unphysical negative components of $P(D)$ are minimized.

The information content of XSI distance distributions depends on several experimental factors. The first factor is the scattering power of the probes relative to the macromolecule. This ratio determines the highest value of S at which $I_\Delta(S)$ can be measured over noise, and sets the distance resolution of the corresponding distance distribution. A second factor is how spherical and uniformly sized the probes are. These properties influence the extent to which macromolecular conformation can be inferred from probe center-to-center distances. A third factor is how rigidly the probes are affixed to the macromolecule. Compact and highly localized probes will provide the finest level of structural detail. In pioneering XSI work, heavy metal probes of just 1–4 atoms were used (Miakelye, Doniach, & Hodgson, 1983; Vainshtein et al., 1980). Alternatively, deuterated ribosomal subunits were used as neutron-scattering probes (Capel, Kjeldgaard, Engelman, & Moore, 1988). The marginal scattering power of these probes and the large size of the deuterated subunits limited the resolution.

The XSI approach detailed below utilizes spherical gold nanocrystals composed of 68 atoms (Fig. 2; Mathew-Fenn, Das, & Harbury, 2008; Shi et al., 2013). These nanocrystals scatter as strongly as a 20 kDa nucleic acid. Larger ellipsoidal nanocrystals composed of ~10,000 gold atoms have been utilized in other work (Hura et al., 2013; Mastroianni, Sivak, Geissler, &

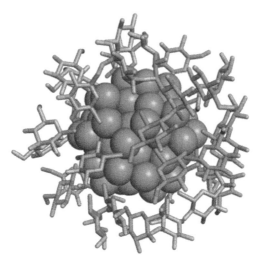

Figure 2 An illustration of the Au nanocrystal (spheres) with its thioglucose protection shell. The nanocrystal coordinates are based on a substructure of the nanocrystal reported in Jadzinsky, Calero, Ackerson, Bushnell, and Kornberg (2007) and the experimental analysis of Dass (2009). *Figure 2 is reproduced from figure 2C of Shi et al. (2013).*

Alivisatos, 2009). The variable size and shape of the ellipsoids reduced the resolution of the XSI distance measurements. Nevertheless, they scatter 15,000-fold more strongly than the 68-atom nanocrystals and are thus compatible with larger macromolecular complexes and enable measurements at sample concentrations down to 100 nM (Hura et al., 2013). At the opposite extreme, it should be possible to implement XSI with very small clusters. Gold atoms rigidly embedded in the side chain of an amino acid, for example, could provide high-resolution information about protein structure and fluctuations. This approach would require accurate matching of the solvent electron density to the protein electron density to minimize scattering by the protein.

Over the last few years, XSI has been fruitfully applied. Initially, distance distribution measurements on short 3′-end-labeled DNA helices revealed a previously unknown cooperative stretching motion (Mathew-Fenn, Das, & Harbury, 2008). A subsequent study analyzed long DNA fragments approaching one persistence length (Mastroianni et al., 2009). XSI was then used to measure the twisting and bending elasticity of DNA helices at short length scales (Shi et al., 2013). More recent work probed the solution conformational ensemble of a DNA bulge (Fig. 3; Shi et al., 2014). This model helix–junction–helix motif adopts heterogeneous structures in solution.

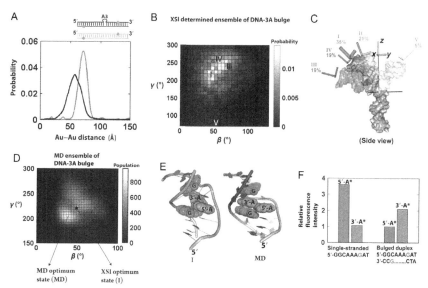

Figure 3 An ensemble model for a three adenosine DNA (DNA-3A) bulge based on XSI data. (A) Comparison of the Au–Au center-to-center distance distributions for a 26-base pair DNA helix (gray) and for DNA-3A, a 26-base pair DNA helix with a three adenosine bulge (black). (B) A heat map showing the conformational ensemble of the DNA-3A bulge determined by XSI. β and γ represent, respectively, the bending angle and the bending direction of the top helix relative to the bottom helix (see panel C). The model is based on six XSI distance distributions with different labeling positions for the Au nanocrystal probes. The distribution in panel A is one of the six data sets. (C) Five representative conformations of the DNA-3A ensemble (I–V) are shown as three-dimensional models. (D) A heat map showing the conformational ensemble of the DNA-3A bulge generated by an molecular dynamics (MD) ensemble simulation. The MD ensemble differs significantly from the XSI ensemble. For example, the most highly populated conformer in the XSI ensemble (state I, black star) is sparsely populated in the MD ensemble. (E) Atomic-level models of the most populated conformer from the XSI ensemble (state I) and from the MD. The XSI model predicts that the 5′ adenosine of the bulge is stacked on a flanking guanine, whereas the MD model predicts that the 3′ adenosine of the bulge is stacked on the flanking guanine. (F) Test of atomic-level models. The 5′ and 3′ adenosines of the bulge were replaced with fluorescent 2-aminopurine analogs. Stacking of 2-aminopurine on a guanine base quenches its fluorescence. Thus, the XSI model predicts that the 5′ 2-aminopurine fluorophore (A*) should be quenched when the bulge is formed from single-stranded oligonucleotide precursors, whereas the MD model predicts that the 3′ 2-aminopurine fluorophore (A*) should be quenched. Experimentally, quenching is observed at the 5′ position, consistent with the XSI model. *Panels (A–F) are reproduced from figures 1D, S6A, 4C, S12, and 5 of Shi et al. (2014).* (See the color plate.)

The DNA bulge experiments established the utility of XSI for studying complex ensemble systems. XSI has also been successfully applied to RNA and RNA-protein complexes (Shi et al., manuscript in preparation). Finally, large structural intermediates in bacterial DNA mismatch repair have been characterized by XSI (Hura et al., 2013).

The future directions and unsolved problems for XSI are numerous and include developing in-depth knowledge of the ensembles of RNA motifs and structured RNAs and RNA-protein complexes across a variety of solution conditions, all problems inaccessible by traditional approaches. Principal among unexplored areas is the application of XSI to proteins, which awaits an advance in probe attachment strategies. Another outstanding problem is to create large, spherical, and monodisperse nanoparticle probes. If they existed, these probes could be used to accurately measure structural rearrangements in biological supercomplexes, such as the ribosome. The scattering power of spherical XSI probes grows as the sixth power of their diameter. Thus, a modest increase in probe size can allow probe scattering to overwhelm the scattering of large biological assemblies. A third possible direction is single-molecule XSI using a free electron laser. This could provide correlated distance measurements between multiple sites on a macromolecule. Finally, new motifs for rigidly attaching scattering probes to nucleic acid and protein secondary structures could simplify the setup and interpretation of XSI experiments.

4. A GENERAL PROTOCOL FOR XSI MEASUREMENT OF NUCLEIC ACID CONSTRUCTS

This protocol is aimed to provide a detailed guide for carrying out an XSI experiment. The protocol is divided into several sections that roughly correspond to the chronological order of a typical project.

4.1 Getting Started: The Choice of Nucleic Acid System and Construct Design

The current protocol uses thioglucose passivated gold nanocrystals with a diameter of 1.2 nm (Fig. 2). These probes are adequate for nucleic acids up to 40 base pairs in size. We are not aware of larger spherical nanocrystals that are monodisperse and suitable for XSI. Making such probes would be a significant technical advance.

Note 1: *Proteins scatter X-rays more weakly than nucleic acids, and can be contrast matched with buffers containing sucrose or heavy salts* (Stuhrmann & Miller,

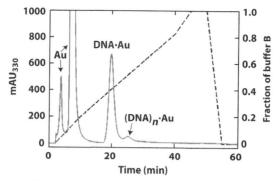

Scheme 1 Two sets of hypothetical 15-mer DNA constructs. The top quartet is labeled with gold nanocrystal probes at its ends, and the bottom quartet is labeled with gold nanocrystals at internal thymidine residues. The quartets are assembled from four single-stranded oligonucleotides. Within each quartet, "W" denotes the Watson strand and "C" denotes the Crick strand. The superscripts "3'-OH" and "3'-Au" indicate that the 3'-terminus of the respective oligonucleotide is a free hydroxyl group or is coupled to a gold nanocrystal. The superscripts "T-H" and "T-Au" indicate that the 5-methyl group of the internal thymine base is coupled to a proton or is coupled to a gold nanocrystal.

1978). *In protein–nucleic acid complexes, protein size is less limiting than nucleic acid size on a per kilo Dalton basis.*

Gold nanocrystals can be attached at the 3'- or 5'-end of a nucleic acid strand (end labeled) or to a modified base within the strand (internally labeled). To avoid undesirable fraying of terminal base pairs, it is advisable to put a G–C cap at the ends of helices. For internally labeled constructs, we recommend that labeling sites be placed at least three base pairs inward from the helix ends.

In the following protocol, we describe two hypothetical 15-mer DNA constructs, one end labeled, and the other internally labeled (Scheme 1).

4.2 Stage 1: Synthesis and Purification of Au Nanocrystals

Steps 1.1 and 1.2 prepare FPLC columns.

Two FPLC columns are used for the purification of gold nanocrystals, a Sephadex G15 column (GE Healthcare) for rapid desalting and a Superdex 30 column (GE Healthcare) for size exclusion purification.

1.1. The G15 column (50 mL in a 26/10 column housing) is first washed with two column volumes of water, followed by four column volumes of cleaning solution (20 mM Tris–HCl, pH 8.0, and 100 mM DTT) at 2 mL/min. The column is then equilibrated with four column volumes of water.

1.2. The Superdex 30 column (120 mL in a 16/60 column housing) is first washed with two column volumes of water at 0.75 mL/min, followed by 4–5 column volumes of cleaning solution and another 4–5 column volumes of water to remove the residual DTT. The column is then equilibrated with two column volumes of 150 mM ammonium acetate, pH 5.6, the running buffer for nanocrystal purification.

Note: The DTT in the cleaning solution removes gold residue left behind by previous samples. The residue causes a brown discoloration of the white chromatography resin. The column should turn white again after it is treated with the cleaning solution.

Steps 1.3–1.9 synthesize Au nanocrystals.

1.3. Prepare a 250-mL round-bottom flask, washed, dried, and cooled. Add a stir bar and cap the flask with a vented septum. Mount the flask above a magnetic stir plate inside a dark fume hood.

1.4. Prepare a 72 mL solution of 5:1 by volume methanol–acetic acid.

1.5. Weigh 0.544 g of hydrogen tetrachloroaurate (III) hydrate (Strem Chemicals 79-0500). The gold salt is light sensitive. Immediately transfer the powder to the round-bottom flask. Add 36 mL of the 5:1 methanol–acetic acid solution to the flask. The stirred solution should be clear with a bright orange color.

Note: The hydrogen tetrachloroaurate (III) hydrate bottle should be prewarmed to room temperature on a bench before it is opened. This minimizes the condensation of water inside the bottle. The bottle should be flushed with nitrogen or argon after use.

1.6. Dissolve 1 g of 1-thio-β-D-glucose (Sigma T6375 or Santa Cruz Biotech sc-216128) in the remaining 36 mL of the 5:1 methanol–acetic acid solution using a vortexer. Add the thioglucose solution to the stirred reaction flask. The mixture in the round-bottom flask should become cloudy.

1.7. Add 0.9 g of sodium borohydride to 20 mL of water in a 50 mL Falcon tube. Dissolve the solid by inverting the tube two to three times, or by vortexing briefly. *How you perform the next step is critical to the quality of the nanocrystal synthesis!* Add the sodium borohydride solution to the reaction flask in a dropwise manner over 3–4 min. Use a 1-mL pipettor.

Note: To maintain a more constant addition speed, a glass addition funnel with a teflon stopcock can be used instead of the pipettor. The funnel needs to be uncapped during the addition and the teflon stopcock needs to preadjusted to ensure a suitable dropwise draining rate.

1.8. Allow the reaction to stir at ambient temperature for another 30 min.

1.9. Reduce the volume of the crude reaction product to about 12 mL using a rotary evaporator.

Steps 1.10–1.15 purify the Au nanocrystals.

1.10. Keep the crude Au nanocrystal preparation on ice or in a cold room. Filter the nanocrystal solution with a 0.22-μm filter to remove particulates. The filtered Au nanocrystals solution is then ready for FPLC purification.

1.11. Desalt the crude preparation on a G15 FPLC column (see step 1.1). Load a maximum of 6 mL of the Au nanocrystals solution, and then pump water over the column at 2 mL/min. The nanocrystal peak can be detected by absorption at any wavelength between 260 and 360 nm. Au nanocrystals of 1.2–1.4 nm in diameter elute in the 9–12 min window. Collect the main nanocrystal peak, and avoid the salt front by monitoring conductivity. Repeat step 1.11 until the entire volume of the crude material is desalted.

Note: As the crude Au nanocrystals are relatively unstable at high salt, high pH, and high temperature, try to minimize unnecessary downtime in steps 1.10 and 1.11. Before loading onto the G15 column, check the solution for visible precipitates and refilter if necessary.

Note: Once the Au nanocrystal solution is desalted, it is safe to store it at 4 °C for days or at −20 °C for months.

1.12. Reduce the volume of the desalted Au nanocrystal solution to about 12 mL by either centrifugal filtration (3K Amicon) or rotary evaporation.

1.13. Inject up to 6 mL of the concentrated Au nanocrystal solution onto the Superdex 30 column. Pump 150 mM ammonium acetate, pH 5.6, over the column at 0.75 mL/min for 205 min. Collect only the center of the largest UV-absorbing peak and avoid the shoulder region to ensure nanocrystal size homogeneity. Repeat step 1.13 until the entire volume of the desalted Au nanocrystal solution is purified. Immediately after each iteration of step 1.13, repeat step 1.12 to reduce the solution volume.

1.14. Immediately repeat step 1.11 to desalt the purified Au nanocrystals.

1.15. The concentration of the final purified and desalted Au nanocrystals is determined by UV absorption using an extinction coefficient of 0.076 μM/cm at 360 nm.

Note: The size distribution of the Au nanocrystals can be determined from their X-ray scattering profile (see step 4.1 below). The purity and homogeneity of the Au nanocrystals can also be checked by 15% denaturing PAGE.

4.3 Stage 2: Generating Au Nanocrystal Labeled Constructs

We describe the procedures below using an end-labeled and an internally labeled duplex (Scheme 1) as examples. Although the examples are in DNA, the same procedure also applies to RNA. Suggested minor modifications to the procedure for RNA are noted at steps 2.9 and 2.17.

Steps 2.1–2.3 synthesize and purify single-stranded oligonucleotides.

2.1. Purchase amine-modified oligonucleotides at the 200-nmol scale from a commercial vendor. Larger quantities may be needed for repeated measurements. Alternatively, practitioners can synthesize these oligonucleotides using a DNA/RNA synthesizer (e.g., ABI 394). As illustrated in Scheme 1, oligonucleotides $W^{3'-OH}$, $C^{3'-OH}$, W^{T-H}, and C^{T-H} are unmodified. Oligonucleotides $W^{3'-Au}$ and $C^{3'-Au}$ incorporate a 3'-thiol modification (Glen Research C3-S-S, 20-2933). Oligonucleotides W^{T-Au} and C^{T-Au} incorporate an amine-modified T base (Glen Research amino-modifier C2 dT, 10-1037) at positions intended for gold labeling.

Note: The 3'-thiol modification (C3-S-S) contains a short three-carbon linker between the terminal phosphate and the sulfur atom. We are not aware of a comparably short 5'-thiol modification. Longer six-carbon thiol modifications are commercially available for both the 3'- and 5'-ends of oligonucleotides.

Note: For the 3'-thiol modification (C3-S-S), leave the disulfide bond intact.

Note: It is advised to order oligonucleotides as "DMT-on." These oligonucleotides retain their 5'-terminal dimethoxytrityl protecting group and can be cleaned up over a reverse phase cartridge (e.g., Glen Research Poly-Pak cartridges).

2.2. Purify the single-stranded oligonucleotides by anion exchange HPLC using a Dionex DNAPac semipreparative column (9 × 250 mm) at 3 mL/min. Oligonucleotides are eluted with a salt gradient formed from buffer A (10 mM NaCl, 20 mM Na-borate, pH 7.8) and buffer B (1500 mM NaCl, 20 mM Na-borate, pH 7.8).

Note: Borate buffer is used instead of an amine-containing buffer, which would interfere with the reaction in step 2.6 below.

Note: Before each HPLC run, particulates should be removed. The samples can be spun in a microfuge for 4 min at 4 °C, followed by transfer of the cleared solution to a new tube. Alternatively, the samples can be passed through a 0.22 μm filter.

2.3. Desalt and concentrate the HPLC-purified oligonucleotides by centrifugal filtration (3K Amicon) or ethanol precipitation. Check the purity by PAGE, analytical HPLC, or capillary electrophoresis. Typical yields from the HPLC purification are 30–60%, depending on the purity of the synthesized oligonucleotides.

Note: One typically gets ~120 nmol of crude oligonucleotide from a 200 nmol scale synthesis (step 2.1). Following step 2.3, ~60 nmol of purified single-stranded oligonucleotide is recovered.

Steps 2.4–2.11 introduce a thiol moiety into oligonucleotides with internal amine modifications.

Note: The internally amino modified W^{T-Au} and C^{T-Au} must be reacted with SPDP (N-succinimidyl 3-[2-pyridyldithio]-propionate) to introduce a disulfide group. This is described in steps 2.4–2.8. The oligonucleotides $W^{3'-Au}$ and $C^{3'-Au}$ arrive from the vendor with 3' disulfide modifications, so they do not require any additional chemical steps.

2.4. Reduce the volume of W^{T-Au} and C^{T-Au} (~60 nmol) to less than 160 μL. Add water to bring the total volume to 160 μL. Then, add 20 μL of 1 M borate buffer (pH 7.8).

2.5. Make an SPDP (N-succinimidyl 3-[2-pyridyldithio]-propionate) solution by dissolving SPDP in DMSO (1 mg SPDP per 10 μL of DMSO). A total of 40 μL of SPDP solution is required for each oligonucleotide (a total of 80 μL of SPDP solution for the W^{T-Au} and C^{T-Au} pair).

2.6. Preheat the oligonucleotide solutions to 37 °C. Add 20 μL of SPDP solution to each tube. Mix by pipetting up and down, and then incubate for 30 min at 37 °C.

2.7. After the 30 min, add another 20 μL of SPDP solution and incubate for another 30 min at 37 °C.

2.8. Add 2 μL of 2 M MgCl$_2$ and 1 mL of cold ethanol. After gentle mixing, incubate the reaction tubes in dry ice for 40 min. Spin the tubes for 30 min at maximum speed using a microfuge at 4 °C, and remove the supernatant. Wash by adding 1 mL of cold ethanol. Spin for another 15 min at 4 °C, and then remove the supernatant.

Note: W^{T-Au} and C^{T-Au} are now internally labeled with a disulfide group. The typical yield for steps 2.4–2.7 is close to 100%.

2.9. Reduce the disulfide bonds in $W^{3'-Au}$, $C^{3'-Au}$, W^{T-Au}, and C^{T-Au} by dissolving them in 150 μL of 200 mM DTT, 50 mM Tris–HCl, pH 9.0. Incubate the solutions at 50–70 °C for 30 min.

Note: 50 °C for RNA.

2.10. Purify the oligonucleotides by ethanol precipitation as in step 2.8.

2.11. Dissolve the pellets in 500 µL of water, and remove residual DTT by centrifugal filtration at 4 °C (3K Amicon, 14,000 × g, 30 min). Determine the concentration of each oligonucleotide by UV absorption (using a Nanodrop for example).

Note: Excess DTT destabilizes nanocrystal-DNA conjugates. If a significant amount of residual DTT is observed, indicated by a strong absorbance at 230 nm and a shoulder peak at ≥300 nm, repeat step 2.11.

Steps 2.12–2.15 label the single-stranded oligonucleotides with Au nanocrystals.

Perform step 2.12 immediately after completing step 2.11.

2.12. This step uses a 6-to-1 molar ratio of purified and desalted Au nanocrystals to oligonucleotide. For example, use 300 nmol of Au nanocrystals for 50 nmol of oligonucleotide from step 2.11. Add the oligonucleotide solution to the Au nanocrystal solution and gently vortex. Add 20 µL of 1 M Tris–HCl, pH 9.0, and gently vortex again. Incubate for 2 h at room temperature.

2.13. Stop the reaction with 15 µL of 2 M ammonium acetate, pH 5.6, and store the samples on ice.

2.14. Purify the Au-labeled oligonucleotides by anion exchange HPLC using a Dionex DNAPac column. The conjugates are eluted with a salt gradient formed from buffer A (10 mM NaCl, 20 mM ammonium acetate, pH 5.6) and buffer B (1500 mM NaCl, 20 mM ammonium acetate, pH 5.6). Monitor the absorbance at 260 nm, where nanocrystals and nucleic acids both absorb. Also monitor absorbance at 330 nm, where only nanocrystals absorb. A sample chromatogram is provided in Fig. 4.

Note: The 1:1 Au-oligonucleotide conjugates elute from the DNAPac column earlier than unlabeled oligonucleotides of the same length. Depending on oligonucleotide length, the conjugates come off of the column in the 25–50%B range. The conjugates absorb at 330 nm whereas the free oligonucleotides do not. There should be almost no free oligo remaining after step 2.12. Unmodified gold nanocrystals do not bind strongly to the column.

Note: There will be by-products consisting of gold nanocrystals coupled to multiple oligonucleotides. These species elute later than the 1:1 nanocrystal-oligonucleotide conjugate.

2.15. Buffer exchange the nanocrystal-oligonucleotide fractions into water by centrifugal filtration at 4 °C (3K Amicon, 3500 × g with a swinging bucket rotor for 30–40 min; repeat three times).

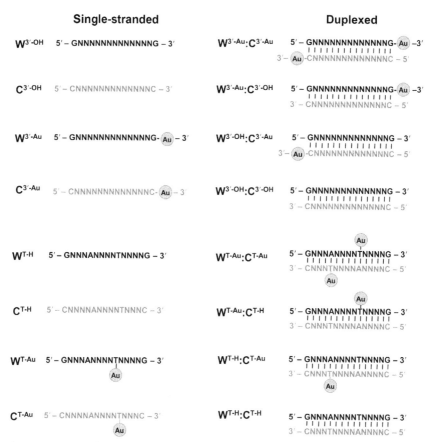

Figure 4 A sample HPLC chromatogram of Au-labeled oligonucleotides. An ion-exchange HPLC chromatogram of a crude coupling reaction between gold nanocrystals and thiol-modified DNA. The different products and biproducts include uncoupled gold nanocrystals (Au), a 1:1 complex of a gold nanocrystal and a 21-mer single-stranded DNA (DNA·Au), and 1:multiple complexes of gold nanocrystals and 21-mer single-stranded DNAs (DNA_n·Au).

2.16. Determine the concentration of the purified conjugates by absorption at 260 nm. Calculate the extinction coefficient of the conjugate as the sum of the extinction coefficient for the oligonucleotide component and an extinction coefficient of 0.215 μM/cm for the gold nanocrystal component.

Note: After desalting, the samples can be stored at $-20\,°C$ for extended periods of time before step 2.17.

Note: A typical yield for steps 2.12–2.16 is 20%. Starting with 50 nmol at step 2.12, roughly 10 nmol are recovered after step 2.16.

Steps 2.17–2.22 prepare the final duplexed constructs for XSI measurements.

2.17. Mix pairs of complementary single-stranded oligonucleotides in a 1:1 molar ratio. Incubate the solutions at room temperature for 30 min.
Note: For RNA, incubate, the solutions at 40 °C for 30 min.

2.18. Purify the annealed samples by anion exchange HPLC as in step 2.14. The duplex DNA elutes roughly 5% Buffer B (See Step 2.2 above for the definition of Buffer B) later than the single-stranded starting materials.

2.19. Buffer exchange the duplex DNA into water by centrifugal filtration at 4 °C (10K Amicon, $3500 \times g$ with a swinging bucket rotor for 15 min, repeat three times).

2.20. Reduce the volume of the desalted samples to about 30–40 μL by centrifugal filtration at 4 °C (10K Amicon, $10,000 \times g$ with a 0.5 mL filter unit for 20 min).

2.21. Determine the concentration of the purified samples by absorption at 260 nm and 360 nm. Calculate the extinction coefficient of the double-stranded conjugate as the sum of the extinction coefficients for the DNA component and the gold nanocrystal component.

2.22. Store the samples in a −20 or a -80 °C freezer. Desalted samples free of DNase and RNase contamination are stable for at least months.
Note: The typical yield for steps 2.17–2.22 is about 50%. Starting with 10 nmol of each single-stranded oligonucleotide at step 2.17, 5 nmol of duplexed product is obtained. Each XSI measurement requires 0.9 nmol of sample.

4.4 Stage 3: Collecting the X-ray Scattering Data

The XSI measurements are carried out at a synchrotron beamline configured for small-angle X-ray scattering. An example is beamline 4-2 of the Stanford Synchrotron Radiation Lightsource (SSRL).

3.1. Use a sample-to-detector distance that covers a q range from 0.01 to 1 Å^{-1}. This corresponds to 1.1 m at 11 keV on SSRL beamline 4-2.

3.2. Make a $10\times$ buffer solution containing Tris–HCl, sodium ascorbate, the desired concentration of additional salt, and any other components required for the experiment. An example of a $1\times$ buffer is 70 mM Tris–HCl, pH 7.4, 10 mM sodium ascorbate, 150 mM NaCl, and 1 mM MgCl$_2$.
Note: Ascorbate functions as radical scavenger, as does Tris to a lesser extent. A radical scavenger is essential; it protects the DNA from oxidative damage.

3.3. Before measurement, the samples are thawed and vortexed. Prepare a 200 μM solution of Au nanocrystals in 1 × buffer. Spin the solution in a microfuge for 2 min at 10,000 × g and 4 °C. This removes potential large particle contaminants that can dominate the scattering.

3.4. Set up the data collection as 10 repeats of a 3 s exposure. In general, it is advisable not to reuse the sample.

Note: Oxidative damage to the sample can be detected by a gradual change in scattering intensity over the ten repeats. Alternatively, samples can be recovered after exposure to X-rays and analyzed by HPLC. Do not use data collected under conditions that cause oxidative damage.

3.5. Measure the scattering of the pure Au nanocrystal solution. Repeat step 3.5 for a solution of the buffer alone.

Note: The scattering profiles of Au nanocrystals under different salt conditions are nearly identical.

Note: It is advisable to measure scattering for a dilution series of the Au nanocrystals, for example, at 200, 100, 50, and 25 μM. The shape of the scattering profiles should be independent of concentration. A concentration dependence of the profile at low q can result from interparticle scattering. Work in a concentration range where interparticle scattering is negligible.

3.6. Prepare a quartet of samples for data collection. The quartet consists of an unlabeled duplex, two singly labeled duplexes, and a doubly labeled duplex (e.g. see Fig. 1 and Scheme 1). The nanocrystal–oligonucleotide conjugates should be at 30 μM concentration in 1 × buffer. Spin the samples as in step 3.3. Measure the scattering profiles of the quartet in a direct sequence, so that the data collection conditions are as closely matched as possible.

Note: Measurements typically require 30 μL of solution (0.9 nmol of sample). In cases, where the concentration of phosphate in the nucleic acid backbone is comparable to the concentration of a counterion in the buffer, for example, at very low concentrations of $MgCl_2$, the samples should be prepared by exchange into 1 × buffer using a centrifugal filter (10K Amicon).

3.7. Repeat step 3.6 with additional quartets, with alternate buffers, and with additional experimental components such as binding partners.

4.5 Stage 4: Analyzing the Data

Data processing is carried out using beamline software and custom MATLAB scripts. The conceptual basis for the processing algorithms is described in detail elsewhere (Mathew-Fenn, Das, Silverman, et al., 2008).

A brief overview of the analysis pipeline is summarized here. Following regular SAXS data processing (step 4.1), the radius distribution of the Au nanocrystals is first determined. This is done by decomposing the measured gold auto-scattering signal into a sum of scattering profiles for solid spheres of varying radius (step 4.2). The radius distribution is then used to calculate IqD, which are basis functions for scattering interference between two nanocrystals separated by varying center-to-center distances (step 4.3). Independently, the scattering profiles of the double-, single-, and non-labeled constructs are each decomposed into a sum of simple cardinal sine functions (these are denoted $P(D)$). These crude transforms are used to optimize scaling coefficients so that the sum $P_\Delta(D) = P_{AB}(D) - c_A P_A(D) - c_B P_B(D) + c_{macro} P_{macro}(D)$ does not include negative values. Following optimization of the scaling coefficients, a final high-resolution probability distribution for nanocrystal-nanocrystal center-to-center distances is obtained by decomposing the $I_\Delta(S)$ interference profile into a sum of the IqD basis functions using a maximum entropy procedure (step 4.5).

Below, we provide the step-by-step guide for obtaining the nanocrystal-nanocrystal scattering interference profile, and the center-to-center distance distribution between a pair of nanocrystals attached to a nucleic acid construct. The custom MATLAB scripts can be downloaded from the open-access website for (Mathew-Fenn, Das, Silverman, et al., 2008). The procedure to compute an ensemble from distance distribution data is described elsewhere (Shi et al., 2014, 2013).

4.1. Reduce the raw data to a one-dimensional X-ray scattering profile. The raw scattering data are a two-dimensional matrix of photon intensities as a function of position on the X-ray detector. Using beamline-specific software, the matrix is integrated into a one-dimensional scattering profile. The data reduction program at SSRL beamline 4-2 is named sastool. The processed data consist of ordered pairs of scattering intensity and magnitude of the scattering vector. There are two conventions for expressing the magnitude of the scattering vector, represented by the symbols q and S. Specifically, q is calculated as $4\pi\sin(\theta/\lambda)$, where θ is half of the scattering angle and λ is the wavelength of the X-ray radiation. S is calculated as $q/(2\pi)$. q and S have units of Å^{-1}. Both are linearly proportional to scattering angle for small angles. Ten separate scattering profiles are measured for each sample (step 3.4). The variance in the measured scattering intensities for each value of q (or S) is included in the data set. The data should be truncated at very low values of q to remove beamline artifacts. This is generally done by visual inspection.

4.2. Obtain a volume-weighted Au nanocrystal radius distribution, denoted PAU6. First, subtract the buffer scattering profile from the pure Au nanocrystal scattering profile (these data are obtained in step 3.5). The buffer-corrected profile is denoted I_Au. Then, obtain PAU6 with the MATLAB function call: PAU6=lsqnonneg(gen_IqR(S, 1:100), I_Au). S is a vector containing the magnitudes of the scattering vectors for each intensity value in I_Au.

4.3. Obtain basis functions for the scattering interference profile of a pair of Au nanocrystals at different center-to-center distances. These are stored in a matrix denoted IqD. Obtain IqD with the MATLAB function call: IqD=gen_IqD(gen_Ic2c0(PAU6, 1:100), S), S, 1:200). The resulting IqD matrix includes interference basis functions for inter-nanocrystal distances of 1-200 Å in 1 Å increments.

4.4. Obtain the nanocrystal-nanocrystal scattering interference profile, denoted I_delta. I_delta is computed by subtracting the two singly labeled scattering profiles from the doubly labeled scattering profile. The profiles must be properly scaled relative to each other. Obtain I_delta with the MATLAB function call: I_delta=gen_Idelta(S, I_U, I_A, I_B, I_AB, I_Au, I_Buf, 1). I_U, I_A, I_B, I_AB, I_Au, and I_Buf are, respectively, the scattering profiles of the unlabeled duplex, the duplexes that are singly labeled at sites A and B, the doubly labeled duplex, the Au nanocrystals and the buffer. The gen_Idelta function also returns the optimal coefficients for scaling the scattering profiles. These are denoted c_U, c_A, c_B, c_AB, c_Au, and c_Buf. The variance in I_delta for each value of S, denoted V_delta, is returned by the script.

Note: The coefficient c_Au was originally included to correct for free Au nanocrystals in the experimental samples. With the improved sample-preparation protocols described above, free Au nanocrystals are negligible. Therefore, c_Au can be set to 0.

4.5. Obtain the center-to-center distance distribution between the pair of nanocrystals, denoted $P(D)$. The script is based on a maximum entropy optimization. Obtain $P(D)$ with the MATLAB function call: [~,P_D]=jacknife(S, I_delta, V_delta, IqD, ICF).

Note: A canonical ICF (intrinsic correlation function matrix) is included with the MATLAB script download. If desired, alternative ICF matrices can be generated by modifying the script gen_ICF.

ACKNOWLEDGMENTS

We thank H. Tsuruta (deceased), T. Matsui, and T. Weiss at beamline 4-2 of the Stanford Synchrotron Radiation Lightsource (SSRL) for their technical support on synchrotron experiments and during the development of the X-ray scattering interferometry technique reported here. This work was supported by NIH grants PO1 GM066275 (D.H.) and DP-OD000429-01 (P.B.H.).

REFERENCES

Bacolla, A., & Wells, R. D. (2004). Non-B DNA conformations, genomic rearrangements, and human disease. *Journal of Biological Chemistry*, 279(46), 47411–47414.

Capel, M. S., Kjeldgaard, M., Engelman, D. M., & Moore, P. B. (1988). Positions of S2, S13, S16, S17, S19 and S21 in the 30-S-ribosomal subunit of escherichia-coli. *Journal of Molecular Biology*, 200(1), 65–87.

Dass, A. (2009). Mass spectrometric identification of Au-68(SR)(34) molecular gold nanoclusters with 34-electron shell closing. *Journal of the American Chemical Society*, 131(33), 11666–11667.

Dethoff, E. A., Petzold, K., Chugh, J., Casiano-Negroni, A., & Al-Hashimi, H. M. (2012). Visualizing transient low-populated structures of RNA. *Nature*, 491(7426), 724–728.

Feldkamp, U., & Niemeyer, C. M. (2006). Rational design of DNA nanoarchitectures. *Angewandte Chemie-International Edition*, 45(12), 1856–1876.

Fenwick, R. B., Esteban-Martin, S., & Salvatella, X. (2011). Understanding biomolecular motion, recognition, and allostery by use of conformational ensembles. *European Biophysics Journal*, 40(12), 1339–1355.

Fischer, N., Konevega, A. L., Wintermeyer, W., Rodnina, M. V., & Stark, H. (2010). Ribosome dynamics and tRNA movement by time-resolved electron cryomicroscopy. *Nature*, 466(7304), 329–333.

Frank, J., & Agrawal, R. K. (2000). A ratchet-like inter-subunit reorganization of the ribosome during translocation. *Nature*, 406(6793), 318–322.

Hura, G. L., Tsai, C. L., Claridge, S. A., Mendillo, M. L., Smith, J. M., Williams, G. J., et al. (2013). DNA conformations in mismatch repair probed in solution by X-ray scattering from gold nanocrystals. *Proceedings of the National Academy of Sciences of the United States of America*, 110(43), 17308–17313.

Jadzinsky, P. D., Calero, G., Ackerson, C. J., Bushnell, D. A., & Kornberg, R. D. (2007). Structure of a thiol monolayer-protected gold nanoparticle at 1.1 A resolution. *Science*, 318(5849), 430–433.

Jares-Erijman, E. A., & Jovin, T. M. (2003). FRET imaging. *Nature Biotechnology*, 21(11), 1387–1395.

Kaplan, N., Moore, I. K., Fondufe-Mittendorf, Y., Gossett, A. J., Tillo, D., Field, Y., et al. (2009). The DNA-encoded nucleosome organization of a eukaryotic genome. *Nature*, 458(7236), 362–366.

Lange, O. F., Lakomek, N.-A., Fares, C., Schroeder, G. F., Walter, K. F. A., Becker, S., et al. (2008). Recognition dynamics up to microseconds revealed from an RDC-derived ubiquitin ensemble in solution. *Science*, 320(5882), 1471–1475.

Mastroianni, A. J., Sivak, D. A., Geissler, P. L., & Alivisatos, A. P. (2009). Probing the conformational distributions of subpersistence length DNA. *Biophysical Journal*, 97(5), 1408–1417.

Mathew-Fenn, R. S., Das, R., & Harbury, P. A. (2008). Remeasuring the double helix. *Science*, 322(5900), 446–449.

Mathew-Fenn, R. S., Das, R., Silverman, J. A., Walker, P. A., & Harbury, P. A. B. (2008). A molecular ruler for measuring quantitative distance distributions. *PLos One, 3*(10), e3229.

Miakelye, R. C., Doniach, S., & Hodgson, K. O. (1983). Anomalous X-ray-scattering from terbium-labeled parvalbumin in solution. *Biophysical Journal, 41*(3), 287–292.

Mittermaier, A., & Kay, L. E. (2006). Review—New tools provide new insights in NMR studies of protein dynamics. *Science, 312*(5771), 224–228.

Olson, W. K., Gorin, A. A., Lu, X. J., Hock, L. M., & Zhurkin, V. B. (1998). DNA sequence-dependent deformability deduced from protein-DNA crystal complexes. *Proceedings of the National Academy of Sciences of the United States of America, 95*(19), 11163–11168.

Palecek, E. (1991). Local supercoil-stabilized DNA structures. *Critical Reviews in Biochemistry and Molecular Biology, 26*(2), 151–226.

Palmer, A. G. (2004). NMR characterization of the dynamics of biomacromolecules. *Chemical Reviews, 104*(8), 3623–3640.

Rohs, R., Jin, X., West, S. M., Joshi, R., Honig, B., & Mann, R. S. (2010). Origins of specificity in protein-DNA recognition. *Annual Review of Biochemistry, 79*, 233–269.

Salmon, L., Yang, S., & Al-Hashimi, H. M. (2014). Advances in the determination of nucleic acid conformational ensembles. *Annual Review of Physical Chemistry, 65*(65), 293–316.

Shi, X. S., Beauchamp, K. A., Harbury, P. B., & Herschlag, D. (2014). From a structural average to the conformational ensemble of a DNA bulge. *Proceedings of the National Academy of Sciences of the United States of America, 111*(15), E1473–E1480.

Shi, X. S., Herschlag, D., & Harbury, P. A. B. (2013). Structural ensemble and microscopic elasticity of freely diffusing DNA by direct measurement of fluctuations. *Proceedings of the National Academy of Sciences of the United States of America, 110*(16), E1444–E1451.

Stuhrmann, H. B., & Miller, A. (1978). Small-angle scattering of biological structures. *Journal of Applied Crystallography, 11*(Oct), 325–345.

Vainshtein, B. K., Feigin, L. A., Lvov, Y. M., Gvozdev, R. I., Marakushev, S. A., & Likhtenshtein, G. I. (1980). Determination of the distance between heavy-atom markers in hemoglobin and histidine-decarboxylase in solution by small-angle X-Ray-scattering. *FEBS Letters, 116*(1), 107–110.

Woodward, C., Simon, I., & Tuchsen, E. (1982). Hydrogen-exchange and the dynamic structure of proteins. *Molecular and Cellular Biochemistry, 48*(3), 135–160.

Zhang, Q., Stelzer, A. C., Fisher, C. K., & Al-Hashimi, H. M. (2007). Visualizing spatially correlated dynamics that directs RNA conformational transitions. *Nature, 450*(7173), 1263–1267.

CHAPTER FOUR

2-Aminopurine Fluorescence as a Probe of Local RNA Structure and Dynamics and Global Folding

Michael J. Rau, Kathleen B. Hall[1]

Department of Biochemistry and Molecular Biophysics, Washington University Medical School, St. Louis, Missouri, USA
[1]Corresponding author: e-mail address: kathleenhal@gmail.com

Contents

1. Introduction — 100
   1.1 Sample preparation — 103
   1.2 Thermal denaturation — 105
   1.3 Circular dichroism spectropolarimetry — 108
   1.4 Selecting fluorescence probe sites — 108
   1.5 Steady-state fluorescence — 109
   1.6 Time-correlated single photon counting/time-resolved anisotropy — 113
   1.7 Stopped-flow fluorescence — 117
2. Conclusions: GAC Folding—Local and Global Events — 121

Acknowledgments — 122
References — 122

Abstract

The biology of an RNA is encoded in its structure and dynamics, whether that be binding to a protein, binding to another RNA, enzymatic catalysis, or becoming a substrate. In solution, most RNA molecules are sampling conformations, and their structures are best described as conformational ensembles. For larger RNAs, experiments that can describe the conformations of their domains can be particularly daunting, especially when the RNA is novel and not well characterized. Here, we explain how we have used site-specific 2-aminopurine as a fluorescent probe of the secondary and tertiary structures of a 60 nucleotide RNA, and what new findings we have about its Mg^{2+}-dependent conformational changes. We focus on this RNA from prokaryotic ribosome as a proof of concept as well as a research project. Its tertiary structure is known from a cocrystal, and its secondary structure is modeled from phylogenetic conservation, but there are virtually no data describing the motions of its nucleotides in solution, or its folding kinetics. It is a perfect system to illustrate the unique information that comes from a comprehensive fluorescence study of this intricate RNA.

1. INTRODUCTION

The 60 nucleotide prokaryotic GTPase center of the 23S rRNA is phylogenetically conserved (Gutell & Woese, 1990) and its tertiary structure is known from cocrystal structures of the RNA together with the L11-binding protein (Conn, Draper, Lattman, & Gittis, 1999; Conn, Gittis, Lattman, Misra, & Draper, 2002; Wimberley, Guymon, McCutcheon, White, & Ramakrishnan, 1999). There is an extensive literature from the Draper group describing the ion dependence of its folding, in particular its requirement for divalent Mg^{2+} ions (Laing, Gluick, & Draper, 1994; Lu & Draper, 1994, and subsequent). However, there are almost no data describing the properties of this essential RNA in solution: in the absence of Mg^{2+}, are its hairpin loops structured or floppy? Is its 3-way junction rigid or flexible? How do its structural elements change their dynamics in the presence of divalent ions? Does the RNA adopt its tertiary fold in bits and pieces or as one global collapse? While these features are specific for this RNA, the questions are general, and the answers are important to know for any RNA that has a biological function in the cell.

Here, we use RNAs that contain 2-aminopurine (2AP) substituted in a single position within the 60 nucleotide strand. Each sample is studied by fluorescence methods: steady-state fluorescence, time-resolved fluorescence, time-resolved fluorescence anisotropy, and stopped-flow fluorescence. Our goal is to understand how each element of the RNA contributes to Mg^{2+}-dependent tertiary structure formation, but each study could stand on its own for what it reveals about specific sites in the RNA. If these methods were to be applied to an RNA for which there was limited structural information, the results would provide insights into the properties of the RNA that otherwise would take years of NMR structure determination (for a large RNA). We note that there is no solution structure of this 60-mer RNA, either free $\pm Mg^{2+}$ or bound to protein, so what we learn about this RNA is novel and exciting.

2AP has historically been used as a probe for nucleic acid structure and structural perturbations (Ward, Reich, & Stryer, 1969). While its quantum yield (0.68) is not high compared to commonly used fluorophores such as Cy3 or fluorescein, it is readily accepted into nucleic acid duplexes to base pair with U or T and can be inserted into unstructured regions without the worry that bulky substituents will interfere with tertiary interactions. Its great disadvantage is that its excitation is in the UV (305 nm), which makes it problematic for single-molecule experiments (but see Alemán, de Silva, Patrick, Musier-Forsyth, & Rueda, 2014). Its real value is its exquisite sensitivity to context: its fluorescence is not sensitive to hydrogen bonding, but is quenched when stacked with nucleobases (Hardman & Thompson, 2006; Jean & Hall, 2001; Rachofsky, Osman, & Ross, 2001; Stivers, 1998). In loops or single-stranded regions, its fluorescence intensity as well as time-resolved fluorescence lifetimes and anisotropy can reveal the structure and dynamics of the nucleic acid (e.g., Guest, Hochstrasser, Sowers, & Millar, 1991; Hall & Williams, 2004; Rai, Cole, Thompson, Millar, & Linn, 2003), with the caveat that because its structure is not identical to that of adenine or guanine, hydrogen bonding patterns required for a tertiary structure could be disrupted by substitution with 2AP. It has recently been used to investigate RNA riboswitch properties (reviewed in St-Pierre, McCluckey, Shaw, Penedo, & LaMontaine, 2014). Using steady-state 2AP fluorescence methods, folding and ligand binding of several riboswitches including SAM-I (Heppell et al., 2011), SAM-II (Haller, Reider, Aigner, Blanchard, & Micura, 2011), TPP (Lang, Rieder, & Micura, 2007), lysine (Garst, Porter, & Batey, 2012), and preQ1 (Rieder, Kreutz, & Micura, 2010; Souliere, Haller, Rieder, & Micura, 2011) were examined. In most examples, the 2AP was introduced into the longer RNA via splint-ligation methods (e.g., Lang et al., 2007). However, here we use chemically synthesized 60 nucleotide RNAs, with a single 2AP substituted for a specific purine. We will use one sample throughout to illustrate the applications and interpretations.

We refer to the 60-mer RNA as the GAC (GTPase center), as shown in Fig. 1. Its tertiary structure has been solved in cocrystals with L11 protein (Conn et al., 1999, 2002; Wimberley et al., 1999), where interactions between nucleotides are dominated by nucleobase stacking and noncanonical hydrogen bonding. In particular, we see that both the 1093–1098 loop and 1082–1086 loop form U-turns, while loop 1065–1073 is a T-loop (Nagaswamy & Fox, 2002). The T-loop has several Mg^{2+} ions localized near its nucleobases and phosphate oxygens and contributes two nucleotides to base triples that form the core. Leipply and Draper (2011) found thermodynamic evidence of a chelated Mg^{2+} ion in the GAC, and tentatively assigned its site to the 1065–1073 loop. The 1093–1098 stemloop appears to be a classical four-nucleotide UNR U-turn; its loop sequence and loop-closing base pair are invariant. Base triples that stabilize the tertiary structure are formed from two nucleotides to the two G:C base pairs in its stem. The global GAC tertiary

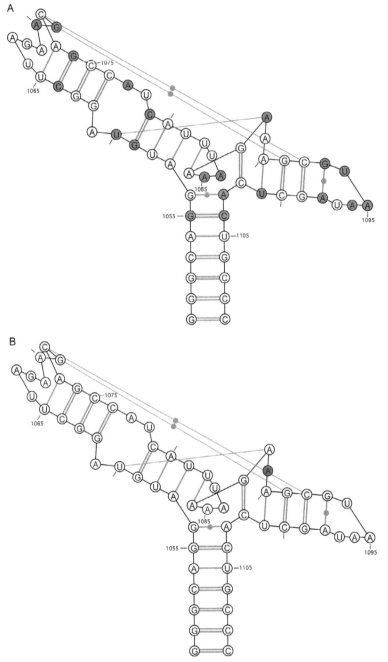

Figure 1 Secondary structure schematic of the *E. coli* U1061A GAC. (A) The invariant sites in the sequence. (B) The A1089 site that was changed to 2-aminopurine.

structure is also stabilized by a key basepair that forms between U1060 and A1088 both of which are conserved.

In the crystal structure (1hc8; Conn et al., 2002) of the folded U1061A GAC variant, A1089 is stacked with G1071, which is part of a base triple and with G1087 in the hinge (Fig. 2). A1089 does not appear to make hydrogen bonds with other residues; its stacking interactions are its sole contributions to the tertiary structure. Several Mg^{2+} ions are seen in this crystal structure to be proximal to A1089, but their role in folding is not apparent. Since A1089 is not required for specific interactions, is not phylogenetically conserved, and is near crystallographic Mg^{2+} ions, we selected it for substitution with the fluorescent base analog 2AP, for use as a probe of tertiary structure formation. Since here we will be asking for details of the folding process, this information about A1089 was important to us, since we needed to maintain normal interactions. However, the absence of such details should not inhibit exploratory uses of 2AP fluorescence investigations.

1.1 Sample preparation

RNAs shown in this work were either transcribed using T7 polymerase run-off transcription from a plasmid (Milligan, Groebe, Witherell, & Uhlenbeck, 1987) or chemically synthesized 2AP-labeled RNAs received from Agilent. Folding large RNAs can be challenging, and developing a protocol that consistently produces a single structure needs to be in place before proceeding with the types of detailed investigations of structure and dynamics described here. Fortunately, the 58 nucleotide GAC adopts its correct secondary structure in a buffer of 10 mM sodium cacodylate, pH 6.5, and 100 mM KCl. The Draper lab developed the folding protocol for the GAC that we use (Leipply & Draper, 2011): heat at 65 °C for 30 min in its buffer (without Mg^{2+}), then shift it to room temperature for about 15 min. High concentrations (1 mM) of the RNA will result in minimal dimer formation, as assayed by native gel electrophoresis.

Before investigating the properties of any RNA, it is crucial to know precisely what ions are associated with it. If you transcribe your own RNAs or purchase chemically synthesized RNAs, never assume that the RNA is fully desalted. We have used two methods to desalt both our transcribed and chemically synthesized RNAs; each method has its pros and cons but they both work. Dialyzing RNAs in Slide-A-Lyzer dialysis cassette (Thermo-Fisher) requires minimal labor and is an efficient way to remove divalent ions because you can dialyze against large concentrations and

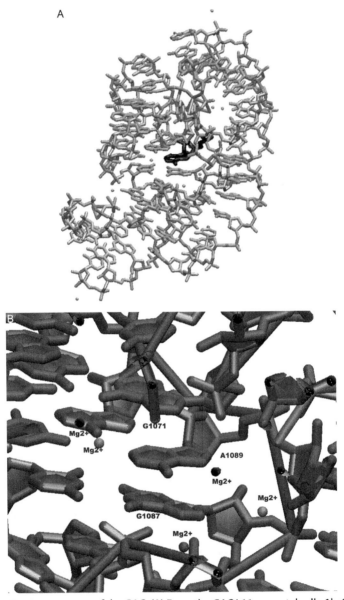

Figure 2 Tertiary structure of the GAC. (A) From the GAC:L11 cocrystal pdb: 1hc8, showing only the RNA. A1089 is in black. (B) Close-up of A1089, showing its stacking partners and the close crystallographic Mg^{2+} ions.

volumes of EDTA. These dialysis cassettes come in various molecular weight cutoffs and volumes. For the 60-mer GAC, we used 0.1–0.5 ml cassette volumes with a 3.5 K molecular weight cutoff, but other varieties are available. The down sides are: dialysis is time-consuming, there is an inherent risk of a leak in the membrane or that the gasket could fail, and the return yield is less because of the way sample is removed from the cassette. The other method is to use Amicon Ultra centrifugal filters (Millipore). These filters are easy to use with great sample recovery. They also come in different molecular weight cutoffs and different volumes. For the GAC, we used 2 ml volume filters with a 3 K molecular weight cutoff. The down sides are: they can take time to remove the salt to a sufficient level and because you are adding volume directly to your sample there is an increased risk of RNase contamination. We favor the dialysis cassettes, and typically dialyze against $500\times$ sample volume using 10 mM sodium cacodylate, pH 6.5, 100 mM KCl, and 20 mM EDTA first for 3 h at room temperature, then twice more for 3 h with fresh buffer and salt but no EDTA. The fourth dialysis is usually overnight. After the first round, you can also dialyze into water if you prefer. We used 10 mM sodium cacodylate, pH 6.5, and 100 KCl for all experiments, unless otherwise noted.

1.2 Thermal denaturation

Thermal denaturation of RNA is typically monitored by its change in absorbance with temperature in the ultraviolet at 260 nm. The increase in absorbance as bases unstack (hypochromicity) is a classic method to assess thermodynamic stability of duplexes and hairpins. "UV melting" can also be very useful in structure determination with large RNAs when there is no crystal structure and the tertiary and secondary structures are only predicted. Applied to the GAC, it can assess both secondary structure and tertiary structure formation. Since the thermal and thermodynamic stability of any RNA structure will be dependent on the concentrations of specific ions, a first series of experiments on a new RNA molecule will be to measure its thermal denaturation in whatever solution condition is appropriate (physiological salts are generally 150 mM K^+, 2 mM Mg^{2+}; prokaryotic anion is glutamate, eukaryotic is often approximated with phosphate, but a note of caution: we use cacodylate rather than phosphate, since heating RNA to 95 °C in phosphate will result in an insoluble precipitate).

GAC has two distinct denaturation (melting) transitions when Mg^{2+} is present which can be assigned to the tertiary and secondary structures

(Lu & Draper, 1994). These transitions can be resolved in a UV melt measured at 260 nm, but using 280 nm only secondary structure melting is observed. (This observation is unique to the GAC and is due to the guanosine absorbance band at 280 nm. Guanosines do not contribute substantially to tertiary stacking interactions in this RNA.) Using this information, the Draper lab was able to monitor the effect of various base substitutions on either the secondary or tertiary structure (Maeder, Conn, & Draper, 2006). Analyzing UV melts with multiple transitions is best done by taking the first derivative of the absorbance with respect to temperature, and plotting versus temperature (Puglisi & Tinoco, 1989).

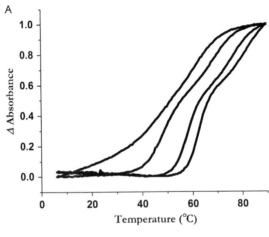

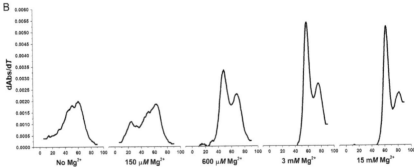

Figure 3 Ultraviolet spectra and circular dichroism spectra of GAC as a function of Mg^{2+}. (A) Thermal denaturation experiments, monitoring absorbance at 260 nm, in 100 mM KCl, +0.6 mM $MgCl_2$, +3 mM $MgCl_2$, +15 mM $MgCl_2$ (left to right curves). (B) First derivative d(Abs)/dT of the data in (A), showing the appearance of the two transitions. The transition at highest temperature is the denaturation of the secondary structure; the transition before it is the melting of the tertiary structure.

(Continued)

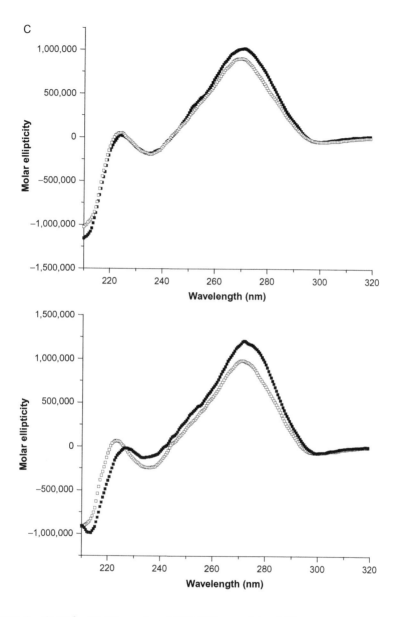

Figure 3—Cont'd (C) CD spectra of GAC RNA-containing 2AP at position 1089 (top) and RNA transcribed *in vitro* (bottom). Addition of 8 mM MgCl$_2$ to either RNA increases the ellipticity at 275 nm, indicating an increase in base stacking.

Denaturation profiles shown (Fig. 3A) were performed on a Gilford 260 fitted with a Gilford thermoprogrammer 2527. RNA concentrations used for these experiments were 1.5 μM and cuvettes path length was 5 mm. Sample temperature was ramped at 1 °C/min from 6 to 90 °C, while absorbance was recorded at 260 nm. $MgCl_2$ titrations showed incremental changes of the melting profile, resulting in stabilization of the tertiary structure. This become clearer when plotted as d(Abs)/dT, where the broad single peak transforms into two well-defined peaks (Fig. 3B).

1.3 Circular dichroism spectropolarimetry

Circular dichroism (CD) spectropolarimetry provides another measure of the extent of base stacking in a nucleic acid. The CD spectrum of an A-form duplex is distinctive, due to the geometry of the nucleobases in the double strand. Single-stranded RNA can also have a CD spectrum if the bases are stacked in a helical structure. It was possible that the GAC RNA would have different CD spectra with and without Mg^{2+}, if addition of Mg^{2+} led to a change in base stacking.

The spectra of the GAC RNA are shown in Fig. 3C, and both display the typical A-form characteristics: a maximum ellipticity at ∼270 nm and a null at ∼245 nm. It was disappointing that there was not a dramatic change in the spectrum upon Mg^{2+} addition, but there is an increase in the ellipticity at 270 nm, indicating more stacking of the bases in the tertiary structure.

CD measurements were carried out on a Jasco J715 spectropolarimeter at room temperature (23 °C). This is an older model, but has excellent optics. For these experiments, a Peltier sample holder is extremely useful. We used an RNA concentration of 4 μM and a path length of 10 mm. Three scans were collected from 320 to 210 nm and then averaged for all samples. We used a scanning speed of 50 nm/min in a continuous scanning mode with the bandwidth set at 1.0 nm. Buffer-only control was subtracted from each spectrum. Mg^{2+} was added to a final concentration of 8 mM.

1.4 Selecting fluorescence probe sites

Using fluorescent base analogs is a very powerful way to monitor the dynamics of specific sites that might be important for structural integrity or accessibility by ions or proteins. The most readily available and well-characterized fluorescent base analog is 2AP. To decide where to make the substitutions, it is best to get as much information about the secondary and tertiary structures of the RNA as possible. Crystal structures are convenient for this purpose, although there can be unpleasant surprises if the structure has trapped an alternative conformation. Using information about phylogenetically conserved bases and thermal denaturation data from RNAs with specific base substitutions is reliable guides for selection of sites to probe. In the case of GAC, we looked for sites that were predicted to be sensitive to tertiary structure formation, and so result in a loss or increase of 2AP fluorescence intensity. In practice, the 2AP substitutions were in hairpin loops and internal bulges where the formation of the tertiary structure is predicted to change the stacking interactions. One 2AP substitution per RNA offers the most unambiguous information.

Our 2AP GAC RNAs were chemically synthesized to contain a single 2AP at a specific site. Agilent provided them, and here we discuss the results from one that is labeled at the position A1089 (A1089AP). This position was chosen based on examination of the crystal structure and prior work from the Draper lab on GAC with specific base substitutions (Lu & Draper, 1994). A1089AP should report on the stacking of G1071 when it forms a base triple with G1091 in the tertiary structure. We compared the Mg^{2+} dependence of thermal denaturation by UV absorbance to that of the wild-type GAC, as well as the CD spectra of A1089AP GAC with and without Mg^{2+}, to determine if the substitution disrupted formation of the tertiary structure or altered its Mg^{2+} requirement. We found that A1089AP GAC requires a higher Mg^{2+} concentration to stabilize its tertiary structure (8 mM) but otherwise there were no significant changes.

1.5 Steady-state fluorescence

The easiest way to evaluate if the site selected for 2AP substitution can monitor tertiary structure formation is to use steady-state fluorescence of the 2AP-RNA molecule. Assuming that the RNA of interest will require a divalent ion cofactor to adopt its tertiary structure, a first experiment on a new 2AP-RNA would be a Mg^{2+} titration. If the 2AP-RNA fluorescence intensity changes upon addition of Mg^{2+}, then that site becomes a reporter on a conformational change (an increase in fluorescence emission intensity is consistent with unstacking, while an intensity decrease indicates more stable nucleobase stacking). The temperature dependence of the fluorescence is an important variable, since we have examples of 2AP sites with identical spectra at 20 °C, with/without Mg^{2+}, but with quite different spectra at either lower or higher temperatures. It is important to appreciate that a change in the 2AP fluorescence does not mean that it "binds" a Mg^{2+} ion in these titration experiments. While there are some examples of specific Mg^{2+} ion binding (Leipply & Draper, 2011), most Mg^{2+} ions are loosely associated with an RNA where they act to neutralize repulsion of proximal phosphates when the RNA folds into a tertiary structure. Many experiments from the Draper lab show quantitatively that the GAC tertiary structure is Mg^{2+} dependent, so we use it here as an example of how we characterize the Mg^{2+} dependence of RNA structure.

2AP in solution. For reference, free 2AP triphosphate fluorescence was monitored versus temperature at the same concentration and buffer as the GAC A1089AP. 2AP at these concentrations (2 μM) does not form aggregates or intermolecular dimers, so a measurement of its fluorescence intensity provides a value of the maximum fluorescence intensity possible in the solution conditions and fluorimeter setup. The fluorescence intensity is temperature dependent, since at high temperatures collisional quenching with solvent will decrease the fluorescence (Lakowicz, 2006). (We do not degas our solutions, so molecular oxygen is also present and acts as a quencher.) Free 2AP triphosphate can be used as an external standard to maintain consistent instrument response.

Mg^{2+} titrations were used to investigate the Mg^{2+} dependence of A1089AP fluorescence. When performing Mg^{2+} titrations it is important to measure the excitation and emission spectra when Mg^{2+} is added, since a different environment of the 2AP can alter its photophysics. If there is a shift of the wavelength maximum of either the excitation or emission, then it will be necessary to acquire full excitation and emission spectra instead of monitoring the titration at a single wavelength. Wavelength shifts usually happened when there is a change in stacking environment of the 2AP. This could be either stacking with different bases or stacking with a different number of bases (Hardman & Thompson, 2006). For this A1089AP RNA, there was no shift in either emission or excitation maximum, so we used 309 nm for excitation and observed at 368 nm.

We first measured the apparent K_D of Mg^{2+} binding using the decrease in fluorescence intensity of A1089AP, shown in Fig. 4. We emphasize that this is only an apparent K_D, since the stoichiometry of binding is not known. For these experiments, the RNA concentration is 2 μM, while the Mg^{2+} concentration is much higher. The isotherm was fit using Origin, taking into account the sloping baselines, to give a single dissociation constant of 348 ± 115 μM. We assign this Mg^{2+} binding to the general association of Mg^{2+} to GAC, without defining where the site(s) are located. Most

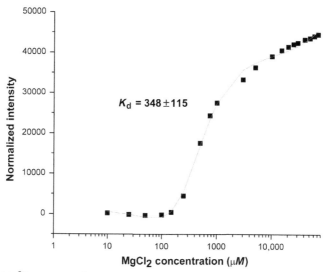

Figure 4 Mg^{2+} titration of GAC A1089AP. The fluorescence intensity decreases with added Mg^{2+}, but here, we plot (1 − Intensity) for fitting. The upper baseline is not well defined for this RNA, and in fact, it appears that there could be two isotherms. [GAC] = 2 μM. $\lambda_{excitation} = 309$ nm; $\lambda_{emission} = 368$ nm.

importantly, this experiment defines concentrations where the Mg^{2+} ions should be saturating: we estimate that >3 mM Mg^{2+} is necessary, based on the shape of the titration curve. This is important information for the design of subsequent stopped-flow fluorescence experiments.

The temperature dependence of 2AP fluorescence (fluorescence melts) provide more information about the stability of 2AP in a specific position in the structure of the RNA. In the absence of Mg^{2+}, the fluorescence intensity of A1089AP is quite constant from 6 to 30 °C, but then becomes progressively reduced (Fig. 5). In this KCl concentration, GAC A1061 does not adopt its tertiary structure (Leipply & Draper, 2011), so the fluorescence intensity changes that we observe must be reporting on the local environment of the 2AP in the secondary structure. The relatively low fluorescence intensity relative to free 2AP indicates that A1089AP is stacked in the secondary structure. UV thermal denaturation data shows that the RNA undergoes some melting of its structure from 6 to 30 °C, but since A1089AP fluorescence intensity remains unchanged in this temperature range, this region of the RNA is not denaturing. When 3 mM Mg^{2+} is added there is a 50% decrease in the fluorescence intensity which is constant from 6 to 40 °C, an indication that the nucleobase stacking has increased. At 40 °C,

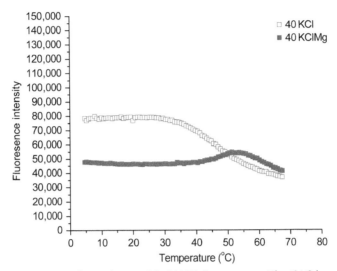

Figure 5 Temperature dependence of A1089AP fluorescence. The GAC has only secondary structure in the top curve, while the lower curve (lower fluorescence intensity) reports on fluorescence in the tertiary structure. [GAC] = 2 μM. $\lambda_{excitation}$ = 309 nm; $\lambda_{emission}$ = 368 nm.

the fluorescence intensity increases until about 50 °C, then decreases with higher temperature. In the tertiary structure, A1089AP is base-stacked in an internal loop, which corresponds to the loss of fluorescence intensity upon addition of Mg^{2+}. The fluorescence change from 40 to 50 °C indicates a conformational change has occurred, probably disruption of the internal loop that leads to exposure of 1089AP. After 55 °C, the effects of collisional quenching (both solvent and neighboring nucleobases "collide" with the 2AP) further reduce the fluorescence intensity as the structure is disrupted and the nucleobase exposed to solvent. In this example, 2AP fluorescence has reported the thermal stability of this local structural element, which will be incorporated into a model of the unfolding pathway.

If this were an RNA for which little structural information was available, would these experiments be valuable? Imagine that these data describe a novel RNA and that there is a secondary structure prediction available. Then, we known from the initial low fluorescence intensity that this nucleobase spends a significant amount of time stacked in KCl, but that when Mg^{2+} is added, its stacking propensity is increased. Now include the thermal denaturation and fluorescence data to the picture, and you have a description of a local structural element that is stable until about 40 °C, while other parts of the RNA are melting. With these data, secondary structure predictions could be modified or validated, and with some imaginative modeling, tertiary interactions could be proposed.

These temperature-dependent fluorescence experiments were performed on a PTI (Photon Technology International) fitted with a Peltier controlled four-cuvette turret. RNA concentration for these experiments was 2 μM. Emission and excitation scans were performed to find the optimal wavelength for $\pm Mg^{2+}$, to give 308 nm for excitation and 368 nm for emission. Lamp power and PMT voltage were kept constant for all experiments at 70 W and 1000 V, respectively. Slit widths were also kept constant throughout the experiments. Temperature was increased with a ramp of 1 °C/min and acquiring for 5 s with an integration time of 1 s. For each 5-s acquisition, the points were averaged then plotted versus temperature.

1.6 Time-correlated single photon counting/time-resolved anisotropy

Time-correlated single photon counting (TCSPC) and time-resolved anisotropy (TRA) are fluorescence methods that provide information about environment and motions of the fluorophore. TCSPC decay curves show the time-dependent loss of fluorescence intensity (the fluorescence lifetime) and are typically multiexponential. Data are interpreted in terms of fraction of time spent in a specific environment with a corresponding fluorescence lifetime. TRA decay curves show the time dependence of fluorescence depolarization. Emitted fluorescence becomes depolarized as the molecule tumbles, either globally if the fluorophore is rigid in the molecular frame and/or locally if the fluorophore has motion independent of the molecular frame. 2AP nucleotide alone has a single fluorescence lifetime of 10–11 ns, and a single TRA decay of 9 ps, but in the context of a nucleic acid, it typically has two or three fluorescence decay lifetimes and two depolarization (anisotropy) decay times.

In the GAC secondary structure, A1089 might be stacked with its neighbors, but this region of the GAC acts as a hinge, so it could be flexible. When GAC adopts its tertiary structure, A1089 is stacked on both sides with other nucleobases, so we anticipated that its conformational mobility would be restricted. The raw data from a TCSPC experiment are shown in Fig. 6 as a function of temperature. Each trace is fit independently to multi-exponential components, and for GAC 1089AP $\pm Mg^{2+}$, those data are given in Table 1. The surprise is that there are three lifetimes with and without Mg^{2+} and that those lifetimes are virtually identical. We do see a change

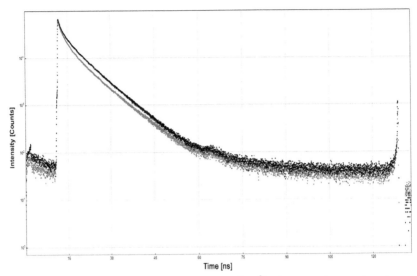

Figure 6 TCSPC trace of 2AP-RNA as a function of Mg^{2+}. These are the raw data that are subsequently fit with multiexponential components in Fluo-Fit (see methods) 30°C. Black is no Mg^{2+}; grey is 8mm Mg^{2+}.

Table 1 Time-resolved fluorescence parameters
TCSPC

Mg^{2+} (mM)	T (°C)	τ_1 (ns)	β_1 (%)	τ_2 (ns)	β_2 (%)	τ_3 (ns)	β_3 (%)	χ^2
0	30	7.6	24	2.3	26	0.30	50	2.2
8	30	7.7	16	2.1	29	0.25	55	2.6

TRA

Mg^{2+} (mM)	T (°C)	R_o	ϕ_1 (ns)	β_1 (%)	ϕ_2 (ns)	β_2 (%)	χ^2
0	30	0.27	3.6	85	0.56	15	1.2
8	30	0.26	5.6	91	0.78	9	1.2

TCSPC, time-correlated single photon counting; TRA, time-resolved anisotropy.

in the proportion (amplitude) of each component, especially a much lower proportion of the longest lifetime (7.7 ns) when the GAC is folded into its tertiary structure in Mg^{2+}. A first approximation is that the environment of A1089 is not very different in the two conformational states of the GAC.

In general, for 2AP fluorescence decays, we assign long lifetimes to an unstacked environment, intermediate lifetimes to a partially or transiently stacked ensemble, and short lifetimes to a predominantly stacked environment (in our instrument, the shortest lifetime we can resolve is ~190 ps).

The three fluorescence lifetimes of A1089AP (Table 1) are thus assigned to these physical environments to provide a structural/dynamical interpretation. When Mg^{2+} is added, the proportion of 1089AP in a stacked environment increases slightly (50–55%), while the fraction of time spent in the unstacked environment is reduced (24–16%). These data indicate that A1089AP in the GAC structures does not have completely unhindered mobility, but neither is it rigidly constrained. If these data were obtained for an RNA where no starting and ending structures were known, one reasonable interpretation would be that addition of Mg^{2+} has no effect on its position in the RNA (at least at the temperature measured).

TRA experiments provide a different picture of the 2AP position in the RNA, and how that position changes with addition of Mg^{2+}. In the GAC, the addition of Mg^{2+} should alter the global tumbling time of the RNA: it is tempting to interpret the secondary structure representation literally, and think of the GAC as three arms extended around a central junction. In the absence of Mg^{2+}, the three arms could be free to move independently. Upon addition of Mg^{2+}, the GAC folds into the compact structure seen in the cocrystal structures. The global tumbling times (rotational correlation time) of the RNA should be detectably different in those two environments, and so the TRA depolarization should have different values.

We first calculated the theoretical rotational correlation time of the GAC in its tertiary structure. For example, if 2AP were incorporated into a 40 kDa globular RNA, with a rotational correlation time of 15 ns, then one of its depolarization decays (anisotropy) would be due to global tumbling (15 ns). However, if the longest 2AP fluorescent lifetime is only 7.7 ns (as it is here), then its fluorescence will be lost long before the rotational correlation time of the molecule; i.e., the fluorescence lifetime is too short to allow the anisotropy to measure global tumbling. However, if the RNA has a rotational correlation time of 3 ns, then 2AP TRA will accurately describe global (and possibly local) tumbling.

To calculate the theoretical rotational correlation time for the global tumbling of the RNA, there has to be some knowledge of its shape, either through modeling of its structure, from SAXS, or using a crystal structure. The equations to calculate the rotational correlation time for different shapes can be found in Lakowicz (2006). GAC RNA can be classified as a prolate ellipsoid when Mg^{2+} is present, based on the dimensions of the crystal structure, with an axial ratio of 1.8. For this shape, three different rotational correlation times are needed: $\Theta_1=5.7$, $\Theta_2=5.2$, $\Theta_3=5.9$ ns at 30 °C. These predicted values assume an axial ratio of 1.8, viscosity of 0.797 cP, molecular weight of 21 kDa, specific volume of 0.6 ml/g, and a hydration of 0.2 ml/g.

Each component contributes differently to the overall anisotropy value making it difficult to get an exact value, but at least this approximation provides a general idea of the tumbling time. An additional caveat is that these are the values used for calculations of protein parameters, and we expect that the specific volume and hydration values are not accurate for RNA. Since the longest A1089AP fluorescence lifetime is greater than 7 ns, and since we predict that the rotational correlation time of GAC in its tertiary fold is ~6 ns, these measurements should report its global rotational correlation time. Since the shape of the unfolded GAC is not known, TRA measurements may or may not report it.

TRA experiments were done at several temperatures, and the data are shown in Table 1 for 30 °C to correspond to our calculations of rotational correlation times. TRA data for A1089AP were best fit with two components, with or without added Mg^{2+}. When GAC has adopted its tertiary fold in 8 mM Mg^{2+}, we interpret the longest depolarization decay (5.6 ns) as global tumbling of the RNA, in surprising agreement with the theoretical calculations. Global tumbling contributes the dominant component to the anisotropy, indicating that A1089 is fixed in position in the folded GAC. There is a small component of apparently local motion: 9% of the anisotropy decay is attributed to local depolarization at 780 ps. Recalling that there are three fluorescence lifetimes of A1089AP, we suggest that the local motion occurs when A1089AP moves from stacked to unstacked environments.

By analogy, the longest depolarization time of 2AP in GAC without Mg^{2+} should also report on global tumbling. Curiously, this decay time is significantly shorter (3.6 ns) than in the tertiary folded GAC. Again by analogy, the shorter depolarization decay time is attributed to local motion on the timescale of 560 ps. In the GAC with only secondary structure, this component contributes 15% of the anisotropy decay, about twice what is observed in the tertiary structure. TRA experiments show clearly that 1089AP inhabits two different contexts in the GAC secondary and tertiary structures.

Several important questions remain from these fluorescence experiments. For example, do the TRA data support the model of a GAC with floppy arms that move independently of each other on this timescale? Alternatively, is the GAC secondary structure rigidly extended with a short and very long axial rotational component? If we want to understand the mechanism of Mg^{2+}-dependent tertiary structure formation, the starting state of the secondary structure is important to know in order to interpret the energetics. Certainly, the solution of GAC structural flexibility is grossly underdetermined with only one site as a reporter, and other sites are needed to model the structure, and other experiments (e.g., SAXS; Grilley, Misra, Caliskan, & Draper, 2007) are required.

All lifetime and anisotropy measurements were performed on a home-built TCSPC instrument (Jean & Hall, 2001; 2004). A Ti:Sapphire laser was pulse-picked and tripled to excite the 2AP using 300 nm. Samples were heated in a water-jacketed cuvette holder using a water bath and the temperature was monitored using a thermistor. Lifetime and IRF decay curves were collected until overflow (65,536 counts) with the polarizer at 55°. IRF was collected with a 1/100 dilution of LUDOX (Aldrich) solution; its FWHM was measured every day and varied from 215 to 222 ps. RNA concentrations for both TCSPC and TRA were 2 μM. Buffer scattering accounted for <5% of the counts in the peak channel; it was not subtracted from decay curves. Fluorescence lifetime decays were best fit with a three component exponential model with reconvolution. Anisotropy decays were collected in separate runs where vertical polarization was collected first to overflow then horizontal polarization was collected with the same acquisition time. Anisotropy decays were fit using a two component exponential model and the G-factor was calculated using tail matching. Fluofit Pro 4.4 (Picoquant) was used to fit both lifetime and anisotropy curves. Goodness of fit was primarily assessed by support plane analysis.

1.7 Stopped-flow fluorescence

The previous experiments were conducted in conditions of structural equilibrium, when the RNA tertiary structure was either absent or present. Now, we can use the measured change in 2AP fluorescence in those two states to monitor the kinetics of Mg^{2+}-dependent tertiary structure formation. For these experiments, we use stopped-flow fluorescence. Stopped-flow measurements should be done in saturating conditions of ligand, which from our equilibrium measurements we know is >3 mM MgCl$_2$ when the GAC concentration is ~2 μM.

The stopped-flow kinetics data reveal the complexity of the folding process; here, we show only one temperature experiment for the 1089AP GAC, but the results are typical. The fluorescence traces and the fitted decay curves are shown in Fig. 7 and summarized in Table 2. Certainly, the data cannot be fit by a simple two-state transition. Instead, at all concentrations of added Mg^{2+}, there is a hidden event that is only discernable from the observation that the traces do not start at 100% signal intensity at $t=1$ ms. The initial loss of signal amplitude (from 10% to 20%) means there is a fast kinetic component that occurs within the 1 ms dead time of the instrument. At the lowest Mg^{2+} concentration tested (3 mM), the kinetic curve can be described by a

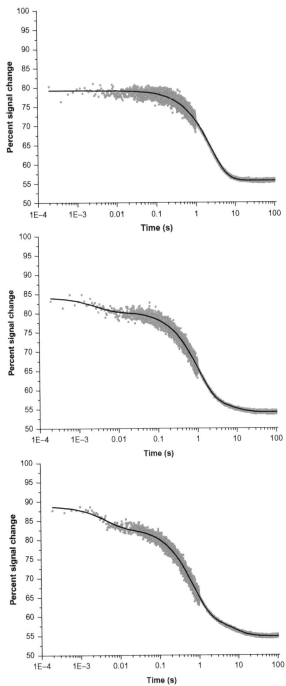

Figure 7 Stopped-flow fluorescence traces and fits for GAC A1089AP at 30 °C. (Top) 3 mM MgCl$_2$ added. (Middle) 8 mM MgCl$_2$ added. (Bottom) 20 mM MgCl$_2$ added. Fits (black line) to an exponential model. The dead time of the instrument is ∼1 ms. [GAC] = 100 nM.

Table 2 Stopped-flow kinetic parameters at 30 °C

Mg^{2+}	Y offset	τ_1 (s)	k_1 (1/s)	β_1 (%)	τ_2 (s)	k_2 (1/s)	β_2 (%)	τ_3 (s)	k_3 (1/s)	β_3 (%)
3	56	2.4	0.42	24						
8	54	7.0	0.14	4	1.0	1.0	22	0.003	333	4
20	55	7.0	0.14	6	0.68	1.47	23	0.004	250	6

100 nM RNA in 100 mM KCl, 10 mM sodium cacodylate, pH 6.5.

single transition with a rate constant of 0.42 s^{-1} (Table 2). However, upon addition of 8 or 20 mM Mg^{2+}, there are three discernable kinetic phases reported by A1089AP, some of which might be local events, but other(s) are global. There is an initial fast phase with a time constant of 3 ms and a rate of $\sim$300 s^{-1}, followed by a slower phase with large amplitude fluorescence intensity change with a time constant of 0.7–1 s, and finally a slow phase, with a time constant of 7 s. The structural interpretation of these kinetic transitions is not immediately apparent: if 1089AP fluorescence reports on overall folding to the tertiary structure, which component describes this global event?

Using data from a single probe position is not sufficient for an unambiguous interpretation of the kinetic parameters. To create a model of the GAC tertiary folding process, we need more probes that encompass other events, such as the juxtaposition of the two hairpins and the burial of specific nucleobases in the stacked core, and formation of the two hairpin loop structures. In that respect, these data are tantalizing, for they provide a glimpse of the intricate rearrangements of the nucleobases that accompany tertiary structure formation in the GAC.

Would analogous experiments be useful for an RNA with only a predicted secondary structure and no tertiary fold? Perhaps, but considering that they report on only one position of the RNA, they leave a lot to imagination. If the timescale of folding events is an important biological parameter (for example, does it depend on specific ions, proteins, cofactors), then the investment could be worthwhile. It is not the first experiment to do, however, since interpretation is challenging. A more general approach is that of time-resolved hydroxyl radical footprinting (Sclavi, Sullivan, Chance, Brenowitz, & Woodson, 1998), which simultaneously monitors all riboses in the RNA as it probes for sites that are protected from cleavage. Descriptions of RNA folding from riboses and nucleobases are complementary, so ideally both experiments should be done.

Our lab used an Applied Photophysics SX-20 Stopped-Flow spectrometer for all measurements. This instrument has two syringes for mixing; one holds the RNA in buffer and KCl, the other contains the $MgCl_2$ in buffer and KCl. We used an excitation wavelength of 305 nm based on steady-state fluorescence excitation and emission spectra of A1089AP. We set both excitation slit widths to 2 mm, which results in a bandwidth of 9.3 nm. We chose a long pass filter from Semrock FF01-341/LP-25 for the emission filter, which has an optimal cutoff for removal of the Raman peak at 305 nm excitation. The Applied Photophysics PMT R6095 has increased sensitivity between 300 and 600 nm. For optimal sensitivity, the PMT voltages were set between 8 and 9 V to give a maximum fluorescence signal. The GAC concentration in the syringe was 200 nM, in a total volume of 1.5 ml; each experiment required 80 µl per syringe for a total of 160 µl shot volume. Experiments at each temperature were repeated multiple times. Temperature was regulated with a water bath and sample temperature was monitored through Applied Photophysics software. The number of points collected per time interval was adjusted to achieve the best resolution; from $t=0$ to $t=1$ s, we sampled 5000 points, while from 1 to 100 s, we collected 10,000 points. The dead time of the instrument is ~1 ms.

Before beginning a series of experiments, we first cleaned the system to ensure no RNAases were present, using a combination of 1 M NaOH, 2 M HNO_3, and lots of Milli-Q water. The NaOH is first loaded into both sample syringes and then at least five shots per syringe were run (injected). After a wait of about 1 min, two additional shots were run. Immediately following, the system is washed with Milli-Q water using both sample syringes and running about 10 shots. Next HNO_3 is used, using the same protocol as per NaOH. Again the system is washed with Mill-Q water, typically about four sample syringe volumes per syringe. Finally, we run one sample syringe volume of buffer through the entire system.

The stopped-flow experiments are simple to perform, but the controls must be rigorously done. We have found that the series of controls needs to be run *every* time the instrument is used or *anytime* a setting is changed. Our controls are RNA mixing with buffer only (no Mg^{2+}) and buffer mixing with buffer; in both cases, multiple mixing experiments were performed. Whenever a solution is changed in the sample syringe, it is important to insure that the pervious solution is flushed out of the system to prevent contamination, which necessitates several initial dummy mixes.

The fluorescence intensity of A1089AP allowed us to use 100 nM RNA final concentration for all experiments. $MgCl_2$ addition was in increments of 3, 8, and 20 mM final concentration. Buffer was the same for all solutions: 10 mM sodium cacodylate, pH 6.5, and 100 mM KCl. The RNA secondary structure was formed as described before it was loaded into the syringe. Buffer with no RNA or MgCl$_2$ was also prepared for control and for system washing. We used BD 3 ml syringes as our loading syringes and had no RNase or other contaminates.

The raw data traces were processed before fitting in the following manner. In each condition, all measurements were first averaged. We averaged between 8 and 12 measurements per Mg^{2+} condition to ensure a high signal/noise. The buffer/buffer background control trace was subtracted from the [RNA without Mg^{2+}] control and each of the [RNA+Mg^{2+}] experiments. The corrected [RNA without Mg^{2+}] trace was then smoothed using a 20-point running average. This trace was essentially flat: there was no change in the fluorescence during the runs and is a measure of 100% of A1089AP fluorescence. To compare the fluorescence change to the steady-state fluorescence, the lower "baseline" is adjusted to the final fluorescence value of A1089AP after addition of Mg^{2+} (56% of the initial fluorescence; this is the Y offset or Y_0 in the fitting equation). Fitting the decay curves to the exponential model in Origin (Eq. 1) is expressed as fluorescence changes with corresponding fraction of the total fluorescence signal (β_n): 56% is the signal is constant, a varying fraction is lost in <1 ms, and the remainder is lost during the folding events.

$$y = y_0 + \beta_n e^{-(x/t_n)} + \beta_{n+1} e^{-(x/t_{n+1})} \tag{1}$$

2. CONCLUSIONS: GAC FOLDING—LOCAL AND GLOBAL EVENTS

We have explored the structure and dynamics of the GAC of 23S rRNA when it has only secondary structure and when it has adopted its tertiary fold. Our methods are centered in optical spectroscopy, and in particular fluorescence of a single substituted 2AP nucleobase. That site is used to report on events during Mg^{2+}-induced folding kinetics. The site we selected is sensitive to GAC conformational changes, based on cocrystal structures, providing a sensitive reporter of the environment at one position in the GAC.

What we observe is that A1089AP is predominantly stacked, with little dynamic motion, independent of the folded state of GAC. However, the data indicate that the local environments are not the same in the unfolded and folded states, and that to form the tertiary structure, this local area of the GAC undergoes a conformational change that rearranges the stacking partners of A1089. This conclusion is based on steady-state fluorescence data showing that the fluorescence intensity of 2AP is reduced when the RNA forms its tertiary structure with additional insight into the rearrangement from the kinetics data: there are multiple events that lead to progressive loss of fluorescence. Each event that we observe is reporting on a dynamic conformational change involving A1089AP, on timescales ranging from microseconds to seconds.

The GAC RNA is well characterized, but a new RNA of biological interest will probably be lacking substantial information on its structural transitions. Are these 2AP fluorescence experiments pertinent to such a new RNA? We would argue "yes." The first experiment to conduct on any new RNA is to predict its secondary structure via thermodynamic rules, which can be done online by several software packages. To validate the secondary structure requires chemical and/or enzymatic probing under different solution conditions, perhaps in the presence or absence of a protein. Thermal UV denaturation experiments can also give some information on structural variations as a function of solution conditions. Alternatively, a single site of particular interest can be substituted with 2AP and probed to provide immediate insight into whether it is stacked or freely rotating, suggesting its possible conformations in the RNA.

ACKNOWLEDGMENTS

This work is supported by NIH R01-GM098102 to K.B.H., and a gift from Agilent. The RNAs were chemically synthesized by Agilent labs, and we thank Dr. Laurakay Bruhn, Dr. Doug Dellinger, Dr. Jeff Sampson, and their colleagues. We thank Professor Roberto Galletto for use of his stopped-flow instrument and Mr. Robb Welty for his stopped-flow experimental data. We thank Mr. W. Tom Stump for his many insights and his excellent instrument maintenance.

REFERENCES

Alemán, E. A., de Silva, C., Patrick, E. M., Musier-Forsyth, K., & Rueda, D. (2014). Single-molecule fluorescence using nucleotide analogs: A proof-of-principle. *Journal of Physical Chemistry Letters, 5*, 777–781.

Conn, G. L., Draper, D. E., Lattman, E. E., & Gittis, A. G. (1999). Crystal structure of a conserved ribosomal protein-RNA complex. *Science, 284*, 1171–1174.

Conn, G. L., Gittis, A. G., Lattman, E. E., Misra, V. K., & Draper, D. E. (2002). A compact RNA tertiary structure contains a buried backbone-K^+ complex. *Journal of Molecular Biology, 318*, 963–973.

Garst, A. D., Porter, E. B., & Batey, R. T. (2012). Insights into the regulatory landscape of the lysine riboswitch. *Journal of Molecular Biology, 423*, 17–33.

Grilley, D., Misra, V., Caliskan, G., & Draper, D. E. (2007). Importance of partially unfolded conformations for Mg^{2+}-induced folding of RNA tertiary structure: Structural models and free energies of Mg^{2+} interactions. *Biochemistry, 46,* 10266–10278.

Guest, C. R., Hochstrasser, R. A., Sowers, L. C., & Millar, D. P. (1991). Dynamics of mismatched base pairs in DNA. *Biochemistry, 30,* 3271–3279.

Gutell, R. R., & Woese, C. C. (1990). Higher order structural elements in ribosomal RNAs: Pseudo-knots and the use of noncanonical pairs. *Proceedings of the National Academy of Sciences of the United States of America, 87,* 663–667.

Hall, K. B., & Williams, D. J. (2004). Dynamics of the IRE RNA hairpin loop probed by 2-aminopurine fluorescence and stochastic dynamics simulations. *RNA, 10,* 34–47.

Haller, A., Reider, U., Aigner, M., Blanchard, S. C., & Micura, R. (2011). Conformational capture of the SAM-II riboswitch. *Nature Chemical Biology, 7,* 393–400.

Hardman, S. J., & Thompson, K. C. (2006). Influence of base-stacking and hydrogen bonding on the fluorescence of 2-aminopurine and pyrrolocytosine in nucleic acids. *Biochemistry, 45,* 9145–9155.

Heppell, B., Blouin, S., Dussault, A. L., Muhlbacher, J., Ennifar, E., Penedo, J. C., et al. (2011). Molecular insights into the ligand-controlled organization of the SAM-I riboswitch. *Nature Chemical Biology, 7,* 384–392.

Jean, J. M., & Hall, K. B. (2001). 2-Aminopurine fluorescence quenching and lifetimes: Role of base stacking. *Proceedings of the National Academy of Sciences of the United States of America, 98,* 37–41.

Jean, J. M., & Hall, K. B. (2004). Stacking-unstacking dynamics of oligodeoxynucleotide trimers. *Biochemistry, 43,* 10277–10284.

Laing, L. G., Gluick, T. C., & Draper, D. E. (1994). Stabilization of RNA structure by Mg ions. Specific and non-specific effects. *Journal of Molecular Biology, 237,* 577–587.

Lakowicz, J. R. (2006). *Principles of fluorescence spectroscopy* (3rd ed.). New York: Springer.

Lang, K., Rieder, R., & Micura, R. (2007). Ligand-induced folding of the thiM TPP riboswitch investigated by a structure-based fluorescence spectroscopic approach. *Nucleic Acids Research, 35,* 5370–5378.

Leipply, D., & Draper, D. E. (2011). Evidence for a thermodynamically distinct Mg^{2+} ion associated with formation of an RNA tertiary structure. *Journal of the American Chemical Society, 133,* 13397–13405.

Lu, M., & Draper, D. E. (1994). Bases defining an ammonium and magnesium ion-dependent tertiary structure within the large subunit ribosomal RNA. *Journal of Molecular Biology, 244,* 572–585.

Maeder, C., Conn, G. L., & Draper, D. E. (2006). Optimization of a ribosomal structural domain by natural selection. *Biochemistry, 45,* 6635–6643.

Milligan, J. F., Groebe, D. R., Witherell, G. W., & Uhlenbeck, O. C. (1987). Oligoribonucleotide synthesis using T7 RNA polymerase and synthetic DNA templates. *Nucleic Acids Research, 15,* 8783–8798.

Nagaswamy, U., & Fox, G. E. (2002). Frequent occurrence of the T-loop RNA folding motif in ribosomal RNAs. *RNA, 8,* 1112–1119.

Puglisi, J. D., & Tinoco, I., Jr. (1989). Absorbance melting curves of RNA. *Methods in Enzymology, 180,* 304–325.

Rachofsky, E. I., Osman, R., & Ross, F. B. (2001). Probing structure and dynamics of DNA with 2-aminopurine: Effects of local environment on fluorescence. *Biochemistry, 40,* 946–956.

Rai, P., Cole, T. D., Thompson, E., Millar, D. P., & Linn, S. (2003). Steady-state and time-resolved fluorescence studies indicate an unusual conformation of 2-aminopurine within ATAT and TATA duplex DNA sequences. *Nucleic Acids Research, 31,* 2323–2332.

Rieder, U., Kreutz, C., & Micura, R. (2010). Folding of a transcriptionally acting PreQ$_1$ riboswitch. *Proceedings of the National Academy of Sciences of the United States of America, 107*, 10804–10809.

Sclavi, B., Sullivan, M., Chance, M. R., Brenowitz, M., & Woodson, S. A. (1998). RNA folding at millisecond intervals by synchrotron hydroxyl radical footprinting. *Science, 279*, 1940–1943.

Souliere, M. F., Haller, A., Rieder, R., & Micura, R. (2011). A powerful approach for the selection of 2-aminopurine substitution sites to investigate RNA folding. *Journal of the American Chemical Society, 133*, 16161–16167.

Stivers, J. T. (1998). 2-Aminopurine fluorescence studies of base stacking interactions at abasic sites in DNA: Metal-ion and base sequence effects. *Nucleic Acids Research, 26*, 3827–3844.

St-Pierre, P., McCluckey, K., Shaw, E., Penedo, J. C., & LaMontaine, D. A. (2014). Fluorescence tools to investigate riboswitch structural dynamics. *Biochimica et Biophysica Acta, 1839*, 1005–1019.

Ward, D. C., Reich, E., & Stryer, L. (1969). Fluorescence studies of nucleotides and polynucleotides. I. Formycin, 2-aminopurine riboside, 2,6-diaminopurine riboside, and their derivatives. *Journal of Biological Chemistry, 244*, 1228–1237.

Wimberley, B. T., Guymon, R., McCutcheon, J. P., White, S. W., & Ramakrishnan, V. (1999). A detailed view of a ribosomal active site: The structure of the L11-RNA complex. *Cell, 97*, 491–502.

CHAPTER FIVE

Mod-seq: A High-Throughput Method for Probing RNA Secondary Structure

Yizhu Lin[1], Gemma E. May[1], C. Joel McManus[2]

Department of Biology, Carnegie Mellon University, Pittsburgh, Pennsylvania, USA
[2]Corresponding author: e-mail address: mcmanus@andrew.cmu.edu

Contents

1. Theory	127
2. Equipment	128
3. Materials	129
3.1 Solutions and buffers	130
3.2 Custom oligonucleotides and adapters	132
3.3 Tip	133
4. Protocol	133
4.1 Duration 3–4 days	133
4.2 Preparation	134
4.3 Caution	135
5. Step 1: RNA Fragmentation	135
5.1 Overview	135
5.2 Duration	135
6. Step 2: Phosphatase and Kinase the RNA	135
6.1 Overview	135
6.2 Duration	136
7. Step 3: Ligate the 3′ Linker	136
7.1 Overview	136
7.2 Duration	137
7.3 Tip	137
8. Step 4: Ligate the 5′ Linker	137
8.1 Overview	137
8.2 Duration	137
9. Step 5: 5′ Linker Selection	138
9.1 Overview	138
9.2 Duration	138

---

[1] Contributed equally to this work.

10. Step 6: Reverse Transcription	140
10.1 Overview	140
10.2 Duration	140
11. Step 7: PAGE Purification	141
11.1 Overview	141
11.2 Duration	141
11.3 Tip	143
12. Step 8: Ethanol Precipitation	143
12.1 Overview	143
12.2 Duration	143
13. Step 9: Circularization	143
13.1 Overview	143
13.2 Duration	144
14. Step 10: Subtractive Hybridization	144
14.1 Overview	144
14.2 Duration	144
15. Step 11: PCR Amplification	146
15.1 Overview	146
15.2 Duration	146
15.3 Tip	147
15.4 Tip	147
16. Step 12: Purification and Analysis	147
16.1 Overview	147
16.2 Duration	147
16.3 Tip	147
17. Step 13: Data Analysis Using Mod-Seeker	148
17.1 Overview	148
17.2 Duration	150
References	151

Abstract

It has become increasingly clear that large RNA molecules, especially long noncoding RNAs, function in almost all gene regulatory processes (Cech & Steitz, 2014). Many large RNAs appear to be structural scaffolds for assembly of important RNA/protein complexes. However, the structures of most large cellular RNA molecules are currently unknown (Hennelly & Sanbonmatsu, 2012). While chemical probing can reveal single-stranded regions of RNA, traditional approaches to identify sites of chemical modification are time consuming. Mod-seq is a high-throughput method used to map chemical modification sites on RNAs of any size, including complex mixtures of RNA. In this protocol, we describe preparation of Mod-seq high-throughput sequencing libraries from chemically modified RNA. We also describe a software package "Mod-seeker," which is a compilation of scripts written in Python, for the analysis of Mod-seq data. Mod-seeker returns statistically significant modification sites, which can then be used to aid in secondary structure prediction.

1. THEORY

Chemical modification is a well-established method to probe RNA secondary structure. RNAs are treated with chemicals that react selectively with nucleotides in single-stranded regions. Several small molecules that modify specific RNA bases have been utilized (Weeks, 2010), including DMS (modifies A and C), Kethoxal (modifies U), and CMCT (modifies G). More recently, small molecules targeting the 2′ OH of RNA have been developed for selective hydroxyl acylation analyzed by primer extension (SHAPE) analysis. These chemicals lead to premature termination of reverse transcription. Thus, sites of modification can be detected by acrylamide or capillary gel electrophoresis of reverse transcription products.

Several RNA structure-probing approaches have been developed which utilize high-throughput sequencing to identify sites of chemical modification. SHAPE-seq (Lucks et al., 2011), the earliest of such techniques, allows probing of short *in vitro* transcripts. More recent methods allow analysis of long RNA structures. These include DMS-seq (Rouskin, Zubradt, Washietl, Kellis, & Weissman, 2014), Structure-seq (Ding et al., 2014), SHAPE-MaP (Siegfried, Busan, Rice, Nelson, & Weeks, 2014), and Mod-seq (Talkish, May, Lin, Woolford, & McManus, 2014). These methods are similar in that chemically probed RNA molecules are reverse transcribed into cDNA for sequencing library preparation. Sites of modifications can then be determined by sequence alignment and statistical analysis. Compared to traditional secondary structure-probing methods, these high-throughput methods allow transcriptome-wide probing without limitations on the lengths of target RNAs.

This chapter details a protocol for Mod-seq mapping of RNA chemical modification sites. Mod-seq is the most useful for mapping structures of large RNAs (>300 nucleotides) *in vitro* and complex mixtures of RNAs (e.g., whole transcriptomes). For probing mixtures of several long RNAs *in vitro*, an Illumina MiSeq run is sufficient to identify sites of modification. To probe entire transcriptomes, much higher read coverage is required. Although the topic has not been quantitatively analyzed, at least one read per nucleotide of RNA has been used as a general rule-of-thumb for transcriptome-wide experiments (Ding et al., 2014). Because mRNA abundance is roughly log-normally distributed, reaching this depth for the majority of the transcriptome requires at least 50 M reads per sample for yeast and other microorganisms and at least 100 M reads per sample for higher eukaryotes. Considerably, more coverage ($\sim 5\times$) would be needed to probe low-abundance transcripts. Mod-seq requires Illumina libraries prepared from each replicate for both chemically modified and untreated (negative control) RNA. Thus, at least several lanes of Illumina HiSeq are necessary for transcriptome-wide RNA structure probing.

Several steps are required for Mod-seq library preparation. After chemical probing, RNA molecules are randomly fragmented and specific linkers are ligated to the 5′ and 3′ ends of fragmented RNA. The RNA molecules are then reverse transcribed. For RNA molecules that contain a chemically modified base, reverse transcription stops one nucleotide before the modification site. For RNAs without modification, reverse transcription continues through the end of the 5′ linker sequence. The reverse transcription products are circularized and cDNA are reduced via subtractive hybridization. Libraries are then PCR amplified for high-throughput sequencing on an Illumina-based platform. Illumina reads initiating with the 5′ linker sequence are identified as full-length reverse transcription products and removed from further analysis. This allows reduced background for higher signal-to-noise ratios and is the primary difference between Mod-seq and other high-throughput chemical modification methods. As with all high-throughput RNA structure-probing methods, ligation, reverse transcription, and PCR amplification steps can introduce bias in the final sequencing library. To account for this bias, libraries are prepared and sequenced in parallel using untreated RNA as a negative control.

We also describe a data analysis package for Mod-seq, named Mod-seeker. Mod-seeker takes high-throughput sequence data in fastq format from both chemically treated and negative control samples, a target sequence for alignment, and an annotation file listing the positions of RNAs on the target sequence as input. Mod-seeker output includes per-nucleotide counts of reverse transcription termination events for each transcript in the annotation file, as well as statistical analysis of modification enrichment in chemically treated samples compared to untreated control samples.

2. EQUIPMENT

Thermocycler
Micropipettors
Micropipettor tips
Microcentrifuge
0.2 ml thin-walled tubes
50 ml conical tube
Serological pipettes

1.5 ml tubes
Heating block (80 °C)
Tube rotator
Vortexer
Plastic wrap
DynaMag™-2 Magnet (Life Technologies)
Polyacrylamide gel electrophoresis equipment—vertical setup
Light box/Dark reader camera
Dark reader
Corning® Costar® Spin-X® centrifuge tube filters, cellulose acetate membrane, pore size 0.22 μM.
EMD Millipore Steriflip™ sterile disposable vacuum filter unit 0.22 µm PVDF
50 ml syringe
18G × 1½ needles
Razor blades
Vacuum source
RNA Clean & Concentrator™-5 (Zymo Research)
DNA Clean & Concentrator™-5 (Zymo Research)
Tapestation or Bioanalyzer (Agilent)

3. MATERIALS

Nuclease free water
Hydrochloric acid (HCl)
Sodium chloride (NaCl)
Sodium acetate (NaOAc)
Sodium hydroxide (NaOH)
EDTA, pH 8.0
Tris, pH 7.4
Tween-20
Sodium citrate
Isopropanol
Ethanol
10 × TBE
Urea
40% Acrylamide/Bis solution 19:1
TEMED (tetramethylethylenediamine)
Ammonium persulfate ($(NH_4)_2S_2O_8$)
Deionized formamide

Bromophenol blue
Manganese chloride ($MnCl_2$)
10 mM dNTP mix
100 mM DTT (dithiothreitol)
DMSO (dimethyl sulfoxide)
10 mM ATP
50% PEG 8000
Ambion RNA fragmentation reagent (Life Technologies)
T4 Polynucleotide Kinase (New England Biolabs)
Superase-In™ RNase Inhibitor (Life Technologies)
Universal miRNA Cloning Linker (New England Biolabs)
T4 RNA Ligase 2, truncated (New England Biolabs)
2× Quick ligase buffer (New England Biolabs)
T4 RNA ligase 1 (ssRNA Ligase) (New England Biolabs)
Dynabeads® MyOne™ Streptavidin C1 beads (Life Technologies)
Yeast tRNA (10 mg/ml)
SuperScript® II Reverse Transcriptase (Life Technologies)
GlycoBlue™ Coprecipitant (Life Technologies)
10 bp DNA ladder (Life Technologies)
CircLigase™ ss DNA Ligase (Epicentre)
Phusion® High-Fidelity DNA Polymerase (New England Biolabs)
SYBR® Gold Nucleic Acid Gel Stain, 10,000× (Life Technologies)

3.1 Solutions and buffers

Step 5: Solution A

Component	Final concentration (*M*)	Stock (*M*)	Amount/50 ml
NaOH	0.1	1	5 ml
NaCl	0.05	1	2.5 ml
Nuclease free water			42.5 ml

Solution B

Component	Final concentration (*M*)	Stock (*M*)	Amount/50 ml
NaCl	0.1	1	5 ml
Nuclease free water			45 ml

Steps 5 and 10 $20\times$ saline sodium citrate (SSC)

Component	Final concentration (M)	Stock	Amount/50 ml
NaCl	3		8.77 g
Sodium citrate	0.3		4.41 g

Add nuclease free water up to 40 ml, adjust the pH of the solution to 7.0 with 14 N HCl. Add nuclease free water up to 50 ml.

$2\times$ Bind/Wash buffer

Component	Final concentration	Stock	Amount/50 ml
NaCl	2 M	5 M	20 ml
EDTA, pH 8	1 mM	0.5 M	100 µl
Tris, pH 7.4	10 mM	1 M	500 µl
Tween-20	0.02%	100%	10 µl
Nuclease free water			29.4 ml

Step 6: RT Precipitation mix

Component	Per reaction (µl)	Stock (M)	Volume for 20 reactions
Nuclease free water	156.5		3.13 ml
NaOAc, pH 5.5	20	3	400 µl

Step 7: 8% Acrylamide 8 M Urea gel

Component	Final concentration	Stock	Amount/40 ml
Urea	8 M		19.4 g
TBE	1$\times$	10$\times$	4 ml
Acrylamide/Bis solution 19:1	8%	40%	8 ml

Add deionized water to 40 ml. Heat solution in warm water bath until the urea has dissolved.

2× Formamide gel loading buffer

Component	Final concentration	Stock	Amount/10 ml
Deionized formamide	95%	100%	9.5 ml
EDTA, pH 8	0.5 mM	0.5 M	100 µl
Bromophenol blue			0.005 g
Nuclease free water			400 µl

DNA gel extraction buffer

Component	Final concentration (mM)	Stock (M)	Amount/50 ml
Nuclease free water			34.4 ml
NaCl	300	1	15 ml
Tris pH 8	10	1	500 µl
EDTA pH 8	1	0.5	100 µl

Step 14: DNA precipitation mix

Component	Per reaction (µl)	Stock (M)	Volume for 20 reactions
Nuclease free water	460		9.2 ml
NaCl	64	3	1.28 µl

3.2 Custom oligonucleotides and adapters

RNA 5′ linker (upper case is DNA, lower case is RNA)	ATCGTaggcaccugaaa
Anti-5′ linker	5′-TTTCAGGTGCCTACGAT-3′—Biotin–TEG
Reverse transcription (RT) primer	5′-(Phos)AGATCGGAAGAGCGTCGTGTAGGG AAAGAGTGTAGATCTCGGTGGTCGC(SpC18) CACTCA(SpC18)TTCAGACGTGTGCTC TTCCGATCTATTGATGGTGCCTACAG-3′
Post-RT sub-hyb	Biotin–TEG—5′-ATCGTAGGCACCTGAAA-3′

Universal forward PCR primer	5'-AATGATACGGCGACCACCGAGATCTACAC-3'
Barcoded reverse PCR primer	5'-CAAGCAGAAGACGGCATACGAGAT-(6 nt Illumina barcode)-TGTACTGGAGTTCAGACGTGTGCTCTTCCG-3'
Blocking forward PCR primer	5'-CGCTCTTCCGATCTATCGTAGGCAC/3ddC/-3'

3.3 Tip

The 6-nucleotide barcodes should be chosen from standard Illumina barcode sequences, which can be found online or in consultation with a high-throughput sequencing service provider. Illumina sequencers require "balanced" sequences to properly deconvolute barcoded libraries. In general, the pool of barcodes needs to have at least one A or C and at least one G or T at each position, e.g., GCCAAT and CTTGTA.

4. PROTOCOL

4.1 Duration 3–4 days

Day 1
    Step 1: RNA fragmentation
    Step 2: Phosphatase and kinase the RNA
    Step 3: Ligate the 3' linker
    Step 4: Ligate the 5' linker
Day 2
    Step 5: 5' Linker selection
    Step 6: Reverse transcription
    Step 7: PAGE purification
Day 3
    Step 8: Ethanol precipitation
    Step 9: Circularization
    Step 10: Subtractive hybridization
    Step 11: PCR amplification
    Step 12: Purification and analysis
See Fig. 1 for a flowchart of sequencing library preparation steps.

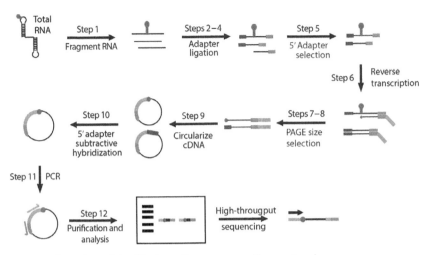

Figure 1 Overview of the Mod-seq library preparation protocol.

4.2 Preparation

1. This protocol begins with chemically modified RNA (either *in vivo* or *in vitro* probed). There are many methods available for probing RNA *in vivo* (Wells, Hughes, Igel, & Ares, 2000) and *in vitro* (Brunel & Romby, 2000). We have found that these give appropriate modification of RNA for Mod-seq. Negative control samples (either untreated or solvent-only treated) are essential for the analysis of Mod-seq data using the Mod-seeker software. We recommend including RNAs of known structure as positive controls in *in vitro* (e.g., a ribozyme) and *in vivo* (e.g., rRNA) Mod-seq experiments.
2. It is crucial to begin with good quality RNA for this protocol. Contamination with impurities such as phenol or ethanol can severely decrease the quality of the resulting sequencing libraries. We suggest that input RNA is purified using a cleanup column (e.g., RNA Clean and Concentrator-5®) prior to beginning the protocol.
3. It is possible to prepare all buffers and solutions in advance. The buffers and solutions can be stored at room temperature, with the exception of 2× formamide gel loading buffer, which is stored at −20 °C.
4. Order all of the custom oligonucleotides and resuspend in 1× TE buffer. All oligonucleotides should be stored at −20 °C as 100 μM stocks. Working stocks are at a 10 μM concentration, with the exception of the RT primer with its working stock being 2.5 μM, and the Blocking Forward PCR primer, with its working stock being 50 μM.

4.3 Caution

Care must be taken to ensure an RNase-free environment while working with RNA. RNase-free materials (tips, tubes, etc.) and gloves should always be used while working with RNA. Furthermore, good molecular biology techniques and caution must be used to prevent outside contamination of libraries with exogenous DNA or RNA sources. Acrylamide and SYBR® gold nucleic acid stain are both toxic and carcinogenic. Use standard biosafety laboratory procedures for working with and disposing of these materials.

5. STEP 1: RNA FRAGMENTATION

5.1 Overview

In this step, modified and unmodified RNA are randomly fragmented to an approximate size of 50–300 bp. It is extremely important to incubate the sample with the fragmentation buffer for precisely the duration indicated. When processing multiple samples, staggering sample fragmentation is recommended. The protocol can accommodate 1 to 2 µg of input RNA. Less input RNA may be used, but library preparation becomes more challenging and may require more PCR cycles in the final step.

5.2 Duration

1 h

1.1 In a 0.2-ml thin-walled tube, bring RNA (2 µg) to a volume of 18 µl with nuclease free water. Keep the samples on ice.

1.2 Add 2 µl of 10× Ambion RNA fragmentation reagent to the sample.

1.3 Mix, spin briefly, and incubate at 70 °C for exactly 2 min on a PCR thermocycler.

1.4 Immediately add 2 µl of Ambion RNA fragmentation stop solution, place on ice.

1.5 Clean up with RNA Clean & Concentrator™-5 (>17 nt protocol, as per manufacturers instructions), eluting in 11.5 µl nuclease free water. Proceed directly to step 2, or store the RNA at −20 °C.

6. STEP 2: PHOSPHATASE AND KINASE THE RNA

6.1 Overview

The subsequent linker ligation steps require RNA templates carrying 5′ phosphate and a 3′ hydroxyl groups. In this step, T4 Polynucleotide Kinase is used in the absence of ATP to remove 3′ terminal cyclic phosphates. ATP is then added to the reaction to phosphorylate RNA 5′ ends.

6.2 Duration

3 h

1.1 In a 0.2-ml thin-walled tube, heat the RNA sample at 80 °C for 2 min and place on ice.

1.2 Add the following to each RNA sample (total volume 15 μl).

Reagent	Per reaction (μl)
10× T4 PNK buffer (no ATP)	1.5
SUPERase-In® (20 U/μl)	1
T4 PNK	1

1.3 Incubate at 37 °C for 1 h. Proceed directly to the next step.

1.4 Add the following to each RNA sample (total volume 30 μl).

Reagent	Per reaction (μl)
Nuclease free water	9.5
10× T4 PNK buffer (no ATP)	1.5
10 mM ATP	3
T4 PNK	1

1.5 Incubate at 37 °C in a thermocycler for 1 h.

1.6 Heat inactivate by incubating the samples on a thermocycler for 10 min at 70 °C.

1.7 Clean up with RNA Clean & Concentrator™-5 (>17 nt protocol, as per manufacturers instructions). Elute in 8.5 μl nuclease free water. Proceed directly to step 3.

7. STEP 3: LIGATE THE 3′ LINKER

7.1 Overview

The 3′ linker is ligated onto fragmented mRNA, and will serve as a universal primer binding site for reverse transcription in step 6.

7.2 Duration

3.5 h

1.1 Transfer the sample to a 0.2-ml thin-walled tube.
1.2 Add 1 µl (375 ng/µl) NEB miRNA cloning linker to each sample.
1.3 In a thermocycler, heat the RNA to 80 °C for 2 min. Place the samples on ice.
1.4 Add 10 µl of reaction mix (below) to each sample.

Reagent	Per reaction (µl)
50% PEG 8000	5
DMSO	2
10 × RNA ligase buffer	2
SUPERase-In® (20 U/µl)	1.0

1.5 Add 1 µl T4 RNA ligase 2, truncated, to each sample.
1.6 Incubate at for 2.5 h at 22 °C. Alternatively, incubate overnight at 16 °C.
1.7 Clean up with RNA Clean & Concentrator™-5 (>200 nt protocol, as per manufacturers instructions). Elute in 6 µl nuclease free water. Proceed directly to step 4.

7.3 Tip

50% PEG 8000 is very viscous. Pipette slowly so that the volume is accurate.

8. STEP 4: LIGATE THE 5' LINKER

8.1 Overview

The 5' linker is ligated onto fragmented RNA. This step marks the 5' ends of fragments in order to differentiate reverse transcription products ending in run-off from those ending due to premature termination.

8.2 Duration

30 min preparation, ~24 h incubation

1.1 Transfer the sample to a 0.2-ml thin-walled tube. Place it on ice.
1.2 Aliquot 1.1 × N µl of the RNA 5' linker into a separate 0.2-ml thin-walled tube, where N is equal to the number of samples.
1.3 In a thermocycler, heat the 5' linker to 70 °C for 2 min. Place the tube of 5' linker on ice.
1.4 Add 14 µl of reaction mix (below) to each sample.

Reagent	Per reaction (μl)
DMSO	2
2 × Quick ligase buffer (NEB)	10
T4 RNA ligase 1 (ssRNA ligase)	1
5′ RNA linker (375 ng/μl)	0.75
SUPERase-In® (20 U/μl)	0.25

1.5 Incubate at 22 °C overnight.

1.6 Clean up with RNA Clean & Concentrator™-5 (>200 nt protocol, as per manufacturers instructions). Elute in 6 μl nuclease free water. Proceed directly to step 5.

9. STEP 5: 5′ LINKER SELECTION

9.1 Overview

5′ Linker ligation is relatively inefficient. To enrich for RNA products that have been ligated to the 5′ linker, the sample is annealed to a complimentary oligo ("anti-5′ linker" oligo). The anti-5′ linker oligo is biotinylated for selection with streptavidin beads. The streptavidin beads are first blocked with yeast tRNA to reduce nonspecific binding. Before use with RNA, the MyOne Streptavidin C1 beads need to be washed with two solutions (Solutions A and B) as described below. This method is outlined in the manufacturers' instructions, but is repeated here for ease of use. All of the steps are at room temperature unless indicated otherwise.

9.2 Duration

2 h

1.1 Transfer the sample to a 0.2-ml thin-walled tube. Place it on ice.

1.2 Add 24 μl of the mix below.

Reagent	Per reaction (μl)
Nuclease free water	20
Anti-5′ linker oligo (10 μM)	2
20 × SSC buffer	2

1.3 In a thermocycler, denature the sample for 90 s at 95 °C and cool to 37 °C by cooling 3 °C/min.

1.4 Meanwhile, for each RNA sample, pipette 40 µl of MyOne Streptavidin C1 beads into a 1.5-ml tube.

1.5 Place the tube on a DynaMag™-2 Magnet and incubate for 1 min. Remove and discard the supernatant.

1.6 Add one volume of solution A to the beads. Remove the tube from the magnet and vortex briefly. Put the tube back on the magnet and incubate for 1 min. Remove and discard the supernatant.

1.7 Repeat step 1.6 above.

1.8 Add one volume of solution B to the beads. Remove the tube from the magnet and vortex briefly. Put the tube back on the magnet and incubate for 1 min. Remove and discard the supernatant.

1.9 Resuspend the beads in 80 µl 2× Bind/Wash buffer per sample.

1.10 Add 8 µl of 10 mg/ml yeast tRNA to each tube of washed beads, then add 72 µl of nuclease free water to each tube.

1.11 Place the tubes of beads on a tube rotator, and incubate for 20 min at room temperature.

1.12 Place the tubes on the DynaMag™-2 Magnet and incubate for 1 min. Remove and discard the supernatant.

1.13 Add 100 µl of 1× Bind/Wash buffer to each tube. Remove the tube from the magnet and vortex briefly. Put the tube back on the magnet and incubate for 1 min. Remove and discard the supernatant.

1.14 Repeat step 1.13 twice.

1.15 Resuspend the beads in 80 µl of 2× Bind/Wash per tube.

1.16 Add 50 µl of nuclease free water to the RNA from the thermocycler and then add 1 µl of SUPERase-in®.

1.17 Add 80 µl of RNA sample to the 80 µl of beads.

1.18 Place the tubes containing the beads and RNA samples on a tube rotator. Incubate at room temperature for 15 min.

1.19 Add 100 µl of 1× Bind/Wash buffer to each tube. Remove the tube from the magnet and vortex briefly. Put the tube back on the magnet and incubate for 1 min. Remove and discard the supernatant.

1.20 Repeat step 1.19 twice.

1.21 In each tube, resuspend the beads in 11 µl of nuclease free water.

1.22 Heat the beads for 2 min at 80 °C in a heat block. After the incubation, immediately place the sample on the DynaMag™-2 Magnet.

1.23 Incubate on the magnet for 20 s and pipette off the supernatant into a new 0.2-ml thin-walled tube. Place it on ice. Proceed directly to step 6.

10. STEP 6: REVERSE TRANSCRIPTION

10.1 Overview

RNAs are reverse transcribed into cDNA in this step.

10.2 Duration

3 h

1.1 Add 1 µl of 2.5 µM RT primer to 11 µl of each RNA sample.
1.2 On a thermocycler, heat the RNA and primer mix by incubating for 2 min at 80 °C followed by 5 min at 65 °C. Place the RNA and primer mix on ice.
1.3 Add 7 µl reaction mix (below):

Reagent	Per reaction (µl)
5 × First strand buffer	4
10 mM dNTPs	1
100 mM DTT	1
SUPERase-In® (20 U/µl)	1

1.4 Add 1 µl Superscript® II to each reaction.
1.5 On a thermocycler, incubate at 50 °C for 30 min.
1.6 Destroy RNA template by adding 2.0 µl of 1 M NaOH and incubating for 20 min at 98 °C on a thermocycler. Place on ice.
1.7 Add 1.5 µl of GlycoBlue and 177.5 µl of RT precipitation mix to each sample.
1.8 Add 300 µl isopropanol to each sample and precipitate cDNA by incubating −80 °C for 30 min. Alternatively, precipitate at −20 °C for 1 h or overnight.
1.9 To pellet the cDNA, centrifuge for 30 min at 4 °C at 20,000 × g.
1.10 Remove and discard supernatant. *Note: at this point, care must be taken to not pipette off the pellet or dislodge the pellet.*
1.11 Wash the pellet by adding 600 µl of 70% ethanol. Centrifuge the samples for 15 min at 4 °C at 20,000 × g.
1.12 Resuspend pellets in 10 µl nuclease free water. The sample can be stored at −20 °C. Alternatively, proceed directly to step 7.

11. STEP 7: PAGE PURIFICATION

11.1 Overview

In order to remove excess RT primer and shorter products, the cDNA is purified and size selected to 150–330 bp. This eliminates unincorporated RT primer (at 100 bp) and also reduces or eliminates size distribution biases. The unmodified samples do not have as many reverse transcription stops as the modified samples and so the distribution of sizes in the unmodified samples is slightly larger than the modified samples (see Section 11.3 and Fig. 2). Any PAGE set up that can accommodate 20 µl of sample and result in an effective size selection of 150–330 bp can be used. The following protocol was designed for 18×16 cm electrophoresis plates.

11.2 Duration

2.5 h, overnight elution

1.1 In a 50 ml conical tube, prepare 40 ml of an 8% polyacrylamide, 8 M Urea gel.

1.2 Once the urea has dissolved, filter the gel using a 0.22 μM Steriflip™.

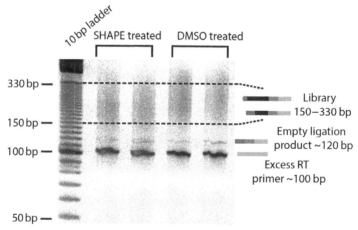

Figure 2 Example of Mod-seq reverse transcription products resolved by denaturing polyacrylamide gel electrophoresis. The bottom two bands in each lane represent excess RT primer and empty ligation product contaminants. Empty ligation cDNA arises from reverse transcription of templates generated by ligation of 5' adapter to 3' adapter sequences and should be excluded from further steps. Chemically modified RNA (SHAPE treated) typically produces slightly shorter cDNA fragments than control RNA (DMSO treated).

1.3 Set up the vertical gel apparatus, glass plates, and gel comb so that it is ready to be poured.

1.4 Add 200 μl of 10% ammonium persulfate and 20 μl of TEMED to the 40 ml of filtered 8% polyacrylamide and urea. Mix well by inversion and immediately pour the gel.

1.5 Add 10 μl 2× formamide load dye to each sample.

1.6 Prepare 10 bp ladder (1 μl ladder, 9 μl nuclease free water, 10 μl 2× formamide load dye).

1.7 Heat all samples at 80 °C for 3 min and place on ice.

1.8 Once the gel has polymerized, place the gel into the gel box and pour 1× TBE into the gel box at the top and at the bottom so that the electrodes are submersed in 1× TBE.

1.9 Fill a syringe with a needle attached with 1× TBE and blow the urea out from each well.

1.10 Pre-run the gel at 300 V for at least 20 min.

1.11 After the pre-run is over, fill a syringe with a needle attached with 1× TBE and blow the urea out from each well again.

1.12 Immediately load the denatured cDNA onto the gel, along with the denatured 10 bp ladder. Load ladder every few lanes (~4) to ensure accurate sizing.

1.13 Run the gel at 300 V for 1 h.

1.14 Prepare a spin extractor for each sample by puncturing a hole at the bottom of a 0.5-ml tube. Place the 0.5-ml tube inside of a 2-ml tube.

1.15 Remove the gel from the apparatus and place it in a container lined with plastic wrap. Add 100 ml of 1× TBE and 10 μl of SYBR® Gold nucleic acid stain and mix briefly. Incubate for 5 min at room temperature.

1.16 Visualize the gel on a dark reader. For most samples, the cDNA can be viewed with the naked eye. For lower concentration samples, the cDNA product may not be visible. See Fig. 2 for an example of cDNA separation by PAGE.

1.17 Using razor blades, excise the extended product (from 150 to 330 bp) and place cut out gel bands in the top tube of spin extractors. Use a new razor blade to cut out each sample.

1.18 Spin the spin extractors for 2 min in a centrifuge at top speed to fragment the gel into the 2-ml tube.

1.19 Add 600 μl DNA extraction buffer to the 2-ml tube containing the fragmented gel.

1.20 Incubate at room temperature on a tube rotator or mixer overnight.

11.3 Tip

The length of cDNA products indicates how effectively the original RNAs were chemically modified. cDNA products from treated samples should be slightly shorter than those from untreated controls. If the treated and untreated are of equal length, it may indicate that additional modification is necessary for your RNA(s) of interest.

12. STEP 8: ETHANOL PRECIPITATION

12.1 Overview

Eluted cDNA products are purified by precipitation with ethanol.

12.2 Duration

2 h

1.1 For each sample, add the gel slurry and liquid to a Costar® Spin-X® centrifuge tube filter.
1.2 In a microcentrifuge, spin tubes at for 2 min at top speed.
1.3 Discard the filter containing the gel slurry.
1.4 Add 1 µl glycoblue to each tube.
1.5 Add one volume of isopropanol.
1.6 Precipitate the cDNA at −80 °C for 30 min, or at −20 °C for 1 h or overnight.
1.7 To pellet the cDNA, centrifuge for 30 min at 4 °C at $20{,}000 \times g$.
1.8 Remove and discard supernatant. *Note: at this point, care must be taken to not pipette off the pellet or dislodge the pellet.*
1.9 Wash the pellet by adding 600 µl of 70% ethanol. Centrifuge the samples for 15 min at 4 °C at $20{,}000 \times g$.
1.10 Resuspend pellets in 15 µl nuclease free water. For low concentration samples (i.e., samples that could not be visualized in step 7.16), resuspend the pellet in 7.5 µl of nuclease free water. The sample can be stored at −20 °C. Alternatively, proceed to step 9.

13. STEP 9: CIRCULARIZATION

13.1 Overview

Single-stranded cDNA products are circularized using the circligase enzyme. This intramolecular ligation reaction is essentially unbiased for short cDNA products (Talkish et al., 2014).

13.2 Duration
2 h
- **1.1** Transfer 7.5 µl of each sample into a 0.2-ml thin-walled tube.
- **1.2** Add 2 µl Reaction mix (below) to each sample and mix well.

Reagent	Per reaction (µl)
10 × Circligase Buffer	1
1 mM ATP	0.5
50 mM MnCl$_2$	0.5

- **1.3** Add 0.5 µl CircLigase to each sample.
- **1.4** In a thermocycler, incubate for 1.5 h at 60 °C.
- **1.5** Heat inactivate by incubating the sample at 80 °C for 10 min. The sample can be stored at −20 °C. Alternatively, proceed to step 10.

14. STEP 10: SUBTRACTIVE HYBRIDIZATION

14.1 Overview

Reverse transcriptase terminates at modified nucleotides, prior to reaching 5′ linker on the RNA template. In this step, full-length cDNA molecules are subtracted from the sample to enrich the cDNA library for premature termination events. A biotinylated oligo complimentary to the 5′ linker sequence is annealed to the cDNA. These molecules are subtracted from the sample using streptavidin beads. The remaining cDNA is enriched for modification stops.

14.2 Duration
3 h
- **1.1** In a 0.2-ml thin-walled tube, combine 5 µl of circularized cDNA with the following reagents (below).

Reagent	Per reaction (µl)
Post-RT sub-hyb 10 µM	2
20 × SSC	2
Nuclease free water	21

1.2 In a thermocycler, heat the sample for 90 s at 95 °C and cool to 37 °C by cooling 3 °C/min.

1.3 Meanwhile, for each sample, pipette 40 µl of MyOne Streptavidin C1 beads into a 1.5-ml tube.

1.4 Place the tube on a DynaMag™-2 Magnet and incubate for 1 min. Remove and discard the supernatant.

1.5 Add one volume of Bind/Wash buffer to the beads. Remove the tube from the magnet and vortex briefly. Put the tube back on the magnet and incubate for 1 min. Remove and discard the supernatant.

1.6 Repeat step 1.5 twice.

1.7 Resuspend the beads in 80 µl 2× Bind/Wash buffer per sample.

1.8 Add 8 µl of 10 mg/ml yeast tRNA to each tube of washed beads, then add 72 µl of nuclease free water to each tube.

1.9 Place the tubes of beads on a tube rotator, and incubate for 20 min at room temperature.

1.10 Place the tubes on the DynaMag™-2 Magnet and incubate for 1 min. Remove and discard the supernatant.

1.11 Add 100 µl of 1× Bind/Wash buffer to each tube. Remove the tube from the magnet and vortex briefly. Put the tube back on the magnet and incubate for 1 min. Remove and discard the supernatant.

1.12 Repeat step 1.11 twice.

1.13 Resuspend the beads in 80 µl of 2× Bind/Wash per tube.

1.14 Add 50 µl of nuclease free water to each sample of cDNA and Post-RT sub-hyb oligo from the thermocycler.

1.15 Add 80 µl of the sample to the 80 µl of MyOne Streptavidin C1 beads.

1.16 Place the tubes containing the beads and samples on a tube rotator. Incubate at room temperature for 15 min.

1.17 For each sample, aliquot 525.5 µl of DNA precipitation mix to a 1.5-ml tube.

1.18 Place the 1.5-ml tubes containing the beads on the DynaMag™-2 Magnet and incubate at room temperature for 1 min.

1.19 Pipette off approximately 160 µl of the supernatant into the 1.5-ml tubes containing the 525.5 µl of DNA precipitation mix.

1.20 Add 700 µl isopropanol to each tube and incubate at −80 °C for 30 min. Alternatively, incubate at −20 °C for 1 h or overnight.

1.21 To pellet the cDNA, centrifuge for 30 min at 4 °C at $20,000 \times g$.

1.22 Remove and discard supernatant. *Note: at this point, care must be taken to not pipette off the pellet or dislodge the pellet.*

- **1.23** Wash the pellet by adding 600 μl of 70% ethanol. Centrifuge the samples for 15 min at 4 °C at 20,000 × g.
- **1.24** Resuspend pellets in 10 μl nuclease free water. The sample can be stored at −20 °C. Alternatively, proceed to step 11.

15. STEP 11: PCR AMPLIFICATION

15.1 Overview

In this step, cDNA is PCR amplified for approximately 12–16 cycles, depending on the amount of starting material present (see Section 15.3). An oligo that blocks the amplification of sequences that contain the 5′ linker is added to the reaction to minimize the number of reads that contain the linker sequence in the final sequencing library (Vestheim & Jarman, 2008). The reverse primer contains a 6-nucleotide barcode that is read during high-throughput sequencing. A different reverse primer containing a unique barcode can be used for each sample, allowing multiple libraries to be sequenced in the same sequencing lane. Barcode sequences should be compatible with Illumina libraries and conform to the rule that the pool of sequences must have at least one A or C and one G or T at every position (see Section 3.3).

15.2 Duration

1 h

- **1.1** In a 0.2-ml thin-walled tube, combine 1 μl of cDNA from step 10 with the following reagents (below).

Reagent	Per reaction (μl)
5× Phusion HF buffer	4
Nuclease free water	12.4
10 mM dNTP mix	0.4
10 μM Universal Forward PCR primer	1
50 μM Blocking Forward PCR primer	1
Phusion® DNA polymerase	0.2

- **1.2** Add 1 μl of barcoded reverse PCR primer to each sample.
- **1.3** In a thermocycler, use following PCR protocol 98 °C (30 s) [98 °C (10 s), 64 °C (10 s), 72 °C (30 s)] × 12–18 to amplify the DNA.

15.3 Tip

For most samples, 1 μl of template is sufficient to generate enough PCR product after 12 cycles of amplification. However, for lower concentration samples, increasing the amount of template to 2–6 μl, and increasing the total reaction volume to 50–60 μl may be required. Interestingly, for very concentrated samples, too much template will result in secondary PCR products and a reduction in the amount of desired PCR product. This phenomenon is likely a result of the polymerase continuing along a circular template, resulting in a "smear" of products larger than the ~200 bp peak of products. If this occurs, reduce the amount of starting template by diluting the original template. A dilution rage in the range of 1:2–1:8 should result in the desired PCR product.

15.4 Tip

The ideal number of PCR cycles will vary depending on amount of original RNA sample. For lower concentration samples, 18 cycles may be necessary for any product to be visualized. For most samples, approximately 12–14 cycles of PCR are sufficient.

16. STEP 12: PURIFICATION AND ANALYSIS
16.1 Overview

In this step, PCR products are purified and analyzed to check the size distribution and quality of the Illumina sequencing library.

16.2 Duration

1 h

1.1 Clean up the PCR product using a DNA Clean and Concentrator™-5 column, according to the manufacturers instructions. Elute in 10 μl TE.

1.2 Determine the size and concentration of the PCR product by running 1 μl on a D1000 ScreenTape on a Tapesation (or Bioanalyzer equivalent). The PCR product ranges in size between 180 and 350 bp, with a peak around 190–200 bp. See Fig. 3 for an example of Agilent Tapestation analysis of Mod-seq libraries.

16.3 Tip

Empty library, which results from amplification of residual RT oligo, is an undesirable product that is 140–145 bp in length. The empty library usually occurs when the amount of input RNA is low (less than 1 μg) or if the cDNA was not properly

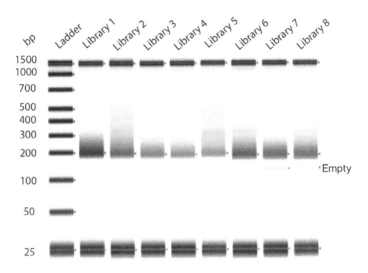

Figure 3 Example of Agilent Tapestation image of Mod-seq libraries. After PCR, libraries usually peak around 200 bp. Shorter "empty" library contaminants are visible in libraries 7 and 8. If empty library contaminants comprise more than 5% of the total sample, we recommend size selection via gel electrophoresis or SPRI beads to remove the contaminants.

resolved during PAGE purification. To avoid sequencing empty library, it is advisable to remove this product if it represents more than 10% of the library. It is possible to size select and purify the PCR products on an agarose gel or by SPRI bead enrichment. Then, reanalyze the purified product for size and concentration.

17. STEP 13: DATA ANALYSIS USING MOD-SEEKER

17.1 Overview

Mod-seq data can be analyzed using the Mod-seeker pipeline (Fig. 4), a python script package that runs from the command line in Linux or Apple OS X. Mod-seeker requires installation of the following software/packages: python 2.6 or later, cutadapt (Martin, 2011), bowtie (Langmead, Trapnell, Pop, & Salzberg, 2009), samtools (Li et al., 2009), and bedtools (Quinlan & Hall, 2010). Also, a reference sequence needs to be built as a bowtie index and a gene annotation file (either in .gff or .bed format) is required. Mod-seeker is available both at the McManus lab website and as supplemental data from the published manuscript (Talkish et al., 2014).

Mod-seeker contains two separate scripts. "Mod-seeker-map.py" is used to count the number of modifications at each position in each gene from

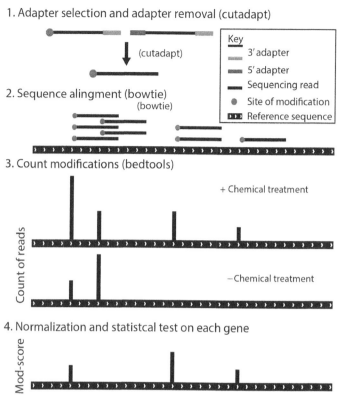

Figure 4 Mod-seeker analysis pipeline summary. Steps 1, 2, and 3 are run using the Mod-seeker-map.py script. This script uses cutadapt, bowtie, samtools, and bedtools for read processing. Step 4 is performed using the Mod-seeker-stats.py script.

each sample. In this script, sequencing reads are first trimmed to remove 3′ and 5′ adapters using cutadapt. During adapter trimming, reads beginning with a 5′ linker are removed from further analysis. The remaining trimmed reads are aligned to the reference sequence using bowtie, and short 5′ mismatches indicating untemplated nucleotides introduced during reverse transcription are removed. Reads are then mapped to annotated genes using samtools and bedtools. Finally, Mod-seeker-map.py counts the number of modifications at each position by tallying reads whose sequence initiates 3′ to each nucleotide. In the final output files ("CountMod" files), each gene with modifications is represented by two lines, where the first line is a summary of the gene and the second line records space-separated counts of modifications at each position.

The second script, "Mod-seeker-stats.py" finds statistically significant sites of modifications by comparing chemically treated samples with no-treatment controls. This script uses the Cochran–Mantel–Haenszel test with two or more replicates, or a chi-squared test for cases with no replicates. In addition to p-values from the statistical tests, the output file will also report odds ratios as a measurement of the modification level. The odds ratio is calculated as shown in Eq. (1), for a gene with length n, the odds ratio of position is:

$$OR_i = \left(\frac{T_i}{\sum T}\right) / \left(\frac{C_i}{\sum C}\right) \qquad (1)$$

where T is the count of modifications at position i in chemical-treated sample and C is the count of modifications at position i in control sample. Additional data processing can be applied for further analysis. For example, odds ratios can be log transformed and rescaled to mimic SHAPE scores as described previously (Cordero, Kladwang, VanLang, & Das, 2012) and serve as input for the program RNAstructure (Hajdin et al., 2013; Reuter & Mathews, 2010) to predict RNA secondary structures.

17.2 Duration

Variable, depending on the amount of data analyzed.

1.1 Place your Illumina fastq files in the "InputFiles" directory.

1.2 Check the settings in map_settings.txt. This file requires users to identify the location (ref_path) and name (ref_genome) of the bowtie target sequence. The user must also set the location of the file describing the position of genes on the genome (GeneAnnotationFile). Additional options allow users to set the sequence of 5′ and 3′ adapters used in the study, with defaults set to the adapters described in this protocol. Two other options set the number of processors used for alignment (bowtie_alignment_threads) and Mod-seeker runs (max_python_$threads). These can be increased to reduce processing time on multiprocessor systems.

1.3 Type "python Mod-seeker-map.py" in the command line. The program will produce a dated Results directory containing the output from your run.

1.4 When Mod-seeker-map.py has finished, copy the CountMod files from the Results directory into the Mod-seeker directory and check the settings on stats_settings.txt. This file allows users to set the false discovery rate (FDR), the minimum fold-enrichment for significant sites (Odds_Ratio_Threshold), and the number of replicates used in the study. Users must also set the names of treatment and control CountMod files in stats_settings.txt.

1.5 Type "python Mod-seeker-stats.py" in the command line. The program will produce a tab-delimited Stats_output file listing the statistically significant sites of chemical modification determined by mod-seeker.

REFERENCES

Brunel, C., & Romby, P. (2000). Probing RNA structure and RNA-ligand complexes with chemical probes. *Methods in Enzymology*, *318*, 3–21.

Cech, T. R., & Steitz, J. A. (2014). The noncoding RNA revolution—Trashing old rules to forge new ones. *Cell*, *157*(1), 77–94.

Cordero, P., Kladwang, W., VanLang, C. C., & Das, R. (2012). Quantitative dimethyl sulfate mapping for automated RNA secondary structure inference. *Biochemistry*, *51*(36), 7037–7039.

Ding, Y., Tang, Y., Kwok, C. K., Zhang, Y., Bevilacqua, P. C., & Assmann, S. M. (2014). In vivo genome-wide profiling of RNA secondary structure reveals novel regulatory features. *Nature*, *505*, 696–700.

Hajdin, C. E., Bellaousov, S., Huggins, W., Leonard, C. W., Mathews, D. H., & Weeks, K. M. (2013). Accurate SHAPE-directed RNA secondary structure modeling, including pseudoknots. *Proceedings of the National Academy of Sciences of the United States of America*, *110*(14), 5498–5503.

Hennelly, S. P., & Sanbonmatsu, K. Y. (2012). Sizing up long non-coding RNAs: Do lncRNAs have secondary and tertiary structure? *BioArchitecture*, *2*(6), 189–199.

Langmead, B., Trapnell, C., Pop, M., & Salzberg, S. (2009). Ultrafast and memory-efficient alignment of short DNA sequences to the human genome. *Genome Biology*, *10*(3), R25.

Li, H., Handsaker, B., Wysoker, A., Fennell, T., Ruan, J., Homer, N., et al. (2009). The sequence alignment/map format and SAMtools. *Bioinformatics (Oxford, England)*, *25*(16), 2078–2079.

Lucks, J. B., Mortimer, S. A., Trapnell, C., Luo, S., Aviran, S., Schroth, G. P., et al. (2011). Multiplexed RNA structure characterization with selective 2′-hydroxyl acylation analyzed by primer extension sequencing (SHAPE-Seq). *Proceedings of the National Academy of Sciences of the United States of America*, *108*(27), 11063–11068.

Martin, M. (2011). Cutadapt removes adapter sequences from high-throughput sequencing reads. *EMBnet. Journal*, *17*(1), 10–12.

Quinlan, A. R., & Hall, I. M. (2010). BEDTools: A flexible suite of utilities for comparing genomic features. *Bioinformatics (Oxford, England)*, *26*(6), 841–842.

Reuter, J. S., & Mathews, D. H. (2010). RNAstructure: Software for RNA secondary structure prediction and analysis. *BMC Bioinformatics*, *11*(1), 129.

Rouskin, S., Zubradt, M., Washietl, S., Kellis, M., & Weissman, J. S. (2014). Genome-wide probing of RNA structure reveals active unfolding of mRNA structures in vivo. *Nature*, *505*, 701–705.

Siegfried, N. A., Busan, S., Rice, G. M., Nelson, J. A. E., & Weeks, K. M. (2014). rnA motif discovery by shAPe and mutational profiling (shAPe-maP). *Nature Methods*, *11*, 959–965.

Talkish, J., May, G., Lin, Y., Woolford, J. L., & McManus, C. J. (2014). Mod-seq: High-throughput sequencing for chemical probing of RNA structure. *RNA*, *20*, 713–720.

Vestheim, H., & Jarman, S. N. (2008). Blocking primers to enhance PCR amplification of rare sequences in mixed samples—A case study on prey DNA in Antarctic krill stomachs. *Frontiers in Zoology, 5*(1), 12.

Weeks, K. M. (2010). Advances in RNA structure analysis by chemical probing. *Current Opinion in Structural Biology, 20*(3), 295–304.

Wells, S. E., Hughes, J. M., Igel, A. H., & Ares, M. (2000). Use of dimethyl sulfate to probe RNA structure in vivo. *Methods in Enzymology, 318*, 479–493.

CHAPTER SIX

Reproducible Analysis of Sequencing-Based RNA Structure Probing Data with User-Friendly Tools

Lukasz Jan Kielpinski, Nikolaos Sidiropoulos, Jeppe Vinther[1]

Section for RNA and Computational Biology, Department of Biology, University of Copenhagen, Copenhagen, Denmark
[1]Corresponding author: e-mail address: jvinther@bio.ku.dk

Contents

1. Introduction — 154
2. Data Analysis — 159
   2.1 Introduction — 159
   2.2 Preprocessing — 159
   2.3 Mapping — 162
   2.4 Summarize unique barcodes — 164
   2.5 Normalization — 167
   2.6 Data export and visualization — 174
3. Concluding Remarks — 177
Acknowledgments — 177
References — 177

Abstract

RNA structure-probing data can improve the prediction of RNA secondary and tertiary structure and allow structural changes to be identified and investigated. In recent years, massive parallel sequencing has dramatically improved the throughput of RNA structure probing experiments, but at the same time also made analysis of the data challenging for scientists without formal training in computational biology. Here, we discuss different strategies for data analysis of massive parallel sequencing-based structure-probing data. To facilitate reproducible and standardized analysis of this type of data, we have made a collection of tools, which allow raw sequencing reads to be converted to normalized probing values using different published strategies. In addition, we also provide tools for visualization of the probing data in the UCSC Genome Browser and for converting RNA coordinates to genomic coordinates and vice versa. The collection is implemented as functions in the R statistical environment and as tools in the Galaxy platform, making them easily accessible for the scientific community. We demonstrate the usefulness of the collection by applying it to the analysis of sequencing-based hydroxyl radical probing data and comparing different normalization strategies.

1. INTRODUCTION

RNA molecules can be probed in different ways to provide information on their structure and interactions with proteins. A good probing reagent reacts differentially with the RNA depending on its structure in a way that makes it possible to read out the probing signal. In most cases, the probing reagent either cleaves the RNA or adds a chemical group to the RNA, which causes reverse transcriptase to terminate. Enzymes such as nuclease S1 and RNAse V1, which preferentially cleave single- and double-stranded RNA regions, respectively, can be used to probe the secondary structure of RNA (Ehresmann et al., 1987). An alternative to enzymatic probing is to use chemical reagents such as dimethyl sulfate (DMS), which reacts with accessible and unpaired cytosines and adenines (Peattie & Gilbert, 1980), and the selective 2′-hydroxyl acylation analyzed by primer extension (SHAPE) reagents, which react with the 2′ position of flexible RNA nucleotides (Merino, Wilkinson, Coughlan, & Weeks, 2005). SHAPE reagents have been used with great success to probe RNA secondary structures and to improve computational predictions of RNA secondary structure (Deigan, Li, Mathews, & Weeks, 2009). Likewise, prediction of RNA tertiary structure can be improved by including hydroxyl radical probing data in predictions (Ding, Lavender, Weeks, & Dokholyan, 2012). Hydroxyl radical probing provides information on the solvent accessibility of the RNA backbone (Latham & Cech, 1989; Tullius & Greenbaum, 2005), which can also be used to identify footprints of proteins on RNA (Powers & Noller, 1995). Compared to data from high-resolution methods, such as X-ray crystallography and NMR, these types of probing data are of much lower resolution, but they have advantages which make them the obvious choice for most studies dealing with RNA structure. First, most RNA molecules are very difficult to analyze with the high-resolution methods, and for these RNAs, probing methods are the preferred and perhaps the only option. Second, structure probing allows dynamic changes to be interrogated and provides a straightforward strategy for identifying RNA positions that undergo structural change when comparing two conditions.

Third, a subset of the probing methods can be applied inside cells, which allows biological hypotheses to be directly investigated in the physiologically relevant environment. Finally, over the last 4–5 years, probing methods have been adapted to the use of massive parallel sequencing as read-out, which has dramatically increased the throughput and allows thousands of RNAs to be investigated simultaneously. These sequencing-based methods have already uncovered important global principles of RNA structure (Rouskin, Zubradt, Washietl, Kellis, & Weissman, 2014; Wan et al., 2014) and promise to accelerate the rate of new discoveries related to RNA structure in the years to come. An example of a sequencing-based probing experiment is outlined in Fig. 1. After probing, a primer containing a 5′ Illumina adapter sequence is extended by reverse transcriptase resulting in termination at the probed position. Next, an Illumina adaptor is ligated to the 3′ end of the cDNA, thereby allowing the sequencing library to be PCR amplified. Index sequences located in the primer used for PCR amplification make it possible to pool samples during sequencing and by using a 3′ adaptor with a degenerate sequence (Barcode), it is possible to correct for biases in the PCR amplification step. After sequencing and mapping of the obtained sequencing reads, information of the identity of the RNA, the position probed, and the position of priming for reverse transcription will be known for each PCR fragment. In addition, the index sequence is used to identify the sample from which a fragment is originating, allowing many samples to be sequenced in the same sequencing lane. As the next step, all the terminations observed at a given position need to be summarized, and if a barcode was used, this termination count (TC) can be corrected for PCR duplicates. In the final steps of data analysis, the counts need to be normalized for differences in coverage, either by estimating the coverage or by using a window-based smoothing of the signal. Finally, depending on the experimental setup, the sample can be normalized with the control to correct for background-induced termination of reverse transcriptase or can be compared to detect relative differences between conditions. With the large amount of data obtained from sequencing experiments, the data analysis is becoming a larger and larger part of the overall study, and although there are variations in the experimental setup of the studies that have applied sequencing-based RNA structure probing, most of the steps required to go from sequencing reads to normalized probing signal are identical. Nevertheless, each of the studies has developed its own computational pipeline for analysis of the data and in many cases also its own normalization strategy (Table 1). Here, we present a set of tools that simplify data processing

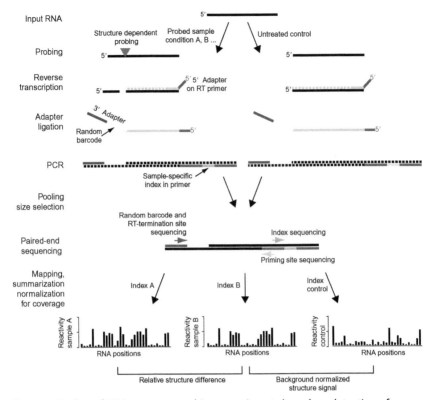

Figure 1 Outline of RNA structure-probing experiments based on detection of reverse transcriptase termination sites by sequencing. RNA is reacted with a reagent, which probe the RNA structure and can lead to termination of reverse transcription. The probed and control RNA samples are reverse transcribed with the random primer carrying a 5′ adapter compatible with the Illumina platform. The 3′ end of a synthesized cDNA carries information on the reverse transcription termination site, which can correspond to the natural RNA 5′ end, random RNA breakage, spontaneous RT stop, or probing reagent-induced termination. The latter carries information useful for structural investigation, while the previous are accounted for by the control sample. Following reverse transcription, an adapter is ligated to the 3′ end of cDNA which introduces another Illumina sequence and a random barcode, which is useful for distinguishing if sequenced reads are PCR duplicates or are derived from separate cDNA molecules. Single-stranded DNA constructs with adapters at both ends are ready for PCR amplification, which elongates the adapters and introduces sample-specific indexes (present in one of the primers). These indexes allow for pooling the libraries intended to be sequenced together which decreases per-sample cost and equalizes size selection and sequencing biases. Multiplex paired-end sequencing provides information on three regions of each DNA cluster. Read 1 starts with the random barcode of known length and follows into the sequence corresponding to the 3′ end of cDNA. Then, the index read reads out the sample index, which allows for classification of a cluster into one of the pooled samples. Finally, read 2 starts by sequencing the fragment corresponding to the random part of the reverse transcription primer and follows into cDNA body, providing information on reverse transcription priming site. Sequencing data from each of the pooled samples are first processed independently, after which information can be combined to pinpoint relative structural differences or give normalized structural signal.

Table 1 Summary of high-throughput sequencing-based methods for probing RNA structure

Method name	Probing reagent	Reagent compatibility	Termini-based signal detection	Detection of reverse transcription termination sites	Data analysis strategy	Applicable in vivo	Signal enrichment	References
PARS	S1 and V1 nucleases	Nucleases leaving 5′ phosphate	Yes	No	Log ratio	No	5′ Phosphate	Kertesz et al. (2010), Wan et al. (2012), and Wan et al. (2014)
FragSeq	P1 nuclease	Nucleases leaving 5′ phosphate	Yes	No	Log ratio	No	5′ Phosphate	Underwood et al. (2010)
ssRNA/dsRNA-Seq	RNase V1/Rnase I	ssRNA or dsRNA degrading nucleases	No	No	Identification of protected stretches	No	Degradation	Li, Zheng, Ryvkin, et al. (2012), Li, Zheng, Vandivier, et al. (2012), and Zheng et al. (2010)
SHAPE-Seq	1M7	Reagents inducing RT stop or RNA break	Yes	Yes	Modification probability	No		Lucks et al. (2011) and Mortimer, Trapnell, Aviran, Pachter, and Lucks (2012)
DMS-Seq	DMS	Reagents inducing RT stop or RNA break	Yes	Yes	Winsorization	Yes	Size selection	Rouskin et al. (2014)
Structure-seq	DMS	Reagents inducing RT stop or RNA break	Yes	Yes	$\Delta(\log(\text{counts}))/\text{mean}(\log(\text{counts}))$	Yes		Ding et al. (2014)

Continued

Table 1 Summary of high-throughput sequencing-based methods for probing RNA structure—cont'd

Method name	Probing reagent	Reagent compatibility	Termini-based signal detection	Detection of reverse transcription termination sites	Data analysis strategy	Applicable in vivo	Signal enrichment	References
Mod-Seq	DMS	Reagents inducing RT stop or RNA break	Yes	Yes	CMH test	Yes	Subtractive hybridization	Talkish, May, Lin, Woolford, and McManus (2014)
MAP-Seq	DMS, CMCT, 1M7, NMIA	Reagents inducing RT stop or RNA break	Yes	Yes	Modification probability	No		Seetin, Kladwang, Bida, and Das (2014)
HRF-Seq	Hydroxyl radicals	Reagents inducing RT stop or RNA break	Yes	Yes	Modification probability	Yes		Kielpinski and Vinther (2014)
SHAPE-MaP	1M7, 1M6, NMIA, DMS	?	No	No	Normalized differential mutation rate	Yes		Homan et al. (2014) and Siegfried, Busan, Rice, Nelson, and Weeks (2014)

of sequencing-based probing experiments and implement different published strategies for using barcodes to estimate unique counts, for normalizing counts across the probed RNA molecules, and for background signal.

2. DATA ANALYSIS

2.1 Introduction

With the dramatically increased throughput resulting from the use of massive parallel sequencing, data analysis is becoming the major bottleneck for RNA probing experiments. In this chapter, we describe a workflow that allows reproducible and convenient analysis of sequencing-based RNA probing data (Fig. 2). The workflow starts from FASTQ files generated from probing experiments, where the termini of the reads corresponds to the RNA positions that have been probed. For each of the required steps (preprocessing, mapping, summarization of unique counts, and normalization), we provide tools that have been implemented in the Galaxy environment (Goecks, Nekrutenko, & Taylor, 2010) to allow researchers without training in bioinformatics to easily perform the data analysis. The output of the workflow is data tables with normalized values corresponding to the probing reactivities at each RNA position, different plot options, and tracks for the UCSC Genome Browser (Kent et al., 2002; Fig. 2). Furthermore, for researchers with more computational experience with data analysis, we provide command line scripts for the preprocessing and mapping steps and a new R package (RNAprobR) for summarization, normalization, and plotting. The R package is equipped with an exhaustive help system and examples explaining each of the functions, as well as a vignette guiding a user through an HRF-Seq analysis workflow. Below, we demonstrate the utility of the tools by reanalyzing the data from our own hydroxyl radical footprinting (HRF-Seq) publication (Kielpinski & Vinther, 2014) and the data from the *in vivo* DMS-Seq publication from the Weissman Lab (Rouskin et al., 2014). For each of the different steps in the workflow, we provide detailed protocols for performing the analysis either in Galaxy or in the command line/R environment.

2.2 Preprocessing

Libraries prepared for sequencing on the Illumina platform consist of sequences derived from the nucleic acid molecules of interest, as well as library-specific elements introduced during sample preparation. The main

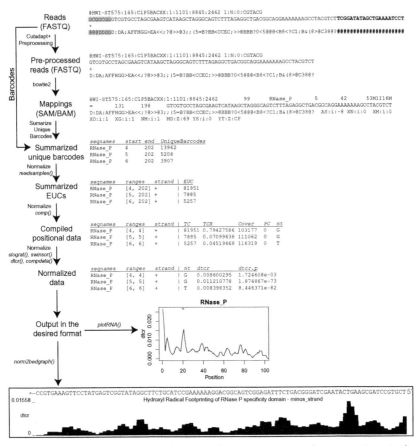

Figure 2 Overview of the data analysis workflow. Data types in the processing pipeline with data examples from RNase P HRF-Seq analysis (right side). Names of functions used to convert between data types printed by the straight arrows (RNAprobR functions in *italic*). Shaded area in the FASTQ file example corresponds to random barcode sequence, bold text to low quality tail. Upper plot was generated in R environment, lower in UCSC Microbial Genome Browser.

purpose of preprocessing is to remove the latter. The first step in our procedure is to remove sequences derived from adapters and low quality stretches using the *Cutadapt* tool (Martin, 2011). The next step, for which we have prepared a *Preprocessing* tool, removes and saves the random barcodes sequences, if they were ligated to 3′ ends of cDNA. The resulting Barcodes file is used for calculating Estimated Unique Counts (EUCs) during the summarization step. In our previous publication

(Kielpinski & Vinther, 2014), we used a seven-nucleotide fully random barcode (5′-PHO-NNNNNNN[...]). However, we have noticed that the adapters in some cases contain indels, which offsets the position recognized as the cDNA termini. We therefore recommend to use an adapter with a partially random sequence 5′-PHO-NWTRYSNNNN[...]. By requiring the barcode part of the first-in-pair reads to match this pattern, it is possible to recognize the majority of barcodes containing indels and remove the affected fragments from the analysis.

In the case of paired-end sequencing, the first nucleotides of second-in-pair reads are derived from the random primer and therefore have a higher risk of carrying mutations and if reads are long, the random primer sequence may also be present in the 3′ end of the first-in-pair read. Likewise, for long reads, the 3′ end of second-in-pair reads may be derived from the random barcode. Thus, to avoid including these sequences in the analysis the preprocessing tool trims: (1) the 5′ end of the second-in-pair reads to remove the reverse transcription primer-derived sequence and (2) 3′ end of both reads to remove possible random barcode incorporation in the second-in-pair read and random primer in first-in-pair read. Running the *Preprocessing* tool is mandatory as it realigns pairs of reads coming from paired-end sequencing.

2.2.1 Galaxy
2.2.1.1 Cutadapt tool
For standard probing data, we use the *Cutadapt* default options, except that (1) the "3′ Adapters" option is set to the sequence of the adapters used in the experiment, (2) the "Quality cutoff" option (found under the "Additional modifications to reads" group) is set to 17, and (3) the "Minimum length" option (under "Output filtering options") is set to a suitable length depending on sample complexity (the more complex RNA mixture, the longer minimal length required). With paired-end sequencing, the *Cutadapt* needs to be run twice, once for each FASTQ file (providing its respective adapter sequence).

2.2.1.2 Preprocessing tool
For the *Preprocessing* tool, the type of sequencing that was performed (paired-end or single-end) has to be specified and relevant adapter-trimmed FASTQ files chosen. Then, the IUPAC DNA code of the random barcode should be specified in the "Barcode sequence" option. If no random barcode was

introduced in the experimental procedure, the "Barcode sequence" option should be left empty. Finally, in the "3′ trimming length" field the length of the random part of the RT primer should be specified.

2.2.2 Command line
2.2.2.1 Cutadapt tool
Typical run:

```
cutadapt -a $ADAPTER_SEQUENCE -q $QUALITY_CUTOFF -m $MINIMAL_LENGTH reads.fastq > reads_trimmed.fastq
```

where ADAPTER_SEQUENCE is a sequence of the adapter to be trimmed, QUALITY_CUTOFF is a cutoff at which low quality stretches should be trimmed (suggested 17), and MINIMAL_LENGTH determines minimal length of reads to be kept.

2.2.2.2 Preprocessing tool
We have prepared a bash script (*preprocessing.sh*) to allow for easy preprocessing of Cutadapt trimmed reads in the command line. For help on its usage type:

```
./preprocessing.sh -h
```

An example run for processing single-end, nonbarcoded reads, trimming 15-nucleotide long random primer sequence:

```
./preprocessing.sh -1 reads.fastq -t 15
```

Processing paired-end, barcoded reads, and writing output to "preprocessing_out" directory:

```
./preprocessing.sh -1 reads1.fastq -2 reads2.fastq -b NNNNNNN -t 15 -o"preprocessing_out"
```

2.3 Mapping
For sequencing-based RNA probing experiments, the preprocessed reads need to be mapped to the investigated RNAs. This requires several specific considerations. First, as the alignment must faithfully report the ends of mapped reads, only end-to-end procedures are suitable. Second, the stringency of mapping must allow for mapping even if mismatched untemplated nucleotides are present at the read's end. Finally, with paired-end sequencing it is necessary to allow a distance between read pairs to vary within the whole range of experimentally determined insert size distribution. Out of several available mapping algorithms (Hatem, Bozdag, Toland, & Catalyurek,

2013), we have chosen Bowtie 2 (Langmead & Salzberg, 2012) for use in our pipeline.

For mapping, a FASTA file containing the sequences to which reads will be mapped is required. For probing studies focusing on a known set of RNA molecules, e.g., RNA from *in vitro* transcription or from specific enrichment, all the sequences of these molecules should be included in the FASTA file. On the other hand, in studies probing a whole transcriptome, annotations such as ENSEMBL (Flicek et al., 2014) or RefSeq (Pruitt et al., 2014) longest isoforms (Rouskin et al., 2014), or a preselected set, such as top expressed RNA molecules, can be used. In any case, our pipeline is designed to study a predefined set of transcripts, and mapping should be performed to transcript sequences and not a genomic reference. The result of mapping is saved to SAM or BAM files, which contain information on mapping coordinates and various alignment properties. Mapping is the most computationally heavy step in our pipeline.

2.3.1 Galaxy
Bowtie2 (version 0.2) is available on the Galaxy main server and is one of the tools included in the standard Galaxy instance that can be installed locally or run as a cloud-based instance of Galaxy. The following parameters should be specified:
- Specify sequencing type (paired-end or single-end).
- Use as input "FASTQ file" (or files, if paired-end sequencing) produced by *Preprocessing* tool (with paired-end sequencing use "Read 1" file in the first field, "Read 2" in the second).
- For the paired-end sequencing, modify "Maximum insert size for valid paired-end alignments" to suitable length, which depends on the size selection performed during library preparation and the sequencer capabilities (for HRF-Seq use 700).
- Change "Will you select a reference genome from your history or use a built-in index?" to "Use one from the history" and choose the uploaded FASTA file in "Select the reference genome."
- Check if the type of alignment under "Parameter Settings" is set to –end-to-end options, advisably –sensitive or –very sensitive.

2.3.2 Command line
When running Bowtie 2 from command line, it is necessary to index the reference FASTA file with Bowtie2-build before mapping:

```
bowtie2-build index.fasta index
```

A typical mapping and subsequent file compression can be performed with the following code:

```
bowtie2 -p $THREADS -X 700 -x index -1 read1.fastq
-2 read2.fastq | gzip > alignedReads.sam.gz
```

where THREADS specifies number of parallel threads. In the case of single-end sequencing, use -U option instead of -1 and -2.

2.4 Summarize unique barcodes

The main function of the *Summarize Unique Barcodes* tool is to count the number of unique random barcodes associated with each sequenced fragment. A fragment is understood as (1) a pair consisting of RT termination site and RT priming site given paired-end sequencing or (2) an RT termination site in single-end sequencing. One of the options of this tool is trimming of untemplated nucleotides. Untemplated nucleotides can be added to cDNA 3' ends via terminal transferase activity of reverse transcriptase (Schmidt & Mueller, 1999). After sequencing and mapping, these nucleotides will offset the location of the read-end mapping by as many nucleotides as were added. In probing experiments this is a problem, because the probing signal will be shifted from the position that were probed in the experiment to positions immediately upstream (Kielpinski & Vinther, 2014; Talkish et al., 2014). In this workflow, before counting unique barcodes, the optional step of trimming untemplated nucleotides from cDNA 3' ends can be performed. It works as previously described (Kielpinski, Boyd, Sandelin, & Vinther, 2013), but here assuming that ambiguous nucleotide ("N") matches a template. We recommend to use trimming for methods based on detecting reverse transcription termination sites and not to use it for methods based on ligating the linker directly to RNA (Table 1). We have applied our trimming procedure to the published HRF-Seq and DMS-Seq data (fibroblasts *in vivo* probing, mapped to RefSeq longest isoforms, top 1000 transcripts by read count) (Rouskin et al., 2014). The fraction of reads that were trimmed is larger for HRF-Seq data than for DMS-Seq (Fig. 3A). This difference is likely caused by the methylated bases in the DMS probing experiment blocking reverse transcription, while the reverse transcriptase can run off the end of the cleaved RNA molecules in the HRF-Seq experiment. In DMS-Seq, only modifications of A or C nucleotides should lead to RT terminations, and as expected, we find that our trimming procedure increases the number of reads terminating before A or C (Fig. 3B). Moreover, for the HRF-Seq

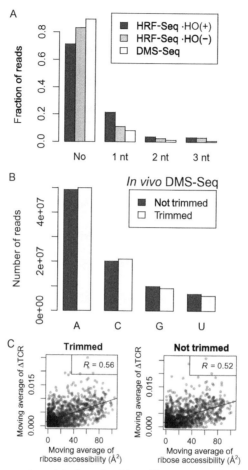

Figure 3 Trimming untemplated nucleotides. (A) Fraction of reads which did not undergo trimming (No) or had one, two, or three nucleotides trimmed from the read's 5′ ends in HRF-Seq treated (•HO(+)), HRF-Seq control (•HO(−)) and *in vivo* DMS-Seq experiments (fibroblasts). (B) Number of reads terminating upon reaching different nucleotides in DMS-Seq probing with or without trimming untemplated nucleotides. (C) Correlation between moving average of ΔTCR and moving average of 16S rRNA ribose accessibility (ball size 3 Å) in HRF-Seq experiment with or without untemplated nucleotides trimming.

data, we find that trimming improves the correlation between our HRF-Seq probing measure ΔTCR and ribose accessibility as calculated from the ribosome crystal structure (Dunkle, Xiong, Mankin, & Cate, 2010; Kielpinski & Vinther, 2014; Fig. 3C).

2.4.1 Galaxy

To run *Summarize Unique Barcodes* tool in Galaxy the following options need to be specified:

- In the "Aligned Reads" option, specify the BAM file produced in the Mapping step.
- In the "Barcodes" option, specify the Barcodes file generated by *Preprocessing* tool.
- In the "Produce k2n file" option, specify if a *k2n file* should be produced. This file is necessary for EUC to be calculated according to the HRF-Seq method.
- In the "Trim untemplated nucleotides" option, choose whether untemplated nucleotides should be trimmed.
- In the "Set priming position" option, choose whether the priming position should be fixed (applicable with fixed priming site).

Summarize Unique Barcodes tool produces the following output:

- *Unique Barcodes*: file with four columns: (1) transcript name (seqnames), (2) RT termination site (start), (3) RT priming site (end), and (4) count of unique barcodes associated with fragments matching first three columns.
- *Read Counts*: similar to *Unique Barcodes* file, but the fourth column is a number of reads matching first three columns. Used for the analysis of nonbarcoded experimental setups.
- *k2n file*: file required for calculating EUCs. It is a sequence of numbers, where nth number informs how many unique cDNA molecules gives rise to observation of n unique barcodes in a given sample. Required for EUC calculation according to HRF-Seq method (Formula 1 in HRF-Seq paper; Kielpinski & Vinther, 2014).
- *Trimming stats*: statistics of trimming untemplated nucleotides from reads' ends.
- A warning message with the highest number of observed unique barcodes.

2.4.2 Command line

We have prepared a bash script (*summarize_unique_barcodes.sh*) to facilitate easy summarization of unique barcodes in the command line. For help on its usage type:

```
./summarize_unique_barcodes.sh -h
```

An example run for nonbarcoded sample with untemplated nucleotides trimming followed by writing output to "summarize_out" directory:

```
./summarize_unique_barcodes.sh -f alignedReads.sam.gz
-t -o "summarize_out"
```

An example run for barcoded sample, with untemplated nucleotides trimming, calculation of *k2n file* and setting the priming to a fixed position (202):

```
./summarize_unique_barcodes.sh -f alignedReads.sam.gz
-b barcodes.txt -t -k -p 202
```

2.5 Normalization

An important step in the analysis of sequencing-based probing data is to summarize counts of observed events (cleavages or modifications) at each position within the studied transcripts. One possibility is to simply count the observed sequenced reads that terminate at the given position (most common approach). In experiments using a large amount of starting material and relatively few PCR cycles, this approach can produce unbiased data. However, in most cases, the PCR amplification of the sequencing library will lead to biases in the TCs observed at different positions. If a random barcode is added to the cDNAs in the ligation step (Fig. 1), it is possible to reduce PCR bias by counting unique barcodes, it is sets of sequenced reads that have the same barcode sequence and position of priming and probing (Casbon, Osborne, Brenner, & Lichtenstein, 2011; Konig et al., 2010). Finally, in cases with high coverage of the investigated RNA, the same barcode may be attached to a particular fragment more than one time. This will cause an underestimation of the number of unique barcodes, which can be corrected by estimating the unique count either by assuming equal barcode frequencies (Fu, Hu, Wang, & Fodor, 2011) or by estimating the barcode frequencies (Kielpinski & Vinther, 2014). Our workflow implements all of the four methods of counting events mentioned above. Whenever possible, we recommend including barcodes in the experimental protocol to allow correction of PCR-induced biases. For RNA molecules with low coverage (highest count per position is lower than the square root of number of different barcodes; Casbon et al., 2011), counting of unique barcodes will be sufficient to correct for PCR-induced biases. For RNAs with high coverage, we recommend using EUC with equal or estimated barcode frequencies, depending on the barcode frequency distribution in the barcode preparation used for the experiments (Kielpinski & Vinther, 2014).

The next fundamental step of data processing in RNA structure probing experiments is the normalization of the observed probing signal. Two different types of noises typically need to be accounted for. First, data from probing experiments will typically contain some background signal derived from background cleavage of RNA or pretermination of reverse transcriptase, which is typically corrected with a mock-treated control sample. Second, the coverage along each RNA molecule will influence the probing signal. Several different methods have been proposed to convert sequencing counts from probing experiments into a score reflecting RNA reactivity with the probing reagents. One of these is the logarithm of the ratio approach, which was applied in the PARS and FragSeq methods (Table 1). In this normalization procedure, at each assayed position, the counts from the treated sample are divided by the counts from the control sample followed by logarithmic transformation of the resulting ratio. Prior to log-ratio calculation, both methods normalized counts to sequencing depth either within sample (Kertesz et al., 2010) or within RNA molecule (Underwood et al., 2010) and increased them by a value of pseudocount. The use of the $\log_2$ ratio brings positions with high background TCs to a range consistent with the rest of positions (Underwood et al., 2010) and the added pseudocount ensures that positions with low counts compared to the pseudocount will obtain a $\log_2$ ratio close to zero. In our workflow, we have implemented a similar Smooth Log_2 Ratio Normalization procedure. Our tool calculates the binary logarithm of the ratio of sum of TCs of treated sample to sum of TCs of the control sample (increased at each nucleotide by a pseudocount value "P") in overlapping windows of length "w" centered at nucleotide "i" (Formula 1) (Wan et al., 2014). Before calculation, counts are adjusted to the sum of sequencing counts within sample or within RNA molecule depending on the chosen option. Additionally, the function calculates p values by comparing the sum of TC's in a specific window to the sum of TC's from an entire transcript for the treated and the control samples using a pooled two-sample Z-test (calculated before adding pseudocount or adjusting for sequencing depth).

$$\text{slograt}_{i-f} = \log_2 \left(\sum_{j=i-((w-1)/2)}^{j=i+((w-1)/2)} \left(\text{TC}_{\text{treated},j} + P \right) \right) \\ - \log_2 \left(\sum_{j=i-((w-1)/2)}^{j=i+((w-1)/2)} \left(\text{TC}_{\text{control},j} + P \right) \right) \quad (1)$$

Another normalization approach is based on estimating the probability of each position in the transcript being modified in the course of probing. It has been applied for SHAPE-Seq, MAP-Seq, and HRF-Seq (Table 1) and relies on estimating the coverage at each position. The coverage is the count of cDNA molecules that either terminates at a given position or spans the position in such a way that a sequencing fragment could have been observed if the cDNA had terminated at the position. The coverage can be understood as a measure for how many reverse transcriptases reached this at a given position and could potentially have terminated there and been detected in the sequencing experiment. The probing occurring at a given position will determine what fraction of the reverse transcriptases that reach the position also terminates. The HRF-Seq and MAP-Seq protocols estimate the reverse transcription termination probability at each position by calculating ratio of TC to coverage (termination-coverage ratio, TCR). By subtracting the TCR observed in a mock-treated (control) sample from the TCR observed in the probed sample (ΔTCR), the specific modification probability at each position can be calculated (Karabiber, McGinnis, Favorov, & Weeks, 2013). In SHAPE-Seq, the background drop-off rate is calculated as TCR in the control sample, while SHAPE reactivities are calculated by likelihood maximization of Poisson-distributed signal mixed with background (Aviran et al., 2011).

In our workflow, we have implemented the ΔTCR method. Coverage can be calculated based on a fixed priming site (Aviran et al., 2011; Seetin et al., 2014). Alternatively, if random priming was used during reverse transcription and the position of priming was obtained by paired-end sequencing, the coverage can be deduced from the termination sites and observed priming sites as the sum of EUCs of inserts spanning each location (Kielpinski & Vinther, 2014). In sequencing-based probing experiments, short inserts (typically <100 nts) are not observed because of size selection. Thus, when calculating coverage, inserts that would not have been observable if terminated at a given position (resulting in a product too short to be detected) should not be used for the coverage calculation of that position.

After having summarized TCs and coverage (Cover) for each location, the TCR (Formula 2) can be calculated. Next, to estimate the probability with which each position in the transcript is modified by the probing reagent, the ΔTCR is calculated based on TCR values of control and treated samples (Formula 3). As a last step, the signal can be smoothened in running windows of length "w" and offset by "f" nucleotides (usually 1) to account for signal being detected before the modified residue/bond (Formula 4).

The implemented ΔTCR function also calculates p values using a strategy similar to the one implemented for Smooth Log$_2$ Ratio Normalization, but using the local coverage instead of the transcript-wide sum of TCs.

$$\text{TCR}_i = \frac{\text{TC}_i}{\text{Cover}_i} \qquad (2)$$

$$\Delta\text{TCR}_i = \max\left(\frac{\text{TCR}_{\text{treated},i} - \text{TCR}_{\text{control},i}}{1 - \text{TCR}_{\text{control},i}}, 0\right) \qquad (3)$$

$$\text{dtcr}_{i-f} = \sum_{j=i-((w-1)/2)}^{j=i+((w-1)/2)} \frac{\Delta\text{TCR}_j}{w} \qquad (4)$$

Several methods for RNA structure investigation aim at reducing the amount of sequenced background at the experimental stage (Table 1). For example, PARS and FragSeq methods utilize the presence of a phosphate left at the RNA 5′ end after hydrolysis with the probing nuclease. In Mod-Seq, background is reduced by subtractive hybridization of cDNA molecules that did not terminate at the modified residue. DMS-Seq and Pseudo-Seq introduce two stringent size selections into the experimental protocol which removes nonprobed molecules from the pool before sequencing (Carlile et al., 2014; Rouskin et al., 2014). In addition to these methods, we have recently developed a SHAPE probing method called SHAPE selection (SHAPES), which efficiently reduces background signal both from preterminated reverse transcriptases and from unprobed molecules using biotin-coupled probing reagents and streptavidin selection (Poulsen, Kielpinski, Salama, Krogh, & Vinther, in press). For the log$_2$ ratio and the ΔTCR normalization methods, an untreated control sample is required to normalize the data for background termination of reverse transcription, but if the background has been efficiently reduced at the experimental stage, it is possible to leave out the mock-treated sample. In this case, the data still need to be normalized for differences in coverage across the RNA, but background normalization is not needed. For example, in DMS-Seq, the counts are normalized in nonoverlapping windows by 90% Winsorization, which replaces counts greater than the 95th percentile with the 95th percentile value followed by linear transformation to [0,1] range. We have implemented a modified Winsorization procedure for experiments which do not use an untreated control, named here Smooth Winsorization. The first step of the calculation is the Winsorization (Hastings, Mosteller, Tukey, & Winsor, 1947) of level "L," which sets all values which are above $(1+L)/2$ quantile or below $(1-L)/2$ quantile to $(1+L)/2$ quantile and $(1-L)/2$ quantile, respectively (Formula 5). This is followed by scaling the values to the $[0,1]$ interval (Formula 6). To avoid introducing artificial boundaries in the analysis and to be able to calculate the standard deviation of the predictions, we calculate scaled Winsorized values in all possible contiguous stretches of length "w" within each transcript and report the mean and standard deviation of prediction for each site (Formula 7):

$$\text{Winsor}_L(TC_{i...j}) = \begin{cases} Q_{\frac{1+L}{2}}(TC_{i...j}), \forall TC \in \left(Q_{\frac{1+L}{2}}(TC_{i...j}), \max(TC_{i...j})\right] \\ Q_{\frac{1-L}{2}}(TC_{i...j}), \forall TC \in \left[\min(TC_{i...j}), Q_{\frac{1-L}{2}}(TC_{i...j})\right) \\ TC, \forall TC \in \left[Q_{\frac{1-L}{2}}(TC_{i...j}), Q_{\frac{1+L}{2}}(TC_{i...j})\right] \end{cases}$$

(5)

$$\text{Scaled Winsor}_L(TC_{i...j}) = \frac{\text{Winsor}_L(TC_{i...j}) - Q_{\frac{1-L}{2}}(TC_{i...j})}{Q_{\frac{1+L}{2}}(TC_{i...j}) - Q_{\frac{1-L}{2}}(TC_{i...j})}$$

(6)

$$\text{swinsor}_{i-f} = \sum_{j=i-((w-1)/2)}^{j=i+((w-1)/2)} \left(\frac{\text{Scaled Winsor}_L(TC_{(j-((w-1)/2))...(j+((w-1)/2))})}{w}\right)$$

(7)

To compare the three different normalization schemes implemented in our workflow, we applied these to the data from our recent HRF-Seq publication. Visual inspection of normalized data shows that overall trends in the data are similar, but also that considerable differences are apparent between the normalized datasets (Fig. 4A).

We have evaluated the three normalization methods by comparing two metrics: (1) the Pearson correlation with ribose accessibility calculated from the crystal structure (Dunkle et al., 2010; Kielpinski & Vinther, 2014) and (2) area under the curve (AUC) of receiver operating characteristic (ROC) curve (Robin et al., 2011) with the response calculated from ribose accessibilities (accessibilities which are below first quartile are set to 0, those above third quartile are set to 1, and remaining values are set to NA). Using both benchmarks, the ΔTCR normalization scheme scored highest, the Smooth Log_2 Ratio Normalization scheme second, and the Smooth Winsorization normalization scheme third (Fig. 4B). This is consistent with input information requirements of the different functions, with ΔTCR requiring both

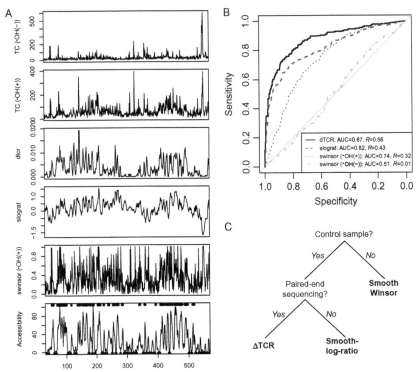

Figure 4 Comparison of normalization methods. (A) Distribution of TCs in the control (•HO(−)) and treated (•HO(+)) samples and of probing signal normalized with different methods (ΔTCR, Smooth Log$_2$ Ratio, and Smooth Winsor) for HRF-Seq probed 16S *Escherichia coli* rRNA (only fragment shown). The bottom plot shows moving average of ribose accessibility (line) and whether nucleotides were categorized as accessible (top quartile of accessibility—top points) or inaccessible (bottom quartile of accessibility—bottom points). (B) ROC curves for the different normalized datasets using the accessible and inaccessible categories as the binary classifier. The AUCs as well as Pearson correlations between moving average of ΔTCR and moving average of 16S rRNA ribose accessibility are indicated in the legend. (C) Decision tree for the choice of normalization method.

paired-end sequencing and a control sample, Smooth Log$_2$ Ratio requiring single-end sequencing and a control sample, and Smooth Winsorization requiring single-end sequencing and no control sample (Fig. 4C).

2.5.1 Galaxy

Normalization in the Galaxy environment is performed with the *Normalize* tool. It uses *Unique Barcodes* or *Read Counts* files produced by *Summarize Unique Barcodes* tool as input. To perform ΔTCR or Smooth Log$_2$ Ratio

Normalization, files for both a treated and a control sample are necessary. First, a method for EUC calculation has to be specified. The possible options are

- "Counts", count from the input file is not transformed—count of unique barcodes or read count (depending on the input) is preserved. As long as the highest observed unique barcode count is lower than the square root of the count of all possible barcode combinations (Casbon et al., 2011), this is the recommended method. The *Summarize Unique Barcodes* tool outputs value of maximum observed barcodes. It is the only suitable method for *Read Counts* file.
- "Fu", EUCs are calculated using a method which corrects for the possibility of specific barcodes to be ligated to the same fragment multiple times assuming equal probability of ligating each barcode sequence (Fu et al., 2011). It requires specifying value of random barcode complexity, e.g., for NWTRYSNNNN barcode, the complexity equals 16384 ($4 \times 2 \times 1 \times 2 \times 2 \times 2 \times 4 \times 4 \times 4 \times 4$).
- "HRF-Seq", EUCs are calculated similarly as in "Fu," but the probability of ligating different barcodes is estimated via observed frequencies of nucleotides at each barcode position (Kielpinski & Vinther, 2014). Requires that *k2n files* from *Summarize Unique Barcodes* tool are provided for both treated and control samples.

Additionally, a FASTA file (the one used for initial read mapping) can be specified to provide sequence information at each position. Next, it can be chosen whether the raw counts should be included (Print raw values option) and a normalization method should be specified. Moreover, the cut-off length used for coverage calculation and nucleotide offset ("f" in Formulas 1, 4, and 7) has to be specified. The nucleotide offset should be set to "1" for probing methods for which RT terminates one nucleotide before modified residue and "0" for methods for which RT terminates at the modified residue. Finally, the user should choose if a BedGraph file should be produced. This requires that a BED file containing the coordinates of the RNA molecules that were used for mapping is uploaded to Galaxy and specified along with the genome build and track name. BedGraph files can be uploaded as custom tracks for display of the probing data in UCSC Genome Browser at single nucleotide resolution.

2.5.2 RNAprobR

The initial step of data analysis with the RNAprobR package is to read-in the *Unique Barcodes* or *Read Counts* files into R environment with the readsamples() function. Depending on the choice of the euc option, EUCs will be calculated according to different methods ("counts," "Fu," or "HRF-Seq"), which may require setting additional parameters (see above). The function readsamples() pools together data from separate samples that are meant to be combined (e.g., replicated controls) and produce GenomicRanges object with an EUC value for each sequenced fragment.

```
control_euc <- readsamples(samples="UniqueBarcodes_control",
euc="HRF-Seq", k2n_files=k2n_control)
```

After reading-in the data, the comp() function can be used to compile information for each position of RNA molecules. The function computes the TCs, coverage (Cover), TCR, and priming counts for each position within studied transcripts and creates a GenomicRanges object. Additionally, one can specify the fasta_file option with the path to FASTA file (the same which was used for mapping) to add the nucleotide identity at each position. In some instances, it may be beneficial to discard fragments shorter than a certain threshold, which can be specified by the option cutoff (influences coverage calculation).

```
control_comp <- comp(euc_GR=control_euc, cutoff=101,
fasta_file = "index.fasta")
```

The compiled data from control and treated samples can be used to perform the normalization with different methods. The code below creates the GenomicRanges object with data normalized with each method and adds the compiled unnormalized data (compdata() function):

```
norm_data <- dtcr(control_GR=control_comp,
treated_GR=treated_comp, window_size=3, nt_offset=1)
norm_data <- slograt(control_GR=control_comp,
treated_GR=treated_comp, add_to=norm_data)
norm_data <- swinsor(Comp_GR=treated_comp, add_to=norm_data)
norm_data <- compdata(Comp_GR=treated_comp,
add_to=norm_data)
```

2.6 Data export and visualization

The final steps supported by our workflow are data export and visualization. The possible output formats are a table with normalized data, plots of probing signal over specific RNA molecules, and BedGraph files for display of data as custom tracks on the UCSC Genome Browser (Fig. 5).

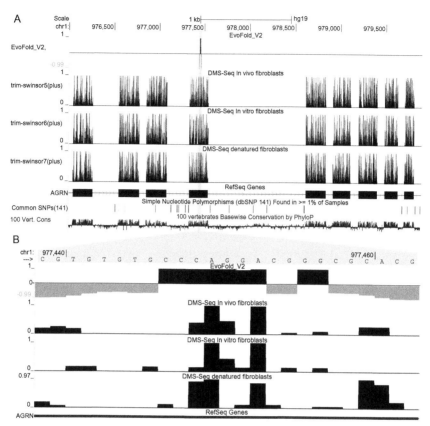

Figure 5 UCSC Genome Browser display of normalized data. BedGraph file with Smooth Winsor normalized DMS-Seq data uploaded to the genome browser. (A) View of the region surrounding an EvoFold v.2 (Parker et al., 2011) predicted structure in the AGRN gene. (B) Zoom-in to the location of the predicted structure. Bars in EvoFold track indicate prediction confidence and base-pairing status (positive values—nucleotide is unpaired and negative values—nucleotide is paired).

2.6.1 Galaxy

In Galaxy, the normalized data in tabular form and a BedGraph file (if the option is chosen in the normalization tool) are produced as output from the *Normalize* tool. The files can be downloaded with the "Download" button. Plots for specific RNA molecules can be generated with the "RNA Plot" tool by specifying the normalized dataset and the transcript identifier for the RNA molecule. Moreover, a threshold level of the p value (ΔTCR or Smooth Log_2 Ratio) or the standard deviation (Smooth Winsorization) below which sites will be marked with asterisks can be specified.

2.6.2 RNAprobR

In the R environment, the normalization functions produce Bioconductor-compatible (Gentleman et al., 2004) GenomicRanges objects (Lawrence et al., 2013), which need to be converted to a tabular format for export. This is achieved with the GR2norm_df() function, which creates a data frame with normalized values for transcripts and normalization methods of choice.

```
norm_df <- GR2norm_df(norm_data, RNAid = "all",
norm_methods = "all")
```

The norm2bedgraph() function converts the transcript coordinates to genomic coordinates and creates a Genome Browser compatible-bedgraph tracks displaying normalized values over genomic positions. The function takes as input a normalized GenomicRanges object and transcript definitions (BED file or TranscriptDb object) and creates a BedGraph file which can be uploaded to the UCSC Genome Browser via "Manage Custom Tracks" tool.

```
norm2bedgraph(norm_data, bed_file="hg19_refseq.bed",
norm_method="slograt", genome_build="hg19",
bedgraph_out_file = "out_file", track_name = "slograt",
track_description = "Smooth Log2 Ratio normalization")
```

To create a plot of normalized values over a transcript of interest the plotRNA() function can be used. This produces a plot with normalized value over transcript positions and adds the asterisks on top of each nucleotide with *p* value or standard deviation below a chosen threshold.

```
plotRNA(norm_data, RNAid="my_RNA", norm_method="dtcr",
stat_cutoff="0.001")
```

2.6.3 VARNA

In many cases, it may be of interest to visualize probing data on a known secondary structure annotation. Given the dot-bracket structural annotation and the sequence, the freely available VARNA applet (Darty, Denise, & Ponty, 2009) draws a secondary structure of RNA and allows to color nucleotides according to specified values. To color the representation of studied molecule with normalized values, prepare a single column file containing the normalized values for relevant region, avoiding scientific notation of numbers or nonnumerical values. Next, paste the RNA sequence and dot-bracket representation of RNA structure into VARNA session, right click at the structure, select "Color map -> Load values..." and choose file with normalized values.

3. CONCLUDING REMARKS

The Galaxy tools presented here are implemented and can be accessed on our local Galaxy server (http://galaxy.bio.ku.dk/). The Galaxy tools and workflows prepared for easy installation on local Galaxy servers, as well as command line scripts, the RNAprobR package and the HRF-Seq dataset are available for download at http://people.binf.ku.dk/jvinther/data/rna_probing. Moreover, the RNAprobR package has been submitted to the open source Bioconductor repository for R-based tools for analysis of high-throughput genomic data and comes with an easy to follow documentation. The Galaxy workflow is based on the scripts and RNAprobR functions and produces exactly the same output as the command line/R package option.

The tools, we describe here, are generic and can be used for the analysis of sequencing data from any type of experiment that depends on mapping ends of sequencing reads to RNA molecules. This includes the types of RNA probing described in Table 1, but also data from other methods, such as the iCLIP method for identification of protein–RNA interactions (Konig et al., 2010) or transcriptome-wide methods for detection of pseudouridines (Carlile et al., 2014; Schwartz et al., 2014). Moreover, in addition to the analysis of experiments comparing treated samples to their controls, our tools are also suitable for relative comparison of probed samples, e.g., RNA pools probed at different ionic conditions. Our tools simplify the analysis of sequencing-based RNA probing data and should make it possible for experimentalist without formal training in bioinformatics to perform this type of analysis. Importantly, this set of predefined tools will also allow users to easily and accurately describe how they performed data analysis and allow published RNA probing results to be reproduced from the raw sequencing reads.

ACKNOWLEDGMENTS

This work was supported by the Danish Council for Strategic Research [Center for Computational and Applied Transcriptomics, DSF-10-092320]. We thank Line D. Poulsen for discussions and helpful suggestions.

REFERENCES

Aviran, S., Trapnell, C., Lucks, J. B., Mortimer, S. A., Luo, S., Schroth, G. P., et al. (2011). Modeling and automation of sequencing-based characterization of RNA structure. *Proceedings of the National Academy of Sciences of the United States of America*, 108(27), 11069–11074.

Carlile, T. M., Rojas-Duran, M. F., Zinshteyn, B., Shin, H., Bartoli, K. M., & Gilbert, W. V. (2014). Pseudouridine profiling reveals regulated mRNA pseudouridylation in yeast and human cells. *Nature*, *515*(7525), 143–146.

Casbon, J. A., Osborne, R. J., Brenner, S., & Lichtenstein, C. P. (2011). A method for counting PCR template molecules with application to next-generation sequencing. *Nucleic Acids Research*, *39*(12), e81.

Darty, K., Denise, A., & Ponty, Y. (2009). VARNA: Interactive drawing and editing of the RNA secondary structure. *Bioinformatics*, *25*(15), 1974–1975.

Deigan, K. E., Li, T. W., Mathews, D. H., & Weeks, K. M. (2009). Accurate SHAPE-directed RNA structure determination. *Proceedings of the National Academy of Sciences of the United States of America*, *106*(1), 97–102.

Ding, F., Lavender, C. A., Weeks, K. M., & Dokholyan, N. V. (2012). Three-dimensional RNA structure refinement by hydroxyl radical probing. *Nature Methods*, *9*(6), 603–608.

Ding, Y., Tang, Y., Kwok, C. K., Zhang, Y., Bevilacqua, P. C., & Assmann, S. M. (2014). In vivo genome-wide profiling of RNA secondary structure reveals novel regulatory features. *Nature*, *505*(7485), 696–700.

Dunkle, J. A., Xiong, L., Mankin, A. S., & Cate, J. H. (2010). Structures of the *Escherichia coli* ribosome with antibiotics bound near the peptidyl transferase center explain spectra of drug action. *Proceedings of the National Academy of Sciences of the United States of America*, *107*(40), 17152–17157.

Ehresmann, C., Baudin, F., Mougel, M., Romby, P., Ebel, J. P., & Ehresmann, B. (1987). Probing the structure of RNAs in solution. *Nucleic Acids Research*, *15*(22), 9109–9128.

Flicek, P., Amode, M. R., Barrell, D., Beal, K., Billis, K., Brent, S., et al. (2014). Ensembl 2014. *Nucleic Acids Research*, *42*(database issue), D749–D755.

Fu, G. K., Hu, J., Wang, P. H., & Fodor, S. P. (2011). Counting individual DNA molecules by the stochastic attachment of diverse labels. *Proceedings of the National Academy of Sciences of the United States of America*, *108*(22), 9026–9031.

Gentleman, R. C., Carey, V. J., Bates, D. M., Bolstad, B., Dettling, M., Dudoit, S., et al. (2004). Bioconductor: Open software development for computational biology and bioinformatics. *Genome Biology*, *5*(10), R80.

Goecks, J., Nekrutenko, A., & Taylor, J. (2010). Galaxy: A comprehensive approach for supporting accessible, reproducible, and transparent computational research in the life sciences. *Genome Biology*, *11*(8), R86.

Hastings, C., Mosteller, F., Tukey, J. W., & Winsor, C. P. (1947). Low moments for small samples—A comparative study of order statistics. *Annals of Mathematical Statistics*, *18*(3), 413–426.

Hatem, A., Bozdag, D., Toland, A. E., & Catalyurek, U. V. (2013). Benchmarking short sequence mapping tools. *BMC Bioinformatics*, *14*, 184.

Homan, P. J., Favorov, O. V., Lavender, C. A., Kursun, O., Ge, X., Busan, S., et al. (2014). Single-molecule correlated chemical probing of RNA. *Proceedings of the National Academy of Sciences of the United States of America*, *111*(38), 13858–13863.

Karabiber, F., McGinnis, J. L., Favorov, O. V., & Weeks, K. M. (2013). QuShape: Rapid, accurate, and best-practices quantification of nucleic acid probing information, resolved by capillary electrophoresis. *RNA*, *19*(1), 63–73.

Kent, W. J., Sugnet, C. W., Furey, T. S., Roskin, K. M., Pringle, T. H., Zahler, A. M., et al. (2002). The human genome browser at UCSC. *Genome Research*, *12*(6), 996–1006.

Kertesz, M., Wan, Y., Mazor, E., Rinn, J. L., Nutter, R. C., Chang, H. Y., et al. (2010). Genome-wide measurement of RNA secondary structure in yeast. *Nature*, *467*(7311), 103–107.

Kielpinski, L. J., Boyd, M., Sandelin, A., & Vinther, J. (2013). Detection of reverse transcriptase termination sites using cDNA ligation and massive parallel sequencing. *Methods in Molecular Biology, 1038*, 213–231.

Kielpinski, L. J., & Vinther, J. (2014). Massive parallel-sequencing-based hydroxyl radical probing of RNA accessibility. *Nucleic Acids Research, 42*(8), e70.

Konig, J., Zarnack, K., Rot, G., Curk, T., Kayikci, M., Zupan, B., et al. (2010). iCLIP reveals the function of hnRNP particles in splicing at individual nucleotide resolution. *Nature Structural & Molecular Biology, 17*(7), 909–915.

Langmead, B., & Salzberg, S. L. (2012). Fast gapped-read alignment with Bowtie 2. *Nature Methods, 9*(4), 357–359.

Latham, J. A., & Cech, T. R. (1989). Defining the inside and outside of a catalytic RNA molecule. *Science, 245*(4915), 276–282.

Lawrence, M., Huber, W., Pages, H., Aboyoun, P., Carlson, M., Gentleman, R., et al. (2013). Software for computing and annotating genomic ranges. *PLoS Computational Biology, 9*(8), e1003118.

Li, F., Zheng, Q., Ryvkin, P., Dragomir, I., Desai, Y., Aiyer, S., et al. (2012). Global analysis of RNA secondary structure in two metazoans. *Cell Reports, 1*(1), 69–82.

Li, F., Zheng, Q., Vandivier, L. E., Willmann, M. R., Chen, Y., & Gregory, B. D. (2012). Regulatory impact of RNA secondary structure across the Arabidopsis transcriptome. *Plant Cell, 24*(11), 4346–4359.

Lucks, J. B., Mortimer, S. A., Trapnell, C., Luo, S., Aviran, S., Schroth, G. P., et al. (2011). Multiplexed RNA structure characterization with selective 2′-hydroxyl acylation analyzed by primer extension sequencing (SHAPE-Seq). *Proceedings of the National Academy of Sciences of the United States of America, 108*(27), 11063–11068. http://dx.doi.org/10.1073/pnas.1106501108.

Martin, M. (2011). Cutadapt removes adapter sequences from high-throughput sequencing reads. *EMBnet. Journal, 17*(1), 10–12.

Merino, E. J., Wilkinson, K. A., Coughlan, J. L., & Weeks, K. M. (2005). RNA structure analysis at single nucleotide resolution by selective 2′-hydroxyl acylation and primer extension (SHAPE). *Journal of the American Chemical Society, 127*(12), 4223–4231.

Mortimer, S. A., Trapnell, C., Aviran, S., Pachter, L., & Lucks, J. B. (2012). SHAPE-Seq: High-throughput RNA structure analysis. *Current Protocols in Chemical Biology, 4*(4), 275–297.

Parker, B. J., Moltke, I., Roth, A., Washietl, S., Wen, J., Kellis, M., et al. (2011). New families of human regulatory RNA structures identified by comparative analysis of vertebrate genomes. *Genome Research, 21*(11), 1929–1943.

Peattie, D. A., & Gilbert, W. (1980). Chemical probes for higher-order structure in RNA. *Proceedings of the National Academy of Sciences of the United States of America, 77*(8), 4679–4682.

Poulsen, L. D., Kielpinski, L. J., Salama, S. R., Krogh, A., & Vinther, J. (in press). SHAPE Selection (SHAPES) enrich for RNA structure signal in SHAPE sequencing based probing data. *RNA (New York, N. Y.)*.

Powers, T., & Noller, H. F. (1995). Hydroxyl radical footprinting of ribosomal proteins on 16S rRNA. *RNA, 1*(2), 194–209.

Pruitt, K. D., Brown, G. R., Hiatt, S. M., Thibaud-Nissen, F., Astashyn, A., Ermolaeva, O., et al. (2014). RefSeq: An update on mammalian reference sequences. *Nucleic Acids Research, 42*(database issue), D756–D763.

Robin, X., Turck, N., Hainard, A., Tiberti, N., Lisacek, F., Sanchez, J. C., et al. (2011). pROC: An open-source package for R and S+ to analyze and compare ROC curves. *BMC Bioinformatics, 12*, 77.

Rouskin, S., Zubradt, M., Washietl, S., Kellis, M., & Weissman, J. S. (2014). Genome-wide probing of RNA structure reveals active unfolding of mRNA structures in vivo. *Nature*, *505*(7485), 701–705.

Schmidt, W. M., & Mueller, M. W. (1999). CapSelect: A highly sensitive method for 5′ CAP-dependent enrichment of full-length cDNA in PCR-mediated analysis of mRNAs. *Nucleic Acids Research*, *27*(21), e31.

Schwartz, S., Bernstein, D. A., Mumbach, M. R., Jovanovic, M., Herbst, R. H., Leon-Ricardo, B. X., et al. (2014). Transcriptome-wide mapping reveals widespread dynamic-regulated pseudouridylation of ncRNA and mRNA. *Cell*, *159*(1), 148–162.

Seetin, M. G., Kladwang, W., Bida, J. P., & Das, R. (2014). Massively parallel RNA chemical mapping with a reduced bias MAP-seq protocol. *Methods in Molecular Biology*, *1086*, 95–117.

Siegfried, N. A., Busan, S., Rice, G. M., Nelson, J. A., & Weeks, K. M. (2014). RNA motif discovery by SHAPE and mutational profiling (SHAPE-MaP). *Nature Methods*, *11*(9), 959–965.

Talkish, J., May, G., Lin, Y., Woolford, J. L., Jr., & McManus, C. J. (2014). Mod-seq: High-throughput sequencing for chemical probing of RNA structure. *RNA*, *20*(5), 713–720.

Tullius, T. D., & Greenbaum, J. A. (2005). Mapping nucleic acid structure by hydroxyl radical cleavage. *Current Opinion in Chemical Biology*, *9*(2), 127–134.

Underwood, J. G., Uzilov, A. V., Katzman, S., Onodera, C. S., Mainzer, J. E., Mathews, D. H., et al. (2010). FragSeq: Transcriptome-wide RNA structure probing using high-throughput sequencing. *Nature Methods*, *7*(12), 995–1001.

Wan, Y., Qu, K., Ouyang, Z., Kertesz, M., Li, J., Tibshirani, R., et al. (2012). Genome-wide measurement of RNA folding energies. *Molecular Cell*, *48*(2), 169–181.

Wan, Y., Qu, K., Zhang, Q. C., Flynn, R. A., Manor, O., Ouyang, Z., et al. (2014). Landscape and variation of RNA secondary structure across the human transcriptome. *Nature*, *505*(7485), 706–709.

Zheng, Q., Ryvkin, P., Li, F., Dragomir, I., Valladares, O., Yang, J., et al. (2010). Genome-wide double-stranded RNA sequencing reveals the functional significance of base-paired RNAs in Arabidopsis. *PLoS Genetics*, *6*(9), e1001141.

CHAPTER SEVEN

Computational Methods for RNA Structure Validation and Improvement

Swati Jain*,†,‡, David C. Richardson†,1, Jane S. Richardson†
*Program in Computational Biology and Bioinformatics, Duke University, Durham, North Carolina, USA
†Department of Biochemistry, Duke University Medical Center, Durham, North Carolina, USA
‡Department of Computer Science, Duke University, Durham, North Carolina, USA
[1]Corresponding author: e-mail address: dcrjsr@kinemage.biochem.duke.edu

Contents

1. Introduction	182
2. General Validation Criteria	186
2.1 Data validation	186
2.2 Model-to-data match	186
2.3 Model validation	188
3. Criteria Specific to RNA Conformation	189
3.1 RNA bases	189
3.2 Ribose pucker	191
3.3 Backbone conformers	193
4. How to Interpret Validation Results	195
4.1 As an end user	195
4.2 As a journal referee	198
4.3 As a structural biologist	200
5. Correcting the Problems	201
5.1 Manual rebuilding	202
5.2 RNABC	203
5.3 Coot and RCrane	203
5.4 PHENIX refinement with pucker-specific targets	204
5.5 ERRASER, with examples	204
6. What's Coming Next	207
6.1 ERRASER for RNA/protein	207
6.2 Better ion identification	208
6.3 Putting conformers into initial fitting	208
Acknowledgments	208
References	209

Abstract

With increasing recognition of the roles RNA molecules and RNA/protein complexes play in an unexpected variety of biological processes, understanding of RNA structure–function relationships is of high current importance. To make clean biological interpretations from three-dimensional structures, it is imperative to have high-quality, accurate RNA crystal structures available, and the community has thoroughly embraced that goal. However, due to the many degrees of freedom inherent in RNA structure (especially for the backbone), it is a significant challenge to succeed in building accurate experimental models for RNA structures. This chapter describes the tools and techniques our research group and our collaborators have developed over the years to help RNA structural biologists both evaluate and achieve better accuracy. Expert analysis of large, high-resolution, quality-conscious RNA datasets provides the fundamental information that enables automated methods for robust and efficient error diagnosis in validating RNA structures at all resolutions. The even more crucial goal of correcting the diagnosed outliers has steadily developed toward highly effective, computationally based techniques. Automation enables solving complex issues in large RNA structures, but cannot circumvent the need for thoughtful examination of local details, and so we also provide some guidance for interpreting and acting on the results of current structure validation for RNA.

1. INTRODUCTION

As any reader of this volume appreciates, the discovery of new RNA biology has exploded in recent decades. Determination of 3D structures for large RNAs and ribonucleoprotein complexes (RNPs) has surged as well to provide the detailed molecular descriptions essential for mechanistic understanding of the diverse biological functions. The primary techniques used for large RNA structures are electron microscopy and X-ray crystallography. The developments described in this chapter have so far been applied almost exclusively to crystallography but should become useful for electron microscopy as it increasingly attains higher resolutions.

The methodology for RNA crystallography has been improving rapidly. However, the tools still lag far behind those for protein crystallography and the problems are inherently more difficult, both in crystallization and phasing and also in accurate model building and refinement. Base pairing can be predicted with quite good, if far from perfect, accuracy, and the expected double helices and their base pairs can be located with good reliability in electron density, even at the 2.5–4 Å resolution typical for attainable crystal structures of large RNAs and RNPs. Most phosphates are also identifiable as peaks of strong, nearly spherical density (see Fig. 1). In contrast, it is very difficult to fit an accurate atomic model for the sugar–phosphate backbone into low-resolution electron density (Murray, Arendall, Richardson, & Richardson, 2003). A nucleotide residue has six torsion-angle parameters (α–ζ), each of which frequently takes on a value widely different from that in A-form. At best, low-resolution density shows a blob for the ribose, with no shape to indicate the ring pucker, and a round tube for the backbone sections between ribose and phosphate, with no way to see which direction the atoms actually zigzag (Fig. 1). The result of this anisotropy in structural clarity is that RNA bases are usually well positioned in deposited crystal structures (e.g., Fig. 2A), while even structures done by expert and careful practitioners often have high rates of physically impossible backbone conformations (e.g., Fig. 2B).

Many important things can be learned from the overall 3D structures of large RNAs, but the details of the backbone conformation also matter, more than one might think, for mechanistic understanding of biological function. Backbone atoms are involved in most types of RNA catalysis (Ferré-D'Amaré & Scott, 2010; Lang, Erlacher, Wilson, Micura, & Polacek, 2008; Zaher, Shaw, Strobel, & Green, 2011). Particular non-A-form backbone conformations are critical to specificity for a wide variety of protein, ligand, aptamer, and drug interactions of RNAs (Klein, Schmeing, Moore, & Steitz, 2001; Warner et al., 2014), as for instance seen in Fig. 3, where one entire side of the specific drug binding is to non-A-form

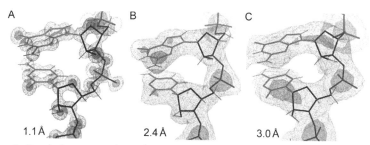

Figure 1 Resolution comparison: three examples, each a backbone conformer **1a** suite in regular A-form double helix. (A) From PDB 2Q1O at 1.1 Å (Li et al., 2007); (B) 3CC2 at 2.4 Å (Blaha, Gural, Schroeder, Moore, & Steitz, 2008), and (C) 3R8T at 3.0 Å (Dunkle et al., 2011). The 2F$_o$-F$_c$ density is shown at 1.8σ as pale gray mesh and at 5.3σ as purple (dark gray in the print version) mesh. Note that the phosphate density is high and round in all cases. Base type is unambiguous at 1.1 Å; the G is still identifiable in this case at 2.4 Å, but only purine-versus-pyrimidine at 3.0 Å. At 2.4 Å, there is a bump for the 2'O, but not at 3.0 Å. The zigzag of backbone atoms between P and ribose is clear at 1.1 Å, but that backbone density is just a round tube in (B) and (C).

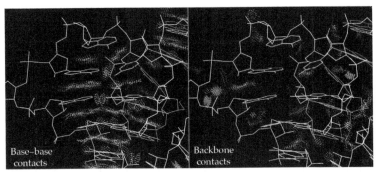

Figure 2 Base versus backbone accuracy as shown by all-atom steric contacts. (A) Well-fit base–base contacts, shown by favorable H-bond and van der Waals contact dots (green and blue). (B) Fairly frequent backbone problems shown by spike clusters of steric clash overlap ≥ 0.4 Å (hotpink). *From PDB 2HOJ at 2.5 Å (Edwards & Ferré-D'Amaré, 2006).* (See the color plate.)

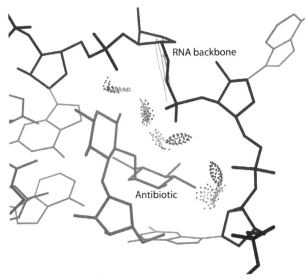

Figure 3 Specific ribostamycin binding by non-A-form RNA backbone. All-atom contacts are shown only between drug and RNA backbone. *From PDB 3C3Z at 1.5 Å (Freisz, Lang, Micura, Dumas, & Ennifar, 2008).*

backbone. RNA–RNA backbone interactions stabilize distant binding sites (Batey, Gilbert, & Montange, 2004) or in general enable the conformational changes that accompany RNA–protein recognition (Williamson, 2000). Other molecules feel the RNA backbone in its full-atomic detail, even if our experimental techniques cannot always see that detail. Therefore, it is important both to pursue better accuracy where feasible and to provide validation measures that allow end users of RNA structures to easily judge reliability of the local features most important to their work.

Fortunately, the addition of explicit hydrogen atoms and calculation of all-atom contacts (Word, Lovell, LaBean, et al., 1999; Word, Lovell, Richardson, & Richardson, 1999) provide sensitive local validation such as the markup shown in Fig. 2, positive for H-bonds and favorable van der Waals and negative for unfavorable steric clashes. Even more importantly, physical factors are now known to limit the accessible combinations of RNA backbone torsions and ribose puckers (Richardson et al., 2008), enabling inferences about detailed backbone conformation to be calculated from the better-observed base and phosphate positions. Complete versions of these tools and their results are available in user-friendly form both online at the MolProbity site (Chen et al., 2010; http://molprobity.biochem.duke.edu) and in the PHENIX crystallographic software system (Adams et al., 2010; http://phenix-online.org). Validation summaries are now available for each X-ray entry in the PDB (Protein Data Bank; Berman et al., 2000), including RNAs, at the RCSB, PDBe, and PDBj sites of the wwPDB (Berman, Henrick, & Nakamura, 2003; Gore, Velankar, & Kleywegt, 2012). Pucker diagnosis and pucker-specific target parameters aid RNA refinement in PHENIX. Highly effective correction of RNA backbone problems is now feasible with the new enumerative real-space refinement assisted by electron density under Rosetta (ERRASER) program (Adams et al., 2013; Chou, Sripakdeevong, Dibrov, Hermann, & Das, 2012), which uses Rosetta's RNA tools (Das & Baker, 2007) and density match (DiMaio, Tyka, Baker, Chiu, & Baker, 2009), MolProbity diagnosis, and crystallographic refinement to do exhaustive local conformational sampling.

A DNA side note: DNA structure can be validated by traditional model-to-data match and covalent geometry, and by all-atom contacts, but the RNA conformational tools are unfortunately not valid for DNA because it is more locally flexible, with broader torsion-angle preferences and more sugar-pucker states accessible (Svozil, Kalina, Omelka, & Schneider, 2008).

This chapter describes the basis and properties of currently available and forthcoming computational tools and visualizations that help structural biologists improve the accuracy of new experimental models for large RNAs. We also discuss how any user can readily interpret the validation results to judge the overall accuracy, or most importantly the local reliability, of those RNA crystal structures.

2. GENERAL VALIDATION CRITERIA

Crystallographic validation includes both global (whole structure) and local (residue level) criteria, and it has three logically distinct components: for the experimental data, for the 3D model, and for the model-to-data match.

2.1 Data validation

Validation of the X-ray data (reviewed in Read et al., 2011) is of concern to expert crystallographer reviewers and of course to the experimenters themselves, since it can determine whether or not a structure solution is possible at all. However, there are several aspects that are understandable and of relevance to an end user. The most obvious is the resolution, the higher the better (lower absolute number), which is the most important global, single-number criterion of structure quality. Better than 1.5 Å is called atomic resolution, since most nonhydrogen atoms are separately visualized in the electron density, and interatomic distances and conformational parameters are determined with reliably high accuracy. The caveat is that in disordered regions with multiple conformations or high B-factors, accuracy can become very poor at any resolution. Historically, and for proteins, a majority of macromolecular crystal structures are in the quite dependable, workhorse 1.5–2.5 Å resolution range. At 2.5–3 Å, the overall chain trace is nearly always correct, although some atomic groups will be misplaced into the wrong piece of density or the wrong orientation. Beyond 3 Å resolution, local details are not reliable, and the incidence of local sequence misalignment increases. Effective resolution is better than the nominal value if there is a lot of noncrystallographic symmetry (such as for viruses) and worse than nominal if the data are less than 85–90% complete in the highest resolution shell or especially if the crystal is "twinned" (with parts at different orientations). Twinning is noted in the wwPDB validation report; it can be dealt with fairly well in refinement (Yeates, 1997; Zwart, Grosse-Kunstleve, & Adams, 2005), but there is still less information content than for an untwinned crystal.

2.2 Model-to-data match

Model-to-data match evaluates how well the modeled atomic coordinates account for the observed diffraction data: the structure-factor amplitudes, or "F"s, measured for each X-ray reflection. The global criterion is called the R-factor, which is the residual disagreement between the F's observed and the F's back-calculated from the model. Even more revealing is the cross-validation residual, R_{free} (Brunger, 1992), calculated for 5–10% of the data kept out of the refinement process. A long-time rule of thumb was that near 2 Å resolution the R should be ≤20% and $R_{free} - R$ should be ≤5%. Now, however, one need not rely on such rules, because wwPDB validation for each structure reports how its R_{free} compares, as a percentile relative to all other deposited structures at similar resolution (see open bar on top "slider" bar in Fig. 4).

Local match of an individual residue's model cannot be evaluated in the "reciprocal space" of the diffraction peaks, because the Fourier transform between the data and the electron density image relates every atom to every reflection and vice versa. Therefore, local match is evaluated in the "real space" of the model and in the electron density map. The usual criterion is either the real-space correlation coefficient or the real-space residual (RSR) measured as the agreement between the "experimental" electron density (usually $2mF_{obs} - DF_{calc}$; Read, 1986) and the purely model-calculated F_{calc} electron density, within a mask around the atoms in that residue (Jones, Zou, Cowan, & Kjeldgaard, 1991). The wwPDB reports percentile scores for RSR-Z, which is a version of the RSR normalized as a Z-score (essentially the number of standard deviations from the mean)

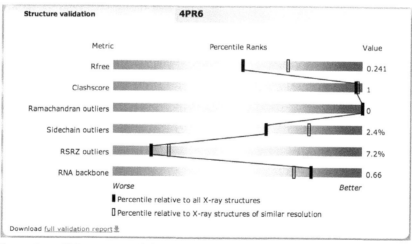

Figure 4 wwPDB summary "slider" validation plot for an RNA/protein complex (4PR6 at 2.3 Å; Kapral et al., 2014). Percentile scores on six validation criteria are plotted versus all PDB X-ray structures (filled bars) and versus the cohort at similar resolution (open bars). (See the color plate.)

compared to values at that resolution for the same amino acid or nucleotide (Gore et al., 2012; Kleywegt et al., 2004; next-to-last slider bar in Fig. 4).

2.3 Model validation

Model validation covers covalent geometry and steric interactions, both applicable to any molecule, and conformational parameters, which are quite distinct for protein and RNA and are therefore covered in detail in Section 3.

Geometry criteria include covalent bond lengths and angles, planarity, and chirality. Target values, including their estimated standard deviations or weights, are derived from the databases of small-molecule crystal structures (Allen, 2002; Grazulis et al., 2009), or perhaps from quantum calculations, especially for unusual bound ligands (e.g., Moriarty, Grosse-Kunstleve, & Adams, 2009). With some exceptions discussed below, geometry validation primarily serves as a sanity check for whether sensible restraints were used in the refinement.

Steric interactions include both favorable hydrogen bonds and van der Waals interactions and also unfavorable or even impossible atomic overlaps, or "clashes." Not every donor or acceptor is H-bonded and not every atom grouping is tightly packed, but that should nearly always be true in the molecule interior and especially in regular secondary structure. If there are two possible conformations consistent with the electron density and one of them has more of the good interactions, then it is much more likely to be correct. Bad steric clashes have been flagged as problems in just about every relevant analysis, but only for non-H atoms until our lab's development of all-atom contacts (Word, Lovell, LaBean, et al., 1999; Word, Lovell, Richardson, et al., 1999), which is the most distinctive contribution of MolProbity validation. Our Reduce program adds all H atoms, now by default in the electron cloud-center positions (Deis et al., 2013) that are most appropriate both for crystallography, where it is the electrons that diffract X-rays, and also for all-atom contact analysis, where van der Waals interactions are between the electron clouds not between the nuclei. Most H atoms lie in directions determined quite closely by the planar or tetrahedral geometry of their parent heavy atoms, and even methyl groups spend almost all their time very close to a staggered orientation. Reduce then optimizes the rotatable OH, SH, and NH_3 positions within entire H-bond networks, including any needed correction of the 180° "flip" orientation for side-chain amides and histidine rings (Word, Lovell, Richardson, et al., 1999). The Probe program analyzes all atom–atom contacts within 0.5 Å of touching van der Waals surfaces, assigning numerical scores and producing visualizations as paired patches of dot surface like those seen in Figs. 2 and 3. A cluster of hotpink clash spikes gives the most telling signal of a serious local problem in the model. Barring a misunderstood atom nomenclature, a flagged steric overlap >0.4 usually, and >0.5 Å nearly always, means that at least one of the clashing atoms must move away. The MolProbity "clashscore" is normalized as the number of clashes per 1000 atoms in the structure and is reported as percentile scores by the wwPDB (Fig. 4, second slider bar). As shown by the example in Fig. 2, all-atom clashes are a valuable diagnostic for RNA backbone conformation.

3. CRITERIA SPECIFIC TO RNA CONFORMATION
3.1 RNA bases

Pairing and stacking of the bases is the most obvious and the most energetically important aspect of RNA conformation. Nearly all bases in large RNAs or RNPs are both stacked and paired, but the stacking may be with intercalated small molecules or with protein aromatic side chains and the pairing is often not canonical Watson–Crick. Base pairs can deviate a fair amount from coplanarity but should maintain good H-bonding. There is simply not enough room for a purine–purine pair within an A-form helix, but occasionally a pyrimidine–pyrimidine singly H-bonded pair can be tolerated, such as the C–U that ends the helix below the S-motif in the rat sarcin–ricin loop (Correll et al., 1998) compared to the G–C pair in that position for *E. coli* (Correll, Beneken, Plantinga, Lubbers, & Chan, 2003). Comparing across phylogenetically related sequences, base pairing is very strongly conserved both in helices and in other structural motifs, as judged by "isosteric" replacements (Leontis & Westhof, 2001; Stombaugh, Zirbel, Westhof, & Leontis, 2009), but there are occasional surprises such as the hole in *E. coli* Phe tRNA next to the nonpaired position 26 base (Byrne, Konevega, Rodnina, & Antson, 2010; Dunkle et al., 2011). The H-bonds of base pairs and the broad, tight van der Waals contacts of base stacking are shown visually by all-atom contacts (as in Fig. 2), but neither MolProbity nor wwPDB validation evaluates them quantitatively. They can be checked out with the MC-Fold/MC-Sym system (Parisien & Major, 2008), for instance, and probably should be added into formal nucleic acid validation criteria.

The bond that connects the base to the ribose is called the glycosidic bond, and the dihedral angle across that bond is called the χ angle, defined by atoms O4'-C1'-N9-C4 for purines (bases A and G) and O4'-C1'-N1-C2 for pyrimidines (bases C and U). The χ angle is known to adopt two major conformations: in the much more common *anti* conformation, the bulkier side of the base (the six-membered ring for purines and the ring substituent oxygen for pyrimidines) points away from the sugar, putting χ near 180°, while in the less common *syn* conformation the bulky group points toward the sugar (Saenger, 1983). A variant of *anti* conformation called high-*anti* (near 270°) is also seen sometimes, which is awkward for the generic chemical definition of *syn* as $0 \pm 90°$ and *anti* as $180 \pm 90°$. An even more serious disconnect, now corrected, is that early parameters for χ were often based on DNA (B-form) rather than on the very different RNA distributions. It is now generally accepted that the spread of χ values is greater for purines than for pyrimidines.

To define updated parameters for RNA χ, we used the RNA11 dataset of low-redundancy ≤ 3 Å resolution chains and then filtered to omit any base with a steric clash, B-factor ≥ 40, chemical modification, or ribose-pucker outlier (see Section 3.2). The original 25,000-residue distribution populates all 360° of χ, but the high-quality 10,000-residue distribution is empty for two ranges of 110–150° and 320–360°. Both base type and pucker state make large differences in what values can occur, in patterns that are highly nonperiodic and not symmetric around zero. Comparing Fig. 5A

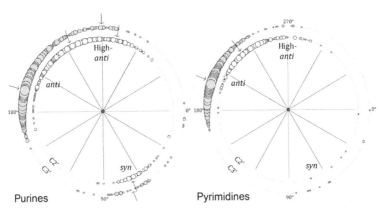

Figure 5 Empirical plot of preferred base χ angle values from the RNA11 dataset, separated by purine (left) versus pyrimidine (right) and by C3'-*endo* pucker (outer ring) versus C2'-*endo* pucker (inner ring). Occurrence frequency is shown by circle radius, on a log scale. Arrows mark the pucker-specific χ-angle parameters now used in PHENIX.

(purines) with B (pyrimidines) shows that pyrimidines have essentially no reliable instances of *syn* χ. Comparing the outer versus inner plots shows a very large shift of the *anti* peak between C3'-*endo* and C2'-*endo* ribose puckers, and also that C3'-*endo* pyrimidines almost never have high-*anti* conformations. Since purine bases are much larger, it seems counterintuitive that they have less conformational restriction. However, close interactions between base and backbone are with the larger six-membered ring for pyrimidines and with the smaller five-membered ring for purines. The end result of this update was a set of RNA preferred χ values for use in PHENIX, specific both for base type and for pucker type, as marked by the arrows in Fig. 5.

3.2 Ribose pucker

The ribose ring in RNA structures is known to adopt two main conformations, C3'-*endo* pucker and C2'-*endo* pucker. There is a clean bimodal distribution in the CSD small-molecule data (Allen, 2002;), echoed closely at high resolution and high quality in the RNA11 dataset, here plotted in Fig. 6 as a function of the closely correlated backbone dihedral angle δ (C5'-C4'-C3'-O3'). The mean value of δ is 84° for C3'-*endo* and 145°

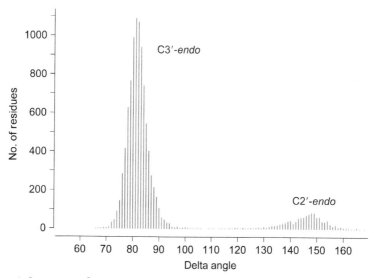

Figure 6 Occurrence frequency at each δ dihedral-angle value. *Data from the RNA11 dataset, with residue-level filters of backbone atom clash, B-factor ≥60 and geometry outliers applied. The distribution is bimodal, with δ angle range of 60°–105° for C3'-endo pucker and 125°–165° for C2'-endo pucker.*

for C2′-endo pucker, with acceptable ranges of 60°–110° and 125°–165°, respectively.

However, as described in Section 1, it is extremely difficult to place the backbone atoms correctly into the electron density and distinguish between the two puckers, especially at the resolutions lower than 2.5 Å that are typical for large RNA structures. Therefore, ribose puckers are often modeled incorrectly in RNA structures, with an understandable but too thorough bias toward the C3′-endo conformation found four times more often for RNA residues. The three large groups of atoms attached to the ribose (the two backbone directions and the base) are pointed in different directions by the different pucker states, which means that fitting and refining those groups around an incorrect pucker nearly always produces further problems such as clashes and bond-angle outliers. Pucker outliers occur even at fairly high resolution and may be in biologically important regions, such as the tRNA/synthetase example in Fig. 7.

Fortunately, there exists a simple and reliable way to detect the pucker of an RNA residue from several atoms of RNA structure models that in practice turn out to be relatively well placed in the electron density. Even at low resolution, the phosphorus atom is well centered in a nearly spherical high peak of the map (see Fig. 1) but flanked by round, featureless tubes of backbone density. The large blob of base density is offset to one side from the smaller ribose blob. Fortunately, initial model construction into this evident configuration and spacing of density usually places the P atom and the C1′–N1/9 glycosidic bond (the line connecting ribose and base) very close to their final refined positions. If a perpendicular is dropped from the 3′ P to

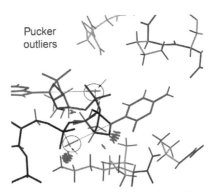

Figure 7 Example of pucker outliers (magenta (highlighted by circle and arrow in the print version) crosses, pointing along the glycosidic bond vector), at the contact of a tRNA CCA end with its synthetase enzyme. *From PDB 1N78 at 2.1 Å (Sekine et al., 2003).*

the extended line of the glycosidic bond vector, the length of that perpendicular (abbreviated as the "Pperp" distance) reliably indicates the pucker of the ribose ring: it is long for C3′-*endo* pucker and short for C2′-*endo* pucker.

This relationship is empirically confirmed by the data shown in Fig. 8, where an initial approximate correlation improves to a clean distinction as additional residue-level filters are applied to the residues in the RNA11 dataset. A simple Pperp distance cutoff can reliably distinguish between the two puckers: ≥ 2.9 Å for C3′-*endo* pucker and <2.9 Å for C2′-*endo* pucker. This relationship can be judged quite easily by eye, and is trivial to automate. It holds well even for misfit models, so validation can tell whether the ribose pucker of an RNA residue is correct or not by testing whether the Pperp distance is paired with the correct range of backbone dihedral angle δ. If not, it is flagged as a pucker outlier. Tools for correcting ribose-pucker outliers are described in Section 5 below.

3.3 Backbone conformers

The RNA backbone is highly flexible with 6 degrees of freedom (dihedral angles α–ζ) and, as described in Section 1, it is extremely difficult to place correctly in the electron density. Fortunately, there is order in this chaos. The individual dihedral angles sample a large range of angle values, but when analyzed together as high-dimensional clusters, they describe a discrete set of

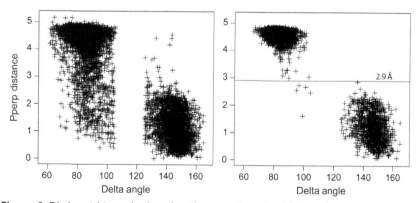

Figure 8 Distinguishing whether the ribose pucker should be C3′-*endo* or C2′-*endo*, from the "Pperp" length of the perpendicular dropped from the 3′ P to the glycosidic bond vector. On the left, Pperp versus δ angle for unfiltered RNA11 data. On the right, RNA11 data filtered on backbone atom clashes, on backbone atom B-factor ≥ 60, and on ε and geometry outliers. The cutoff used to distinguish between the two pucker conformations is 2.9 Å, with Pperp value ≥ 2.9 Å for C3′-*endo* pucker and <2.9 Å for C2′-*endo* pucker.

conformations that the RNA backbone can adopt. RNA backbone has been shown to be rotameric (Murray et al., 2003), a study that was extended by the RNA Ontology Consortium to describe 54 distinct conformations commonly adopted by the RNA backbone (Richardson et al., 2008).

These backbone conformers are described in terms of the sugar-to-sugar unit of the RNA backbone called a "suite", because the dihedral angles within a suite are better correlated than the dihedral angles in a traditional nucleotide (Fig. 9). Each suite i consists of seven dihedral angles: δ, ε, and ζ from the heminucleotide i-1, and α, β, γ, and δ from the heminucleotide i. The 54 backbone conformers were identified using the data from a quality-conscious, nonredundant dataset of RNA crystal structures (RNA05), with each distinct cluster in the seven dimensions of the suite identified as an individual conformer. Most conformers are well separated in the seven-dimensional space but some are not, such as the "satellites" around the big A-form cluster. Close cluster pairs can usually be distinguished on the

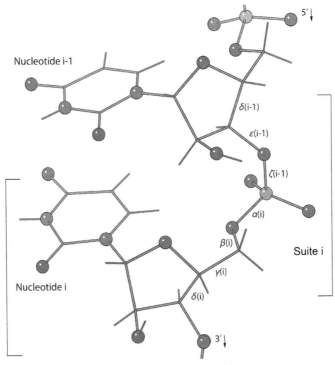

Figure 9 Definition of the "suite" division of RNA backbone that spans from sugar to sugar, with its seven backbone dihedral angles: δ, ε, and ζ from heminucleotide i-1, and α, β, γ, and δ from heminucleotide i.

basis of their different structural roles, such as the base-stacked **1b** versus the intercalated **1[**.

Each conformer is given a two-character name with first a number describing the conformation of heminucleotide i-1 and second a letter (or letter-related character) describing the conformation of heminucleotide i. For example, **1a** represents the RNA backbone conformation in a standard A-form helix, **1[** is the most common intercalation conformation, **1g** is the starting conformer for the GNRA tetraloop motif, and **5z** and **4s** are two of the three conformers that describe the backbone of an S-motif. For automated backbone conformer assignment in PHENIX and MolProbity, a program called Suitename was developed. Suitename takes as input the seven backbone dihedral angles, assigns the suite one of the 12 δ–γ–δ bins (two ranges for δ angles and three ranges for γ angle) and then assigns a backbone conformer to the suite within that bin by a rather complex process (Richardson et al., 2008). It also provides a value called the "suiteness" for each suite, which indicates how far from the center of the backbone cluster this particular suite is. Any suite which does not belong to the 54 backbone conformers is flagged as a backbone-conformer outlier, denoted by !!.

One of the advantages of using backbone conformers and their two-character names is that any three-dimensional RNA structure can be represented as a string of two-character backbone conformers' names, called a "suitestring." Many commonly occurring structural motifs have consensus suite strings, for example, kink-turns (**7r6p2[0a** on the kinked strand), S-motifs (**5z4s#a** on the "S" strand, **1e** opposite), or GNRA tetraloops (**1g1a1a1c** above a Watson–Crick pair). These suitestrings are useful to identify new occurrences of known structural motifs and to find new structural motifs. In the present context, they form the basis for validating RNA backbone conformation and are especially important to the correction process (described in Section 5). This is a two-way relationship, as the recognized suites are found by ERRASER even though they are not explicitly used in its scoring function, and we now use ERRASER to verify possible new conformers that have too few examples to define empirically.

4. HOW TO INTERPRET VALIDATION RESULTS

4.1 As an end user

For a general orientation to the 3D structure of your RNA of interest, resolution or validation scores do not matter, and primarily you just want to take into account the state of processing, binding partners, ligands, or alternative conformational forms for the molecule in a particular deposited structure as well as how closely related it is to your own research subject. However, when specific details matter, you can benefit greatly from considering validation information, both global and local.

If many relevant structures are available, you can choose the best one or ones to use overall by starting with the "slider plots" (such as the one in Fig. 4) on each structure's page at the wwPDB Web sites. Compare the absolute values of the various validation criteria, or just visually judge the positions of the filled bars on each slider, which show the percentile score within PDB entries at all resolutions. If the most relevant structure is low quality, then you should look both at that one and also at the best generic structure of the set.

Most important are local validation criteria in the regions you especially care about for your own work. Local accuracy and reliability often vary widely within a single structure, mostly because some parts are much more mobile than others. Where the map is unclear, the model can be influenced by subjective judgments or by software idiosyncrasies. Even high resolution and excellent quality scores cannot protect you from a serious local modeling error; conversely, even in a poor-quality structure, if the local region of interest has no validation outliers at all, then it is very probably correct.

At the wwPDB sites (RCSB, PDBe, PDBj), local validation is in the "full validation report" pdf for each entry; a download link is just above or below the summary "slider" report (see Fig. 4). In the "Overall quality at a glance" section, the last table lists any ligands whose geometry or electron density quality is an outlier; further details are in later sections. An outlier in electron density quality (LLDF) is certainly not bound at full occupancy and risks having a component of wishful thinking. In the "Residue property plots" section, you can see where there are concentrations of geometry or density match outliers along the sequence, in each RNA or protein chain. Note: the wwPDB uses "geometry" to include dihedral-angle conformation as well as covalent geometry. Look for the residue numbers you care about; if they show up, then you probably want to check the more detailed information sections, or better yet consult a representation of validation markup on the 3D structure.

The most powerful way to evaluate local model quality is interactively in the three-dimensional structure, where you can see clustering in space as well as in sequence, and even consult the electron density map (usually what you want is called a $2F_o-F_c$ map). At the wwPDB and elsewhere, there are now many interactive 3D viewers for a quick and easy look at overall molecular structure, but they are not as good for comprehending local detail, and none of them show validation markup. The Electron Density Server (EDS; Kleywegt et al., 2004; eds.bmc.uu.se/eds) has two different online viewers available for looking at the model and electron density map in 3D. It also has excellent per-residue sequence plots of the various model-to-data validation metrics such as RSR-Z, with nothing omitted and mouse-over details, and it is the easiest source to download maps for viewing in other software. PyMol (www.pymol.org), still free to academics, is probably the most full featured of the nonexpert-accessible macromolecular viewers, both for interactive use and for making 2D presentation images. It can show electron density maps but does not have an easy way to import validation markup. The only user-friendly, high-comprehension software we know of for viewing models, maps, and validation markup in 3D is KiNG, which was used for producing most of the figures here. It is a central feature of the MolProbity Web site, for viewing the "multi-criterion kinemage," doable directly online without any installation needed as long as you can run Java, or KiNG can be run on your own computer. From the overview, either locate the worst clusters of problems visually and zoom in on them, or find and center on a residue of interest and see how good that region is. If it is in a cluster of problems, you should treat its information as unreliable in detail. On the other hand, if that residue and its neighbors are free of major outliers (true for most parts of most crystal structures), then the local model and inferences from it are trustworthy.

Single outliers need a more nuanced approach. A bad steric clash really does mean that one or both of the clashing atoms must move, although perhaps not by much. A ribose-pucker outlier is almost certainly wrong. A backbone-conformer outlier (!!), however, is fairly often valid, especially if the bases of the suite are far apart such as in a one-residue bulge or a helix junction. Bond-angle outliers are not necessarily serious for their own sake, but in RNA they are most often a symptom of a locally misfit conformation. As in any macromolecule, disorder at chain termini or elsewhere means there is more than one conformation, which makes fitting even more difficult and almost always produces outliers. Although that casts no doubt on the overall quality of the structure, it still means that you cannot know the local conformation well.

For the proteins in RNP structures at low resolution, keep in mind that there may be significant problems such as local out-of-register sequence, unjustified Ramachandran outliers or *cis* peptides, or mismodeled interactions at the protein/RNA interface. For instance, an Arg side chain ending near a phosphate but forming no H-bonds is very likely to have its rotamer fit backward. In such a case, you might benefit from consulting the version on the PDB Redo Web site (www.cmbi.ru.nl/pdb_redo; Joosten, Joosten, Murshudov, & Perrakis, 2012), which re-refines all PDB crystal structure entries, including automated side-chain and peptide-flip rebuilding for proteins but not yet local corrections for RNA.

In this process, when you have discovered a potential local problem of concern in the most relevant structure, it is often quite helpful to check out that same region in other related structures. See whether they agree or disagree, and consider that in the context of how they differ both in information content and in biological context.

4.2 As a journal referee

First of all, feel free to ask for coordinates and structure factors—you may well get them, and are then positioned to do the best job of fairly evaluating the structure. Failing that, however, you should certainly require access to the wwPDB validation report. Here are some suggestions on what to look for in such reports: what's good, or at least OK; what needs an explanation; and what is unacceptable.

In the context of journal review, the technical issue that matters most is how well the authors handled the data that could be collected from the system in question. Therefore, the percentile scores versus the resolution cohort (the open bars on the wwPDB summary slider plots, as in Fig. 4) matter much more than absolute or all-PDB scores. A high-resolution structure should be held to higher absolute standards of accuracy than a low-resolution one. The validation criteria have been chosen to be as independent as feasible, so they do not necessarily all have similar values; it is somewhat unusual for a structure to score near the 60th percentile, say, on all measures. It is not particularly disturbing, for instance, to have very good Ramachandran scores and quite poor rotamer scores; that presumably just means the crystallographer, or the software, concentrated much more on the backbone. Some combinations are disturbing, however. Top scores for R_{free} and/or RSR-Z coupled with near-zero percentiles for clash and conformation criteria might either indicate some poor methodology that produced extreme model bias (possible, for instance, if twinning is handled incorrectly) or even possible fraud (Janssen, Read, Brunger, & Gros, 2007), or else might just mean that the methodology was complex enough to be treated poorly by the automated routines that calculate those crystallographic scores. Such cases need to be looked at carefully, and probably questions should be asked of the authors. And of course the issue still remains of the near-zero conformational percentiles.

Note that RSR-Z is reported for values >2, which is only a 2σ outlier, compared to the 4σ level or more for most other outlier definitions. The overall RSR-Z score can look bad just from having some mobile regions in the structure (e.g., for the 4PR6 of Fig. 4), because the RSR is even more sensitive to low electron density than to the exact shape of that density. This effect will change dramatically depending on whether the depositor believes in making a complete model, or believes in leaving out anything not seen clearly; there is no good answer to that dilemma, so neither side should be penalized for their decision. RNA crystallographers are generally on the side of more complete models. Locally, poor (high) per-residue RSR-Z scores just mean that the density in that region was weak and that part of the molecule presumably mobile but still modeled, not that anything is wrong with the structure or the methodology in general. But if poor RSR-Z scores occur in a place important to the conclusions of the paper, then they must be explained or discounted, and in general necessitate explicit defense of any strong claims.

An especially important aspect of validating model-to-data match is whether or not there is good evidence in the electron density to support the modeled presence of a bound ligand, inhibitor, or drug. If a bound inhibitor, drug, or other ligand is discussed in the paper, then check its ligand geometry and especially its match to data in the detailed tables for the presence of yellow flags. Highly deviant ligand geometry is always suspect, especially at lower resolution or in weak density, and may mean something else is locally wrong besides that geometry. The "LLDF" score estimates the electron density quality of the ligand relative to that of the surrounding macromolecule. If that score is an outlier, the ligand is certainly not bound at full occupancy, and no strong conclusions should have been drawn about its conformation or even its binding. The request to make in such cases is for an omit map covering the region of the ligand, to see whether there is unbiased evidence supporting its presence and conformation.

Not every structure needs to be, or can be, in the top 10%, so scores that are on average typical for the resolution should be considered acceptable in most cases. A structure far worse than average should be questioned, however. If it is not an especially important structure, then it may not be worth reporting in an inaccurate state. If it is really important, then as a referee it would be worth pushing for improved quality to better support the presumed important uses of it; if it is not corrected now, then neither the depositor nor anyone else will be able to get funding to improve it later.

4.3 As a structural biologist

Besides a better report card from the wwPDB, you can benefit from better refinement behavior, clarify the map in other places through phase improvement, and especially can locate the places that are wrong in ways that can often be successfully corrected but will occasionally be valid and interesting, such as the well-established case in Fig. 10. In general, MolProbity validation aims for conservative identification of problems, to avoid wasting user effort on chasing false alarms.

Backbone clashes are especially diagnostic in RNA, and some types are easy to correct, such as when the 5′ H atoms are turned inward, putting them too close to the ribose (as for the manual correction example in Section 5.1). A ribose-pucker outlier is essentially always incorrect; they can sometimes be fixed in Coot, or even by refinement with pucker-specific targets, but their repair may require running ERRASER (see Section 5 below). Suite outliers are less serious, since not all valid conformers have yet been identified. Aim for 80–90% known conformers, but not for 100% unless your structure is pure A-form helix. The wwPDB slider reports percentile scores

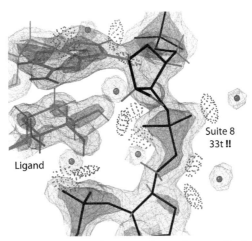

Figure 10 A valid !! "outlier" suite conformer, with excellent confirming electron density and favorable interactions. The $2F_o$-F_c density is shown at 1.2σ as pale gray mesh and at 3.2σ as purple. The water molecules are shown as separate balls and hydrogen bonds as pillows of dots. From PDB 3C3Z, HIV-1 RNA/antibiotic complex at 1.5 Å (Freisz et al., 2008), near residue B 8. (See the color plate.)

for the average "suiteness" per residue, which is useful but not the best indication of outright mistakes. Using MolProbity, or consulting the detailed tables in the wwPDB validation report, you can find specifically which nucleotides or suites are pucker or conformer outliers.

A few types of base modification, such as the dihydro-uridine common in tRNAs (Dalluge, Hashizume, Sopchik, McCloskey, & Davis, 1996), break aromaticity of the ring and should indeed be strongly nonplanar. For other bases, deviations from base-ring planarity great enough to notice by eye are a warning sign of missing or poorly balanced refinement restraints—although not necessarily of an otherwise inaccurate structure. In contrast, a spatial cluster of bad bond-angle outliers is usually caused by a serious misfitting of the local conformation and is worth worrying about, in either protein or RNA. Historically, achieving a low level of geometry outliers has been quite difficult for RNA structures, and simply tightening the restraints is not very effective. However, now that we have access to better tools for correcting local errors in RNA conformation, such as ERRASER, we find that geometry improves as an additional result and persists robustly in further refinement.

As a crystallographer, the most helpful form of validation and correction is to work within the PHENIX GUI (Echols et al., 2012), where MolProbity-style validation is reported at the end of each macrocycle in tables, plots, and a multi-criterion kinemage. Each entry in a table can with one click take you to the relevant place in Coot to work on the problem. Even if you prefer another program for refinement, it might well be worth an occasional excursion into PHENIX, to refine briefly with pucker-specific targets and then use the RNA-specific validation as coupled with Coot. Tools for RNA backbone correction and examples of their use will be found in Section 5.

The take-home message (Richardson & Richardson, 2013) is what we like to call:

The zen of model anomalies:
Consider each outlier and correct most.
Treasure the valid, meaningful few.
Do not fret over a small inscrutable remainder.

5. CORRECTING THE PROBLEMS

Detection of modeling errors in RNA and RNP structures is valuable, especially for the hard-to-fit backbone. However, it is more satisfying and very much more useful if that diagnosis can be followed by correction. This section describes the development of increasingly powerful and user-friendly tools to correct RNA modeling errors.

5.1 Manual rebuilding

Manual rebuilding of individual RNA backbone suites, even with graphics tools developed specifically for the task, is tedious and limited in scope, but quite educational. First the SuiteFit tool in Mage (Richardson & Richardson, 2001) graphics and then the RNA Rotator tool available in KiNG (Chen, Davis, & Richardson, 2009) were developed for manual rebuilding of individual RNA backbone suites. They allow the user to interactively adjust the seven backbone dihedral angles and two χ angles associated with the suite, or to select a new starting point from a list of all 54 backbone conformers. As well as the essential reference of the electron density contours, display of all-atom contact dots is updated in real time as the user changes dihedral angles, providing essential feedback. In Mage, the user controls overall rotation and translation of the suite explicitly. In KiNG, the user controls suite positioning by specifying a set of atoms to keep optimally superimposed on the original as angles are changed, and additional feedback is provided by real-time update of the "suiteness" conformer-quality parameter (see Section 3.3). This process is only sometimes successful even for expert users.

Figure 11 shows before and after states for a manual correction done using RNA Rotator on suite 542 in the 50S ribosomal subunit of

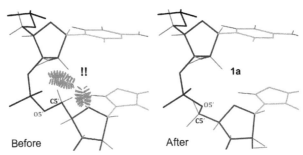

Figure 11 Example of a manual RNA backbone correction. At left, the originally modeled backbone of suite 542 clashes in both directions, with backbone of 541 and base of 542 (clusters of hotpink (dark gray in the print version) spikes), and suite 542 is a backbone-conformer outlier (!!). At right, after clash correction using RNA Rotator, suite 542 adopts a valid **1a** conformer. *From PDB 3CC2, Hm 50S ribosomal subunit at 2.4 Å (Blaha et al., 2008).*

H. marismortui (PDB ID: 3CC2). Changing the α, β, and γ backbone dihedral angles and superimposing the new suite on the 5' and 3' ends and the base atoms, suite 542 was changed from a **!!** to a **1a** conformation, which then refined successfully.

5.2 RNABC

RNABC is an automated tool developed to help fix all-atom clashes in RNA backbone (Wang et al., 2008). As previously noted, the base and the phosphate are most easily visible in the density, hence RNABC keeps the base and P atom fixed and searches for a better rebuild of the rest of a specified dinucleotide using forward kinematics. The bond lengths and angles are held fixed, and the pucker of the two ribose rings in the dinucleotide is either specified by the user or determined by the Pperp test (described in Section 3.2). Possible conformations are scored based on steric clashes, pucker, and geometry terms, and the best-scoring set of nonredundant possible conformations are output for the user to choose from. However, RNABC does not take the electron density or the known RNA backbone conformers into account during the rebuilding process.

5.3 Coot and RCrane

RCrane (Keating & Pyle, 2012) is a plug-in available in the Coot model-building software (Emsley, Lohkamp, Scott, & Cowtan, 2010), to do semiautomated building of RNA structure into electron density. It is partially based on the coarse-grained RNA backbone parameters η and θ (the two pseudo-dihedral angles defined by the P and C4' atoms), used successfully to describe and locate RNA structural motifs (Duarte & Pyle, 1998). For this purpose, they are modified to a θ', η' form that uses the more reliably fit C1' atom rather than C4', and the suite rather than the nucleotide division. These dihedrals are related to the all-dihedral RNA backbone suite conformers (see Section 3.3), though the relationship is not one-to-one. The tool takes as input the electron density and builds a trace for the user-selected nucleotides. This is done by marking the highest intensity peaks as crosses (see Fig. 12 screen shot) and allowing the user to choose the probable P and C1' atoms from these density peaks and adjust their positions (within 10 Å) as needed. They are joined alternately into a proposed virtual backbone trace. An automated selection is then made of the few most probable backbone conformers (based on θ', η', Pperp, C1'–C1', and P–P distances), followed by individual coordinate minimization within the

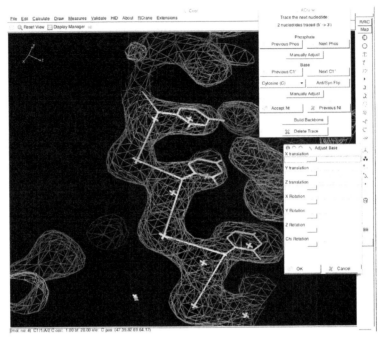

Figure 12 Screen shot of the RCrane tool in Coot. Peaks of electron density are marked with crosses. RCrane allows the user to choose and manually adjust the position of the P and C1' atoms. The resulting virtual-bond backbone trace is shown, from which probable backbone conformers will then be proposed.

electron density. The user chooses from the list of possible conformers and their scores, for each built nucleotide. This technique can also be used for correction of errors in the RNA structure, given the input model and the electron density, and rebuilding selected nucleotides.

5.4 PHENIX refinement with pucker-specific targets

Now that the right pucker state can be identified even for misfit nucleotides using the Pperp test, that simple function allows the use of pucker-specific dihedral-angle and bond-angle targets to be used in refinement, a functionality available in PHENIX (Adams et al., 2010).

5.5 ERRASER, with examples

All the correction and rebuilding tools described above require major input from the user, either in identifying errors or in the actual correction or building process, and none of them have nearly as high a success rate as outlier correction in protein structures achieved some time ago

(Arendall et al., 2005). Now, however, we have software that has proven to be truly effective in automated error correction of RNA backbone, called ERRASER (Adams et al., 2013; Chou et al., 2012). Figure 13 shows the quite revolutionary level of cleanup that ERRASER can achieve. It utilizes capabilities in PHENIX and MolProbity for identification of modeling errors, and a stepwise assembly (SWA) procedure to rebuild each residue by enumerating many conformations covering all build up paths, taking into account the fit of the model to the electron density. ERRASER can be used to rebuild whole RNA structures (where error detection is done automatically), or single residues as specified by the user, in which case ERRASER returns its top 10 distinct conformations for the user to choose among. One can also specify that a particular set of residues remain fixed.

The rebuilding process in ERRASER consists of three steps: First, ERRASER minimizes all torsion angles and all backbone bond lengths and bond angles using the Rosetta energy function for RNA (Das & Baker, 2007), including an electron density correlation score (DiMaio et al., 2009). Second, PHENIX's MolProbity-style RNA validation tools are used to identify errors (geometry, pucker, and unrecognized backbone conformations) in the minimized model. These residues, as well as residues with large rms deviation (>2 Å) between their original position and the

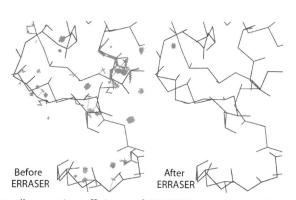

Figure 13 Overall correction efficiency of ERRASER, supplemented with other tools. The active-site region of the uncleaved HDV ribozyme structure, with the RNA backbone in black. At left, the original structure (PDB ID: 1VC7; Ke, Zhou, Ding, Cate, & Doudna, 2004), with many clashes (hotpink spikes), bond-length outliers (red and blue spirals), bond-angle outliers (red and blue fans), and ribose-pucker outliers (magenta crosses). At right, the rebuilt structure (PDB ID: 4PRF; Kapral et al., 2014), with essentially all validation outliers corrected. (See the color plate.)

minimized position, are identified as residues to be rebuilt. The second step is skipped if the user has specified a particular residue to be rebuilt. Third, the residues from step two are rebuilt one at a time with the SWA procedure, and then minimized again. This process is carried out usually for three cycles.

Figure 14 shows an example of a ribose-pucker correction done using ERRASER. Residue 152 in the uncleaved HDV ribozyme structure (PDB ID: 1VC7; Ke et al., 2004) is incorrectly modeled as a C3′-endo pucker, leading to deviations in geometry and steric clashes with the phosphate group of the next residue. O2′ is modeled out of the $2F_o$-F_c density, and there is a large peak of positive difference density (blue mesh) near the residue. ERRASER was run on the entire HDV ribozyme structure, which resulted in this residue being remodeled as C2′-endo pucker and correction of all other validation outliers (PDB ID: 4PRF; Kapral et al., 2014). The O2′ moves into $2F_o$-F_c density, getting rid of the positive difference density peak.

ERRASER can also be used to correct a single residue at a time. Figure 15 shows an incorrectly modeled kink-turn in the 50S ribosomal subunit of *H. marismortui* (PDB ID: 3CC2; Blaha et al., 2008), with the suite identities shown in the figure. The first suite in the kink-turn is a !! conformer outlier, its backbone clashing with other residues in the kink-turn. ERRASER was run on that residue (1603), and the top-scoring conformation it returned fixed the steric clash and changed suite 1603 from !! to 7r,

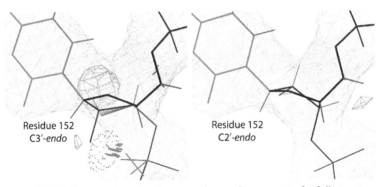

Figure 14 ERRASER correction of a ribose-pucker outlier, as part of a full-structure run. At left, residue 152 modeled incorrectly as $C3_0$-endo pucker (PDB ID: 1VC7), shown in $2F_o$-F_c density at 1.2σ (pale gray mesh). It has a too-tight, clashing hydrogen bond (hotpink (dark gray in the print version) spikes within the green (dark gray in the print version) dot pillow), a bond-angle outlier (red (dark gray in the print version) fan), ribose-pucker outlier (magenta (light gray in the print version) cross), and a large positive difference density peak at 3.5σ (blue (light gray in the print version) mesh). At right, the residue 152 rebuilt by ERRASER as $C2_0$-endo pucker, and all other outliers gone (PDB ID: 4PRF).

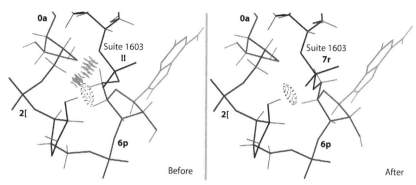

Figure 15 ERRASER single-residue correction of a kink-turn suite. Suites 1603–1606 form a kink-turn motif in the Hm 50S ribosomal subunit (PDB ID: 3CC2), but suite 1603 was modeled as a backbone-conformer outlier (!!), badly clashing with the 1605 backbone. At right, ERRASER has made large rotations in the backbone between ribose and 5′ P to resolve the clash and rebuild suite 1603 as a **7r** conformer.

making the kink-turn backbone conformation consistent with the consensus suitestring for a kink-turn motif (**7r6p2[0a**).

ERRASER has proven to be very powerful for correcting backbone errors in RNA structures. However, it requires installation of Rosetta, and because of its comprehensive search for different possible conformations, running ERRASER is compute intensive, especially for large RNA or RNP structures. As with any rebuilding procedure, crystallographic refinement should be run both before and after ERRASER. Compared with the time and effort required for any other RNA correction method, and with their relatively limited success rate, ERRASER is still a very worthwhile bargain.

6. WHAT'S COMING NEXT
6.1 ERRASER for RNA/protein

The current version of ERRASER recognizes only nonmodified RNA residues and removes all other residue types (protein residues, waters, ions, modified residues, DNA, ligands) from the PDB file before starting the structure optimization. This is a problem for rebuilding RNA residues that interact with the neglected residue types, such as in large RNA/protein complexes such as ribosomes. ERRASER can be misled into moving the new RNA conformation into protein or ligand density, which creates impossible steric clashes during the final step of merging the rebuilt structure with the original PDB file. We are working on the thorough rewrite of

ERRASER that will allow it to recognize the currently neglected residue types, and thus take into account their interaction with RNA residues during the rebuilding process. This should greatly improve modeling and correction of both RNA–small molecule and RNP structures.

6.2 Better ion identification

Because of the high negative charge on nucleic acid backbones, ions are a crucial factor for RNA folding, stability, and function. However, it is tricky to locate them in a structure, and even more so to identify their molecular type, since their coordination to the RNA surface is seldom neat and complete, and either partial occupancy or small ions such as Mg are hard to distinguish from waters. Tools are being developed (Echols et al., 2014) and will be expanded, to make use of anomalous-scattering X-ray data that can be diagnostic of ion presence and identity, and of contact information that distinguishes waters from + or − ions, or unmodeled alternate conformations, based on distances and nature of the interacting atoms (Headd & Richardson, 2013).

6.3 Putting conformers into initial fitting

The set of 54 recognized RNA backbone suite conformers (Richardson et al., 2008) are utilized for structure validation and correction, but could also be very valuable in initial fitting of RNA models, especially in non-helical regions. Some initial trials have been done, and work is in progress to take advantage of that possibility. The conformer set is similar in nature to protein side-chain rotamers, but perhaps even more useful, because they apply to the continuous backbone rather than just to an individual side branch. In implementing such a tool, the philosophy is to work from the features best seen in electron density, a P and its two flanking glycosidic bonds, but using at least seven parameters, so as to encapsulate the full information contained in those positions. Such a system would be integrated into the automated procedures in PHENIX.

ACKNOWLEDGMENTS

We would like to acknowledge the contributions of former lab members Laura Murray and Gary Kapral and of our collaborators Rhiju Das, Jack Snoeyink, Kevin Keating, the Phenix developers, and the validation team at the wwPDB. This work was supported by the National Institutes of Health (previously GM088674 and GM074127, currently GM063210 and GM073919).

REFERENCES

Adams, P. D., Afonine, P. V., Bunkóczi, G., Chen, V. B., Davis, I. W., Echols, N., et al. (2010). PHENIX: A comprehensive Python-based system for macromolecular structure solution. *Acta Crystallographica, D66,* 213–221.

Adams, P. D., Baker, D., Brunger, A. T., Das, R., DiMaio, F., Read, R. J., et al. (2013). Advances, interactions, and future developments in the CNS, Phenix, and Rosetta structural biology software systems. *Annual Review of Biophysics, 42,* 265–287.

Allen, F. H. (2002). The Cambridge Structural Database: A quarter of a million crystal structures and rising. *Acta Crystallographica, B58,* 380–388.

Arendall, B. W., III, Tempel, W., Richardson, J. S., Zhou, W., Wang, S., Davis, I. W., et al. (2005). A test of enhancing model accuracy in high-throughput crystallography. *Journal of Structural and Functional Genomics, 6,* 1–11.

Batey, R. T., Gilbert, S. D., & Montange, R. K. (2004). Structure of a natural guanine-responsive riboswitch complexed with the metabolite hypoxanthine. *Nature, 432,* 411–415, 1U8D.

Berman, H. M., Henrick, K., & Nakamura, H. (2003). Announcing the worldwide Protein Data Bank. *Nature Structural Biology, 10,* 980.

Berman, H. M., Westbrook, J., Feng, Z., Gilliland, G., Bhat, T. N., Weissig, H., et al. (2000). The Protein Data Bank. *Nucleic Acids Research, 28,* 235–242.

Blaha, G., Gural, G., Schroeder, S. J., Moore, P. B., & Steitz, T. A. (2008). Mutations outside the anisomycin-binding site can make ribosomes drug-resistant. *Journal of Molecular Biology, 379,* 505–519, 3CC2.

Brunger, A. T. (1992). Free R-value: a novel statistical quantity for assessing the accuracy of crystal structures. *Nature, 355,* 472–475.

Byrne, R. T., Konevega, A. L., Rodnina, M. V., & Antson, A. A. (2010). The crystal structure of unmodified tRNA Phe from *Escherichia coli. Nucleic Acids Research, 38,* 4154–4162, 4L0U.

Chen, V. B., Arendall, W. B., III, Headd, J. J., Keedy, D. A., Immormino, R. M., Kapral, G. J., et al. (2010). MolProbity: All-atom structure validation for macromolecular crystallography. *Acta Crystallographica, D66,* 12–21.

Chen, V. B., Davis, I. W., & Richardson, D. C. (2009). KiNG (Kinemage, Next Generation): A versatile interactive molecular and scientific visualization program. *Protein Science, 18,* 2403–2409.

Chou, F.-C., Sripakdeevong, P., Dibrov, S. M., Hermann, T., & Das, R. (2012). Correcting pervasive errors in RNA crystallography through enumerative structure prediction. *Nature Methods, 10,* 74–76.

Correll, C. C., Beneken, J., Plantinga, M. J., Lubbers, M., & Chan, Y. L. (2003). The common and distinctive features of the bulged-G motif based on a 1.04 Å resolution RNA structure. *Nucleic Acids Research, 31,* 6806–6818, 1Q9A.

Correll, C. C., Munishkin, A., Chan, Y. L., Ren, Z., Wool, I. G., & Steitz, T. A. (1998). Crystal structure of the ribosomal RNA domain essential for binding elongation factors. *Proceedings of the National Academy of Sciences of the United States of America, 95,* 13436–13441, 430D.

Dalluge, J. J., Hashizume, T., Sopchik, A. E., McCloskey, J. A., & Davis, D. R. (1996). Conformational flexibility in RNA: The role of dihydrouridine. *Nucleic Acids Research, 24,* 1073–1079.

Das, R., & Baker, D. (2007). Automated *de novo* prediction of native-like RNA tertiary structures. *Proceedings of the National Academy of Sciences of the United States of America, 104,* 14664–14669.

Deis, L. N., Verma, V., Videau, L. L., Prisant, M. G., Moriarty, N. W., Headd, J. J., et al. (2013). Phenix/MolProbity hydrogen parameter update. *The Computational Crystallography Newsletter, 4,* 9–10.

DiMaio, F., Tyka, M., Baker, M., Chiu, W., & Baker, D. (2009). Refinement of protein structures into low-resolution density maps using Rosetta. *Journal of Molecular Biology, 392*, 181–190.

Duarte, C. M., & Pyle, A. M. (1998). Stepping through an RNA structure: A novel approach to conformational analysis. *Journal of Molecular Biology, 284*, 1465–1478.

Dunkle, J. A., Wang, L., Feldman, M. B., Pulk, A., Chen, V. B., Kapral, G. J., et al. (2011). Structures of the bacterial ribosome in classical and hybrid states of tRNA binding. *Science, 332*, 981–984, 4GD1,2, 3R8S,T.

Echols, N., Grosse-Kunstleve, R. W., Afonine, P. V., Bunkoczi, G., Chen, V. B., Headd, J. J., et al. (2012). Graphical tools for macromolecular crystallography in Phenix. *Journal of Applied Crystallography, 45*, 581–586.

Echols, N., Morshed, N., Afonine, P. V., McCoy, A. J., Miller, M. D., Read, R. J., et al. (2014). Automated identification of elemental ions in macromolecular crystal structures. *Acta Crystallographica, D70*, 1104–1114.

Edwards, T. E., & Ferré-D'Amaré, A. R. (2006). Crystal structures of the Thi-Box riboswitch bound to thiamine pyrophosphate analogs reveal adaptive RNA-small molecule recognition. *Structure, 14*, 1459–1468, 2HOJ.

Emsley, P., Lohkamp, B., Scott, W. G., & Cowtan, K. (2010). Features and development of Coot. *Acta Crystallographica, D66*, 486–501.

Ferré-D'Amaré, A. R., & Scott, W. G. (2010). Small self-cleaving ribozymes. *Cold Spring Harbor Perspectives in Biology, 2*, a003574.

Freisz, S., Lang, K., Micura, R., Dumas, P., & Ennifar, E. (2008). Binding of aminoglycoside antibiotics to the duplex form of the HIV-1 genomic RNA dimerization initiation site. *Angewandte Chemie (International Edition in English), 47*, 4110–4113, 3C3Z.

Gore, S., Velankar, S., & Kleywegt, G. J. (2012). Implementing an X-ray validation pipeline for the Protein Data Bank. *Acta Crystallographica, D68*, 478–483.

Grazulis, S., Chateigner, D., Downs, R. T., Yokochi, A. T., Quiros, M., Lutterotti, L., et al. (2009). Crystallography Open Database—An open-access collection of crystal structures. *Journal of Applied Crystallography, 42*, 726–729.

Headd, J., & Richardson, J. (2013). Fitting Tips #5: What's with water? *The Computational Crystallography Newsletter, 4*, 2–5.

Janssen, B. J. C., Read, R. J., Brunger, A. T., & Gros, P. (2007). Crystallographic evidence for deviating C3b structure? *Nature, 448*, E1–E2.

Jones, T. A., Zou, J. Y., Cowan, S. W., & Kjeldgaard, M. (1991). Improved methods for building protein models in electron density maps and the location of errors in these models. *Acta Crystallographica, A47*, 110–119.

Joosten, R. P., Joosten, K., Murshudov, G. N., & Perrakis, A. (2012). PDB_REDO: Cobstructive validation, more than just looking for errors. *Acta Crystallographica, D68*, 484–496.

Kapral, G. J., Jain, S., Noeske, J., Doudna, J. A., Richardson, D. C., & Richardson, J. S. (2014). New tools provide a second look at HDV ribozyme structure, dynamics, and cleavage. *Nucleic Acids Research, 42*, 12833–12846, 4PR6, 4PRF.

Ke, A., Zhou, K., Ding, F., Cate, J. H. D., & Doudna, J. A. (2004). A conformational switch controls hepatitis delta virus ribozyme catalysis. *Nature, 429*, 201–205, 1VC7.

Keating, K. S., & Pyle, A. M. (2012). RCrane: Semi-automated RNA model building. *Acta Crystallographica, D68*, 985–995.

Klein, D. J., Schmeing, T. M., Moore, P. B., & Steitz, T. A. (2001). The kink-turn: A new RNA secondary structure motif. *The EMBO Journal, 20*, 4214–4221.

Kleywegt, G. J., Harris, M. R., Zou, J. Y., Taylor, T. C., Wahlby, A., & Jones, T. A. (2004). The Uppsala Electron Density Server. *Acta Crystallographica, D60*, 2240–2249.

Lang, K., Erlacher, M., Wilson, D. N., Micura, R., & Polacek, N. (2008). The role of 23S ribosomal RNA residue A2451 in peptide bond synthesis revealed by atomic mutagenesis. *Chemistry & Biology, 15*, 485–492.

Leontis, N. B., & Westhof, E. (2001). Geometric nomenclature and classification of RNA base pairs. *RNA, 7*, 499–512.

Li, F., Pallan, P. S., Maier, M. A., Rajeev, K. G., Mathieu, S. L., Kreutz, C., et al. (2007). Crystal structure, stability, and in vitro RNAi activity of oligoribonucleotides containing the ribo-difluorotoluyl nucleotide: Insights into substrate requirements by the human RISC Ago2 enzyme. *Nucleic Acids Research, 35*, 6424–6438, 2Q1O.

Moriarty, N. W., Grosse-Kunstleve, R. W., & Adams, P. D. (2009). electronic Ligand Builder and Optimization Workbench (eLBOW): A tool for ligand coordinate and restraint generation. *Acta Crystallographica, D65*, 1074–1080.

Murray, L. W., Arendall, W. B., III, Richardson, D. C., & Richardson, J. S. (2003). RNA backbone is rotameric. *Proceedings of the National Academy of Sciences of the United States of America, 100*, 13904–13909.

Parisien, M., & Major, F. (2008). The MC-Fold and MC-Sym pipeline infers RNA structure from sequence data. *Nature, 452*, 51–55.

Read, R. J. (1986). Improved Fourier coefficients for maps using phases from partial structures with errors. *Acta Crystallographica, A42*, 140–149.

Read, R. J., Adams, P. D., Arendall, W. B., III, Brunger, A. T., Emsley, P., Joosten, R. P., et al. (2011). A new generation of crystallographic validation tools for the protein data bank. *Structure, 19*, 1395–1412.

Richardson, D. C., & Richardson, J. S. (2001). MAGE, PROBE, and Kinemages. In M. G. Rossmann & E. Arnold (Eds.), *Crystallography of biological macromolecules: Vol. F. IUCr's international tables of crystallography* (pp. 727–730). Dortrecht: Kluwer Academic Press, (chapter 25.2.8).

Richardson, J. S., & Richardson, D. C. (2013). Doing molecular biophysics: Finding, naming, and picturing signal within complexity. *Annual Review of Biophysics, 42*, 1–28.

Richardson, J. S., Schneider, B., Murray, L. W., Kapral, G. J., Immormino, R. M., Headd, J. J., et al. (2008). RNA backbone: Consensus all-angle conformers and modular string nomenclature (an RNA Ontology Consortium contribution). *RNA, 14*, 465–481.

Saenger, W. (1983). *Principles of nucleic acid structure.* New York: Springer.

Sekine, S., Nureki, O., Dubois, D. Y., Bernier, S., Chenevert, R., Lapointe, J., et al. (2003). ATP binding by glutamyl-tRNA synthetase is switched to the productive mode by tRNA binding. *The EMBO Journal, 22*, 676–688, 1N78.

Stombaugh, J., Zirbel, C. L., Westhof, E., & Leontis, N. B. (2009). Frequency and isostericity of RNA base pairs. *Nucleic Acids Research, 37*, 2294–2312.

Svozil, D., Kalina, J., Omelka, M., & Schneider, B. (2008). DNA conformations and their sequence preferences. *Nucleic Acids Research, 36*, 3690–3706.

Wang, X., Kapral, G. J., Murray, L. W., Richardson, D. C., Richardson, J. S., & Snoeyink, J. (2008). RNABC: Forward kinematics to reduce all-atom steric clashes in RNA backbone. *Journal of Mathematical Biology, 56*, 253–278.

Warner, K. D., Chen, M. C., Song, W., Straek, R. L., Thorn, A., Jaffery, S. R., & Ferre-D'Amare, A. R. (2014). Structural basis for activity of highly efficient RNA mimics of green fluorescent protein. *Nature Structural & Molecular Biology, 21*, 658–663.

Williamson, J. R. (2000). Induced fit in RNA-protein recognition. *Nature Structural Biology, 7*, 834–837.

Word, J. M., Lovell, S. C., LaBean, T. H., Taylor, H. C., Zalis, M. E., Presley, B. K., et al. (1999). Visualizing and quantifying molecular goodness-of-fit: Small-probe contact dots with explicit hydrogen atoms. *Journal of Molecular Biology, 285*, 1711–1733.

Word, J. M., Lovell, S. C., Richardson, J. S., & Richardson, D. C. (1999). Asparagine and glutamine: Using hydrogen atom contacts in the choice of sidechain amide orientation. *Journal of Molecular Biology, 285*, 1735–1747.

Yeates, T. O. (1997). Detecting and overcoming crystal twinning. *Methods in Enzymology, 276*, 344–358.

Zaher, H. S., Shaw, J. J., Strobel, S. A., & Green, R. (2011). The 2′-OH group of the peptidyl-tRNA stabilizes an active conformation of the ribosomal PTC. *The EMBO Journal, 30*, 2445–2453.

Zwart, P. H., Grosse-Kunstleve, R. W., & Adams, P. D. (2005). Xtriage and Fest: Automatic assessment of X-ray data and substructure structure factor estimation. *CCP4 Newsletter, 43*, Contribution 7.

CHAPTER EIGHT

Structures of Large RNAs and RNA–Protein Complexes: Toward Structure Determination of Riboswitches

Jason C. Grigg, Ailong Ke[1]
The Department of Molecular Biology and Genetics, Cornell University, Ithaca, New York, USA
[1]Corresponding author: e-mail address: ailong.ke@cornell.edu

Contents

1. Introduction 214
2. Rational Construct Design for Structure Determination 215
   2.1 Sequence Gazing, Devil Is in the Details 216
   2.2 Do Your Homework, Do Not Skip the Biochemistry 218
   2.3 Which Sequence to Choose for Crystallization? 221
   2.4 Maximize Your Chance of Success by Rational Construct Engineering 222
   2.5 Use of RNA-Binding Proteins as Crystallization Aid 223
3. Crystallization and Structure Determination 223
4. Structure–Function Validation 225
   4.1 Mutagenesis Assayed by *In Vivo* and *In Vitro* Experiments 225
   4.2 Guide the Design and Interpretation of Single-Molecule Measurement and Molecular Simulations 226
   4.3 Capturing Ligand-Free and Other Important Functional States 226
5. Conclusions 227
Acknowledgments 227
References 227

Abstract

Riboswitches are widespread and important regulatory elements. They are typically present in the mRNA of the gene under their regulation, where they form complex three-dimensional structures that can bind an effector and regulate either transcription or translation of the mRNA. Structural biology has been essential to our understanding of their ligand recognition and conformational switching mechanisms, but riboswitch determination presents several important complications. Overcoming these challenges requires a synergistic approach using rational design of the constructs and supporting methods to biochemically validate the designs and resulting structures.

ABBREVIATIONS
SAM S-adenosyl-L-methionine
tRNA transfer RNA

1. INTRODUCTION

Structured RNAs perform numerous important roles in biology, with historically well-characterized examples in protein synthesis by RNA components of the ribosome and transfer RNA (tRNA) (Ban, Nissen, Hansen, Moore, & Steitz, 2000; Muth, Ortoleva-Donnelly, & Strobel, 2000; Nissen, Hansen, Ban, Moore, & Steitz, 2000). The breadth of other activities performed by RNA has begun to be understood, including everything from structural scaffolding to catalysis and regulation of transcription or translation. Riboswitches are an important type of structured RNA; they are broadly distributed regulatory sequences present in mRNA that sense and respond to an effector to control gene expression (Breaker, 2012). They can act at the levels of transcription, translation, and even splicing, independent of protein cofactors. Riboswitches were first described in 2002 (Mironov et al., 2002; Nahvi et al., 2002; Winkler, Cohen-Chalamish, & Breaker, 2002), but have been intensely studied over the past 12 years and are now appreciated to perform regulation in many bacteria, archaea, and even higher organisms with recent examples from plants, fungi, and algae (for reviews, see Breaker, 2012).

Riboswitches are generally comprised of separate aptamer and effector domains (Fig. 1). The aptamer domain is the ligand-binding region, where binding induces structural changes that are transmitted to the effector domain. The effector domain then converts the input signal into a regulatory output. In the case of translational riboswitches, the effector typically exposes or sequesters the ribosome binding site (Shine–Dalgarno sequence) to turn translation on or off. In transcriptional riboswitches, the effector enables or disrupts the formation of a premature transcription terminator that dissociates the RNA polymerase, producing a truncated mRNA. Alternative regulatory mechanisms, such as ligand-induced mRNA self-cleavage, have been reported (Winkler, Nahvi, Roth, Collins, & Breaker, 2004). Riboswitches in eukaryotes can regulate alternative splicing (Cheah, Wachter, Sudarsan, & Breaker, 2007; Li & Breaker, 2013) or bacterial riboswitches can possess additional levels of complexity, such as controlling

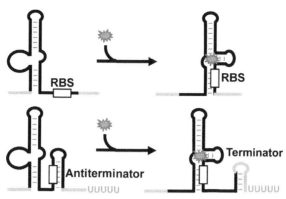

Figure 1 Two popular mechanisms for riboswitches to regulate gene expression in bacteria. The general principle of translational (top) and transcriptional (bottom) riboswitches is the same. Ligand binding drives formation of alternate structures that either restrict or expose necessary sequences for ribosome binding (top), or to terminate RNA polymerase extension (bottom).

mRNA turnover by sequestering RNase cleavage sites (Caron et al., 2012). Regardless of their gene regulatory mechanism, the same general principles apply to most of the described riboswitches, that they are inherently unstable switches that alter their conformation upon binding of a ligand to turn gene expression on or off depending on their environment.

Our understanding of riboswitch mechanisms has greatly expanded since their discovery, with structures now determined for most validated riboswitches classes (Serganov & Nudler, 2013). This is encouraging considering that riboswitches offer their own particular challenges for structural characterization, being flexible switches by design. In this chapter, we examine strategies used to overcome the challenges of working with riboswitches and outline strategies for solving new structures as well as validating the results. Where possible, we will draw from our experience characterizing the SAM-I (Lu et al., 2010), S_{MK} box (Lu et al., 2011, 2008; Price, Grigg, & Ke, 2014), and T-box riboswitches (Grigg et al., 2013; Grigg & Ke, 2013a, 2013b) to illustrate these points.

2. RATIONAL CONSTRUCT DESIGN FOR STRUCTURE DETERMINATION

A major obstacle in studying riboswitches is that they are inherently dynamic structures that often fold into alternatively base-paired conformers as part of their switching mechanism. The switching mechanism can also vary, being kinetically driven or with stable alternative conformers existing in thermodynamic equilibrium (Garst, Heroux, Rambo, & Batey, 2008; Lemay et al., 2011; Wickiser, Cheah, Breaker, & Crothers, 2005; Wickiser, Winkler, Breaker, & Crothers, 2005). Both mechanisms add their own complications when studying samples, *in vitro*, and neither X-ray crystallography nor NMR fair well with flexible structures or a mixture of multiple conformers. Accordingly, constructs for structure determination are designed with structural stability (or conformation homogeneity) in mind by accurately defining the nucleotides involved in alternative conformations, choosing inherently stable sequences to study, and by making rational modifications to enhance stability.

2.1 Sequence Gazing, Devil Is in the Details

Accurately, defining the core structural elements is one of the most important initial steps for any successful *in vitro* riboswitch work. Extraneous regions that do not contribute to the key function of the riboswitch are not under the same level of selection pressure to adopt the most concise and stable conformation. As a result, they may possess unwanted flexibility that may impede structure determination, or form alternate structures. Most published riboswitch structures represent the aptamer domain, in isolation. By working with the aptamer domain alone, the possibility to form the alternative structure with the effector domain is eliminated. The same general characteristics that can be used to rationally identify riboswitches can be used to define the aptamer and the expression platform. For instance, transcription terminators are defined by a hairpin preceding a poly-U stretch and Shine–Dalgarno sequences (ribosome binding sites) consist of the AGGAGG consensus, approximately eight nucleotides upstream of the start codon. Identifying either feature in alternatively pairing helices could suggest the presence of a riboswitch. By including these regions in secondary structure predictions, with tools such as RNAfold (Lorenz et al., 2011) or Mfold (Zuker, 2003), the different folding states can often be identified. Since riboswitches have typically been identified by sequence first, multiple sequence alignments, phylogenetic covariation analysis, and secondary structure predictions provide a rich source of information to define secondary (and sometimes even tertiary) structure and core regions (Griffiths-Jones et al., 2005). For example, distal stemloops that are less important for riboswitch function may exhibit a higher degree of variation in loop length and sequence in multiple sequence alignments. Helices that only participate in secondary structure formation may show obvious signs of sequence covariation in base-paired residues. On the other hand, nucleotides that mediate either secondary and tertiary structures or RNA–ligand interactions, may display a higher degree of sequence conservation and less of covariation. For instance, in the S_{MK} box riboswitch, the stretch of residues between nucleotides 68 and 74, CUUGUAA according to numbering in Fuchs, Grundy, and Henkin (2006), is not only part of the ligand-binding site in the off state but also part of the P0 helix in the on-state (Fig. 2). With these dual roles, not surprisingly, the CUUGUAA nucleotides are nearly completely conserved (Fuchs et al., 2006; Fuchs, Grundy, & Henkin, 2007) and as a result, so is the complementary strand in the P0 helix. In contrast, the P4 helix is only 2-bp long with a large, highly degenerate loop and was therefore not noticed in the initial studies. We noticed the high degree of sequence covariation in the two base pair region in preparation for the structural studies and designed our crystallization constructs with the goals of preserving the P4 structure, while reducing unwanted conformational

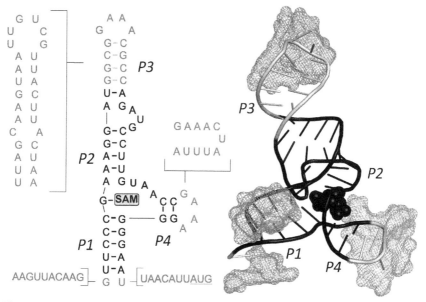

Figure 2 Construct engineering to obtain the S_{MK} box riboswitch structure. The secondary and tertiary structures of the S_{MK} box riboswitch are showing with unmodified regions (black) and modified regions (gray) indicated. The wild-type sequences are shown beside their corresponding element for comparison. SAM is drawn in the binding pocket as black spheres. The crystal packing contacts (gray mesh) are primarily mediated by engineered regions.

flexibility in this stemloop. The effort enabled us to successfully capture the S-adenosyl-L-methionine (SAM)-bound structure (Fig. 2; Lu et al., 2008). More careful analysis later revealed that the conformational flexibility is built into P4 as a mechanism to maintain an energetic balance between alternative conformers (Lu et al., 2011). Sequence conservation versus covariation is well utilized in databases such as Rfam to determine RNA consensus structures with great success (Griffiths-Jones et al., 2005). This information is readily available to anyone proceeding with structural work on structured RNAs and is crucial in the initial design stages.

2.2 Do Your Homework, Do Not Skip the Biochemistry

Biochemical validation is indispensible at two different stages in characterizing riboswitch structures: during the construct design/optimization stages, and also in the function validation stage following successful structure determination. Prior to structural work, rational approaches are routinely used to modify the RNA sequences, but great care must be taken to modify regions without impeding biological functions under investigation. No matter how modest the changes, the onus is on the experimenter to demonstrate that the modifications do not appreciably alter function and relevance. For riboswitches, this typically entails validating both folding and maintenance of ligand binding (or whatever target feature is being examined).

To illustrate this point, we will use a typical protocol from our lab. Numerous permutations exist, but in our lab, we primarily use T7 polymerase systems to transcribe RNA from a 5′ T7 promoter and a 3′ HDV ribozyme to ensure homogeneous blunt ends are produced (Ke & Doudna, 2004). The RNA of interest is subsequently separated from the HDV ribozyme and precursors on preparative denaturing polyacrylamide gel electrophoresis (PAGE), and the RNA band of interest is visualized by UV shadow, excised, and crushed. RNA is eluted into water and buffer-exchanged into the refolding buffer. A purity of over 95% is typically achieved following this procedure (Ke & Doudna, 2004). Despite every care to engineer stable structures, each time a new RNA is produced, it is important to assess and optimize folding to ensure homogeneity. Native PAGE is an easy and inexpensive screening tool for folding. Typically, the gel is supplemented with the stabilizing cations, such as Mg^{2+}. A single band in the lane is suggestive of one well-formed conformation, whereas multiple bands could indicate discrete conformations or poor folding (Grigg et al., 2013). If native PAGE calls for optimization of the refolding procedure, we typically test heating with quick (on ice) and slow cooling (removing heat block from heater) first and when necessary, move to ultraslow cooling (thermocycler) or equilibrium dialysis to slowly removed denaturant such as urea (Batey, Sagar, & Doudna, 2001; Ke & Doudna, 2004). Once folding has been optimized, other methods, such as chemical probing are ideal to demonstrate that the RNA is achieving a defined (and most often, predicted) structure. In our lab, we prefer to use selective 2′-hydroxyl acylation analyzed by primer extension (SHAPE) over traditional chemical probing tools (Wilkinson, Merino, & Weeks, 2006). Although SHAPE requires adding sequences to the 5′ and 3′ ends of the RNA, we always take care to avoid potential alternative structures, and in our hands, the modifications have not interfered with riboswitch function or folding (Grigg et al., 2013; Lu et al., 2011). In cases where primer extension is to be avoided, advancements to the technique, such as RNase detected SHAPE allow analysis of the native structure (Steen, Siegfried, & Weeks, 2011). Not only is SHAPE useful for folding analysis, but because it is a sensitive technique amenable to analysis by capillary electrophoresis, it has also proven very useful in monitoring ligand binding and subsequent structural changes.

While structural probing is critical, function should also be investigated. For many riboswitches, this could simply entail using a ligand-binding assay specific to the system. This would be the ultimate assay for the fold of the riboswitch, and in many cases prove essential in choosing appropriate riboswitch homologs for structural study. As an example, our investigation of the T-box riboswitch (an atypical riboswitch that senses a macromolecule tRNA rather than a small molecule) proceeded in this fashion. The T-box is a large, multistem structure that binds to tRNA by recognizing its codon and its aminoacylation site to regulate various components of amino acid state. We began working with the full version (>180 nts) of several T-box riboswitches, but generally they folded into multiple conformations or failed to robustly bind tRNA, *in vitro*, as shown by native PAGE. We screened several homologs of *Bacillus subtilis glyQS*, including the thermophilic bacteria: *Geobacillus kaustophilus* and *G. thermodenitrificans*. While the *B. subtilis* construct displayed some folding heterogeneity, the *G. kaustophilus glyQS* folded homogeneously and bound tRNA robustly *in vitro* (Grigg et al., 2013). We further questioned what constituted the minimal functional unit of the *G. kaustophilus glyQS* riboswitch. Previous studies established that tRNA stably bound to the riboswitch regardless of its aminoacylation status, which is sensed by Stem IV (Green, Grundy, & Henkin, 2010). Further, it was established that a codon–anticodon interaction in Stem I determined tRNA specificity (Grundy & Henkin, 1993; Grundy, Hodil, Rollins, & Henkin, 1997; Rollins, Grundy, & Henkin, 1997). We used two approaches to characterize the interaction: (1) we tried to crystallize several smaller fragments throughout the riboswitch in the absence of tRNA, regardless of their ability to maintain tRNA binding and (2) we used mobility shift assays to determine the minimal fragment of the T-box riboswitch that retained its ability to bind tRNA (Fig. 3; Grigg et al., 2013; Grigg & Ke, 2013b). Both approaches proved fruitful because we successfully determined the distal region of Stem I and the structure alone provided significant insight, which allowed us to demonstrate that it forms an intricate inverted T-loop motif that mediates RNA–RNA interactions, and guided our design of biochemical experiments to validate the hypothesis. This work also revealed that the region forms the second, requisite contact to tRNA outside of its anticodon (Grigg et al., 2013; Lehmann, Jossinet, & Gautheret, 2013). This was primarily supported by our biochemical validation, since the minimal fragment for robust complex formation between Stem I and tRNA included the anticodon and distal inverted T-loop (Grigg et al., 2013). Establishing the minimal stable structure for tRNA binding enabled us to crystallize and solve the 3.2 Å resolution crystal structure of an 85-nt fragment of the T-box

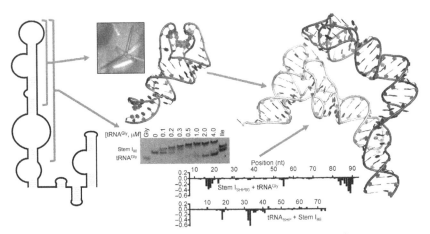

Figure 3 Strategy to determine the T-box–tRNA complex structure. Deletion constructs were generated based on previous publications and secondary structure analysis of the *Geobacillus kaustophilus* GlyQS T-box (left). These constructs allowed structure determination of the distal T-box domain (top middle) and the biochemical identification of key tRNA interactions and the minimum tRNA-binding fragment (bottom middle). The combined efforts led to a robust reconstitution of the T-box–tRNA ternary complex and successful structure determination (right).

riboswitch in complex with a 75 nt tRNA (Grigg & Ke, 2013b), a feat that would not have been possible without using the synergistic approach of rational design with biochemical validation.

2.3 Which Sequence to Choose for Crystallization?

A given riboswitch family may contain thousands of homologs, but as structural analysis is quite expensive, usually only a handful can be chosen for crystallization trials. How does one choose the optimal structural targets? The following criteria have worked well for us and colleagues in the field: (1) choose the well-characterized sequences, as the published biochemical data will provide insight in construct design. (2) Choose the more concise sequences and those with less possibility of forming alternative structures qualitatively estimated by tools like Mfold (Zuker, 2003) and RNAfold (Lorenz et al., 2011). (3) Include some targets from thermophiles, as they generally possess less conformation flexibility at ambient temperature. (4) Diversify your targets by combining candidates from the above selection strategies and carry out a broad crystallization screening approach.

(1) Choosing the well-characterized target. If foundational biochemistry has been done prior to crystallization, these data usually provide valuable insights in helping the design or modification of crystallization constructs. A prime example of this approach is our *B. subtilis yitJ* SAM-I riboswitch structure (Lu et al., 2010). A strong foundation had been built for this riboswitch from genetic and biochemical characterizations (Grundy & Henkin, 1998; McDaniel, Grundy, Artsimovitch, & Henkin, 2003; McDaniel, Grundy, & Henkin, 2005; Tomsic, McDaniel, Grundy, & Henkin, 2008; Winkler, Nahvi, Sudarsan, Barrick, & Breaker, 2003). In this case, we opted to work directly with this sequence by producing the aptamer domain (117 nt) in a largely unmodified form and fortunately, using established strategies, we were able to determine the ~3.0 Å X-ray crystal structure (Lu et al., 2010).

(2) Choosing more concise structures is also a common approach in selecting starting constructs. Even closely related riboswitches can display a large variability in length of helices and loops at the periphery. The S_{MK} box riboswitch family maintains the basic structural core with absolutely conserved SAM-binding sites, but the sequences display enormous diversity in length of insertions at both P3 and P4 varying by anywhere up to 200 nts (Fuchs et al., 2006). The *Enterococcus faecalis* S_{MK} box riboswitch has a short extension at P3 relative to some others, but it also has the

shortest P4 helix, making it one of the more compact homologs and a good starting point for engineering. For this reason, it was chosen as a prime target for subsequent crystallization (Lu et al., 2008).

(3) Riboswitches are also often chosen for their predicted inherent stability. For instance, many structures have been determined for riboswitches from thermophilic bacteria. These organisms grow at elevated temperatures and their genomes require higher G–C contents for stability. These characters are reflected in their riboswitches as well. Several successful examples of this approach include: the SAM-I riboswitch from *Thermoanaerobacter tengcongensis* (Montange & Batey, 2006), the fluoride riboswitch was from *Thermatoga petrophila* (Ren, Rajashankar, & Patel, 2012), the lysine riboswitch from *Thermatoga maritima* (Garst, Porter, & Batey, 2012), and our T-box structures from *G. kaustophilus* (Grigg et al., 2013; Grigg & Ke, 2013b), to name a few.

2.4 Maximize Your Chance of Success by Rational Construct Engineering

Even after selecting the most stable constructs and defining the minimal structures of the riboswitch, the construct will almost always need to be further engineered to enhance its stability, minimize flexible regions, and optimize crystallization. There are numerous strategies to achieve this, including stabilizing the structure by mutating base pairs to G–C, lengthening helices, removing or reducing loops, and adding highly stable turns. Several of these approaches were used in our S_{MK} box riboswitch structures (Fig. 2): in this structure, the first base pair in P1 was swapped with a G–C pair for T7 transcription, P3 was largely replaced by a short run of four G–C base pairs and capped by a GAAA tetraloop, and P4 was shortened and replaced by two G–C pairs and another GAAA tetraloop (Lu et al., 2008). The resulting construct maintains the wild-type core with a highly stabilized periphery.

All forms of nucleic acid crystallography share the same challenge, that structured nucleic acids have a negatively charged surface, which hinders crystal packing. To counter this, a common approach is to introduce crystallization modules to facilitate crystal contacts. Again, care must be taken to maintain the structure and ligand-binding abilities. Some common modifications include introducing GNRA- or UNR-turns into hairpins that not only facilitate a stable turn but also display the Watson–Crick edges of the bases in the turn, available to form base pairs at symmetry contacts, illustrated by both the P3 and P4 helices of our S_{MK} box structure (Fig. 2). Crystal contacts are frequently observed across the base of helices where coaxial stacking leads to the formation of a pseudo-continuous superhelix. This interaction can form across blunt ends, but can often be further stabilized by single- or double-nucleotide overhangs. Therefore, the length and composition of major helices are often systematically screened in crystallization constructs. Ultimately, most successful attempts have utilized some combination of selecting more stable versions of riboswitches and engineering.

2.5 Use of RNA-Binding Proteins as Crystallization Aid

Another crystallization strategy that has been successfully applied to RNA structure determination is to insert protein-binding motifs into RNA and use the bound protein to facilitate crystallization. The U1SnRNP U1A protein binds a specific loop sequence in RNA and the complex structure has been well characterized (Allain, Howe, Neuaus, & Varani, 1997; Oubridge, Ito, Evans, Teo, & Nagai, 1994). The motif has been used to successfully cocrystallize several riboswitch structures, including the c-di-GMP riboswitch (Kulshina, Baird, & Ferre-D'Amare, 2009; Smith et al., 2009; Smith, Lipchock, Livingston, Shanahan, & Strobel, 2010; Smith, Lipchock, & Strobel, 2012), the TPP riboswitch (Kulshina, Edwards, & Ferre-D'Amare, 2010), the glycine riboswitch (Butler, Xiong, Wang, & Strobel, 2011), and an artificial tetracycline aptamer (Xiao, Edwards, & Ferre-D'Amare, 2008). In these cases, the U1A protein is indispensible because it mediates important crystal contacts. The K-turn binding proteins have also been successfully used in crystallization (Hamma & Ferre-D'Amare, 2004). The advantage is that they tightly bind a common motif (the K-turn) in many structured RNAs and not only mediate crystal contacts but also rigidify the K-turn conformation that may otherwise be quite flexible. This strategy was used for crystallization of the SAM-I riboswitch (Baird, Zhang, Hamma, & Ferre-D'Amare, 2012) and for the T-box Stem I–tRNA ternary complex (Zhang & Ferre-D'Amare, 2013, 2014), again with YbxF involved in crystal contacts in both examples. Given the success of these techniques, they will undoubtedly continue to be a prominent strategy for crystallization. Another strategy coming to the horizon has been the use of RNA-binding antibodies to facilitate crystallization (Huang et al., 2014; Shechner et al., 2009). The large size of an antibody creates more opportunity for mediating crystal contacts, and this method also does not introduce nonnative sequences into the RNA prior to crystallization.

3. CRYSTALLIZATION AND STRUCTURE DETERMINATION

Significant advancements have been made for structure determination of nucleic acids by both crystallography and NMR. NMR structures of the preQ1 class I (Kang, Peterson, & Feigon, 2009) and preQ1 class II (Kang, Eichhorn, & Feigon, 2014) riboswitches, and small portions of the T-box (Chang & Nikonowicz, 2013), the Mg^{2+} riboswitch antiterminator (PDB ID: 2IIL) and an engineered neomycin-sensing riboswitch (Duchardt-Ferner et al., 2010) have provided valuable information about their function. The majority of riboswitches and all large riboswitches have been determined using X-ray crystallography, so that will be the focus of this section. Many commercial crystallization screen kits specific to DNA or RNA are now available, so initial screening has become straightforward; however, an important part in crystallizing the modified RNAs is the feedback between crystal trays and constructs. Using iterations between crystallization screening, hit optimization, diffraction analysis, and tweaking rational design approaches listed above, optimized crystals can be obtained. Further optimization for diffraction follows general protocols for dehydration and optimizing cryoprotection. Due to the negative charge and large solvent channels, RNA crystals are particularly sensitive to cations and dehydration and optimizing these variables can significantly affect diffraction (Ben-Shem, Jenner, Yusupova, & Yusupov, 2010; Zhang & Ferre-D'Amare, 2014).

Phasing riboswitch structures still primarily relies on heavy metal soaking or cocrystallization with heavy atoms. Cobalt hexamine is often used as an additive in crystallizations and tends to bind in the groove of helices and loops, behaving similarly to the typical binding of hydrated magnesium. Two heavy atom alternatives that have proven extremely useful have been iridium hexamine and osmium hexamines, either by cobalt hexamine replacement for cocrystallization or by crystal soaking. Tantalum cluster compound is also gaining population in phasing large RNA structures (Ding et al., 2011; Johnson, Reyes, Polaski, & Batey, 2012). In our hands, it has often worked when standard heavy metal compounds failed to yield interpretable solutions and has proven successful at moderate resolutions. Selenium phasing methods have been developed for RNA crystallization and have proven quite successful (Sheng & Huang, 2010), but do not yet share the broad adoption analogous to the use of selenomethionine in proteins crystallography, so soaking methods are still often favored.

Once the structure has been phased, several tools now exist to optimize RNA model building and refinement, but only within the last few years, these have been implemented into common software packages used in crystallography. Two additions that we have found particularly useful are RCrane (Keating & Pyle, 2010, 2012) and ERRASER (Chou, Sripakdeevong, Dibrov, Hermann, & Das, 2013). RCrane was developed to aid in building RNA chains into density or improve current models. It aids with phosphate and bases/nucleoside placement and uses consensus backbone conformer libraries to assist with modeling (Keating & Pyle, 2010, 2012). RCrane has now become widely used and is implemented within the molecular visualization tool, Coot (Emsley, Lohkamp, Scott, & Cowtan, 2010). Another major development in RNA structure refinement is ERRASER (Enumerative Real-space Refinement Assisted by Electron-density under Rosetta) (Chou et al., 2013) that has now been integrated into the PHENIX suite of software (Adams et al., 2010). ERRASER is extremely powerful, using Rosetta RNA structure optimization with a weighted electron density restraint. ERRASER is generally useful throughout resolution ranges, but is particularly powerful for low-resolution structures, where density is often unclear and tight nucleotide geometry restraints are necessary for initial building. It functions by performing successive rounds of identifying geometric problems, rebuilding, and refinement. We have found it particularly useful in our work for correcting initial models at mid- to low resolution (Grigg et al., 2013; Grigg & Ke, 2013b).

4. STRUCTURE–FUNCTION VALIDATION

Even with a riboswitch structure in hand, it is important to validate the model biochemically for several reasons: (1) many riboswitch structures are determined to low or moderate resolutions, leading to some uncertainty in the final model. (2) The crystal structure of a riboswitch represents a single snapshot of a dynamic molecule. (3) Given the limited possibility for crystal contacts, it is not always clear if the crystallized form is a biologically relevant form.

4.1 Mutagenesis Assayed by *In Vivo* and *In Vitro* Experiments

Structures provide the basis to make key mutants to perturb secondary or tertiary structure contacts or RNA–ligand interactions or to test hypothesized switching mechanisms. The methods used to assess riboswitch function using mutagenesis will vary with the system to be tested, but we have generally used a combination of techniques to support structural data. For instance, in our S_{MK} box work, we used a combination of SHAPE and equilibrium dialysis to probe nucleotides involved in SAM binding (Lu et al., 2010, 2011, 2008). In our T-box work, the initial structures revealed a putative docking interface and there, we used electrophoretic mobility shift assays to probe various interface mutants for loss of binding (Grigg et al., 2013; Grigg & Ke, 2013b). Alternatively, the mutagenesis can also be performed on the ligand side of the interaction. Ligand analogs have proven extremely useful in probing the S_{MK} box structure to validate the unique observation that the amino end of the methionine moiety in SAM is not recognized by the S_{MK} box riboswitch, unlike all other known SAM riboswitches (Lu et al., 2008).

4.2 Guide the Design and Interpretation of Single-Molecule Measurement and Molecular Simulations

Single-molecule fluorescence resonance energy transfer (smFRET) studies have proven extremely useful for probing the dynamic switching behaviour of riboswitches (St-Pierre, McCluskey, Shaw, Penedo, & Lafontaine, 2014). Structures enable rational placement of fluorophores for FRET, as well as determining hypothetical FRET changes corresponding to a particular motion. smFRET studies have provided significant insight into the purine (Lemay, Penedo, Mulhbacher, & Lafontaine, 2009; Lemay, Penedo, Tremblay, Lilley, & Lafontaine, 2006), lysine (Fiegland, Garst, Batey, & Nesbitt, 2012), and TPP (Haller, Altman, Souliere, Blanchard, & Micura, 2013) riboswitches. Another particularly powerful single-molecule method in riboswitch study is the use of optical tweezers (Savinov, Perez, & Block, 2014). In these excellent studies of the adenine riboswitch, the folding pathways of the riboswitch can be discerned and recent advancements have enabled direct observation of the cotranscriptional folding process of this riboswitch (Frieda & Block, 2012; Greenleaf, Frieda, Foster, Woodside, & Block, 2008).

4.3 Capturing Ligand-Free and Other Important Functional States

A major hurdle in structural biology of riboswitches is capturing the spectrum of functional states. Since the majority of riboswitch structures represent the ligand bound structure, the structure themselves leave several open questions, including: Does ligand binding induce structural change? Are the changes induced by binding or by conformational selection? Is the switching driven by thermodynamic or kinetic principles? (Price et al., 2014). To begin to answer these deeper mechanistic questions, only a synergistic approach combining structure and biochemistry tools can provide key insights.

5. CONCLUSIONS

The field of riboswitch structural biology has made major strides in the last decade with improved strategies and techniques, making structure determination approachable for these dynamic structures. Future studies will undoubtedly see new riboswitch structures determined as regulatory structured RNAs continue to be revealed as a prominent strategy. Another area lacking major structural characterization has been alternate ligand-free structures and many of the mechanistic details involved in the switch remain unknown.

ACKNOWLEDGMENTS

This work was supported by NIH operating grants GM-086766 and GM-102543 to A.K. J.C.G. was supported by a Postdoctoral Fellowship from the Canadian Institute for Health Research.

REFERENCES

Adams, P. D., Afonine, P. V., Bunkoczi, G., Chen, V. B., Davis, I. W., Echols, N., et al. (2010). PHENIX: A comprehensive Python-based system for macromolecular structure solution. *Acta Crystallographica Section D: Biological Crystallography*, 66(Pt. 2), 213–221. http://dx.doi.org/10.1107/s0907444909052925.

Allain, F. H., Howe, P. W., Neuhaus, D., & Varani, G. (1997). Structural basis of the RNA-binding specificity of human U1A protein. *EMBO Journal*, 16(18), 5764–5772. http://dx.doi.org/10.1093/emboj/16.18.5764.

Baird, N. J., Zhang, J., Hamma, T., & Ferre-D'Amare, A. R. (2012). YbxF and YlxQ are bacterial homologs of L7Ae and bind K-turns but not K-loops. *RNA*, 18(4), 759–770. http://dx.doi.org/10.1261/rna.031518.111.

Ban, N., Nissen, P., Hansen, J., Moore, P. B., & Steitz, T. A. (2000). The complete atomic structure of the large ribosomal subunit at 2.4 A resolution. *Science*, 289(5481), 905–920.

Batey, R. T., Sagar, M. B., & Doudna, J. A. (2001). Structural and energetic analysis of RNA recognition by a universally conserved protein from the signal recognition particle. *Journal of Molecular Biology*, 307(1), 229–246. http://dx.doi.org/10.1006/jmbi.2000.4454.

Ben-Shem, A., Jenner, L., Yusupova, G., & Yusupov, M. (2010). Crystal structure of the eukaryotic ribosome. *Science*, 330(6008), 1203–1209. http://dx.doi.org/10.1126/science.1194294.

Breaker, R. R. (2012). Riboswitches and the RNA world. *Cold Spring Harbor Perspectives in Biology*, 4(2). http://dx.doi.org/10.1101/cshperspect.a003566.

Butler, E. B., Xiong, Y., Wang, J., & Strobel, S. A. (2011). Structural basis of cooperative ligand binding by the glycine riboswitch. *Chemistry & Biology*, 18(3), 293–298. http://dx.doi.org/10.1016/j.chembiol.2011.01.013.

Caron, M. P., Bastet, L., Lussier, A., Simoneau-Roy, M., Masse, E., & Lafontaine, D. A. (2012). Dual-acting riboswitch control of translation initiation and mRNA decay. *Proceedings of the National Academy of Sciences of the United States of America*, 109(50), E3444–3453. http://dx.doi.org/10.1073/pnas.1214024109.

Chang, A. T., & Nikonowicz, E. P. (2013). Solution NMR determination of hydrogen bonding and base pairing between the glyQS T box riboswitch Specifier domain and the anticodon loop of tRNA(Gly.). *FEBS Letters, 587*(21), 3495–3499. http://dx.doi.org/10.1016/j.febslet.2013.09.003.

Cheah, M. T., Wachter, A., Sudarsan, N., & Breaker, R. R. (2007). Control of alternative RNA splicing and gene expression by eukaryotic riboswitches. *Nature, 447*(7143), 497–500. http://dx.doi.org/10.1038/nature05769.

Chou, F. C., Sripakdeevong, P., Dibrov, S. M., Hermann, T., & Das, R. (2013). Correcting pervasive errors in RNA crystallography through enumerative structure prediction. *Nature Methods, 10*(1), 74–76. http://dx.doi.org/10.1038/nmeth.2262.

Ding, F., Lu, C., Zhao, W., Rajashankar, K. R., Anderson, D. L., Jardine, P. J., et al. (2011). Structure and assembly of the essential RNA ring component of a viral DNA packaging motor. *Proceedings of the National Academy of Sciences of the United States of America, 108*(18), 7357–7362. http://dx.doi.org/10.1073/pnas.1016690108.

Duchardt-Ferner, E., Weigand, J. E., Ohlenschlager, O., Schmidtke, S. R., Suess, B., & Wohnert, J. (2010). Highly modular structure and ligand binding by conformational capture in a minimalistic riboswitch. *Angewandte Chemie International Edition in English, 49*(35), 6216–6219. http://dx.doi.org/10.1002/anie.201001339.

Emsley, P., Lohkamp, B., Scott, W. G., & Cowtan, K. (2010). Features and development of Coot. *Acta Crystallographica Section D: Biological Crystallography, 66*(Pt. 4), 486–501. http://dx.doi.org/10.1107/s0907444910007493.

Fiegland, L. R., Garst, A. D., Batey, R. T., & Nesbitt, D. J. (2012). Single-molecule studies of the lysine riboswitch reveal effector-dependent conformational dynamics of the aptamer domain. *Biochemistry, 51*(45), 9223–9233. http://dx.doi.org/10.1021/bi3007753.

Frieda, K. L., & Block, S. M. (2012). Direct observation of cotranscriptional folding in an adenine riboswitch. *Science, 338*(6105), 397–400. http://dx.doi.org/10.1126/science.1225722.

Fuchs, R. T., Grundy, F. J., & Henkin, T. M. (2006). The S(MK) box is a new SAM-binding RNA for translational regulation of SAM synthetase. *Nature Structural and Molecular Biology, 13*(3), 226–233. http://dx.doi.org/10.1038/nsmb1059.

Fuchs, R. T., Grundy, F. J., & Henkin, T. M. (2007). S-adenosylmethionine directly inhibits binding of 30S ribosomal subunits to the SMK box translational riboswitch RNA. *Proceedings of the National Academy of Sciences of the United States of America, 104*(12), 4876–4880. http://dx.doi.org/10.1073/pnas.0609956104.

Garst, A. D., Heroux, A., Rambo, R. P., & Batey, R. T. (2008). Crystal structure of the lysine riboswitch regulatory mRNA element. *Journal of Biological Chemistry, 283*(33), 22347–22351. http://dx.doi.org/10.1074/jbc.C800120200.

Garst, A. D., Porter, E. B., & Batey, R. T. (2012). Insights into the regulatory landscape of the lysine riboswitch. *Journal of Molecular Biology, 423*(1), 17–33. http://dx.doi.org/10.1016/j.jmb.2012.06.038.

Green, N. J., Grundy, F. J., & Henkin, T. M. (2010). The T box mechanism: tRNA as a regulatory molecule. *FEBS Letters, 584*(2), 318–324. http://dx.doi.org/10.1016/j.febslet.2009.11.056.

Greenleaf, W. J., Frieda, K. L., Foster, D. A., Woodside, M. T., & Block, S. M. (2008). Direct observation of hierarchical folding in single riboswitch aptamers. *Science, 319*(5863), 630–633. http://dx.doi.org/10.1126/science.1151298.

Griffiths-Jones, S., Moxon, S., Marshall, M., Khanna, A., Eddy, S. R., & Bateman, A. (2005). Rfam: Annotating non-coding RNAs in complete genomes. *Nucleic Acids Research, 33*(Database issue), D121–124. http://dx.doi.org/10.1093/nar/gki081.

Grigg, J. C., Chen, Y., Grundy, F. J., Henkin, T. M., Pollack, L., & Ke, A. (2013). T box RNA decodes both the information content and geometry of tRNA to affect gene expression. *Proceedings of the National Academy of Sciences of the United States of America, 110*(18), 7240–7245. http://dx.doi.org/10.1073/pnas.1222214110.

Grigg, J. C., & Ke, A. (2013a). Sequence, structure, and stacking: Specifics of tRNA anchoring to the T box riboswitch. *RNA Biology, 10*(12), 1761–1764. http://dx.doi.org/10.4161/rna.26996.

Grigg, J. C., & Ke, A. (2013b). Structural determinants for geometry and information decoding of tRNA by T box leader RNA. *Structure (London, England: 1993), 21*(11), 2025–2032. http://dx.doi.org/10.1016/j.str.2013.09.001. http:/dx/doi.org/10.1016/j.str.2013.1009.1001.

Grundy, F. J., & Henkin, T. M. (1993). tRNA as a positive regulator of transcription antitermination in B. subtilis. *Cell, 74*(3), 475–482.

Grundy, F. J., & Henkin, T. M. (1998). The S box regulon: A new global transcription termination control system for methionine and cysteine biosynthesis genes in gram-positive bacteria. *Molecular Microbiology, 30*(4), 737–749.

Grundy, F. J., Hodil, S. E., Rollins, S. M., & Henkin, T. M. (1997). Specificity of tRNA-mRNA interactions in Bacillus subtilis tyrS antitermination. *Journal of Bacteriology, 179*(8), 2587–2594.

Haller, A., Altman, R. B., Souliere, M. F., Blanchard, S. C., & Micura, R. (2013). Folding and ligand recognition of the TPP riboswitch aptamer at single-molecule resolution. *Proceedings of the National Academy of Sciences of the United States of America, 110*(11), 4188–4193. http://dx.doi.org/10.1073/pnas.1218062110.

Hamma, T., & Ferre-D'Amare, A. R. (2004). Structure of protein L7Ae bound to a K-turn derived from an archaeal box H/ACA sRNA at 1.8 A resolution. *Structure, 12*(5), 893–903. http://dx.doi.org/10.1016/j.str.2004.03.015.

Huang, H., Suslov, N. B., Li, N. S., Shelke, S. A., Evans, M. E., Koldobskaya, Y., et al. (2014). A G-quadruplex-containing RNA activates fluorescence in a GFP-like fluorophore. *Nature Chemical Biology, 10*(8), 686–691. http://dx.doi.org/10.1038/nchembio.1561.

Johnson, J. E., Jr., Reyes, F. E., Polaski, J. T., & Batey, R. T. (2012). B12 cofactors directly stabilize an mRNA regulatory switch. *Nature, 492*(7427), 133–137. http://dx.doi.org/10.1038/nature11607.

Kang, M., Eichhorn, C. D., & Feigon, J. (2014). Structural determinants for ligand capture by a class II preQ1 riboswitch. *Proceedings of the National Academy of Sciences of the United States of America, 111*(6), E663–671. http://dx.doi.org/10.1073/pnas.1400126111.

Kang, M., Peterson, R., & Feigon, J. (2009). Structural Insights into riboswitch control of the biosynthesis of queuosine, a modified nucleotide found in the anticodon of tRNA. *Molecular Cell, 33*(6), 784–790. http://dx.doi.org/10.1016/j.molcel.2009.02.019.

Ke, A., & Doudna, J. A. (2004). Crystallization of RNA and RNA-protein complexes. *Methods, 34*(3), 408–414. http://dx.doi.org/10.1016/j.ymeth.2004.03.027.

Keating, K. S., & Pyle, A. M. (2010). Semiautomated model building for RNA crystallography using a directed rotameric approach. *Proceedings of the National Academy of Sciences of the United States of America, 107*(18), 8177–8182. http://dx.doi.org/10.1073/pnas.0911888107.

Keating, K. S., & Pyle, A. M. (2012). RCrane: Semi-automated RNA model building. *Acta Crystallographica Section D: Biological Crystallography, 68*(Pt. 8), 985–995. http://dx.doi.org/10.1107/s0907444912018549.

Kulshina, N., Baird, N. J., & Ferre-D'Amare, A. R. (2009). Recognition of the bacterial second messenger cyclic diguanylate by its cognate riboswitch. *Nature Structural and Molecular Biology, 16*(12), 1212–1217. http://dx.doi.org/10.1038/nsmb.1701.

Kulshina, N., Edwards, T. E., & Ferre-D'Amare, A. R. (2010). Thermodynamic analysis of ligand binding and ligand binding-induced tertiary structure formation by the thiamine pyrophosphate riboswitch. *RNA, 16*(1), 186–196. http://dx.doi.org/10.1261/rna.1847310.

Lehmann, J., Jossinet, F., & Gautheret, D. (2013). A universal RNA structural motif docking the elbow of tRNA in the ribosome, RNAse P and T-box leaders. *Nucleic Acids Research, 41*(10), 5494–5502. http://dx.doi.org/10.1093/nar/gkt219.

Lemay, J. F., Desnoyers, G., Blouin, S., Heppell, B., Bastet, L., St-Pierre, P., et al. (2011). Comparative study between transcriptionally- and translationally-acting adenine riboswitches reveals key differences in riboswitch regulatory mechanisms. *PLoS Genetics*, 7(1), e1001278. http://dx.doi.org/10.1371/journal.pgen.1001278.

Lemay, J. F., Penedo, J. C., Mulhbacher, J., & Lafontaine, D. A. (2009). Molecular basis of RNA-mediated gene regulation on the adenine riboswitch by single-molecule approaches. *Methods in Molecular Biology*, 540, 65–76. http://dx.doi.org/10.1007/978-1-59745-558-9_6.

Lemay, J. F., Penedo, J. C., Tremblay, R., Lilley, D. M., & Lafontaine, D. A. (2006). Folding of the adenine riboswitch. *Chemistry & Biology*, 13(8), 857–868. http://dx.doi.org/10.1016/j.chembiol.2006.06.010.

Li, S., & Breaker, R. R. (2013). Eukaryotic TPP riboswitch regulation of alternative splicing involving long-distance base pairing. *Nucleic Acids Research*, 41(5), 3022–3031. http://dx.doi.org/10.1093/nar/gkt057.

Lorenz, R., Bernhart, S. H., Honer Zu Siederdissen, C., Tafer, H., Flamm, C., Stadler, P. F., et al. (2011). ViennaRNA Package 2.0. Algorithms. *Molecular Biology*, 6, 26. http://dx.doi.org/10.1186/1748-7188-6-26.

Lu, C., Ding, F., Chowdhury, A., Pradhan, V., Tomsic, J., Holmes, W. M., et al. (2010). SAM recognition and conformational switching mechanism in the Bacillus subtilis yitJ S box/SAM-I riboswitch. *Journal of Molecular Biology*, 404(5), 803–818. http://dx.doi.org/10.1016/j.jmb.2010.09.059.

Lu, C., Smith, A. M., Ding, F., Chowdhury, A., Henkin, T. M., & Ke, A. (2011). Variable sequences outside the SAM-binding core critically influence the conformational dynamics of the SAM-III/SMK box riboswitch. *Journal of Molecular Biology*, 409(5), 786–799. http://dx.doi.org/10.1016/j.jmb.2011.04.039.

Lu, C., Smith, A. M., Fuchs, R. T., Ding, F., Rajashankar, K., Henkin, T. M., et al. (2008). Crystal structures of the SAM-III/S(MK) riboswitch reveal the SAM-dependent translation inhibition mechanism. *Nature Structural and Molecular Biology*, 15(10), 1076–1083. http://dx.doi.org/10.1038/nsmb.1494.

McDaniel, B. A., Grundy, F. J., Artsimovitch, I., & Henkin, T. M. (2003). Transcription termination control of the S box system: Direct measurement of S-adenosylmethionine by the leader RNA. *Proceedings of the National Academy of Sciences of the United States of America*, 100(6), 3083–3088. http://dx.doi.org/10.1073/pnas.0630422100.

McDaniel, B. A., Grundy, F. J., & Henkin, T. M. (2005). A tertiary structural element in S box leader RNAs is required for S-adenosylmethionine-directed transcription termination. *Molecular Microbiology*, 57(4), 1008–1021. http://dx.doi.org/10.1111/j.1365-2958.2005.04740.x.

Mironov, A. S., Gusarov, I., Rafikov, R., Lopez, L. E., Shatalin, K., Kreneva, R. A., et al. (2002). Sensing small molecules by nascent RNA: A mechanism to control transcription in bacteria. *Cell*, 111(5), 747–756.

Montange, R. K., & Batey, R. T. (2006). Structure of the S-adenosylmethionine riboswitch regulatory mRNA element. *Nature*, 441(7097), 1172–1175. http://dx.doi.org/10.1038/nature04819.

Muth, G. W., Ortoleva-Donnelly, L., & Strobel, S. A. (2000). A single adenosine with a neutral pKa in the ribosomal peptidyl transferase center. *Science*, 289(5481), 947–950.

Nahvi, A., Sudarsan, N., Ebert, M. S., Zou, X., Brown, K. L., & Breaker, R. R. (2002). Genetic control by a metabolite binding mRNA. *Chemistry & Biology*, 9(9), 1043.

Nissen, P., Hansen, J., Ban, N., Moore, P. B., & Steitz, T. A. (2000). The structural basis of ribosome activity in peptide bond synthesis. *Science*, 289(5481), 920–930.

Oubridge, C., Ito, N., Evans, P. R., Teo, C. H., & Nagai, K. (1994). Crystal structure at 1.92 A resolution of the RNA-binding domain of the U1A spliceosomal protein complexed with an RNA hairpin. *Nature*, *372*(6505), 432–438. http://dx.doi.org/10.1038/372432a0.

Price, I. R., Grigg, J. C., & Ke, A. (2014). Common themes and differences in SAM recognition among SAM riboswitches. *Biochimica et Biophysica Acta*, *1839*(10), 931–938. http://dx.doi.org/10.1016/j.bbagrm.2014.05.013.

Ren, A., Rajashankar, K. R., & Patel, D. J. (2012). Fluoride ion encapsulation by Mg2+ ions and phosphates in a fluoride riboswitch. *Nature*, *486*(7401), 85–89. http://dx.doi.org/10.1038/nature11152.

Rollins, S. M., Grundy, F. J., & Henkin, T. M. (1997). Analysis of cis-acting sequence and structural elements required for antitermination of the Bacillus subtilis tyrS gene. *Molecular Microbiology*, *25*(2), 411–421.

Savinov, A., Perez, C. F., & Block, S. M. (2014). Single-molecule studies of riboswitch folding. *Biochimica et Biophysica Acta, Protein Structure and Molecular Enzymology*, *1839*(10), 1030–1045. http://dx.doi.org/10.1016/j.bbagrm.2014.04.005.

Serganov, A., & Nudler, E. (2013). A decade of riboswitches. *Cell*, *152*(1–2), 17–24. http://dx.doi.org/10.1016/j.cell.2012.12.024.

Shechner, D. M., Grant, R. A., Bagby, S. C., Koldobskaya, Y., Piccirilli, J. A., & Bartel, D. P. (2009). Crystal structure of the catalytic core of an RNA-polymerase ribozyme. *Science*, *326*(5957), 1271–1275. http://dx.doi.org/10.1126/science.1174676.

Sheng, J., & Huang, Z. (2010). Selenium derivatization of nucleic acids for X-ray crystal-structure and function studies. *Chemistry & Biodiversity*, *7*(4), 753–785. http://dx.doi.org/10.1002/cbdv.200900200.

Smith, K. D., Lipchock, S. V., Ames, T. D., Wang, J., Breaker, R. R., & Strobel, S. A. (2009). Structural basis of ligand binding by a c-di-GMP riboswitch. *Nature Structural and Molecular Biology*, *16*(12), 1218–1223. http://dx.doi.org/10.1038/nsmb.1702.

Smith, K. D., Lipchock, S. V., Livingston, A. L., Shanahan, C. A., & Strobel, S. A. (2010). Structural and biochemical determinants of ligand binding by the c-di-GMP riboswitch. *Biochemistry*, *49*(34), 7351–7359. http://dx.doi.org/10.1021/bi100671e.

Smith, K. D., Lipchock, S. V., & Strobel, S. A. (2012). Structural and biochemical characterization of linear dinucleotide analogues bound to the c-di-GMP-I aptamer. *Biochemistry*, *51*(1), 425–432. http://dx.doi.org/10.1021/bi2016662.

Steen, K. A., Siegfried, N. A., & Weeks, K. M. (2011). Selective 2′-hydroxyl acylation analyzed by protection from exoribonuclease (RNase-detected SHAPE) for direct analysis of covalent adducts and of nucleotide flexibility in RNA. *Nature Protocols*, *6*(11), 1683–1694. http://dx.doi.org/10.1038/nprot.2011.373.

St-Pierre, P., McCluskey, K., Shaw, E., Penedo, J. C., & Lafontaine, D. A. (2014). Fluorescence tools to investigate riboswitch structural dynamics. *Biochimica et Biophysica Acta, Protein Structure and Molecular Enzymology*, *1839*(10), 1005–1019. http://dx.doi.org/10.1016/j.bbagrm.2014.05.015.

Tomsic, J., McDaniel, B. A., Grundy, F. J., & Henkin, T. M. (2008). Natural variability in S-adenosylmethionine (SAM)-dependent riboswitches: S-box elements in bacillus subtilis exhibit differential sensitivity to SAM in vivo and in vitro. *Journal of Bacteriology*, *190*(3), 823–833. http://dx.doi.org/10.1128/jb.01034-07.

Wickiser, J. K., Cheah, M. T., Breaker, R. R., & Crothers, D. M. (2005). The kinetics of ligand binding by an adenine-sensing riboswitch. *Biochemistry*, *44*(40), 13404–13414. http://dx.doi.org/10.1021/bi051008u.

Wickiser, J. K., Winkler, W. C., Breaker, R. R., & Crothers, D. M. (2005). The speed of RNA transcription and metabolite binding kinetics operate an FMN riboswitch. *Molecular Cell*, *18*(1), 49–60. http://dx.doi.org/10.1016/j.molcel.2005.02.032.

Wilkinson, K. A., Merino, E. J., & Weeks, K. M. (2006). Selective 2′-hydroxyl acylation analyzed by primer extension (SHAPE): Quantitative RNA structure analysis at single nucleotide resolution. *Nature Protocols*, *1*(3), 1610–1616. http://dx.doi.org/10.1038/nprot.2006.249.

Winkler, W. C., Cohen-Chalamish, S., & Breaker, R. R. (2002). An mRNA structure that controls gene expression by binding FMN. *Proceedings of the National Academy of Sciences of the United States of America*, *99*(25), 15908–15913. http://dx.doi.org/10.1073/pnas.212628899.

Winkler, W. C., Nahvi, A., Roth, A., Collins, J. A., & Breaker, R. R. (2004). Control of gene expression by a natural metabolite-responsive ribozyme. *Nature*, *428*(6980), 281–286. http://dx.doi.org/10.1038/nature02362.

Winkler, W. C., Nahvi, A., Sudarsan, N., Barrick, J. E., & Breaker, R. R. (2003). An mRNA structure that controls gene expression by binding S-adenosylmethionine. *Nature Structural Biology*, *10*(9), 701–707. http://dx.doi.org/10.1038/nsb967.

Xiao, H., Edwards, T. E., & Ferre-D'Amare, A. R. (2008). Structural basis for specific, high-affinity tetracycline binding by an in vitro evolved aptamer and artificial riboswitch. *Chemistry & Biology*, *15*(10), 1125–1137. http://dx.doi.org/10.1016/j.chembiol.2008.09.004.

Zhang, J., & Ferre-D'Amare, A. R. (2013). Co-crystal structure of a T-box riboswitch stem I domain in complex with its cognate tRNA. *Nature*, *500*(7462), 363–366. http://dx.doi.org/10.1038/nature12440.

Zhang, J., & Ferre-D'Amare, A. R. (2014). Dramatic improvement of crystals of large RNAs by cation replacement and dehydration. *Structure*, *22*(9), 1363–1371. http://dx.doi.org/10.1016/j.str.2014.07.011.

Zuker, M. (2003). Mfold web server for nucleic acid folding and hybridization prediction. *Nucleic Acids Research*, *31*(13), 3406–3415.

SECTION II

RNA-Protein Interactions

CHAPTER NINE

One, Two, Three, Four! How Multiple RRMs Read the Genome Sequence

Tariq Afroz*,[1], Zuzana Cienikova*,[1], Antoine Cléry*,[2], Frédéric H.T. Allain*,[2]

*Institute of Molecular Biology and Biophysics, ETH Zürich, Zürich, Switzerland
[2]Corresponding authors: e-mail address: aclery@mol.biol.ethz.ch; allain@mol.biol.ethz.ch

Contents

1. Introduction — 236
2. Structural Investigations of RRMs: How Many? — 236
3. The RRM Fold and Its Numerous Variations — 237
4. Single RRMs Recognize a Large Repertoire of RNA Sequences: Is There a Recognition Code? — 239
    4.1 Use of elements outside the β-sheet of RRMs to achieve sequence-specificity — 246
    4.2 Noncanonical RRMs: How to bind RNA without the aromatic side-chains of the RNP motifs — 249
5. Increasing Specificity and Affinity by Recognition of Longer RNA Sequences — 251
6. RNA Recognition by Tandem RRMs — 254
7. Assembly of Higher Order Ribonucleoprotein Complexes by Multi RRM–RNA and Multi Protein–RNA Interactions — 258
    7.1 RNA recognition by multi-RRM proteins — 258
    7.2 RRM–RNA interactions in the higher order RNP structures — 259
8. Structural and Genome-Wide Approaches: How do They Match? — 260
9. Conclusions and Perspectives — 267
Acknowledgments — 268
References — 268

Abstract

RRM-containing proteins are involved in most of the RNA metabolism steps. Their functions are closely related to their mode of RNA recognition, which has been studied by structural biologists for more than 20 years. In this chapter, we report on high-resolution structures of single and multi RRM–RNA complexes to explain the numerous strategies used by these domains to interact specifically with a large repertoire of RNA sequences. We show that multiple variations of their canonical fold can be used to adapt to

---

[1] These authors contributed equally to this work.

different single-stranded sequences with a large range of affinities. Furthermore, we describe the consequences on RNA binding of the different structural arrangements found in tandem RRMs and higher order RNPs. Importantly, these structures also reveal with very high accuracy the RNA motifs bound specifically by RRM-containing proteins, which correspond very often to consensus sequences identified with genome-wide approaches. Finally, we show how structural and cellular biology can benefit from each other and pave a way for understanding, defining, and predicting a code of RNA recognition by the RRMs.

1. INTRODUCTION

RNA-recognition motif (RRM), also known as RNA-binding domain (RBD) or ribonucleoprotein (RNP) domain is the most abundant RBD in higher vertebrates (Venter et al., 2001). It is ubiquitously found in all kingdoms of life, including prokaryotes and viruses. RRM-containing proteins are involved in most of the posttranscriptional gene expression processes (such as mRNA and rRNA processing, RNA export, translation, localization, stability, and turnover) and therefore are still a topic of extensive research. Since the identification of RRMs (more than 30 years ago), biochemical and structural studies continue to reveal plasticity in their structures and the versatility of their interactions with single-stranded nucleic acids, proteins (Cléry, Blatter, & Allain, 2008), and most recently with lipids (Clingman et al., 2014). In this chapter, we first report on the important structural discoveries published within the last two decades on RRM–RNA interactions and emphasize their importance for the understanding of the mechanism of action of RRM-containing proteins. Furthermore, we discuss how the molecular knowledge revealed by high-resolution structures correlates to the genomic protein–RNA interaction profiles, paving the way for functional interpretation of RNA regulatory networks.

2. STRUCTURAL INVESTIGATIONS OF RRMs: HOW MANY?

Although the number of structures of protein–RNA complexes solved in the past years has grown exponentially, it still remains difficult to predict a general code of RNA recognition by RRMs (Auweter, Oberstrass, & Allain, 2006). NMR and X-ray crystallography have shown to be two very efficient methods for determination of high-resolution structures (Daubner, Cléry, & Allain, 2013). Since their discovery and the first

RRM structure (Nagai, Oubridge, Jessen, Li, & Evans, 1990), 248 structures of RRMs have been solved in their free form (169 solution and 79 crystal structures) and 70 bound to RNA (38 solution structures and 32 crystal structures). These two methods are complementary and provide a detailed description of the system under investigation (Daubner et al., 2013). In contrast to NMR, X-ray crystallography provides the position of water molecules and therefore allows the prediction of water-mediated hydrogen bonds, which in some cases are important to understand the specificity of protein–RNA interactions (Jenkins, Malkova, & Edwards, 2011). On the other hand, solution NMR has the advantage of studying biomolecules closer to their physiological states. In addition, NMR allows the identification of flexible parts of proteins, typically unobservable in the crystal structures. Furthermore, solution NMR studies allowed to study conformational changes associated with RRM–RNA interactions involving multi-RRM proteins like U2AF (Mackereth et al., 2011) or more recently CPEB (Afroz et al., 2014). Even though X-ray crystallography is still preferentially used to determine structures of large RNA–protein complexes (>30 kDa), their investigation in solution is nevertheless possible using a combination of methods such as paramagnetic relaxation enhancement (Mackereth, Simon, & Sattler, 2005), residual dipolar coupling, and/or electron paramagnetic resonance (EPR) (Duss, Michel, et al., 2014; Duss, Yulikov, Jeschke, & Allain, 2014).

3. THE RRM FOLD AND ITS NUMEROUS VARIATIONS

RRMs are approximately 90 amino acids long and contain a $\beta1–\alpha1–\beta2–\beta3–\alpha2–\beta4$ arrangement of secondary structures, which form a characteristic four-stranded β-sheet packed against the two α-helices (Fig. 1A and B). Using this scaffold, several structural variations have been observed. The loops located between the secondary structure elements (loops 1 to 5, Fig. 1B) can have very different lengths. For example, an unusually long Lys-rich loop2 ("KK" loop) was reported in FUS RRM (Fig. 1C) (Liu et al., 2013). In addition, contrary to loops 1 to 4, which are primarily disordered in their free form, loop5 often forms a short β-hairpin (Fig. 1A). In RRM1 of CPEB1 and CPEB4, this structural element can even be part of the β-sheet extending the surface to six β-strands (Fig. 1D) (Afroz et al., 2014). Furthermore, the length of the α-helices and β-strands can vary as well. U2AF^{35} RRM is a representative example, because it contains an α1-helix three times longer than in canonical RRMs (Fig. 1E) (Kielkopf

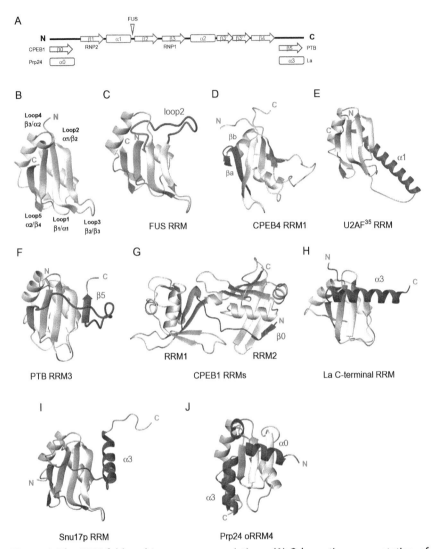

Figure 1 The RRM fold and its numerous variations. (A) Schematic representation of secondary structure elements found in RRMs. Canonical elements are in blue (light gray in the print version) and additional secondary structure elements that can be found in RRMs are in red (dark gray in the print version). (B) hnRNP A1 RRM2 as a model of the typical RRM fold (Ding et al., 1999). (C–J) Structures of single RRMs that contain unusually long or additional secondary structure elements (in red (dark gray in the print version)). (C) FUS RRM (Liu et al., 2013), (D) CPEB4 RRM1 (Afroz et al., 2014), (E) U2AF^{35} RRM (Kielkopf, Rodionova, Green, & Burley, 2001), (F) PTB RRM3 (Oberstrass et al., 2005), (G) CPEB1 RRM1 and RRM2 (Afroz et al., 2014), (H) La C-terminal RRM (Jacks et al., 2003), (I) Snu17p RRM (Wysoczanski et al., 2014), and (J) Prp24 oRRM4 (Martin-Tumasz, Richie, Clos, Brow, & Butcher, 2011).

et al., 2001). Moreover, structural investigations revealed the frequent presence of additional secondary structure elements (Fig. 1A). In PTB RRM2 and 3 (Oberstrass et al., 2005) and hnRNP L RRM3 (Zhang et al., 2013), the C-terminus extends the surface of the β-sheet by forming an additional β5-strand, which is antiparallel to β2 (Fig. 1F) while in CPEB1 RRM2, it is the region N-terminal to the RRM1 that extends the surface (Fig. 1G) (Afroz et al., 2014). Additional α-helices are also frequently found like in La C-terminal RRM (Jacks et al., 2003), U1A N-terminal RRM (Avis et al., 1996) and CstF-64 RRM (Perez-Canadillas & Varani, 2003) where the region C-terminal to the RRM forms an α-helix that lies on the β-sheet surface (Fig. 1H). More unusual, in Snu17p RRM, a C-terminal α-helix folds upon binding a protein partner and lies antiparallel to the β2-strand (Fig. 1I) (Wysoczanski et al., 2014). In the so-called occluded RRM4 (oRRM4) of Prp24 (Martin-Tumasz et al., 2011), both RRM extremities adopt a helical conformation in their free form that masks the β-sheet surface (Fig. 1J). However, contrary to other RRM–RNA complexes for which the additional C-terminal α-helix unfolds (Perez-Canadillas & Varani, 2003) or is displaced (Allain, Howe, Neuhaus, & Varani, 1997; Avis et al., 1996) upon RNA binding, both α-helices of the oRRM still interact with the RRM β-sheet surface even in the presence of RNA (Montemayor et al., 2014). The discovery of additional secondary structure elements at RRM extremities and their functional importance changed our understanding of RRM boundaries (Fig. 1A), which remain crucial when cloning and studying such RRMs in isolation or bound to RNA. After prediction of the characteristic secondary elements β1–α1–β2–β3–α2–β4, it is typically recommended to include 10 to 15 residues on each side of the domain. However, this might not even be sufficient since in the case of Snu17p, 25 additional amino acids would have been needed at the C-terminal extremity of the RRM to include the C-terminal α3-helix (Fig. 1I) (Wysoczanski et al., 2014). A good strategy to predict the presence of long additional regions at the RRM extremities consists if possible in looking into the sequence conservation of these regions.

4. SINGLE RRMs RECOGNIZE A LARGE REPERTOIRE OF RNA SEQUENCES: IS THERE A RECOGNITION CODE?

Typically, RRMs interact with single-stranded RNA targets using their β-sheet surface. A single RRM binds a variable number of nucleotides ranging from a minimum of two for CBP20 (Calero et al., 2002; Mazza,

Segref, Mattaj, & Cusack, 2002) to a maximum of eight for U2B″ (Price, Evans, & Nagai, 1998) (Table 1). The RNA bases are usually spread on the protein surface, while the RNA phosphates point away toward the solvent. The RNA molecule most always runs across the β-sheet from β4 to β2 strand in the 5′ to 3′ direction (Fig. 2A). Although nanomolar affinity can be observed for a few single RRMs, RRMs typically bind their single-stranded RNA targets with K_d in the micromolar range (Table 2). At the primary sequence level, RRMs are characterized by the presence of two consensus sequences: RNP1 (Lys/Arg-Gly-Phe/Tyr-Gly/Ala-Phe/Tyr-Val/Ile/Leu-X-Phe/Tyr) and RNP2 (Ile/Val/Leu-Phe/Tyr-Ile/Val/Leu-X-Asn-Leu) located on the β3 and β1 strands, respectively. The amino acids of these motifs which are exposed on the surface of the β-sheet are crucial for RRM–RNA interactions. Most commonly, three of the aromatic side-chains (RNP2 position 2 and RNP1 positions 3 and 5) accommodate two nucleotides as follows: the bases of the 5′-nucleotide (referred as the N1 position) and of the 3′-nucleotide (referred as the N2 position) stack on the aromatic rings located in position 2 of RNP2 and position 5 of RNP1, respectively (Fig. 2A). The third aromatic ring (RNP1 position 3) is inserted between the two sugar rings of the dinucleotide. Finally, a positively charged side-chain (RNP1 position 1) forms a salt bridge with the phosphate located between the two nucleotides (Fig. 2A). These conserved contacts provide affinity but do not explain the sequence-specificity reported for most RRMs. Also, despite being present in a vast majority of the RRM–RNA complexes, there are some complexes where some of these contacts are missing and some where this canonical binding surface is not used at all (see Section 4.2).

It is remarkable that such small protein domains appear to have the ability to bind a huge diversity of RNA sequences. Every type of nucleotide (A, C, G, or U) can be found in the five main binding pockets of RRMs (Table 1) although some clear nucleotide preferences can be observed (Table 1). It was previously proposed that the N1 and N2 RRM pockets would be readily shaped for a C or A and a G with a *syn* conformation, respectively, since four intermolecular hydrogen bonds can be formed with only the main-chain of two residues after the end of the β4-strand (Fig. 2B) (Auweter, Oberstrass, et al., 2006). In good agreement with this statement, cytosines and adenines are enriched at position N1 (as 52% of the bound Cs and 35% of the bound As are found in N1) and almost 30% of the bound Gs are found at position N2 in a *syn* conformation. More surprisingly, it appears from the analysis of Table 1 that

Table 1 Register of the RNA or DNA sequences in complex structures of RRM domain-containing proteins

Position on the RRM domain	N_{-3}	N_{-2}	N_{-1}	N_0	N_1	N_2	N_3	N_4	N_5
Musashi1 (Ohyama et al., 2012)			G	U	A	G*	U		
hnRNP G (Moursy, Allain, & Clery, 2014)			U	C	A*	A	A		
SUP-12 (Amrane, Rebora, Zniber, Dupuy, & Mackereth, 2014; Kuwasako et al., 2014)		G	U	G*	U	G	C		
SRSF2 (Daubner, Cléry, Jayne, Stevenin, & Allain, 2012)				U	C	C	A	G	U
SRSF2 (Daubner et al., 2012)				U	G*	G*	A	G	U
DAZL (Jenkins et al., 2011)		U	U	G	U	U	C	U	U
Nab3 (Lunde, Horner, & Meinhart, 2011)				U	C	U	U		
Tra2-β1 (Cléry et al., 2011; Tsuda et al., 2011)			A	A	G	A* A	C		
Rna15 (Leeper, Qu, Lu, Moore, & Varani, 2010)				A	A	U	A*	A*	
Prp24 RRM2 (Montemayor et al., 2014)		U	A*	C	A	G*	A*	G	
Prp24 RRM3 (Montemayor et al., 2014)				A	C	A	A*	A*	
CUGBP1 RRM1/2 (Teplova, Song, Gaw, Teplov, & Patel, 2010)				U G*	U	U			
CUGBP1 RRM3 (Tsuda et al., 2009)		U	G	U	G	U	G		
RBMY (Skrisovska et al., 2007)				C	A*	A			
SRSF3 (SRp20) (Hargous et al., 2006)					C	A*	U	C	
hnRNP C (Cienikova, Damberger, Hall, Allain, & Maris, 2014)				U	U	U	U		
U1A (Allain et al., 1997; Oubridge, Ito, Evans, Teo, & Nagai, 1994)	A	U	U	G	C	A	C		
Sxl RRM1 (Handa et al., 1999)	U	U	U	U	U				
Sxl RRM2 (Handa et al., 1999)				U	G	U			
PABP RRM1 (Deo, Bonanno, Sonenberg, & Burley, 1999)				A	A	A A			
PABP RRM2 (Deo et al., 1999)				A	A	A	A		
U2B" (Price et al., 1998)	A	U	U	G	C	A	G*	U	
hnRNP A1 RRM1 (Ding et al., 1999)				T	A	G*	G		
hnRNP A1 RRM2 (Ding et al., 1999)			T	T	A	G*	G		
Nucleolin RRM1 (Allain, Bouvet, Dieckmann, & Feigon, 2000; Johansson et al., 2004)					C	G*	A*		

Continued

Table 1 Register of the RNA or DNA sequences in complex structures of RRM domain-containing proteins—cont'd

Position on the RRM domain	N₋₃	N₋₂	N₋₁	N₀	N₁	N₂	N₃	N₄	N₅
Nucleolin RRM2 (Allain, Bouvet, et al., 2000; Johansson et al., 2004)				U		C	C		
HuD RRM1 (Wang & Tanaka Hall, 2001)	U	U	A*	U	U	U			
HuD RRM2 (Wang & Tanaka Hall, 2001)					U	U			
HuD RRM2 (Wang & Tanaka Hall, 2001)				U	A	U			
HuC RRM1 (Inoue, Hirao, et al., 2000)				U	U	U			
HuC RRM2 (Inoue, Hirao, et al., 2000)					A	U			
CBP20 (Calero et al., 2002; Mazza et al., 2002)					G		N		
PTB RRM1 (Oberstrass et al., 2005)				U	C		U		
PTB RRM2 (Oberstrass et al., 2005)					C		U		
PTB RRM3 (Oberstrass et al., 2005)				U	C		U		
PTB RRM4 (Oberstrass et al., 2005)				U	C		N		
Fox-1 (Auweter, Fasan, et al., 2006)	U	G	C	A	U	G*	U		
hnRNP D (Enokizono et al., 2005)				T	A	G*	G		
Hrp1 RRM1 (Perez-Canadillas, 2006)			A	U	A	U			
Hrp1 RRM2 (Perez-Canadillas, 2006)				U	A	U			
U2AF⁶⁵ RRM1 (Sickmier et al., 2006)				U	U	U	U		
U2AF⁶⁵ RRM2 (Sickmier et al., 2006)			U	U	U	U			
HuR RRM1 (Wang et al., 2013)		U	U	A	U	U	U		
HuR RRM2 (Wang et al., 2013)					U	U			
TDP-43 RRM1 (Lukavsky et al., 2013)		G	U	G*	U	G			
TDP-43 RRM2 (Lukavsky et al., 2013)			A	A	U	G*			
CPEB4 RRM1 (Afroz et al., 2014)			C	U	U	U			
CPEB4 RRM2 (Afroz et al., 2014)					A				
			19% A	22% A	31% A	24% A	17% A		
			8% C	7% C	23% C	5% C	17% C		
			8% G	15% G	12.5% G	26% G	31% G		
			65% U	56% U	33.5% U	45% U	35% U		

The star indicates that the nucleotide adopts a *syn* conformation.

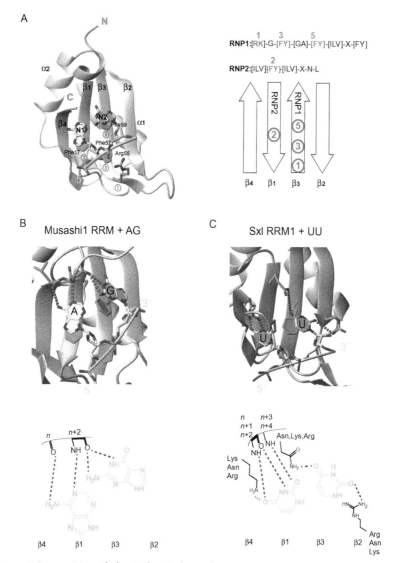

Figure 2 Recognition of 5′-AG-3′ and 5′-UU-3′ dinucleotides at positions N1 and N2 of RRMs. (A) Structure of hnRNP A1 RRM2 bound to 5′-AG-3′ dinucleotide as a model of single-stranded nucleic acid binding (Ding et al., 1999). (B) Sequence-specific contacts involved in the recognition of the 5′-AG-3′ dinucleotide by Musashi1 RRM (32). (C) Sequence-specific contacts involved in the recognition of the 5′-UU-3′ dinucleotide by Sxl RRM1 (Handa et al., 1999). Side chains and hydrogen bonds (dashed lines) that were previously found to be involved in the UU dinucleotide recognition (Afroz et al., 2014; Cienikova et al., 2014; Jenkins et al., 2011; Sickmier et al., 2006; Teplova et al., 2010; Wang et al., 2013; Wang & Tanaka Hall, 2001) are schematically shown in the lower part of each panel. One hydrogen bond that is not observed in the Sxl–RNA complex but found in other RRM–RNA complexes (Afroz et al., 2014; Cienikova et al., 2014; Jenkins et al., 2011; Sickmier et al., 2006; Teplova et al., 2010; Wang et al., 2013; Wang & Tanaka Hall, 2001) is shown in blue (light gray in the print version) instead of red (dark gray in the print version).

Table 2 Dissociation constant (K_d) of RRM–RNA complexes

Protein	Complex	Method	K_d (M)
Single RRM–RNA complexes			
U1A	RRM1 + 23 nt U1 hairpin II RNA (Law, Rice, Lin, & Laird-Offringa, 2006)	SPR	$(3.4 \pm 0.2) \times 10^{-11}$
Fox-1	RRM + 5′-CUCUGCAUGU-3′ (Auweter, Fasan, et al., 2006)	SPR	0.49×10^{-9}
SRSF3	RRM + 5′-CAUC-3′ (Hargous et al., 2006)	NMR	18×10^{-6}
RBMY	RRM + 21 nt stem loop RNA (Skrisovska et al., 2007)	EMSA	1×10^{-9}
YxiN (Dead box helicase)	RRM + 72 nt segment of 23S rRNA fragment (Hardin, Hu, & McKay, 2010)	EMSA	0.054×10^{-9}
Tra2-β1	RRM + 5′-AAGAAC-3′ (Cléry et al., 2011)	ITC	$(2.25 \pm 0.05) \times 10^{-6}$
Nab3	RRM + 5′-UCUU-3′ (Lunde et al., 2011)	FA	$70 (+20/-10) \times 10^{-6}$
TAP	RRM and LRR (NTD) + 62 nt segment (hCTE) (Teplova, Wohlbold, Khin, Izaurralde, & Patel, 2011)	ITC	$(90 \pm 20) \times 10^{-6}$
DAZL	RRM + 5′-UUGUUCUU-3′ (Jenkins et al., 2011)	FA	38×10^{-9}
SRSF2	RRM + 5′-UCCAGU-3′ (Daubner et al., 2012)	ITC	$(0.27 \pm 0.02) \times 10^{-6}$
p65	RRM + 45 nt (stem IV TER RNA) (Singh et al., 2012)	ITC	$(36.4 \pm 3.6) \times 10^{-9}$
hnRNP G	RRM + 5′-AUCAAA-3′ (Moursy et al., 2014)	ITC	18×10^{-6}
SRSF1	Pseudo-RRM + 5′-UGAAUUAC-3′ (Cléry et al., 2013)	ITC	0.8×10^{-6}
hnRNP C	RRM + 5′-AUUUUUC-3′ (Cienikova et al., 2014)	ITC	$(2.0 \pm 0.3) \times 10^{-6}$
Hera (Dead box helicase)	RRM + 5′-GGGC-3′ (Steimer et al., 2013)	FA	$(0.13 \pm 0.7) \times 10^{-6}$
hnRNP F	qRRM1 + 5′-AGGGAU-3′ (Dominguez, Fisette, Chabot, & Allain, 2010)	ITC	0.4×10^{-6}
	qRRM1 + 5′-AGGGAU-3′ (Dominguez et al., 2010)	ITC	4.6×10^{-6}
	qRRM1 + 5′-AGGGAU-3′ (Dominguez et al., 2010)	ITC	0.4×10^{-6}
Multi RRM–RNA complexes			
Sxl	RRM1 + 20 nt (U)$_4$GUUG(U)$_8$CUAG (Kanaar, Lee, Rudner, Wemmer, & Rio, 1995)	EMSA	4.8×10^{-8}
	RRM2 + 20 nt (U)$_4$GUUG(U)$_8$CUAG (Kanaar et al., 1995)	EMSA	4.0×10^{-6}
	RRM12 + 20 nt (U)$_4$GUUG(U)$_8$CUAG (Kanaar et al., 1995)	EMSA	6.7×10^{-11}
CUGBP1	RRM1 + 5′-UUGUU-3′ (Teplova et al., 2010)	ITC	$(29 \pm 0.3) \times 10^{-6}$
	RRM2 + 5′-UUGUU-3′ (Teplova et al., 2010)	ITC	$(45 \pm 5) \times 10^{-6}$
	RRM12 + 5′-GUUGUUUUGUUU-3′ (Teplova et al., 2010)	ITC	$(0.65 \pm 0.07) \times 10^{-6}$
	RRM3 + (UG)$_3$ (Tsuda et al., 2009)	ITC	1.9×10^{-6}
	Full-length protein + (UG)$_{15}$ (Mori, Sasagawa, Kino, & Ishiura, 2008)	SPR	$(0.25 \pm 0.1) \times 10^{-6}$

Table 2 Dissociation constant (K_d) of RRM–RNA complexes—cont'd

Protein	Complex	Method	K_d (M)
Nucleolin	RRM12 + 22 nt SNRE RNA (Allain, Gilbert, Bouvet, & Feigon, 2000)	NMR	$(10-100) \times 10^{-9}$
PTB	RRM1 + 5'-CUCUCU-3' (Auweter, Oberstrass, & Allain, 2007)	NMR	$(3.2 \pm 1.8) \times 10^{-6}$
	RRM2 + 5'-CUCUCU-3' (Auweter et al., 2007)	NMR	$(0.93 \pm 0.29) \times 10^{-6}$
	RRM34, mutation in RRM3 + 5'-CUCUCU-3' (Auweter et al., 2007)	NMR	$(2.1 \pm 1.3) \times 10^{-6}$
	RRM34 + (poly-pyrimidine tract)$_{15}$ (Lamichhane et al., 2010)	FRET	$(11 \pm 5) \times 10^{-9}$
U2AF^{65}	RRM1 + (U)$_9$ (Mackereth et al., 2011)	ITC	800×10^{-6}
	RRM2 + (U)$_9$ (Mackereth et al., 2011)	ITC	$(7 \pm 0.1) \times 10^{-6}$
	RRM12 + (U)$_9$ (Mackereth et al., 2011)	ITC	$(1.32 \pm 0.04) \times 10^{-6}$
HuC	RRM1 + 5'-AUUUA-3' (Inoue, Muto, et al., 2000)	NMR	40×10^{-6}
	RRM2 (Inoue, Muto, et al., 2000)		n.d
	RRM12 (Inoue, Muto, et al., 2000)		n.d
HuD	RRM1 + 38 nt (UUAU)$_4$U (Park, Myszka, Yu, Littler, & Laird-Offringa, 2000)	EMSA	$>1 \times 10^{-4}$
	RRM2 + 38 nt (UUAU)$_4$U (Park et al., 2000)	EMSA	$>1 \times 10^{-3}$
	RRM12 + 38 nt (UUAU)$_4$U (Park et al., 2000)	EMSA	$(3.9 \pm 0.2) \times 10^{-8}$
Prp24	RRM2 + 5'-AGAGAU-3' (Martin-Tumasz, Reiter, Brow, & Butcher, 2010)	NMR	$(90 \pm 10) \times 10^{-6}$
	RRM23 + 5'-AGAGAU-3' (Martin-Tumasz et al., 2010)	NMR	$(30 \pm 10) \times 10^{-6}$
TDP-43	RRM1 + (UG)$_6$ (Kuo, Doudeva, Wang, Shen, & Yuan, 2009)	NFBS	$(6.5 \pm 0.2) \times 10^{-8}$
	RRM2 + (UG)$_6$ (Kuo et al., 2009)	NFBS	$(4.2 \pm 0.4) \times 10^{-6}$
	RRM12 + (UG)$_6$ (Kuo et al., 2009)	NFBS	$(1.4 \pm 0.1) \times 10^{-8}$
Ternary RRM–RNA complexes			
ASD-1, SUP-12	SUP-12 RRM1 + 12 nt RNA (Kuwasako et al., 2014)	ITC	$(5.65 \pm 0.87) \times 10^{-7}$
	ASD-1 RRM1 + 12 nt RNA + SUP-12 RRM1 (Kuwasako et al., 2014)	ITC	$(4.24 \pm 0.34) \times 10^{-8}$
Sxl, UNR	UNR CSD1 + 5'-GAGCAC (Hennig et al., 2014)	ITC	$(1.4 \pm 0.8) \times 10^{-6}$
	Sxl RRM12 + 5'-(U)$_7$GAGCACGUGAA-3' (Hennig et al., 2014)	ITC	$(200 \pm 30) \times 10^{-9}$
	UNR CSD1 + Sxl RRM12 + 5'-(U)$_7$GAGCACGUGAA-3' (Hennig et al., 2014)	ITC	$(15 \pm 5) \times 10^{-9}$

SPR, surface plasmon resonance; NMR, nuclear magnetic resonance; FA, fluorescence anisotropy; ITC, isothermal titration calorimetry; EMSA, electrophoretic mobility shift assay; NFBA, nitrocellulose filter binding assay.

uracil is the most frequently found nucleotide that is sequence-specifically recognized by RRMs (based of the structures already solved). Uracils can be recognized sequence-specifically in all nucleotide-binding pockets found in RRMs, even the N1 and N2 pockets. Indeed, recognition of a dinucleotide UU at these positions is found quite frequently and present common features among the different structures (Afroz et al., 2014; Cienikova et al., 2014; Jenkins et al., 2011; Sickmier et al., 2006; Teplova et al., 2010; Wang et al., 2013; Wang & Tanaka Hall, 2001). Typically, the 5′-uracils in N1 form one hydrogen bond with a side-chain (Lys, Asn, or Arg) present on the β4-strand and one to three hydrogen bonds with the main-chains of residues up to four residues after β4. The 3′-uracils in N2 are not recognized by the main-chain but via two side-chains located on the β1 and β2 strands, which contact their carbonyl oxygen O4 and O2, respectively (Fig. 2C). Finally, despite some clear sequence bias, RRMs have shown the capacity to bind almost any dinucleotide sequences using their canonical binding platform, although cytosines are rare except in the N1 position.

Importantly, the unusual versatility of RNA recognition observed with RRMs is also due to the formation of additional contacts involving supplementary secondary structure elements, loops, N- and C-terminal regions of the domain. Here also, it implies the formation of hydrogen bonds between atoms of RNA bases and main-chains and side-chains of RRMs. In Section 4, we report on the different strategies used by isolated RRMs to accommodate different RNA sequences. Subsequently, we discuss the various modes of RNA recognition by multi-RRM proteins, which add an additional layer of diversity in protein–RNA interaction by RRM proteins.

4.1 Use of elements outside the β-sheet of RRMs to achieve sequence-specificity

We mentioned above the pronounced plasticity of RRM structures in terms of sequence-specificity and protein folds. The presence of additional secondary structure elements is largely used by RRMs to interact with longer RNA sequences in a sequence-specific fashion. The fifth β-strand present in RRM2 and RRM3 of PTB results in an extension of the β-sheet, which allows the binding of one (RRM2) or two (RRM3) additional nucleotides (Fig. 3A) (Oberstrass et al., 2005). Similarly, RRM1 of CPEB1 and CPEB4 harbors two additional antiparallel β-strands (β^a and β^b)

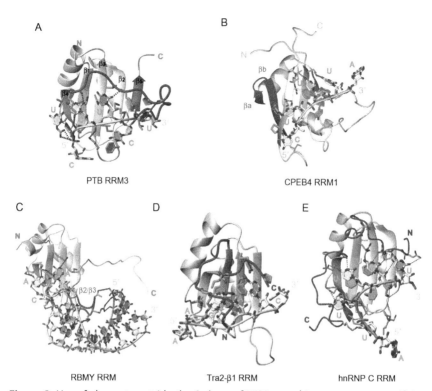

Figure 3 Use of elements outside the β-sheet of RRMs to achieve sequence-specificity. (A) Structure of PTB RRM3 in complex with the 5′-CUCUCU-3′ RNA (Oberstrass et al., 2005). The additional β5-strand, which is involved in the RRM–RNA interaction, is shown in red. (B) Solution structure of CPEB4 RRMs bound to 5′-CUUUA-3′ (Afroz et al., 2014). For clarity, only RRM1 domain is shown. The insertion of β-strands is shown in red. (C) Structure of RBMY RRM in complex with a stem-loop RNA capped by a 5′-CACAA-3′ pentaloop (Skrisovska et al., 2007). The β2/β3 loop that binds the RNA stem is in red. (D) Solution structure of Tra2-β1 RRM bound to 5′-AAGAAC-3′ RNA (Cléry et al., 2011). The N- and C-terminal extremities of the RRM are in red. (E) Solution structure of hnRNP C RRM bound to 5′-AUUUUU-3′ RNA (Cienikova et al., 2014) (PDB ID: 2MXY). The N- and C-terminal extremities of the RRM are in red. In all the figures, the ribbon of the RRM is shown in gray, the RNA nucleotides are in yellow, and the protein side-chains in green. The N, O, and P atoms are in blue, red, and orange, respectively. The N- and C-terminal extensions of the RRM and 5′- and 3′-end of RNA are indicated. Hydrogen bonds are represented by purple dashed lines. All the figures were generated by the program MOLMOL (Koradi, Billeter, & Wuthrich, 1996). (See the color plate.)

between the α2-helix and the β4 strand, extending the β-sheet surface. The β^b strand exposes a conserved aromatic residue, which makes sequence-specific hydrogen bonds with the U_2 (Fig. 3B). In the RRM of RBMY, several residues located in the β-sheet specifically interact with a CAA triplet, whereas the β2–β3 loop is crucial for the recognition of the major groove of the RNA stem-loop (Skrisovska et al., 2007) (Fig. 3C). Finally, regions N- and C-terminal of RRMs are also almost always involved in RNA-sequence recognition, acting like a clamp that sandwiches the RNA molecule lying on the β-sheet surface. In addition to stabilize the RNA molecule, they provide a large set of additional side-chain and main-chain contacts that are used to increase the repertoire of sequences recognized by RRMs. The N-terminus (CUGBP1 (Tsuda et al., 2009), CBP20 (Mazza et al., 2002)), the C-terminus (PTB (Oberstrass et al., 2005), PABP (Deo et al., 1999), RBMY (Skrisovska et al., 2007), DAZL (Jenkins et al., 2011), SUP-12 (Amrane et al., 2014; Kuwasako et al., 2014), hnRNP G (Moursy et al., 2014), and Musashi1 (Ohyama et al., 2012)) or both extremities (Tra2-β1 (Cléry et al., 2011; Tsuda et al., 2011), SRSF2 (Daubner et al., 2012), and hnRNP C (Cienikova et al., 2014)) were used for this purpose in the respective complexes. Importantly, these domain extensions do not need to adopt a secondary structure to interact with RNA and the number of residues involved in RNA binding can be up to 10 amino acids as shown for CUGBP1 RRM3 N-terminus (Tsuda et al., 2009). In the case of Tra2-β1 and hnRNP C proteins, both the N- and C-terminal extensions participate in RNA binding. They cross each other in Tra2-β1 complex (Fig. 3D) (Cléry et al., 2011; Tsuda et al., 2011), whereas they are perpendicular in hnRNP C RRM bound to RNA (Fig. 3E) (Cienikova et al., 2014). Except for the pre-positioned extremities of PTB RRMs and CUGBP1 RRM3, N- and C-terminal parts that are involved in RNA binding are flexible in the RRM free form and become rigid upon RNA binding. In addition to revealing the RNA-binding modes of RRMs, these structural investigations allowed the characterization of residues that are crucial for RRM–RNA interactions and the identification of RNA sequences that are recognized by these domains. As described in Section 8, this information has been essential to identify natural binding sites of these proteins and understand their mode of action *in vivo*. For example, it was proposed for Tra2-β1 that the RNA-induced positioning of both extremities of the domain was functionally important (Cléry et al., 2011). It could explain the recruitment by Tra2-β1 of two additional proteins, hnRNP G and SRSF9, which together regulate SMN exon 7 splicing (Cléry et al., 2011; Hofmann & Wirth, 2002; Young et al., 2002).

4.2 Noncanonical RRMs: How to bind RNA without the aromatic side-chains of the RNP motifs

4.2.1 The xRRM binds 5′-GA-3′

This domain was initially found in the Tetrahymena p65 protein, a telomerase holoenzyme essential for telomerase RNA accumulation *in vivo* (Singh et al., 2012). Structural investigations revealed the presence of an additional C-terminal α3-helix, which masks the β-sheet surface. However, contrary to Prp24 oRRM, two conserved residues present in the β2 strand (Tyr-X-Asp referred as a new RNP3 sequence) are still accessible to solvent and participate in single-stranded RNA binding (Fig. 4A) (Singh, Choi, & Feigon, 2013; Singh et al., 2012). The structure of p65 xRRM2 bound to RNA shows that a GA bulge is extruded out of the RNA helix and is specifically recognized by these residues (Singh et al., 2012). The guanine base stacks on Tyr407, whereas Asp409 side-chain forms hydrogen bonds with the Watson–Crick edge of the base (Fig. 4A). In addition, a conserved Arg465 in β3 interacts with the Hoogsteen edges of the guanine and the adenine bases (Fig. 4A). Interestingly, the C-terminal α3-helix is extended upon RNA binding and inserts into the major groove of the telomerase RNA stem IV leading to a 105° bend of the RNA backbone (Singh et al., 2013; Singh et al., 2012). This conformational change induced by p65 xRRM2 is required for the recruitment of the telomerase reverse transcriptase and explains why this domain is sufficient to direct hierarchical assembly of the catalytic core *in vitro* (Singh et al., 2012).

4.2.2 The quasi-RRM binds sequence-specifically 5′-GGG-3′

This domain was first identified in the hnRNP F protein (Dominguez & Allain, 2006). Due to the lack of conserved aromatic residues in RNP1 and RNP2, the three RRMs of hnRNP F have been renamed as quasi-RRMs (qRRMs) (Dominguez & Allain, 2006). Structures of the three hnRNP F qRRMs in free and RNA-bound states were solved by NMR in our lab and revealed that they all specifically interact with a G-tract in an identical and very unusual manner (Dominguez et al., 2010). Indeed, the β1/α1, β2/β3, and α2/β4 loops form the main surface of interaction instead of the β-sheet. The three guanines adopt a compact conformation surrounded by three conserved residues (two aromatics and one arginine) located in these loops (Fig. 4B) (Dominguez et al., 2010). These side-chains stack on each guanine base encaging the G-tract. This mode of binding appears conserved among qRRMs, because amino acids involved in

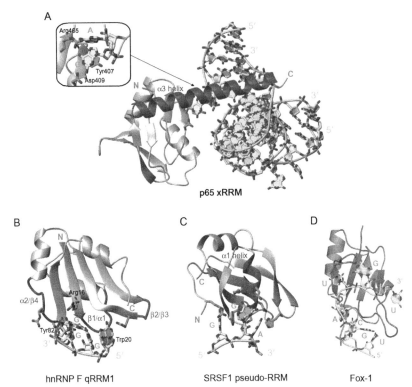

Figure 4 Structures of single RRMs using an unusual mode of interaction with RNA. (A) Crystal structure of p65 xRRM bound to the telomerase RNA stem IV (Singh et al., 2012). The α3-helix is shown in red. (B) Solution structure of the hnRNP F qRRM1 bound to 5′-AGGGAU-3′ RNA (Dominguez et al., 2010). Loops involved in RNA binding are in red. (C) Solution structure of SRSF1 pseudo-RRM bound to RNA (Cléry et al., 2013). The α1-helix, which is primarily involved in RNA binding, is in red. (D) Structure of Fox-1 RRM in complex with RNA (Auweter, Fasan, et al., 2006). For all figures, the color code is the same as in Fig. 3. (See the color plate.)

sequence-specific recognition of this G-repeat sequence (R-G-L-P-[W/F/Y] in loop1 and [R/K]-[X5]-R-Y-[V/I/L]-E-[V/I/L]-F in loop5) are conserved among identified qRRMs (Dominguez et al., 2010). Based on these structures, it was proposed that hnRNP F could regulate splicing by maintaining G-rich sequences in a single-stranded conformation, preventing RNA to form secondary structures (Dominguez et al., 2010).

4.2.3 The pseudo-RRM binds 5′-GGA-3′

Unlike canonical RRMs, pseudo-RRMs (ΨRRMs) have a conserved distinctive heptapeptide (SWQDLKD) on their α1-helix (Birney, Kumar, &

Krainer, 1993). In addition, their β-sheet surface lacks the set of conserved aromatic residues usually involved in RNA binding suggesting that this domain might use an unusual mode of RNA recognition (Tintaru et al., 2007). We recently solved by NMR the first structure of this RRM type (the ΨRRM of the SR protein SRSF1) bound to RNA and indeed, it revealed a completely unexpected mode of interaction (Cléry et al., 2013). ΨRRMs do not use the β-sheet or loops to interact with RNA, but the side of the domain where α1-helix packs against β2 (Fig. 4C). This interface was previously reported as interacting with the kinase SRPK1, which is responsible for the phosphorylation of the C-terminal RS domain of SRSF1. This structural information indicated that the protein could not bind RNA and be phosphorylated at the same time. Importantly, this surface is primarily negatively charged and was not predicted as a RNA-binding surface. Remarkably, all residues of the SWQDLKD conserved motif (except the leucine that anchors α1-helix into the hydrophobic core) are involved in the recognition of a GGA motif (Fig. 4C) (Cléry et al., 2013). This explains their strong conservation and would predict that all ΨRRMs use a similar mode of interaction (Cléry et al., 2013). Using the structure to design a series of mutants that specifically affect SRSF1 ΨRRM interaction with RNA, we showed that the isolated domain can regulate alternative splicing events in cells by competing with the recruitment of the splicing repressor hnRNP A1 on overlapping binding sites (Cléry et al., 2013). Interestingly, the structures of two bacterial RRMs were also recently shown to interact with RNA using the α1-helix (Hardin et al., 2010; Steimer et al., 2013). Although these RRMs are not defined as ΨRRMs, because they lack the SWQDLKD motif, this suggests that this unusual mode of interaction for an RRM might originate from a common RRM precursor.

Altogether, structures of these nonanonical RRMs bound to RNA showed that RRMs can bind RNA sequence-specifically outside the canonical β-sheet surface. Importantly, their structure determination revealed a sequence motif corresponding to conserved residues directly involved in the RNA-binding specificity. Therefore, by identifying proteins containing such motifs, one can predict the sequence they recognize, i.e., GA, GGG, and GGA for xRRMs, qRRM, and ΨRRM, respectively.

5. INCREASING SPECIFICITY AND AFFINITY BY RECOGNITION OF LONGER RNA SEQUENCES

As discussed in Section 4, single RRMs can employ diverse modes of binding for sequence-specific recognition of short RNA motifs. Although some single RRMs can bind seven to eight nucleotides with nanomolar affinity, most RRMs recognize two to four nucleotides with a low micromolar affinity (Table 2). Therefore, it is difficult to explain how RRMs can efficiently find their RNA targets *in vivo*. Hence, RRM-containing proteins developed different strategies to achieve this goal. The simplest way to increase the RNA-binding specificity and affinity of a protein is to bind the longest possible RNA sequence. As mentioned above, this is observed in the case of alternative splicing regulator Fox-1, whose RRM employs an unusually large surface of interaction to bind a longer RNA sequence (seven nucleotides) (Fig. 4D) (Auweter, Fasan, et al., 2006). In the corresponding structure, the last three nucleotides are bound in a canonical fashion by the β-sheet surface of the RRM, while the first four nucleotides are bound by two loops (loop1 and loop3) in a manner reminiscent of the way qRRMs bind RNA. This extended interaction of Fox-1 RRM with RNA results in a subnanomolar RNA-binding affinity (K_d of 0.49 nM, Table 2) (Auweter, Fasan, et al., 2006).

A second strategy adopted by RRM proteins to accommodate longer RNA sequences consists in employing several RRMs. Indeed, most RRM-containing proteins harbor multiple juxtaposed RRMs, which often have high sequence similarity suggesting that they resulted from gene duplication events during evolution. As a consequence, each RRM copy tends to bind a similar sequence explaining why repeat sequences are often targets of proteins containing multiple RRMs. The majority of the RNA-recognition principles of multi-RRM-containing proteins have been elucidated from the structures of tandem RRM–RNA complexes, where specific arrangement of the two RRMs provides an extended RNA-binding surface. This results in the sequence-specific recognition of longer RNA sequences compared to single RRMs. Structures of tandem RRMs characterized in their free states show extensive variations. In most cases, a flexible interdomain linker separates the two RRMs and no inter-RRM interactions are present as seen for Sxl, Nucleolin, PTB RRM12, Hrp1, Npl3, and HuR (Allain, Gilbert, Bouvet, & Feigon, 2000; Crowder, Kanaar, Rio, & Alber, 1999; Perez-Canadillas, 2006; Skrisovska & Allain, 2008; Vitali et al., 2006; Wang et al., 2013). In contrast, a few tandem RRMs show stable interactions between the RRMs in the free state as shown for PTB, nPTB, Prp24, FIR, U2AF[65], hnRNP A1, hnRNP L, and more recently CPEB1 and 4 (Afroz et al., 2014; Bae et al., 2007; Barraud & Allain, 2013; Cukier et al., 2010; Mackereth et al., 2011; Vitali et al., 2006; Zhang et al., 2013). Superposition of the structures of tandem RRMs interacting in the free state reveals a surprisingly large diversity in the positioning of the second RRM with respect to the first (Fig. 5A). The interaction between

RRM–RNA Interactions 253

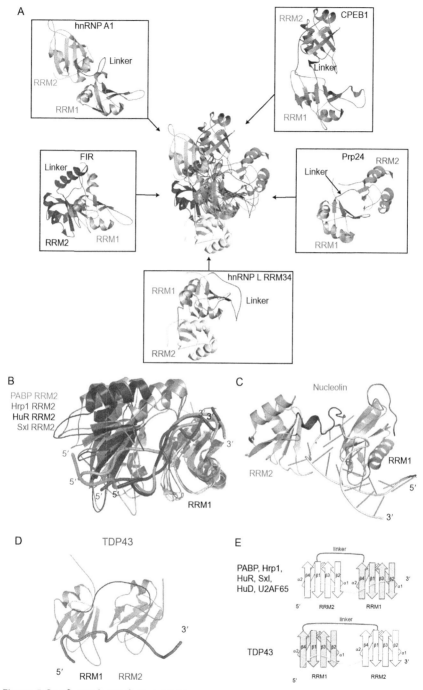

Figure 5 See figure legend on next page.

the RRMs can be directly between the two domains like in hnRNP A1, hnRNP L, or PTB/nPTB where the α-helices of the domains interact, or like Prp24, FIR, or U2AF[65] where the β-sheet of one RRM interacts with the α-helices of the other. The mode of interaction in CPEB1/4 is very different, as the interdomain linker, and the region N-terminal to RRM1 interact with each other and fix the orientation of the two RRMs relative to each other (Fig. 1G). Altogether, this great variety of structural arrangements in tandem RRMs has direct impact on their target RNAs.

6. RNA RECOGNITION BY TANDEM RRMs

Structures of tandem RRMs bound to RNA have revealed enormous diversity in the mode of RNA recognition adding an additional layer of complexity to RRM–RNA interactions. This is mainly achieved by the diversity in the arrangement of tandem RRMs, which has direct implications on the RNA-binding mode. In Section 7, we discuss these diverse modes of RNA recognition resulting from cooperative and dynamic interactions between the RRMs and their implications on RNA orientation, RNA-binding affinity, and specificity.

RNA recognition by tandem RRMs can be classified into three categories based on the mode of RNA binding. In the first category (the most frequently found to date), the tandem RRMs are independent of each other in the free state and adopt a rigid structure upon RNA binding. In all cases except TDP-43, the directionality of RNA binding is conserved with the

Figure 5—Cont'd (A) Overlay of CPEB1 RRM12 with four previously characterized tandem RRMs in the free state—hnRNP L (3S01), FIR (2KXF), hnRNP A1 (2LYV), and Prp24 (2GO9). All the structures were overlaid on the RRM1 domain (shown in gray for all structures). RRM2 domain is shown in yellow, blue, cyan, orange, and green for hnRNP L, FIR, hnRNP A1, CPEB1, and Prp24, respectively. For clarity, the loops in the structures have been smoothened. (B) Overlay of PABP (Deo et al., 1999), Hrp1 (Perez-Canadillas, 2006), HuR (Wang et al., 2013), and Sxl (Handa et al., 1999) tandem RRMs in complex with RNA. Structures have been overlaid on the first RRM of each structure (shown in gray). RRM2 of PABP, Hrp1, HuR, and Sxl are shown in orange, magenta, blue, and green, respectively. RNA is shown in tube representation in the same color as RRM2 for individual structures. (C) Structure of Nucleolin RRMs bound to RNA (Allain, Gilbert, Bouvet, & Feigon, 2000). (D) Structure of TDP-43 tandem RRMs in complex with RNA (Lukavsky et al., 2013). (E) Schematic representation to illustrate the two modes of RNA binding based on the directionality of RNA binding by tandem RRMs. (See the color plate.)

5′-end of the RNA being bound by the second RRM and the 3′-end by the first RRM (found for Sxl, PABP, Hrp1, HuR, HuD, and Nucleolin) (Allain, Bouvet, Dieckmann, & Feigon, 2000; Deo et al., 1999; Handa et al., 1999; Mackereth et al., 2011; Perez-Canadillas, 2006; Wang et al., 2013; Wang & Tanaka Hall, 2001). In all these structures except for Nucleolin, the two RRMs lie side-by-side forming an extended platform upon which the RNA binds in a linear and extended manner. Such inter-RRM interactions further stabilize the complex. Despite binding very different sequences, the similarity in the mode of RNA binding by PABP, Hrp1, HuR, and Sxl tandem RRMs is striking (Fig. 5B). The mode of binding of Nucleolin differs as the two RRMs do not lie side-by side but rather face-to-face in order to sandwich the RNA (Fig. 5C) (Allain, Bouvet, Dieckmann, & Feigon, 2000). Table 2 clearly illustrates the increase in affinity of the protein for RNA achieved by employing tandem RRMs to bind RNA rather than a single one. In all this cases, even though the binding sites on the two RRMs are independent, the increase in affinity is 1000-fold compared to the single RRMs.

The exception to this frequent arrangement is found in the recently determined structure of TDP-43 tandem RRMs bound to a UG-rich sequence (Lukavsky et al., 2013). The single-stranded RNA adopts an extended conformation and the RRMs lie side-by-side. However, in this structure the RNA 5′-end is bound by RRM1 and the 3′-end by RRM2 (Fig. 5D). The fixed orientation between the RRMs is again mediated via inter-RRM interactions as well as interactions to one guanine which is contacted by side-chains from both domains. This different topology for TDP-43 compared to the others allowed us to speculate on why most tandem RRMs bind RNA with RRM2 at the 5′-end and RRM1 at the 3′-end. The answer we believe lies in the topology of the RRM fold and the directionality that the RNA adopts on the surface of the β-sheet surface running 5′ to 3′ from β4 to β2. When two RRMs lie side-by-side to target six to eight consecutive nucleotides, the interdomain linker that connects β4 of RRM1 to β1 of RRM2 can be much shorter if RRM1 is on the right side of the RNA binding the 3′-end and RRM2 on the left-side binding the 5′-end (Fig 5E). In the other topology as seen in TDP-43, to position the RRM2 on the right, the linker must be longer (Fig. 5E). In agreement with this hypothesis, the interdomain linker of PABP, Hrp1, HuR, Sxl, and Nucleolin is less than 10-amino acid long while the one of TDP-43 is 15-amino acid long. The smaller entropic cost required to order a short linker compared to a longer one might explain that a topology with

RRM2 binding the 5′-end and RRM1 the 3′-end has been found more frequently so far. Furthermore, although the affinity of TDP-43 tandem RRMs increases compared to single RRM binding (Table 2), only fivefold affinity is gained compared to the 1000-fold increase for the tandem RRMs with the other arrangement.

In the second category of complexes, the RRMs have a fixed orientation in the free state that is found invariant in the RNA-bound form. HnRNP L and PTB RRM34 are the two tandem RRMs belonging to this category and we expect that findings similar to PTB will apply to nPTB (Joshi et al., 2014) and hnRNP L RRM34 (Zhang et al., 2013). In both hnRNP L and PTB RRM34, the position of the two RRMs relative to each other is such that the binding of a continuous sequence by both RRM surface is topologically impossible (Fig. 6A). The tandem RRMs of both proteins are therefore ideal scaffolds to promote RNA looping in single-stranded RNA. Indeed, RNA binding of noncontiguous RNA sequences resulting from inter-RRM interactions has been shown to be functionally important in the context of splicing regulation (Lamichhane et al., 2010). In this mode of binding also, there is a gain of affinity for the tandem RRMs compared to single RRMs (Table 2).

In the third category of tandem RRM–RNA complexes, a conformational change in the structure of the tandem RRMs is observed upon RNA binding. Two recent reports utilizing solution techniques have unequivocally highlighted the importance of structural changes in RNA recognition by multi-RRM proteins. In the structure of U2AF^{65}, RRM12 tandem domains can adopt two distinct structures in solution depending on the presence or absence of high-affinity RNA ligand (Mackereth et al., 2011). In free state, the tandem RRMs adopt a closed conformation occluding the RNA binding β-sheet surface of RRM1 due to its interaction with the α-helices of RRM2. However, in presence of high-affinity RNA ligand (U)$_9$, the tandem RRMs adopt an open conformation, where the β-sheet surfaces of the two RRMs lie side-by-side forming an extended RNA-binding surface (Fig. 6B) (Mackereth et al., 2011). Such a conformational change might be physiologically relevant because it allows the protein to differentiate long from short poly-U sequences, splicing being productive only with recognition of long poly-U sequences (Mackereth et al., 2011). A related scanning mode of RNA recognition has also been suggested for CPEB proteins for finding their RNA target in the 3′-UTRs (Afroz et al., 2014). In CPEB1 and CPEB4, the tandem RRMs adopts a rigid V-shaped structure in the free form. This positively

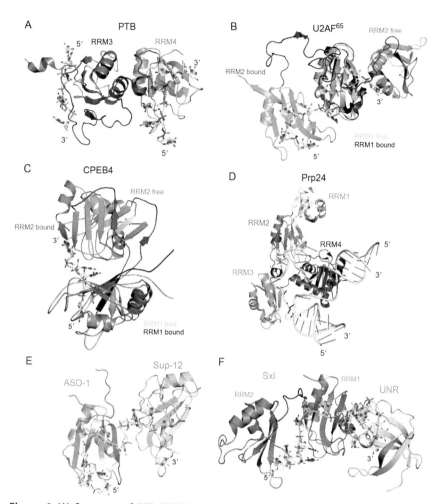

Figure 6 (A) Structure of PTB RRM34 in complex with RNA (Oberstrass et al., 2005). RRM3 (dark gray), RRM4 (green) with RNA (in yellow) binding on individual RRMs. (B and C) Overlay of structures of U2AF^{65} (B) (Mackereth et al., 2011) and CPEB4 (C) (Afroz et al., 2014) tandem RRMs in their free form and in complex with RNA. The structures have been overlaid on RRM1 (shown in gray). RRM2 in the free state is depicted in orange, while in green in complex with RNA. RNA is shown in stick representation in yellow. (D) Crystal structure of yeast U6-Prp24 complex (Montemayor et al., 2014). Prp24 RRM1–4 are shown in ribbon representation in yellow, gray, green, and blue, respectively. RNA is shown in tube representation in yellow with base pairs as ladders. (E) Structure of ternary complex of ASD-1 and SUP-12 RRMs in complex with RNA (Kuwasako et al., 2014). (F) Crystal structure of ternary complex of Sxl tandem RRMs (in gray), UNR CSD1 (in cyan) and RNA (shown as yellow sticks) (Hennig et al., 2014). (See the color plate.)

charged groove favors the binding of a penta-nucleotide RNA in a sandwich fashion with two nucleotides U_3 and U_4 interacting with the protein sidechains of both domains. This configuration allows binding of the RNA similarly to TDP-43 in the unusual orientation with the 5′-end bound by RRM1 and the 3′-end by RRM2 (Fig. 6C). When comparing the structures of the free and bound forms of CPEB1 and 4, a conformational change reminiscent of a Venus fly trap closure is observed (Afroz et al., 2014) (Fig. 6C). In both CPEB1/4 and U2AF^{65}, the conformational change observed allows the RRM proteins to achieve high sequence-specificity despite an overall modest affinity (Table 2).

7. ASSEMBLY OF HIGHER ORDER RIBONUCLEOPROTEIN COMPLEXES BY MULTI RRM–RNA AND MULTI PROTEIN–RNA INTERACTIONS

7.1 RNA recognition by multi-RRM proteins

As mentioned previously, several proteins harbor more than two copies of RRMs generating even higher level of complexity and diversity than in tandem RRM–RNA complexes. However, to our knowledge, the only structure of protein–RNA complex involving more than two RRMs reported so far is the structure of Prp24 RRMs in complex with U6 RNA (Montemayor et al., 2014). This structure reveals a unique RNA architecture, with three RRMs encircling a large RNA loop to form an interlocked protein–RNA ring (Fig. 6D) and highlights the importance of cooperativity in RNA recognition by multi-RRM proteins. RRM2 and RRM3 form the main RNA-binding site, which recognizes the single-stranded RNA loop in a classical fashion, the 5′-end binding RRM3 and the 3′-end RRM2. RRM1 contacts the RNA belonging to a neighboring complex in the crystal. Interestingly, the occluding RRM4 (oRRM4) makes noncanonical contacts with the double-stranded RNA and fixes the angle between the RNA helices (Fig. 6D). Comparison with the crystal structure of Prp24 RRMs in free state suggests that the RRM3 undergoes a rotation of 180° and 20 Å displacement relative to RRM2 similarly to U2AF^{65} (Montemayor et al., 2014), making RRM2 accessible for RNA binding and highlighting once again the importance of conformational changes in RNA recognition by multi-RRM proteins. Furthermore, the linker between RRM3 and RRM4 passes through an RNA stem-loop of U6 snRNA and literally locks the RNP structure, the protein not being able to dissociate unless the RNA unfolds. However, the crystal structures reveal only the static picture and further studies employing solution techniques are essential to understand the dynamics of association and dissociation of this complex.

7.2 RRM–RNA interactions in the higher order RNP structures

RNA-binding proteins (RBPs) often harbor other types of domains in addition to the RRMs. These domains impart additional functions to the proteins such as increasing their RNA-binding affinity and specificity, mediating protein–protein interactions or even specific enzymatic functions. Other RBDs accompanying RRMs include K homology domains, zinc fingers, cold-shock domains (CSD), and sterile alpha motifs domains, which may independently bind to target RNAs or might even cooperate with the RRMs to potentially increase the RNA-binding specificity and/or affinity for the target RNA. However, to our knowledge, there is no structural data of an RRM protein harboring any of the above-mentioned domains in complex with RNA.

The recently reported structure of the ternary complex between the RRMs of Fox-1 homolog in C. *elegans* (ASD-1) and SUP-12 illustrates intermolecular cooperativity in RNA recognition (Kuwasako et al., 2014). SUP-12 RRM lacks a tight consensus for RNA, and therefore a cooperative recognition is essential for targeting specific functions. The structure of ASD-1 RRM and SUP-12 RRM in the ternary complex with RNA reveals a side-by-side arrangement of the RRMs forming a positively charged surface. The first six nucleotides are bound by ASD-1 RRM, while the SUP-12 RRM binds the last five nucleotides (Fig. 6E) (Kuwasako et al., 2014). The cooperativity originates from the interaction with the 7th nucleotide, which is sandwiched by the RRM of each protein. Indeed, the apparent affinity of SUP-12 for RNA is increased by a factor of 10 in the presence of ASD-1 RRM (Table 2).

Another recently reported structure is a ternary complex between the Sxl tandem RRMs and the CSD1 of UNR bound to msl2 RNA. It also reveals intermolecular cooperativity between the Sxl RRMs and the UNR CSD1 in RNA binding (Hennig et al., 2014). In this ternary complex, the Sxl RRMs bind to the 5′-half of the 18-mer RNA, while CSD1 interacts with the 3′-end of the RNA (Fig. 6F). The RNA is sandwiched between the Sxl tandem RRMs and UNR CSD by a set of unique, intertwined protein–RNA–protein interactions resembling a "triple zipper," which mediates cooperative RNA recognition by Sxl RRM1 and UNR CSD1 (Hennig et al., 2014).

Although these structures represent RRM–RRM (in ASD-1–SUP-12–RNA) and RRM–CSD (Sxl–UNR–RNA) intermolecular cooperativity, the principles might as well extend to intramolecular cooperativity between the RRMs and additional RBDs. In other cases, biochemical assays have convincingly shown that the RRMs cooperate with additional domains like with the binuclear zinc-binding domain of CPEB1 that increases the RNA-binding affinity (Afroz et al., 2014).

Hence, it becomes increasingly clear that the scope of structural studies surpasses the atomic level of information. Most recent investigations provide, beyond the basic understanding of nucleic acid or protein-binding specificities, also knowledge about intraprotein domain coordination as well as protein interplay within RNP complexes. The focus shifts gradually toward understanding the posttranscriptional RNA processing at the cellular scale. In Section 8, we discuss potential strategies to exploit structural information to elucidate RNA regulatory networks in the genomic context.

8. STRUCTURAL AND GENOME-WIDE APPROACHES: HOW DO THEY MATCH?

As shown above, structural investigations allow decoding the molecular determinants of RRM–RNA interaction. As discussed earlier as well, follow-up *in vivo* studies support the functional relevance of the structural information. Mutational analyses allowed to validate the intermolecular contacts observed in the complex structures and confirmed the role of RRMs in specific recognition of conserved sites regulating alternative splicing (Cléry et al., 2013; Daubner et al., 2012; Moursy et al., 2014) or translation (Afroz et al., 2014) . The conformational constraints in protein domain orientation found in the structures of Tra2-β1 RRM (Cléry et al., 2011) or PTB RRM34 (Lamichhane et al., 2010; Oberstrass et al., 2005) bound to RNA allowed to hypothesize about interdomain or interprotein interaction and to suggest models for splicing regulation mechanism.

These assays, performed on isolated protein/mRNA systems, contribute valuable functional insights but warrant caution when attempting broader generalizations. With the advent of the "–omics" era and increasing availability of genome-wide *in vivo* protein–RNA interaction datasets, this obstacle can potentially be lifted. An early in-depth profiling of RNPs associated with the human spliceosome by liquid chromatography coupled with mass spectrometry (LC-MS (Rappsilber, Ryder, Lamond, & Mann, 2002)) showed the potential of the high-throughput (HT) interactome analysis approach. RNP immunoprecipitation followed by microarray hybridization (RIP-Chip (Keene, Komisarow, & Friedersdorf, 2006)) or HT sequencing (RIP-Seq (Zhao et al., 2010)), as well as HT UV-cross-linking and immunoprecipitation (HiTS-CLIP or CLIP-Seq, reviewed in Darnell, 2010) or affinity precipitation (CRAC (Granneman, Kudla, Petfalski, & Tollervey, 2009)) methods are now widely used for transcriptome-wide RNA–RBP interaction site mapping.

The number of RRM-carrying proteins with both the structural and genomic interaction data available is steadily increasing (Table 3). The comparison of these two approaches shows a rather good agreement between the base preferences of an RBP derived from molecular contacts observed in structural studies and the consensus motifs of binding sites overrepresented in precipitated transcripts. Strongly selective proteins exhibiting a preference for complex motifs (Fox-1/2, hnRNP A1, SRSF1, or Tra2-β1 (Auweter, Fasan, et al., 2006; Cléry et al., 2011; Cléry et al., 2013; Ding et al., 1999)) show a particularly good match between the molecular and the genomic data. Such proteins are however in minority among RBPs. Many other RNA *trans*-acting factors recognize binding sites that are less well defined. This is the case of RBPs recognizing degenerate tracts or repeats (CUGBP1, hnRNP C, Hu family members, TDP-43, U2AF^{65} (Cienikova et al., 2014; Inoue, Hirao, et al., 2000; Lukavsky et al., 2013; Sickmier et al., 2006; Teplova et al., 2010; Wang et al., 2013; Wang & Tanaka Hall, 2001)); their binding sites have typically imperfectly delimited boundaries with an increased length that compensates for deviations from the ideal consensus. Similar situation arises for promiscuous proteins with few or weakly specific binding pockets (Rna15, SRSF2, and SRSF3 (Daubner et al., 2012; Hargous et al., 2006; Leeper et al., 2010)), recognizing sites sharing low degree of similarity. Finally, for proteins with several independent RBD copies such as PTB (Oberstrass et al., 2005), each domain with a slightly different specificity contributes to a blurring of the binding site consensus. The task of localizing such binding sites using the over-representation and conservation criteria becomes challenging on transcripts that are several tens or hundreds of nucleotides long, which is the typical setup of classical CLIP or RIP experiments. Identified RNA-recognition elements (RRE) are either divergent (several variants for SRSF2 (Pandit et al., 2013)) or have low information content. Often, only global enrichment in certain bases or dinucleotides rather than a proper consensus motif can be defined, as seen for CUGBP1, HuC/D/R, and hnRNP F (Daughters et al., 2009; Huelga et al., 2012; Ince-Dunn et al., 2012). In these cases, *a priori* knowledge about

Table 3 List of RRM-containing proteins for which protein:RNA structures and *in vivo* interaction data are available

Complex structure	Structural motif	*In vivo* method	*In vivo* motif	*In vitro* method	*In vitro* motif
CUGBP1 RRM12 3NNH, 3NMR (Teplova et al., 2010)	UGUU repeats	RIP-Chip (Lee, Lee, Wilusz, Tian, & Wilusz, 2010), HiTS-CLIP (Daughters et al., 2009)	GU repeats in G, U-rich background	SELEX (Marquis et al., 2006)	UGU in U-rich background
HuR RRM12 4ED5 (Wang et al., 2013)	UUUAUUU	PAR-CLIP (Lebedeva et al., 2011)	UUUDUUU	RNAcmp (Ray et al., 2009)	UUUNUUU
		PAR-CLIP (Mukherjee et al., 2011)	WUUUW	RNAcmp (Ray et al., 2013)	UUNNuUU
HuC RRM12 1FNX (Inoue, Hirao, et al., 2000)	UUUAUUU	HiTS-CLIP (Ince-Dunn et al., 2012)	U/g rich	SELEX (Abe, Sakashita, Yamamoto, & Sakamoto, 1996)	$(U)_{2-5}(R)_{1-2}(U)_{2-5}$
HuD RRM12 1FXL, 1G2E (Wang & Tanaka Hall, 2001)	UUUAUUU	HiTS-CLIP (Ince-Dunn et al., 2012)	U/g rich		
Fox-1/2 RRM 2ERR (Auweter, Fasan, et al., 2006)	UGCAUGN	HiTS-CLIP (Yeo et al., 2009)	UGCAUG	SELEX (Jin et al., 2003)	GCAUG
				SELEX (Ponthier et al., 2006)	UGCAUG
hnRNP A1 RRM12 2UP1 (Ding et al., 1999), 1PGZ (Myers, Moore, & Shamoo, 2003)	TAGG resp. YAGG repeats	HiTS-CLIP (Huelga et al., 2012)	UAG resp. YAG repeats	SELEX (Burd & Dreyfuss, 1994)	UAGGGW
				RNAcmp (Ray et al., 2013)	UAGGGA
hnRNP C RRM (Cienikova et al., 2014)	₄uUUU	iCLIP (König et al., 2010), (Zarnack et al., 2013)	uUUUU in U-rich background	SELEX (Görlach, Burd, & Dreyfuss, 1994), RNAcmp (Ray et al., 2013)	UUUUU
hnRNP F/H RRM1 2KFY, RRM2 2KG0, RRM3 2KG1 (Dominguez et al., 2010)	NGGGN	HiTS-CLIP (Huelga et al., 2012)	gGU-rich	RNAcmp of H2 (Ray et al., 2013)	gGGaGGg
Hrp1 RRM12 2CJK (Perez-Canadillas, 2006)	auAUAU	CRAC (Tuck & Tollervey, 2013)	$(UA)_3$	SELEX (Valentini, Weiss, & Silver, 1999)	$(UA)_3$
				RNAcmp (Ray et al., 2013)	UAGGGA
Nab3 RRM 2L41 (Hobor et al., 2011), 2XNR (Lunde, et al., 2011)	UCUN	PAR-CLIP (Webb, Hector, Kudla, & Granneman, 2014)	uGUAG		
		PAR-CLIP (Creamer et al., 2011)	GNUCUUGU		
		RIP-Chip (Hogan, Riordan, Gerber, Herschlag, & Brown, 2008)	WUUCUUGU		
PTB1 RRM1 2AD9, RRM2 2ADB, RRM34 2ADC (Oberstrass et al., 2005)	YCU CUNN YCUNN, YCN	HiTS-CLIP (Xue et al., 2009)	UYUYY	RNAcmp (Ray et al., 2009)	CUNUC
				SELEX (Pérez, Lin, McAfee, & Patton, 1997)	UCUUC in Y-rich background
				RNAcmp (Ray et al., 2013)	YUUUY

Table 3 List of RRM-containing proteins for which protein:RNA structures and *in vivo* interaction data are available—cont'd

Complex structure	Structural motif	In vivo method	In vivo motif	In vitro method	In vitro motif
Rna15 RRM 2X1F (Pancevac, Goldstone, Ramos, & Taylor, 2010), 2KM8 (Leeper et al., 2010)	gu aaua	PAR-CLIP (Baejen et al., 2014)	UUUUCUUU	SELEX (Takagaki & Manley, 1997)	UGYGUAUUYUCC
SRSF1 RRM2 2M8D (Cléry et al., 2013)	GGA	HiTS-CLIP (Sanford et al., 2009)	GAAGA	SELEX (Tacke & Manley, 1995)	GaagaaC
		HiTS-CLIP with deletion analysis (Pandit et al., 2013)	GGAGA, C(GGA)$_2$	FSELEX (Liu, Zhang, & Krainer, 1998)	SRSASGA
				RNAcmp (Ray et al., 2013)	(GGA)$_2$
SRSF2 RRM 2LEB, 2LEC (Daubner et al., 2012)	SSNG	HiTS-CLIP with deletion analysis (Pandit et al., 2013)	gaGGA; CUUC, CNGG, CGNNg	SELEX (Cavaloc, Bourgeois, Kister, & Stévenin, 1999)	Heterogenous, contains SSNG
				SELEX (Tacke & Manley, 1995)	agNaG
				FSELEX (Schaal & Maniatis, 1999)	UGCNGYY
				FSELEX (Liu, Chew, Cartegni, Zhang, & Krainer, 2000)	GRYYCSYR
				RNAcmp (Ray et al., 2013)	AGGAG
SRSF3 RRM 2I2Y (Hargous et al., 2006)	Cauc	iCLIP (Anko et al., 2012)	YHWCH	SELEX (Cavaloc et al., 1999)	WCAUCGAY, YWCUUCAU, CUWCAAC
				FSELEX (Schaal & Maniatis, 1999)	CUCKUCY
TDP-43 RRM12 4BS2 (Lukavsky et al., 2013)	GNGUGNNUGN	iCLIP (Tollervey et al., 2011)	(GU)$_7$	RNAcmp (Ray et al., 2013)	GaAUGa
		HiTS-CLIP (Polymenidou et al., 2011)	(GU)$_4$		
RRM1 4IUF (Kuo, Chiang, Wang, Doudeva, & Yuan, 2014)	GNGCG	RIP-Chip (Colombrita et al., 2012)	(GU)$_{10}$		
		RRIP-Chip (Sephton et al., 2011)	(GU)$_6$		
Tra2-β1 RRM 2KXN (Cléry et al., 2011), 2RRA (Tsuda et al., 2011)	NaGAAN	HiTS-CLIP (Grellscheid et al., 2011a)	aGAAga	SELEX (Tacke, Tohyama, Ogawa, & Manley, 1998)	AA(GAA)$_2$
U2AF^{65} RRM12 2G4B (Sickmier et al., 2006), 3VAH (Jenkins, Agrawal, Gupta, Green, & Kielkopf, 2013)	UUUUYYY	iCLIP (Zarnack et al., 2013)	YUUyUYYYY in Y-rich background	SELEX (Singh, Valcarcel, & Green, 1995; Singh, Banerjee, & Green, 2000)	(U)$_6$CCCuuu(U)$_6$
				RNAcmp (Ray et al., 2013)	UUUUYy

N = A, C, G or U; D = A, G or U; H = A, C or U; R = A or G; S = C or G; Y = C or U; W = A or U.
Upper case: strong base preference; lower case: weak preference; protein names are in bold.
SELEX, systematic evolution of ligands by exponential enrichment; FSELEX, functional SELEX; RNAcmp, RNAcompete.

the base specificity of an RBP can be useful or even necessary to guide the RRE search in the large-scale *in vivo* datasets. If available, the use of *in vitro* structural information is preferable over selection/amplification (SELEX) data, as the latter suffers from ambiguities due to its lower resolution and aggregate nature. The importance of the *a priori* information for binding site identification within CLIP datasets is illustrated by the case of Nab3. The top ranking binding motif found in one of its PAR-CLIP experiments (Webb et al., 2014) does not match the known Nab3 consensus (Hobor et al., 2011); instead, it corresponds to sequences recognized by its hetero-dimerization partner Nrd1 (Carroll, Ghirlando, Ames, & Corden, 2007). Similarly, the immediate neighborhood of Tra2-β1 *in vivo* binding sites was found to be enriched in adenines, leading to the extended consensus motif identified within sequences cross-linked by this protein (Grellscheid et al., 2011). In agreement with a proposed co-recruitment model of Tra2-β1 and hnRNP G regulating factors in alternative splicing (Hofmann & Wirth, 2002), a recent structural work with hnRNP G RRM (Moursy et al., 2014) indeed established this protein's preference for adenine repeats.

Recognition of neighboring regulatory sites by *trans*-acting factors in a coordinated fashion, nicely illustrated by the recent ternary complex structures of Sxl RRM–UNR CSD1–RNA (Hennig et al., 2014) and ASD-1–SUP-12–RNA (Kuwasako et al., 2014), is in fact a recurrent phenomenon in eukaryotic posttranscriptional control. It is therefore clear that the interpretation of genomic data requires a certain level of precaution and it may fail to identify the exact binding position of the precipitated protein. In this respect, high-resolution interactome data offer marked advantages. Three variants of the CLIP method provide high level of accuracy: individual-nucleotide resolution CLIP (iCLIP) (König et al., 2010), photoactivatable ribonucleoside-enhanced (PAR)-CLIP (Ascano, Hafner, Cekan, Gerstberger, & Tuschl, 2012; Hafner et al., 2010), and cDNA damage analysis in standard CLIP (Zhang & Darnell, 2011). The analysis of hnRNP C and U2AF^{65} interactions illustrates the remarkable match between structural and cell-wide, single-nucleotide resolution-binding properties. Recent structural investigation of hnRNP C (Cienikova et al., 2014) showed that its single RRM contains five nucleotide pockets for uridines with selectivity increasing gradually in the 5′ to 3′ direction of the poly-U tract. This pattern was mirrored by uridine conservation frequencies of full-length hnRNP C iCLIP sites aligned on the cross-linked nucleotide (König et al., 2010) (Fig. 7A). Moreover, the authors found a very good agreement between the *in vitro* RRM affinities for uridine oligomers of variable length and

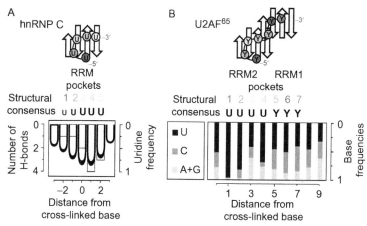

Figure 7 Comparison between the structurally derived sequence-specificity of hnRNP C RRM (A) and U2AF^{65} RRM12 (B) (Jenkins et al., 2013), and their transcriptome-wide CLIP consensus sequences (König et al., 2010; Zarnack et al., 2013). (A) The number of intermolecular hydrogen bonds observed for each RRM pocket is compared to the uridine frequency observed in aligned cross-linked sites. (B) The frequencies of the nucleotides U, C, and A+G in aligned cross-linked sites are matched against the base specificities in the individual RRM12 pockets. *Panel (A): Accepted with permission from Cienikova et al. (2014). Copyright 2014, American Chemical Society.*

the transcriptome-wide short uridine tract saturation by hnRNP C, giving a strong support for a quantitative interpretation of HT cross-linking data, as proposed earlier by Kishore et al. (2011) and Sugimoto et al. (2012). In the case of U2AF^{65}, the molecular features driving the recognition of poly-pyrimidine tracts by its RRM12 subdomain were extensively examined by X-ray crystallography (Jenkins et al., 2013). The structures in complex with multiple pyrimidine-rich RNA heptamers report a strong uridine preference of the 5′-end binding RRM2, while the downstream binding RRM1 seems to adapt readily to both uridines and cytidines, in particular in the most 3′-end pockets. This observation nicely agrees with transcriptome-wide single-nucleotide precision consensus of the sites cross-linked by full-length U2AF^{65} protein (Zarnack et al., 2013), where a similar 5′ to 3′ pattern or uridine versus pyrimidine conservation can be recognized (Fig. 7B). The side-by-side cross-validation of the structural and the iCLIP data hints that the two RRMs of U2AF^{65} indeed bind in tandem on their cellular target sites, with their relative orientation preserved.

Together, the examples of these two RBPs suggest that combining high-resolution structural and genomic data allows to describe cellular RBP:RNA interaction with high confidence, creating a foundation for further thermodynamic and mechanistic understanding of posttranscriptional regulation, well beyond binding site matching. The structural perspective focuses on the protein and gives very detailed insights about the molecular basis of its RNA recognition; the information content of these elementary interaction rules is however insufficient *per se* for identification of cellular targets (Bulyk, 2004; Zhang, Leslie, & Chasin, 2005). On the other hand, large-scale interaction studies produce a wealth of information on the RNA side of the equation. This data, once stringently clustered and filtered of transient or nonspecific complexes, is free of false positives or biases inherent to small-scale assays and allows for robust generalizations. High-resolution CLIP variants are particularly interesting since they capture only direct (short-range) contacts with—ideally—single-nucleotide precision. The complexity of these transcriptome-wide datasets hinders however straightforward interpretation. An approach combining structural and high-resolution cellular perspectives will allow analyzing the content-rich HT datasets through the lens of the molecular interaction rules, as they operate under biologically relevant constraints, with a potential to provide new insights into the mechanisms of protein–RNA recognition. Combining these two approaches will therefore allow to understand better the combinatorial action of multiple RBDs or proteins (competition or cooperativity) and even provide quantitative information such as the affinities or the stoichiometry of the protein–RNA complexes.

The analysis of hnRNP C transcriptome-wide binding behavior using a structure-based model (Cienikova et al., 2014) is, to our knowledge, the first study that endeavored to bridge the gap between molecular and cellular levels of understanding of protein–RNA interactions. Further venues for investigating and interpreting RNP interaction networks will certainly present themselves as the amount of published structural and HT interaction data grows. As new methodologies and analytical tools are developed, we imagine that future studies will incorporate, in addition to structural and cellular binding data, also additional genomic knowledge. In particular, the information from HT RNA-Seq data will allow accounting for RNA expression levels, abundance of splicing variants, and other posttranscriptional modifications (Cloonan et al., 2008; Nagalakshmi et al., 2008). Similarly, methods probing RNA structural context and binding site accessibility (Fukunaga et al., 2014; Li, Quon, Lipshitz, & Morris, 2010) could help refining modeling algorithms and improve quantitative and mechanistical interpretation of interaction data.

9. CONCLUSIONS AND PERSPECTIVES

In this chapter, we gave an overview of the numerous modes of interaction used by RRMs to bind RNA and showed that they can specifically interact with a large repertoire of RNA sequences. In addition to be crucial for the understanding of these RRM–RNA interactions, we also show that structural investigations are needed to better characterize RBP-binding sites and functions *in vivo*. Interestingly, structures of RRM–RNA complexes suggest that during their evolution, single RRMs modulated their fold, enriched their mode of RNA recognition involving different networks of hydrogen bonds and started to use additional parts of the domain outside the canonical β-sheet interface to interact with RNA (loops, α1-helix, and additional secondary structure elements). Consequently, RRMs acquired the ability to bind diverse target RNA sequences (in terms of length and nucleotide content) in a specific fashion and with a large range of affinities (K_d from subnanomolar to micromolar). This allows RRMs to bind almost all single-stranded RNA sequences and it most likely explains their abundance in RBPs. However, it also increases the difficulty to predict a recognition code for these domains.

Although single RRMs can use very diverse and sophisticated modes of RNA recognition, we showed that the use of multiple RRMs in a single protein increases the complexity of RBP interactions with their RNA targets. This observation tempted us to make an analogy between a protein containing one, two, three, or four RRMs and a musical ensemble of one, two, three, or four instruments interpreting genomic information and musical score, respectively. Indeed, similar to individual instruments that provide specific tunes, each RRM brings its part of RNA-binding specificity. Therefore, in proteins containing multiple RRMs, the domains can cooperate to achieve specific functions that cannot be achieved by individual domains. In addition to binding RNA with more affinity and specificity, they can modulate RNA architecture, mask or stabilize longer RNA sequences and cooperate with additional domains which bring new functions. As a result, the information that is generated out of a single RNA molecule becomes more sophisticated in the presence of multi-RRM-containing proteins.

Despite progress in the last decade in this growing field, more high-resolution structures of multi RRM–RNA complexes are still required in order to derive general principles. However, determination of these structures is challenging for both X-ray crystallography (obtaining crystals) and NMR (size limitation). The use of recent solution techniques such as SAXS and EPR in conjunction with NMR spectroscopy should facilitate structure determination of high-molecular-weight complexes. The increasing resolution of structures solved by electron microscopy should also start to give more and more details about these intermolecular interactions. Importantly, we also need to have a more dynamic view of protein–RNA interactions, which reflects *in vivo* situation. Most often, these interactions are reversible and depend on competitive binding of RBPs on overlapping sequences. Finally, several additional questions still need to be answered. How do several RBPs assemble or multimerize on RNA? How do posttranslational modifications (e.g., phosphorylation) and posttranscriptional modifications (e.g., methylation and pseudouridylation) influence this dynamic? As illustrated in Section 8, interdisciplinary work is now required to find the corresponding answers, which are needed to fully understand posttranscriptional gene regulation.

ACKNOWLEDGMENTS

Research of F. H. -T. A. is supported by the Swiss National Science Foundation (SNF) and the National Center for Competence in Research (NCCR) in Structural Biology. A. C. is additionally supported by the European Molecular Biology Organization for a postdoctoral fellowship, Fondation Suisse de Recherche sur les Maladies Musculaires (FSRMM), and Spinal Muscular Atrophy (SMA)-Europe. T. A. and Z. C. are supported by the SNF.

REFERENCES

Abe, R., Sakashita, E., Yamamoto, K., & Sakamoto, H. (1996). Two different RNA binding activities for the AU-rich element and the poly(A) sequence of the mouse neuronal protein mHuC. *Nucleic Acids Research*, 24(24), 4895–4901. http://dx.doi.org/10.1093/nar/24.24.4895.

Afroz, T., Skrisovska, L., Belloc, E., Guillen-Boixet, J., Mendez, R., & Allain, F. H. (2014). A fly trap mechanism provides sequence-specific RNA recognition by CPEB proteins. *Genes & Development*, 28(13), 1498–1514. http://dx.doi.org/10.1101/gad.241133.114.

Allain, F. H., Bouvet, P., Dieckmann, T., & Feigon, J. (2000). Molecular basis of sequence-specific recognition of pre-ribosomal RNA by nucleolin. *The EMBO Journal*, 19(24), 6870–6881. http://dx.doi.org/10.1093/emboj/19.24.6870.

Allain, F. H., Gilbert, D. E., Bouvet, P., & Feigon, J. (2000). Solution structure of the two N-terminal RNA-binding domains of nucleolin and NMR study of the interaction with its RNA target. *Journal of Molecular Biology*, 303(2), 227–241. http://dx.doi.org/10.1006/jmbi.2000.4118.

Allain, F. H., Howe, P. W., Neuhaus, D., & Varani, G. (1997). Structural basis of the RNA-binding specificity of human U1A protein. *The EMBO Journal*, 16(18), 5764–5772. http://dx.doi.org/10.1093/emboj/16.18.5764.

Amrane, S., Rebora, K., Zniber, I., Dupuy, D., & Mackereth, C. D. (2014). Backbone-independent nucleic acid binding by splicing factor SUP-12 reveals key aspects of

molecular recognition. *Nature Communications*, 5, 4595. http://dx.doi.org/10.1038/ncomms5595.
Anko, M.-L., Muller-McNicoll, M., Brandl, H., Curk, T., Gorup, C., Henry, I., et al. (2012). The RNA-binding landscapes of two SR proteins reveal unique functions and binding to diverse RNA classes. *Genome Biology*, 13(3), R17.
Ascano, M., Hafner, M., Cekan, P., Gerstberger, S., & Tuschl, T. (2012). Identification of RNA–protein interaction networks using PAR-CLIP. *Wiley Interdisciplinary Reviews RNA*, 3(2), 159–177. http://dx.doi.org/10.1002/wrna.1103.
Auweter, S. D., Fasan, R., Reymond, L., Underwood, J. G., Black, D. L., Pitsch, S., et al. (2006). Molecular basis of RNA recognition by the human alternative splicing factor Fox-1. *The EMBO Journal*, 25(1), 163–173. http://dx.doi.org/10.1038/sj.emboj.7600918.
Auweter, S. D., Oberstrass, F. C., & Allain, F. H. (2006). Sequence-specific binding of single-stranded RNA: Is there a code for recognition? *Nucleic Acids Research*, 34(17), 4943–4959. http://dx.doi.org/10.1093/nar/gkl620.
Auweter, S. D., Oberstrass, F. C., & Allain, F. H. (2007). Solving the structure of PTB in complex with pyrimidine tracts: An NMR study of protein-RNA complexes of weak affinities. *Journal of Molecular Biology*, 367(1), 174–186. http://dx.doi.org/10.1016/j.jmb.2006.12.053.
Avis, J. M., Allain, F. H., Howe, P. W., Varani, G., Nagai, K., & Neuhaus, D. (1996). Solution structure of the N-terminal RNP domain of U1A protein: The role of C-terminal residues in structure stability and RNA binding. *Journal of Molecular Biology*, 257(2), 398–411. http://dx.doi.org/10.1006/jmbi.1996.0171.
Bae, E., Reiter, N. J., Bingman, C. A., Kwan, S. S., Lee, D., Phillips, G. N., Jr., et al. (2007). Structure and interactions of the first three RNA recognition motifs of splicing factor prp24. *Journal of Molecular Biology*, 367(5), 1447–1458. http://dx.doi.org/10.1016/j.jmb.2007.01.078.
Baejen, C., Torkler, P., Gressel, S., Essig, K., Söding, J., & Cramer, P. (2014). Transcriptome maps of mRNP biogenesis factors define Pre-mRNA recognition. *Molecular Cell*, 55(5), 745–757. http://dx.doi.org/10.1016/j.molcel.2014.08.005.
Barraud, P., & Allain, F. H. (2013). Solution structure of the two RNA recognition motifs of hnRNP A1 using segmental isotope labeling: How the relative orientation between RRMs influences the nucleic acid binding topology. *Journal of Biomolecular NMR*, 55(1), 119–138. http://dx.doi.org/10.1007/s10858-012-9696-4.
Birney, E., Kumar, S., & Krainer, A. R. (1993). Analysis of the RNA-recognition motif and RS and RGG domains: Conservation in metazoan pre-mRNA splicing factors. *Nucleic Acids Research*, 21(25), 5803–5816.
Bulyk, M. L. (2004). Computational prediction of transcription-factor binding site locations. *Genome Biology*, 5(1), 201.
Burd, C. G., & Dreyfuss, G. (1994). RNA binding specificity of hnRNP A1: Significance of hnRNP A1 high-affinity binding sites in pre-mRNA splicing. *The EMBO Journal*, 13(5), 1197–1204.
Calero, G., Wilson, K. F., Ly, T., Rios-Steiner, J. L., Clardy, J. C., & Cerione, R. A. (2002). Structural basis of m7GpppG binding to the nuclear cap-binding protein complex. *Nature Structural Biology*, 9(12), 912–917. http://dx.doi.org/10.1038/nsb874.
Carroll, K. L., Ghirlando, R., Ames, J. M., & Corden, J. L. (2007). Interaction of yeast RNA-binding proteins Nrd1 and Nab3 with RNA polymerase II terminator elements. *RNA*, 13(3), 361–373. http://dx.doi.org/10.1261/rna.338407.
Cavaloc, Y., Bourgeois, C. F., Kister, L., & Stévenin, J. (1999). The splicing factors 9G8 and SRp20 transactivate splicing through different and specific enhancers. *RNA*, 5(3), 468–483.
Cienikova, Z., Damberger, F., Hall, J., Allain, F., & Maris, C. (2014). Structural and mechanistic insights into poly(uridine) tract recognition by the hnRNP C RNA recognition

motif. *Journal of the American Chemical Society*, *136*, 14536–14544. http://dx.doi.org/10.1021/ja507690d.

Cléry, A., Blatter, M., & Allain, F. H. (2008). RNA recognition motifs: Boring? Not quite. *Current Opinion in Structural Biology*, *18*(3), 290–298. http://dx.doi.org/10.1016/j.sbi.2008.04.002.

Cléry, A., Jayne, S., Benderska, N., Dominguez, C., Stamm, S., & Allain, F. H. (2011). Molecular basis of purine-rich RNA recognition by the human SR-like protein Tra2-beta1. *Nature Structural & Molecular Biology*, *18*(4), 443–450. http://dx.doi.org/10.1038/nsmb.2001.

Cléry, A., Sinha, R., Anczuków, O., Corrionero, A., Moursy, A., Daubner, G., et al. (2013). Isolated pseudo–RNA-recognition motifs of SR proteins can regulate splicing using a noncanonical mode of RNA recognition. *Proceedings of the National Academy of Sciences of the United States of America*, *110*(30), 2802–2811. http://dx.doi.org/10.1073/pnas.1303445110.

Clingman, C. C., Deveau, L. M., Hay, S. A., Genga, R. M., Shandilya, S. M. D., Massi, F., et al. (2014). Allosteric inhibition of a stem cell RNA-binding protein by an intermediary metabolite. *eLife*, *3*, 1–26. http://dx.doi.org/10.7554/Elife.02848.

Cloonan, N., Forrest, A. R. R., Kolle, G., Gardiner, B. B. A., Faulkner, G. J., Brown, M. K., et al. (2008). Stem cell transcriptome profiling via massive-scale mRNA sequencing. *Nature Methods*, *5*(7), 613–619.

Colombrita, C., Onesto, E., Megiorni, F., Pizzuti, A., Baralle, F. E., Buratti, E., et al. (2012). TDP-43 and FUS RNA-binding proteins bind distinct sets of cytoplasmic messenger RNAs and differently regulate their post-transcriptional fate in motoneuron-like cells. *Journal of Biological Chemistry*, *287*(19), 15635–15647. http://dx.doi.org/10.1074/jbc.M111.333450.

Creamer, T. J., Darby, M. M., Jamonnak, N., Schaughency, P., Hao, H., Wheelan, S. J., et al. (2011). Transcriptome-wide binding sites for components of the *Saccharomyces cerevisiae* non-poly(A) termination pathway: Nrd1, Nab3, and Sen1. *PLoS Genetics*, *7*(10), e1002329. http://dx.doi.org/10.1371/journal.pgen.1002329.

Crowder, S. M., Kanaar, R., Rio, D. C., & Alber, T. (1999). Absence of interdomain contacts in the crystal structure of the RNA recognition motifs of sex-lethal. *Proceedings of the National Academy of Sciences of the United States of America*, *96*(9), 4892–4897.

Cukier, C. D., Hollingworth, D., Martin, S. R., Kelly, G., Diaz-Moreno, I., & Ramos, A. (2010). Molecular basis of FIR-mediated c-myc transcriptional control. *Nature Structural & Molecular Biology*, *17*(9), 1058–1064. http://dx.doi.org/10.1038/nsmb.1883.

Darnell, R. (2010). HITS-CLIP: Panoramic views of protein–RNA regulation in living cells. *Wiley Interdisciplinary Reviews RNA*, *1*(2), 266–286. http://dx.doi.org/10.1002/wrna.31.

Daubner, G. M., Cléry, A., & Allain, F. H. (2013). RRM-RNA recognition: NMR or crystallography...and new findings. *Current Opinion in Structural Biology*, *23*(1), 100–108. http://dx.doi.org/10.1016/j.sbi.2012.11.006.

Daubner, G. M., Cléry, A., Jayne, S., Stevenin, J., & Allain, F. H. (2012). A syn-anti conformational difference allows SRSF2 to recognize guanines and cytosines equally well. *The EMBO Journal*, *31*(1), 162–174. http://dx.doi.org/10.1038/emboj.2011.367.

Daughters, R. S., Tuttle, D. L., Gao, W., Ikeda, Y., Moseley, M. L., Ebner, T. J., et al. (2009). RNA gain-of-function in spinocerebellar ataxia type 8. *PLoS Genetics*, *5*(8), e1000600. http://dx.doi.org/10.1371/journal.pgen.1000600.

Deo, R. C., Bonanno, J. B., Sonenberg, N., & Burley, S. K. (1999). Recognition of polyadenylate RNA by the poly(A)-binding protein. *Cell*, *98*(6), 835–845.

Ding, J., Hayashi, M. K., Zhang, Y., Manche, L., Krainer, A. R., & Xu, R.-M. (1999). Crystal structure of the two-RRM domain of hnRNP A1 (UP1) complexed with single-stranded telomeric DNA. *Genes & Development*, *13*(9), 1102–1115.

Dominguez, C., & Allain, F. H. (2006). NMR structure of the three quasi RNA recognition motifs (qRRMs) of human hnRNP F and interaction studies with Bcl-x G-tract RNA: A novel mode of RNA recognition. *Nucleic Acids Research, 34*(13), 3634–3645.

Dominguez, C., Fisette, J. F., Chabot, B., & Allain, F. H. (2010). Structural basis of G-tract recognition and encaging by hnRNP F quasi-RRMs. *Nature Structural & Molecular Biology, 17*(7), 853–861. http://dx.doi.org/10.1038/nsmb.1814.

Duss, O., Michel, E., Yulikov, M., Schubert, M., Jeschke, G., & Allain, F. H. (2014). Structural basis of the non-coding RNA RsmZ acting as a protein sponge. *Nature, 509*(7502), 588–592. http://dx.doi.org/10.1038/nature13271.

Duss, O., Yulikov, M., Jeschke, G., & Allain, F. H. (2014). EPR-aided approach for solution structure determination of large RNAs or protein-RNA complexes. *Nature Communications, 5*, 3669. http://dx.doi.org/10.1038/ncomms4669.

Enokizono, Y., Konishi, Y., Nagata, K., Ouhashi, K., Uesugi, S., Ishikawa, F., et al. (2005). Structure of hnRNP D complexed with single-stranded telomere DNA and unfolding of the quadruplex by heterogeneous nuclear ribonucleoprotein D. *The Journal of Biological Chemistry, 280*(19), 18862–18870. http://dx.doi.org/10.1074/jbc.M411822200.

Fukunaga, T., Ozaki, H., Terai, G., Asai, K., Iwasaki, W., & Kiryu, H. (2014). CapR: Revealing structural specificities of RNA-binding protein target recognition using CLIP-seq data. *Genome Biology, 15*(1), R16.

Görlach, M., Burd, C. G., & Dreyfuss, G. (1994). The determinants of RNA-binding specificity of the heterogeneous nuclear ribonucleoprotein C proteins. *Journal of Biological Chemistry, 269*(37), 23074–23078.

Granneman, S., Kudla, G., Petfalski, E., & Tollervey, D. (2009). Identification of protein binding sites on U3 snoRNA and pre-rRNA by UV cross-linking and high-throughput analysis of cDNAs. *Proceedings of the National Academy of Sciences of the United States of America, 106*(24), 9613–9618. http://dx.doi.org/10.1073/pnas.0901997106.

Grellscheid, S., Dalgliesh, C., Storbeck, M., Best, A., Liu, Y., Jakubik, M., et al. (2011). Identification of evolutionarily conserved exons as regulated targets for the splicing activator Tra2β in development. *PLoS Genetics, 7*(12), e1002390. http://dx.doi.org/10.1371/journal.pgen.1002390.

Hafner, M., Landthaler, M., Burger, L., Khorshid, M., Hausser, J., Berninger, P., et al. (2010). Transcriptome-wide identification of RNA-binding protein and MicroRNA target sites by PAR-CLIP. *Cell, 141*(1), 129–141. http://dx.doi.org/10.1016/j.cell.2010.03.009.

Handa, N., Nureki, O., Kurimoto, K., Kim, I., Sakamoto, H., Shimura, Y., et al. (1999). Structural basis for recognition of the tra mRNA precursor by the sex-lethal protein. *Nature, 398*(6728), 579–585. http://dx.doi.org/10.1038/19242.

Hardin, J. W., Hu, Y. X., & McKay, D. B. (2010). Structure of the RNA binding domain of a DEAD-box helicase bound to its ribosomal RNA target reveals a novel mode of recognition by an RNA recognition motif. *Journal of Molecular Biology, 402*(2), 412–427. http://dx.doi.org/10.1016/j.jmb.2010.07.040.

Hargous, Y., Hautbergue, G. M., Tintaru, A. M., Skrisovska, L., Golovanov, A. P., Stevenin, J., et al. (2006). Molecular basis of RNA recognition and TAP binding by the SR proteins SRp20 and 9G8. *The EMBO Journal, 25*(21), 5126–5137. http://dx.doi.org/10.1038/sj.emboj.7601385.

Hennig, J., Militti, C., Popowicz, G. M., Wang, I., Sonntag, M., Geerlof, A., et al. (2014). Structural basis for the assembly of the Sxl-Unr translation regulatory complex. *Nature, 515*, 287–290. http://dx.doi.org/10.1038/nature13693.

Hobor, F., Pergoli, R., Kubicek, K., Hrossova, D., Bacikova, V., Zimmermann, M., et al. (2011). Recognition of transcription termination signal by the nuclear polyadenylated RNA-binding (NAB) 3 protein. *Journal of Biological Chemistry, 286*(5), 3645–3657. http://dx.doi.org/10.1074/jbc.m110.158774.

Hofmann, Y., & Wirth, B. (2002). hnRNP-G promotes exon 7 inclusion of survival motor neuron (SMN) via direct interaction with Htra2-beta1. *Human Molecular Genetics*, *11*(17), 2037–2049.

Hogan, D. J., Riordan, D. P., Gerber, A. P., Herschlag, D., & Brown, P. O. (2008). Diverse RNA-binding proteins interact with functionally related sets of RNAs, suggesting an extensive regulatory system. *PLoS Biology*, *6*(10), e255. http://dx.doi.org/10.1371/journal.pbio.0060255.

Huelga, Stephanie C., Vu, Anthony Q., Arnold, Justin D., Liang, Tiffany Y., Liu, Patrick P., Yan, Bernice Y., et al. (2012). Integrative genome-wide analysis reveals cooperative regulation of alternative splicing by hnRNP proteins. *Cell Reports*, *1*(2), 167–178. http://dx.doi.org/10.1016/j.celrep.2012.02.001.

Ince-Dunn, G., Okano, Hirotaka J., Jensen, K. B., Park, W.-Y., Zhong, R., Ule, J., et al. (2012). Neuronal elav-like (Hu) proteins regulate RNA splicing and abundance to control glutamate levels and neuronal excitability. *Neuron*, *75*(6), 1067–1080. http://dx.doi.org/10.1016/j.neuron.2012.07.009.

Inoue, M., Hirao, M., Kasashima, K., Kim, I. S., Kawai, G., Kigawa, T., et al. (2000). Solution structure of mouse HuC RNA-binding domains complexed with an AU-rich element reveals determinants of neuronal differentiation (unpublished).

Inoue, M., Muto, Y., Sakamoto, H., & Yokoyama, S. (2000). NMR studies on functional structures of the AU-rich element-binding domains of Hu antigen C. *Nucleic Acids Research*, *28*(8), 1743–1750.

Jacks, A., Babon, J., Kelly, G., Manolaridis, I., Cary, P. D., Curry, S., et al. (2003). Structure of the C-terminal domain of human La protein reveals a novel RNA recognition motif coupled to a helical nuclear retention element. *Structure*, *11*(7), 833–843.

Jenkins, J., Agrawal, A., Gupta, A., Green, M., & Kielkopf, C. (2013). U2AF65 adapts to diverse pre-mRNA splice sites through conformational selection of specific and promiscuous RNA recognition motifs. *Nucleic Acids Research*, *41*(6), 3859–3873. http://dx.doi.org/10.1093/nar/gkt046.

Jenkins, H. T., Malkova, B., & Edwards, T. A. (2011). Kinked beta-strands mediate high-affinity recognition of mRNA targets by the germ-cell regulator DAZL. *Proceedings of the National Academy of Sciences of the United States of America*, *108*(45), 18266–18271. http://dx.doi.org/10.1073/pnas.1105211108.

Jin, Y., Suzuki, H., Maegawa, S., Endo, H., Sugano, S., Hashimoto, K., et al. (2003). A vertebrate RNA-binding protein Fox-1 regulates tissue-specific splicing via the pentanucleotide GCAUG. *The EMBO Journal*, *22*(4), 905–912. http://dx.doi.org/10.1093/emboj/cdg089.

Johansson, C., Finger, L. D., Trantirek, L., Mueller, T. D., Kim, S., Laird-Offringa, I. A., et al. (2004). Solution structure of the complex formed by the two N-terminal RNA-binding domains of nucleolin and a pre-rRNA target. *Journal of Molecular Biology*, *337*(4), 799–816. http://dx.doi.org/10.1016/j.jmb.2004.01.056.

Joshi, A., Esteve, V., Buckroyd, A. N., Blatter, M., Allain, F. H., & Curry, S. (2014). Solution and crystal structures of a C-terminal fragment of the neuronal isoform of the polypyrimidine tract binding protein (nPTB). *PeerJ*, *2*, e305. http://dx.doi.org/10.7717/peerj.305.

Kanaar, R., Lee, A. L., Rudner, D. Z., Wemmer, D. E., & Rio, D. C. (1995). Interaction of the sex-lethal RNA binding domains with RNA. *The EMBO Journal*, *14*(18), 4530–4539.

Keene, J. D., Komisarow, J. M., & Friedersdorf, M. B. (2006). RIP-Chip: the isolation and identification of mRNAs, microRNAs and protein components of ribonucleoprotein complexes from cell extracts. *Nature Protocols*, *1*(1), 302–307.

Kielkopf, C. L., Rodionova, N. A., Green, M. R., & Burley, S. K. (2001). A novel peptide recognition mode revealed by the X-ray structure of a core U2AF35/U2AF65 heterodimer. *Cell*, *106*(5), 595–605.

Kishore, S., Jaskiewicz, L., Burger, L., Hausser, J., Khorshid, M., & Zavolan, M. (2011). A quantitative analysis of CLIP methods for identifying binding sites of RNA-binding proteins. *Nature Methods*, *8*(7), 559–564.

König, J., Zarnack, K., Rot, G., Curk, T., Kayikci, M., Zupan, B., et al. (2010). iCLIP reveals the function of hnRNP particles in splicing at individual nucleotide resolution. *Nature Structural & Molecular Biology*, *17*(7), 909–915. http://dx.doi.org/10.1038/nsmb.1838.

Koradi, R., Billeter, M., & Wuthrich, K. (1996). MOLMOL: A program for display and analysis of macromolecular structures. *Journal of Molecular Graphics*, *14*(1), 51–55, 29–32.

Kuo, P.-H., Chiang, C.-H., Wang, Y.-T., Doudeva, L. G., & Yuan, H. S. (2014). The crystal structure of TDP-43 RRM1-DNA complex reveals the specific recognition for UG- and TG-rich nucleic acids. *Nucleic Acids Research*, *42*(7), 4712–4722. http://dx.doi.org/10.1093/nar/gkt1407.

Kuo, P.-H., Doudeva, L. G., Wang, Y.-T., Shen, C.-K. J., & Yuan, H. S. (2009). Structural insights into TDP-43 in nucleic-acid binding and domain interactions. *Nucleic Acids Research*, *37*(6), 1799–1808. http://dx.doi.org/10.1093/nar/gkp013.

Kuwasako, K., Takahashi, M., Unzai, S., Tsuda, K., Yoshikawa, S., He, F., et al. (2014). RBFOX and SUP-12 sandwich a G base to cooperatively regulate tissue-specific splicing. *Nature Structural & Molecular Biology*, *21*(9), 778–786. http://dx.doi.org/10.1038/nsmb.2870.

Lamichhane, R., Daubner, G. M., Thomas-Crusells, J., Auweter, S. D., Manatschal, C., Austin, K. S., et al. (2010). RNA looping by PTB: Evidence using FRET and NMR spectroscopy for a role in splicing repression. *Proceedings of the National Academy of Sciences of the United States of America*, *107*(9), 4105–4110. http://dx.doi.org/10.1073/Pnas.0907072107.

Law, M. J., Rice, A. J., Lin, P., & Laird-Offringa, I. A. (2006). The role of RNA structure in the interaction of U1A protein with U1 hairpin II RNA. *RNA*, *12*(7), 1168–1178. http://dx.doi.org/10.1261/rna.75206.

Lebedeva, S., Jens, M., Theil, K., Schwanhäusser, B., Selbach, M., Landthaler, M., et al. (2011). Transcriptome-wide analysis of regulatory interactions of the RNA-binding protein HuR. *Molecular Cell*, *43*(3), 340–352. http://dx.doi.org/10.1016/j.molcel.2011.06.008.

Lee, J. E., Lee, J. Y., Wilusz, J., Tian, B., & Wilusz, C. J. (2010). Systematic analysis of cis-elements in unstable mRNAs demonstrates that CUGBP1 is a key regulator of mRNA decay in muscle cells. *PLoS One*, *5*(6), e11201. http://dx.doi.org/10.1371/journal.pone.0011201.

Leeper, T. C., Qu, X., Lu, C., Moore, C., & Varani, G. (2010). Novel protein–protein contacts facilitate mRNA 3′-processing signal recognition by Rna15 and Hrp1. *Journal of Molecular Biology*, *401*(3), 334–349. http://dx.doi.org/10.1016/j.jmb.2010.06.032.

Li, X., Quon, G., Lipshitz, H., & Morris, Q. (2010). Predicting in vivo binding sites of RNA-binding proteins using mRNA secondary structure. *RNA*, *16*(6), 1096–1107. http://dx.doi.org/10.1261/rna.2017210.

Liu, H.-X., Chew, S. L., Cartegni, L., Zhang, M. Q., & Krainer, A. R. (2000). Exonic splicing enhancer motif recognized by human SC35 under splicing conditions. *Molecular and Cellular Biology*, *20*(3), 1063–1071. http://dx.doi.org/10.1128/mcb.20.3.1063-1071.2000.

Liu, X., Niu, C., Ren, J., Zhang, J., Xie, X., Zhu, H., et al. (2013). The RRM domain of human fused in sarcoma protein reveals a non-canonical nucleic acid binding site. *Biochimica et Biophysica Acta*, *1832*(2), 375–385. http://dx.doi.org/10.1016/j.bbadis.2012.11.012, S0925-4439(12)00271-2 [pii].

Liu, H.-X., Zhang, M., & Krainer, A. R. (1998). Identification of functional exonic splicing enhancer motifs recognized by individual SR proteins. *Genes & Development*, *12*(13), 1998–2012. http://dx.doi.org/10.1101/gad.12.13.1998.

Lukavsky, P. J., Daujotyte, D., Tollervey, J. R., Ule, J., Stuani, C., Buratti, E., et al. (2013). Molecular basis of UG-rich RNA recognition by the human splicing factor TDP-43. *Nature Structural & Molecular Biology*, *20*(12), 1443–1449. http://dx.doi.org/10.1038/nsmb.2698.

Lunde, B. M., Horner, M., & Meinhart, A. (2011). Structural insights into cis element recognition of non-polyadenylated RNAs by the Nab3-RRM. *Nucleic Acids Research*, *39*(1), 337–346. http://dx.doi.org/10.1093/Nar/Gkq751.

Mackereth, C. D., Madl, T., Bonnal, S., Simon, B., Zanier, K., Gasch, A., et al. (2011). Multi-domain conformational selection underlies pre-mRNA splicing regulation by U2AF. *Nature*, *475*(7356), 408–411. http://dx.doi.org/10.1038/nature10171.

Mackereth, C. D., Simon, B., & Sattler, M. (2005). Extending the size of protein-RNA complexes studied by nuclear magnetic resonance spectroscopy. *Chembiochem*, *6*(9), 1578–1584. http://dx.doi.org/10.1002/cbic.200500106.

Marquis, J., Paillard, L., Audic, Y., Cosson, B., Danos, O., Le bec, C., et al. (2006). CUG-BP1/CELF1 requires UGU-rich sequences for high-affinity binding. *Biochemical Journal*, *400*(2), 291–301. http://dx.doi.org/10.1042/bj20060490.

Martin-Tumasz, S., Reiter, N. J., Brow, D. A., & Butcher, S. E. (2010). Structure and functional implications of a complex containing a segment of U6 RNA bound by a domain of Prp24. *RNA*, *16*(4), 792–804. http://dx.doi.org/10.1261/rna.1913310.

Martin-Tumasz, S., Richie, A. C., Clos, L. J., Brow, D. A., & Butcher, S. E. (2011). A novel occluded RNA recognition motif in Prp24 unwinds the U6 RNA internal stem loop. *Nucleic Acids Research*, *39*(17), 7837–7847. http://dx.doi.org/10.1093/Nar/Gkr455.

Mazza, C., Segref, A., Mattaj, I. W., & Cusack, S. (2002). Large-scale induced fit recognition of an m(7)GpppG cap analogue by the human nuclear cap-binding complex. *The EMBO Journal*, *21*(20), 5548–5557.

Montemayor, E. J., Curran, E. C., Liao, H. H., Andrews, K. L., Treba, C. N., Butcher, S. E., et al. (2014). Core structure of the U6 small nuclear ribonucleoprotein at 1.7-A resolution. *Nature Structural & Molecular Biology*, *21*(6), 544–551. http://dx.doi.org/10.1038/nsmb.2832.

Mori, D., Sasagawa, N., Kino, Y., & Ishiura, S. (2008). Quantitative analysis of CUG-BP1 binding to RNA repeats. *Journal of Biochemistry*, *143*(3), 377–383. http://dx.doi.org/10.1093/Jb/Mvm230.

Moursy, A., Allain, F. H., & Clery, A. (2014). Characterization of the RNA recognition mode of hnRNP G extends its role in SMN2 splicing regulation. *Nucleic Acids Research*, *42*(10), 6659–6672. http://dx.doi.org/10.1093/nar/gku244.

Mukherjee, N., Corcoran, David L., Nusbaum, Jeffrey D., Reid, David W., Georgiev, S., Hafner, M., et al. (2011). Integrative regulatory mapping indicates that the RNA-binding protein HuR couples pre-mRNA processing and mRNA stability. *Molecular Cell*, *43*(3), 327–339. http://dx.doi.org/10.1016/j.molcel.2011.06.007.

Myers, J. C., Moore, S. A., & Shamoo, Y. (2003). Structure-based incorporation of 6-methyl-8-(2-deoxy-β-ribofuranosyl)isoxanthopteridine into the human telomeric repeat DNA as a probe for UP1 binding and destabilization of G-tetrad structures. *Journal of Biological Chemistry*, *278*(43), 42300–42306. http://dx.doi.org/10.1074/jbc.M306147200.

Nagai, K., Oubridge, C., Jessen, T. H., Li, J., & Evans, P. R. (1990). Crystal structure of the RNA-binding domain of the U1 small nuclear ribonucleoprotein A. *Nature*, *348*(6301), 515–520. http://dx.doi.org/10.1038/348515a0.

Nagalakshmi, U., Wang, Z., Waern, K., Shou, C., Raha, D., Gerstein, M., et al. (2008). The transcriptional landscape of the yeast genome defined by RNA sequencing. *Science*, *320*(5881), 1344–1349. http://dx.doi.org/10.1126/science.1158441.

Oberstrass, F. C., Auweter, S. D., Erat, M., Hargous, Y., Henning, A., Wenter, P., et al. (2005). Structure of PTB bound to RNA: Specific binding and implications for splicing regulation. *Science*, *309*(5743), 2054–2057. http://dx.doi.org/10.1126/science.1114066.

Ohyama, T., Nagata, T., Tsuda, K., Kobayashi, N., Imai, T., Okano, H., et al. (2012). Structure of Musashi1 in a complex with target RNA: The role of aromatic stacking interactions. *Nucleic Acids Research*, *40*(7), 3218–3231. http://dx.doi.org/10.1093/nar/gkr1139.

Oubridge, C., Ito, N., Evans, P. R., Teo, C. H., & Nagai, K. (1994). Crystal structure at 1.92 A resolution of the RNA-binding domain of the U1A spliceosomal protein complexed with an RNA hairpin. *Nature*, *372*(6505), 432–438. http://dx.doi.org/10.1038/372432a0.

Pancevac, C., Goldstone, D. C., Ramos, A., & Taylor, I. A. (2010). Structure of the Rna15 RRM–RNA complex reveals the molecular basis of GU specificity in transcriptional 3′-end processing factors. *Nucleic Acids Research*, *38*(9), 3119–3132. http://dx.doi.org/10.1093/nar/gkq002.

Pandit, S., Zhou, Y., Shiue, L., Coutinho-Mansfield, G., Li, H., Qiu, J., et al. (2013). Genome-wide analysis reveals SR protein cooperation and competition in regulated splicing. *Molecular Cell*, *50*(2), 223–235. http://dx.doi.org/10.1016/j.molcel.2013.03.001.

Park, S., Myszka, D. G., Yu, M., Littler, S. J., & Laird-Offringa, I. A. (2000). HuD RNA recognition motifs play distinct roles in the formation of a stable complex with AU-rich RNA. *Molecular and Cellular Biology*, *20*(13), 4765–4772.

Pérez, I., Lin, C. H., McAfee, J. G., & Patton, J. G. (1997). Mutation of PTB binding sites causes misregulation of alternative 3′ splice site selection in vivo. *RNA*, *3*(7), 764–778.

Perez-Canadillas, J. M. (2006). Grabbing the message: Structural basis of mRNA 3′UTR recognition by Hrp1. *The EMBO Journal*, *25*(13), 3167–3178. http://dx.doi.org/10.1038/sj.emboj.7601190.

Perez-Canadillas, J. M., & Varani, G. (2003). Recognition of GU-rich polyadenylation regulatory elements by human CstF-64 protein. *The EMBO Journal*, *22*(11), 2821–2830. http://dx.doi.org/10.1093/emboj/cdg259.

Polymenidou, M., Lagier-Tourenne, C., Hutt, K. R., Huelga, S. C., Moran, J., Liang, T. Y., et al. (2011). Long pre-mRNA depletion and RNA missplicing contribute to neuronal vulnerability from loss of TDP-43. *Nature Neuroscience*, *14*(4), 459–468.

Ponthier, J. L., Schluepen, C., Chen, W., Lersch, R. A., Gee, S. L., Hou, V. C., et al. (2006). Fox-2 splicing factor binds to a conserved intron motif to promote inclusion of protein 4.1R alternative exon 16. *Journal of Biological Chemistry*, *281*(18), 12468–12474. http://dx.doi.org/10.1074/jbc.M511556200.

Price, S. R., Evans, P. R., & Nagai, K. (1998). Crystal structure of the spliceosomal U2B″-U2A′ protein complex bound to a fragment of U2 small nuclear RNA. *Nature*, *394*(6694), 645–650. http://dx.doi.org/10.1038/29234.

Rappsilber, J., Ryder, U., Lamond, A. I., & Mann, M. (2002). Large-scale proteomic analysis of the human spliceosome. *Genome Research*, *12*(8), 1231–1245. http://dx.doi.org/10.1101/gr.473902.

Ray, D., Kazan, H., Chan, E. T., Castillo, L. P., Chaudhry, S., Talukder, S., et al. (2009). Rapid and systematic analysis of the RNA recognition specificities of RNA-binding proteins. *Nature Biotechnology*, *27*(7), 667–670.

Ray, D., Kazan, H., Cook, K. B., Weirauch, M. T., Najafabadi, H. S., Li, X., et al. (2013). A compendium of RNA-binding motifs for decoding gene regulation. *Nature*, *499*(7457), 172–177.

Sanford, J. R., Wang, X., Mort, M., VanDuyn, N., Cooper, D. N., Mooney, S. D., et al. (2009). Splicing factor SFRS1 recognizes a functionally diverse landscape of RNA transcripts. *Genome Research*, *19*(3), 381–394. http://dx.doi.org/10.1101/gr.082503.108.

Schaal, T. D., & Maniatis, T. (1999). Selection and characterization of Pre-mRNA splicing enhancers: Identification of novel SR protein-specific enhancer sequences. *Molecular and Cellular Biology*, *19*(3), 1705–1719.

Sephton, C. F., Cenik, C., Kucukural, A., Dammer, E. B., Cenik, B., Han, Y., et al. (2011). Identification of neuronal RNA targets of TDP-43-containing ribonucleoprotein complexes. *Journal of Biological Chemistry*, *286*(2), 1204–1215. http://dx.doi.org/10.1074/jbc.M110.190884.

Sickmier, E. A., Frato, K. E., Shen, H., Paranawithana, S. R., Green, M. R., & Kielkopf, C. L. (2006). Structural basis for polypyrimidine tract recognition by the essential pre-mRNA splicing factor U2AF65. *Molecular Cell*, *23*(1), 49–59. http://dx.doi.org/10.1016/j.molcel.2006.05.025.

Singh, R., Banerjee, H., & Green, M. R. (2000). Differential recognition of the polypyrimidine-tract by the general splicing factor U2AF65 and the splicing repressor sex-lethal. *RNA*, *6*(6), 901–911.

Singh, M., Choi, C. P., & Feigon, J. (2013). XRRM: A new class of RRM found in the telomerase La family protein p65. *RNA Biology*, *10*(3), 353–359. http://dx.doi.org/10.4161/rna.23608.

Singh, R., Valcarcel, J., & Green, M. (1995). Distinct binding specificities and functions of higher eukaryotic polypyrimidine tract-binding proteins. *Science*, *268*(5214), 1173–1176. http://dx.doi.org/10.1126/science.7761834.

Singh, M., Wang, Z., Koo, B. K., Patel, A., Cascio, D., Collins, K., et al. (2012). Structural basis for telomerase RNA recognition and RNP assembly by the Holoenzyme La family protein p65. *Molecular Cell*, *47*(1), 16–26. http://dx.doi.org/10.1016/j.molcel.2012.05.018.

Skrisovska, L., & Allain, F. H. (2008). Improved segmental isotope labeling methods for the NMR study of multidomain or large proteins: Application to the RRMs of Npl3p and hnRNP L. *Journal of Molecular Biology*, *375*(1), 151–164. http://dx.doi.org/10.1016/j.jmb.2007.09.030.

Skrisovska, L., Bourgeois, C. F., Stefl, R., Grellscheid, S. N., Kister, L., Wenter, P., et al. (2007). The testis-specific human protein RBMY recognizes RNA through a novel mode of interaction. *EMBO Reports*, *8*(4), 372–379.

Steimer, L., Wurm, J. P., Linden, M. H., Rudolph, M. G., Wohnert, J., & Klostermeier, D. (2013). Recognition of two distinct elements in the RNA substrate by the RNA-binding domain of the T. thermophilus DEAD box helicase Hera. *Nucleic Acids Research*, *41*(12), 6259–6272. http://dx.doi.org/10.1093/nar/gkt323.

Sugimoto, Y., Konig, J., Hussain, S., Zupan, B., Curk, T., Frye, M., et al. (2012). Analysis of CLIP and iCLIP methods for nucleotide-resolution studies of protein-RNA interactions. *Genome Biology*, *13*(8), R67.

Tacke, R., & Manley, J. L. (1995). The human splicing factors ASF/SF2 and SC35 possess distinct, functionally significant RNA binding specificities. *The EMBO Journal*, *14*(14), 3540–3551.

Tacke, R., Tohyama, M., Ogawa, S., & Manley, J. L. (1998). Human Tra2 proteins are sequence-specific activators of pre-mRNA splicing. *Cell*, *93*(1), 139–148.

Takagaki, Y., & Manley, J. L. (1997). RNA recognition by the human polyadenylation factor CstF. *Molecular and Cellular Biology*, *17*(7), 3907–3914.

Teplova, M., Song, J., Gaw, H. Y., Teplov, A., & Patel, D. J. (2010). Structural insights into RNA recognition by the alternate-splicing regulator CUG-binding protein 1. *Structure*, *18*(10), 1364–1377. http://dx.doi.org/10.1016/J.Str.2010.06.018.

Teplova, M., Wohlbold, L., Khin, N. W., Izaurralde, E., & Patel, D. J. (2011). Structure-function studies of nucleocytoplasmic transport of retroviral genomic RNA by mRNA export factor TAP. *Nature Structural & Molecular Biology*, *18*(9), 990–998. http://dx.doi.org/10.1038/nsmb.2094.

Tintaru, A. M., Hautbergue, G. M., Hounslow, A. M., Hung, M. L., Lian, L. Y., Craven, C. J., et al. (2007). Structural and functional analysis of RNA and TAP binding to SF2/ASF. *EMBO Reports*, *8*(8), 756–762.

Tollervey, J. R., Curk, T., Rogelj, B., Briese, M., Cereda, M., Kayikci, M., et al. (2011). Characterizing the RNA targets and position-dependent splicing regulation by TDP-43. *Nature Neuroscience*, *14*(4), 452–458.

Tsuda, K., Kuwasako, K., Takahashi, M., Someya, T., Inoue, M., Terada, T., et al. (2009). Structural basis for the sequence-specific RNA-recognition mechanism of human CUG-BP1 RRM3. *Nucleic Acids Research*, *37*(15), 5151–5166. http://dx.doi.org/10.1093/nar/gkp546.

Tsuda, K., Someya, T., Kuwasako, K., Takahashi, M., He, F., Unzai, S., et al. (2011). Structural basis for the dual RNA-recognition modes of human Tra2-beta RRM. *Nucleic Acids Research*, *39*(4), 1538–1553. http://dx.doi.org/10.1093/nar/gkq854.

Tuck, Alex C., & Tollervey, D. (2013). A transcriptome-wide atlas of RNP composition reveals diverse classes of mRNAs and lncRNAs. *Cell*, *154*(5), 996–1009. http://dx.doi.org/10.1016/j.cell.2013.07.047.

Valentini, S. R., Weiss, V. H., & Silver, P. A. (1999). Arginine methylation and binding of Hrp1p to the efficiency element for mRNA 3'-end formation. *RNA*, *5*(2), 272–280.

Venter, J. C., Adams, M. D., Myers, E. W., Li, P. W., Mural, R. J., Sutton, G. G., et al. (2001). The sequence of the human genome. *Science*, *291*(5507), 1304–1351. http://dx.doi.org/10.1126/science.1058040.

Vitali, F., Henning, A., Oberstrass, F. C., Hargous, Y., Auweter, S. D., Erat, M., et al. (2006). Structure of the two most C-terminal RNA recognition motifs of PTB using segmental isotope labeling. *The EMBO Journal*, *25*(1), 150–162. http://dx.doi.org/10.1038/sj.emboj.7600911.

Wang, X., & Tanaka Hall, T. M. (2001). Structural basis for recognition of AU-rich element RNA by the HuD protein. *Nature Structural & Molecular Biology*, *8*(2), 141–145. http://dx.doi.org/10.1038/84131.

Wang, H., Zeng, F., Liu, Q., Liu, H., Liu, Z., Niu, L., et al. (2013). The structure of the ARE-binding domains of Hu antigen R (HuR) undergoes conformational changes during RNA binding. *Acta Crystallographica. Section D: Biological Crystallography*, *69*(Pt. 3), 373–380. http://dx.doi.org/10.1107/S0907444912047828.

Webb, S., Hector, R., Kudla, G., & Granneman, S. (2014). PAR-CLIP data indicate that Nrd1-Nab3-dependent transcription termination regulates expression of hundreds of protein coding genes in yeast. *Genome Biology*, *15*(1), 1–15. http://dx.doi.org/10.1186/gb-2014-15-1-r8.

Wysoczanski, P., Schneider, C., Xiang, S., Munari, F., Trowitzsch, S., Wahl, M. C., et al. (2014). Cooperative structure of the heterotrimeric pre-mRNA retention and splicing complex. *Nature Structural & Molecular Biology*, *21*(10), 911–918. http://dx.doi.org/10.1038/nsmb.2889.

Xue, Y., Zhou, Y., Wu, T., Zhu, T., Ji, X., Kwon, Y.-S., et al. (2009). Genome-wide analysis of PTB-RNA interactions reveals a strategy used by the general splicing repressor to modulate exon inclusion or skipping. *Molecular Cell*, *36*(6), 996–1006. http://dx.doi.org/10.1016/j.molcel.2009.12.003.

Yeo, G. W., Coufal, N. G., Liang, T. Y., Peng, G. E., Fu, X.-D., & Gage, F. H. (2009). An RNA code for the FOX2 splicing regulator revealed by mapping RNA-protein interactions in stem cells. *Nature Structural & Molecular Biology*, *16*(2), 130–137.

Young, P. J., DiDonato, C. J., Hu, D., Kothary, R., Androphy, E. J., & Lorson, C. L. (2002). SRp30c-dependent stimulation of survival motor neuron (SMN) exon 7 inclusion is facilitated by a direct interaction with hTra2 beta 1. *Human Molecular Genetics*, *11*(5), 577–587.

Zarnack, K., König, J., Tajnik, M., Martincorena, I., Eustermann, S., Stévant, I., et al. (2013). Direct competition between hnRNP C and U2AF65 protects the transcriptome from the exonization of Alu elements. *Cell, 152*(3), 453–466. http://dx.doi.org/10.1016/j.cell.2012.12.023.

Zhang, C., & Darnell, R. B. (2011). Mapping in vivo protein-RNA interactions at single-nucleotide resolution from HITS-CLIP data. *Nature Biotechnology, 29*(7), 607–614.

Zhang, X. H. F., Leslie, C. S., & Chasin, L. A. (2005). Computational searches for splicing signals. *Methods, 37*(4), 292–305. http://dx.doi.org/10.1016/j.ymeth.2005.07.011.

Zhang, W., Zeng, F., Liu, Y., Zhao, Y., Lv, H., Niu, L., et al. (2013). Crystal structures and RNA-binding properties of the RNA recognition motifs of heterogeneous nuclear ribonucleoprotein L: Insights into its roles in alternative splicing regulation. *Journal of Biological Chemistry, 288*(31), 22636–22649. http://dx.doi.org/10.1074/jbc.M113.463901.

Zhao, J., Ohsumi, T. K., Kung, J. T., Ogawa, Y., Grau, D. J., Sarma, K., et al. (2010). Genome-wide identification of polycomb-associated RNAs by RIP-seq. *Molecular Cell, 40*(6), 939–953. http://dx.doi.org/10.1016/j.molcel.2010.12.011.

CHAPTER TEN

Combining NMR and EPR to Determine Structures of Large RNAs and Protein–RNA Complexes in Solution

Olivier Duss*,[1], Maxim Yulikov[†,1], Frédéric H.T. Allain*, Gunnar Jeschke[†,2]

*Institute of Molecular Biology and Biophysics, ETH Zürich, CH-8093 Zürich, Switzerland
[†]Laboratory of Physical Chemistry, ETH Zürich, CH-8093 Zürich, Switzerland
[2]Corresponding author: e-mail address: gunnar.jeschke@phys.chem.ethz.ch

Contents

1. Introduction — 280
2. Limitations of Current Techniques to Solve Structures of Large RNAs and Protein–RNA Complexes in Solution — 282
3. EPR Theory — 285
   3.1 Introduction and Description of Measurement Settings — 285
   3.2 Dipolar Interaction and Analysis of the DEER Time Traces — 287
4. EPR-Aided Approach for Structure Determination of Large RNAs and Protein–RNA Complexes in Solution — 292
   4.1 Overview — 292
   4.2 Site-Specific Labeling of RNA and Proteins — 297
   4.3 Measurement of DEER EPR Distance Distributions — 300
   4.4 Modeling Spin-Label Conformations — 308
   4.5 Simple EPR Distance Constraints — 310
   4.6 Protocol for CYANA Structure Calculation Using Combined NMR/EPR Data — 313
   4.7 Structure Validation — 319
5. Future Challenges and Implications — 322
Acknowledgments — 324
References — 324

Abstract

Although functional significance of large noncoding RNAs and their complexes with proteins is well recognized, structural information for this class of systems is very scarce. Their inherent flexibility causes problems in crystallographic approaches, while their

---

[1] These authors contributed equally to this work.

typical size is beyond the limits of state-of-the-art purely NMR-based approaches. Here, we review an approach that combines high-resolution NMR restraints with lower resolution long-range constraints based on site-directed spin labeling and measurements of distance distribution restraints in the range between 15 and 80 Å by the four-pulse double electron–electron resonance (DEER) EPR technique. We discuss sample preparation, the basic assumptions behind data analysis in the EPR-based distance measurements, treatment of the label-based constraints in generation of the structure, and the back-calculation of distance distributions for structure validation. Step-by-step protocols are provided for DEER distance distribution measurements including data analysis and for CYANA based structure calculation using combined NMR and EPR data.

1. INTRODUCTION

Although RNA is overshadowed in public perception by proteins and DNA, it is one of the key components of living cells. Besides simply transferring the information coded in the DNA to the ribosomes, which are the protein-producing machines in the cell, RNA is central in gene regulation and early organisms might even have consisted solely of RNA (Szathmáry, 1999). Nowadays, it is assumed that 98% of the transcribed RNA in humans is not translated into a protein (Mattick & Makunin, 2005). Previously considered to be simply transcriptional junk, these untranslated RNAs (noncoding RNAs, ncRNAs) can have catalytic activity, serve as protein scaffolds, guide the cellular machinery, and are crucial in all aspects of gene expression, such as regulating chromatin remodeling, transcription, and many posttranscriptional events, both in prokaryotes and eukaryotes (Baker, 2011; Guttman & Rinn, 2012; Mattick et al., 2005; Morris & Mattick, 2014; Ponting, Oliver, & Reik, 2009; Storz, Altuvia, & Wassarman, 2005; Wiedenheft, Sternberg, & Doudna, 2012). Some ncRNAs have 100,000 nucleotides such as the long ncRNA Air present in mammals, whereas silencing RNAs may be only 20 nucleotides long (Storz et al., 2005) and are considered as a potential new class of drugs (Fichou & Férec, 2006). While the number of newly discovered ncRNAs is dramatically increasing and many new functions are annotated to these RNAs, structural information of these RNAs and the complexes they form with proteins is almost negligible.

High-resolution structures of large macromolecular systems are usually obtained by X-ray crystallography, a method which, in principle, has no size limitation. However, the flexibility of local structural elements such as loops, linkers, and protein or RNA termini but also global large-scale motions of RNA helices or protein domains in macromolecular complexes can lead to incompletely resolved structures or even impede crystallization. The problem can sometimes be circumvented by truncation of flexible parts, although such parts may be critical for function. Even when RNA or its macromolecular complexes crystallize, the domain arrangement or even stoichiometry can be different from that observed in solution (Nam, Chen, Gregory, Chou, & Sliz, 2011; Sickmier et al., 2006; Simon, Madl, Mackereth, Nilges, & Sattler, 2010; Teplova et al., 2011). For instance, the crystal structure of the tandem RRM domains of U2AF^{65} in complex with a polyuridine RNA (Sickmier et al., 2006) has a different RRM domain orientation than in solution (Simon et al., 2010). In the crystal structure, the polyuridine RNA is bound to RRM1 and RRM2 of two distinct protein molecules indicating that crystal packing is influencing the relative orientation of the two RRM domains (Sickmier et al., 2006).

Macromolecules and their complexes are not rigid but can adopt multiple conformations in solution (Baldwin & Kay, 2009; Clore & Iwahara, 2009; Dethoff, Chugh, Mustoe, & Al-Hashimi, 2012; Mackereth & Sattler, 2012). Recently, we found that the bacterial noncoding RNA RsmZ sequesters several RsmE proteins (translational repressor) resulting into two structurally different native protein–RNA complexes (Duss, Michel, Yulikov, et al., 2014). Similarly, U2AF^{65} in complex with RNA is present in at least two distinct conformations, with the relative populations of both conformations being dependent on relative affinity of the target RNA sequence to the two RRMs (Mackereth et al., 2011). RNA in isolation also is very dynamic as shown for the HIV-1 TAR RNA that can sample in its free state all the conformations that it adopts when bound to different ligands (Zhang, Stelzer, Fisher, & Al-Hashimi, 2007).

Transient macromolecular interactions, low populated intermediates (Clore, Tang, & Iwahara, 2007), or minor conformations of macromolecular complexes might be overlooked due to the crystallization process that often traps a single conformation. It is, therefore, crucial to develop methods to study large RNAs and RNPs (ribonucleoprotein complexes) in solution.

In this chapter, we will start by discussing limitations of current techniques available to determine structures of large RNAs and protein–RNA complexes in solution (Section 2). We continue by introducing the theoretical background of electron paramagnetic resonance (EPR) (Section 3). The main part of this review consists of a detailed presentation of an EPR-aided approach for structure determination of large RNAs or protein–RNA complexes in solution (Section 4). The structure determination protocol will be showcased by means of the 70 kDa complex of the bacterial noncoding RNA RsmZ bound to three homodimeric RsmE proteins, a complex which is important in fine-tuning bacterial virulence (Duss, Michel, Diarra dit Konte, Schubert, & Allain, 2014; Duss, Michel, Yulikov, et al., 2014; Duss, Yulikov, Jeschke, & Allain, 2014). We will conclude this chapter, by discussing future challenges and implications of combining NMR with EPR for structure determination of large RNAs and protein–RNA complexes.

2. LIMITATIONS OF CURRENT TECHNIQUES TO SOLVE STRUCTURES OF LARGE RNAs AND PROTEIN–RNA COMPLEXES IN SOLUTION

Large macromolecules or their complexes can be studied in liquid or frozen solution by spectroscopic techniques, scattering techniques, and electron microscopy (EM). Among the spectroscopic techniques, NMR spectroscopy is the only one that can provide a sufficiently large number of conformational restraints to define a structure at atomic resolution. However, unfavorable relaxation properties and limitations in spectral resolution limit the molecular weight of biomolecular complexes that can be studied solely by NMR.

During the past decade, the rather small molecular weight limit (about 25 kDa) imposed by classical NMR approaches relying only on the use of nuclear Overhauser effects (NOEs)-derived distance constraints was surpassed by the use of additional measurements such as residual dipolar couplings, which provide global orientation information (Hass & Ubbink, 2014; Lipsitz & Tjandra, 2004; Schwieters et al., 2010), and paramagnetic relaxation enhancements (PREs) (Clore & Iwahara, 2009; Lapinaite et al., 2013; Madl, Guttler, Gorlich, & Sattler, 2011; Simon et al., 2010), which provide medium-range distance restraints up to 20–25 Å.

More recently, NMR restraints have been combined with lower resolution global shape information resulting from small angle X-ray (SAXS) and neutron scattering (SANS) measurements (Burke, Sashital, Zuo, Wang, & Butcher, 2012; Grishaev, Wu, Trewhella, & Bax, 2005; Grishaev, Ying, Canny, Pardi, & Bax, 2008; Lapinaite et al., 2013; Madl, Gabel, & Sattler, 2011) or envelops resulting from EM (Miyazaki et al., 2010).

Overall, these combined approaches proved to be useful for determining the structures of intermediate-size RNAs (Lukavsky, Kim, Otto, & Puglisi, 2003) or protein–RNA complexes (Dominguez, Schubert, Duss, Ravindranathan, & Allain, 2011; Duss, Michel, Diarra dit Konte, et al.,

2014; Hudson, Martinez-Yamout, Dyson, & Wright, 2004; Johansson et al., 2004; Lee et al., 2006; Leeper, Qu, Lu, Moore, & Varani, 2010; Mackereth et al., 2011; Stefl et al., 2010; Varani et al., 2000). However, only two solution structures of macromolecular complexes of over 50 kDa molecular weight containing an RNA component have been determined so far at high resolution (Duss, Michel, Yulikov, et al., 2014; Lapinaite et al., 2013). While for the smaller 70 kDa RsmE-RsmZ protein–RNA complex (Duss, Michel, Yulikov, et al., 2014; Duss, Yulikov, et al., 2014), both the RNA and the protein components could be determined at high resolution, no high-resolution information could be obtained for the RNA in the 390 kDa box C/D ribonucleoprotein enzyme (Lapinaite et al., 2013).

In large proteins or protein complexes, NOEs between amide protons or protonated methyl groups can be resolved by deuterating all the other protons thereby reducing the fast relaxation by nearby proton spins (Religa, Sprangers, & Kay, 2010; Sprangers & Kay, 2007; Sprangers, Velyvis, & Kay, 2007). Observation of NOEs in large RNAs or RNA macromolecular complexes is more difficult than in large proteins. Very sophisticated RNA-labeling schemes are required that involve segmental isotope labeling (Duss, Lukavsky, & Allain, 2012; Duss, Maris, von Schroetter, & Allain, 2010; Lu, Miyazaki, & Summers, 2010) and selective deuteration/protonation of specific nucleotides (Davis et al., 2005; D'Souza, Dey, Habib, & Summers, 2004; Duss et al., 2012). Even then, the observation of NOEs within RNA or between RNA and proteins is very difficult, if not impossible for macromolecular complexes larger than 40 kDa. Faster relaxation at larger molecular weight leads to line broadening and the narrow chemical shift dispersion in RNA leads to even more rapid spectral crowding than for proteins at equivalent molecular weight.

For the same reason, RDCs are more difficult to measure with RNA than with proteins. Segmental labeling of RNA dramatically helps to reduce such spectral overlap (Duss et al., 2012, 2010; Lu et al., 2010). However, production of the many RNA samples needed to measure a sufficient number of RDCs can be time consuming and very costly. Also, due to particularly fast relaxation of the sugar protons and of the pyrimidine base H5 and H6 protons, only RDC from one-bond couplings of purine H8, adenine H2, and imino protons of uracil and guanine can be used in large RNAs or protein–RNA complexes, therefore reducing the number of measurable RDCs in the RNA part (Dominguez et al., 2011; Duss, Michel, Yulikov, et al., 2014).

Attaching a spin label on the RNA and detecting the PRE effect on the isotopically labeled protein has been shown to provide rich intermediate-range spatial information between protein and RNA (Edwards, Long, de Moor, Emsley, & Searle, 2013; Leeper et al., 2010; Ramos & Varani, 1998). However, the presence of more than one conformation of the protein–RNA macromolecular complex or the spin-label side chain significantly complicates the interpretation of the PRE data, because minor conformations (<5%) can significantly contribute to the observed PRE effect (Clore et al., 2007; Iwahara, Schwieters, & Clore, 2004; Jeschke, 2013; Ramos et al., 2000).

In contrast to attaching the spin label on the RNA and observing the PRE on many protein resonances, the opposite (attaching the spin label on the protein and observing the PRE effect on many RNA resonances) would provide valuable spatial high-resolution information on the RNA component. However, this requires the observation of a sufficient number of well-separated RNA resonances of good spectral quality, which also necessitates the time consuming and costly production of isotopically labeled RNAs with particular isotope-labeling schemes or segmental labeling (see above). Finally, the medium-range distances from PRE (up to 20–25 Å) might not provide enough long-range information in large protein–RNA complexes, for which distances between domains might be much longer (30–100 Å).

Low-resolution techniques such as SAXS, SANS, or EM provide valuable global shape information. The possibility of solvent contrast matching allows SANS to visualize separate domains in protein–RNA complexes (i.e., only the RNA or only a domain within a multi-domain protein if segmentally labeled; Madl, Gabel, et al., 2011). However, the resolution is so low that interpretation of the data could become ambiguous. Hence, these methods should be supplemented by additional medium-range or long-range constraints that can be unambiguously assigned, such as RDCs, PREs, or distance constraints between spin labels obtained by EPR measurements (see below). Moreover, the presence of more than a single conformation significantly complicates the interpretation of such low-resolution measurements.

Compared to NMR spectroscopy, EPR spectroscopy can access much longer distances due to the larger magnetic moment of the electron spin. Such EPR distance measurements rely on the dipole–dipole coupling that is revealed as a broadening in continuous-wave EPR spectra of nitroxide spin labels up to distances of about 20 Å (Rabenstein & Shin, 1995), by relaxation enhancement up to distances of about 30–40 Å (Jäger, Koch, Maus, Spiess, & Jeschke, 2008; Lueders, Razzaghi, et al., 2013; Voss et al., 2001), and by pulse EPR techniques up to distances of about 80 Å (Jeschke, 2012; Schiemann & Prisner, 2007). The data obtained by pulse techniques can be converted to distance distributions (Jeschke et al., 2006; Jeschke, Koch, Jonas, & Godt, 2002), which is particularly useful in cases where the macromolecule or complex exists in several distinct conformations. The accessible distance ranges match typical dimensions between domains in large macromolecules and their complexes. If complications from orientation selection are avoided (see below) and the label is rigidly attached to the macromolecule, measurements by double electron–electron resonance (DEER) are more precise than measurements in the same distance range by Förster resonance energy transfer techniques (Klostermeier & Millar, 2001). Furthermore, EPR distance measurements between two site-selectively introduced spin labels do not suffer from signal assignment problems and can be performed irrespective of the size of the macromolecule or macromolecular complex. Limitations arise from the necessity to perform the measurements at cryogenic temperature, the production of a separate sample for each distance to be measured and from conformational distribution of the label that must be considered in structure modeling. This limits resolution.

In contrast, by combining long-range distance constraints from EPR with short-range but high-resolution restraints from NMR, high-resolution structures of large RNAs and protein–RNA complexes can be obtained in solution. In the following, we will first give a theoretical introduction on EPR spectroscopy and then review the combined NMR–EPR approach for structure determination of large RNAs and protein–RNA complexes focusing on experimental details.

3. EPR THEORY

3.1 Introduction and Description of Measurement Settings

Measurement of distances in pairs of nitroxide radicals is based on selective observation of the static magnetic dipole–dipole interaction between the unpaired electron spins of the two radicals. Other interactions are either refocused by suitable pulse sequences or made small by assuring certain necessary experimental conditions (*vide infra*). The most commonly used experiment to detect such an interaction is the four-pulse DEER experiment

(Pannier, Veit, Godt, Jeschke, & Spiess, 2000), which is sometimes also referred to as pulse electron–electron double resonance (PELDOR). The DEER pulse sequence (Fig. 1) is a dead-time free modification of the previous three-pulse version of PELDOR (Milov, Salikhov, & Shchirov, 1981).

In DEER, one frequency is used for echo detection of observer spins A around frequency ω_{obs} and the other one for inverting polarization of pumped spins B around frequency ω_{pump}. As the observer pulses and the pump pulse both act on the nitroxide radicals, the pulse spectral positions and excitation bandwidths need to be properly set to avoid excitation of the A spins by the pump pulse (Pannier et al., 2000; Polyhach et al., 2012). At Q-band frequencies around 34 GHz, pump and detection pulses of 12 ns are recommended with the pump-detection frequency offset of 100 MHz (Polyhach et al., 2012), while for the narrower nitroxide spectrum at X-band with frequencies around 9.5 GHz, optimum sensitivity is obtained with a smaller frequency offset of 65 MHz and reduced excitation bandwidths of observer pulses (32 ns) and the pump pulse (12 ns).

Alternative single-frequency experiments, such as double-quantum coherence buildup (Borbat & Freed, 1999) or SIFTER (Jeschke, Pannier, Godt, & Spiess, 2000), have higher requirements on excitation bandwidth and suffer from the lack of an analytical model for signal decay caused by intermolecular dipole–dipole interaction. They are thus less frequently used, but may have a sensitivity advantage if sufficient excitation bandwidth is available (Borbat & Freed, 2007). The following analysis of dipolar evolution traces is analogous for all mentioned experiments, with some differences on the stage of decomposition into the intramolecular form factor and intermolecular background.

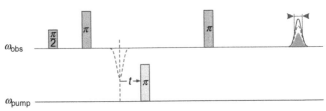

Figure 1 Four-pulse DEER sequence. Time t is varied and the echo integral in a gate with full width at half height (blue; light gray color in the print version) is recorded.

3.2 Dipolar Interaction and Analysis of the DEER Time Traces

We first consider the dipole–dipole interaction between an observer spin A, excited by the pulses at microwave frequency ω_{obs}, and a pumped spin B, excited by the π pulse at frequency ω_{pump}. If the spins are well localized, as generally applies to spin labels, and if the interspin distance r exceeds 15 Å, the exchange interaction is negligible compared to the dipole–dipole interaction. Furthermore, at such distances, the dipole–dipole interaction is much smaller than the difference of the Larmor frequencies of spins A and B, which is of the order of $\omega_{obs} - \omega_{pump}$. For all common labels, g anisotropy is less than 10%, so that to a good approximation the two magnetic moments are aligned parallel to the static magnetic field. Under all these conditions, the dipolar frequency is given by

$$\nu_{dd} = g_A g_B \mu_B^2 / (4\pi\mu_0 h) \left(3\cos^2\theta - 1\right)/r^3 \quad (1)$$

where g_A and g_B are the g values of the A and B spin, respectively, μ_B is the Bohr magneton, μ_0 is the permeability of vacuum, h is Planck's quantum of action, and θ is the angle between the spin–spin vector and the external magnetic field. For common labels, g_A and g_B can be approximated by the g value of the free electron ($g_e \approx 2.002319$), which gives

$$g_A g_B \mu_B^2 / (4\pi\mu_0 h) = 52.04\,\text{GHz}\,\text{Å}^3 \quad (2)$$

Now, we consider the signal for a sample consisting of isolated pairs of a spin A and a spin B with the same A–B distance. The contribution for a differentially small angle range $d\theta$ is given by

$$f(t)\,d\theta = \{1 - \lambda(\theta)[1 - \cos(\omega_{dd}t)]\}\sin\theta\,d\theta \quad (3)$$

where the inversion efficiency $\lambda(\theta)$ is the fraction of B spins for given value θ that are inverted by the selective pump pulse and $\omega_{dd} = 2\pi\nu_{dd}$. The inversion efficiency λ does not depend on θ if the molecular frame orientations of A and B spins are uncorrelated, which is often a good approximation. Then, Eq. (3) can be integrated assuming constant inversion efficiency λ. The normalized form factor

$$F(t) = 1 - \lambda + \lambda \int_0^{\pi/2} \cos(\omega_{dd}t)]\sin\theta\,d\theta \quad (4)$$

depends only on distance and inversion efficiency, but no longer on orientation θ of the spin–spin vector. Integration assumes a macroscopically isotropic sample where angles θ are populated with weightings $\sin\theta$.

If the A and B spin orientations are correlated, averaging of the signal over a certain range of magnetic fields largely suppresses the dependence of λ on θ (Godt, Schulte, Zimmermann, & Jeschke, 2006).

After removing the constant part of $F(t)$ and another renormalization, we obtain the dipolar evolution function

$$D(t) = \int_0^{\pi/2} \cos(\omega_{dd} t)] \sin\theta \, d\theta \tag{5}$$

A distribution $P(r)$ of distances r in the isolated spin pairs causes a distribution of dipolar frequencies ω_{dd}. We shall see below that careful data analysis can recover $P(r)$ from $D(t)$.

In practice, samples usually cannot be diluted to an extent that makes the assumption of isolated spin pairs a good approximation. Inversion of spins C_i in neighboring macromolecules contributes to the signal as a background decay $B(t)$. In three-pulse and four-pulse DEER experiments, the signal of a system with multiple pumped spins factorizes into spin pair contributions. Accordingly, the total normalized primary DEER signal is given by

$$V(t)/V(0) = F(t) B(t) \tag{6}$$

Hence, the form factor $F(t)$ can be recovered from the primary data, if a good approximation for $B(t)$ can be found. The factorization described by Eq. (6) does not strictly apply to single-frequency experiments or two-frequency experiments with more than one pump pulse.

For a homogeneous distribution of C_i spins in three-dimensional space, $B(t)$ is an exponentially decaying function $B(t) = \exp(-kt)$, with the decay constant k being proportional to C_i spin concentration and inversion efficiency λ_C for the C spins. Spatial confinement of the macromolecules can be approximated by a homogeneous distribution in space with fractional dimension $D < 3$ (Milov & Tsvetkov, 1997). In this case, the background decay takes the form $B(t) = \exp(-k_D t^{D/3})$. This is observed for membrane proteins reconstituted in liposomes (Hilger, Polyhach, Padan, Jung, & Jeschke, 2007). Fast-relaxing C_i spins lead to contributions corresponding to an apparent dimension $D = 6$ and exclusion of C spins in the volume of the macromolecule can also be modeled by a stretched exponential background function (Kattnig, Reichenwallner, & Hinderberger, 2013). Hence, it is usually sufficient to fit intramolecular background by a function

$$B(t) = \exp\left(-k_\xi t^\xi\right) \tag{7}$$

with variable stretch exponent ξ and a variable decay constant k_ξ, which is not usually further interpreted.

Such background fitting to primary data $V(t)/V(0)$ requires identification of a time interval $t_{\min,\text{bckg}} \leq t \leq t_{\max,\text{bckg}}$, where the fit is not significantly biased by the contribution from the form factor $F(t)$. In the form factor described by Eq. (4), the first half oscillation has much larger amplitude than subsequent oscillations. Accordingly, $t_{\min,\text{bckg}}$ should exceed one half of the dipolar oscillation period for the mean distance. Whenever possible, data should be measured to a time corresponding to two oscillation periods of the mean distance, even if the distance distribution is broad and the dipolar oscillation is thus overdamped.

We now assume that the dipolar evolution function $D(t)$ is defined at discrete times t_i, which we denote by a time vector $\boldsymbol{t}$. If the distance distribution $\boldsymbol{P}$ is given at discrete distances r_j, we can compute the expected dipolar evolution function as $\boldsymbol{D} = \boldsymbol{K} \cdot \boldsymbol{P}$, where the matrix elements of kernel matrix $\boldsymbol{K}$ at t_i and r_j are computed by Eqs. (5) and (1). If $\boldsymbol{D}$ were known with infinite precision, the matrix equation could be inverted to compute $\boldsymbol{P}$ from $\boldsymbol{D}$ and $\boldsymbol{K}$. This inverse problem is, however, ill-posed, meaning that even small deviations of the experimental function $\boldsymbol{D}$, such as noise, can cause large deviations of $\boldsymbol{P}$. This problem can be somewhat alleviated, but not solved, by least-squares fitting with the nonnegativity constraint for the probability distribution $P(r) \geq 0$.

The inverse problem can be further stabilized by smoothing of the distance distribution (Jeschke et al., 2002), which is best achieved by Tikhonov regularization (Chiang, Borbat, & Freed, 2005; Jeschke, Panek, Godt, Bender, & Paulsen, 2004). Tikhonov regularization minimizes the function

$$G_\alpha = \|\boldsymbol{K}\boldsymbol{P} - \boldsymbol{D}\|^2 + \alpha \|d^2/dr^2 \boldsymbol{P}\|^2 = \rho + \alpha\eta \qquad (8)$$

where ρ is the mean square deviation between the simulated and experimental dipolar evolution functions, η is the square norm of the second derivative of the distance distribution, and α is the regularization parameter. For larger α the distance distribution becomes smoother, but the deviation between simulated and experimental data also becomes larger. For very small α, the experimental data, including the noise, are almost perfectly fitted, but at the expense of an unrealistically ragged distance distribution. A good compromise value for α can be found by the L-curve method. A parametric plot of $\log \eta$ versus $\log \rho$ is L-shaped (Fig. 2B) with a steep decay for small α where smoothing strongly decreases η without causing much deviation of the

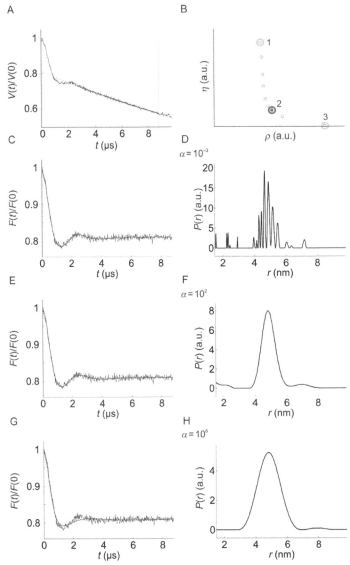

Figure 2 DEER measurements of the RsmE protein dimer bound to two singly labeled RNA stem-loops and their analysis with the DeerAnalysis package: (A) primary DEER data and best fit of the intermolecular background as obtained by the package by default optimization routines; (B) L-curve with the cases 1–3 marked (see subplots C–H), values of η and ρ as in Eq. (8); (C and D) case of the least smoothness ($\alpha = 0.001$), intramolecular form factor along with its best fit (C) as well as the corresponding distance distribution (D) are presented; (E and F) optimal smoothness of the distance distribution as judged by the L-curve criterion ($\alpha = 100$)—form factor and its best fit (E) and the corresponding distance distribution (F); (G and H) over-smoothed distance distribution case ($\alpha = 1{,}000{,}000$), form factor time trace (G) and distance distribution (H). *Experimental data from Duss, Yulikov, et al. (2014).*

simulation from experiment (undersmoothing region, Fig. 2D) and moderate decay for large α where the smoothing artificially broadens $\boldsymbol{P}$ (oversmoothing region, Fig. 2H). The two branches meet at the corner of the L-shape and the regularization parameter α corresponding to this corner is the best compromise (Fig. 2F).

The data analysis procedure described above is implemented in the open-source software package DeerAnalysis (Jeschke et al., 2006, 2004), which can be downloaded at www.epr.ethz.ch/software/index. DeerAnalysis can automatically optimize $t_{\min,\mathrm{bckg}}$ by assuming that the isolated pair distance distribution should approach zero at the longest accessible distances. This is a safe assumption if the maximum dipolar evolution time $t_{\max}$ is long enough to cover at least one full oscillation of the longest expected distance. Otherwise, background correction tends to suppress contributions from long distances.

Furthermore, DeerAnalysis provides some guidance on the reliability of features of the distance distribution (Fig. 3). This guidance refers to a base distance $r_0 = 10\,\text{Å}\cdot(t_{\max}/2\mu\text{s})^{1/3}$. The range where contributions are likely to be suppressed or strongly shifted is colored pale red. It extends from $5r_0$ to $6r_0$. For a dipolar evolution function extending to $t_{\max} = 2\,\mu\text{s}$, distances longer than 50 Å cannot be reliably determined and contributions from distances longer than 60 Å may be completely unnoticed. In the range between $4r_0$ and $5r_0$, the mean distance can be safely determined, but not the width of the distance distribution, because the width is encoded in the decay of the dipolar oscillation. This range is colored pale orange. In the pale yellow-colored range between $3r_0$ and $4r_0$, the mean and the width can be reliably estimated, but the shape of the distribution (asymmetry, shoulders) is unreliable. The shape can be trusted up to $3r_0$ (pale green range). In other words, experimental data must extend to 16 μs in order to determine shapes up to 60 Å, distribution widths up to 80 Å, and mean distances up to 100 Å.

Tikhonov regularization enforces the same smoothness of the distribution throughout the whole distance range. This may lead to distortions if both very narrow and very broad features are present, for instance features from a folded and an unfolded fraction of the protein. In such a case, the largest regularization parameter α that is capable of properly reproducing the shape of the narrowest peaks splits broader peaks in the distance distribution into sub-peaks of approximately the same width as the narrowest peak. Conversely, the smallest regularization parameter that is appropriate for reproducing the broad features leads to extra broadening of the narrowest

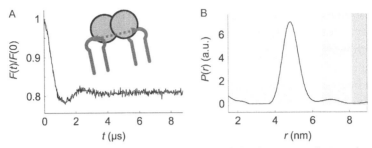

Figure 3 Form factor time trace along with its best fit by the DeerAnalysis package (A), as well as the best smoothness distance distribution (B) for a complex of the RsmE protein homodimer bound to two singly labeled RNA stem-loops, RNA-labeling site "A". The inset in (A) is a schematic of the labeled complex. The program marks the distance ranges for the precise determination of the shape of the distance distribution (green; light gray in the print version), only the mean distance and the width of the distribution (yellow; white color in the print version), only the mean distance (pale orange; light gray color in the print version), and the range where distance determination is ambiguous (red; gray color in the print version) are marked. In the case at hand, color coding indicates that mean distance and width can be safely interpreted, while the shape of the long-distance wing may be distorted. *Experimental data from Duss, Yulikov, et al. (2014).*

features. Therefore, one should analyze several distance distributions and the associated form factor fits, corresponding to different regularization parameters close to the corner of the L-curve. If in doubt, it is better to assume one broad feature rather than a multimodal distribution. Distance domain peak splittings can be safely interpreted only if they can also be recognized in the form factor as a superposition of oscillations with significantly different periods.

4. EPR-AIDED APPROACH FOR STRUCTURE DETERMINATION OF LARGE RNAs AND PROTEIN–RNA COMPLEXES IN SOLUTION

4.1 Overview

Recently, we introduced a combined NMR and EPR-aided approach for the solution structure determination of a 70 kDa protein–RNA complex simultaneously present in two conformations (Duss, Michel, Yulikov, et al., 2014; Duss, Yulikov et al., 2014). The complex consists of a 72 nucleotides RNA with four stem-loops (SLs) and a single-stranded linker, which all bind one binding site of a homodimeric RsmE protein (see Fig. 4, step 1). SL2 and SL3 bind the first RsmE dimer, SL1 and SL4 the second and the linker nucleotides bind one side of the third RsmE dimer.

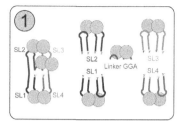
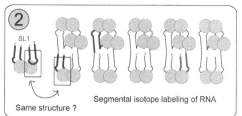
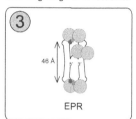
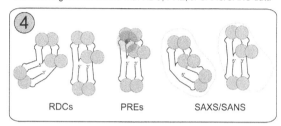

Figure 4 Procedure for elucidating the structures of large RNAs or protein–RNA complexes in solution. (1) The structures of the isolated domains are determined. (2) The chemical shifts of the isolated domains and the domains in context of the whole macromolecular complex are compared. Because of severe spectral overlap, the RNA (and potentially also multidomain proteins) has to be segmentally labeled in the full macromolecular complex. The labeled part is shown in color, the unlabeled in gray. Those parts having identical or very similar chemical shifts are considered to have the same structure and the restraints defining the isolated domains can then be used to determine the structure of the full macromolecular complex. (3) The global structure is determined using long-range EPR distance constraints. The spin label (orange star; dark gray color in the print version) can be attached on the RNA to a modified 4-thiouridine residue (red ball; black color in the print version) or onto the protein to a cysteine residue. (4) The global structure can be refined with RDC, PRE, and/or SAXS/SANS data if available.

To determine the solution structure of this complex, we used a modular and combinatorial approach, which generally consists of four steps (Fig. 4). First, the structures of the isolated domains (SL or linker nucleotides bound to one RsmE homodimeric protein) are solved. Second, the NMR chemical shifts of the isolated domains are compared with those in the full complex to verify that the domains have the same structure in isolation as in the fully assembled complex. Third, the global structure of the whole complex is determined using the isolated domains as building blocks and long-range distance distribution constraints (up to 80 Å) obtained by EPR to determine the relative arrangement of the domains (Fig. 5). Fourth, the obtained structural ensemble can be further refined or independently validated using complementary methods such as RDCs, PREs, SAXS/SANS, or EM data. If such additional data are not available, a cross-validation analysis can be performed by omitting a set of EPR constraints and investigating its effect on the resulting structural ensemble.

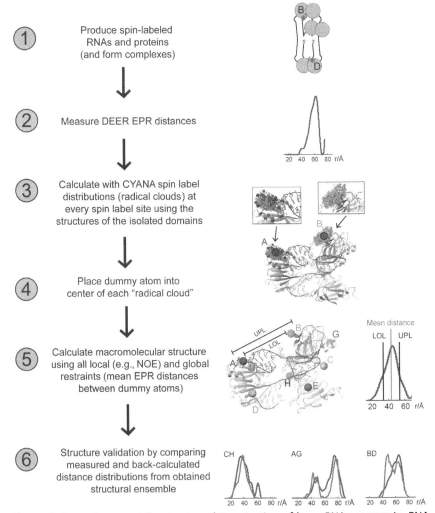

Figure 5 General protocol for structure determination of large RNAs or protein–RNA complexes using EPR-derived distance constraints. Abbreviations LOL and UPL correspond to the lower and upper limit of the distance, respectively, that is assumed in structure calculation. *This figure has been taken from Duss, Yulikov et al. (2014).*

4.1.1 Structures of Isolated Domains (Fig. 4, Step 1)

First, the structures of the isolated domains have to be determined. In some cases, the structures might already be available in the protein data bank. Isolated protein domains might be obtained using homology modeling and A-form helical RNA structures can be built using nucleic acid modeling software (Lu & Olson, 2008). X-ray crystallography is the method of choice, if a rigid subdomain is crystallizing and the phase problem can be solved. However, extended RNA structures free or in complex with proteins are often difficult to crystallize and NMR spectroscopy has proven to be a powerful method to determine the structures of small- to medium-size RNAs and protein–RNA complexes (Dominguez et al., 2011).

4.1.2 Verification that the Local Structure is the Same in Isolation and in the Full Complex (Fig. 4, Step 2)

After having determined the structures of the separated domains, it has to be verified if their structure is also the same in the context of the full complex. Chemical shifts of nuclei are very sensitive to their chemical environment (Aeschbacher et al., 2013; Aeschbacher, Schubert, & Allain, 2012). It has been shown that chemical shifts are sensitive to report on specific RNA helical conformations (Bullock, Ringel, Ish-Horowicz, & Lukavsky, 2010). Therefore, comparing the chemical shifts of the resonances in the isolated domains and the domains in the full complex is a valid approach to identify if their structures are identical in both environments.

Due to severe resonance overlap in large RNAs or protein–RNA complexes, specific isotope labeling strategies have to be applied.

4.1.2.1 Protein Resonances

Amide protons can be observed in large proteins when the protein is deuterated and TROSY-type experiments are recorded (Tzakos, Grace, Lukavsky, & Riek, 2006). Methyl group protons in an otherwise perdeuterated background can even be observed in Mega-Dalton protein complexes (Religa et al., 2010; Sprangers & Kay, 2007; Sprangers et al., 2007). Segmental isotope labeling of protein domains significantly reduces signal overlap (Skrisovska, Schubert, & Allain, 2010).

4.1.2.2 RNA Resonances

Compared to proteins, the helical nature of RNA leads to more severe resonance overlap and mainly the imino and aromatic protons can be observed in larger RNAs and protein–RNA complexes (Dominguez et al., 2011;

Duss, Michel, Yulikov, et al., 2014). Therefore, segmental isotope labeling of RNA should be done in order to reduce spectral overlap (Duss et al., 2012, 2010; Lu et al., 2010). Furthermore, selective deuteration/protonation of specific nucleotides could improve the line-width (by reducing relaxation by nearby protons) and reduce spectral overlap (Davis et al., 2005; D'Souza et al., 2004; Duss et al., 2012). When determining the structure of RsmZ bound to three RsmE dimers, five segmentally labeled RNAs were prepared in which each SL or one linker is labeled, while the rest of molecule is unlabeled (Duss et al., 2010; Duss, Michel, Yulikov, et al., 2014).

4.1.3 Global Structure Determination, Refinement, and Validation
(Fig. 4, Steps 3 and 4)
After having defined the isolated domains, they have to be positioned with respect to each other to obtain a model of the entire complex. In large macromolecular complexes, in which the "isolated domains" are simply different macromolecules, the different macromolecules can be treated as rigid bodies for which their relative spatial arrangement is best-fitted to the experimental EPR DEER data (Hilger et al., 2007; Jeschke, 2012; Park et al., 2006; Schiemann & Prisner, 2007).

However, the different domains in large RNAs are not separate macromolecules. Rather the domains consist of RNA secondary structures (free or bound to protein domains) connected by (short) single-stranded RNA linkers. This requires a protocol that determines the relative positions of the domains but still allows for restricted flexibility on the lower parts of the stems and the single-stranded linker RNA nucleotides. Therefore, our structure calculation protocol uses all the local restraints which determine the structure of the isolated domains (NOEs, dihedral angles, hydrogen-bond restraints), together with the long-range DEER EPR constraints to provide sufficient information on the global arrangement of the domains. The whole complex is folded from a random coil using the torsion angle molecular dynamics (MD) simulated annealing program CYANA (Güntert, 2004). This generates an ensemble of structures, for which the global arrangement of the domains is close to the global minimum.

The structures are further refined with a simulated annealing protocol using an RNA-optimized AMBER force field (e.g., ff99; Case et al., 2005; Dominguez et al., 2011). If additional data can be measured, the structures can be refined against RDC, PRE, or SAXS data (Dominguez et al., 2011).

In the following, we provide detailed explanations on how to use EPR for solution structure determination of large RNAs or protein–RNA complexes, which is summarized in Fig. 5.

4.2 Site-Specific Labeling of RNA and Proteins (Fig. 5, Step 1)
4.2.1 How to Choose Good Spin-Labeling Sites?
Careful selection of the spin-labeling sites is a crucial factor for a successful structure determination. The following criteria should be considered:

1. Introduction of the spin label should not disturb the complex formation nor the global structure of the complex. Successful complex formation can be verified by comparing the binding affinities of the macromolecules in the spin-labeled versus unlabeled states. Native gels can provide information on the global shape of the complex and can, therefore, indicate if spin-labeling influences the global structure of the complex. Although, it is more time consuming and requires much more sample, SAXS data can be acquired to provide information on the dimensions of the macromolecular complex with and without spin labels.

2. The structure of the RNA or protein at the labeling site should be known. If the structure at the labeling site is not known *a priori*, significant uncertainty is introduced into modeling the spin-label distributions. This uncertainty is translated into a complicated correlation between the measured EPR data and the applied structural constraint during structure calculation.

3. The site to which the spin label is attached should not be too flexible. A rigid site results into a minimization of the space that the spin label can sample in respect to the fixed domain to which it is attached. Besides reducing the disturbance of the macromolecular structure by the tag, a rigid spin-label attachment site also allows for a better modeling of the spin-label distribution. Direct experimental information on the mobility of the spin-label attachment site can be obtained by introducing two spin labels on the same rigid domain. The width of the distance distributions can provide rich information on the dynamical characteristics of the spin-label sites (Duss, Yulikov, et al., 2014).

4. The spin label should be able to move relatively freely and should not have preferred orientations. Very tight sites may result in selection of a single conformation of the spin-label side chain, but often this conformation is hard to predict (Polyhach & Jeschke, 2010). Furthermore, very tight sites increase the risk that the structure or conformational ensemble of the macromolecule is distorted or that the site cannot be labeled or

only incompletely. In the computational rotamer library approach for prediction of the conformational distribution of the label (Polyhach, Bordignon, & Jeschke, 2011), very tight sites are recognized by a small number of populated rotamers and by a small value for the partition function (Polyhach & Jeschke, 2010). Note that the narrow distance distributions that are encountered with very tightly packed labels can come at the expense of strong orientation selection, i.e., a significant dependence of inversion efficiency λ on angle θ, which in turn can cause distortions in the distance distribution.

Mild spatial restrictions of the label by the protein backbone and neighboring amino acid side chains can narrow spatial distribution of the electron spin without compromising protein structure, labeling efficiency, and predictability.

The implementation of the rotamer library approach in the open-source software MMM, which can also be downloaded at www.epr.ethz.ch/software/index, allows for scanning all possible labeling sites in a protein within a few minutes and reports the number of populated rotamers and coordinate uncertainty of the electron spin for each site. MMM version 2014 extends this approach to RNA and DNA. Alternatively, calculating the conformational distribution of the spin label at each labeling site can be performed with the CYANA structure calculation program (Duss, Yulikov, et al., 2014; see Section 4.4).

Note that in contrast to flexible spin labels, rigid spin labels can be introduced (Schiemann, Cekan, Margraf, Prisner, & Sigurdsson, 2009). The orientation selection that ensues requires more elaborate data analysis but can provide additional information that can be used in structure determination of macromolecules and has even provided information on flexibility of DNA (Marko et al., 2011).

4.2.2 Production of Large Doubly Spin-Labeled RNAs

Measuring long-range distance constraints with pulsed EPR requires the site-specific attachment of two spin labels into the RNA containing modified nucleotides. Small RNAs containing modified nucleotides can be chemically synthesized. The most commonly used modified nucleotide that is incorporated into RNA by chemical synthesis is the 4-thiouridine. A 3-(2-iodoacetamido)-proxyl (IA-proxyl) spin label can be attached to the sulfur of the 4-thiouridine base using a simple protocol (Ramos & Varani, 1998). A variety of other spin labels in RNA have also been used and is reviewed elsewhere (Zhang, Cekan, Sigurdsson, & Qin, 2009).

While chemical synthesis is the method of choice for introducing modified nucleotides into smaller RNAs (<30–50 nucleotides), many biologically relevant RNAs are significantly larger and cannot be obtained by the same strategy. The introduction of spin labels at the 5′- or 3′-ends of larger RNAs have been described and reviewed (Sowa & Qin, 2008). However, internal labeling of larger RNAs has only been possible recently.

While two elegant methods have been introduced by the Hobartner lab performing ligations of smaller RNA fragments using deoxyribozymes (Buttner, Javadi-Zarnaghi, & Hobartner, 2014; Buttner, Seikowski, Wawrzyniak, Ochmann, & Hobartner, 2013), these approaches are currently limited by strong sequence requirements at the sites of ligation.

To overcome the sequence requirements at the ligation site and to access RNAs of virtually no upper size limitation, we have recently used splinted enzymatic ligation of shorter fragments using T4 DNA ligase (Duss, Yulikov, et al., 2014). Some of these fragments contain a 4-thiouridine to which a nitroxide spin label can be attached.

Two strategies are possible: the 4-thiouridine modified and unmodified fragments can be ligated first and then labeled with a nitroxide. Alternatively, the spin label can be first attached to the fragments containing the modified 4-thiouridine followed by ligation of the spin-labeled and unlabeled fragments. We found that only the second approach is successful (Duss, Yulikov, et al., 2014). The separation of the doubly modified 4-thiouridine RNA of interest from the different unreacted RNA fragments, the RNA side-products from ligation and the long DNA splint requires a harsh denaturing purification at high temperature (85 °C) on an anion exchange HPLC (Duss et al., 2010). This results in partial hydrolysis of the 4-thiocarbonyl groups. Note also that the second approach has the advantage of being able to quantify the spin-labeling efficiency at every spin-labeling site independently, which would not be the case if the 4-thiouridine modified RNA fragments were ligated before spin-labeling. The harsh HPLC purification of the spin-labeled RNA does not lead to cleavage of the spin label. The labeling efficiency after purification determined by CW EPR is generally in the range of 85–100% for all doubly spin-labeled RNAs. Labeling of the fragments and subsequent ligation has the further advantage that different labels can be easily introduced into the same long RNA. This advantage could become important with increasing complexity of the systems under study, which may require application of orthogonal spin-labeling techniques. A step-by-step protocol will be provided elsewhere (O.D., Nana Diarra dit Konté, F.H.-T.A.).

4.2.3 Production of Spin-Labeled Proteins

Site-directed spin labeling of proteins (Hubbell, Lopez, Altenbach, & Yang, 2013) is usually performed by removing native cysteine residues (cys-less mutant), introducing cysteine residues at the desired labeling sites by site-directed mutagenesis, and attachment of thiol-reactive labels that bear methanethiosulfonate, iodoacetamide, or maleimide functionalities.

The methanethiosulfonate spin label (MTSL) is often preferred for its good compromise between flexibility and narrow spatial distribution of the spin-label site and for its high selectivity for thiol groups (Jeschke, 2013). The main disadvantage of MTSL is the chemical lability of the disulfide bond that is formed on attachment.

For proteins that have to be reconstituted under mildly reducing conditions, iodoacetamido or maleimide chemistry is preferable, although it is less selective and the more flexible side chains lead to broader distance distributions. Maleimide labels have the additional disadvantage of introducing new stereocenters, which compromises predictability of their conformational distribution.

For some proteins, not all native cysteines can be removed without distorting structure or function of the protein. Such proteins could be labeled via unnatural amino acids (Fleissner et al., 2009; Schmidt, Borbas, & Drescher, 2014), with a range of such approaches being under current development. If measurements need to be performed under reducing conditions, for instance in living bacterial cells, the redox lability of the nitroxide group also poses problems. In such cases, Gd(III) labels are advantageous (Martorana et al., 2014; Qi, Gross, Jeschke, Godt, & Drescher, 2014).

The specific labeling protocol varies between proteins and labeling sites, depending mainly on possibilities for separation of unreacted label from the protein, protein stability, and accessibility of the labeling site. Typically, cysteines are labeled with twofold to 10-fold excess of MTSL at 4 °C with incubation times between an hour and 12 h. Separation of unreacted label from the protein is particularly difficult for membrane proteins, since the hydrophobic labels tend to partition into detergent micelles and lipid bilayers. If the membrane protein bears a His-tag for purification, this problem can be solved by carrying out the labeling reaction on the purification column (Hilger, Polyhach, Jung, & Jeschke, 2009).

4.3 Measurement of DEER EPR Distance Distributions
(Fig. 5, Step 2)

Here, we will concentrate on DEER-based distance measurements with conventional nitroxide spin labels. Some comments on other types of spin

labels will be given in the later sections. The main concerns in the DEER-based distance measurements are:
- The S/N ratio in the form factor time trace.
- The largest reliably detectable distance and the precision of the shape of the computed distance distribution.
- The interference between spin–spin distances and local molecular frame orientations in the measured dipolar frequency patterns.
- The possibility of partial sample aggregation or agglomeration that would affect the measured distance distributions.

Nearly, every experimental parameter or sample preparation condition influences at least two items in the above list. As a result, optimization of the DEER measurements faces the problem of balancing multiple contradictory requirements. The typically considered optimization parameters are:
- EPR detection frequency
- Microwave power/lengths of the pulses
- Microwave resonator mode bandwidth and the pump-detection frequency separation
- Sample volume
- Concentration of spin-labeled molecules
- Length of the DEER time trace
- Type of the spin labels

Out of the detection frequencies commercially available, the best concentration sensitivity can be achieved at Q-band (34–36 GHz) with high-power mw amplifier (150–200 W) and broad-band resonator (Polyhach et al., 2012), whereas the best absolute sensitivity is rather achieved at W-band (94 GHz; Goldfarb et al., 2008). For samples with correlated orientations of the two spin labels, it is important to keep in mind that orientation selection, which is detrimental for obtaining precise distance distributions, increases with decreasing ratio between excitation bandwidth of the microwave pulses and spectral width of the spin label. Since g anisotropy leads to increasing spectral width with increasing magnetic field and available microwave power tends to decrease with increasing frequency, orientation selection may become a serious problem at high field and frequency. The problem can be overcome with dedicated high-power spectrometers (Cruickshank et al., 2009) or using Gd(III) labels, which do not exhibit orientation selection (Goldfarb, 2014). If these approaches cannot be taken and if there is no prior knowledge about the absence of preferred nitroxide mutual orientations, distance determination at W-band requires multiple DEER measurements with pump and detection frequencies set at different positions in the nitroxide EPR spectrum (Polyhach et al., 2007). In contrast, at Q-band a single DEER measurement with nonselective pulses is typically sufficient for proper distance determination (Polyhach et al., 2012). For hard nonselective microwave pulses, it is also important to have a broadband (or bimodal) resonator, so that separation between pump and detection frequency can be set sufficiently large to reduce interference between microwave pulses at these two frequencies.

With a given resonator type, the maximum possible sample volume scales with the cube of the wavelength and thus with the inverse cube of microwave frequency. Sample volumes of 50–200 µL can be used with conventional X-band resonators optimized for pulse EPR, whereas the volume decreases to 8 µL at Q-band and to less than 1 µL at W-band. This volume decrease counteracts the sensitivity increase due to higher Boltzmann polarization of the EPR transitions and the larger energy of the microwave quantum and the possible use of higher quality factor resonators at the same bandwidth. Up to Q-band frequencies, concentration sensitivity nevertheless increases, but when going to W-band, noise figures of microwave components increase and a loss in concentration sensitivity is encountered. Sample volumes up to about 30 µL can be used at Q-band when sacrificing intrinsic quality factor up to the point where this factor is optimal for the DEER experiment (Polyhach et al., 2012). An increase in concentration sensitivity by about an order of magnitude compared to X-band can be realized that way without serious reduction in microwave field homogeneity. At W-band with sufficient microwave power of 1 kW, similar sample volumes can be used by working with a nonresonant sample holder (Cruickshank et al., 2009), which then leads to similar concentration sensitivity. As the concentration of biomacromolecules is an issue more often than the available sample amount (for instance, some proteins tend to aggregate above certain concentrations), it is advisable to rely on such "oversized sample" concepts at Q- and W-band. With 30 µL sample volume at Q-band, biomolecule concentrations down to ~5 μM are safely accessible by DEER measurements (Duss, Yulikov, et al., 2014; Polyhach et al., 2012). Note that the lowest accessible concentration strongly depends on the distance to be measured, i.e., the required t_{max}, as well as on transverse relaxation time of the label, which differs strongly between different environments. Hence, lower concentrations have been reported elsewhere and can be achieved for soluble proteins in deuterated matrices when distances below 40 Å are measured.

The S/N achievable in DEER measurements increases with the increase of spin concentration. On the other hand, the stability of biomolecule solutions often reduces above a certain concentration range, which might lead to precipitation or formation of ordered or randomly formed aggregates.

The maximum measurable length of DEER traces depends on the transverse relaxation of the used spin labels, on their surrounding, and on the local spin concentration. Provided that the local spin concentration is sufficiently low, typically below 200 µM, transverse relaxation time of spin labels is longer in deuterated solvents and it further increases if deuterated biomacromolecules are prepared for spin labeling. Considering nitroxide radicals in protonated surrounding, a DEER trace with a length of 2–6 µs can be measured, which corresponds to the upper detectable distance limit of 5–7 nm. For protein–RNA complexes with protonated RNA molecules, and protein molecules expressed in D_2O but with protonated sugar source, and with EPR samples prepared in D_2O/glycerol-d_8 mixture, DEER traces with lengths of up to 20 µs could be measured. This allowed for detecting distances up to ~10.5 nm. Note that the upper limits reduce to 80% of the stated values if also the width of the distribution needs to be estimated. With more extensive deuteration schemes, samples can be prepared for detection of distances probably in excess of 10 nm, with good resolution having been proved at 7 nm (Ward et al., 2010).

4.3.1 Protocol

The following protocol assumes the use of a high-power Q-band setup, described in Gromov et al. (2001) and Polyhach et al. (2012), which is now also commercially available, together with a broad-band resonator for oversized samples (Polyhach et al., 2012; Tschaggelar et al., 2009).

1. If previously lyophilized, the sample is dissolved in either protonated or deuterated buffer to achieve twice the desired final concentration. For measurement of distances above ~6 nm, the use of deuterated buffer and final concentration of 100 µM is recommended. For distances below 6 nm, protonated buffer may be sufficient and concentrations up to ~400 µM can be used to speed up measurements. Deuterated (recommended) or protonated glycerol is added to make a 1/1 (v/v) mixture at final concentration of the biomacromolecules. A volume of 20–30 µL is filled into an EPR quartz tube (3 mm o.d., Wilmad). The sample is frozen either by immersion into liquid nitrogen or, to increase freezing speed, by immersion into a mixture of isopentane with dry ice ($T = -80\ °C$). It is kept frozen until measurement.

2. The cryostat of the microwave resonator is precooled to 50 K with liquid helium, and the sample is transferred into the resonator without warming it up. If liquid nitrogen is used for cooling, the lowest possible temperature should be set. Usually, with a sufficiently strong nitrogen pump, temperatures in the range between 75 and 80 K can be achieved. Note that use of higher measurement temperatures somewhat reduces the maximum accessible length of DEER time trace and thus the accessible distance range. The temperature of sample and resonator needs to equilibrate for 20–30 min before the DEER measurement is started to reduce phase and frequency drifts. Except for setting the final signal phase, DEER setup can be performed during equilibration time.

3. If a single-mode resonator is used, it is over-coupled to a quality factor that enables π-pulses of 12 ns over a range of 80–100 MHz. The microwave frequency of the main frequency source is then set to the upper limit of this range. At first, an echo-detected EPR spectrum is recorded. The magnetic field is then set to the position of maximum intensity in this spectrum. The frequency of the second microwave source for the pump pulse is set equal to the frequency of the main source (observer frequency). Microwave power of the pump pulse is adjusted with an inversion recovery sequence consisting of an inversion pulse at the pump frequency and a Hahn echo at the observer frequency by maximizing echo inversion. Somewhat more precise settings can be achieved by variation of the duration of the pump pulse in an echo-detected nutation experiment (Schweiger & Jeschke, 2001). In that case, the nutation experiment is repeated with different pump pulse power settings until the first minimum of the nutation trace is encountered exactly at 12 ns.

4. The observer frequency is set 80–100 MHz below the frequency of the pump source. If available, two microwave channels should be used to set up two-step [(+x)-(−x)] phase cycling on the first $\pi/2$ pulse (pulse duration 12 ns), and a third channel at higher microwave power level to set up the π-pulses of 12 ns duration. If only two channels are available, 6 ns $\pi/2$ pulses should be used. For pulses of 12 ns duration, a separation between the pump and detection frequency of 100 MHz is sufficient to reduce excitation band overlap of pump and detection pulses to an acceptable level. In case of pulse interference, a so called "2+1" effect occurs (Kurshev, Raitsimring, & Tsvetkov, 1989). This leads to a small additional dipolar oscillation signal that attains its maximum at the end of the trace, which disturbs the data. If such a raise in signal at the end of the DEER trace is observed, the trace needs to be cut at a point where

the "2 + 1" contamination is still below noise level. This cutting can be done in postprocessing after the measurement. During experiment setup, the length of the DEER trace should be set such that about two full dipolar oscillations could be observed for the longest expected distance. If no significant refocused echo signal can be detected at such a long time, sequence length is reduced until such a signal is detected. Before the measurement is started, echo phase should be readjusted with the pump pulse being present, as this pulse induces a phase shift (Bowman & Maryasov, 2007). Accumulation time needed for a sufficient signal-to-noise ratio depends on the required length of the DEER trace, on spin concentration and labeling efficiency and ranges from a few minutes to 40–50 h.

5. Distance analysis can be performed with the DeerAnalysis software (Jeschke et al., 2006). After loading, the data phase is automatically corrected and the zero time automatically determined. Good data have near zero imaginary part and a clear maximum at the determined zero time. It is recommended to optimize the background fit range automatically by clicking on the exclamation mark button, first for a 3D background (exponential decay function) and then with variable background dimensionality (stretched exponential function). The latter correction should be applied only if the former correction does not lead to a form factor trace that is flat in its last quarter and only if background dimension $2 \leq D \leq 6$ is found. Otherwise, 3D background correction is preferable. If the intramolecular distance distribution is sufficiently narrow, one or more full oscillations are visible on top of the smoothly decaying intermolecular background. In such cases, separation to the intra- and intermolecular contributions is straightforward. If the intramolecular distance distribution is broad, the dipolar signal is overdamped and essentially consists of only half an oscillation, or even resembles a decay function, without oscillatory character. In such situations, the separation of intra- and intermolecular contributions relies on the fact that at sufficiently low concentration the typical distances between neighboring molecules in the frozen solution are much longer than the longest possible intramolecular distance. If this is the case, a clear kink is visible around the time, where the intramolecular dipolar signal has decayed and the DEER trace starts to be dominated by the intermolecular background. Processing data that do not even exhibit such a kink is unreliable, as inter- and intramolecular contributions cannot be separated. After the background function has been fitted, the

primary DEER trace is automatically divided by this function resulting in the form factor trace that encodes only the intramolecular distance distribution.

6. At this point, DeerAnalysis displays a rough estimate of the distance distribution that is obtained by approximate Pake transformation (Jeschke et al., 2002). To improve the estimate, Tikhonov regularization with full L-curve computation should be selected. After the computation, distance distributions and form factor fits are available for several regularization parameters α and the L-curve is displayed. Several distance distributions, corresponding to different regularization parameters close to the corner on the L-curve, should be examined in order to assess possible shapes of the distance distribution.

A nontrivial case of a distance analysis is presented in Fig. 6. In this case, the DEER time trace encodes a bimodal distance distribution (Duss, Michel, Yulikov, et al., 2014). The primary DEER time trace has two visible kinks in the time range between $t=0$ and $t\sim5$ μs. While the first kink is automatically recognized by the DeerAnalysis routines, the second kink is partially incorporated into the default background function as shown in Fig. 6A. The problem with the background fit in this case can be recognized from the fact that the intermolecular background function (red curve (light gray color in the print version) in Fig. 6A) does not perfectly overlap with the primary data in the fitted range. The difference between the background function and the primary DEER data appears at the beginning of the range, automatically assigned for the background fit, as well as at the end of this range, the latter caused by the above mentioned "2+1" effect (Kurshev et al., 1989). The imperfections in the background correction can be seen in the resulting form factor time trace (Fig. 6B, black), which shows a gradual increase in the time range after $t\sim5$ μs. Shifting the beginning of the time range for the background fitting to the second kink position (Fig. 6D) makes most of the form factor range $t>5$ μs flat, except of the end of the region, still influenced by the "2+1" effect. The corresponding form factor time trace (black) and its fit are shown in Fig. 6E and the resulting distance distribution for the best smoothness case is presented in Fig. 6F. Cutting out the "2+1" range results in almost perfect fit of the background function to the primary DEER data in the remaining time range (Fig. 6G). The quality of the background fit in this case demonstrates that in the selected time range, indeed, the intermolecular interactions dominate the DEER data. Since the intramolecular DEER signal has completely decayed by the time $t\sim5$ μs, this indicates

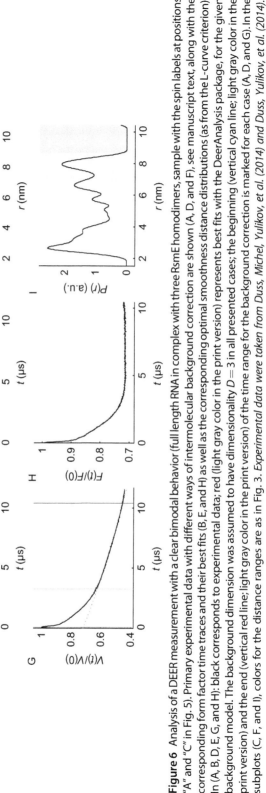

Figure 6 Analysis of a DEER measurement with a clear bimodal behavior (full length RNA in complex with three RsmE homodimers, sample with the spin labels at positions "A" and "C" in Fig. 5). Primary experimental data with different ways of intermolecular background correction are shown (A, D, and F), see manuscript text, along with the corresponding form factor time traces and their best fits (B, E, and H) as well as the corresponding optimal smoothness distance distributions (as from the L-curve criterion). In (A, B, D, E, G, and H): black corresponds to experimental data; red (light gray color in the print version) represents best fits with the DeerAnalysis package, for the given background model. The background dimension was assumed to have dimensionality $D = 3$ in all presented cases; the beginning (vertical cyan line; light gray color in the print version) and the end (vertical red line; light gray color in the print version) of the time range for the background correction is marked for each case (A, D, and G). In the subplots (C, F, and I), colors for the distance ranges are as in Fig. 3. *Experimental data were taken from Duss, Michel, Yulikov, et al. (2014) and Duss, Yulikov, et al. (2014).*

that distance distribution broadening due to the time trace truncation (in this case at about $t \sim 12$ μs) does not strongly influence the width of the distance distribution as compared to its native width. This means that the shape of the distance distribution is predicted with a better quality as compared to what is indicated by the colored ranges in Fig. 6I. For the case of the proper background correction without disturbing artifacts (Fig. 6G), the obtained distance distribution is, indeed, rather broad with remaining "wavy" features most probably originating from the Tikhonov-enforced homogeneous smoothness. As Tikhonov regularization procedure results in a distance distribution with the same smoothness over all distance ranges, it is possible that the actual distance distribution in the presented case has steeper shoulders at the lowest distances range as well at the edge of the longest distances, but in the central region, it could be smoother than presented in Fig. 6I. The distinction between such a case of a smoother central range of the distance distribution and the multiple sub-peak case predicted by the Tikhonov regularization cannot be done based on the presented analysis. In such a case, one has to assume the simplest situation, namely a smooth distance distribution. Accordingly, in the published data analysis, the distance distribution with yet higher regularization parameter that enforced smooth distance distribution in the central region was taken (Duss, Michel, Yulikov, et al., 2014).

The two main peaks in the ranges of 2–4 and 6–8 nm are safely visible in all three cases presented in Fig. 6. It is important to note that neither the most probable distance nor the range of distances that covers, e.g., 70% of the area of the distance distribution around the mean distance is strongly dependent on the background model. This feature of the analysis of the DEER data has made the determination of the long-distance constraints from the DEER data rather robust and only weakly dependent on the particular way of analyzing the experimental data (see Section 4.5.2 for how to convert EPR distance distributions into structural constraints for structure calculation). The "soft constraints," just restricting the average spin–spin distances to be within the main areas of the distance distributions were used in structure determination (Duss, Michel, Yulikov, et al., 2014; Duss, Yulikov, et al., 2014) in order to account for possible errors in the spin–spin distance predictions as well as for the possible complex flexibility.

4.4 Modeling Spin-Label Conformations (Fig. 5, Step 3)

Extracting structural information from an experimental DEER distance distribution requires knowledge about the conformational space of the spin label attached to a specific site in the macromolecule. Several approaches have been discussed how to model the spin-label conformations.

Long (4 ns) MD simulations of the R5 nitroxide label have been performed and yielded mean distances that agreed well with the experimental distance (Price, Sutch, Cai, Qin, & Haworth, 2007).

Because MD simulations require a large computational effort, rotamer libraries have been constructed for which 200 rotamers represent the conformational space of the spin label (Polyhach et al., 2011). Libraries were constructed for the MTSSL and the IA-proxyl spin labels attached to a cysteine residue. The simulated and experimental DEER signals are in an overall good agreement. Although some simulated distance distribution widths are overestimated, the mean distances are well reproduced (Jeschke, 2013; Polyhach et al., 2011). However, until recently rotamer libraries were only available for MTSL and iodoacetamido-Proxyl attached to cysteine residues in proteins. Version 2014 of the MMM program offers libraries for additional labels, among them iodoacetamido-Proxyl attached to 4-thiouridine.

Simpler conformer search algorithms have also been introduced. A simple conformer search algorithm (NASNOX) that is purely based on steric contacts was used to eliminate disallowed conformers for the R5 nitroxide spin label attached to the RNA or DNA phosphate backbone by a phosphorothioate (Cai et al., 2007, 2006; Price et al., 2007; Qin et al., 2007). The program varies the torsion angles of the spin label in a stepwise fashion and finds those spin-label conformers that have no steric clashes between the nitroxide and the rigid parent molecule. Interestingly, this simple algorithm predicts almost equally well the experimental distances as the time-consuming MD simulations (Price et al., 2007). Unfortunately, the NASNOX algorithm is currently only applicable for the R5 nitroxide probe attached to a phosphorothioate in DNA or RNA (Price et al., 2007; Qin et al., 2007; Zhang et al., 2012). The corresponding PRONOX algorithm is available for MTSL attached to cysteine residues in proteins (Hatmal et al., 2012). A similar accessible-volume algorithm is used by MtsslWizard (Hagelueken, Ward, Naismith, & Schiemann, 2012). These algorithms provide similar prediction accuracy as first-generation rotamer libraries (Jeschke, 2013), but none of the existing approaches is easy to interface with existing programs for generation of atomic resolution structures from NMR data.

In contrast, the well-established torsion angle dynamics simulations program CYANA (Güntert, 2004) allows for a straightforward computation of the spin-label conformations and is applicable to a broad range of spin label types (Duss, Yulikov, et al., 2014). Because the local structure is known at each spin-labeling site, a structure calculation can be performed in which all NMR restraints (NOEs, dihedral angles, hydrogen bonds) used to determine the structure of the nonspin-labeled macromolecular complex (including the residue to which the spin label is attached) are applied but the spin label is left unrestrained. This allows for a simple and fast sampling of the side chain conformations at every spin label attachment site. Distance distributions between two spin labels for which their side chain conformations were modeled using this simple approach represent a very good approximation of the experimental DEER distance distribution (Duss, Yulikov, et al., 2014). If the structures of the isolated domains (local structures at the spin-labeling sites) were not determined by NMR or the corresponding NMR restraints are unknown, artificial restraints extracted from the corresponding PDB file can be used.

This approach was also successful for modeling well the distance distribution of spin labels attached to partially mobile nucleotides by simply increasing the NOEs upper limits restraining such nucleotides and by that, simulating an increased flexibility of the spin-labeling site (Duss, Yulikov, et al., 2014).

4.5 Simple EPR Distance Constraints (Fig. 5, Step 4)

4.5.1 Introduction

Once two spin labels are attached at two specific sites in a macromolecule and interspin distances have been measured, distance information reporting on the macromolecular structure has to be extracted from the experimental data. However, the experimental information does not directly report on the distance between the two sites on the macromolecule to which the spin label is attached but on the distance between the radical positions on the spin label.

To include EPR restraints into structure calculation programs, several approaches have been used. Measured DEER data were directly translated as distance constraints between the Cα positions (Park et al., 2006) or the Cβ positions (Yang et al., 2010) of the cysteine residues to which the spin label was attached. Because the measured data reports on the distances between the radical positions of the two spin labels and not the Cα or Cβ positions to which the spin label is attached, this approach leads to uncertainties in the conversion of the experimental data into structural information. In another study, the interelectron distance constraints were assigned between the O atoms of the nitroxide spin label (Rao, Jao, Hegde, Langen, & Ulmer, 2010). Using this approach, it is crucial to use a densely interconnected network of constraints between the spin labels in order to reduce the degrees of freedom of each spin label during structure calculation.

Overall, all these approaches are obstructed by uncertainties and inaccuracies of the exact radical position of the flexible spin label leading to wrong correlations between the measured distance and the applied distance constraint during structure calculation.

4.5.2 Restraining Geometric Centers of Spin Label Distributions with Experimental Mean Distance

It has been shown that the mean distances are stable parameters of broad distances distributions and that there is a strong correlation between the measured and predicted mean distance (Cai et al., 2007, 2006; Duss, Yulikov, et al., 2014; Hilger et al., 2007; Jeschke et al., 2004). Furthermore, in contrast to a distance distribution, the mean distances are not much affected by the ill-posedness of the data conversion from time to distance domain (Jeschke, 2012; Jeschke et al., 2004), see also Fig. 6 and corresponding discussion at the end of Section 4.3.

The experimental mean distance is reporting in a very good approximation on the distance between the geometrical centers of the spin-label distribution (radical clouds) of two spin-labels (Duss, Yulikov, et al., 2014). Thus, a simple way to convert the experimental distance distributions into simple constrains for structure calculation is to restrain the geometrical centers of two "radical clouds" to each other using the experimental mean distance measured in DEER data (Fig. 5, steps 3–5).

To account for uncertainties resulting from distortions of the structure by the spin label, the limited precision of the experimental data (Fig. 5, step 2), and uncertainties in the prediction of the spin-label conformations (Fig. 5, step 3), we found convenient to define DEER-based distance constraints using lower and upper limit distance limits flanking the corresponding experimental mean distance similarly to NMR distance restraints in structure calculation. For the structure calculation of the RsmZ/RsmE complex, we used broad distance ranges covering about 70% of the distributions around their maxima. These very broad ranges were required because the complex is simultaneously present in two conformations resulting mainly in broad distance distributions in which the two conformations were not resolved. We expect that significantly narrower distance constraint ranges could be used if the macromolecular complex exists in a single well-defined conformation. Note that after calculating the ensembles of structures from such sets of constraints, one needs to validate the stability of the resulting ensembles by loosening or tightening the constraints or by even dropping a fraction of them.

Because the center of a "radical cloud" is not represented by a specific atom position, a dummy atom is fixed to the center of each "radical cloud" (see Fig. 5, step 4). For simplicity, the dummy atom is defined as the CA of a glycine residue, which is treated as separate molecule during structure calculation (e.g., using CYANA) and refinement (e.g., using AMBER). For each spin-label position in the full complex, another glycine CA dummy atom is defined. These dummy atoms are fixed to the geometrical centers of their corresponding "radical clouds" using tight lower and upper limit constraints between the CA of the dummy atoms and several atoms of the protein–RNA complex. Preferentially, atoms on the protein backbone are chosen to restrain the dummy atom in order to minimize distortions of local structure of the protein–RNA complex. Fixing the dummy atoms on the RNA moiety might lead to distortion of the RNA structure because the RNA has a less dense ensemble of restraints than a protein. Alternatively, the local structure of the RNA (or protein) can be fixed artificially during structure calculation such that the structure does not get distorted.

Overall, determining the relative position of these dummy atoms in space directly reports on the spatial arrangement of the different domains of the RNA– or protein–RNA complex and therefore defines the three-dimensional structure.

4.5.3 Required Number of Long-Range Distance Distribution Constraints

Let us consider an RNA or protein–RNA complex consisting of N relatively rigid domains that are joined by flexible linkers: the first domain fixes the global coordinate frame. Each additional domain introduces three rotational and three translational degrees of freedom. Hence, in an ideal system, $6(N-1)$ long-range distance distribution constraints would be required to unambiguously determine the system. To compensate for the limited precision of the experimental data and the uncertainty in translating the experimental data into distance constraints, it has been suggested to overdetermine the problem at least twofold for structure determination of multiprotein complexes (Hilger et al., 2007). In cases where the RNA or protein–RNA complex exists in two or more distinct conformations, assignment problems may arise in the distance distributions. In this case, additional distance distribution measurements may be required to resolve these ambiguities.

However, this theoretically relatively high number of required distance constraints for unambiguous structure determination is significantly decreased by steric and topological constraints imposed by the different domains onto each other. Besides simple steric hindrance between different domains, the RNA and protein linkers connecting the different domains significantly reduce the degrees of freedom. For the structure determination of the RsmZ/RsmE complex consisting of three independent domains connected by several short RNA linkers, less than 21 long-range distance constraints (each one defined by a lower and higher distance limit) were sufficient to unambiguously determine the structure of the complex. This was the case even though the complex simultaneously consists of two different conformations in solution (Duss, Michel, Yulikov, et al., 2014; Duss, Yulikov, et al., 2014). The number of required long-range distance constraints depends on the system under study. To determine if enough long-range distance constraints have been measured, a cross-validation analysis is crucial. This can be done by performing structure calculations in which a single or several distance constraints are omitted followed by verification of the generated solutions.

4.6 Protocol for CYANA Structure Calculation Using Combined NMR/EPR Data (Fig. 5, Step 5)

The entire structure calculation is based on a standard CYANA structure calculation protocol (Güntert, 2004). However, the following modifications are required.

4.6.1 Structure Calculation Protocol

A CYANA structure calculation is performed by simultaneously including all the local NMR restraints consisting of NOE distance, torsion angle and hydrogen-bond restraints, and the global EPR constraints, with all restraints having the same weight during structure calculation (Fig. 7). The number of structures calculated and the number of simulated annealing steps used depends on the system under study and has to be optimized by careful inspection of the quality of the output structural ensemble. Using a too small number of simulated annealing steps or calculating too little structures might result into higher CYANA energy target functions and can also lead to distortions of regions containing only few or even

```
# ------ read input files ------
###
      read upl      NMR_NOEs.upl              append
###
      read upl      NMR_Hbonds.upl            append
      read lol      NMR_Hbonds.lol            append
###
      read aco      NMR_torsion_angles.aco    append
###
      read upl      EPR.upl                   append
      read lol      EPR.lol                   append
###
      read upl      dummy_residue_1_GLY80.upl    append
      read lol      dummy_residue_1_GLY80.lol    append
      read upl      dummy_residue_2_GLY280.upl   append
      read lol      dummy_residue_2_GLY280.lol   append
#                                             # as an example only two dummy residues shown
# ------ structure calculation ------
seed=5671                                                       # random number generator seed
calc_all structures=500 command=anneal steps=200000             # calculate conformers
overview StructureX.ovw structures=50 pdb details               # write overview file and 50 best conformers
```

Figure 7 Typical CALC.cya file. The different input files are discussed further below.

no restraints such as the RNA phosphate backbone. For structure calculation of the 70 kDa RsmZ/RsmE protein–RNA complex (Duss, Michel, Yulikov, et al., 2014; Duss, Yulikov, et al., 2014), 2500 preliminary structures of the full complex were calculated using 200,000 simulated annealing steps. Note that several CYANA runs were required because the maximum number of structures per calculation is 999. The 50 lowest energy structures (based on the CYANA target function) were selected for further refinement (e.g., AMBER). Note that for initial structure calculations, a significantly lower number of simulated annealing steps (e.g., 30,000) and of structures (e.g., 200–500) are sufficient in order to save time and computational resources.

4.6.2 Sequence File

The protein and RNA chains and each dummy residue are connected by virtual linkers resulting into a single chain (Fig. 8). Each center of a spin-

```
MET      1
LEU      2
. . .
HIS      65
PL       66
LL5      67
. . .
LL5      78
LP       79
GLY      80
PL       81
LL5      82
. . .
LL5      99
LN       100
URA      101
. . .
RADE     172
```

Figure 8 StructureX.seq example file: This example sequence file consists of a protein (residues 1–65), an RNA (residues 101–172), and one dummy residue (GLY 80). Each of the three molecules (the dummy residue is defined as a separate molecule) is separated by virtual linkers. Note that in a real sequence file, more than a single dummy residue would be present (at least two in order to define one interspin distance).

label conformational ensemble is represented by a different GLY residue (defining a separate molecule). The CA atoms of these GLY residues are used to include the EPR lower and upper limit distance constraints into structure calculation (see Section 4.5.2 and Fig. 5, step 4). Each GLY residue and all the protein and RNA chains must be separated by long enough linkers, the length depending on the size of the macromolecular complex and its structure. For example, a LL5 linker residue has a virtual bond length of 5 Å. Thus, calculating the structure of a macromolecular complex with a maximal dimension of 100 Å would require at least 20 LL5 linker residues between the different chains (protein, RNA, and dummy residues).

4.6.3 Input Restraint Files
4.6.3.1 NMR NOE Restraints
The NOEs are semiquantitatively classified according to their intensities in the 2D- and 3D-NOESY spectra and are typically represented as upper limit constraints.

4.6.3.2 NMR Hydrogen-Bond Restraints
For the RNA helical regions: the hydrogen-bonding distance restraints are based on the observation of imino resonances of the corresponding base pairs. For protein secondary structures, while not used at initial stages of structure calculation, hydrogen-bond restraints can help for structure convergence if experimentally supported. Experimental support is, for example, provided by slowly exchanging amides (amide protection in D_2O) or substantial chemical shift perturbations of the carbonyl groups.

4.6.3.3 NMR Torsion Angles
Protein backbone angles: the φ- and ψ-angles are determined using TALOS + (Shen, Delaglio, Cornilescu, & Bax, 2009). RNA backbone angles are based on A-form geometry derived from high-resolution crystal structures: $\alpha = 270°-330°$, $\beta = 150°-210°$, $\gamma = 30°-60°$, $\delta = 50°-110°$, $\varepsilon = 180°-240°$, $\zeta = 260°-320°$ (Dominguez et al., 2011). The restraints are only used for double-stranded regions, which were identified by the presence of a protected imino resonance and confirmed by NOEs. δ-Angles in the loop residues: $50°-110°$ for C3′-endo conformation and $130°-190°$ for C2′-endo conformation.

4.6.3.4 EPR Constraints

Each pair of dummy atoms is constrained by an upper and lower distance limit, which is derived from a DEER distance distribution to confine 70% of the peak area from each side of the peak maximum (see Section 4.5.2). These broad 70% distance ranges cover most of the distance distributions with reasonable probability but also account for uncertainties in the simulations of the spin-label distributions and in DEER data analysis, such as broadened wings in distance distributions due to Tikhonov regularization or possible low amplitude artifacts at the ends of the distance distributions (for illustration, see Fig. 9). Note that narrower distance ranges might be used, especially if only a single macromolecular conformation is present in solution. However, it has to be verified that narrower distance ranges do not result into increased energy target functions.

4.6.3.5 Restraints to Fix the GLY-dummy Residues (CA atoms) to the Centers of the Spin-label Distributions

The CA of the GLY-dummy residues is fixed to the geometrical centers of the modeled spin-label distributions (Fig. 5, step 4) using several tight upper and lower distance limit restraints (Fig. 10) between the CA of the GLY-dummy residues and backbone atoms of protein or RNA residues to which the spin-label is attached to (see also Section 4.5.2).

The CA of the GLY-dummy residues are fixed to the geometrical centers of the modeled spin-label distributions (Fig. 5, step 4) using several tight

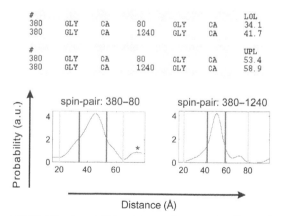

Figure 9 Typical EPR.lol (top) and EPR.upl (center) constraint files with corresponding EPR DEER distance distributions (bottom) for distance 380–80 (left) and distance 380–1240 (right). The red bars (dark gray color in the print version) in the DEER spectra represent the lower and upper distance limits confining the 70% distance range.

dummy atom			backbone of protein			LOL	UPL
80	GLY	CA	44	ARG	CA	14.7	16.2
80	GLY	CA	36	ALA	CA	25.3	26.1
80	GLY	CA	14	ILE	CA	22.8	23.5
80	GLY	CA	34	ILE	CA	22.4	22.9
80	GLY	CA	4	LEU	CA	17.3	17.9
80	GLY	CA	10	GLU	CA	25.5	26
80	GLY	CA	24	GLY	CA	24.1	24.8
80	GLY	CA	26	SER	CA	22.4	22.8
80	GLY	CA	244	ARG	CA	24.7	25.5
80	GLY	CA	236	ALA	CA	16.7	17.5
80	GLY	CA	214	ILE	CA	18.1	18.6
80	GLY	CA	234	ILE	CA	14	14.3
80	GLY	CA	204	LEU	CA	20.4	20.6
80	GLY	CA	210	GLU	CA	9.8	10.1
80	GLY	CA	224	GLY	CA	12.3	13.2
80	GLY	CA	226	SER	CA	19.2	19.9

Figure 10 Typical restraint file for fixing the dummy atom CA of a dummy residue (here GLY80) to the center of the corresponding spin-label distribution (here fixed to the protein backbone). Note that two files are required per dummy residue to be fixed, one including the lower limit restraints (LOL, e.g., dummy_residue_1_GLY80.lol, see Fig. 7) and one including the upper limit restraints (UPL, e.g., dummy_residue_1_GLY80.upl).

upper and lower distance limit restraints (Fig. 10) between the CA of the GLY-dummy residues and backbone atoms of protein or RNA residues to which the spin-label is attached to (see also Section 4.5.2).

4.6.3.6 Additional Restraints
Optional restraints can be used to enforce the RNA phosphate backbone of regions for which the structure is known to be identical or very similar to the structure of the isolated domains but is distorted in the structure calculation of the full complex.

4.6.4 Simple Extension If More than One Conformation is Present in Solution
When separated peaks (corresponding to different global macromolecular conformations present in solution) are present in certain DEER distance distributions, then two or more structure calculations should be performed in parallel by including into the structure calculation the distance information from the different isolated peaks corresponding to the different global

conformations present in solution. For the structure calculation of the 70 kDa RsmZ/RsmE complex (Duss, Michel, Yulikov, et al., 2014), only two of the 21 measured distance distributions (AC and AG) consisted of two well-separated peaks one of each belonging to one conformation, whereas the other distance distributions were not resolving such peak pairs but rather were represented by broad single-peak distributions. Two separate structure calculations were performed in which only the two resolved distance distributions are providing distinct distance constraints, whereas all broad unresolved distance distributions are used as same constraints for both conformations. Note that due to geometrical reasons only one of the two distance combinations was possible. In Fig. 11, the resulting structural ensembles and the structural statistics for 70 kDa RsmZ/RsmE complex are shown.

4.7 Structure Validation (Fig. 5, Step 6)

When a structural ensemble is calculated with the above described approach, the quality of the obtained solution can be inspected by overlapping the experimentally measured spin–spin distance distributions with the ones calculated from all the individual structures in the ensemble with explicitly adding all possible conformations of the spin labels to each individual structure. To back-calculate the distance distribution for a specific spin-label pair, the two "radical clouds" (spin-label distributions, see Fig. 5, step 3) corresponding to the two spin labels are superimposed onto one structure and all the possible distances from each radical position of spin label 1 to each radical position of spin label 2 are measured (Fig. 12) using a home-written Matlab script composed of several routines from the MMM package (www.epr.ethz.ch/software/index). The same is repeated for all the structures of the structural ensemble and the distances are plotted as a normalized histogram resulting into the back-calculated distance distribution. These back-calculated distance distributions can then be superimposed onto the measured DEER data (Fig. 12). Note that if more than a single structural conformation is present in solution, the same can be performed by simply weighting the structural ensembles of both conformations according to their population in solution.

Besides comparing measured and back-calculated distance distributions, it is also advisable to verify if all conformations of the spin labels (radical clouds: Fig. 5, step 3), predicted from the structures of the individual domains, can also be adopted in the structure of the full complex.

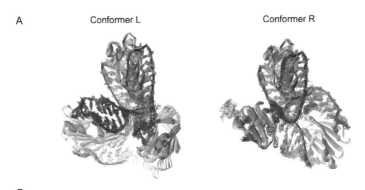

	Conformer L protein	RNA	Conformer R protein	RNA
NMR distance and dihedral constraints				
Distance restraints				
Total NOE (intramolecular)	4664	837	4664	873
Intra-residue	451	368	451	385
Inter-residue	4213	469	4213	488
Sequential (|i-j| = 1)	2091	353	2091	369
Non-sequential (|i-j| > 1)	2122	116	2122	119
Hydrogen bonds	96	46	96	48
Protein–protein or protein–nucleic acid intermolecular NOEs	1000	904	1000	904
Total dihedral angle restraints	580	205	580	217
Protein				
phi	288		288	
psi	292		292	
Nucleic acid				
Base pair				
Sugar pucker		29		29
Backbone				
Based on A-form geometry		176		188
EPR restraints (upper and lower limits)		42		42
Structure statistics				
Violations (mean and s.d.)				
Distance constraints (>0.2 Å)	1.2 ± 1.0		2.5 ± 1.5	
Dihedral angle constraints (>5 °)	0.0 ± 0.0		0.1 ± 0.3	
Max. dihedral angle violation (°)	0.61 ± 0.30		1.22 ± 2.14	
Max. distance constraint violation (Å)	0.25 ± 0.07		0.30 ± 0.06	
Deviations from idealized geometry				
Bond lengths (Å)	0.009		0.01	
Bond angles (°)	1.6		1.6	
Average pairwise r.m.s.d.* (Å)				
Protein (1-53, all six chains)				
Heavy	1.59 ± 0.57		1.24 ± 0.14	
Backbone	1.30 ± 0.66		0.83 ± 0.18	
RNA				
All RNA heavy (1-72)	1.83 ± 0.26		1.26 ± 0.25	
RNA binding site†	0.90 ± 0.17		0.97 ± 0.26	
Complex				
All complex heavy (C, N, O, P)‡	1.48 ± 0.51		1.19 ± 0.17	

*Pairwise r.m.s.d. was calculated among 20 refined structures.
†Conformer L: 1–16, 19–36, 38–42, 44–56, 59–71 and conformer R: 1–16, 19–36, 38–42, 44–56, 58–72.
‡Protein: 1–53 and RNA same as in †.

Figure 11 Seventy kilo Dalton complex of the ncRNA RsmZ bound to three RsmE protein homo-dimers. (A) Structural ensemble (20 lowest energy structures) and (B) structural statistics after refinement. *This figure has been modified from Duss, Michel, Yulikov, et al. (2014).*

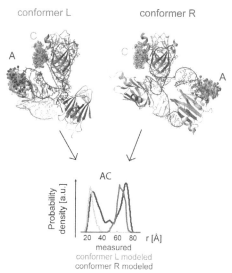

Figure 12 How to obtain the back-calculated (modeled) distance distribution for a single spin label pair (A–C as an example). The modeled distance distribution is calculated by plotting the occurrence of distances between all the radical positions of spin label A to all the radical positions of spin label C of one specific structure. This is repeated for all the structures in the conformational ensemble. On the top, the "radical clouds" (spin-label distributions) for spin labels A and C are superimposed onto one structure of each conformation. Bottom, the measured (blue, black color in the print version) DEER distance distribution is superimposed onto the modeled ones of both conformations present in solution, which is conformation L (cyan, light gray color in the print version) and conformation R (magenta, dark gray color in the print version). The population weight for both structural conformations was assumed to be 2:1, conformation R being more populated. *This figure has been taken from Duss, Michel, Yulikov, et al. (2014).*

If it is not the case, the back-calculated distance distributions are less accurate and differences between measured and back-calculated distance distributions might be larger.

To make sure that the structures are not over-restrained using too tight distance ranges to confine the centers of the "radical clouds" to each other, it should be checked by how much the ensemble of structures changes upon slight variations of the distance ranges. Violations in the EPR distance constraints during structure calculation are also an indication of too tight distance ranges.

Finally, a structure validation should be performed by omitting some DEER constraints and computing the structural ensemble with the reduced set of DEER-based constraints.

5. FUTURE CHALLENGES AND IMPLICATIONS

Many noncoding RNAs and their complexes with proteins are larger and of greater structural complexity than the ones approached to date and they may exist as equilibrium of several conformations that are distinct in long-range arrangement of their domains. Combining long-range EPR distance constraints with short-range (NOEs) and intermediate-range NMR distance (PREs) or orientational (RDCs) restraints and with data from other biophysical methods (SAXS, SANS, EM) has high potential to provide high-resolution structural information on such large and dynamic RNAs and protein–RNA assemblies.

EPR spectroscopy can measure distances independently of the size of a macromolecule and RNA and protein NMR resonances can be observed in high-molecular-weight complexes with appropriate isotope labeling schemes. It is therefore expected that in the future also significantly larger RNAs or protein–RNA complexes of more than 70 kDa will be at reach (Duss, Michel, Yulikov, et al., 2014; Duss, Yulikov, et al., 2014).

Compared to other methods, DEER also allows a less complicated observation and evaluation of different conformations simultaneously present in solution (Duss, Michel, Yulikov, et al., 2014; Duss, Yulikov, et al., 2014). However, future improvement in methods dealing with several structural conformations simultaneously present in solution will be essential. A second population present by less than 5–10% cannot be safely assigned or excluded with the current accuracy of DEER distance distributions, mainly because of the ill-posedness of the conversion of noisy time-domain data to distance distributions. Also, structure determination of two globally similar structures, which differ slightly in domain positions or orientations, will not be successful, but will result in an average structure.

Larger and dynamic systems require the measurement of an increasing amount of long-range distance distribution constraints as well as strategies for assigning peaks in the distance distributions to the distinct conformations. Orthogonal spin labeling, as reviewed recently (Yulikov, 2015), can help to distinguish protein–protein, RNA–RNA, and RNA–protein distances. Furthermore, orthogonal labeling can be used to recognize aggregation (Lueders, Jäger, et al., 2013) and thus avoid misinterpretation of distance distributions. Unambiguous assignment of peaks in distance distributions to distinct conformations will require two-dimensional correlation experiments that could be most easily realized with orthogonal labeling schemes with three distinct labels. The trityl radical (Kunjir, Reginsson, Schiemann, & Sigurdsson, 2013; Reginsson, Kunjir, Sigurdsson, & Schiemann, 2012; Shevelev et al., 2014; Yang et al., 2012) is the most promising option as the third spin label for such an approach.

With increasing molecular weight and an increasing number of required restraints, sensitivity of the DEER distance measurements becomes even more important. Large macromolecular assemblies are often difficult to obtain in high enough concentrations required for EPR long-range distance measurements. Absolute sensitivity can be increased by applying W-band Gd(III)–Gd(III) DEER (Goldfarb, 2014) or RIDME (Razzaghi et al., 2014) experiments. However, to avoid aggregation, it may also be necessary to improve concentration sensitivity. Arbitrary waveform excitation (Spindler, Glaser, Skinner, & Prisner, 2013; Spindler et al., 2012), and in particular ultrawideband EPR technology (Doll & Jeschke, 2014; Doll, Pribitzer, Tschaggelar, & Jeschke, 2013) is expected to lead to an increase in concentration sensitivity by up to one order of magnitude, depending on the type of label. With current technology, DEER distance distributions can be obtained at concentrations down to 5–20 μM of spin pairs in sample volumes of 10–50 µL, with the lower values corresponding to distances up to 4 nm and requiring deuteration of at least the solvent and cryoprotectant. It may be expected that sensitivity can be improved to 1–2 μM in sample volumes of 2–5 µL during the next years.

In contrast, when high-concentrated samples can be produced, EPR should also allow obtaining structural information of low-affinity transient complexes (intermediate μM to low mM range), which are difficult to study by NMR or other methods. Low-affinity protein–RNA complexes often result in bad spectral line shapes and intermolecular NOEs between RNA and protein are often absent (Dominguez et al., 2011). When long-range EPR distances can be measured between different domains of the same macromolecule in complex (different RNA secondary structural elements within a large RNA or different protein domains within the same protein), then the total concentration of the RNA or protein can be increased by supplementing the spin-labeled macromolecule with unlabeled one. Increasing the concentrations of the macromolecules forming a complex results in shifting the equilibrium toward the bound form, which then allows detecting structural information on the low-affinity complex by EPR.

Finally, improvement in the precision of the DEER distance distribution restraints may help to obtain more precise structural ensembles and to reduce the number of restraints that are required for structure determination. When RNA or DNA is labeled with an MTSL or iodoacetamido-Proxyl label at a thiouridine, the resulting distribution of nitroxide positions is more compact than in the case of labeling a cysteine residue in a protein. The reason for that is that the linker between the RNA base and the spin label ring is one bond shorter as compared to the length of the chain between a protein backbone and the spin label. Therefore, further optimization of the length of the flexible linkers of different types of spin labels could improve the precision of the distance constraints obtained in DEER measurements.

Overall, a perfect correlation between measured and back-calculated distance distributions would allow refining the structures directly against the experimental DEER distance distributions. Such improvements would simplify determining structures of protein–RNA complexes having multiple conformations in solution paving the way for exploring the complex universe of dynamic ribonucleoprotein assemblies.

ACKNOWLEDGMENTS

This work was supported by the Swiss National Science Foundation (SNF) Grant no. 3100A0-118118, 31003ab-133134, and 31003A-149921 to F.H.-T.A.

REFERENCES

Aeschbacher, T., Schmidt, E., Blatter, M., Maris, C., Duss, O., Allain, F. H. T., et al. (2013). Automated and assisted RNA resonance assignment using NMR chemical shift statistics. *Nucleic Acids Research*, 41(18), e172.

Aeschbacher, T., Schubert, M., & Allain, F. H. T. (2012). A procedure to validate and correct the 13C chemical shift calibration of RNA datasets. *Journal of Biomolecular NMR*, 52(2), 179–190.

Baker, M. (2011). Long noncoding RNAs: The search for function. *Nature Methods*, 8, 379–383.

Baldwin, A. J., & Kay, L. E. (2009). NMR spectroscopy brings invisible protein states into focus. *Nature Chemical Biology*, 5(11), 808–814.

Borbat, P. P., & Freed, J. H. (1999). Multiple-quantum ESR and distance measurements. *Chemical Physics Letters*, 313, 145–154.

Borbat, P. P., & Freed, J. H. (2007). Measuring distances by pulsed dipolar ESR spectroscopy: Spin-labeled histidine kinases. *Methods in Enzymology*, 423, 52–116.

Bowman, M. K., & Maryasov, A. G. (2007). Dynamic phase shifts in nanoscale distance measurements by double electron electron resonance (DEER). *Journal of Magnetic Resonance*, 185(2), 270–282.

Bullock, S. L., Ringel, I., Ish-Horowicz, D., & Lukavsky, P. J. (2010). A′-form RNA helices are required for cytoplasmic mRNA transport in Drosophila. *Nature Structural & Molecular Biology*, 17(6), 703–709.

Burke, J. E., Sashital, D. G., Zuo, X., Wang, Y. X., & Butcher, S. E. (2012). Structure of the yeast U2/U6 snRNA complex. *RNA*, 18(4), 673–683.

Buttner, L., Javadi-Zarnaghi, F., & Hobartner, C. (2014). Site-specific labeling of RNA at internal ribose hydroxyl groups: Terbium-assisted deoxyribozymes at work. *Journal of the American Chemical Society, 136*(22), 8131–8137.

Buttner, L., Seikowski, J., Wawrzyniak, K., Ochmann, A., & Hobartner, C. (2013). Synthesis of spin-labeled riboswitch RNAs using convertible nucleosides and DNA-catalyzed RNA ligation. *Bioorganic & Medicinal Chemistry, 21*(20), 6171–6180.

Cai, Q., Kusnetzow, A. K., Hideg, K., Price, E. A., Haworth, I. S., & Qin, P. Z. (2007). Nanometer distance measurements in RNA using site-directed spin labeling. *Biophysical Journal, 93*(6), 2110–2117.

Cai, Q., Kusnetzow, A. K., Hubbell, W. L., Haworth, I. S., Gacho, G. P. C., Van Eps, N., et al. (2006). Site-directed spin labeling measurements of nanometer distances in nucleic acids using a sequence-independent nitroxide probe. *Nucleic Acids Research, 34*(17), 4722–4730.

Case, D. A., Cheatham, T. E., 3rd., Darden, T., Gohlke, H., Luo, R., Merz, K. M., Jr., et al. (2005). The Amber biomolecular simulation programs. *Journal of Computational Chemistry, 26*(16), 1668–1688.

Chiang, Y., Borbat, P. P., & Freed, J. H. (2005). The determination of pair distance distributions by pulsed ESR using Tikhonov regularization. *Journal of Magnetic Resonance, 172*, 279–295.

Clore, G. M., & Iwahara, J. (2009). Theory, practice, and applications of paramagnetic relaxation enhancement for the characterization of transient low-population states of biological macromolecules and their complexes. *Chemical Reviews, 109*(9), 4108–4139.

Clore, G. M., Tang, C., & Iwahara, J. (2007). Elucidating transient macromolecular interactions using paramagnetic relaxation enhancement. *Current Opinion in Structural Biology, 17*(5), 603–616.

Cruickshank, P. A. S., Bolton, D. R., Robertson, D. A., Hunter, R. I., Wylde, R. J., & Smith, G. M. (2009). A kilowatt pulsed 94 GHz electron paramagnetic resonance spectrometer with high concentration sensitivity, high instantaneous bandwidth, and low dead time. *The Review of Scientific Instruments, 80*, 103102.

Davis, J. H., Tonelli, M., Scott, L. G., Jaeger, L., Williamson, J. R., & Butcher, S. E. (2005). RNA helical packing in solution: NMR structure of a 30 kDa GAAA tetraloop-receptor complex. *Journal of Molecular Biology, 351*(2), 371–382.

Dethoff, E. A., Chugh, J., Mustoe, A. M., & Al-Hashimi, H. M. (2012). Functional complexity and regulation through RNA dynamics. *Nature, 482*(7385), 322–330.

Doll, A., & Jeschke, G. (2014). Fourier-transform electron spin resonance with bandwidth-compensated chirp pulses. *Journal of Magnetic Resonance, 246*, 18–26.

Doll, A., Pribitzer, S., Tschaggelar, R., & Jeschke, G. (2013). Adiabatic and fast passage ultra-wideband inversion in pulsed EPR. *Journal of Magnetic Resonance, 230*, 27–39.

Dominguez, C., Schubert, M., Duss, O., Ravindranathan, S., & Allain, F. H. T. (2011). Structure determination and dynamics of protein-RNA complexes by NMR spectroscopy. *Progress in Nuclear Magnetic Resonance Spectroscopy, 58*(1–2), 1–61.

D'Souza, V., Dey, A., Habib, D., & Summers, M. F. (2004). NMR structure of the 101-nucleotide core encapsidation signal of the Moloney murine leukemia virus. *Journal of Molecular Biology, 337*(2), 427–442.

Duss, O., Lukavsky, P. J., & Allain, F. H. T. (2012). Isotope labeling and segmental labeling of larger RNAs for NMR structural studies. *Advances in Experimental Medicine and Biology, 992*, 121–144.

Duss, O., Maris, C., von Schroetter, C., & Allain, F. H. T. (2010). A fast, efficient and sequence-independent method for flexible multiple segmental isotope labeling of RNA using ribozyme and RNase H cleavage. *Nucleic Acids Research, 38*(20), e188.

Duss, O., Michel, E., Diarra dit Konte, N., Schubert, M., & Allain, F. H. T. (2014). Molecular basis for the wide range of affinity found in Csr/Rsm protein-RNA recognition. *Nucleic Acids Research, 42*(8), 5332–5346.

Duss, O., Michel, E., Yulikov, M., Schubert, M., Jeschke, G., & Allain, F. H. T. (2014). Structural basis of the non-coding RNA RsmZ acting as a protein sponge. *Nature*, *509*(7502), 588–592.

Duss, O., Yulikov, M., Jeschke, G., & Allain, F. H. T. (2014). EPR-aided approach for solution structure determination of large RNAs or protein-RNA complexes. *Nature Communications*, *5*, 3669.

Edwards, J. M., Long, J., de Moor, C. H., Emsley, J., & Searle, M. S. (2013). Structural insights into the targeting of mRNA GU-rich elements by the three RRMs of CELF1. *Nucleic Acids Research*, *41*(14), 7153–7166.

Fichou, Y., & Férec, C. (2006). The potential of oligonucleotides for therapeutic applications. *Trends in Biotechnology*, *24*(12), 563–570.

Fleissner, M. R., Brustad, E. M., Kalai, T., Altenbach, C., Cascio, D., Peters, F. B., et al. (2009). Site-directed spin labeling of a genetically encoded unnatural amino acid. *Proceedings of the National Academy of Sciences of the Unites States of America*, *106*, 21637–21642.

Godt, A., Schulte, M., Zimmermann, H., & Jeschke, G. (2006). How flexible are oligo(paraphenyleneethynylene)s? *Angewandte Chemie, International Edition*, *45*, 7560–7564.

Goldfarb, D. (2014). Gd3+ spin labeling for distance measurements by pulse EPR spectroscopy. *Physical Chemistry Chemical Physics*, *16*, 9685–9699.

Goldfarb, D., Lipkin, Y., Potapov, A., Gorodetsky, Y., Epel, B., Raitsimring, A. M., et al. (2008). HYSCORE and DEER with an upgraded 95 GHz pulse EPR spectrometer. *Journal of Magnetic Resonance*, *194*, 8–15.

Grishaev, A., Wu, J., Trewhella, J., & Bax, A. (2005). Refinement of multidomain protein structures by combination of solution small-angle X-ray scattering and NMR data. *Journal of the American Chemical Society*, *127*(47), 16621–16628.

Grishaev, A., Ying, J., Canny, M. D., Pardi, A., & Bax, A. (2008). Solution structure of tRNA(Val) from refinement of homology model against residual dipolar coupling and SAXS data. *Journal of Biomolecular NMR*, *42*(2), 99–109.

Gromov, I., Shane, J., Forrer, J., Rakhmatoullin, R., Rozentzwaig, Y., & Schweiger, A. (2001). A Q-band pulse EPR/ENDOR spectrometer and the implementation of advanced one- and two-dimensional pulse EPR methodology. *Journal of Magnetic Resonance*, *149*, 196–203.

Güntert, P. (2004). Automated NMR structure calculation with CYANA. *Methods in Molecular Biology*, *278*, 353–378.

Guttman, M., & Rinn, J. L. (2012). Modular regulatory principles of large non-coding RNAs. *Nature*, *482*(7385), 339–346.

Hagelueken, G., Ward, R., Naismith, J. H., & Schiemann, O. (2012). mtsslSuite: In silico spin labelling, trilateration and distance-constrained rigid body docking in PyMOL. *Applied Magnetic Resonance*, *42*, 377–391.

Hass, M. A., & Ubbink, M. (2014). Structure determination of protein-protein complexes with long-range anisotropic paramagnetic NMR restraints. *Current Opinion in Structural Biology*, *24*, 45–53.

Hatmal, M. M., Li, Y. Y., Hegde, B. G., Hegde, P. B., Jao, C. C., Langen, R., et al. (2012). Computer modeling of nitroxide spin labels on proteins. *Biopolymers*, *97*, 35–44.

Hilger, D., Polyhach, Y., Jung, H., & Jeschke, G. (2009). Backbone structure of transmembrane domain IX of the Na+/proline transporter PutP of Escherichia coli. *Biophysical Journal*, *96*, 217–225.

Hilger, D., Polyhach, Y., Padan, E., Jung, H., & Jeschke, G. (2007). High-resolution structure of a Na+/H+ antiporter dimer obtained by pulsed EPR distance measurements. *Biophysical Journal*, *93*, 3675–3683.

Hubbell, W. L., Lopez, C. J., Altenbach, C., & Yang, Z. Y. (2013). Technological advances in site-directed spin labeling of proteins. *Current Opinion in Structural Biology*, *23*, 725–733.

Hudson, B. P., Martinez-Yamout, M. A., Dyson, H. J., & Wright, P. E. (2004). Recognition of the mRNA AU-rich element by the zinc finger domain of TIS11d. *Nature Structural & Molecular Biology*, *11*(3), 257–264.
Iwahara, J., Schwieters, C. D., & Clore, G. M. (2004). Ensemble approach for NMR structure refinement against H-1 paramagnetic relaxation enhancement data arising from a flexible paramagnetic group attached to a macromolecule. *Journal of the American Chemical Society*, *126*(18), 5879–5896.
Jäger, H., Koch, A., Maus, V., Spiess, H. W., & Jeschke, G. (2008). Relaxation-based distance measurements between a nitroxide and a lanthanide spin label. *Journal of Magnetic Resonance*, *194*, 254–263.
Jeschke, G. (2012). DEER distance measurements on proteins. *Annual Review of Physical Chemistry*, *63*, 419–446.
Jeschke, G. (2013). Conformational dynamics and distribution of nitroxide spin labels. *Progress in Nuclear Magnetic Resonance Spectroscopy*, *72*, 42–60.
Jeschke, G., Chechik, V., Ionita, P., Godt, A., Zimmermann, H., Banham, J., et al. (2006). DeerAnalysis2006—A comprehensive software package for analyzing pulsed ELDOR data. *Applied Magnetic Resonance*, *30*, 473–498.
Jeschke, G., Koch, A., Jonas, U., & Godt, A. (2002). Direct conversion of EPR dipolar time evolution data to distance distributions. *Journal of Magnetic Resonance*, *155*, 72–82.
Jeschke, G., Panek, G., Godt, A., Bender, A., & Paulsen, H. (2004). Data analysis procedures for pulse ELDOR measurements of broad distance distributions. *Applied Magnetic Resonance*, *26*, 223–244.
Jeschke, G., Pannier, M., Godt, A., & Spiess, H. W. (2000). Dipolar spectroscopy and spin alignment in electron paramagnetic resonance. *Chemical Physics Letters*, *331*, 243–252.
Johansson, C., Finger, L. D., Trantirek, L., Mueller, T. D., Kim, S., Laird-Offringa, I. A., et al. (2004). Solution structure of the complex formed by the two N-terminal RNA-binding domains of nucleolin and a pre-rRNA target. *Journal of Molecular Biology*, *337*(4), 799–816.
Kattnig, D. R., Reichenwallner, J., & Hinderberger, D. (2013). Modeling excluded volume effects for the faithful description of the background signal in double electron–electron resonance. *The Journal of Physical Chemistry. B*, *117*(51), 16542–16557.
Klostermeier, D., & Millar, D. P. (2001). RNA conformation and folding studied with fluorescence resonance energy transfer. *Methods*, *23*(3), 240–254.
Kunjir, N. C., Reginsson, G. W., Schiemann, O., & Sigurdsson, S. T. (2013). Measurements of short distances between trityl spin labels with CW EPR, DQC and PELDOR. *Physical Chemistry Chemical Physics*, *15*, 19673–19685.
Kurshev, V. V., Raitsimring, A. M., & Tsvetkov, Y. D. (1989). Selection of dipolar interaction by the "2 + 1" pulse train ESE. *Journal of Magnetic Resonance*, *81*, 441–454.
Lapinaite, A., Simon, B., Skjaerven, L., Rakwalska-Bange, M., Gabel, F., & Carlomagno, T. (2013). The structure of the box C/D enzyme reveals regulation of RNA methylation. *Nature*, *502*(7472), 519–523.
Lee, B. M., Xu, J., Clarkson, B. K., Martinez-Yamout, M. A., Dyson, H. J., Case, D. A., et al. (2006). Induced fit and "lock and key" recognition of 5S RNA by zinc fingers of transcription factor IIIA. *Journal of Molecular Biology*, *357*(1), 275–291.
Leeper, T. C., Qu, X., Lu, C., Moore, C., & Varani, G. (2010). Novel protein-protein contacts facilitate mRNA 3′-processing signal recognition by Rna15 and Hrp1. *Journal of Molecular Biology*, *401*(3), 334–349.
Lipsitz, R. S., & Tjandra, N. (2004). Residual dipolar couplings in NMR structure analysis. *Annual Review of Biophysics and Biomolecular Structure*, *33*, 387–413.
Lu, K., Miyazaki, Y., & Summers, M. F. (2010). Isotope labeling strategies for NMR studies of RNA. *Journal of Biomolecular NMR*, *46*(1), 113–125.

Lu, X. J., & Olson, W. K. (2008). 3DNA: A versatile, integrated software system for the analysis, rebuilding and visualization of three-dimensional nucleic-acid structures. *Nature Protocols, 3*(7), 1213–1227.

Lueders, P., Jäger, H., Hemminga, M. A., Jeschke, G., & Yulikov, M. (2013). Distance measurements on orthogonally spin-labeled membrane spanning WALP23 polypeptides. *The Journal of Physical Chemistry. B, 117*, 2061–2068.

Lueders, P., Razzaghi, S., Jäger, H., Tschaggelar, R., Hemminga, M. A., Yulikov, M., et al. (2013). Distance determination from dysprosium induced relaxation enhancement: A case study on membrane-inserted WALP23 polypeptides. *Molecular Physics, 111*, 2824–2833.

Lukavsky, P. J., Kim, I., Otto, G. A., & Puglisi, J. D. (2003). Structure of HCV IRES domain II determined by NMR. *Nature Structural Biology, 10*(12), 1033–1038.

Mackereth, C. D., Madl, T., Bonnal, S., Simon, B., Zanier, K., Gasch, A., et al. (2011). Multi-domain conformational selection underlies pre-mRNA splicing regulation by U2AF. *Nature, 475*(7356), 408–411.

Mackereth, C. D., & Sattler, M. (2012). Dynamics in multi-domain protein recognition of RNA. *Current Opinion in Structural Biology, 22*(3), 287–296.

Madl, T., Gabel, F., & Sattler, M. (2011). NMR and small-angle scattering-based structural analysis of protein complexes in solution. *Journal of Structural Biology, 173*(3), 472–482.

Madl, T., Guttler, T., Gorlich, D., & Sattler, M. (2011). Structural analysis of large protein complexes using solvent paramagnetic relaxation enhancements. *Angewandte Chemie (International Ed. in English), 50*(17), 3993–3997.

Marko, A., Denysenkov, V. P., Margraf, D., Cekan, P., Schiemann, O., Sigurdsson, S. T., et al. (2001). Conformational flexibility of DNA. *Journal of the American Chemical Society, 133*, 13375–13379.

Martorana, A., Bellapadrona, G., Feintuch, A., Di Gregorio, E., Aime, S., & Goldfarb, D. (2014). Probing protein conformation in cells by EPR distance measurements using Gd(3+) spin labeling. *Journal of the American Chemical Society, 136*, 13458–13465.

Mattick, J. S., & Makunin, I. V. (2005). Small regulatory RNAs in mammals. *Human Molecular Genetics, 14*(1), R121–R132.

Milov, A. D., Salikhov, K. M., & Shchirov, M. D. (1981). Use of the double resonance in electron spin echo method for the study of paramagnetic center spatial distribution in solids. *Fizika Tverdogo Tela (Leningrad), 23*, 975–982.

Milov, A. D., & Tsvetkov, Y. D. (1997). Double electron–electron resonance in electron spin echo: Conformations of spin-labeled poly-4-vinilpyridine in glassy solutions. *Applied Magnetic Resonance, 12*, 495–504.

Miyazaki, Y., Irobalieva, R. N., Tolbert, B. S., Smalls-Mantey, A., Iyalla, K., Loeliger, K., et al. (2010). Structure of a conserved retroviral RNA packaging element by NMR spectroscopy and cryo-electron tomography. *Journal of Molecular Biology, 404*(5), 751–772.

Morris, K. V., & Mattick, J. S. (2014). The rise of regulatory RNA. *Nature Reviews. Genetics, 15*(6), 423–437.

Nam, Y., Chen, C., Gregory, R. I., Chou, J. J., & Sliz, P. (2011). Molecular basis for interaction of let-7 microRNAs with Lin28. *Cell, 147*(5), 1080–1091.

Pannier, M., Veit, S., Godt, A., Jeschke, G., & Spiess, H. W. (2000). Dead-time free measurement of dipole-dipole interactions between electron spins. *Journal of Magnetic Resonance, 142*, 331–340.

Park, S. Y., Borbat, P. P., Gonzalez-Bonet, G., Bhatnagar, J., Pollard, A. M., Freed, J. H., et al. (2006). Reconstruction of the chemotaxis receptor-kinase assembly. *Nature Structural & Molecular Biology, 13*(5), 400–407.

Polyhach, Y., Bordignon, E., & Jeschke, G. (2011). Rotamer libraries of spin labelled cysteines for protein studies. *Physical Chemistry Chemical Physics, 13*(6), 2356–2366.

Polyhach, Y., Bordignon, E., Tschaggelar, R., Gandra, S., Godt, A., & Jeschke, G. (2012). High sensitivity and versatility of the DEER experiment on nitroxide radical pairs at Q-band frequencies. *Physical Chemistry Chemical Physics, 14*, 10762–10773.

Polyhach, Y., Godt, A., Bauer, C., & Jeschke, G. (2007). Spin pair geometry revealed by high-field DEER in the presence of conformational distributions. *Journal of Magnetic Resonance, 185,* 118–129.

Polyhach, Y., & Jeschke, G. (2010). Prediction of favourable sites for spin labelling of proteins. *Spectroscopy, 24,* 651–659.

Ponting, C. P., Oliver, P. L., & Reik, W. (2009). Evolution and functions of long noncoding RNAs. *Cell, 136*(4), 629–641.

Price, E. A., Sutch, B. T., Cai, Q., Qin, P. Z., & Haworth, I. S. (2007). Computation of nitroxide-nitroxide distances in spin-labeled DNA duplexes. *Biopolymers, 87*(1), 40–50.

Qi, M., Gross, A., Jeschke, G., Godt, A., & Drescher, M. (2014). Gd(III)-PyMTA label is suitable for in-cell EPR. *Journal of the American Chemical Society, 136,* 15366–15378.

Qin, P. Z., Haworth, I. S., Cai, Q., Kusnetzow, A. K., Grant, G. P., Price, E. A., et al. (2007). Measuring nanometer distances in nucleic acids using a sequence-independent nitroxide probe. *Nature Protocols, 2*(10), 2354–2365.

Rabenstein, M. D., & Shin, Y. K. (1995). Determination of the distance between 2 spin labels attached to a macromolecule. *Proceedings of the National Academy of Sciences of the United States of America, 92*(18), 8239–8243.

Ramos, A., Grunert, S., Adams, J., Micklem, D. R., Proctor, M. R., Freund, S., et al. (2000). RNA recognition by a Staufen double-stranded RNA-binding domain. *The EMBO Journal, 19*(5), 997–1009.

Ramos, A., & Varani, G. (1998). A new method to detect long-range protein-RNA contacts: NMR detection of electron-proton relaxation induced by nitroxide spin-labeled RNA. *Journal of the American Chemical Society, 120*(42), 10992–10993.

Rao, J. N., Jao, C. C., Hegde, B. G., Langen, R., & Ulmer, T. S. (2010). A combinatorial NMR and EPR approach for evaluating the structural ensemble of partially folded proteins. *Journal of the American Chemical Society, 132*(25), 8657–8668.

Razzaghi, S., Qi, M., Nalepa, A. I., Godt, A., Jeschke, G., Savitsky, A., et al. (2014). RIDME spectroscopy with Gd(III) centers. *Journal of Physical Chemistry Letters, 5,* 3970–3975.

Reginsson, G. W., Kunjir, N. C., Sigurdsson, S. T., & Schiemann, O. (2012). Trityl radicals: Spin labels for nanometer-distance measurements. *Chemistry: A European Journal, 18,* 13580–13584.

Religa, T. L., Sprangers, R., & Kay, L. E. (2010). Dynamic regulation of archaeal proteasome gate opening as studied by TROSY NMR. *Science, 328*(5974), 98–102.

Schiemann, O., Cekan, P., Margraf, D., Prisner, T. F., & Sigurdsson, S. T. (2009). Relative orientation of rigid nitroxides by PELDOR: Beyond distance measurements in nucleic acids. *Angewandte Chemie, International Edition, 48,* 3292–3295.

Schiemann, O., & Prisner, T. F. (2007). Long-range distance determinations in biomacromolecules by EPR spectroscopy. *Quarterly Reviews of Biophysics, 40*(1), 1–53.

Schmidt, M. J., Borbas, J., & Drescher, M. (2014). A genetically encoded spin label for electron paramagnetic resonance distance measurements. *Journal of the American Chemical Society, 136,* 1238–1241.

Schweiger, A., & Jeschke, G. (2001). *Principles of pulse electron paramagnetic resonance.* Oxford: Oxford University Press.

Schwieters, C. D., Suh, J. Y., Grishaev, A., Ghirlando, R., Takayama, Y., & Clore, G. M. (2010). Solution structure of the 128 kDa enzyme I dimer from Escherichia coli and its 146 kDa complex with HPr using residual dipolar couplings and small- and wide-angle X-ray scattering. *Journal of the American Chemical Society, 132*(37), 13026–13045.

Shen, Y., Delaglio, F., Cornilescu, G., & Bax, A. (2009). TALOS+ is a hybrid method for predicting protein backbone torsion angles from NMR chemical shifts. *Journal of Biomolecular NMR, 44,* 213–223.

Shevelev, G. Y., Krumkacheva, O. A., Lomzov, A. A., Kuzhelev, A. A., Rogozhnikova, O. Y., Trukhin, D. V., et al. (2014). Physiological-temperature distance

measurement in nucleic acid using triarylmethyl-based spin labels and pulsed dipolar EPR spectroscopy. *Journal of the American Chemical Society, 136*, 9874–9877.
Sickmier, E. A., Frato, K. E., Shen, H., Paranawithana, S. R., Green, M. R., & Kielkopf, C. L. (2006). Structural basis for polypyrimidine tract recognition by the essential pre-mRNA splicing factor U2AF65. *Molecular Cell, 23*(1), 49–59.
Simon, B., Madl, T., Mackereth, C. D., Nilges, M., & Sattler, M. (2010). An efficient protocol for NMR-spectroscopy-based structure determination of protein complexes in solution. *Angewandte Chemie (International Ed. in English), 49*(11), 1967–1970.
Skrisovska, L., Schubert, M., & Allain, F. H. T. (2010). Recent advances in segmental isotope labeling of proteins: NMR applications to large proteins and glycoproteins. *Journal of Biomolecular NMR, 46*(1), 51–65.
Sowa, G. Z., & Qin, P. Z. (2008). Site-directed spin labeling studies on nucleic acid structure and dynamics. *Progress in Nucleic Acid Research and Molecular Biology, 82*, 147–197.
Spindler, P. E., Glaser, S. J., Skinner, T. E., & Prisner, T. F. (2013). Broadband inversion PELDOR spectroscopy with partially adiabatic shaped pulses. *Angewandte Chemie, International Edition, 52*, 3425–3429.
Spindler, P. E., Zhang, Y., Endeward, B., Gershernzon, N. A., Skinner, T. E., Glaser, S. J., et al. (2012). Shaped optimal control pulses for increased excitation bandwidth in EPR. *Journal of Magnetic Resonance, 218*, 49–58.
Sprangers, R., & Kay, L. E. (2007). Quantitative dynamics and binding studies of the 20S proteasome by NMR. *Nature, 445*(7128), 618–622.
Sprangers, R., Velyvis, A., & Kay, L. E. (2007). Solution NMR of supramolecular complexes: Providing new insights into function. *Nature Methods, 4*(9), 697–703.
Stefl, R., Oberstrass, F. C., Hood, J. L., Jourdan, M., Zimmermann, M., Skrisovska, L., et al. (2010). The solution structure of the ADAR2 dsRBM-RNA complex reveals a sequence-specific readout of the minor groove. *Cell, 143*(2), 225–237.
Storz, G., Altuvia, S., & Wassarman, K. M. (2005). An abundance of RNA regulators. *Annual Review of Biochemistry, 74*, 199–217.
Szathmáry, E. (1999). The origin of the genetic code: Amino acids as cofactors in an RNA world. *Trends in Genetics, 15*(6), 223–229.
Teplova, M., Malinina, L., Darnell, J. C., Song, J., Lu, M., Abagyan, R., et al. (2011). Protein-RNA and protein-protein recognition by dual KH1/2 domains of the neuronal splicing factor Nova-1. *Structure, 19*(7), 930–944.
Tschaggelar, R., Kasumaj, B., Santangelo, M. G., Forrer, J., Leger, P., Dube, H., et al. (2009). Cryogenic 35 GHz pulse ENDOR probehead accommodating large sample sizes: Performance and applications. *Journal of Magnetic Resonance, 200*, 81–87.
Tzakos, A. G., Grace, C. R. R., Lukavsky, P. J., & Riek, R. (2006). NMR techniques for very large proteins and RNAs in solution. *Annual Review of Biophysics and Biomolecular Structure, 35*, 319–342.
Varani, L., Gunderson, S. I., Mattaj, I. W., Kay, L. E., Neuhaus, D., & Varani, G. (2000). The NMR structure of the 38 kDa U1A protein–PIE RNA complex reveals the basis of cooperativity in regulation of polyadenylation by human U1A protein. *Nature Structural Biology, 7*, 329–335.
Voss, J., Wu, J., Hubbell, W. L., Jacques, V., Meares, C. F., & Kaback, H. R. (2001). Helix packing in lactose permease of Escherichia coli: Distances between site-directed nitroxides and a lanthanide. *Biochemistry, 40*, 3184–3188.
Ward, R., Bowman, A., Sozudogru, E., El-Mkami, H., Owen-Hughes, T., & Norman, D. G. (2010). EPR distance measurements in deuterated proteins. *Journal of Magnetic Resonance, 207*, 164–167.
Wiedenheft, B., Sternberg, S. H., & Doudna, J. A. (2012). RNA-guided genetic silencing systems in bacteria and archaea. *Nature, 482*(7385), 331–338.

Yang, Z., Liu, Y., Borbat, P., Zweier, J. L., Freed, J. H., & Hubbell, W. L. (2012). Pulsed ESR dipolar spectroscopy for distance measurements in immobilized spin labeled proteins in liquid solution. *Journal of the American Chemical Society, 134*, 9950–9952.

Yang, Y., Ramelot, T. A., McCarrick, R. M., Ni, S., Feldmann, E. A., Cort, J. R., et al. (2010). Combining NMR and EPR methods for homodimer protein structure determination. *Journal of the American Chemical Society, 132*(34), 11910–11913.

Yulikov, M. (2015). Spectroscopically orthogonal spin labels and distance measurements in biomolecules. *Specialist Periodic Reports: Electron Paramagnetic Resonance, 24*, 1–31. http://dx.doi.org/10.1039/9781782620280-00001.

Zhang, X., Cekan, P., Sigurdsson, S. T., & Qin, P. Z. (2009). Studying RNA using site-directed spin-labeling and continuous-wave electron paramagnetic resonance spectroscopy. *Methods in Enzymology, 469*, 303–328.

Zhang, Q., Stelzer, A. C., Fisher, C. K., & Al-Hashimi, H. M. (2007). Visualizing spatially correlated dynamics that directs RNA conformational transitions. *Nature, 450*(7173), 1263–1267.

Zhang, X., Tung, C. S., Sowa, G. Z., Hatmal, M. M., Haworth, I. S., & Qin, P. Z. (2012). Global structure of a three-way junction in a phi29 packaging RNA dimer determined using site-directed spin labeling. *Journal of the American Chemical Society, 134*(5), 2644–2652.

CHAPTER ELEVEN

Structural Analysis of Protein–RNA Complexes in Solution Using NMR Paramagnetic Relaxation Enhancements

Janosch Hennig[*,†], Lisa R. Warner[*,†], Bernd Simon[‡], Arie Geerlof[*,†], Cameron D. Mackereth[§,¶], Michael Sattler[*,†,1]

[*]Institute of Structural Biology, Helmholtz Zentrum München, Oberschleißheim, Germany
[†]Center for Integrated Protein Science Munich at Biomolecular NMR Spectroscopy, Department Chemie, Technische Universität München, Garching, Germany
[‡]European Molecular Biology Laboratory, Heidelberg, Germany
[§]Institut Européen de Chimie et Biologie, IECB, Univ. Bordeaux, Pessac, France
[¶]Inserm, U869, ARNA Laboratory, Bordeaux, France
[1]Corresponding author: e-mail address: sattler@helmholtz-muenchen.de

Contents

1. Introduction — 334
2. Spin Labeling of Protein–RNA Complexes — 336
3. Choice of Spin Label and Attachment Sites — 337
4. Spin Labeling Protocol and Sample Handling — 340
5. Measurement and Analysis of PRE Data — 342
6. Structure Calculation Using PRE-Derived Distance Restraints — 345
7. Case Study I: Structure of the U2AF65 RRM1,2–U9 RNA Complex — 347
8. Case Study II: Conformational Dynamics of RRM Domains in Apo U2AF65-RRM1,2 — 351
9. Case Study III: Use of NMR and PRE Data for Validation of the SxI–Unr mRNA Complex — 353
10. Outlook — 355
Acknowledgments — 355
References — 356

Abstract

Biological activity in the cell is predominantly mediated by large multiprotein and protein–nucleic acid complexes that act together to ensure functional fidelity. Nuclear magnetic resonance (NMR) spectroscopy is the only method that can provide information for high-resolution three-dimensional structures and the conformational dynamics of these complexes in solution. Mapping of binding interfaces and molecular interactions along with the characterization of conformational dynamics is possible for very large protein complexes. In contrast, *de novo* structure determination by NMR becomes very time consuming and difficult for protein complexes larger than 30 kDa as data are noisy and sparse.

Fortunately, high-resolution structures are often available for individual domains or subunits of a protein complex and thus sparse data can be used to define their arrangement and dynamics within the assembled complex. In these cases, NMR can therefore be efficiently combined with complementary solution techniques, such as small-angle X-ray or neutron scattering, to provide a comprehensive description of the structure and dynamics of protein complexes in solution. Particularly useful are NMR-derived paramagnetic relaxation enhancements (PREs), which provide long-range distance restraints (ca. 20 Å) for structural analysis of large complexes and also report on conformational dynamics in solution.

Here, we describe the use of PREs from sample production to structure calculation, focusing on protein–RNA complexes. On the basis of recent examples from our own research, we demonstrate the utility, present protocols, and discuss potential pitfalls when using PREs for studying the structure and dynamic features of protein–RNA complexes.

1. INTRODUCTION

Obtaining the structural details of protein–RNA recognition is fundamental to understanding the molecular mechanisms that direct the regulation of gene expression, e.g., transcription, splicing, and translation. Recent advances have elevated our understanding of how RNA-binding proteins specifically recognize their cognate sequences (Clery, Boudet, & Allain, 2013; Daubner, Clery, & Allain, 2013; Mackereth & Sattler, 2012). However, many RNA *cis* regulatory elements that are important for translation or splicing regulation are found in low complexity regions. For example, poly(U) sequences are abundant in untranslated regions and splice sites (Ray et al., 2013). In theory, many different RNA-binding proteins can thus recognize and compete with one another for these sites. As has been shown, the concerted action of several different RNA-binding domains within a single multidomain protein or the combination of multiple RNA-binding proteins can increase specificity and affinity for low complexity RNA targets (Clery et al., 2013; Hennig et al., 2014; Lunde, Moore, & Varani, 2007; Mackereth et al., 2011; Mackereth & Sattler, 2012). Consequently, structural biology is challenged by the structure and dynamics of high-molecular-weight protein complexes. X-ray crystallography and cryo-EM techniques are capable of handling large complexes (e.g., Cramer et al., 2000; Gagnon, Lin, Bulkley, & Steitz, 2014;

Groll et al., 1997; Makino, Baumgartner, & Conti, 2013; Medalia et al., 2002). However, not all samples are amenable to these methods due to flexible regions, transient binding, and conformational changes (see e.g., Golas et al., 2010).

Advanced biochemical and nuclear magnetic resonance (NMR) spectroscopy methods are available for studying high-molecular-weight proteins and complexes by using deuteration and specific isotope labeling (Gardner, Rosen, & Kay, 1997; Tugarinov, Kanelis, & Kay, 2006) with transverse relaxation optimized spectroscopy (TROSY) or cross-correlated relaxation enhanced polarization transfer (CRINEPT) experiments (Pervushin, Riek, Wider, & Wuthrich, 1997; Riek, Pervushin, & Wuthrich, 2000). These methods enable NMR analysis of large complexes (Mund, Overbeck, Ullmann, & Sprangers, 2013; Religa, Sprangers, & Kay, 2010; Rosenzweig, Moradi, Zarrine-Afsar, Glover, & Kay, 2013; Sprangers, Velyvis, & Kay, 2007). For structural analysis of high-molecular-weight protein complexes, where NMR data are sparse, it is especially powerful to combine NMR with information from complementary methods such as X-ray crystallography, small-angle scattering (SAS), or electron paramagnetic resonance (EPR) experiments (see e.g., Duss, Michel, et al., 2014; Jeschke, 2012; Lapinaite et al., 2013; Morgan et al., 2011; Prisner, Rohrer, & MacMillan, 2001; Takayama, Schwieters, Grishaev, Ghirlando, & Clore, 2011). In particular, NMR residual dipolar couplings (RDCs) and paramagnetic relaxation enhancement (PRE) data provide powerful information about the orientation and long-range distances, respectively, between domains or subunits of a complex.

This chapter focuses on the use of PREs to characterize the structure and dynamics of large multidomain and/or multisubunit protein–RNA complexes, which cannot be crystallized or where crystallization may trap a nonnative conformation. Chemical shift perturbations (CSPs) can report on binding interfaces, i.e., protein–protein and protein–RNA interfaces, but the information obtained is usually insufficient to determine a reliable structure of the complex (Hennig, Wang, Sonntag, Gabel, & Sattler, 2013). PREs can instead provide numerous long-range distance restraints for determining the structure of such complexes (Battiste & Wagner, 2000; Clore & Iwahara, 2009; Gobl, Madl, Simon, & Sattler, 2014; Madl, Felli, Bertini, & Sattler, 2010; Pintacuda, John, Su, & Otting, 2007). We describe experimental aspects, data analysis, and structure calculations for the use of NMR PREs in the structural analysis of protein–RNA complexes.

2. SPIN LABELING OF PROTEIN–RNA COMPLEXES

There are several considerations when using PREs for structural analysis of protein–RNA complexes, including the choice of spin label and the attachment site to the protein or RNA components. It must be considered whether the protein or RNA is spin labeled and to ensure that the spin-labeling site does not alter the native state of the protein–RNA complex (*vide infra*). These requirements are strict if PREs will be used for *de novo* structure calculation of the domain arrangements, as compared to cases where PRE data are merely used for validation of an existing protein–RNA structure, for example that has been determined using X-ray crystallography.

The most common and efficient way of spin labeling is either the covalent attachment of the spin label to a reactive thiol group in a protein harboring a single accessible cysteine residue (Battiste & Wagner, 2000; Gobl et al., 2014; Su & Otting, 2010) or to an RNA containing a single thiouridine (Ramos, Bayer, & Varani, 1999). A variety of nitroxide spin labels or paramagnetic metal-binding tags are available. While nitroxide spin labels provide only distance-dependent PRE effects, paramagnetic metal or lanthanide-binding tags (LBTs) have the advantage in that they also provide distance- and orientation-dependent pseudo-contact shifts and orientation-dependent RDC data (Barthelmes et al., 2011; Bertini, Luchinat, & Parigi, 2002; Bertini, Luchinat, Parigi, & Pierattelli, 2005; Keizers, Saragliadis, Hiruma, Overhand, & Ubbink, 2008; Lu, Berry, & Pfister, 2001; Pintacuda et al., 2007; Su & Otting, 2010; Wohnert, Franz, Nitz, Imperiali, & Schwalbe, 2003). The choice of spin label should be made based on the type of information needed (Gobl et al., 2014). In principle, it is desirable to restrict the conformational flexibility of a paramagnetic tag. For this purpose, a rigid attachment with two appropriately spaced cysteine residues, for example, in a helix, can be used (Keizers et al., 2008). Alternatively, genetically encoded LBTs can be engineered into loop regions of a protein domain as long as the structure of the domain is already known (Barthelmes et al., 2011). An elegant but more elaborate approach is based on the incorporation of unnatural amino acids during protein synthesis, which has the advantage of site-specific placement of the spin label and avoids the requirement of redox-sensitive cysteine residues. This approach involves the site-specific incorporation of either a spin-labeled amino acid or a chemically reactive unnatural amino acid, which can subsequently be spin labeled using click chemistry (Jones et al., 2010; Schmidt, Borbas, Drescher, & Summerer, 2014; Wang, Xie, & Schultz, 2006; Xie & Schultz, 2005).

Protocols for paramagnetic labeling of RNA oligonucleotides are also available (Ramos et al., 1999; Shelke & Sigurdsson, 2012; Su & Otting, 2010). Several methods have been developed to attach a paramagnetic tag to nucleic acids (Keyes & Bobst, 2002; Nguyen & Qin, 2012; Qin & Dieckmann, 2004; Shelke & Sigurdsson, 2012): these include replacing structural Mg^{2+} with paramagnetic Mn^{2+} (Bonneau & Legault, 2014; Kisseleva, Khvorova, Westhof, & Schiemann, 2005), attaching paramagnetic groups to $2'$-amino modified nucleotides or 4-thiouridines (Piton et al., 2007; Qin, Feigon, & Hubbell, 2005), and the use of paramagnetic phosphoramidites during RNA synthesis (Wunderlich et al., 2013). In the case of large RNAs, annealing and ligating short spin-labeled oligonucleotides to a larger RNA scaffold (Helmling et al., 2014), and enzymatically, using T7 RNA polymerase (Lebars et al., 2014) have been reported. A combination of 4-thiouridine spin labeling and segmental labeling of RNA enables double spin labeling of RNA for EPR-based distance measurements (Duss, Maris, von Schroetter, & Allain, 2010; Duss, Michel, et al., 2014; Duss, Yulikov, Jeschke, & Allain, 2014). The choice of method depends on whether the spin label should be placed in a region of the RNA with tertiary structure, double-helical, or single-stranded conformation, and whether the RNA is prepared by chemical synthesis or by *in vitro* transcription. As with protein spin labels, the intrinsic flexibility of the spin label when attached to the RNA needs to be carefully considered. RNA spin labeling can be used to analyze the conformation and dynamics of larger RNA molecules, as has been shown with HIV-1 TAR RNA (Wunderlich et al., 2013), riboswitches, and spliceosomal U6 snRNA (Buttner, Javadi-Zarnaghi, & Hobartner, 2014; Buttner, Seikowski, Wawrzyniak, Ochmann, & Hobartner, 2013). Furthermore, the RNA-binding orientation with respect to an RNA-binding protein can be determined (Mackereth et al., 2011), and protein–RNA distances can be obtained and used for structure calculation and validation (Allen, Varani, & Varani, 2001). Another chapter in this book describes the labeling of RNA and its use in EPR measurements to derive restraints for structure calculations (Duss, Yulikov, Allain, & Jeschke, 2015).

3. CHOICE OF SPIN LABEL AND ATTACHMENT SITES

The preparatory steps for spin-labeling of proteins are cloning, protein expression, purification, and covalent attachment of the spin label. Once this has been achieved, obtaining long-range distance restraints is rather straightforward. Here, we focus on the most efficient protocol, in our hands, based on covalent attachment of nitroxide spin labels to single cysteine variants of the protein. The two most commonly used spin labels are IPSL (3-(2-Iodoacetamido)-PROXYL free radical) and MTSL (S-(1-oxyl-2,2,5,5-tetramethyl-2,5-dihydro-1H-pyrrol-3-yl)methyl methanesulfonothioate) (Berliner, Grunwald, Hankovszky, & Hideg, 1982; Hankovszky, Hideg, & Lex, 1980; Qin & Dieckmann, 2004; Fig. 1). For nitroxides, the correlation time of the electron–nuclear interaction τ_{PRE} is dominated by the rotational correlation time and allows for the measurement of distances up to 20–25 Å for larger molecules. The MTSL spin label contains a reducible disulfide bond linkage to the protein (Fig. 1A), which makes it possible to remove and fully reconstitute the native state of the protein. The disadvantage is the inability to add reducing agents such as dithiothreitol (DTT) or tris(2-carboxyethyl)phosphine (TCEP) to the NMR sample buffer due to the redox-sensitive disulfide bond, which may limit its application when other thiols are present in the complex. In contrast, IPSL is covalently linked to the protein via a stable thioether bond (Fig. 1B) that cannot be reduced by DTT or TCEP. The covalent attachment of either spin label leaves a number of rotatable bonds, and the effects of the spin label flexibility have to be considered for the quantitative interpretation of the PRE data (*vide infra*). As a further practical advantage, nitroxide radical spin labels are easily reduced to the diamagnetic state, which allows the measurement of the paramagnetic and diamagnetic states on a single sample.

To take full advantage of PRE data, it is recommended to have as complete as possible the NMR chemical shift assignments for the components to which PRE effects are to be analyzed. PRE effects of unassigned resonances cannot be included in structure calculations, and the analysis of ambiguously assigned resonances is not recommended. In principle, a preliminary structural model can be created based on CSPs upon complex formation coupled with data-driven docking of the complex (e.g., using HADDOCK; de Vries, van Dijk, & Bonvin, 2010). Such a model can help with choosing appropriate spin label attachment sites by visualizing the possible interaction interfaces and avoiding sites where the spin label attachment could perturb complex formation. On one hand, the spin label needs to be close to the binding interface to provide a sufficient number of distance restraints, but on the other hand it must not be located within the interface and prevent proper assembly of the complex.

Another important consideration is the presence of native cysteines. The accessibility of native cysteine residues can be first assessed with Ellman's test for free thiols (Ellman, 1959). Solvent accessible and/or reactive cysteines

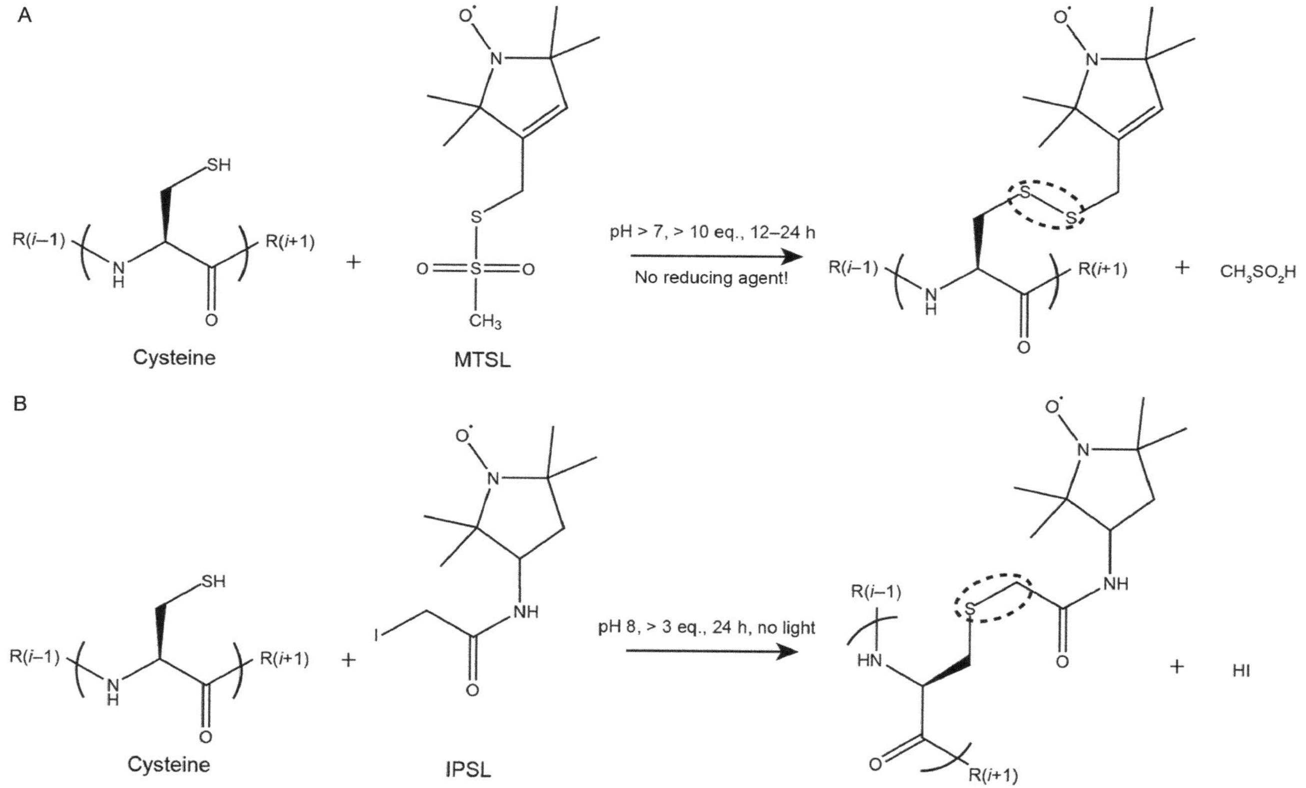

Figure 1 Chemical structures of IPSL and MTSL spin label compounds. Spin labeling of cysteines in a protein with (A) MTSL and (B) IPSL. Using MTSL, a disulfide bond is formed. After the measurements, the reaction can be inverted and the free cysteine thiol recovered by using a reducing agent. With IPSL, a thioether bond is formed and the spin label side chain has only three bonds with rotational freedom. The MTSL-label harbors four bonds with rotational freedom.

need to be replaced so that only a single reactive native cysteine is left, or if a new reactive cysteine is to be added then all native accessible cysteines must be mutated. Native cysteine residues are normally replaced by serine when solvent exposed, or by alanine if they are partially buried. Structurally important disulfide bonds that form part of the native fold should remain untouched, but their compatibility with redox conditions required for the attachment of spin label needs to be considered and tested. In the next step, a number of protein construct variants are designed with single residues mutated to a cysteine for spin label attachment. As a general guideline, solvent-exposed residues in α-helices or β-strands are recommended for the introduction of cysteine mutants whereas flexible linkers should be avoided. NMR ^{15}N relaxation data can be used to identify flexible regions within the protein backbone. It is also possible to infer the presence of ordered or disordered regions from high-resolution structures that may be already available for the single components. It is crucial to ensure that the native structure of the complex is not perturbed by the cysteine mutation or by the attachment of the spin label compounds. Comparison of NMR fingerprint spectra, i.e., $^{1}H,^{15}N$ or $^{1}H,^{13}C_{methyl}$ correlation NMR spectra of the complex with and without spin label attachment, is necessary to confirm that the spin labeling does not affect the structure of the biomolecule or the assembled complex.

Another important consideration for spin label design is the number of different sites to be tested, as the number of distance restraints that can be obtained scales with the number of spin label sites. It is recommended to consider at least five sites for spin labeling in order to adequately cover the surface of each individual domain or subunit (assuming a 10–15 kDa molecular weight for the subunits). The design of additional sites is advisable as some of the mutations may not be usable due to destabilization of the protein fold or the assembled complex, and redundant information can be used for cross-validation of the PRE data.

4. SPIN LABELING PROTOCOL AND SAMPLE HANDLING

Effective spin label attachment depends on the cysteine protonation state. Prior to spin labeling, it is therefore necessary to adjust the pH of the protein buffer to pH 8, i.e., above the sulfhydryl pK_a, preferably using Tris buffer as it has its optimal buffer capacity near pH 8. The labeling reaction is most efficient between pH 7.5 and 8.5, and although some proteins might not be stable at this pH, salt may be added or the sample diluted in order to maintain the protein in solution. pH values above pH 8.5 are also not advisable as other amino acid sidechains may become deprotonated and thus become targets for modification. For optimal labeling efficiency, the cysteine must be fully reduced prior to spin labeling, and it is recommended to place the protein in a strong reducing agent (such as 10 mM DTT) overnight. However, since DTT or other reducing agents inhibit the spin-labeling reaction, complete removal of any reducing agent immediately prior to the addition of the spin label reagent is important. Dialysis for buffer exchange is possible, but it is important to use degassed buffer in order to prevent oxidation of the protein. A faster way is to remove the reducing agent with desalting columns. Here, it is recommended to use disposable columns and perform the desalting step twice, with each elution using 15% less buffer than the recommended amount as a compromise between sufficient protein recovery and complete removal of the reducing agent. A concentrated stock (10–100 mM) of IPSL dissolved in methanol should be added in at least a 3 × molar excess to the protein solution. To prevent spin label insolubility when adding concentrated IPSL directly to a buffer with high ionic strength, it may be necessary to first lower the methanol to 25% (v/v) by adding small amounts of the final buffer just before adding the aliquot to the protein. The stock must be protected from light to prevent degradation, and the labeling process should also be performed in the dark. At a pH of 8.0, the reaction is rapid and generally occurs within 1 h, even at the usual 4 °C; however, the attachment process should be as complete as possible in order to derive accurate information and therefore a longer incubation time is recommended (typically overnight or 24 h). When the reaction is complete, the excess spin label must be completely removed with dialysis, a desalting column, or ideally size exclusion chromatography, depending on the long-term stability of the spin-labeled protein. If required, the sample can be concentrated and returned to the NMR buffer and the final complex can be assembled.

NMR experiments of the paramagnetic sample (*vide infra*) should be acquired soon after the spin label is attached. It is important to retain some spin-labeled protein for quantifying the completeness of the spin labeling by mass spectrometry and to check for free thiols using Ellman's reagent (Ellman, 1959). Additionally, an aliquot of the spin-labeled protein should be set aside for reducing and nonreducing (no reducing agent added to the loading buffer) SDS–PAGE analysis to detect potentially formed dimers by cross-linking via the free cysteine. PRE measurements are only meaningful if the spin labeling is close to 100%, otherwise the PRE effect is underestimated and the distance restraints derived are longer than they should be. If the spin-labeled subunit is also NMR isotope-labeled, a crude quantification of the spin-labeling efficiency is possible by monitoring spins that are within 10 Å radius of the spin label, as NMR signals of these residues are expected to be completely bleached by the PRE effect. In the case of *trans* PRE effects, i.e., where one subunit is spin labeled and PREs are detected on another NMR isotope-labeled subunit, quantification of the spin-labeling efficiency by mass spectrometry is possible and is an advisable step. Following data collection on the paramagnetic protein sample, the spin label must be reduced in order to record NMR data for the diamagnetic state of the protein with the same sample. Reduction of nitroxide spin labels is usually performed by adding 10-fold molar excess of ascorbic acid. Here, it is important that the pH of the ascorbic acid solution is adjusted to match the NMR sample buffer, as otherwise pH-dependent CSPs can be observed or the protein fold may be affected. The reduction is usually complete within 1 h, but care must be taken to have complete reduction prior to collecting data on the diamagnetic sample. The extent of reduction can be followed by measuring 1D ^{1}H spectra until the reaction is complete and no further increase in signal occurs.

5. MEASUREMENT AND ANALYSIS OF PRE DATA

Structural analysis based on PRE data exploits the distance dependence of the paramagnetic relaxation effects. Note that the derivation of distance restraints from PRE measurements as discussed below is applicable for rigid complexes, where little or no conformational dynamics is present. Due to the exquisite sensitivity of the PRE to short distances, even weakly populated (<1%) conformations can give rise to a strong observable PRE. For example, in the presence of conformational dynamics and interconverting species, short distances present in a minor conformation can efficiently report on the minor species and encounter-like contacts (Clore & Iwahara, 2009; Iwahara & Clore, 2006; Tang, Iwahara & Clore, 2006; Volkov, Worrall, Holtzmann, & Ubbink, 2006), or on the conformational behavior of intrinsically disordered proteins (Salmon et al., 2010), but their analysis depends on whether the exchange kinetics between minor and major species is fast or slow compared to PRE rates (Clore & Iwahara, 2009; Iwahara, Schwieters, & Clore, 2004).

PRE data are efficiently measured for amide and methyl protons (Battiste & Wagner, 2000; Clore & Iwahara, 2009; Gelis et al., 2007; Gottstein, Reckel, Dotsch, & Guntert, 2012; Kato et al., 2011; Lapinaite et al., 2013; Liang, Bushweller, & Tamm, 2006; Simon, Madl, Mackereth, Nilges, & Sattler, 2010) or ^{13}C spins (Madl et al., 2010). For proton PREs, it is preferable to measure transverse PREs ($R_{2,PRE}$) as they are less sensitive to internal motion and additional relaxation pathways (Clore & Iwahara, 2009) than the longitudinal $R_{1,PRE}$ rates. Transverse $R_{2,PRE}$ relaxation rates are measured most precisely by sampling the exponential NMR signal decay in the oxidized (paramagnetic) and reduced (diamagnetic) state (Iwahara, Anderson, Murphy, & Clore, 2003). To reduce measurement times, the exponential decay can be sampled with two time points without compromising data quality (Clore & Iwahara, 2009; Gobl et al., 2014; Iwahara, Tang, & Marius Clore, 2007). In the simple approach originally proposed by Battiste and Wagner, amide proton PREs are measured using ^{1}H,^{15}N HSQC experiments (Battiste & Wagner, 2000), where signal intensities in the paramagnetic (oxidized, I_{para}) and diamagnetic (reduced, I_{dia}) states are compared to measure the PRE contribution during a fixed delay. Methyl proton $R_{2,PRE}$ on high-molecular-weight complexes has been measured using ^{1}H,^{13}C HMQC experiments (Gelis et al., 2007; Kato et al., 2011; Lapinaite et al., 2013). Distance restraints for side chains can also be obtained from measuring longitudinal $R_{1,PRE}$ rates for ^{13}C spins in saturation recovery experiments (Madl et al., 2010).

In practice, it is critical that the same experimental settings, i.e., spectrometer and temperature, are used for recording NMR data for the paramagnetic and diamagnetic states of the protein to avoid systematic errors. It is also important to obtain sufficient signal-to-noise for experiments recorded for the paramagnetic state. For all approaches, NMR spectra should be analyzed in a conservative way, i.e., neglecting overlapping signals or simulating such signals so that reliable peak integrations can be obtained. A semiquantitative analysis of PREs from only a few spin label positions may be sufficient for the purposes of validating an existing structure or for qualitative measurements of conformational changes of a specific region in a complex. In contrast, for *de novo* structure calculation of protein folds or domain arrangements, a large number of PREs should be measured with several spin label positions that preferably screen the complete accessible surface of the structure and also allow for the collection of repelling restraints for regions where no PRE is observed (*vide infra*).

In the following, we discuss how distance restraints are obtained from proton PREs derived from signal intensities in NMR correlation spectra

(Battiste & Wagner, 2000), where transverse relaxation is active in constant time delays during the pulse sequences (Battiste & Wagner, 2000; Gobl et al., 2014; Simon et al., 2010). Distance restraints between the electron and the nuclear spin can be derived from the ratio of the cross-peak intensities in the paramagnetic and diamagnetic states. The derivation of Eq. (1) has been described elsewhere (Battiste & Wagner, 2000; Clore & Iwahara, 2009; Gobl et al., 2014; Simon et al., 2010):

$$\frac{I_{para}}{I_{dia}} = \frac{R_{2,dia}e^{-R_{2,PRE}t}}{R_{2,dia} + R_{2,PRE}} \quad (1)$$

$$R_{2,PRE} = \frac{1}{15}\gamma_I^2 g_e^2 \mu_B^2 S(S+1)\left(\frac{\mu_0}{4\pi}\right)^2 \frac{1}{r^6}\left(4\tau_{PRE} + \frac{3\tau_{PRE}}{1+(\omega_H \tau_{PRE})^2}\right) \quad (2)$$

The transverse $R_{2,PRE}$ rate is determined by numerical solution of Eq. (1) based on the experimental intensity ratios I_{para}/I_{dia} (sometimes also referred to as I_{ox}/I_{red} for the oxidized/paramagnetic and reduced/diamagnetic states, respectively) and the diamagnetic relaxation rates, $R_{2,dia}$, which are obtained from relaxation measurements of the sample in the reduced state or of the native protein. The latter method for measuring $R_{2,dia}$ does not take into account potential effects of the spin label on the diamagnetic relaxation rates. As an alternative to recording $R_{2,dia}$ rates for each spin, an average value can be estimated from average amide proton line widths determined using a water-flip-back spin-echo experiment (Anglister, Grzesiek, Ren, Klee, & Bax, 1993) or estimated from the molecular weight and temperature of the measurement (Daragan & Mayo, 1997). This approximation will introduce small errors in the derived distance restraints, which may be considered and compensated for by using larger error bounds, measuring a large number of PRE data from multiple spin labels, and critical evaluation of violated restraints during structure calculations.

The time t in Eq. (1) is the total time of magnetization transfer in the HSQC experiment. Using Eq. (2) and $R_{2,PRE}$ from Eq. (1) the distance r from each nuclear spin to the paramagnetic center can be derived. Here, S is the electron spin quantum number (½), γ is the gyromagnetic ratio, g is the electron g-factor, μ_B is the electron Bohr magneton, μ_0 is the permeability of vacuum, and ω_H is the Larmor frequency of the nuclear spin. In principle, τ_{PRE} is the PRE correlation time which depends on the rotational correlation time of the spin-labeled protein, τ_c, and the electron spin relaxation time of the spin label, τ_S ($\tau_{PRE}^{-1} = \tau_c^{-1} + \tau_S^{-1}$), which is large for nitroxide spin labels ($>10^{-7}$ s) and can therefore be neglected (Kosen, 1989). However, internal dynamics can influence τ_{PRE} and an apparent experimental τ_{PRE}^{app} can be estimated from longitudinal and transverse relaxation rates measured for the diamagnetic and paramagnetic states for each nuclear spin–electron vector (Clore & Iwahara, 2009). Internal dynamics is clearly present due to the rotational degrees of freedom of the spin label and thus the apparent correlation time, τ_{PRE}^{app}, should be smaller than τ_c. Indeed, a τ_{PRE}^{app} for MTSL and IPSL has been reported to be 2.6 and 4.2 ns, respectively, for spin-labeled phospholamban (Kirby, Karim, & Thomas, 2004). However, as can be seen from Eq. (2), variations in this low nanosecond range have a negligible effect on calibrated distance restraints and, in practice, a uniform τ_{PRE} can be assumed for all other spin labels as long as spin labels are attached to rigid regions (i.e., secondary structure) of the protein.

6. STRUCTURE CALCULATION USING PRE-DERIVED DISTANCE RESTRAINTS

PRE data can be incorporated into standard NMR structure calculation by extending existing software packages, such as ARIA/CNS (Madl et al., 2010; Simon et al., 2010), XPLOR-NIH (Clore & Schwieters, 2003), or CYANA (Gottstein et al., 2012). A structure calculation protocol that incorporates PREs and RDCs for multidomain proteins (with semirigid domain structures) and based on the ARIA/CNS structure calculation software (Brunger et al., 1998; Linge, Habeck, Rieping, & Nilges, 2003) is available from the authors (Simon et al., 2010). A number of important aspects should be considered:

1. It is crucial to realize that the PRE effect observed will report on all potentially existing conformations. Due to the r^{-6} dependence, PRE is strongly biased to short distances and thus will report on the existence of minor populations in a dynamic mixture (Clore & Iwahara, 2009; Iwahara et al., 2004). Thus, structure calculations based on PRE data should be applied only to rigid complexes, with a single predominant species existing in solution. To consider the internal motion of the covalently attached spin label, ensemble-averaged distance restraints should be applied using multiple copies (typically four) for each spin label position (Clore & Iwahara, 2009; Iwahara et al., 2004; Simon et al., 2010).

2. For converting the experimental PRE data into distance restraints, the following protocol is recommended: in a first step, distances are

calculated with an initial estimate of τ_{PRE}, i.e., corresponding to rotational correlation time τ_c of the molecule. After a first structure calculation using these distance restraints, τ_{PRE} is refined by a grid search that minimizes the error function Q_{PRE} Eq. (3). In the case of a multidomain protein, only intradomain PREs are used for the initial calculation, so that the distances obtained can be compared with an available structure of the domain on which the spin label is attached. Using the optimized value for τ_{PRE} after the grid search, final distance restraints are calculated and applied with upper and lower bounds calculated from the experimental uncertainties of the intensity ratios. Alternatively, uniform distance bounds for residues with sizable PREs (e.g., ±2 Å) and only lower/upper bounds for residues with small/large intensity ratios can be used (Gobl et al., 2014; Simon et al., 2010). Very small or zero PREs, i.e., for spins that are unaffected by the spin label, are used as repelling restraints to help increase the convergence of the structure calculation.

3. A quality factor, Q_{PRE}, is used to assess the quality of the structure by comparing the experimental PRE data with those back-calculated from the structure according to:

$$Q_{PRE} = \sqrt{\frac{\sum\left(PRE_{backcalc} - PRE_{exp}\right)^2}{\sum\left(PRE_{exp}\right)^2}} \qquad (3)$$

Wherever possible, additional restraints should be provided and combined with the PRE data. A number of examples have recently demonstrated the utility of combining PRE data with RDC data, which provide useful and important complementary structural information that help to define the relative domain orientation in the complex (Blackledge, 2005; Prestegard, Bougault, & Kishore, 2004; Tjandra & Bax, 1997). Vice versa, the PRE-derived distance restraints can help to resolve ambiguities of domain orientations inherent to RDC data (Clore & Schwieters, 2003; Simon et al., 2010).

Additional complementary information can be obtained from combining NMR with small-angle X-ray or neutron scattering (SAXS/SANS) (Bertini et al., 2008; Grishaev, Tugarinov, Kay, Trewhella, & Bax, 2008; Grishaev, Wu, Trewhella, & Bax, 2005; Hennig et al., 2014; Hennig & Sattler, 2014; Hennig et al., 2013; Huang et al., 2014; Lapinaite et al., 2013; Schwieters et al., 2010) and EPR data (Duss, Michel, et al., 2014; Duss, Yulikov, et al., 2014).

7. CASE STUDY I: STRUCTURE OF THE U2AF65 RRM1,2-U9 RNA COMPLEX

The U2 snRNP auxiliary factor 65 (U2AF65) (Zamore, Patton, & Green, 1992) has been shown to play a key role in pre-mRNA splicing (Wahl, Will, & Luhrmann, 2009) by specifically recognizing the 3′ splice-site-associated poly-pyrimidine tract RNA in target pre-mRNAs (Banerjee, Rahn, Davis, & Singh, 2003). Its specificity is based partly on conformational selection between closed and open states adopted by its two N-terminally positioned RNA recognition motifs (RRM), herein named RRM1,2 (Mackereth et al., 2011; Mackereth & Sattler, 2012). The structures of both states have been assessed thoroughly by PRE and RDC measurements, using altogether 10 different spin label mutants in the absence and presence of a high affinity nine uracil (U9) RNA (Mackereth et al., 2011; Simon et al., 2010; Fig. 2). The motivation of using PREs in this system was to monitor the domain rearrangements of both RRMs and the conformational changes that are induced upon RNA binding. To illustrate a favorable case of using PREs for structure calculations of a protein–RNA complex, we walk through the entire process by looking at the U2AF65-N155C mutant in the U9 RNA-bound form.

Considering the suitability of this position, ^{15}N relaxation data have been used to assess the flexibility of the region, where N155 is located. Since the spin label was also used to probe the structural rearrangement upon RNA binding, it was important to make sure that N155 is located far enough away from the β-sheet surface, which forms the canonical RNA-binding site (Fig. 2A). The spin labeling was performed as described, and buffer exchange into spin label buffer (50 mM Tris, 200 mM NaCl, pH 8.0, degassed) was done using dialysis. The HSQC spectrum of the paramagnetic state was acquired directly after exchange into NMR buffer (20 mM NaP$_i$, 50 mM NaCl, 1 mM EDTA, 0.1% NaN$_3$, pH 6.5) at 295 K and at a concentration of around 200 µM. After confirming complete disappearance of cross peaks in ^{1}H,^{15}N HSQC spectra for residues in spatial proximity of residue 155, the sample was either directly reduced by addition of six equivalents of ascorbic acid to measure the signal intensities in the diamagnetic state of the free protein or 1.5 molar equivalents of U9 RNA was added and the HSQC spectrum of the complex was recorded. Immediately afterward, six equivalents of ascorbic acid were added and following a delay of 1 h, the HSQC of the diamagnetic state was acquired for the protein–RNA complex. By

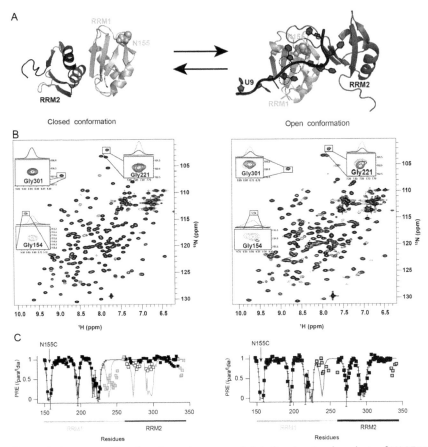

Figure 2 Paramagnetic relaxation enhancement data for structural analysis of U2AF65 RRM1,2 free and bound to U9 RNA. (A) NMR solution structures of RRM1,2 free (left, RRM1: purple; RRM2: yellow, PDB ID: 2YH0) and in complex with U9 RNA (right, RRM1: purple; RRM2: yellow; RNA, black; PDB ID: 2YH1) are shown with the RNP1 and RNP2 of RRM1 indicated with dotted circles and the N155 side chain carbons as a sphere. N155 was chosen as a spin label site because it is attached to a rigid backbone and solvent exposed. N155 is located far enough from the RNA-binding β-sheet surface of RRM1 and thus does not interfere with RNA binding. (B) $^1H,^{15}N$ HSQC spectra of the oxidized (black) and reduced (red) N155C IPSL labeled RRM1,2, left: free and right: bound to U9 RNA. NMR signals of three residues that show a maximal (G154), intermediate (G221), and minimal (G301) PRE effect are shown as zoomed insets together with the ^1H 1D traces (paramagnetic: black line, diamagnetic: red dashed line) at the ^{15}N peak maximum. (C) Experimentally measured amide proton PRE effects plotted as I_{para}/I_{dia} for RRM1,2 with a proxyl spin label attached to residue 155 (SL155), left: free; right: bound to U9 RNA. Filled blue and red boxes are experimental data in rigid and flexible regions of the protein backbone, respectively. Open white boxes are data indicate the presence of the open conformation. Only blue filled boxes were used for structure calculations. (See the color plate.)

comparing both HSQC spectra, signal attenuation in the paramagnetic state of amide protons close in space to the spin label could be observed for the free and U9 RNA-bound protein (Fig. 2B).

The intensity ratios were then plotted versus the residue number to visualize which regions of the protein are affected by the spin label and which regions are far away and unaffected (Fig. 2C). As the high-resolution structures of both domains are available (Sickmier et al., 2006), the distance restraints derived from the PREs could be used for structure calculations. For this particular structure calculation protocol, it was important to keep the linker region flexible in all steps in order to allow movement and reorientation of the domains with respect to each other. Evidence that the linker region is indeed flexible has been obtained from NMR ^{15}N relaxation data (Mackereth et al., 2011). It would be reasonable to predict that with a flexible linker the domains would not interact with each other within the complex. However, the observed PREs clearly show that the domains are in close contact in the protein–RNA complex (Fig. 2C) and measurement of nine additional spin-labeled samples confirmed this observation and allowed for structure calculation of the protein–RNA complex. The structures obtained can thus be considered as hybrid NMR and X-ray-derived structures as NMR data were used along with the available crystal structures of the RRM1 and RRM2 domains bound to oligo-U (Sickmier et al., 2006) in order to obtain the solution conformation of the protein–RNA complex and for the characterization of a large conformational change upon RNA binding (Mackereth et al., 2011).

The rotational correlation time τ_{PRE} for the spin label has been estimated from experimentally measured ^{1}H R_1 and R_2 relaxation rates in the paramagnetic and diamagnetic states (Simon et al., 2010; Fig. 3A). The τ_{PRE} for this spin label is between 2.5 and 3 ns and thus comparable to the value of 2.6 ns which has been determined for IPSL attached to phospholamban (Kirby et al., 2004). The smaller value of τ_{PRE} compared to the rotational correlation time of the protein $\tau_c \approx 10$ ns is likely due to internal dynamics of the spin label which is flexibly attached to a cysteine side chain. A grid search has been performed over all intradomain PREs for this spin label (Fig. 3B), and an optimum has been found at 3.7 ns. As variations on this scale have only a minor effect on the calibrated distance restraints, a value of 4 ns has been used for calibration of all distance restraints derived for all the 10 spin labels used for the structural analysis of the RRM1,2/U9 RNA complex (Simon et al., 2010). Figure 3C shows the distance restraints, derived from the ratio of the experimentally determined signal intensities of

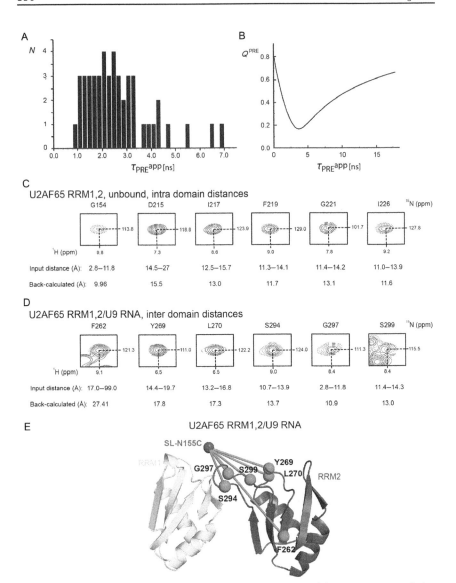

Figure 3 Assessment of spin label dynamics. Determination of the apparent correlation time, τ_{PRE}^{app}, for the electron–nuclear spin vectors in RRM1,2, spin labeled at N155C. (A) τ_{PRE}^{app} derived from the ratio of amide proton R_2 and R_1 relaxation rates in the oxidized (paramagnetic) and reduced (diamagnetic) state. (B) Grid search for the optimal correlation time by structure calculations using only intradomain PRE data (using the known structure of RRM1 and RRM2 as rigid bodies) with intensity ratios between the paramagnetic and the diamagnetic states are between 0.2 and 0.8. Calculation of Q_{PRE} is shown in Eq. (3). (C and D) Zoomed view of NMR signals in the diamagnetic (red (gray in the print version), dashed line) and paramagnetic (blue (dark gray in

the oxidized and reduced states and the corresponding back-calculated average distances for a number of residues in the free and U9-bound U2AF65-RRM1,2 N155C spin-labeled protein.

8. CASE STUDY II: CONFORMATIONAL DYNAMICS OF RRM DOMAINS IN APO U2AF65-RRM1,2

As mentioned above, the first two RRMs of U2AF65 (RRM1,2) are connected by a flexible ~25 residue linker. In the absence of RNA, it was observed that interdomain PRE effects were still present signifying an interaction between the two RRMs. Detailed analyses revealed that although a small proportion of the PREs could be explained by the arrangement of the two domains corresponding to the RNA-bound conformation, the majority of the PREs must arise from an alternative conformation. With a conservative use of only PREs that were distinct from those of the RNA-bound arrangement and also excluding PREs from the linker residues (Fig. 2C, black squares), a single conformation was calculated that represents the closed domain arrangement of the two RRM domains (Mackereth et al., 2011). In addition to the two observed conformations (open and closed arrangements), SAXS data recorded for the free protein (Huang et al., 2014; Jenkins, Laird, & Kielkopf, 2012) demonstrated that the apo state includes the presence of detached conformations, where the domains are not in close contact (Huang et al., 2014). This observation indicated that a more complex conformational ensemble was necessary to describe the apo form of the protein in solution.

In the case of apo RRM1,2, an ensemble approach was therefore employed to describe the conformational state of the protein in the absence of the RNA ligand. The PRE data from seven spin labels were further combined with RDC data from steric alignment and SAXS data by using the

---

the print version), solid line) states is shown for correlations of intradomain distances in free U2AF65 RRM1,2 (C) and for correlations of interdomain distances in U9-bound U2AF65 (D). For each signal, the distance restraint (lower–upper bound) calculated from the peak intensity ratios of the paramagnetic and diamagnetic states and the distances back-calculated from the structure are shown. (E) The distances derived from the PRE effects seen for the signals shown for the bound U2AF65 RRM1,2 in (D) are indicated by green (gray in the print version) lines. The red (dark gray in the print version) sphere indicates the position of the nitroxyl group for the spin label at residue 155, green (gray in the print version) spheres the amides of the residue for which the PRE is observed. *Panel (B) Reproduced with permission from Simon et al. (2010).*

ASTEROIDS method (Guerry et al., 2013; Huang et al., 2014; Salmon et al., 2010) to derive an ensemble of structures that are fully consistent with the NMR and SAS data. The observation of only small chemical shift changes and a single set of NMR signals when comparing the NMR spectra of RRM1 and RRM2 with RRM1,2 suggests that the domain arrangements are dynamically interconverting at time scales that are fast compared to the PRE rates. An analysis of the resulting ensemble shows a highly anisotropic distribution of the RRM domains, with a cluster of interdomain contacts that are mainly electrostatic in nature. To confirm the suggested role of electrostatic interactions, PREs were measured with a sample spin labeled at position 318, where substantial interdomain PREs are observed for the free RRM1,2 protein. As expected, a reduction of interdomain PREs is observed with increasing salt concentration in the buffer as shown in Fig. 4A. The combination of PRE and SAXS therefore provides a more complex picture for the mechanism of RNA recognition by U2AF65 RRM1,2 and involves an ensemble of inactive, closed conformations that also includes a preexisting domain arrangement that resembles the open

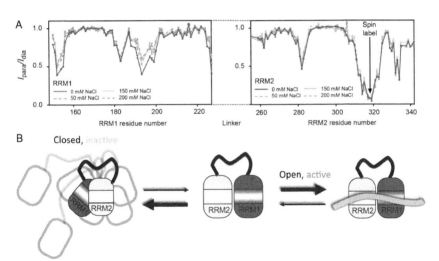

Figure 4 PREs of apo U2AF65 and its RNA-binding mechanism. (A) Salt dependence of interdomain PRE data observed for the free RRM1,2 protein with a spin label at residue 318. The left and right panels are derived from the same sample, and the linker region is indicated with a dashed line. (B) Complex mechanism of RNA binding by U2AF65 RRM1,2 based on PRE and small-angle X-ray scattering (SAXS) data. The free protein is a dynamic ensemble of closed, inactive states, which already includes a small population of preexisting domain arrangements that resemble the open state as seen in the RNA-bound complex.

conformation already in the absence of RNA. This is consistent with a conformational selection mechanism for binding to the U9 RNA ligand (Mackereth et al., 2011; Fig. 4B).

9. CASE STUDY III: USE OF NMR AND PRE DATA FOR VALIDATION OF THE Sxl–Unr mRNA COMPLEX

The third example illustrates how NMR data, i.e., PREs, CSPs, and RDCs, can be used to evaluate if the domain arrangements observed in a crystal structure indeed represent the solution conformation. The ternary Sxl–Unr mRNA complex consists of the tandem RRM domains of the protein Sex-lethal (Sxl), where both RRM domains are connected by a flexible linker and whose structure bound to RNA has been previously solved (Handa et al., 1999) as well as the first cold shock domain (CSD1) of the protein Upstream-of-N-Ras (Unr). We have determined the structure of the ternary complex by using X-ray crystallography and combined this with rigorous validation of the structure using NMR and SAS data (Hennig et al., 2014).

The CSPs of each protein observed upon RNA binding and ternary complex formation fully corroborate the contacts seen in the crystal structure (Hennig et al., 2014, 2013). Moreover, RDC data confirm that the domain orientation observed in the crystal structure represents the solution conformation. Initially, a structural model of this complex had been generated based on the observed CSPs to help design spin label sites for PRE measurements. The resulting models were scored using SAXS and SANS data (Hennig et al., 2013; Fig. 5A). In all lowest energy models, a protein–protein interface was formed between Sxl and CSD1. In contrast, SANS-based *ab initio* bead models suggested that the RNA is sandwiched in between the two proteins (Fig. 5B). Sixteen Sxl and nine CSD1 residues were selected for mutation to cysteine based on the initial NMR CSP-derived model at locations which are close to, but not within the interface, according to the best docking model. One position in each protein was chosen to obtain intradomain PREs for distance calibration.

In parallel, we could solve a crystal structure of the complex, which was indeed found to be consistent with the model derived from the SANS data with the RNA being sandwiched by both proteins. The Sxl RRM and CSD1 domains are therefore further apart than the initial docking model suggested. As a consequence, only 3 out of the 25 spin label sites designed are expected to induce significant PRE effects. From these three spin labels

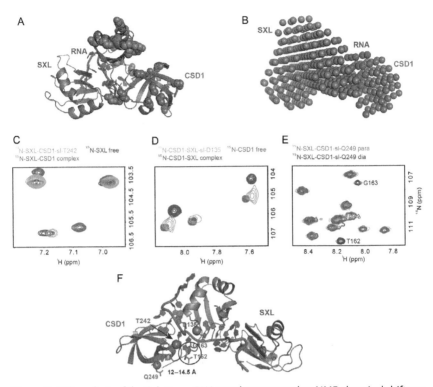

Figure 5 PRE analysis of the Sxl–Unr mRNA regulatory complex. NMR chemical shift perturbations (CSPs), PREs, and small-angle scattering data for the SXL–CSD1 RNA complex. (A) HADDOCK model based on CSPs. The SXL–CSD1 interface residues, determined by CSPs, are shown as spheres (SXL: green, CSD1: blue). The RNA binds along a continuous surface provided by both proteins, which is stabilized by protein–protein contacts. (B) Small-angle neutron scattering of the complex reveals that the RNA is sandwiched by the two proteins. (C and D) Spin labels CSD1-T242 (C) and SXL-D135 (D) could not form the ternary complex as can be seen by the lack of CSP of CSD1 residues in sample with RNA and SXL-D135-spin label (yellow), compared to CSD1 chemical shifts in complex with RNA and SXL (blue). (E) Spin label CSD1-Q249 yields significant PRE effect for SXL residues T162 and G163. (F) The side chains of the three spin labels are indicated as a sphere in the crystal structure of the ternary complex (PDB ID: 4QQB). PREs observed for the CSD1-Q249 spin label to Sxl T162 and G163 (black circle) are indicated by dotted lines and the corresponding distances are annotated. (See the color plate.)

(Sxl-D135C, CSD1-T242C, and Q249C), Sxl-D135C and CSD1-T242C were found to interfere with the complex formation as the spin label attachment is too close to the interface (Fig. 5C and D). The third site (CSD1-Q249C) did indeed provide observable PRE effects for two residues, thus validating the crystal structure (Fig. 5E and F). For all other spin label sites, the absence of significant PREs is consistent with the domain distances observed in the crystal structure. Thus, although the number of observed PRE restraints (combined with RDC, CSP, and SAS data) was insufficient to allow *de novo* structure determination, the information was useful in the validation of the crystal structure and to demonstrate that the observed ternary arrangement accurately represents the solution state. Structural analysis based on NMR data alone would have required the design of additional spin label positions to obtain a sufficient number of interdomain distance restraints. It is noteworthy that the SANS data in particular revealed interesting features of the complex and facilitated the structure determination of the complex [see also Chapter "Small-Angle Neutron Scattering for Structural Biology of Protein–RNA Complexes" by Gabel in this book].

10. OUTLOOK

This chapter reviewed the utility of PRE data for the structural analysis of protein–RNA complexes: (1) for structure determination of multidomain protein–RNA complexes based on available structures of individual domains or subunits, (2) for characterizing dynamic ensembles that involve a range of dynamically interconverting domain arrangements which can be defined by combination of NMR PRE and SAXS data, and (3) for validating the solution conformation of crystal structures or models. Many biologically important protein–RNA interactions involve multidomain RNA-binding proteins where individual domains are connected by flexible linkers. Structural analysis of these proteins and their complexes may not be possible using crystallography or may yield structures that are affected by crystal packing. The combination of NMR data, specifically PREs, but also the combination with RDCs, SAS, and other complementary techniques, is a powerful approach to be used in the characterization of these complex structures and by which to describe their conformational dynamics in solution.

ACKNOWLEDGMENTS

We thank Gerd Gemmecker for the help with NMR measurements and members of the Sattler lab for the discussions.

Funding: J.H. acknowledges the European Molecular Biology Organization (EMBO, grant ALTF-276-2010) and the Swedish Research Council (Vetenskapsrådet) for postdoctoral fellowships. C.D.M. and L.R.W. acknowledge postdoctoral fellowships from EMBO, grants ALTF 74-2003 and ALTF-1520-2011, respectively. C.D.M. also acknowledges EMBO short-term fellowship ASTF 479-2010. Research in the Sattler lab is supported by the Helmholtz Zentrum München and the *Deutsche Forschungsgemeinschaft* (EXC114, SFB1035, and GRK1721 to M.S.).

REFERENCES

Allen, M., Varani, L., & Varani, G. (2001). Nuclear magnetic resonance methods to study structure and dynamics of RNA-protein complexes. *Methods in Enzymology*, *339*, 357–376.

Anglister, J., Grzesiek, S., Ren, H., Klee, C. B., & Bax, A. (1993). Isotope-edited multidimensional NMR of calcineurin B in the presence of the non-deuterated detergent CHAPS. *Journal of Biomolecular NMR*, *3*(1), 121–126.

Banerjee, H., Rahn, A., Davis, W., & Singh, R. (2003). Sex lethal and U2 small nuclear ribonucleoprotein auxiliary factor (U2AF65) recognize polypyrimidine tracts using multiple modes of binding. *RNA*, *9*(1), 88–99.

Barthelmes, K., Reynolds, A. M., Peisach, E., Jonker, H. R., DeNunzio, N. J., Allen, K. N., et al. (2011). Engineering encodable lanthanide-binding tags into loop regions of proteins. *Journal of the American Chemical Society*, *133*(4), 808–819. http://dx.doi.org/10.1021/ja104983t.

Battiste, J. L., & Wagner, G. (2000). Utilization of site-directed spin labeling and high-resolution heteronuclear nuclear magnetic resonance for global fold determination of large proteins with limited nuclear overhauser effect data. *Biochemistry*, *39*(18), 5355–5365.

Berliner, L. J., Grunwald, J., Hankovszky, H. O., & Hideg, K. (1982). A novel reversible thiol-specific spin label: Papain active site labeling and inhibition. *Analytical Biochemistry*, *119*(2), 450–455.

Bertini, I., Calderone, V., Fragai, M., Jaiswal, R., Luchinat, C., Melikian, M., et al. (2008). Evidence of reciprocal reorientation of the catalytic and hemopexin-like domains of full-length MMP-12. *Journal of the American Chemical Society*, *130*(22), 7011–7021. http://dx.doi.org/10.1021/ja710491y.

Bertini, I., Luchinat, C., & Parigi, G. (2002). Magnetic susceptibility in paramagnetic NMR. *Progress in Nuclear Magnetic Resonance Spectroscopy*, *40*, 249–273.

Bertini, I., Luchinat, C., Parigi, G., & Pierattelli, R. (2005). NMR spectroscopy of paramagnetic metalloproteins. *Chembiochem*, *6*(9), 1536–1549. http://dx.doi.org/10.1002/cbic.200500124.

Blackledge, M. (2005). Recent progress in the study of biomolecular structure and dynamics in solution from residual dipolar couplings. *Progress in Nuclear Magnetic Resonance Spectroscopy*, *46*, 23–61.

Bonneau, E., & Legault, P. (2014). Nuclear magnetic resonance structure of the III–IV–V three-way junction from the Varkud satellite ribozyme and identification of magnesium-binding sites using paramagnetic relaxation enhancement. *Biochemistry*, *53*, 6264–6275. http://dx.doi.org/10.1021/bi500826n.

Brunger, A. T., Adams, P. D., Clore, G. M., DeLano, W. L., Gros, P., Grosse-Kunstleve, R. W., et al. (1998). Crystallography & NMR system: A new software suite for macromolecular structure determination. *Acta Crystallographica. Section D, Biological Crystallography*, *54*(Pt. 5), 905–921.

Buttner, L., Javadi-Zarnaghi, F., & Hobartner, C. (2014). Site-specific labeling of RNA at native ribose hydroxyl groups: Terbium-assisted deoxyribozymes at work. *Journal of the American Chemical Society*, *136*(22), 8131–8137. http://dx.doi.org/10.1021/ja503864v.

Buttner, L., Seikowski, J., Wawrzyniak, K., Ochmann, A., & Hobartner, C. (2013). Synthesis of spin-labeled riboswitch RNAs using convertible nucleosides and DNA-catalyzed RNA ligation. *Bioorganic & Medicinal Chemistry*, *21*(20), 6171–6180. http://dx.doi.org/10.1016/j.bmc.2013.04.007.

Clery, A., Boudet, J., & Allain, F. H. (2013). Single-stranded nucleic acid recognition: Is there a code after all? *Structure*, *21*(1), 4–6. http://dx.doi.org/10.1016/j.str.2012.12.006.

Clore, G. M., & Iwahara, J. (2009). Theory, practice, and applications of paramagnetic relaxation enhancement for the characterization of transient low-population states of biological macromolecules and their complexes. *Chemical Reviews*, *109*(9), 4108–4139. http://dx.doi.org/10.1021/cr900033p.

Clore, G. M., & Schwieters, C. D. (2003). Docking of protein-protein complexes on the basis of highly ambiguous intermolecular distance restraints derived from ^{1}H/^{15}N chemical shift mapping and backbone ^{15}N-^{1}H residual dipolar couplings using conjoined rigid body/torsion angle dynamics. *Journal of the American Chemical Society*, *125*(10), 2902–2912. http://dx.doi.org/10.1021/ja028893d.

Cramer, P., Bushnell, D. A., Fu, J., Gnatt, A. L., Maier-Davis, B., Thompson, N. E., et al. (2000). Architecture of RNA polymerase II and implications for the transcription mechanism. *Science*, *288*(5466), 640–649.

Daragan, V. A., & Mayo, K. H. (1997). Motional model analyses of protein and peptide dynamics using ^{13}C and ^{15}N NMR relaxation. *Progress in Nuclear Magnetic Resonance Spectroscopy*, *31*, 63–105.

Daubner, G. M., Clery, A., & Allain, F. H. (2013). RRM-RNA recognition: NMR or crystallography...and new findings. *Current Opinion in Structural Biology*, *23*(1), 100–108. http://dx.doi.org/10.1016/j.sbi.2012.11.006.

de Vries, S. J., van Dijk, M., & Bonvin, A. M. (2010). The HADDOCK web server for data-driven biomolecular docking. *Nature Protocols*, *5*(5), 883–897. http://dx.doi.org/10.1038/nprot.2010.32.

Duss, O., Maris, C., von Schroetter, C., & Allain, F. H. (2010). A fast, efficient and sequence-independent method for flexible multiple segmental isotope labeling of RNA using ribozyme and RNase H cleavage. *Nucleic Acids Research*, *38*(20), e188. http://dx.doi.org/10.1093/nar/gkq756.

Duss, O., Michel, E., Yulikov, M., Schubert, M., Jeschke, G., & Allain, F. H. (2014). Structural basis of the non-coding RNA RsmZ acting as a protein sponge. *Nature*, *509*(7502), 588–592. http://dx.doi.org/10.1038/nature13271.

Duss, O., Yulikov, M., Allain, F. H. -T., & Jeschke, G. (2015). Combining NMR and EPR to determine structures of large RNAs and protein–RNA complexes in solution. *Methods in Enzymology*, *558*, 279–331.

Duss, O., Yulikov, M., Jeschke, G., & Allain, F. H. (2014). EPR-aided approach for solution structure determination of large RNAs or protein-RNA complexes. *Nature Communications*, *5*, 3669. http://dx.doi.org/10.1038/ncomms4669.

Ellman, G. L. (1959). Tissue sulfhydryl groups. *Archives of Biochemistry and Biophysics*, *82*(1), 70–77.

Gagnon, M. G., Lin, J., Bulkley, D., & Steitz, T. A. (2014). Ribosome structure. Crystal structure of elongation factor 4 bound to a clockwise ratcheted ribosome. *Science*, *345*(6197), 684–687. http://dx.doi.org/10.1126/science.1253525.

Gardner, K. H., Rosen, M. K., & Kay, L. E. (1997). Global folds of highly deuterated, methyl-protonated proteins by multidimensional NMR. *Biochemistry*, *36*(6), 1389–1401. http://dx.doi.org/10.1021/bi9624806.

Gelis, I., Bonvin, A. M. J. J., Keramisanou, D., Koukaki, M., Gouridis, G., Karamanou, S., et al. (2007). Structural basis for signal-sequence recognition by the translocase motor SecA as determined by NMR. *Cell*, *131*(4), 756–769. http://dx.doi.org/10.1016/j.cell.2007.09.039.

Gobl, C., Madl, T., Simon, B., & Sattler, M. (2014). NMR approaches for structural analysis of multidomain proteins and complexes in solution. *Progress in Nuclear Magnetic Resonance Spectroscopy*, *80*, 26–63. http://dx.doi.org/10.1016/j.pnmrs.2014.05.003.

Golas, M. M., Sander, B., Bessonov, S., Grote, M., Wolf, E., Kastner, B., et al. (2010). 3D cryo-EM structure of an active step I spliceosome and localization of its catalytic core. *Molecular Cell*, *40*(6), 927–938. http://dx.doi.org/10.1016/j.molcel.2010.11.023.

Gottstein, D., Reckel, S., Dotsch, V., & Guntert, P. (2012). Requirements on paramagnetic relaxation enhancement data for membrane protein structure determination by NMR. *Structure*, *20*(6), 1019–1027. http://dx.doi.org/10.1016/j.str.2012.03.010.

Grishaev, A., Tugarinov, V., Kay, L. E., Trewhella, J., & Bax, A. (2008). Refined solution structure of the 82-kDa enzyme malate synthase G from joint NMR and synchrotron SAXS restraints. *Journal of Biomolecular NMR*, *40*(2), 95–106. http://dx.doi.org/10.1007/s10858-007-9211-5.

Grishaev, A., Wu, J., Trewhella, J., & Bax, A. (2005). Refinement of multidomain protein structures by combination of solution small-angle X-ray scattering and NMR data. *Journal of the American Chemical Society*, *127*(47), 16621–16628. http://dx.doi.org/10.1021/ja054342m.

Groll, M., Ditzel, L., Lowe, J., Stock, D., Bochtler, M., Bartunik, H. D., et al. (1997). Structure of 20S proteasome from yeast at 2.4 Å resolution. *Nature*, *386*(6624), 463–471. http://dx.doi.org/10.1038/386463a0.

Guerry, P., Salmon, L., Mollica, L., Ortega Roldan, J. L., Markwick, P., van Nuland, N. A., et al. (2013). Mapping the population of protein conformational energy sub-states from NMR dipolar couplings. *Angewandte Chemie (International Ed. in English)*, *52*(11), 3181–3185. http://dx.doi.org/10.1002/anie.201209669.

Handa, N., Nureki, O., Kurimoto, K., Kim, I., Sakamoto, H., Shimura, Y., et al. (1999). Structural basis for recognition of the tra mRNA precursor by the Sex-lethal protein. *Nature*, *398*(6728), 579–585. http://dx.doi.org/10.1038/19242.

Hankovszky, O. H., Hideg, K., & Lex, L. (1980). Nitroxyls; VII. Synthesis and reactions of highly reactive 1-oxyl-2,2,5,5-tetramethyl-2,5-dihydropyrrol-3-ylmethyl sulfonates. *Synthesis*, *11*, 914–916.

Helmling, C., Bessi, I., Wacker, A., Schnorr, K. A., Jonker, H. R. A., Richter, C., et al. (2014). Noncovalent spin labeling of riboswitch RNAs to obtain long-range structural NMR restraints. *ACS Chemical Biology*, *9*(6), 1330–1339. http://dx.doi.org/10.1021/cb500050t.

Hennig, J., Militti, C., Popowicz, G. M., Wang, I., Sonntag, M., Geerlof, A., et al. (2014). Structural basis for the assembly of the Sxl-Unr translation regulatory complex. *Nature*, *515*, 287–290. http://dx.doi.org/10.1038/nature13693.

Hennig, J., & Sattler, M. (2014). The dynamic duo: Combining NMR and small angle scattering in structural biology. *Protein Science*, *23*(6), 669–682. http://dx.doi.org/10.1002/pro.2467.

Hennig, J., Wang, I., Sonntag, M., Gabel, F., & Sattler, M. (2013). Combining NMR and small angle X-ray and neutron scattering in the structural analysis of a ternary protein-RNA complex. *Journal of Biomolecular NMR*, *56*, 17–30. http://dx.doi.org/10.1007/s10858-013-9719-9.

Huang, J. R., Warner, L. R., Sanchez, C., Gabel, F., Madl, T., Mackereth, C. D., et al. (2014). Transient electrostatic interactions dominate the conformational equilibrium sampled by multidomain splicing factor U2AF65: A combined NMR and SAXS study. *Journal of the American Chemical Society*, *136*(19), 7068–7076. http://dx.doi.org/10.1021/ja502030n.

Iwahara, J., Anderson, D. E., Murphy, E. C., & Clore, G. M. (2003). EDTA-derivatized deoxythymidine as a tool for rapid determination of protein binding polarity to DNA by intermolecular paramagnetic relaxation enhancement. *Journal of the American Chemical Society*, *125*(22), 6634–6635.

Iwahara, J., & Clore, G. M. (2006). Detecting transient intermediates in macromolecular binding by paramagnetic NMR. *Nature*, *440*(7088), 1227–1230. http://dx.doi.org/10.1038/nature04673.

Iwahara, J., Schwieters, C. D., & Clore, G. M. (2004). Ensemble approach for NMR structure refinement against H-1 paramagnetic relaxation enhancement data arising from a flexible paramagnetic group attached to a macromolecule. *Journal of the American Chemical Society, 126*(18), 5879–5896.

Iwahara, J., Tang, C., & Marius Clore, G. (2007). Practical aspects of (1)H transverse paramagnetic relaxation enhancement measurements on macromolecules. *Journal of Magnetic Resonance, 184*(2), 185–195. http://dx.doi.org/10.1016/j.jmr.2006.10.003.

Jenkins, J. L., Laird, K. M., & Kielkopf, C. L. (2012). A Broad range of conformations contribute to the solution ensemble of the essential splicing factor U2AF(65). *Biochemistry, 51*(26), 5223–5225. http://dx.doi.org/10.1021/bi300277t.

Jeschke, G. (2012). DEER distance measurements on proteins. *Annual Review of Physical Chemistry, 63*, 419–446. http://dx.doi.org/10.1146/annurev-physchem-032511-143716.

Jones, D. H., Cellitti, S. E., Hao, X., Zhang, Q., Jahnz, M., Summerer, D., et al. (2010). Site-specific labeling of proteins with NMR-active unnatural amino acids. *Journal of Biomolecular NMR, 46*(1), 89–100. http://dx.doi.org/10.1007/s10858-009-9365-4.

Kato, H., van Ingen, H., Zhou, B. R., Feng, H., Bustin, M., Kay, L. E., et al. (2011). Architecture of the high mobility group nucleosomal protein 2-nucleosome complex as revealed by methyl-based NMR. *Proceedings of the National Academy of Sciences of the United States of America, 108*(30), 12283–12288. http://dx.doi.org/10.1073/pnas.1105848108.

Keizers, P. H., Saragliadis, A., Hiruma, Y., Overhand, M., & Ubbink, M. (2008). Design, synthesis, and evaluation of a lanthanide chelating protein probe: CLaNP-5 yields predictable paramagnetic effects independent of environment. *Journal of the American - Chemical Society, 130*(44), 14802–14812. http://dx.doi.org/10.1021/ja8054832.

Keyes, R. S., & Bobst, A. M. (2002). Spin-labeled nucleic acids. In L. Berliner (Ed.), *Biological magnetic resonance: Vol. 14.* (pp. 283–338). USA: Springer.

Kirby, T. L., Karim, C. B., & Thomas, D. D. (2004). Electron paramagnetic resonance reveals a large-scale conformational change in the cytoplasmic domain of phospholamban upon binding to the sarcoplasmic reticulum Ca-ATPase. *Biochemistry, 43*(19), 5842–5852. http://dx.doi.org/10.1021/bi035749b.

Kisseleva, N., Khvorova, A., Westhof, E., & Schiemann, O. (2005). Binding of manganese (II) to a tertiary stabilized hammerhead ribozyme as studied by electron paramagnetic resonance spectroscopy. *RNA, 11*(1), 1–6.

Kosen, P. A. (1989). Spin labeling of proteins. *Methods in Enzymology, 177*, 86–121.

Lapinaite, A., Simon, B., Skjaerven, L., Rakwalska-Bange, M., Gabel, F., & Carlomagno, T. (2013). The structure of the box C/D enzyme reveals regulation of RNA methylation. *Nature, 502*, 519–523. http://dx.doi.org/10.1038/nature12581.doi.

Lebars, I., Vileno, B., Bourbigot, S., Turek, P., Wolff, P., & Kieffer, B. (2014). A fully enzymatic method for site-directed spin labeling of long RNA. *Nucleic Acids Research, 42*(15), e117. http://dx.doi.org/10.1093/nar/gku553.

Liang, B., Bushweller, J. H., & Tamm, L. K. (2006). Site-directed parallel spin-labeling and paramagnetic relaxation enhancement in structure determination of membrane proteins by solution NMR spectroscopy. *Journal of the American Chemical Society, 128*(13), 4389–4397.

Linge, J. P., Habeck, M., Rieping, W., & Nilges, M. (2003). ARIA: Automated NOE assignment and NMR structure calculation. *Bioinformatics, 19*(2), 315–316.

Lu, Y., Berry, S. M., & Pfister, T. D. (2001). Engineering novel metalloproteins: Design of metal-binding sites into native protein scaffolds. *Chemical Reviews, 101*(10), 3047–3080.

Lunde, B. M., Moore, C., & Varani, G. (2007). RNA-binding proteins: Modular design for efficient function. *Nature Reviews. Molecular Cell Biology, 8*(6), 479–490. http://dx.doi.org/10.1038/nrm2178.

Mackereth, C. D., Madl, T., Bonnal, S., Simon, B., Zanier, K., Gasch, A., et al. (2011). Multi-domain conformational selection underlies pre-mRNA splicing regulation by U2AF. *Nature, 475*(7356), 408–411. http://dx.doi.org/10.1038/nature10171.

Mackereth, C. D., & Sattler, M. (2012). Dynamics in multi-domain protein recognition of RNA. *Current Opinion in Structural Biology, 22*(3), 287–296. http://dx.doi.org/10.1016/j.sbi.2012.03.013.

Madl, T., Felli, I. C., Bertini, I., & Sattler, M. (2010). Structural analysis of protein interfaces from ^{13}C direct-detected paramagnetic relaxation enhancements. *Journal of the American Chemical Society, 132*(21), 7285–7287. http://dx.doi.org/10.1021/ja1014508.

Makino, D. L., Baumgartner, M., & Conti, E. (2013). Crystal structure of an RNA-bound 11-subunit eukaryotic exosome complex. *Nature, 495*(7439), 70–75. http://dx.doi.org/10.1038/nature11870.

Medalia, O., Weber, I., Frangakis, A. S., Nicastro, D., Gerisch, G., & Baumeister, W. (2002). Macromolecular architecture in eukaryotic cells visualized by cryoelectron tomography. *Science, 298*(5596), 1209–1213. http://dx.doi.org/10.1126/science.1076184.

Morgan, H. P., Schmidt, C. Q., Guariento, M., Blaum, B. S., Gillespie, D., Herbert, A. P., et al. (2011). Structural basis for engagement by complement factor H of C3b on a self surface. *Nature Structural & Molecular Biology, 18*(4), 463–470. http://dx.doi.org/10.1038/nsmb.2018.

Mund, M., Overbeck, J. H., Ullmann, J., & Sprangers, R. (2013). LEGO-NMR spectroscopy: A method to visualize individual subunits in large heteromeric complexes. *Angewandte Chemie (International Ed. in English), 52*(43), 11401–11405. http://dx.doi.org/10.1002/anie.201304914.

Nguyen, P., & Qin, P. Z. (2012). RNA dynamics: Perspectives from spin labels. *Wiley Interdisciplinary Reviews. RNA, 3*(1), 62–72. http://dx.doi.org/10.1002/wrna.104.

Pervushin, K., Riek, R., Wider, G., & Wuthrich, K. (1997). Attenuated T2 relaxation by mutual cancellation of dipole-dipole coupling and chemical shift anisotropy indicates an avenue to NMR structures of very large biological macromolecules in solution. *Proceedings of the National Academy of Sciences of the United States of America, 94*(23), 12366–12371.

Pintacuda, G., John, M., Su, X. C., & Otting, G. (2007). NMR structure determination of protein–ligand complexes by lanthanide labeling. *Accounts of Chemical Research, 40*(3), 206–212. http://dx.doi.org/10.1021/ar050087z.

Piton, N., Mu, Y., Stock, G., Prisner, T. F., Schiemann, O., & Engels, J. W. (2007). Base-specific spin-labeling of RNA for structure determination. *Nucleic Acids Research, 35*, 3128–3143.

Prestegard, J. H., Bougault, C. M., & Kishore, A. I. (2004). Residual dipolar couplings in structure determination of biomolecules. *Chemical Reviews, 104*(8), 3519–3540. http://dx.doi.org/10.1021/cr030419i.

Prisner, T., Rohrer, M., & MacMillan, F. (2001). Pulsed EPR spectroscopy: Biological applications. *Annual Review of Physical Chemistry, 52*, 279–313. http://dx.doi.org/10.1146/annurev.physchem.52.1.279.

Qin, P. Z., & Dieckmann, T. (2004). Application of NMR and EPR methods to the study of RNA. *Current Opinion in Structural Biology, 14*(3), 350–359.

Qin, P. Z., Feigon, J., & Hubbell, W. L. (2005). Site-directed spin labeling studies reveal solution conformational changes in a GAAA tetraloop receptor upon Mg^{2+}-dependent docking of a GAAA tetraloop. *Journal of Molecular Biology, 351*(1), 1–8.

Ramos, A., Bayer, P., & Varani, G. (1999). Determination of the structure of the RNA complex of a double-stranded RNA-binding domain from Drosophila Staufen protein. *Biopolymers, 52*(4), 181–196. http://dx.doi.org/10.1002/1097-0282(1999)52:4<181::AID-BIP1003>3.0.CO;2-5.

Ray, D., Kazan, H., Cook, K. B., Weirauch, M. T., Najafabadi, H. S., Li, X., et al. (2013). A compendium of RNA-binding motifs for decoding gene regulation. *Nature, 499*(7457), 172–177. http://dx.doi.org/10.1038/nature12311.

Religa, T. L., Sprangers, R., & Kay, L. E. (2010). Dynamic regulation of archaeal proteasome gate opening as studied by TROSY NMR. *Science, 328*(5974), 98–102. http://dx.doi.org/10.1126/science.1184991.

Riek, R., Pervushin, K., & Wuthrich, K. (2000). TROSY and CRINEPT: NMR with large molecular and supramolecular structures in solution. *Trends in Biochemical Sciences, 25*(10), 462–468.

Rosenzweig, R., Moradi, S., Zarrine-Afsar, A., Glover, J. R., & Kay, L. E. (2013). Unraveling the mechanism of protein disaggregation through a ClpB-DnaK interaction. *Science, 339*(6123), 1080–1083. http://dx.doi.org/10.1126/science.1233066.

Salmon, L., Nodet, G., Ozenne, V., Yin, G., Jensen, M. R., Zweckstetter, M., et al. (2010). NMR characterization of long-range order in intrinsically disordered proteins. *Journal of the American Chemical Society, 132*(24), 8407–8418. http://dx.doi.org/10.1021/ja101645g.

Schmidt, M. J., Borbas, J., Drescher, M., & Summerer, D. (2014). A genetically encoded spin label for electron paramagnetic resonance distance measurements. *Journal of the American Chemical Society, 136*(4), 1238–1241. http://dx.doi.org/10.1021/ja411535q.

Schwieters, C. D., Suh, J. Y., Grishaev, A., Ghirlando, R., Takayama, Y., & Clore, G. M. (2010). Solution structure of the 128 kDa enzyme I dimer from Escherichia coli and its 146 kDa complex with HPr using residual dipolar couplings and small- and wide-angle X-ray scattering. *Journal of the American Chemical Society, 132*(37), 13026–13045. http://dx.doi.org/10.1021/ja105485b.

Shelke, S. A., & Sigurdsson, S. T. (2012). Site-directed spin labelling of nucleic acids. *European Journal of Organic Chemistry, 2012*(12), 2291–2301. http://dx.doi.org/10.1002/ejoc.201101434.

Sickmier, E. A., Frato, K. E., Shen, H., Paranawithana, S. R., Green, M. R., & Kielkopf, C. L. (2006). Structural basis for polypyrimidine tract recognition by the essential pre-mRNA splicing factor U2AF65. *Molecular Cell, 23*(1), 49–59. http://dx.doi.org/10.1016/j.molcel.2006.05.025.

Simon, B., Madl, T., Mackereth, C. D., Nilges, M., & Sattler, M. (2010). An efficient protocol for NMR-spectroscopy-based structure determination of protein complexes in solution. *Angewandte Chemie (International Ed. in English), 49*(11), 1967–1970. http://dx.doi.org/10.1002/anie.200906147.

Sprangers, R., Velyvis, A., & Kay, L. E. (2007). Solution NMR of supramolecular complexes: Providing new insights into function. *Nature Methods, 4*(9), 697–703. http://dx.doi.org/10.1038/nmeth1080.

Su, X. C., & Otting, G. (2010). Paramagnetic labelling of proteins and oligonucleotides for NMR. *Journal of Biomolecular NMR, 46*(1), 101–112. http://dx.doi.org/10.1007/s10858-009-9331-1.

Takayama, Y., Schwieters, C. D., Grishaev, A., Ghirlando, R., & Clore, G. M. (2011). Combined use of residual dipolar couplings and solution X-ray scattering to rapidly probe rigid-body conformational transitions in a non-phosphorylatable active-site mutant of the 128 kDa enzyme I dimer. *Journal of the American Chemical Society, 133*(3), 424–427. http://dx.doi.org/10.1021/ja109866w.

Tang, C., Iwahara, J., & Clore, G. M. (2006). Visualization of transient encounter complexes in protein-protein association. *Nature, 444*(7117), 383–386. http://dx.doi.org/10.1038/nature05201.

Tjandra, N., & Bax, A. (1997). Direct measurement of distances and angles in biomolecules by NMR in a dilute liquid crystalline medium. *Science, 278*(5340), 1111–1114.

Tugarinov, V., Kanelis, V., & Kay, L. E. (2006). Isotope labeling strategies for the study of high-molecular-weight proteins by solution NMR spectroscopy. *Nature Protocols, 1*(2), 749–754. http://dx.doi.org/10.1038/nprot.2006.101.

Volkov, A. N., Worrall, J. A., Holtzmann, E., & Ubbink, M. (2006). Solution structure and dynamics of the complex between cytochrome c and cytochrome c peroxidase determined by paramagnetic NMR. *Proceedings of the National Academy of Sciences of the United States of America, 103*(50), 18945–18950. http://dx.doi.org/10.1073/pnas.0603551103.

Wahl, M. C., Will, C. L., & Luhrmann, R. (2009). The spliceosome: Design principles of a dynamic RNP machine. *Cell, 136*(4), 701–718. http://dx.doi.org/10.1016/j.cell.2009.02.009.

Wang, L., Xie, J., & Schultz, P. G. (2006). Expanding the genetic code. *Annual Review of Biophysics and Biomolecular Structure, 35*, 225–249. http://dx.doi.org/10.1146/annurev.biophys.35.101105.121507.

Wohnert, J., Franz, K. J., Nitz, M., Imperiali, B., & Schwalbe, H. (2003). Protein alignment by a coexpressed lanthanide-binding tag for the measurement of residual dipolar couplings. *Journal of the American Chemical Society, 125*(44), 13338–13339. http://dx.doi.org/10.1021/ja036022d.

Wunderlich, C. H., Huber, R. G., Spitzer, R., Liedl, K. R., Kloiber, K., & Kreutz, C. (2013). A novel paramagnetic relaxation enhancement tag for nucleic acids: A tool to study structure and dynamics of RNA. *ACS Chemical Biology, 8*(12), 2697–2706. http://dx.doi.org/10.1021/cb400589q.

Xie, J., & Schultz, P. G. (2005). Adding amino acids to the genetic repertoire. *Current Opinion in Chemical Biology, 9*(6), 548–554. http://dx.doi.org/10.1016/j.cbpa.2005.10.011.

Zamore, P. D., Patton, J. G., & Green, M. R. (1992). Cloning and domain structure of the mammalian splicing factor U2AF. *Nature, 355*(6361), 609–614. http://dx.doi.org/10.1038/355609a0.

CHAPTER TWELVE

Resolving Individual Components in Protein–RNA Complexes Using Small-Angle X-ray Scattering Experiments

Robert P. Rambo[1]

Diamond Light Source Ltd., Harwell Science & Innovation Campus, Didcot, United Kingdom
[1]Corresponding author: e-mail address: robert.rambo@diamond.ac.uk

Contents

1. Introduction	364
2. Technical Requirements	365
2.1 Specialized Software	365
2.2 Scripts	365
3. Theoretical Considerations	365
4. Assessing SAXS Sample Quality	374
5. Case Studies	375
5.1 Case I: A Binary Complex: SF1–U2AF65 Splicing Complex	375
5.2 Case II: Visualizing Conformational Changes with SMARCAL1–DNA Complex	379
5.3 Case III: Cautionary Notes from Simulations with the Ternary T-box•tRNA•YbxF Complex	383
6. Considerations	387
Acknowledgments	388
References	388

Abstract

Small-angle X-ray scattering (SAXS) of protein–RNA complexes has developed into an efficient and economical approach for determining low-resolution shapes of particles in solution. Here, we demonstrate a mutliphase volumetric modeling approach capable of resolving individual components within a low-resolution shape. Through three case studies, we describe the SAXS data collecting strategies, premodeling analysis, and computational methods required for deconstructing complexes into their respective components. This chapter presents an approach using the programs ScÅtter and MONSA and custom scripts for averaging and aligning of multiple independent modeling runs. The method can image small (7 kDa) masses within the context of complex and is capable of visualizing ligand-induced conformational changes. Nevertheless, computational algorithms are not without error, and we describe specific considerations during SAXS data reduction and modeling to mitigate possible false positives.

1. INTRODUCTION

Structural biology has been, for a longtime, a solid-state endeavor achieved largely through macromolecular X-ray crystallography and now cryo-electron microscopy. Together, these two techniques cover a wide range of scattering resolutions and macromolecular sizes. The high-resolution information is a consequence of the biophysical measurements being performed on macromolecules trapped in unique, cryo-stablized solid-state conformations. Nonetheless, biology is a solution state game and many interesting biological problems such as delineating the structure–function relationships in the noncoding, untranslated regions of mRNA may not be conducive to the technical requirements of high-resolution studies. Here, small-angle X-ray scattering (SAXS) can complement the high-resolution methods by providing a resolution-limited, structural picture of the thermodynamic state of the biological particle in solution (Putnam, Hammel, Hura, & Tainer, 2007; Rambo & Tainer, 2013a,b).

A solution X-ray scattering signal can be measured to very high spatial resolutions (~2 Å) (Schwieters & Clore, 2007). At this resolution, the scattering signal will contain information regarding the local motions of individual residues (such as base-pair dynamics), but the signal will be inherently weak as it is measured against the immense background scattering of the solvent. At resolutions greater than 10 Å, the scattering signal falls within the SAXS regime and will contain not only information regarding domain motions but also their relative spatial arrangements (Engelman & Moore, 1975; Gabel et al., 2008; Pelikan, Hura, & Hammel, 2009; Svergun, Petoukhov, & Koch, 2001). Furthermore, at sufficiently low resolution, the observed SAXS curve of a particle will be significantly stronger and can be approximated from the particles shape (volume element) assuming a homogenous electron density, ρ (Stuhrmann, 1970, 1973). The difference, $\Delta\rho$, between the particle and solvent defines the particles contrast with $\rho_{nucleic} > \rho_{protein}$. Therefore, in a protein–RNA complex, these contrast differences between $\rho_{nucleic}$, $\rho_{protein}$, and $\rho_{solvent}$ provide a means to model the volume element of each component independently but within the context of the complex (Svergun, 1999).

Resolving the individual components of a complex within a SAXS-based *ab initio* volumetric model requires a proper set of SAXS experiments that are outlined in this chapter. The analysis produces a three-dimensional picture that illustrates the spatial arrangement of the components

(Svergun & Nierhaus, 2000). This method of SAXS analysis known as multiphase volumetric modeling and implemented in the program MONSA (Svergun, 1999), can be used to visualize conformational changes such as the action of a molecular motor on a nucleic acid structure (Mason et al., 2014), and has aided in the understanding of DEAD-box RNA chaperones (Mallam et al., 2011), and binding of PKR protein to HIV TAR and Adenovirus VA_I RNA (Dzananovic et al., 2013). Since SAXS is a thermodynamic measurement that provides information on the available structural states in a single measurement, the resolution of the *ab initio* model will be restricted not only by the observed resolution limit of the SAXS signal but also by the sample quality (Rambo & Tainer, 2010, 2011, 2013a,b). Here, we provide examples and guidelines on how to design SAXS experiments for imaging separate component of protein–nucleic acid complexes.

2. TECHNICAL REQUIREMENTS

The multiphase volumetric analysis described in this chapter requires the software listed in Table 1. MONSA (Svergun, 1999) can be downloaded as a stand-alone application; however, it is recommended that novice users begin with the online version (Petoukhov et al., 2012). The processing of multiple MONSA modeling runs and subsequent alignment of models can be achieved with the scripts detailed in Tables 2 and 3.

2.1 Specialized Software
See Table 1

2.2 Scripts
A typical workflow for multiphase modeling is outlined in Fig. 1. In step 5 (Fig. 1), the averaged bead density model (damfilt.pdb) can be converted to a SITUS density map using pdb2vol (Chacon & Wriggers, 2001). Voxel spacings are defined to be 6 Å with a resolution of 10 Å (kernel width $= -10$). During visualization, the contour of the volumetric density map should be scaled to the same volume as the input damfilt.pdb model.

3. THEORETICAL CONSIDERATIONS

The structural investigations outlined here will describe solution state SAXS experiments of macromolecules under dilute conditions. Dilute conditions refers to the range of macromolecular concentrations and buffer conditions that promote monodispersity or the minimization of any type of

Table 1 Required Software for Multiphase Volumetric Modeling

Software	Web Information	Application Type
ScÅtter	www.bioisis.net/tutorial	Stand-alone application (open source)
Cross-platform JAVA application for data reduction and basic analysis of SAXS data. ScÅtter provides a limited interface to ATSAS programs DAMMIN/F and DATGNOM.		
MONSA	www.embl-hamburg.de/biosaxs/atsas-online/	Web-application (closed source)
Multiphase bead modeling algorithm for use with SAXS and SANS datasets where contrasts for each component (phase) can be specified independently. The software is an extension of DAMMIN.		
DAMAVER	www.embl-hamburg.de/biosaxs/damaver.html	Stand-alone application (closed source)
Program suite for averaging multiple runs of MONSA. Each component (phase) is averaged separately and then realigned to produce a model of the complex.		
VMD	www.ks.uiuc.edu/Research/vmd/	Downloadable binaries (open source)
Molecular viewer that can readily load multiple PDB files from command line. VMD is excellent tool for viewing bead models and SITUS maps.		
Ruby	www.rubyinstaller.org/downloads/	Downloadable binaries (open source)
Complete object-oriented scripting language. Can be interfaced with the GNU Scientific Library (GSL) for advanced computation using ruby-gsl.		
GSL	www.gnu.org/software/gsl/	Downloadable source code (open source)
GNU scientific library for C and C++. Contains useful vector and linear algebra routines.		

The required software (in bold) can be obtained online. The applications are fully compatible with Windows, Mac OS X, and LINUX operating systems. GSL is machine specific and will require compilation using an appropriate compiler.

intermolecular interactions between the particles. Under dilution conditions, the observed SAXS intensity, $I(q)$, will be the direct sum of the scattered intensities from many independent and identical particles. The set of $I(q)$ over the observed q-range describes the SAXS profile of sample and it should be noted that q is measured in inverse Angstroms (Å^{-1}), thus defining the SAXS profile in reciprocal space.

Table 2 Script for Collating MONSA PDB Phase Files into Separate Directories

§	statement
1	`#!/usr/bin/ruby`
2	`require 'find'` `require 'fileutils'`

3	`module Find`
	`  def match(*paths)`
	`    matched = []`
	`    find(*paths) {|path| matched << path if yield path }`
	`    return matched`
	`  end`
	`  module_function :match`
	`end`

4	`# read the number of components from standard input` `phases = (ARGV[0].to_i < 2) ? abort("phases must be greater than 1") : ARGV[0].to_i`
5	`# make a list of all sub-directories, each sub-directory is a separate MONSA run` `subDirs = Dir.glob("**/")`

6	`# make new sub-directories, one for each phase`
	`for i in 0...phases`
	`  name = "phase#{i + 1}"`
	`  FileUtils.mkdir(name)`
	`end`

7	`# initialize parameters`
	`current_working_directory = Dir.pwd`
	`pdb_files = Array.new`

8	`subDirs.each do |sub|`
	`  changeTo = current_working_directory + "/" + sub # change to directory, make list of pdb files`
	`  Dir.chdir(changeTo)`
	`  pdb_files.clear`
	`  Find.match("./"){|p| ext = p[-4...p.size]; if ext && ext == ".pdb"; pdb_files << p; end}`
	
	`  if pdb_files.size >= phases`
	`    # go through each file, copy/rename to appropriate phase directory`
	`    pdb_files.each do |file|`
	`      for i in 1..phases # igore -0 phase`
	`        if (file =~ Regexp.new("-#{i}.pdb"))`

Continued

Table 2 Script for Collating MONSA PDB Phase Files into Separate Directories—cont'd
§ statement

```
                newName = "#{sub.chop}-#{i}.pdb"
                copyTo = current_working_directory + "/phase#{i}/#{newName}"
                FileUtils.cp(file, copyTo)
                break
              end
          end
        end
      else

    end
  end
```

Multiple MONSA runs must be combined to produce a final set of averaged and aligned components. Assuming the modeling runs were performed online, move all uncompressed directories to a common directory. Each directory must have a unique name. From standard input (command line), this script requires the number of modeled phases (§ 4). The script will create new subdirectories named phaseX, where X is an integer corresponding to a phase and will copy to each phaseX directory the appropriate phase from each MONSA modeling run. The copied phases will be named after their corresponding unique directory and appended with "-X." After the script finishes change to the reference phase from the command line and execute "damaver -a *.pdb." This will perform the averaging of the collated phase producing damfilt.pdb and damsup.log. The first file in the damsup.log list is the reference PDB file to which all other PDB files will be aligned.

The SAXS profile of the macromolecule is determined by subtracting the SAXS profile of the buffer (background) from the SAXS profile of the sample (particle + buffer). Therefore, the actual macromolecular SAXS profile is a difference measurement and this difference includes scattering from the particle's hydration and excluded volume (Glatter & Kratky, 1982). For atomistic modeling, corrections must be made to include these differences. However, for the type of analysis presented here, we can exclude these considerations due to the resolution limit imposed by this type of SAXS modeling (Chacon, Moran, Diaz, Pantos, & Andreu, 1998; Svergun, 1999).

In general, the SAXS profile of a protein or nucleic acid macromolecule is due to the spatial distribution of the atoms. For each atom, this spatial distribution determines the location of the atomic scattering form factor, a function that describes the interaction of the incident X-rays with the atom (Fig. 2). At sufficiently low resolution, the atomic scattering form factors appear constant and likewise, at low resolution, the overall scattering of a particle can be approximated by a homogenous collection of volume elements (coarse graining). Therefore, in the low-resolution regime, the difference between the collection of scattering form factors for a protein (F_{protein}) and RNA (F_{RNA}) can be approximated simply by a constant value with $F_{\text{RNA}} > F_{\text{protein}}$.

Table 3 Align Script for Transforming Non-reference Phases Using SUPCOMB Transformation Matrices

§	statement
1	`#!/usr/bin/ruby`

2	`require 'find'`
	`require 'fileutils'`
	`require 'gsl'`

3	`module Find`
	`def match(*paths)`
	`    matched = []`
	`    find(*paths) {|path| matched << path if yield path }`
	`    return matched`
	`  end`
	`  module_function :match`
	`end`

4	`# read the number of components from standard input`
	`number_of_phases = (ARGV[0].to_i < 2) ? abort("must be > 1") :`
	`ARGV[0].to_i source_directory = ARGV[1]`

5	`# initialize parameters`
	`current_working_directory = Dir.pwd`
	`look_in = current_working_directory + "/" + source_directory`
	`pdb_files = Array.new`

6	`# load pdb files from reference phase`
	`Find.match("#{look_in}/"){|p| ext = p[-5...p.size]; if ext && ext ==`
	`"r.pdb"; pdb_files << p; end}`

7	`matrix = GSL::Matrix.alloc(4,4)`
	`# keep only supcomb aligned pdb models`
	`pdb_files.select!{|x| x[-8...x.size] =~ /-[0-9]+r\.pdb/ }`

8	`# loop through each aligned PDB file`
	`pdb_files.each do |file|`
	`  name = file.split(/\//).last`
	`  phase = name.split(/[-r\.pdb]/).last.to_i`
	`  lines =[]`

9	`# Open SUPCOMB aligned PDB file in reference phase directory`
	`open(file){|x| lines = x.readlines}`
	`start_at = lines.index{|l| l =~ /Transformation/}`
	`# extract transformation matrix from header`
	`#REMARK 265 Transformation matrix : -0.2924 0.0701 0.9537 0.0000`
	`#REMARK 265 -16.7828 -24.3914 -8.9793 1.0000`
	`#0123456789012345678901234567890123456789012345678901234567890`

Continued

Table 3 Align Script for Transforming Non-reference Phases Using SUPCOMB Transformation Matrices—cont'd

§	statement
	``
	for i in 0...4
	matrix[i,0] = lines[start_at + i][38..46].to_f
	matrix[i,1] = lines[start_at + i][47..54].to_f
	matrix[i,2] = lines[start_at + i][55..62].to_f
	matrix[i,3] = lines[start_at + i][63..70].to_f
	end
	# must transpose matrix to put into standard convention
	transformation = matrix.transpose
10	# loop through non-reference phases and apply transformation matrix to PDB files
	for i in 1..number_of_phases
	if (i != phase)
	# go into phaseX sub directory
	old_file = "phase#{i}/#{name.split(/\-/).first}-#{i}.pdb"
	if File.exists?(old_file)
	# read PDB file in as an array (atomLines) of atom objects
	atomLines = openPDBFile(old_file)
	# apply transformation matrix to each atom
	atomLines.each do |atom|
	old = GSL::Vector.alloc(atom.xpos, atom.ypos, atom.zpos, 1)
	new_pos = transformation*old
	atom.xpos = new_pos[0]
	atom.ypos = new_pos[1]
	atom.zpos = new_pos[2]
	end
	# create new file
	toPDBFile(atomLines, "#{name.split(/\-/).first}-#{i}m.pdb")
	end
	end
	end
	end # end pdb_files loop

The align script will apply to each of the remaining nonreference phases the rotation and translation transformation matrices determined from SUPCOMB averaging of the reference phases. Two inputs (number of phases and name of the reference phase) are required from standard input (command line) described in §4. In § 10, atomLines is an array of a data structure that holds the atom index, type, and (X,Y,Z) position. Transformed files will be appended with "m." All files can be loaded from command line using VMD with the -m flag.

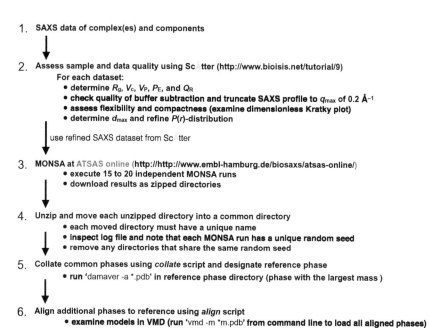

Figure 1 SAXS data analysis workflow for multiphase volumetric modeling using ScÅtter and MONSA. After data collection of relevant complexes and components, the SAXS invariants listed in step 2 can be used to evaluate complex formation and data quality (see http://www.bioisis.net/tutorial). Most importantly, the data can be truncated to 0.2 Å^{-1} to help mitigate poor subtraction. However, datasets exhibiting poor background subtraction should be discarded or recollected. Following assessment of buffer subtraction, determination of d_{max} can proceed using an iterative procedure evaluating several d_{max} values. In ScÅtter, a proper d_{max} for the SAXS dataset will have a χ^2 near 1 and an S_{k2} statistic near 0. Poor buffer subtractions often lead to a difficulty in d_{max} determinations during the real-space transform. Possible d_{max} values should be tested using the "refine" feature in ScÅtter. This feature tests how self-consistent the dataset is with the $P(r)$-distribution defined by the chosen d_{max}. $I(q)$ data points that fall outside a defined cutoff, equivalent to 2.5 × standard deviation are discarded. This refined dataset should be used for all modeling steps. After completion of a table similar to Table 5, 15–20 independent MONSA modeling runs can be instantiated online. Each run should be given a unique name. The results will be returned as compressed zip file. All runs should be unzipped into a common directory for further analysis using scripts described in Tables 2 and 3. It is important to inspect the log file in each directory, noting that the random seed for each run is different. In addition, the χ^2, R_g, and volume for each phase should be recorded to a spreadsheet for basic statistical analysis.

The unique spatial distribution of atomic scattering form factors fundamentally describes the spatial distribution of electrons (electron charge density). The scattering of X-rays is due entirely to the interaction of the X-ray photon with the macromolecule's electrons. Therefore, we can describe

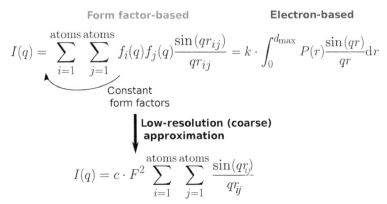

Figure 2 SAXS Debye equations describing relationship between intensity and internal distance. The Debye factor is the sinc function of the product of the scattering vector, q, times the distant r between two points in the particle. In the form factor-based approach, the Debye factor is weighted by the atomic scattering form factors of each atom that defines the paired distance within the particle, whereas in the electron-based approach, the Debye factor is weighted by the frequency of r within the particle. The form factor-based approach can be approximated in the low-resolution regime by assuming an identical constant form factor for each atom. Here, the overall shape of the model will be entirely determined by the Debye factor modulated by the radius of the bead.

SAXS purely from the perspective of the electron using the macromolecule's electron–electron pair distance, $P(r)$, distribution function (Fig. 2). The $P(r)$-distribution is a histogram, binned by the Nyquist–Shannon number ($n_s = 2\pi/q_{max}$) of the observed SAXS profile (Moore, 1980). The distribution is unique to the shape of the macromolecule and is the real-space representation of the SAXS profile. In fact, the $P(r)$-distribution is considered to be the band-limited signal that is being sampled during a SAXS measurement (Rambo & Tainer, 2013a).

In a binary protein–RNA complex, the $P(r)$-distribution is the sum of the distribution for each of the components plus the $P(r)$-distribution describing the distances between the two components (Fig. 3). The set of distances between the two components is a correlation, cross-term and determines the relative orientation of the two components. For a small binary complex consisting of an RNA with 1,389 atoms (~964,000 interatomic distances) and a protein with 2,942 atoms (~4,326,000 interatomic distances), the complex will contain ~4,000,000 additional interatomic distances due to the cross-term alone (Fig. 3).

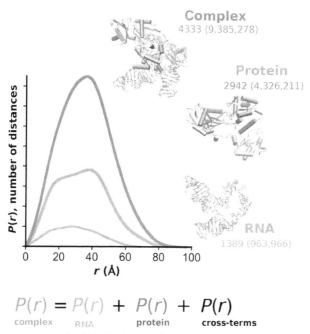

$$P(r) = P(r) + P(r) + P(r)$$
$$\text{complex} \quad \text{RNA} \quad \text{protein} \quad \text{cross-terms}$$

Figure 3 Construction of a P(r)-distribution for a binary complex. The P(r)-distribution for a binary protein–RNA complex is the sum of three unique terms. In this example using PDB: 4N0T, the P(r)-distribution is calculated using the atomic coordinates and scales with the number of atoms, n. The total number of unique distances is given by $n*(n-1)/2$. The RNA component (cyan (light gray shade in the print version)) has the least number of atoms and has the smallest distribution of distances by area. The protein (gray) has nearly double the number of atoms but four times the number of distances. If a complex forms (red (dark gray shade in the print version)), additional distances are available giving a cross-term defining how the two components assemble in the complex.

If the SAXS profiles of each component and complex are measured separately, then there is the opportunity to model each component independently and then assemble them into the complex constrained by the correlation term. This method of *ab initio* modeling separates the components into unique phases or scattering contrasts and is implemented in the program MONSA (Svergun, 1999). MONSA attempts to model the shape of each phase constrained by component's volume. MONSA is an extension of the popular DAMMIN/F (Franke & Svergun, 2009; Svergun, 1999) algorithms for *ab initio* volumetric modeling of SAXS data. These programs favor the lower resolution scattering region ($q < 0.18\,\text{Å}^{-1}$) by assuming a constant scattering form factor and weighting the observed $I(q)$ as a function of q.

4. ASSESSING SAXS SAMPLE QUALITY

Solution state SAXS experiments of biological samples are highly sensitive to contaminants (Jacques & Trewhella, 2010; Putnam et al., 2007). $I(q)$ is directly proportional to the squared mass of the scattering particle; therefore, small amounts of contaminants from aggregates will contribute significantly to the observed scattered intensities at low q-values. Because MONSA or DAMMIN/F favor the lower resolution scattering region (Franke & Svergun, 2009; Svergun, 1999), SAXS samples used for *ab initio* modeling must be made with the highest regard to sample monodispersity that optimizes for conformational homogeneity (Rambo & Tainer, 2010). We refer the reader to several publications for improving sample quality for SAXS (Grishaev, 2012; Rambo & Tainer, 2010; Reyes, Schwartz, Tainer, & Rambo, 2014).

The most basic complex to image by SAXS would be a binary complex containing a protein and RNA component. Here, three separate SAXS measurements would be made corresponding to the complex, and each of the components (phases). MONSA analysis is predicated on the assumption that the structure of each of the components remains largely unchanged in the free and bound states. This assumption is conditionally true, as at very low resolution, a small conformational difference between two states may appear negligible. However, there are many cases where the conformational differences are significant thereby nullifying this assumption. Here, the multiphase modeling can exclude the SAXS data of the unbound component that is suspected of a large conformational change. The structural information for this component will be derived from the SAXS data of the complex and constrained by the specified volume of the phase.

It will be important during the preliminary analysis of the SAXS data to (1) establish the folded state of each of the components and (2) validate formation of a complex. If a component is determined to be flexible or unfolded (e.g., short single-stranded RNA), the SAXS data may be excluded from the collection of datasets used during MONSA analysis. Using a table (Table 5) and a few plotting features in ScÅtter, we can evaluate complex formation and assess the quality of the SAXS data for multiphase *ab initio* modeling.

5. CASE STUDIES

5.1 Case I: A Binary Complex: SF1–U2AF65 Splicing Complex

The SF1–U2AF65 splicing complex consists of two proteins, SF1 (29 kDa), U2AF65 (36 kDa), and a small 3′ splice site RNA AdML (7 kDa) (Wang et al., 2013). SF1 and U2AF65 can form a binary protein assembly that recognizes and binds a 3′ splice site RNA sequence. Here, SAXS data on a small 3′ splice site RNA would be nonsensical as the RNA is a small, 25-nucleotide (nt) single-stranded species that is likely to adopt a grossly different structure in the bound state. However, preliminary analysis of the SAXS data collected on the binary protein assembly (BID: 1PSUKP) and complex (BID: 2PSUKX) can be used to infer properties of the bound RNA.

First, comparison of the radius-of-gyration, R_g, shows a decrease when RNA is present. This suggests the RNA is inducing compaction of the particle or that the RNA is binding near the particle's center-of-mass. Second, comparing the volume-of-correlation, V_c, shows an increase in the bound state suggesting a larger volume for the particle when RNA is present. Changes in both R_g and V_c strongly suggest the RNA is binding near the particle's center-of-mass and confirms assembly of a complex. However, more corroborating evidence can be obtained by determining the particle's Porod Volume, V_P, and mass ratio, Q_R. Here, we see that in the presence of the RNA, the particle's volume increases by 22,000 Å^3. If no binding occurred, V_P for the complex would be less than V_P for the binary protein assembly. This is due to the apparent SAXS-based volume being a weighted average of the smaller and larger particles. In addition, Q_R indicates the particle mass increases within the range of 8.5–15.3 kDa. A simultaneous decrease in all the values (R_g, V_c, V_P, and Q_R) when RNA is present would undoubtedly suggest the lack of a complex.

We can characterize the assembly further by visual inspection of the SAXS data. A plot of the ratio of the two SAXS datasets (Fig. 4A) shows strong features throughout the curve suggesting the molecular form factors are different. Likewise, examination of the dimensionless Kratky plot (Fig. 4B) reveals a significant shift in the principal peak toward the Guinier–Kratky point (√3, 1.104) indicating the protein assembly is

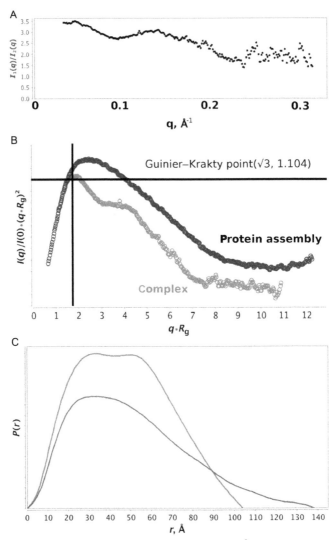

Figure 4 Visual assessment of complex formation using ScÅtter. Changes in the molecular form factor due to complex formation or a conformation change can be readily detected by plotting the ratio of the SAXS profile of the apo-state (black) to the putative bound state (red (dark gray shade in the print version)). A flat ratio indicates no change in the molecular form factor. (A) Ratio of the SAXS profiles for apo-SF1-U2AF[65] assembly to the apo-SF1-U2AF[65]-RNA complex. (B) Semiquantitative analysis using dimensionless Kratky plot. The plot normalizes for mass distribution and concentration and can be used to assert globularity changes to a particle. Globular particles will have a peak coincidence at the Guinier–Kratky point ($\sqrt{3}$, 1.104). The data for the apo-state suggest a nonglobular particle that becomes nearly globular in the presence of the ssRNA. (C) The $P(r)$-distribution is the resolution-limited, real-space transform of the observed SAXS

becoming more globular when bound to the small RNA. Furthermore, comparison of the $P(r)$-distribution (Fig. 4C) shows the maximum dimension, d_{max}, decreasing from 140 to 107 Å indicating a significant compaction of the particle occurs when RNA is present. We can establish firmly that a complex has formed and that the RNA is binding toward the particle's center-of-mass promoting a more compact, globular state. As MONSA and DAMMIN/F are undiscriminating, the *ab initio* modeling must be, at the very least consistent, with this level of understanding.

The available SAXS profiles of the splicing complex consisted of the protein assembly and protein–RNA complex (Wang et al., 2013). MONSA analysis of these datasets can proceed without the measured SAXS data of the RNA but requires constraints that can be inferred from the available information. Tables 4 and 5 list the required information to complete a MONSA modeling run (Fig. 5). The critical information are volume estimates for each of the components that will be modeled independently. In this case of the splicing factor complex, we need volume estimates for the protein assembly (61 kDa) and RNA (7 kDa). For the protein assembly, the SAXS-derived volume would be appropriate (Table 5). However, for the RNA component, the volume is estimated using either the volume difference between the protein assembly and complex or estimated using the partial specific volume for RNA (Table 4). The estimate from the partial specific volume is a nonhydrated estimate thus setting a lower limit on the component volume.

Similar to DAMMIN/F, several independent modeling runs must be performed and averaged (see Section 2.2). Here, the averaging is performed on a reference phase that is typically the component with the largest mass. Averaging the reference phase will determine for each nonreference phase within a single modeling run, a transformation (rotation and translation) matrix that will be applied to the corresponding nonreference components. Ideally, all the aligned components over the independent MONSA modeling runs should converge to the same general region and poor modeling results can be identified by their lack of spatial convergence.

---

data and is the recommended method for comparing SAXS datasets. From the dimensionless Kratky plot, the presence of the RNA suggested a significant change in the particle's globularity. This change can be readily illustrated by the significant decrease in d_{max} (black to red (dark gray shade in the print version)). Furthermore, the profile of the distribution changes from a nearly monomodal distribution to a bimodal distribution. The $P(r)$-distributions were normalized to their respective integrated areas.

Table 4 Modeling Inputs for MONSA Analysis

For Each Component (Phase)	Contrast	d_{max} (Å) Estimated from $P(r)$-Distribution Volume (×10³, Å³)	R_g
Protein	1.0	Use SAXS volume or estimated from $V_{protein}$	This is an optional constraint. The values represent the component within the context of the complex.
Nucleic acid	2.0	Use SAXS volume or estimated from V_{RNA}	
Pseudocomponent	[a]$C_{fractional}$	Use SAXS volume or estimate as sum of the relevant components	

[a]Defined in text.

Each modeled component in a MONSA analysis is referred to as a "phase." A phase will have an assigned contrast that is relative to a protein. By convention, the protein contrast is 1 and nucleic acid contrast is 2. During pseudocomponent analysis, the contrast of a mixed protein–nucleic acid component is estimated using the following equation:

$$C_{fractional} = \frac{V_{RNA}}{V_{total}} \cdot C_{RNA} + \frac{V_{protein}}{V_{total}} \cdot C_{protein}.$$

$C_{fractional}$ requires volume estimates for each of the components. The volume of the pseudocomponent can be measured from a SAXS experiment or estimated from the masses of the respective components using the following equations:

$$V_{protein} = \frac{mass}{1.35} \cdot 1.66$$

$$V_{RNA} = \frac{mass}{1.74} \cdot 1.66.$$

These volumes use the standard density for proteins and RNA derived from crystal structures and represent the dry mass of the component. If the volume cannot be reliably determined from the SAXS data, then the calculated volumes can be used.

In this example, 11 independent runs were performed with the protein assembly designated as the reference phase. The alignment of each run showed that all RNA phases aligned to the same region (Fig. 6B). However, there is a chi-squared, χ^2, difference between the fits of the protein alone (χ^2: 1.07) SAXS data and the complex (χ^2: 2.08) suggesting the conformation of the protein assembly in the bound state is different than the apo-state (Fig. 6C and D). The difference is driven by a poor fit in the high q-region (Fig. 6B, $q > 0.2$ Å^{-1}). If there was no conformational change upon RNA binding, then we would expect the protein model, derived from the complex, to optimally explain the protein alone SAXS data at high resolution. The poorer fit is expected as the $P(r)$-distributions (Fig. 4C) illustrated a

Table 5 Summary of SAXS-Derived Parameters for MONSA Analysis

	Component 1 (SF1-U2AF[65])		Component 2 (25 nt ssRNA)		Complex
	Constraints	Model$_{ave}$	Constraints	Model$_{ave}$	
R_g (Å)	39.5	40.4(±5)	N/a	18(±4)	38.0
V_c (Å^2)	565		N/a		615
Volume (×10^3, Å^3)	113	121(±2)	22^a (7.4)^b	11.1(±0.2)	135
Q_R mass (calc) (kDa)	65.7 (65)		(7.9 kDa)		74.2–81 (73)^c
d_{max} (Å)	140		N/a		107

^{a}Calculated as the difference between component 1 and the complex.
^{b}Calculated dry volume.
^{c}Mass range corresponds to estimates based on if all RNA or all protein using ScÅtter.
The analysis was performed on SAXS datasets deposited in BioIsis.net using BID: 2PSUKX and BID: 1PSUKP for the complex and protein assembly, respectively. This table summarizes the information required for performing a MONSA analysis using a binary complex. For multicomponent complexes, add additional columns as necessary. The R_g estimate is optional but should correspond to the component in the bound state. The search volume is defined by half-d_{max}.
N/a, not available.

significant change. The protein phase is constrained by the expected volume, scattering contrast, and SAXS datasets of the free and bound states, whereas the RNA component is constrained by volume, scattering contrast, and only the SAXS data of the bound state. The differing scattering contrast and volume restraints allows the RNA to be visualized at low resolution ($q < 0.2$ Å^{-1}), where we can now image the protein and RNA components separately. As expected, the RNA component occurs on one side, near the particle's center-of-mass. It should be noted that the final assembled model will have an enantiomorphic equivalent.

The postanalysis should summarize the modeling results by calculating the average and variance for the R_g and volume for each of the respective components (Table 5). Large variances and deviations from expected values may suggest the samples are polydisperse and not suitable for this type of analysis. In this example, the averaged volume of the protein assembly is within 10% of the measured V_P and likewise, for the RNA, the averaged V_P is within the expected range.

5.2 Case II: Visualizing Conformational Changes with SMARCAL1–DNA Complex

SAXS is a powerful tool for detecting conformational changes. As illustrated for the SF1–U2AF[65] splicing complex, conformational changes can be

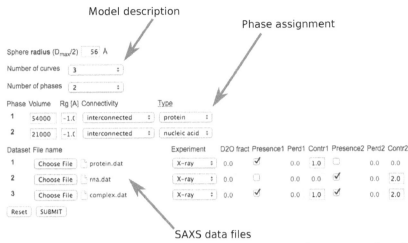

Figure 5 ATSAS online MONSA web interface. The online interface requires the information in Table 4. The description of the model is determined by the number of input SAXS curves and phases (model description red (dark gray shade in the print version) arrow). In this example, MONSA is set to model a binary complex using 3 SAXS curves: protein alone (protein.dat), RNA alone (rna.dat) and complex (complex.dat). From the $P(r)$-distribution, d_{max} is specified as the radius in units of Angstrom. The binary complex contains two unique volume elements (phases) that must be assigned to a phase type (protein, nucleic acid, or custom, see red (dark gray shade in the print version) arrow). For the protein component, it is assigned a volume of 54,000 Å^3 and contrast of 1 whereas the nucleic acid phase has a volume of 21,000 Å^3 and a contrast of 2. R_g for each phase is unknown and will be determined during the modeling run. Each SAXS data file is assigned to either the protein phase, RNA phase, or both and must be assigned in the last section of the web interface. Here, protein.dat is assigned to phase 1 (Presence 1) and rna.dat is assigned to phase 2 (Presence 2) by the checked boxes. Finally, complex.dat is assigned both phases. This completes the web interface and can be submitted.

readily detected by comparing the $P(r)$-distributions of a particle under different conditions. Using ATP analogs, SAXS experiments on SMARCAL1 (64 kDa), an ATP-dependent, DNA-binding motor protein revealed distinct structural states accessible to the SMARCAL1–DNA complex (Mason et al., 2014).

SMARCAL1 is a protein involved in regressing stalled DNA replication forks. Using size-exclusion chromatography (SEC), complexes were formed with 25-bp, 10-nt single-stranded DNA substrates, and purified in the presence of ATP analogs such as AMP-PNP and ADP-BeF$_x$ (representing the pre- and post-ATP hydrolysis states, respectively). The $P(r)$-distributions clearly demonstrated significant conformational differences between the

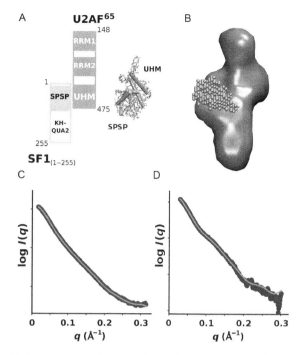

Figure 6 Multiphase volumetric modeling of the SF1-U2AF^{65} splicing complex. (A) Cartoon representation of the SF1 (residues 1–255) and U2AF^{65} (residues 148–475) proteins. The complex forms through an interaction of the N-terminal SF1, SPSP domain with the C-terminal U2AF^{65} UHM domain. (B) Averaged and aligned MONSA model of the protein–RNA splicing complex. Average was determined from 11 independent MONSA runs. Protein phase (purple (dark gray shade in the print version)) forms a shallow cleft for binding the RNA (orange (light gray shade in the print version)). RNA is not extended and, shape of the orange phase suggests the ssRNA loops back. (C) Final MONSA fits to apo-state SAXS data (BID: 1PSUKP) and (D) complex (BID: 2PSUKX).

(+/−) nucleotide states. In the absence of nucleotide, the complex was fully extended with a d_{max} of 124 Å. Upon AMP-PNP binding, the complex became more compact with a significant reduction in R_g (35.4–31.7 Å) and a d_{max} of 108 Å. Finally, trapping the complex in the post-ATP hydrolysis state, the $P(r)$-distribution illustrated a further compaction of the complex (R_g from 31.7 to 29.9 Å) and a smaller d_{max} (101 Å). Since the nucleotides have negligible mass relative to the complex, the R_g decreases suggest that mass is being relocated toward the particle's center-of-mass. Visually, this is illustrated in the $P(r)$-distribution as a shift toward smaller distances of the main peak (Fig. 7A).

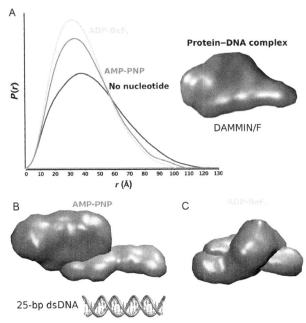

Figure 7 Multiphase volumetric modeling of the SMARCAL1 protein–DNA complex in different ATP hydrolysis states. (A) Normalized P(r)-distributions show significant changes in the presence of different ATP analogs. In the absence of ATP (no nucleotide), the complex is maximally extended. Addition of AMP-PNP (red (dark gray shade in the print version)) or ADP-BeF$_x$ (cyan (light gray shade in the print version)) shows a compaction of the complex. No nucleotide state DAMMIN/F model illustrates a large asymmetric particle with a protrusion consistent with the width of a dsDNA helix. (B) Averaged and aligned MONSA model of the AMP-PNP state shows the 25-bp, 10-nt ssDNA substrate. The length and width of the DNA phase (orange (gray shade in the print version)) is consistent with a 25-bp dsDNA fragment. Protein phase (purple (dark gray shade the in print version)) shows a protein bound preferentially to one end of the DNA, likely the 10-nt ssDNA extension. (C) Averaged and aligned MONSA model of the ADP-BeF$_x$ state showing translocation of the protein along the DNA substrate in the post-ATP hydrolysis state.

Examination of a DAMMIF *ab initio* model suggested the protein was binding asymmetrically to the DNA substrate (Fig. 7B). The DNA component could be inferred from physical dimensions of the model; nevertheless, the model was derived from the assumption of a constant and homogenous form factor for the complex. In this experiment, the DNA substrate had significant mass (19.8 kDa) and structure in the form of a 25-bp double helix that was assumed to be rigid at low resolution. Therefore, MONSA modeling was performed using SAXS data for the complex and unbound DNA. The protein alone SAXS profile was excluded as it was assumed

SMARCAL1 would likely undergo significant conformational changes upon DNA binding. The volume constraints for the modeling were derived from the experimental SAXS datasets consisting of unbound SMARCAL1, DNA, and the complex.

Unlike the protein assembly in the splicing factor complex, the χ^2 for the fits were less than 1.5 suggesting the structure of the DNA was similar in the bound (χ^2: 1.41) and unbound states (χ^2: 1.08). The multiphase model clearly illustrated a protein assembling preferentially to one side and one end of the DNA substrate in the pre-ATP hydrolysis state. In the post-ATP hydrolysis, the protein appears to translocate (Fig. 7C) along the DNA demonstrating the expected motor function of the protein.

These experiments required highly pure and homogenous samples that were purified at the X-ray source during data collection. The complexes were optimized for monodispersity using quasi-elastic and multi-angle light scattering coupled to analytical SEC. Here, an appropriate SEC column (typical choices are Superdex 75 and 200 PC 3.2 from GE Healthcare or KW-402.5, KW-403, and KW-404 from Shodex) was chosen that provided clear separation of the complex from unbound protein and DNA. Complex formation and monodispersity were assessed by the mass and radius-of-hydration distribution across the SEC elution profile during light scattering analysis. To prevent multiple proteins from binding a single DNA molecule, it was essential the DNA were in slight excess at the start of the SEC purification. Furthermore, multimerization of the protein in the absence of nucleic acids may occur and it is suggested that addition of 2% sucrose or 20 mM phosphate be used as a stabilizing additive. SAXS data collection for a specific nucleotide state, over a range of particle concentrations, required ~30 min with subsequent analysis and modeling requiring ~3 days of computational time. SAXS experiments are highly efficient where most of the time is expended on sample preparation. It can be anticipated that inline SEC-SAXS at synchrotron sources will greatly facilitate these types of experiments by ensuring optimal sample homogeneity during data collection (Brookes et al., 2013).

5.3 Case III: Cautionary Notes from Simulations with the Ternary T-box•tRNA•YbxF Complex

Complexes containing more than two components are common in biology and consequently a common sample type for SAXS. While MONSA may seem suitable for analyzing multicomponent complexes (i.e., two RNAs and one protein or two proteins, and one RNA), the algorithm has limitations that are relevant for discussion.

Using a crystal structure of a protein•RNA$_2$ ternary complex (PDB ID: 4TZP) (Zhang & Ferre-D'Amare, 2014), four SAXS curves were simulated consisting of tRNA (22 kDa), YbxF protein (7.2 kDa), T-box(RNA)•YbxF binary complex, and ternary complex (62.2 kDa) (Fig. 8A and B). The simulations were limited to a q_{max} of 0.25 Å^{-1} with noise sampled from an empirical dataset (Fig. 8C) (Rambo & Tainer, 2013a,b). Ideally, MONSA should be able to place each component correctly as the position of the tRNA is represented once and protein twice in the SAXS dataset collection.

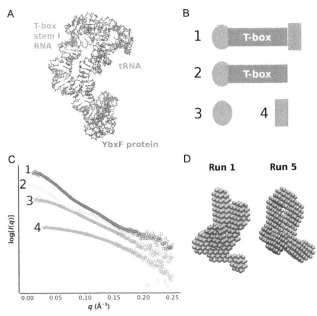

Figure 8 Simulations with the T-box stem I•tRNA•YbxF ternary complex. (A) Crystal structure (PDB: 4TZP) shows the tRNA (orange) is parallel along one arm of the bent T-box stem I RNA (green). YbxF protein (cyan) binds distal to the tRNA arm away from the particle's center-of-mass. (B) Cartoon representation of the ternary complex (1), binary T-box stem I•YbxF complex (2), YbxF protein (3), and tRNA (4). (C) SAXS profiles were simulated of the four components in B using coordinates derived from PDB code: 4TZP. Simulations were performed with CRYSOL using default parameters. Noise was added using a transformation method from an empirical datasets. (D) Two example MONSA models. Fifteen independent runs were performed and each failed to reconstruct the true particle. In each case, the small protein was found near the particle's center-of-mass with the tRNA flipping between two sides of the T-box arm. (See the color plate.)

However, 15 independent MONSA runs failed to correctly reconstitute the complex. Both the YbxF protein and tRNA were positioned around the T-box center-of-mass (Fig. 8D) with the corresponding χ^2 values for all the model data fits less than 1.5 (Fig. 8E). The χ^2 values are unreliable as the position of the protein is fitting the noise. Additional modeling runs that included the tRNA•T-binary complex failed to correctly position YbxF.

In the absence of a prior structure, we can validate a MONSA modeling result by applying the parallel axis theorem (Damaschun, Muller, & Purschel, 1968; Moore, Engelman, & Schoenborn, 1974). For each component, i, given the mass, radius-of-gyration of the ith component ($R_{g,i}$), and center-of-mass coordinate then the total R_g of the complex can be calculated using

$$R_g^2 = \sum_{i=1}^{n} f_i \cdot R_{g,i}^2 + \sum_{i}^{n} \sum_{j \neq i}^{n} f_i f_j \cdot r_{ij}^2,$$

where f_i is the fractional mass of the component and r_{ij} is the Euclidean distance between components i and j. Conveniently, the MONSA log file contains the R_g and center-of-mass coordinate for each modeled component. Therefore, we can compare the modeled and empirical, real-space R_g values of the complex to validate a multiphase volumetric model prior to averaging. For the complex above, there was a mean R_g error of 9.3%, whereas for a correctly modeled complex, the error can be expected to be <5%, in this case 3.4%.

By scattering contrast, the complex is 94% RNA leaving ~6% for the YbxF protein. It is likely that the combination of noise, low scattering mass, use a constant form factor, and q-weighting of the intensities contributed to the false placement of the protein. The modeling accuracy can be improved significantly by restricting the analysis to binary-like complexes (Fig. 9), such that the MONSA analysis is restricted to resolving only two components. The SAXS data for the T-box•YbxF complex is information for a single particle of mixed contrast, i.e., protein and RNA. This can be made into a pseudocomponent for MONSA analysis by specifying the protein•RNA particle as single volume element and assigning a volume-weighted contrast (Table 5). The analysis using SAXS data for the T-box•YbxF pseudocomponent, ternary complex, and tRNA enables the resolution of the tRNA within the ternary complex. Likewise, using SAXS dataset collection consisting of the T-box•tRNA pseudocomponent, ternary complex,

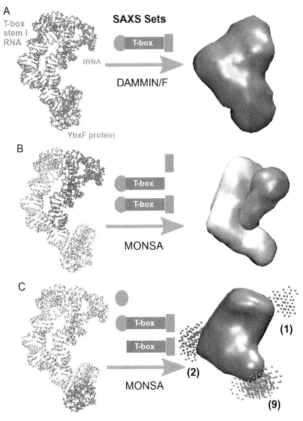

Figure 9 MONSA modeling runs using pseudocomponents. Cartoon above each arrow indicates the simulated SAXS data of the complex or components used in the multiphase volumetric modeling. (A) Averaged DAMMIN/F model calculated from 13 independent runs. (B) MONSA modeling treating the binary T-box stem I•YbxF complex as a single component of mixed phase with a volume-weighted contrast of 1.8. The pseudocomponent is the reference phase during averaging and alignment. This arrangement of SAXS data determines the orientation of the tRNA within the complex. (C) MONSA modeling to determine the position of the YbxF protein. T-box•tRNA binary complex is treated as a single RNA particle (reference phase). Therefore, the MONSA modeling should position the protein. Alignment of the 12 protein phases after averaging the reference phase shows three possible placements of the protein. Most likely position (cyan) is consistent with the true position and can be inferred from the median position (region with a frequency >50%). False positives (red) are artifacts of the modeling algorithm. (See the color plate.)

and YbxF protein in a MONSA analysis will resolve the position of the YbxF protein. Thus, restricting the analysis to binary-like complexes reduces the effective degrees-of-freedom and consequently reduces the false positive rate.

The analysis using the T-box•YbxF pseudocomponent correctly resolved the placement of the tRNA in every independent MONSA run (Fig. 9B). The tRNA volume element associates along one side, within the convex face of the T-box•YbxF component. Likewise, for analysis using the T-box•tRNA pseudocomponent, MONSA now models the protein away from the particle's center-of-mass but in three spatially distinct locations with frequencies of 8.3%, 16.7%, and 75% (Fig. 9C). Given the potential for false positives, we can identify the correct position of the protein by noting that the median of the placements is the position with a frequency >50%. The median is a robust metric that offers resistance to outliers in a set and we recommend noting the median location of the phase in each alignment.

The example with the T-box•tRNA•YbxF complex illustrates limitations inherent to the algorithm and not the method of multiphase volumetric modeling. At the observed resolution, the limitation of the method will be related to how small of a difference in the respective $P(r)$-distributions can be effectively measured, whereas the limitation of the algorithm will be related to how well the coarse graining approximates the scattering of the component. The latter limitation will contribute to a false positive rate during modeling. Nonetheless, these multiphase models are hypotheses that must be validated through an interpretation that enables experimental testing.

6. CONSIDERATIONS

In the early days of ribosome structural investigations, multiphase modeling enabled construction of a three-dimensional map that positioned each of the ribosomal proteins within the respective subunits (Langer, Engelman, & Moore, 1978; Laughrea, Engelman, & Moore, 1978; Moore, Capel, Kjeldgaard, & Engelman, 1986). Here, isotopic labeling of specific ribosomal proteins coupled with small-angle neutron scattering experiments generated the requisite information for placement of the proteins. Likewise, similar studies in the 1970s on the nucleosome established the canonical arrangement of DNA wrapped around a protein core (Hjelm et al., 1977). Modeling these scattering experiments required methodical and restrained interpretations that did not rely on any type of Monte Carlo or simulated annealing algorithms.

Today, advancements in X-rays sources, detectors, and computational methods have extended multiphase volumetric modeling methods to SAXS experiments (Svergun, 1999). These experiments do not require specific isotopic labeling methods but require SAXS datasets to be varied by scattering mass. We showed how multiphase modeling can be applied to resolving components within volumetric density maps. However, the method can be applied to a single protein or RNA species where variations are achieved by domain deletions or extension of specific helical elements. These experiments can highlight specific macromolecular features thus extending the information content of SAXS-derived *ab initio* models.

Improvements in multiphase volumetric modeling will be possible by using methods that better approximate the molecular form factor, and do not preferentially emphasize the low q-range SAXS data during modeling. From information theory, we can establish that the SAXS signal exists error-free in high-noise environments such that the noise can be $\sim 20 \times$ the signal (Rambo & Tainer, 2013a,b). Here, application of the Shannon–Hartley noisy coding channel theorem can help extract the signal thereby allowing an error-free SAXS signal to be modeled. More importantly, with advanced chemical labeling methods that can site specifically conjugate electron dense tags, it may be possible to resolve the location of individual residues as suggested by Kratky and Worthmann (1947). This will require significant improvements in the approximation of the scattering form factors, and likewise new computational approaches to be developed.

ACKNOWLEDGMENTS
I would like to thank James Doutch, Mark Tully, and Katsuaki Inoue for helpful discussions.

REFERENCES
Brookes, E., Perez, J., Cardinali, B., Profumo, A., Vachette, P., & Rocco, M. (2013). Fibrinogen species as resolved by HPLC-SAXS data processing within the () enhanced SAS module. *Journal of Applied Crystallography, 46*(Pt. 6), 1823–1833.

Chacon, P., Moran, F., Diaz, J. F., Pantos, E., & Andreu, J. M. (1998). Low-resolution structures of proteins in solution retrieved from X-ray scattering with a genetic algorithm. *Biophysical Journal, 74*(6), 2760–2775.

Chacon, P., & Wriggers, W. (2001). Using situs for the registration of protein structures with low-resolution bead models from X-ray scattering. *Journal of Applied Crystallography, 34*, 4.

Damaschun, G., Muller, J. J., & Purschel, H. V. (1968). small angle X-ray scattering of objects built of subunits. I. The scattering mass radius. *Acta Biologica et Medica Germanica, 20*(3), 379–392.

Dzananovic, E., Patel, T. R., Deo, S., McEleney, K., Stetefeld, J., & McKenna, S. A. (2013). Recognition of viral RNA stem-loops by the tandem double-stranded RNA binding domains of PKR. *RNA*, *19*(3), 333–344.

Engelman, D. M., & Moore, P. B. (1975). Determination of quaternary structure by small angle neutron scattering. *Annual Review of Biophysics and Bioengineering*, *4*(00), 219–241.

Franke, D., & Svergun, D. I. (2009). DAMMIF, a program for rapid ab-initio shape determination in small-angle scattering. *Journal of Applied Crystallography*, *42*(2), 342–346.

Gabel, F., Simon, B., Nilges, M., Petoukhov, M., Svergun, D., & Sattler, M. (2008). A structure refinement protocol combining NMR residual dipolar couplings and small angle scattering restraints. *Journal of Biomolecular NMR*, *41*(4), 199–208.

Glatter, O., & Kratky, O. (1982). *Small angle X-ray scattering*. London, New York: Academic Press.

Grishaev, A. (2012). Sample preparation, data collection, and preliminary data analysis in biomolecular solution X-ray scattering. *Current protocols in protein science/editorial board, John E. Coligan ... [et al.]*, Chapter 17, Unit 17.

Hjelm, R. P., Kneale, G. G., Sauau, P., Baldwin, J. P., Bradbury, E. M., & Ibel, K. (1977). Small angle neutron scattering studies of chromatin subunits in solution. *Cell*, *10*(1), 139–151.

Jacques, D. A., & Trewhella, J. (2010). Small-angle scattering for structural biology—expanding the frontier while avoiding the pitfalls. *Protein Science: A Publication of the Protein Society*, *19*(4), 642–657.

Kratky, O., & Worthmann, W. (1947). Über die Bestimmbarkeit der Konfiguration gelöster organischer Moleküle durch interferometrische Vermessung mit Röntgenstrahlen. *Monatshefte für Chemie und verwandte Teile anderer Wissenschaften*, *76*(3–5), 263–281.

Langer, J. A., Engelman, D. M., & Moore, P. B. (1978). Neutron-scattering studies of the ribosome of Escherichia coli: A provisional map of the locations of proteins S3, S4, S5, S7, S8 and S9 in the 30 S subunit. *Journal of Molecular Biology*, *119*(4), 463–485.

Laughrea, M., Engelman, D. M., & Moore, P. B. (1978). X-ray and neutron small-angle scattering studies of the complex between protein S1 and the 30-S ribosomal subunit. *European Journal of Biochemistry/FEBS*, *85*(2), 529–534.

Mallam, A. L., Jarmoskaite, I., Tijerina, P., Del Campo, M., Seifert, S., Guo, L., et al. (2011). Solution structures of DEAD-box RNA chaperones reveal conformational changes and nucleic acid tethering by a basic tail. *Proceedings of the National Academy of Sciences of the United States of America*, *108*(30), 12254–12259.

Mason, A. C., Rambo, R. P., Greer, B., Pritchett, M., Tainer, J. A., Cortez, D., et al. (2014). A structure-specific nucleic acid-binding domain conserved among DNA repair proteins. *Proceedings of the National Academy of Sciences of the United States of America*, *111*(21), 7618–7623.

Moore, P. (1980). Small-angle scattering. Information content and error analysis. *Journal of Applied Crystallography*, *13*(2), 168–175.

Moore, P. B., Capel, M., Kjeldgaard, M., & Engelman, D. M. (1986). Quaternary Organization of the 30S Ribosomal Subunit of Escherichia Coli. *Biophysical Journal*, *49*(1), 13–15.

Moore, P. B., Engelman, D. M., & Schoenborn, B. P. (1974). Asymmetry in the 50S ribosomal subunit of Escherichia coli. *Proceedings of the National Academy of Sciences of the United States of America*, *71*(1), 172–176.

Pelikan, M., Hura, G. L., & Hammel, M. (2009). Structure and flexibility within proteins as identified through small angle X-ray scattering. *General Physiology and Biophysics*, *28*(2), 174–189.

Petoukhov, M. V., Franke, D., Shkumatov, A. V., Tria, G., Kikhney, A. G., Gajda, M., et al. (2012). New developments in the ATSAS program package for small-angle scattering data analysis. *Journal of Applied Crystallography*, *45*(2), 342–350.

Putnam, C. D., Hammel, M., Hura, G. L., & Tainer, J. A. (2007). X-ray solution scattering (SAXS) combined with crystallography and computation: Defining accurate macromolecular structures, conformations and assemblies in solution. *Quarterly Reviews of Biophysics*, *40*(3), 191–285.

Rambo, R. P., & Tainer, J. A. (2010). Improving small-angle X-ray scattering data for structural analyses of the RNA world. *RNA*, *16*(3), 638–646.

Rambo, R. P., & Tainer, J. A. (2011). Characterizing flexible and intrinsically unstructured biological macromolecules by SAS using the Porod-Debye law. *Biopolymers*, *95*(8), 559–571.

Rambo, R. P., & Tainer, J. A. (2013a). Accurate assessment of mass, models and resolution by small-angle scattering. *Nature*, *496*(7446), 477–481.

Rambo, R. P., & Tainer, J. A. (2013b). Super-resolution in solution X-ray scattering and its applications to structural systems biology. *Annual Review of Biophysics*, *42*, 415–441.

Reyes, R. E., Schwartz, C. R., Tainer, J. A., & Rambo, R. P. (2014). Methods for using new conceptual tools and parameters to assess RNA structure by small-angle X-ray scattering. *Methods in Enzymology*, *549*(11), 235–265.

Schwieters, C. D., & Clore, G. M. (2007). A physical picture of atomic motions within the Dickerson DNA dodecamer in solution derived from joint ensemble refinement against NMR and large-angle X-ray scattering data. *Biochemistry*, *46*(5), 1152–1166.

Stuhrmann, H. (1970). Interpretation of small-angle scattering functions of dilute solutions and gases. A representation of the structures related to a one-particle scattering function. *Acta Crystallographica. Section A*, *26*(3), 297–306.

Stuhrmann, H. B. (1973). Comparison of the three basic scattering functions of myoglobin in solution with those from the known structure in crystalline state. *Journal of Molecular Biology*, *77*(3), 363–369.

Svergun, D. I. (1999). Restoring low resolution structure of biological macromolecules from solution scattering using simulated annealing. *Biophysical Journal*, *76*(6), 2879–2886.

Svergun, D. I., & Nierhaus, K. H. (2000). A map of protein-rRNA distribution in the 70 S Escherichia coli ribosome. *The Journal of Biological Chemistry*, *275*(19), 14432–14439.

Svergun, D. I., Petoukhov, M. V., & Koch, M. H. (2001). Determination of domain structure of proteins from X-ray solution scattering. *Biophysical Journal*, *80*(6), 2946–2953.

Wang, W., Maucuer, A., Gupta, A., Manceau, V., Thickman, K. R., Bauer, W. J., et al. (2013). Structure of phosphorylated SF1 bound to U2AF(6)(5) in an essential splicing factor complex. *Structure*, *21*(2), 197–208.

Zhang, J., & Ferre-D'Amare, A. R. (2014). Dramatic improvement of crystals of large RNAs by cation replacement and dehydration. *Structure*, *22*(9), 1363–1371.

CHAPTER THIRTEEN

Small-Angle Neutron Scattering for Structural Biology of Protein–RNA Complexes

Frank Gabel*,†,‡,§,1

*Université Grenoble Alpes, Institut de Biologie Structurale, Grenoble Cedex, France
†Commissariat à l'Energie Atomique et aux Energies Alternatives, Direction des Sciences du Vivant, Institut de Biologie Structurale, Grenoble Cedex, France
‡Centre National de la Recherche Scientifique, Institut de Biologie Structurale, Grenoble Cedex 9, France
§Large Scale Structures Group, Institut Laue-Langevin, Grenoble Cedex, France
1Corresponding author: e-mail address: frank.gabel@ibs.fr

Contents

1. Small-Angle Neutron Scattering: A Powerful Tool for Structural Biology of Protein–RNA Complexes — 392
2. Principles of SANS — 393
3. Practical Aspects of SANS and Requirements on the Sample State — 396
4. Model-Free Parameters: Molecular Mass and Radius of Gyration — 397
5. Neutron Scattering Length Densities of Proteins and RNAs and Implications on Model Building — 399
6. Possibilities and Limits of *Ab Initio* Models (Shape Envelopes) — 402
7. Present State-of-the-Art Examples — 405
8. Conclusions and Future Perspectives — 409
Acknowledgments — 411
References — 411

Abstract

This chapter deals with the applications of small-angle neutron scattering (SANS) for the structural study of protein–RNA complexes in solution. After a brief historical introduction, the basic theory and practical requirements (e.g., sample state) for SANS experiments will be treated. Next, model-free parameters, such as the molecular mass and the radius of gyration, which can be obtained without *a priori* structural information, will be introduced. A more detailed section on the specific properties of SANS (with respect to its sister technique, small-angle X-ray scattering), and their implications on possibilities and limits of model building and interpretation will be discussed with a focus on protein–RNA systems. A practical illustration of the information content of SANS data will be given by applying *ab initio* modeling to a tRNA-synthetase system of known high-resolution structure. Finally, two present state-of-the-art examples that combine SANS data with complementary structural biology techniques (NMR and crystallography) will be presented and possible future developments and applications will be discussed.

1. SMALL-ANGLE NEUTRON SCATTERING: A POWERFUL TOOL FOR STRUCTURAL BIOLOGY OF PROTEIN–RNA COMPLEXES

Small-angle neutron scattering (SANS) has been increasingly used since the 1970s (with the advent of high-flux neutron sources) to study a diversity of protein–RNA complexes. It is sensitive to nuclear scattering length density (SLD) fluctuations in the nanometer range, and therefore probes the conformational state of macromolecules. Earliest results include studies on protein/RNA organization in viruses (Jacrot, Chauvin, & Witz, 1977), interaction and stoichiometry of tRNA-synthetase complexes (Dessen, Blanquet, Zaccai, & Jacrot, 1978; Giegé, Jacrot, Moras, Thierry, & Zaccai, 1977), and a mapping of protein positions within the ribosome (Engelman, Moore, & Schoenborn, 1975; Hoppe et al., 1975; Stuhrmann et al., 1976) based on methods of contrast variation and triangulation developed previously for X-rays and neutrons in solution (Engelman & Moore, 1972; Hoppe, 1972; Stuhrmann & Kirste, 1965). In these early studies, the relative arrangement and topology of the protein and RNA subunits were determined decades before the respective atomic-resolution models became available. In passing, it is worth noting that at the same time low-resolution models of nucleosomes were determined by SANS, again many years before atomic models became available (Baldwin, Boseley, Bradbury, & Ibel, 1975; Bram, Butler-Browne, Baudy, & Ibel, 1975; Pardon et al., 1975). The interested reader can find a more complete overview of early studies in appropriate biological SANS reviews (Jacrot, 1976; Zaccai & Jacrot, 1983).

The increase of computing power and the advent of high-throughput software gave rise to new modeling approaches in the 1990s and allowed direct and rapid calculation of SANS (and small-angle X-ray scattering (SAXS)) data from atomic models and their fits against experimental data (Svergun, Barberato, & Koch, 1995; Svergun et al., 1998). At the same time, *ab initio* envelope models could be derived from (multiple) SANS datasets (Svergun, 1999; Svergun & Nierhaus, 2000). In recent years, SANS has been increasingly combined with other structural biology techniques (Petoukhov & Svergun, 2007), with nuclear magnetic resonance (NMR) being a particularly efficient and complementary method (Carlomagno, 2014; Hennig & Sattler, 2014; Madl, Gabel, & Sattler, 2011). In the present chapter, I will provide a short overview of the basics of the SANS technique and discuss its potential and limitations for state-of-the-art modeling with a focus on protein–RNA complexes.

2. PRINCIPLES OF SANS

Neutrons for biological research are produced by nuclear fission or by spallation (Svergun, Koch, Timmins, & May, 2014) at a number of neutron centers worldwide (http://www.neutronsources.org/neutron-centres.html). In general, neutrons from both types of sources need to be moderated (slowed down) in order to display wavelengths in the Angstrom range and guided to an appropriate SANS instruments. A typical SANS experimental setup contains the following main elements (see Fig. 1A): a monochromator (for selection of a specific range of neutron wavelengths), a collimation device (for aligning the neutron beam), a sample holder device (often with the possibility to regulate environment variables such as temperature and pressure), and a detector (for counting the neutrons scattered by the sample). Usually, large parts of the instrument are under vacuum in order to avoid parasitic scattering and scattering/loss of neutrons traveling though air. A more detailed description of SANS instrumentation can be found elsewhere (http://www.ncnr.nist.gov/staff/hammouda/the_SANS_toolbox.pdf).

In the case of a dilute solution of isotropically oriented biological macromolecules, the two-dimensional scattering pattern is symmetric around the direct beam and the scattered intensity I is only a function of the modulus of the momentum transfer $\vec{Q}$ (see Fig. 1A; Feigin & Svergun, 1987; Glatter & Kratky, 1982; Guinier & Fournet, 1955):

$$I(Q) \propto N \left| \int_V \Delta\rho(\vec{r}) e^{i\vec{Q}\cdot\vec{r}} dV \right|^2 \quad (1)$$

The modulus of the momentum transfer $\vec{Q}$ is defined as $Q = \frac{4\pi}{\lambda}\sin\theta$ and the resulting one-dimensional scattering curve $I(Q)$ contains the same information on isotropically oriented particles in solution as their two-dimensional diffraction pattern measured by the detector. Here, 2θ is the scattering angle (Fig. 1A), λ is the neutron wavelength, and N is the number of particles dissolved in solution. The integral runs over the whole (solvent-excluded)

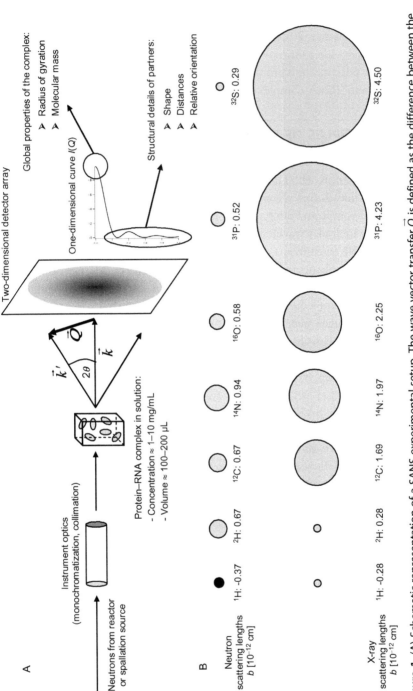

Figure 1 (A) Schematic representation of a SANS experimental setup. The wave vector transfer $\vec{Q}$ is defined as the difference between the outgoing and incoming wave vectors $\vec{k}'$ and $\vec{k}$ (with $|\vec{k}| = |\vec{k}'| = \frac{2\pi}{\lambda}$, λ being the neutron wavelength). (B) Schematic representation of (coherent) neutron scattering lengths (SLDs) and cross sections as compared to the respective ones for X-rays. Please note that the black circle that represents hydrogen (^{1}H) signifies that this isotope has a negative scattering length whereas all other isotopes shown have positive scattering lengths (both for X-rays and neutrons). ^{2}H is the hydrogen isotope deuterium. The values of the SLDs were taken from literature Jacrot (1976).

volume V of a particle with $\vec{r}$ being the vector (of arbitrary origin) pointing to a volume element dV of scattering contrast $\Delta\rho(\vec{r})$ within the particle:

$$\Delta\rho(\vec{r}) = \frac{\sum_i b_i}{dV} - \rho_S \quad (1')$$

b_i are the (coherent) scattering lengths of the atomic nuclei (in the case of neutrons) within the (solvent-excluded) volume element dV in the particle, and ρ_S is the average SLD of the solvent used, e.g., of a certain $H_2O:D_2O$ ratio (see Fig. 2). Values b_i of individual atoms can be retrieved from

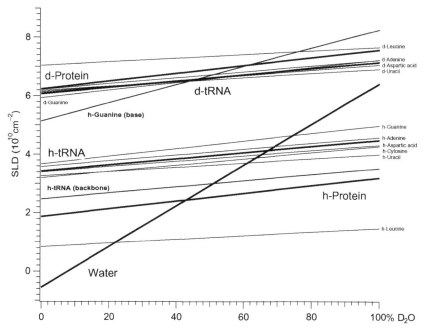

Figure 2 Neutron scattering length densities (SLDs) for several RNA nucleotides and protein residues in the natural, hydrogenated ("h"), and in the perdeuterated ("d") states. "tRNA" and "Protein" refer to the RNA and protein structures used in the section on *ab initio* models (PDB entry 1GTR) and were calculated according to their nucleotide and amino acid sequences. All SLDs were calculated from the chemical composition of the individual molecules and moieties using their atomic scattering lengths and exchangeable hydrogens according to Jacrot (1976). Solvent-excluded volumes of nucleotides and amino acid residues were taken from Jacrot (1976) and Voss and Gerstein (2005), respectively, and also contain the backbone atoms (if not stated otherwise). For reasons of clarity, only two extreme amino acid outliers are shown here. A complete overview of hydrogenated protein amino acid SLDs is provided by Jacrot (1976).

literature Jacrot (1976). Figure 1B shows the ones for the most commonly occurring atoms in proteins and RNA molecules, compared to the respective ones for X-rays. Knowing the solvent-excluded volumes of protein residues (Jacrot, 1976) or RNA nucleotides (Voss & Gerstein, 2005) therefore allows calculating their contrast. Using appropriate software tools, it is possible to calculate the contrast of protein–RNA complexes directly from their primary sequences (Sarachan, Curtis, & Krueger, 2014).

3. PRACTICAL ASPECTS OF SANS AND REQUIREMENTS ON THE SAMPLE STATE

In contrast to SAXS, SANS has certain advantages, including: (1) more freedom in the choice of the buffer, e.g., solutions at high salt concentrations will generally not diminish the signal/noise significantly, (2) the contribution of a hydration shell of different density as the bulk solvent is less pronounced in SANS, in particular when working in H_2O (Kim & Gabel, 2015; Svergun et al., 1998), and (3) no radiation damage of biological samples is usually observed, e.g., bacterial cell growth rates remain unaffected after neutron exposure (Jasnin, Moulin, Haertlein, Zaccai, & Tehei, 2008). The most powerful feature of SANS with respect to SAXS, however, is the possibility to use contrast variation to focus on structural features of specific subunits in a macromolecular complex formed by several partners (see Sections 5, 6 and 7).

In order to have sufficient signal/noise on present-day neutron sources and SANS instruments, sample concentrations of some mg/mL and about 100–200 µL volumes are required. At high concentrations, however, interparticle effects are often observed, in particular, for charged proteins or for RNA and DNA molecules (Pollack, 2011). Optimization of the buffer, e.g., increase of salt content, can often decrease this effect. Usually, concentration series, typically in the range from 1 to 10 mg/ml, should be carried out in order to assess the contribution of interparticle effects to the final SANS curve and to extract the form factor (i.e., the signal corresponding to the shape of an isolated particle in solution) of individual macromolecular complexes (Putnam, Hammel, Hura, & Tainer, 2007).

A paramount issue is the monodispersity of the sample if sophisticated structural modeling of an individual particle in solution (such as a protein–RNA complex) is envisaged. It refers here mainly to the oligomeric state (stoichiometry) of the complex. It should be stressed that SANS is sensitive to all macromolecular particles present in solution that display contrast with respect to the bulk solvent (Eq. 1). Therefore, when protein–RNA complexes are studied, free (unbound) protein, and/or RNA partners will also contribute to the measured SANS signal, in proportion to their volume fraction. Moreover, in mixtures of complexes with different stoichiometries, the higher oligomeric states will scatter the strongest, a fact that is inverse to, e.g., the situation in NMR where the resonances of smaller molecules are strongest. Apart from native gel analysis and light scattering, analytical ultracentrifugation is a method of choice to characterize the monodispersity of a sample. A clear, single peak should be observed in the sedimentation velocity profile. More details on sample preparation, good experimental practice, and pitfalls can be found in recent reviews (Gabel, 2012b; Jacques & Trewhella, 2010).

4. MODEL-FREE PARAMETERS: MOLECULAR MASS AND RADIUS OF GYRATION

While a multitude of sophisticated modeling approaches have become available for biological macromolecules from SANS (and SAXS) data in solution (Putnam et al., 2007; Spinozzi, Ferrero, Ortore, De Maria Antolinos, & Mariani, 2014; Svergun et al., 2014), several important structural parameters can be extracted without *a priori* structural knowledge (such as, e.g., atomic models). The most important ones are the molecular mass and the radius of gyration of particles. A more detailed, model-free, yet quantitative, description of the distribution of contrast within a particle can be obtained by a so-called Stuhrmann plot.

The *molecular mass* can be inferred from the intensity scattered in the forward direction, $I(0)$. Indeed, for $Q=0$, Eq. (1) simplifies and yields:

$$I(Q=0) \propto N \left| \int_V \Delta\rho(\vec{r}) \mathrm{d}V \right|^2 \qquad (2)$$

i.e., the $I(0)$ intensity is proportional to the number N of particles in solution and to the square of the particle contrast $\Delta\rho$, integrated over the solvent-excluded particle volume. For protein–RNA complexes, care needs to be taken since $\Delta\rho$ varies between both partners due to their different chemical composition (see next section). The proportionality constant in Eq. (2) depends on several particle-independent parameters, including the neutron wavelength and geometric parameters of the instrument (Jacrot & Zaccai, 1981). Since both V and N can be expressed in terms of the molecular mass of a particle in solution, a measurement of $I(0)$ can be used to determine the molecular mass directly on an absolute scale by measuring a water (H_2O) sample in the same instrumental setup (Jacrot & Zaccai, 1981). Relative calibration of the molecular mass with respect to a reference sample such as BSA, a procedure commonly used in SAXS for isolated proteins (Mylonas & Svergun, 2006), is not straightforward in SANS for protein–RNA complexes since the relative contrast conditions between a composite particle (protein + RNA) and the reference need to be taken into account (see next section).

Finally and importantly, Eq. (2) implies that $I(0)$ is very sensitive on the oligomeric state of a complex and its interaction with ligand partners in solution since it is proportional to the square of the particle volume. Therefore, a relatively modest increase of 10% in the molecular mass of a complex by binding of a small partner will increase its $I(0)$ signal by more than 20% (assuming the same contrast for both partners), a value that can easily be monitored by SANS.

The *radius of gyration* R_G is a parameter, similar to the moments of inertia in classical mechanics, that describes the spatial distribution of contrast within a given particle of volume V in solution (Guinier, 1939; Guinier & Fournet, 1955):

$$R_G^2 = \frac{\int_V \Delta\rho(\vec{r}) r^2 dV}{\int_V \Delta\rho(\vec{r}) dV} \tag{3}$$

$\Delta\rho$ is the neutron contrast distribution within the particle and depends on the specific amino acid and nucleotide composition (see next section). r is the distance of a given volume element (e.g., an atom, an amino acid, or a nucleotide) from the center of gravity of the contrast $\Delta\rho$ of the particle, $\vec{O}$, which can be determined by the following expression:

$$\vec{O} = \frac{\int_V \Delta\rho(\vec{r}) \vec{r} dV}{\int_V \Delta\rho(\vec{r}) dV} \tag{4}$$

It should be noted that $\vec{O}$ does not necessarily coincide with the (geometrical) particle center (or lie within the particle volume) but can be located at its outside (similar to the center of mass of a hollow sphere).

As the square in Eq. (3) suggests, particles with an anisometric (e.g., elongated) shape have R_Gs that are strongly increased with respect to compact, isometric shapes (an extreme case are spheres which are the particles with minimal R_G for a given volume). Importantly, in contrast to SAXS (where $\Delta\rho$ describes the electronic density and is always positive for proteins and RNAs in aqueous solutions), in SANS $\Delta\rho$ can be positive or negative in $H_2O:D_2O$ mixtures (see next section). As a consequence, R_G^2 can be of either positive or negative sign, i.e., R_G is in general a complex number in SANS. In practice, R_G can be extracted using the so-called Guinier approximation (Guinier, 1939)

$$I(Q) \approx I(0)\exp\left(-\frac{1}{3}R_G^2 Q^2\right) \qquad (5)$$

and using a linear fit at very low angles in a plot of $\ln[I(Q)]$ versus Q^2. Likewise, $I(0)$ is usually extracted by extrapolating this Guinier fit to $Q=0$ (which is experimentally inaccessible since it is contaminated by parasitic scattering from the direct beam and usually hidden by a beam stop). When using the Guinier approximation, the criterion $R_G Q_{max} < 1.0\ldots1.3$ must be respected for globular particles (where Q_{max} is the upper limit of the fit range).

A more detailed (yet model-free) approach than the Guinier analysis is a so-called *Stuhrmann plot* which is a quantitative way to describe the spatial distribution of contrast in a particle composed of several partners of different SLDs based on variations of the R_G as a function of contrast with the solution (Ibel & Stuhrmann, 1974; Stuhrmann & Kirste, 1967). Stuhrmann plots are very useful to determine, in a rapid and model-free way, if an RNA (or a protein) is placed more at the center or at the outside of their complex and if their centers of mass (or, more accurately, their centers of SLD) are displaced with respect to each other.

5. NEUTRON SCATTERING LENGTH DENSITIES OF PROTEINS AND RNAs AND IMPLICATIONS ON MODEL BUILDING

The principle of contrast variation in SANS is based on the difference in the scattering lengths between hydrogen and deuterium (Fig. 1B). This allows distinguishing between deuterium-labeled macromolecules and their unlabeled counterparts, on the one hand, and the introduction of contrast by using solvent mixtures of H_2O and D_2O, on the other hand. Figure 2 displays calculated neutron SLDs of several RNA nucleotides and protein amino acid residues as a function of the percentage of D_2O in an H_2O:D_2O solvent (=water reference line). The difference between the protein and RNA SLDs and the water reference line represents the contrast $\Delta\rho$ measured by a SANS experiment. It is a measure on how strongly a particle is "seen" by neutrons with respect to the solvent. $\Delta\rho$ can be of positive and negative sign and depends on the chemical composition (in particular on the hydrogen density) and the solvent-excluded volume of a given macromolecule. Several important observations can be implied directly from Fig. 2:

(1) Hydrogenated RNAs and proteins have different natural contrast with respect to an aqueous solvent. Their (average) contrast match points (CMPs; i.e., D_2O percentage where $\overline{\Delta\rho} = 0$) are about 68% and 42% D_2O, respectively. This means that both can be made "invisible" (on average) at different $H_2O:D_2O$ ratios (intersection of bold lines of h-tRNA and h-protein with the water line in Fig. 2).

(2) The contrast at a given $H_2O:D_2O$ ratio (and in particular, the exact location of the CMP) depends on the exact nucleotide or amino acid sequence of a given RNA or protein. In the case of hydrogenated nucleotides, the variations between individual nucleotides are relatively moderate: e.g., the CMPs of h-Uracil and h-Guanine are 61% and 74% D_2O, respectively. In the case of amino acids, the spread is much larger; e.g., h-Leucine and h-Aspartic acid have CMPs of 22% and 65% D_2O, respectively.

(3) In general, deuteration homogenizes the respective SLDs of RNAs and proteins, both on average and for individual nucleotides and amino acids (bold d-tRNA and d-protein lines in Fig. 2). However, per-deuterated (i.e., fully deuterated) proteins and RNAs can no longer be contrast-matched, even at 100% D_2O.

(4) For hydrogenated RNAs and proteins, significant SLD differences are observed between the backbone moieties on the one hand and the bases and side chains on the other hand.

The above points are illustrated by a schematic graphical representation of the spatial distributions of contrast in two protein–RNA complexes at the bottom of Fig. 4. $\Delta\rho$ was calculated for several representative protein residues and RNA nucleotides using Eq. (1′) with neutron scattering lengths and solvent-excluded volumes from literatures Jacrot (1976) and Voss and Gerstein (2005), and are plotted as cross sections of the Sxl–Unr–*msl2* and the box C/D complexes (discussed in more detail below).

Observations (1)–(4) have several important implications on SANS experiments and modeling of protein–RNA complexes: first, d-RNA molecules (and to a lesser degree h-RNAs) can in general be considered relatively homogeneous in nonatomic *ab initio* modeling approaches (see next section) while h-proteins are very heterogeneous at a residue level (the heterogeneity for RNAs is mainly between the backbone and the bases). However, protein SLD heterogeneity can be reduced by (per-)deuteration. Second, contrast between a protein and RNA molecule in a complex can (and, if possible, should) be enhanced using (per-)deuteration which will yield a better contrast between the partners and improve the signal/noise level. In order to be matched at 100% D_2O (which yields optimal signal and lowest incoherent background (Gabel et al., 2002) from hydrogens in an aqueous buffer), proteins and RNAs require an average deuteration level of about 75% and 65%, respectively. Third, for modeling of protein–RNA complexes (discussed in more detail in the next two sections), more than one contrast condition should be used in order to improve the uniqueness of the modeling. For a two-partner system (one protein, one RNA), at least four solvent contrasts should be used: the two match points of the partners (i.e., about 42% and 68% D_2O for a nondeuterated protein–RNA complex), a solvent condition where both partners have high contrast (e.g., 0% D_2O), and an intermediate point where both partners have opposite contrast with respect to the solvent (i.e., about 55% D_2O for a nondeuterated protein–RNA complex). As the number of partners in the complex increases, more contrasts and labeling are required; e.g., a system consisting of an RNA and two protein partners would need deuteration of one of the protein partners in order to distinguish it from the other and at least five or six contrast conditions, chosen in a similar way as in the two-partner system (i.e., a combination of a high contrast point, the match points of the individual partners, and one or more intermediate points). In multipartner systems, using (and measuring) multiple labeling schemes (e.g., deuterated protein A, hydrogenated protein B plus the inverse combination) will further increase the quality of the modeling process.

Finally, it should be noted that contrast variation has also been applied by using X-rays (SAXS) (Feigin & Svergun, 1987; Ibel & Stuhrmann, 1974; Zipper, 1982). However, it requires high concentrations of small, electron-rich additives (e.g., salt, sugar, or glycerol) to match the electronic density of proteins, while the electronic densities of RNA (and DNA) are virtually impossible to match. Moreover, high salt concentrations may influence/destabilize delicate molecular interactions as those in protein–

RNA complexes and the more gentle way of using $H_2O:D_2O$ replacement in the solution in a SANS experiment is preferable. However, it should be noted that often higher oligomeric states are favored at high D_2O ratios and special care needs to be taken regarding the sample monodispersity (see above) in SANS experiments using this kind of solvent contrast variation.

6. POSSIBILITIES AND LIMITS OF *AB INITIO* MODELS (SHAPE ENVELOPES)

Several nonatomic modeling techniques have been developed to produce envelopes of macromolecules in solution from SAXS or SANS data (Chacón, Morán, Díaz, Pantos, & Andreu, 1998; Franke & Svergun, 2009; Svergun, 1999), similar to envelopes provided by electron microscopy (EM). However, programs that use a single phase (i.e., a single type of scattering unit or dummy atom) to represent protein–RNA complexes in solution will yield unreliable results in the form of a general, biased, envelope since they do not respect the heterogeneity in contrast between the protein and RNA partners (Fig. 2). Therefore, multiphase modeling, so-called *ab initio* approaches, must be used with at least two phases representing the protein and RNA moieties, respectively.

One program capable of this approach is MONSA (Svergun, 1999; Svergun & Nierhaus, 2000). We illustrate the power and limits of *ab initio* modeling by using MONSA to calculate a low-resolution envelope of a tRNA-synthetase complex using SANS data back-calculated from the known crystal structure (PDB entry 1GTR) (Rould, Perona, & Steitz, 1991). We assume that SANS data at 0% D_2O are available of the unlabeled complex as well as 0%, 42%, 70%, and 100% D_2O of the complex with per-deuterated (d) tRNA. In addition, we assume that a single SAXS dataset of the unlabeled complex has been recorded (at 0% D_2O). SAXS and SANS curves were back-calculated from the PDB atomic complex using the programs CRYSOL and CRYSON (Svergun et al., 1995, 1998) in default mode. Artificial noise was added to all SAXS and SANS datasets, based on experimentally observed noise levels as described elsewhere (Gabel, 2012a). All scattering length densities used in MONSA (Table 1) were calculated from the chemical compositions and solvent-excluded volumes according to the RNA and protein partners as described elsewhere (Jacrot, 1976; Voss & Gerstein, 2005), and as discussed in the previous section. The initial volume chosen was a sphere with a radius of 50 Å, containing 18,780 dummy beads of radius 1.7 Å.

Table 1 Contrast Parameters and Solvent-Excluded Volumes Used for the MONSA Runs (Fig. 3)

Compound	Neutron $\Delta\rho$ (0% D$_2$O) (10^{10}/cm^2)	Neutron $\Delta\rho$ (42% D$_2$O) (10^{10}/cm^2)	Neutron $\Delta\rho$ (70% D$_2$O) (10^{10}/cm^2)	Neutron $\Delta\rho$ (100% D$_2$O) (10^{10}/cm^2)	X-ray $\Delta\rho$ (0% D$_2$O) (e$^-$/Å^3)	Solvent-Excluded Volume (Imposed) (10^3Å^3)
h–Protein	2.42	0.05	−1.53	−3.22	0.09	79.5
h-tRNA	3.99	1.51	−0.14	−1.92	0.21	22.1
d-tRNA	6.63	4.21	2.50	0.72	0.21	22.1

Figure 3 illustrates how faithfully the known 3D crystal structure of the tRNA-synthetase complex can be reconstructed using two phases as a function of the available sets of SAXS/SANS data. Importantly, it should be noted that in each case, the fits from the final shape models against the available SAXS/SANS dataset(s) are very good. This implies that the specific dummy atom models in each case are equivalent to and cannot be distinguished from the original 3D atomic model based on the respective 1D SAXS/SANS dataset(s) available in each scenario. However, the respective information content in the *ab initio* models differs significantly in the individual cases. When only SAXS data (or only SANS data) of the unlabeled complex at 0% D$_2$O are available (Fig. 3B and C), the protein and the RNA are modeled as two elongated partners that are aligned in parallel along their long axes. Alternative configurations, e.g., globular shapes of the protein and tRNA, arranged in a dumbbell shape, can be excluded which is already a wealth of information considering the limited input (a single 1D SAXS/SANS curve).

When several SANS datasets are used, in combination with contrast-enhanced, deuterium-labeled tRNA (d-tRNA), the structural information becomes more detailed and accurate (Fig. 3D): the tRNA assumes the typical kinked, L-shape found in the crystal structure and its overall extension is limited toward the center of the protein density. The mass distribution of the protein itself is more asymmetric along its long axis, in agreement with its crystal structure. This improvement is due to the fact that MONSA is forced to fit several datasets simultaneously. In particular, the 70% D$_2$O dataset, where the h-protein and d-tRNA moieties are of opposite contrast (Fig. 2; Table 1) and close to the overall match point of the complex, yields strong restraints for their relative positions. However, the latter case illustrates also the limits of *ab initio* SANS, even in such a favorable case with

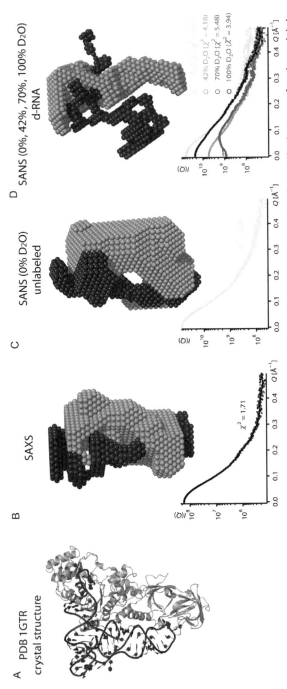

Figure 3 MONSA envelopes of the tRNA-synthetase complex using different SAXS/SANS input. The plots at the bottom of each model show the available SAXS/SANS curves and the back-calculated fits of the models against them, including the respective least χ^2 which is a measure of the deviation of the back calculated from the experiment curve (Svergun et al., 1995).

multiple datasets and the use of labeling schemes that reinforce the signal of the smaller tRNA: in particular, the distribution of mass within the tRNA shape is not very homogeneous and shows an excess toward the hairpin loop with respect to its crystal structure.

In conclusion, *ab initio* approaches can provide very useful information on the shape and relative topology of partners within a complex without *a priori* structural knowledge (e.g., crystal or NMR models of the individual partners). As illustrated in the tRNA-synthetase case in this section, while SANS (or SAXS) at a single contrast might already provide a general idea of the overall shape of the complex and the relative arrangement of both partners, it is in general very beneficial to measure a multitude of SANS data at different solution contrasts ($H_2O:D_2O$ ratios) and/or use deuteration of one (or several) partners in order to increase their signal/noise in the scattering curves and improve the accuracy of the modeling process. However, it is important to stress that *ab initio* approaches should not be used with SANS (or SAXS) data at too high Q values (>0.4–0.5 Å) since at larger angles, the internal heterogeneities of proteins and RNAs (Figs. 2 and 4, bottom) start to contribute to the SANS profiles. Particular care needs to be taken if the dimensions of the molecules studied are of the order of their internal heterogeneity (residues, nucleotides). In the latter case and at higher angles, using directly atomic models as building blocks should be privileged and models built from them directly scored against available SANS (and/or SAXS) data.

7. PRESENT STATE-OF-THE-ART EXAMPLES

While SANS remains a very popular and powerful technique on its own for structural studies of protein–RNA complexes and in particular for viruses (Comellas-Aragones et al., 2011; He et al., 2012), its joint application with other structural biology techniques, including crystallography, EM, and mass spectrometry is gaining momentum (Kim & Gabel, 2015). A particularly useful and powerful complementary technique is NMR. Indeed, both NMR and SANS sample conditions (volumes and concentrations) and states (deuteration labeling) are comparable and the information content is very complementary (Madl et al., 2011). We illustrate this powerful combination by two recent examples, the Sxl–Unr–RNA complex (Hennig et al., 2014) and the C/D box complex (Lapinaite et al., 2013). For a more complete overview of examples combining SANS (SAXS) and NMR on protein–RNA complexes, the reader is referred to recent reviews (Carlomagno, 2014; Hennig & Sattler, 2014; Madl et al., 2011).

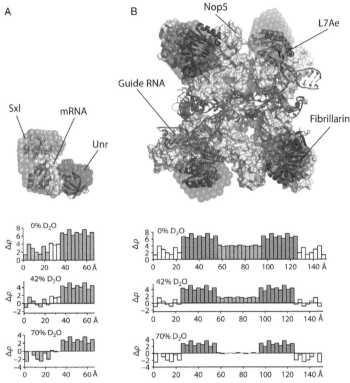

Figure 4 (A) Sxl–Unr–*msl2* complex. The two protein partners and the mRNA are shown in cartoon representation, superposed with their respective MONSA envelopes. The neutron contrasts $\Delta\rho$ of hydrogenated Sxl, hydrogenated mRNA, and deuterated Unr are shown schematically along an imaginary cross section of the complex for several $H_2O:D_2O$ ratios. (B) Box C/D apo-complex (in the absence of substrate rRNA). Fibrillarin, L7Ae, and the guide RNA are depicted as cartoons, Nop5 in space-filled mode. The MONSA envelopes of fibrillarin are superposed with their respective atomic-resolution copies. The neutron contrasts $\Delta\rho$ of hydrogenated Nop5, hydrogenated guide RNA, and deuterated fibrillarin are shown schematically along an imaginary cross section for several $H_2O:D_2O$ ratios. For simplicity, L7Ae is not shown and overlapping regions with contributions from several molecules are equally simplified. (See the color plate.)

In a first example, Hennig et al. determined the structure of a ternary 34 kDa Sxl–Unr–*msl2* ribonucleoprotein complex by a combination of crystallography, NMR, SAXS, and SANS (Hennig et al., 2014). This complex regulates the translation of the *msl2* mRNA transcript (which encodes for a key protein of the dosage compensation complex) in female *Drosophila melanogaster* fruit flies by inhibiting its binding with the small ribosomal subunit. In this work, several labeling schemes (deuterated Sxl and hydrogenated Unr and vice versa) were applied and SANS datasets were recorded of the respective reconstituted complexes at 42% and 70% D_2O. These combinations allowed focusing on the individual partners (Sxl, Unr, and *msl2* RNA) within the reconstituted complex *in situ* and extracting information on their relative positions, whereas the SAXS data provided an overall envelope of the complex. At a first time, the multiphase *ab initio* program MONSA (Petoukhov & Svergun, 2006; Svergun, 1999; Svergun & Nierhaus, 2000) was applied to obtain low-resolution shapes of the three partners within the complex from the SANS data, without using atomic-resolution models as an input. Surprisingly, the results suggested that the RNA is sandwiched between the two protein partners (Fig. 4A). While the low-resolution envelopes were in perfect agreement with all available SAXS and SANS datasets (in analogy to the procedure and data shown in Fig. 3), it should be noted that their major contribution to the structural refinement was to provide the respective positions of the individual partners within the complex. Certain degrees of freedom such as the directionality of the RNA within its envelope, the order of the two SXL domains, or the exact orientation of the three protein domains within their envelopes could not be resolved using the SAXS and SANS data alone. However, this unexpected topological arrangement allowed to guide the application of available NMR residual dipolar couplings and chemical shift perturbations by excluding models with direct protein–protein interactions. The atomic-resolution model was provided by crystallography and cross-checked against the NMR and SAXS/SANS data. In particular, the SAXS and the multiple SANS datasets allowed, through back calculation (Svergun et al., 1995, 1998), to confirm that the crystal structure was in excellent agreement with the data in solution and that crystal-packing artifacts could be excluded.

In a second example (Lapinaite et al., 2013), a combination of NMR, SAXS, and SANS was used to obtain the structure of a 390 kDa box C/D ribonucleoprotein enzyme from *Pyrococcus furiosus*. This complex, which consists of four copies of three protein partners (fibrillarin, L7Ae, and Nop5) and two copies of a structural guide RNA, methylates ribosomal RNA (rRNA) at specific positions via base pairing between the rRNA substrate and the guide RNA. In this structural refinement approach, a number of atomic-resolution subunits (previously determined by crystallography) consisting of single proteins (or domains) and RNA domains or small clusters of 2–3 subunits were assumed as rigid building blocks and, in some cases, connected by flexible linkers. When more than one subunit was present in a building block, the crystallographic interaction surfaces were validated by chemical shift perturbations measured by NMR in solution.

Subsequently, a large number of NMR restraints (paramagnetic relaxation enhancements and chemical shift perturbations) were applied to atomic-resolution building blocks in a classical ARIA-CNS approach (Brünger et al., 1998). However, given the size and the complexity of the assembly, NMR restraints on their own provided only limited information on the RNA conformation and its interaction with the protein partners as well as restraints of partners very distant from one another (up to 200 Å). Therefore, a multitude of reconstituted complexes with one of the protein or RNA partners perdeuterated (in some cases two) were prepared and measured at 0%, 42%, and 70% D_2O by SANS. These data were essential for several aspects of the refinement process: (1) determination of the overall molecular mass, stoichiometry, and oligomeric state; (2) providing an envelope of the guide RNA within the complex in order to restrict its conformational space for the NMR refinement; and (3) pinpointing the approximate positions of the (distant) protein partners as a starting conformations for the rigid body models in the NMR ARIA-CNS refinement process. Indeed, SANS data alone (without any atomic-resolution information) were able to clearly identify the positions and the stoichiometry of the protein partners as illustrated by the low-resolution envelopes (obtained by MONSA) of the four fibrillarin copies within the Box C/D apo-complex (in the absence of rRNA substrate) (Fig. 4B). The final ensemble of atomic structures determined by the ARIA-CNS protocol using NMR restraints only was scored against all SAXS and SANS curves through back calculation (Svergun et al., 1995, 1998). A small family of structures displaying the best fits was found to be grouped together with root mean square displacements of the order of about 5 Å.

Both examples illustrate the power of combining SANS with atomic-resolution techniques to gain insight into the internal topology of challenging and complex protein–RNA systems, even in the case of multiple protein or RNA partners. It can be anticipated that combinations of SANS (and SAXS), NMR, crystallography, and EM will be increasingly applied on such systems in the coming years. Given the complexity of the biomacromolecular assemblies and the multiple datasets involved, it is difficult to provide a single, unique strategy or protocol for such refinement approaches and the application to a given system will in general depend on the *a priori* structural information (high-resolution building blocks, contact information, etc.) and the number and detail of experimental data available. The ultimate goal would be to activate all restraints from SAXS/SANS, NMR, and other techniques simultaneously in an annealing process using full-atom building blocks. Recent progress of incorporating SAXS/SANS restraints into existing annealing protocols (Schwieters & Clore, 2014) is very promising, and the important issue of the uniqueness of models provided by such combined approaches is also being actively discussed (Kim & Gabel, 2015, and references therein).

8. CONCLUSIONS AND FUTURE PERSPECTIVES

SANS is perfectly complementary to its sister technique SAXS (see chapter 'Resolving Individual Components in Protein–RNA Complexes Using Small-Angle X-ray Scattering Experiments' by R.P. Rambo in this volume). It is particularly suited for complexes composed of several partners that can be distinguished in terms of their neutron scattering length densities, either by using natural or artificial (deuterium labeling) contrast. Since its early days in the 1970s, SANS has been providing crucial structural information on a multitude of biomacromolecular complexes, in particular on protein–RNA systems such as the tRNA-synthetase complex, the ribosome, and RNA viruses. With the progress in computational model-building software in the 1990s, the technique became available to a larger scientific community, including nonspecialists. Nowadays, SANS is experiencing a second renaissance by helping to solve the structure of large and challenging protein–RNA complexes in combination with other structural biology techniques such as crystallography and NMR.

What are future applications of SANS and what are its practical and theoretical limitations for structural biology of protein–RNA complexes? The tRNA-synthetase example (Fig. 3) illustrates the present-day limit of structural information (i.e., envelopes of protein or RNA partners in a complex) that can be obtained in a favorable case (several SANS contrast curves) by using only SANS data and no *a priori* structural information (e.g., high-resolution building blocks of the partners). The recent experimental examples in Fig. 4 demonstrate how such low-resolution envelopes compare to models based on actual atomic-resolution data (NMR or crystallography).

An interesting question is at what resolution level coarse-grain RNA models (Jonikas et al., 2008; Mustoe, Al-Hashimi, & Brooks, 2014) could be reliably and uniquely determined from SANS data, e.g., by introducing a granulometry on a base/backbone level and capitalizing on the neat difference in SLDs between them (Fig. 2). This approach would be equivalent to representing the tRNA shape in Fig. 3 by two distinct phases, one for the bases and one for the backbone. This approach, while tempting, has severe limits related to the overall, three-dimensional distribution of bases/backbone and the length scale probed by SANS: (1) maximum distances d between atoms in the sugar/phosphate backbone and the RNA bases are about 10 Å, corresponding to Q values of about 0.6 Å in reciprocal space (using the relationship $Q=2\pi/d$), a range where SANS coherent signal is low and incoherent background high. (2) In contrast to DNA, RNAs display more complex 3D folds yielding a more homogeneous SLD for any given cross section (Fig. 4, bottom) of the molecules. This usually precludes an analysis of the RNA in terms of, e.g., fiber models (Zipper, 1982).

The combination of SANS with multiple complementary structural biology techniques, including crystallography, NMR, and EM, will certainly gain momentum in the coming years for the evaluation of a number of important protein–RNA complexes involved in the regulation of, e.g., gene expression. In principle, any complex in the size range from about 10 kDa to several MDa is amenable to sophisticated structural modeling considering a number of restrictions: (1) most importantly, the complex must be monodisperse in solution (cf. section on sample quality), (2) if large-scale conformational motions are present (e.g., flexible protein domains), a single model will not be able to fit the SANS data and ensemble approaches (Bernado, Mylonas, Petoukhov, Blackledge, & Svergun, 2007; Bernado et al., 2005) should be considered, (3) the protein and RNA moieties should not be too different in size and shape if structural information on both is to be determined, e.g., a 10:1 molecular mass ratio is less favorable for structural data from the small partner than, e.g., a 2:1 ratio.

Innovative labeling schemes (in particular deuteration) that highlight specific moieties in protein–RNA complexes or that allow establishing pairwise distance restraints constitute very promising approaches. For instance, segmental protein labeling (i.e., deuteration), already practiced in NMR (Michel, Skrisovska, Wüthrich, & Allain, 2013) but not yet applied in SANS, would allow, in combination with contrast variation, to access domain-specific structural information of multidomain protein/RNA complexes. The same holds for RNA segmental labeling (Duss, Lukavsky, & Allain, 2012). Other possibilities include the addition of fusion proteins as additional scattering densities (Compton et al., 2014) in order to better localize the partner attached to them within the overall SANS envelope(s). The addition of smaller molecular moieties with strong contrast, as practiced for a long time in SAXS (Hoppe, 1972; Kratky & Worthmann, 1946) is technically more difficult in SANS (Seeger et al., 1997).

While SANS and SAXS at cryotemperatures allow a certain gain in signal/noise (Meisburger et al., 2013; Willumeit et al., 2001), they are very limiting regarding the manipulation of biological samples. Several advances in sample environment such as pressure cells (Winter, 2002), stopped-flow devices (Grillo, 2009), and high temperature baths (Appolaire et al., 2014) have so far mainly been applied to protein systems. Their increased application to protein–RNA systems should yield equally exciting results. While there are some limitations in terms of signal/noise in SANS regarding these devices and the general possibility to carry out high-throughput or time-resolved experiments as practiced in SAXS (Hura et al., 2009; Roh et al., 2010), new developments in SANS instrumentation might help to improve this situation over the coming years (Kynde et al., 2014).

In conclusion, a multitude of recent developments in software, instrumentation, methodology, and biochemistry have been contributing to a renaissance of biological applications of SANS over the last couple of years. It can be hoped that, as a larger biological community is becoming aware of the high potential of this technique for (large) protein–RNA complexes, it will be possible to address a number of exciting structural questions related to the function of these systems.

ACKNOWLEDGMENTS

I would like to thank Giuseppe Zaccai for helpful comments on the historical introduction. Financial support from the "Agence Nationale de la Recherche" (Grant "HYDROSAS" ANR-11-JSV5-003-01) is acknowledged.

REFERENCES

Appolaire, A., Girard, E., Colombo, M., Dura, M. A., Moulin, M., Haertlein, M., et al. (2014). Small-angle neutron scattering reveals the assembly mode and oligomeric architecture of TET, a large, dodecameric aminopeptidase. *Acta Crystallographica. Section D, Biological Crystallography, 70*, 2983–2993.

Baldwin, J. P., Boseley, P. G., Bradbury, E. M., & Ibel, K. (1975). The subunit structure of the eukaryotic chromosome. *Nature, 253*, 245–249.

Bernado, P., Blanchard, L., Timmins, P., Marion, D., Ruigrok, R. W., & Blackledge, M. (2005). A structural model for unfolded proteins from residual dipolar couplings and small-angle X-ray scattering. *Proceedings of the National Academy of Sciences of the United States of America, 102*(47), 17002–17007.

Bernado, P., Mylonas, E., Petoukhov, M. V., Blackledge, M., & Svergun, D. I. (2007). Structural characterization of flexible proteins using small-angle X-ray scattering. *Journal of the American Chemical Society, 129*(17), 5656–5664.

Bram, S., Butler-Browne, G., Baudy, P., & Ibel, K. (1975). Quaternary structure of chromatin. *Proceedings of the National Academy of Sciences of the United States of America, 72*(3), 1043–1045.

Brünger, A. T., Adams, P. D., Clore, G. M., DeLano, W. L., Gros, P., Grosse-Kunstleve, R. W., et al. (1998). Crystallography & NMR system: A new software suite for macromolecular structure determination. *Acta Crystallographica. Section D, Biological Crystallography, 54*(Pt. 5), 905–921.

Carlomagno, T. (2014). Present and future of NMR for RNA-protein complexes: A perspective of integrated structural biology. *Journal of Magnetic Resonance, 241*, 126–136.

Chacón, P., Morán, F., Díaz, J. F., Pantos, E., & Andreu, J. M. (1998). Low-resolution structures of proteins in solution retrieved from X-ray scattering with a genetic algorithm. *Biophysical Journal, 74*(6), 2760–2775.

Comellas-Aragones, M., Sikkema, F. D., Delaittre, G., Terry, A. E., King, S. M., Visser, D., et al. (2011). Solution scattering studies on a virus capsid protein as a building block for nanoscale assemblies. *Soft Matter, 7*(24), 11380–11391.

Compton, E. L., Page, K., Findlay, H. E., Haertlein, M., Moulin, M., Zachariae, U., et al. (2014). Conserved structure and domain organization among bacterial Slc26 transporters. *Biochemical Journal, 463*(2), 297–307.

Dessen, P., Blanquet, S., Zaccai, G., & Jacrot, B. (1978). Antico-operative binding of initiator transfer RNAMet to methionyl-transfer RNA synthetase from *Escherichia coli*: Neutron scattering studies. *Journal of Molecular Biology, 126*, 293–313.

Duss, O., Lukavsky, P. J., & Allain, F. H. (2012). Isotope labeling and segmental labeling of larger RNAs for NMR structural studies. *Advances in Experimental Medicine and Biology, 992*, 121–144.

Engelman, D. M., & Moore, P. B. (1972). A new method for the determination of biological quaternary structure by neutron scattering. *Proceedings of the National Academy of Sciences of the United States of America, 69*, 1997–1999.

Engelman, D. M., Moore, B. P., & Schoenborn, B. P. (1975). Neutron scattering measurements of separation and shape of proteins in 30S ribosomal subunit of *Escherichia coli*: S2-S5, S5-S8, S3-S7. *Proceedings of the National Academy of Sciences of the United States of America, 72*(10), 3888–3892.

Feigin, L. A., & Svergun, D. I. (1987). *Structure analysis by small-angle X-ray and neutron scattering*. New York: Plenum Press.

Franke, D., & Svergun, D. I. (2009). DAMMIF, a program for rapid ab-initio shape determination in small-angle scattering. *Journal of Applied Crystallography, 42*, 342–346.

Gabel, F. (2012a). A simple procedure to evaluate the efficiency of bio-macromolecular rigid-body refinement by small-angle scattering. *European Biophysics Journal, 41*(1), 1–11.

Gabel, F. (2012b). Small angle neutron scattering for the structural study of intrinsically disordered proteins in solution: A practical guide. In N. Uversky & A. K. Dunker (Eds.), *Intrinsically disordered protein analysis* (pp. 123–136). New York: Humana Press.

Gabel, F., Bicout, D., Lehnert, U., Tehei, M., Weik, M., & Zaccai, G. (2002). Protein dynamics studied by neutron scattering. *Quarterly Reviews of Biophysics, 35*(4), 327–367.

Giegé, R., Jacrot, B., Moras, D., Thierry, J.-C., & Zaccai, G. (1977). A neutron investigation of yeast valyl-tRNA synthetase interaction with tRNAs. *Nucleic Acids Research, 7*(7), 2421–2427.

Glatter, O., & Kratky, O. (1982). *Small angle X-ray scattering*. London: Academic Press.

Grillo, I. (2009). Applications of stopped-flow in SAXS and SANS. *Current Opinion in Colloid & Interface Science, 14*(6), 402–408.

Guinier, A. (1939). La diffraction des Rayons X aux Tres Faibles Angles: Applications a l'Etude des Phenomenes Ultra-microscopiques. *Annales de Physique (Paris), 12*, 161–236.

Guinier, A., & Fournet, G. (1955). *Small angle scattering of X-rays*. New York: John Wiley & Sons.

He, L., Piper, A., Meilleur, F., Hernandez, R., Heller, W. T., & Brown, D. T. (2012). Conformational changes in Sindbis virus induced by decreased pH are revealed by small-angle neutron scattering. *Journal of Virology, 86*(4), 1982–1987.

Hennig, J., Militti, C., Popowicz, G. M., Wang, I., Sonntag, M., Geerlof, A., et al. (2014). Structural basis for the assembly of the Sxl-Unr translation regulatory complex. *Nature, 515*, 287–290.

Hennig, J., & Sattler, M. (2014). The dynamic duo: Combining NMR and small angle scattering in structural biology. *Protein Science, 23*, 669–682.

Hoppe, W. (1972). A new X-ray method for the determination of the quaternary structure of protein complexes. *Israel Journal of Chemistry, 10*, 321–333.

Hoppe, W., May, R., Stoeckel, P., Lorenz, S., Erdmann, V. A., Wittmann, H. G., et al. (1975). Neutron scattering measurements with the label triangulation method on the 50S subunit of *E. coli* ribosomes. In *Paper presented at the neutron scattering for the analysis of biological structures, Brookhaven*.

Hura, G. L., Menon, A. L., Hammel, M., Rambo, R. P., Poole, F. L. N., Tsutakawa, S. E., et al. (2009). Robust, high-throughput solution structural analyses by small angle X-ray scattering (SAXS). *Nature Methods, 6*(8), 606–612.

Ibel, K., & Stuhrmann, H. (1974). Comparison of neutron and X-ray scattering of dilute myoglobin solutions. *Journal of Molecular Biology, 93*(2), 255–265.

Jacques, D. A., & Trewhella, J. (2010). Small-angle scattering for structural biology—Expanding the Frontier while avoiding the pitfalls. *Protein Science, 19*, 642–657.

Jacrot, B. (1976). The study of biological structures by neutron scattering from solution. *Reports on Progress in Physics, 39*(10), 911–953.

Jacrot, B., Chauvin, C., & Witz, J. (1977). Comparative neutron small-angle scattering study of small spherical RNA viruses. *Nature, 266*, 417–421.

Jacrot, B., & Zaccai, G. (1981). Determination of molecular weight by neutron scattering. *Biopolymers, 20*, 2413–2426.

Jasnin, M., Moulin, M., Haertlein, M., Zaccai, G., & Tehei, M. (2008). Down to atomic-scale intracellular water dynamics. *EMBO Reports, 9*, 543–547.

Jonikas, M. A., Radmer, R. J., Laederbach, A., Das, R., Pearlman, S., Herschlag, D., et al. (2008). Coarse-grained modeling of large RNA molecules with knowledge-based potentials and structural filters. *RNA, 15*, 189–199.

Kim, H. S., & Gabel, F. (2015). Uniqueness of quasi-atomic bio-macromolecular models derived from small angle scattering data. *Acta Crystallographica. Section D, Biological Crystallography, 71*, 57–66.

Kratky, O., & Worthmann, W. (1946). Ueber die Bestimmabrkeit der Konfiguration geloester organischer Molekule durch interferometrische Vermessung von Roentgenstrahlen. *Monatshefte fuer Chemie, 76*, 263–281.

Kynde, S., Hewitt Klenø, K., Nagy, G., Mortensen, K., Lefmann, K., Kohlbrecher, J., et al. (2014). A compact time-of-flight SANS instrument optimised for measurements of small sample volumes at the European Spallation Source. *Nuclear Instruments and Methods in Physics Research A, 764*, 133–141.

Lapinaite, A., Simon, B., Skjaerven, L., Rakwalska-Bange, M., Gabel, F., & Carlomagno, T. (2013). The structure of the box C/D enzyme reveals regulation of RNA methylation. *Nature, 502*(7472), 519–523.

Madl, T., Gabel, F., & Sattler, M. (2011). NMR and small-angle scattering-based structural analysis of protein complexes in solution. *Journal of Structural Biology, 173*(3), 472–482.

Meisburger, S. P., Warkentin, M., Chen, H., Hopkins, J. B., Gillilan, R. E., Pollack, L., et al. (2013). Breaking the radiation damage limit with Cryo-SAXS. *Biophysical Journal, 104*(1), 227–236.

Michel, E., Skrisovska, L., Wüthrich, K., & Allain, F. H. (2013). Amino acid-selective segmental isotope labeling of multidomain proteins for structural biology. *Chembiochem, 14*(4), 457–466.

Mustoe, A. M., Al-Hashimi, H. M., & Brooks, C. L., III. (2014). Coarse grained models reveal essential contributions of topological constraints to the conformational free energy of RNA bulges. *The Journal of Physical Chemistry B, 118*(10), 2615–2627.

Mylonas, E., & Svergun, D. I. (2006). Accuracy of molecular mass determination of proteins in solution by small-angle X-ray scattering. *Journal of Applied Crystallography, 40*(Suppl.), s245–s249.
Pardon, J. F., Worcester, D. L., Wooley, J. C., Tatchell, K., Van Holde, K. E., & Richards, B. M. (1975). Low-angle neutron scattering from chromatin subunit particles. *Nucleic Acids Research, 2*(11), 2163–2176.
Petoukhov, M. V., & Svergun, D. I. (2006). Joint use of small-angle X-ray and neutron scattering to study biological macromolecules in solution. *European Biophysics Journal, 35*(7), 567–576.
Petoukhov, M. V., & Svergun, D. I. (2007). Analysis of X-ray and neutron scattering from biomacromolecular solutions. *Current Opinion in Structural Biology, 17,* 562–571.
Pollack, L. (2011). SAXS studies of ion–nucleic acid interactions. *Annual Review of Biophysics, 40,* 225–242.
Putnam, C. D., Hammel, M., Hura, G. L., & Tainer, J. A. (2007). X-ray solution scattering (SAXS) combined with crystallography and computation: Defining accurate macromolecular structures, conformations and assemblies in solution. *Quarterly Reviews of Biophysics, 40*(3), 191–285.
Roh, J. H., Guo, L., Kilburn, J. D., Briber, R. M., Irving, T., & Woodson, S. A. (2010). Multistage collapse of a bacterial ribozyme observed by time-resolved small-angle X-ray scattering. *Journal of the American Chemical Society, 132*(29), 10148–10154.
Rould, M. A., Perona, J. J., & Steitz, T. A. (1991). Structural basis of anticodon loop recognition by glutaminyl-tRNA synthetase. *Nature, 352,* 213–218.
Sarachan, K. L., Curtis, J. E., & Krueger, S. (2014). Small-angle scattering contrast calculator for protein and nucleic acid complexes in solution. *Journal of Applied Crystallography, 46,* 1889–1893.
Schwieters, C. D., & Clore, G. M. (2014). Using small angle solution scattering data in Xplor-NIH structure calculations. *Progress in Nuclear Magnetic Resonance Spectroscopy, 80,* 1–11.
Seeger, P. A., Rokop, S. E., Palmer, P. D., Henderson, S. J., Hobart, D. E., & Trewhella, J. (1997). Neutron resonance scattering shows specific binding of plutonium to the calcium-binding sites of the protein calmodulin and yields precise distance information. *Journal of the American Chemical Society, 119,* 5118–5125.
Spinozzi, F., Ferrero, C., Ortore, M. G., De Maria Antolinos, A., & Mariani, P. (2014). GENFIT: Software for the analysis of small-angle X-ray and neutron scattering data of macro-molecules in solution. *Journal of Applied Crystallography, 47*(Pt. 3), 1132–1139.
Stuhrmann, H. B., Haas, J., Ibel, K., Wolf, B., Koch, M. H., Parfait, R., et al. (1976). New low resolution model for 50S subunit of *Escherichia coli* ribosomes. *Proceedings of the National Academy of Sciences of the United States of America, 73*(7), 2379–2383.
Stuhrmann, H. B., & Kirste, R. G. (1965). Elimination der intrapartikulaeren Untergrundstreuung bei der Roentgenkleinwinkelstreuung an kompakten Teilen (Proteinen). *Zeitschrift fuer Physikalische Chemie, 46*(3–4), 247–250.
Stuhrmann, H. B., & Kirste, R. G. (1967). Elimination der intrapartikulaeren Untergrundstreuung bei der Roentgenkleinwinkelstreuung an kompakten Teilen. II. *Zeitschrift fuer Physikalische Chemie, 56,* 334–337.
Svergun, D. I. (1999). Restoring low resolution structure of biological macromolecules from solution scattering using simulated annealing. *Biophysical Journal, 76,* 2879–2886.
Svergun, D. I., Barberato, C., & Koch, M. H. J. (1995). CRYSOL—A program to evaluate X-ray solution scattering of biological macromolecules from atomic coordinates. *Journal of Applied Crystallography, 28,* 768–773.
Svergun, D. I., Koch, M. H. J., Timmins, P. A., & May, R. P. (2014). *Small angle X-ray and neutron scattering from solutions of biological molecules,* Vol. 19. Oxford: Oxford University Press.

Svergun, D. I., & Nierhaus, K. H. (2000). A map of protein–rRNA distribution in the 70S *Escherichia coli* ribosome. *The Journal of Biological Chemistry, 275*(19), 14432–14439.

Svergun, D. I., Richard, S., Koch, M. H. J., Sayers, Z., Kuprin, S., & Zaccai, G. (1998). Protein hydration in solution: Experimental observation by X-ray and neutron scattering. *Proceedings of the National Academy of Sciences of the United States of America, 95*(5), 2267–2272.

Voss, N. R., & Gerstein, M. (2005). Calculation of standard atomic volumes for RNA and comparison with proteins: RNA is packed more tightly. *Journal of Molecular Biology, 346*, 477–492.

Willumeit, R., Diedrich, G., Forthmann, S., Beckmann, J., May, R. P., Stuhrmann, H. B., et al. (2001). Mapping proteins of the 50S subunit from *Escherichia coli* ribosomes. *Biochimica et Biophysica Acta, 1520*(1), 7–20.

Winter, R. (2002). Synchrotron X-ray and neutron small-angle scattering of lyotropic lipid mesophases, model biomembranes and proteins in solution at high pressure. *Biochimica et Biophysica Acta, 1595*(1–2), 160–184.

Zaccai, G., & Jacrot, B. (1983). Small angle neutron scattering. *Annual Review of Biophysics and Biomolecular Structure, 12*, 139–157.

Zipper, P. (1982). Nucleic acids and nucleoproteins. In O. Glatter & O. Kratky (Eds.), *Small angle X-ray scattering* (pp. 295–328). London: Academic Press.

CHAPTER FOURTEEN

Studying RNA–Protein Interactions of Pre-mRNA Complexes by Mass Spectrometry

Saadia Qamar[*,†], Katharina Kramer[*,†,2], Henning Urlaub[*,†,1]

[*]Bioanalytical Mass Spectrometry Group, Department of Cellular Biochemistry, Max Planck Institute for Biophysical Chemistry, Göttingen, Germany
[†]Bioanalytics Research Group, Institute for Clinical Chemistry, University Medical Center, Göttingen, Germany
[1]Corresponding author: e-mail addresses: henning.urlaub@mpibpc.mpg.de; henning.urlaub@med.uni-goettingen.de

Contents

1. Background	419
2. Equipment	421
3. Materials	423
3.1 Solutions and Buffers	424
4. Protocol	433
4.1 Preparation	433
4.2 Duration	434
5. Step 1—*In Vitro* Transcription	434
5.1 Overview	434
5.2 Duration	434
5.3 Tip	438
5.4 Tip	438
5.5 Tip	438
5.6 Tip	438
6. Step 2—*HeLa* Nuclear Extract Preparation	438
6.1 Overview	438
6.2 Duration	439
6.3 Tip	441
6.4 Tip	441
6.5 Tip	441
6.6 Tip	441
7. Step 3—RNA–Protein Complex Assembly and Purification	441
7.1 Overview	441
7.2 Duration	441

---

[2] Present address: Plant Proteomics Group, Max Planck Institute for Plant Breeding Research, Köln, Germany

	7.3	Tip	442
	7.4	Tip	443
	7.5	Tip	443
	7.6	Tip	443
	7.7	Tip	443
8.	Step 4—UV Cross-Linking		443
	8.1	Overview	443
	8.2	Duration	443
	8.3	Tip	443
	8.4	Tip	444
	8.5	Tip	444
9.	Step 5—Dissociation and Hydrolysis of Proteins		445
	9.1	Overview	445
	9.2	Duration	445
	9.3	Tip	446
10.	Step 6—Isolation of Cross-Links		446
	10.1	Overview	446
	10.2	Duration	446
	10.3	Tip	447
	10.4	Tip	447
11.	Step 7—Hydrolysis of RNA and Proteins		450
	11.1	Overview	450
	11.2	Duration	450
	11.3	Tip	451
12.	Step 8—Desalting and Removal of RNA Fragments		451
	12.1	Overview	451
	12.2	Duration	451
	12.3	Tip	453
	12.4	Tip	453
	12.5	Tip	453
13.	Step 9—Enrichment of RNA–Protein Cross-Links		453
	13.1	Overview	453
	13.2	Duration	453
	13.3	Tip	455
14.	Step 10—Analysis by LC-Coupled ESI MS		455
	14.1	Overview	455
	14.2	Duration	455
	14.3	Tip	457
15.	Step 11—MS Data Analysis		457
	15.1	Overview	457
	15.2	Duration	459
16.	Pros and Cons of the Presented Protocol		459
Acknowledgments		460
References		460
Source Reference		463

Abstract

RNA–protein interactions play a crucial role in gene expression. These interactions take place in so-called ribonucleoprotein (RNP) complexes. To investigate which proteins interact with RNA in these complexes, and how they do so, UV-light-induced cross-linking has proven to be a valuable, yet straightforward technique. UV irradiation induces a covalent bond between the RNA and the proteins, whereafter cross-linked proteins can be identified by mass spectrometric (MS) approaches. Moreover, the cross-linked region of the protein, and often the actual cross-linked amino acid, can be identified by state-of-the-art MS, as can the cross-linked RNA moiety. This protocol describes in detail how to isolate peptide–RNA oligonucleotide cross-links from UV-irradiated human pre-mRNA RNPs and to perform the subsequent MS investigation of these peptide–RNA conjugates in combination with a dedicated computational analysis, in order to obtain sequence information about the cross-linked peptide and oligoribonucleotide. The described workflow can be applied to any RNP, irrespective of its origin, e.g., RNPs assembled *in vitro* (as described here) or RNPs isolated from UV-irradiated cells, either *ex vivo* or *in vivo*.

1. BACKGROUND

Together, nucleic acids and proteins play a vital role in the major biological processes, such as transcription (Castro-Roa & Zenkin, 2011), DNA replication, recombination and repair (Antony et al., 2013), and RNA processing, e.g., mRNA splicing (Blencowe, 2006), localization (Lecuyer, Yoshida, & Krause, 2009), degradation (Garneau, Wilusz, & Wilusz, 2007; Tadros et al., 2007; Walser & Lipshitz, 2011), and translation (Wilhelm & Smibert, 2005). About 23% of the human genes so far functionally annotated correspond to nucleic acid-binding proteins (Ule, Jensen, Mele, & Darnell, 2005) and more than 800 RNA-binding proteins (RBPs) have been reported to be encoded by the mammalian genome (Castello et al., 2012). Messenger RNA (mRNA) interacts with more than 30 RBPs during its life cycle (Hogan, Riordan, Gerber, Herschlag, & Brown, 2008). Moreover, these RBPs may possess one or several RNA-binding domains (RBDs) including RNA recognition motif (RRM), K-homology (KH) domain, Arg–Gly–Gly (RGG) box, Sm domain, DEAD/DEAH box, zinc finger, and PAZ and PUF domains (Glisovic, Bachorik, Yong, & Dreyfuss, 2008).

Computational approaches are used to predict RBPs on the basis of their conserved RNA-binding motifs (RBMs) (Chen, Sargsyan, Wright, Huang, & Lim, 2014). Conversely, RBDs can also be predicted by computational analysis when the protein that binds to RNA has been identified (Castello et al., 2013; Landthaler et al., 2008). Experimentally, RBMs are identified by structural approaches, such as X-ray crystallography, nuclear magnetic resonance (NMR), mutational analysis combined with an RNA-binding assay, and protein–RNA cross-linking approaches combined with mapping of the cross-linking site within the protein (Gleghorn & Maquat, 2014; Henkin, 2014; Kramer et al., 2014; Wang et al., 2014). To identify the corresponding RNAs that interact with RBDs, several techniques have been established, including immunoprecipitation of RBP along with interacting RNA, followed by microarray analysis or "next-generation" sequencing (Tenenbaum, Carson, Lager, & Keene, 2000; Zhao et al., 2010), cross-linking immunoprecipitation (CLIP) (Ule et al., 2003), photoactivatable-ribonucleoside-enhanced cross-linking and immunoprecipitation (PAR-CLIP) (Hafner et al., 2010). Further techniques include electrophoretic mobility-shift assay (EMSA) (Hellman & Fried, 2007) and yeast three-hybrid system (SenGupta et al., 1996). However, all these methods have some limitations.

UV-induced cross-linking followed by mass spectrometry (MS) has proven to be a promising technique for the global identification of proteins that interact with RNA (Castello et al., 2013; Hoell et al., 2014; Landthaler et al., 2008) and to determine the exact site(s) of cross-linking, such that either experimental evidence for computationally predicted RBMs is obtained or proteins with hitherto unknown RBMs are identified (Kramer et al., 2014). On the molecular level, UV-induced cross-linking at, e.g., 254 *nm* generates a covalent bond between the amino acid's side chain and the nucleotide base of the RNA, when these are in sufficiently close proximity (König, Zarnack, Luscombe, & Ule, 2012). In this manner, cross-linked proteins resemble proteins that carry a modification, and can, consequently, be investigated by state-of-the-art MS upon digestion of the protein and RNA moieties and subsequent enrichment of the cross-linked peptide–RNA oligonucleotides. Since the cross-linked RNA moiety can have different compositions (ranging from one nucleotide to four or more, depending on the nuclease used for digestion of the intact RNA), and since nucleotides may be detected as derivatives (following loss of water and/or phosphate), this poses a major challenge for the algorithms used for the identification of the cross-linked peptide moiety with its cross-linked oligonucleotides by database searching. We have recently described a computational approach that allows the identification of cross-linked peptides with their cross-linked RNA oligonucleotides derived from complex RNP mixtures, or even from UV-cross-linked yeast cells, by searching the experimental MS data against entire databases, such as the UniprotKB/SwissProt protein sequence database (Kramer et al., 2014). In this way, we have found that with the exceptions of

Asn, Glu, and Gln all amino acids can form UV-induced cross-links to RNA (Kramer et al., 2014; R. Hofele, K. Sharma, S. Qamar, U. Zaman, K. Kramer, unpublished results) and have identified a variety of protein regions that had not previously been described as interacting with RNA.

In the following protocol, we describe in detail the entire approach that lead to the MS-based sequencing of cross-linked peptide–RNA oligonucleotides on purified and subsequently UV-irradiated human pre-mRNA RNPs, which have been obtained by reconstitution of PM5 pre-mRNA transcribed *in vitro* and incubated with *HeLa* nuclear extract. The PM5 pre-mRNA had been used previously for the isolation of splicing complexes (Bessonov et al., 2010). It contains a 60-nucleotide long polypyrimidine tract without 3' exon and terminal AG dinucleotide. As a result, it can undergo 5' splice site cleavage and lariat formation without exon ligation under splicing conditions (Bessonov et al., 2010). In this study, the PM5 pre-mRNA is used to assemble the H complex (Michaud & Reed, 1991), primarily consisting of various hnRNP proteins like hnRNP A, B, C, and L, etc., that under appropriate conditions can be chased to spliceosomal A, B, and C complexes (Wahl, Will, & Lührmann, 2009). Briefly, the protocol comprises: (i) the reconstitution and isolation of pre-mRNA RNPs, (ii) UV-induced cross-linking, (iii) the hydrolysis of proteins and RNA, (iv) the isolation and enrichment of cross-linked peptide–RNA oligonucleotides, (v) MS analysis of cross-linked species, and (vi) database search to identify the cross-linked peptide with its cross-linked nucleotides.

2. EQUIPMENT

1 ml Combitip (Eppendorf)
0.5 ml Microcentrifuge tubes (Eppendorf)
1.5 ml Microcentrifuge tubes (Eppendorf)
2 ml Microcentrifuge tubes (Eppendorf)
1 ml Syringe (Braun)
20 ml Syringe (Braun)
50 ml Syringe (Braun)
NuPAGE® gel electrophoresis system (Invitrogen)
12% Bis–Tris precast gels (1.0 mm × 10 well; Invitrogen)
15 ml Falcon tubes (Sarstedt)
50 ml Falcon tubes (Sarstedt)

Bottle-top filters (0.20 µm and 500 ml; Sarstedt)
Coffee filter paper (Melitta)
Corex glass tubes (Kendro Laboratory Products)
Dialysis tubes (MWCO 6-8000; Spectrapor)
Filtropur S (0.20 µm; Sarstedt)
Glass dishes with a planar surface and an inner diameter of 3.5 cm (constructed in-house)
Glass Dounce homogenizer (100 ml; Kontes)
Glass pipettes (5, 10, and 25 ml; Blaubrand® Eterna)
Gravity flow disposable chromatography columns (Bio-Rad)
LC sample vials (Waters)
LTQ Orbitrap Velos mass spectrometer (Thermo) coupled with nano-LC system (Agilent 1100 series; Agilent)
Microcentrifuge (Eppendorf 5415R)
Heraeus Megafuge/General purpose centrifuge (1.0R; Thermo)
Heraeus Megafuge swing bucket rotor 2704
Microfuge tube shaker (IKA-VIBRAX-VXR)
Micropipette tips (1000, 200, and 10 µl; Starlab)
Micropipettes (Gilson)
NanoDrop ND-1000 spectrophotometer (Thermo Scientific)
Parafilm (Bemis)
pH meter (Mettler Toledo)
Pipette aid (Hirschmann Laborgerate)
Polyacrylamide gel electrophoresis equipment (constructed in-house)
S75 HR size-exclusion (SE) column (3.1 × 300 mm; GE Healthcare)
SMART System for size-exclusion chromatography (SEC) (GE Healthcare)
Sonicator (BANDELIN Electronic)
Sorvall Evolution RC centrifuge
Sorvall Hb-6 rotor
Sorvall SS-34 rotor
SpeedVac (Thermo Scientific)
Syringe filters (0.20 µm; Sarstedt)
Magnetic stirrer (IKA)
Stopwatch (Roth)
Thermomixer (Eppendorf)
UV cross-linking equipment with four 8-W germicidal lamps (G8T5, Herolab, Germany) mounted in parallel (constructed in-house)

Stratalinker UV cross-linker 2400 (Stratagene)
UV/vis spectrophotometer (Amersham Pharmacia Biotech)
Vacuum filtration apparatus (PIAB; Sartorius)
Vortex mixer (Scientific Industries, Inc.)
Weighing balance (Sartorius)

3. MATERIALS

10× NuPAGE MOPS SDS running buffer (Invitrogen)
10× NuPAGE Sample reducing agent (Invitrogen)
4× NuPAGE LDS sample buffer (Invitrogen)
NuPAGE antioxidant (Invitrogen)
152 mM m^{7}GpppG cap (Kedar)
2,5-Dihydroxybenzoic acid (DHB) (Sigma-Aldrich)
4-(2-Hydroxyethyl) piperazine-1-ethanesulfonic acid (HEPES) (Merck)
5× Transcription buffer (Promega)
Acetonitrile (LiChrosolv; Merck)
Ammonium hydroxide solution (28–30% (v/v); Merck)
Ammonium persulfate (APS) (Merck)
Amylose resin (New England Biolabs)
Autoclaved deionized water
Benzonase (25 U; Novagen)
Boric acid (Merck)
Bovine serum albumin (BSA) (Sigma-Aldrich)
Bradford solution, Bio-Rad Protein Assay (Bio-Rad)
Bromophenol blue (Sigma-Aldrich)
C18 reversed-phase chromatography material (C18 AQ 120 Å 5 μm; Dr. Maisch GmbH)
Chloroform (Merck)
Dithiothreitol (DTT) (Roth)
EDTA disodium salt dihydrate (Roth)
EDTA-free protease inhibitor cocktail (Roche)
Ethanol (Merck)
Formamide (Merck)
Formic acid (Fluka)
Glycerol (85% (w/v); Merck)
GlycoBlue (15 mg/ml; Ambion)

Magnesium acetate (Mg(OAc)$_2$·4H$_2$O) (Merck)
Magnesium chloride (MgCl$_2$·4H$_2$O) (Merck)
Maltose (Merck)
Methanol (LiChrosolv; Merck)
MS2-MBP fusion protein (MPI-BPC)
Phenol/chloroform/isoamyl alcohol (PCI) (Roth)
Phenylmethylsulfonyl fluoride (PMSF) (Roche)
Potassium acetate (KOAc) (Merck)
Potassium chloride (KCl) (Merck)
Potassium hydrogen phosphate (K$_2$HPO$_4$) (Merck)
RNase A (1 µg/µl; Ambion, Applied Biosystems)
RNase T1 (1 U/µl; Ambion, Applied Biosystems)
RNasin (40 U/µl; Promega)
rNTPs (0.1 M ATP, 0.1 M CTP, 0.1 M GTP, 0.1 M UTP; Promega)
Rotiphorese gel 40 (38% acrylamide, 2% bis-acrylamide; Roth)
RQ1 DNase (Promega)
Sodium acetate (NaOAc·4H$_2$O) (Merck)
Sodium chloride (NaCl) (Merck)
Sodium dodecyl sulfate (SDS) (Merck)
SP6 RNA polymerase (Promega)
Tetramethylethylenediamine (TEMED) (Sigma-Aldrich)
TiO$_2$ titansphere 5 µm (GL Sciences)
Trifluoroacetic acid (TFA) (Roth)
Tris base (Roth)
Trypsin sequencing grade (Promega)
Urea (Merck)
Water (LiChrosolv; Merck)
Xylene cyanol (Fluka)

3.1 Solutions and Buffers

3.1.1 Step 1: In Vitro Transcription

1 M MgCl$_2$:

9.9 g of MgCl$_2$·4H$_2$O in 50 ml of autoclaved deionized water

1 M DTT:

1.54 g of DTT in 10 ml of autoclaved deionized water

10 mg/ml BSA:

10 mg of BSA in 1 ml of autoclaved deionized water

10% (w/v) APS:

0.1 g of APS in 1 ml of autoclaved deionized water

Always prepare fresh.

3 M NaOAc:

122.47 g of NaOAc·4H$_2$O in 300 ml of autoclaved deionized water (adjust to pH 5.3 with acetic acid)

Filter-sterilize the solution by vacuum filtration apparatus using bottle-top filter.

0.5 M EDTA (pH 8):

9.306 g of Na$_2$EDTA·2H$_2$O in 50 ml of autoclaved deionized water (adjust to pH-8 with NaOH)

1 M Tris–Cl (pH 7.5):

2.43 g of Tris base in 20 ml of autoclaved deionized water (adjust to pH 7.5 with HCl)

10 × TBE:

Component	Amount
Tris base	61.83 g
Boric acid	121.14 g
EDTA	7.44 g
Autoclaved deionized water up to 1000 ml	

1 × TBE:

Component	Stock	Final concentration	Amount
TBE	10×	1×	100 ml
Autoclaved deionized water up to 1000 ml			

5% (w/v) Polyacrylamide gel solution:

Component	Stock	Final concentration	Amount
Rotiphorese gel 40	N/A	N/A	12.5 ml
Urea	N/A	8 M	42 g
TBE	10×	1×	10 ml
Autoclaved deionized water up to 100 ml			
TEMED	N/A	N/A	100 µl
APS	10% (w/v)	1% (w/v)	1000 µl

Filter-sterilize the gel solution before adding TEMED and APS by vacuum filtration apparatus using bottle-top filter.

Always add TEMED and freshly prepared APS just before casting the gel.

RNA extraction buffer:

Component	Stock	Final concentration	Amount
Tris–Cl (pH 7.5)	1 M	20 mM	2 ml
NaCl	N/A	150 mM	0.88 g
SDS	N/A	0.5% (w/v)	0.5 g
EDTA (pH 8)	0.5 M	0.2 mM	40 µl
Autoclaved deionized water up to 100 ml			

Filter-sterilize the buffer before adding SDS by vacuum filtration apparatus using bottle-top filter.

RNA sample-loading buffer:

Component	Stock	Final concentration	Amount
Formamide	N/A	80% (w/v)	8 g
EDTA	N/A	1 mM	0.00373 g
Bromophenol blue	N/A	0.05% (w/v)	0.005 g
Xylene cyanol	N/A	0.05% (w/v)	0.005 g
Autoclaved deionized water up to 10 ml			

80% (v/v) Ethanol:

Component	Stock	Final concentration	Amount
Ethanol	100% (v/v)	80% (v/v)	80 ml
Autoclaved deionized water up to 100 ml			

Store at −20 °C.

3.1.2 Step 2: HeLa Nuclear Extract Preparation

1 M KPO$_4$:

182.584 g of K$_2$HPO$_4$·3H$_2$O in 1000 ml of autoclaved deionized water (adjust to pH 7.4 with 1 M KH$_2$PO$_4$)

1 M HEPES (pH 7.6):

11.92 g of HEPES in 50 ml of autoclaved deionized water (adjust to pH 7.6 with KOH)

1 M KOAc:

4.9 g of KOAc in 50 ml of autoclaved deionized water

1 M Mg(OAc)$_2$:

2.15 g of Mg(OAc)$_2$·4H$_2$O in 10 ml of autoclaved deionized water

0.25 M DTT:

1.92 g of DTT in 50 ml of autoclaved deionized water

Protease inhibitor:

Two tablets of EDTA-free protease inhibitor cocktail in 1 ml of MC buffer

Always prepare fresh before use.

3 M KCl:

44.7 g of KCl in 200 ml of autoclaved deionized water

0.1 M PMSF:

1.74 g of PMSF in 100% (v/v) ethanol to 100 ml

Always prepare fresh PMSF just before use.

$10 \times$ PBS:

Component	Stock	Final concentration	Amount
NaCl	N/A	1300 mM	76 g
KPO$_4$ (pH 7.4)	1 M	200 mM	200 ml
Autoclaved deionized water up to 1000 ml			

$1 \times$ PBS:

Component	Stock	Final concentration	Amount
PBS	$10 \times$	$1 \times$	100 ml
Autoclaved deionized water up to 1000 ml			

Filter-sterilize the buffer by vacuum filtration apparatus using bottle-top filter.

$1 \times$ MC buffer:

Component	Stock	Final concentration	Amount
KOAc	1 M	10 mM	5 ml
HEPES-KOH (pH 7.6)	1 M	10 mM	5 ml
Mg(OAc)$_2$	1 M	0.5 mM	0.25 ml
Autoclaved deionized water to 500 ml			
DTT	0.25 M	0.5 mM	1 ml
Protease inhibitor	N/A	N/A	5 ml

Add DTT and protease inhibitor just before using MC buffer.

Roeder C buffer:

Component	Stock	Final concentration	Amount
Glycerol	85% (w/v)	25% (w/v)	294 ml
HEPES-KOH (pH 7.9)	1 M	20 mM	20 ml
NaCl	N/A	420 mM	24.5 g
MgCl$_2$	1 M	1.5 mM	1.5 ml
EDTA (pH 8)	0.5 M	0.2 mM	0.4 ml

Autoclaved deionized water to 1000 ml

DTT	0.25 M	0.5 mM	2 ml
PMSF	0.1 M	0.5 mM	5 ml

Filter-sterilize the buffer before adding glycerol by vacuum filtration apparatus using bottle-top filter.

Always add DTT and PMSF just before using Roeder C buffer.

Roeder D buffer:

Component	Stock	Final concentration	Amount
Glycerol	85% (w/v)	10% (w/v)	235 ml
HEPES-KOH (pH 7.9)	1 M	20 mM	40 ml
KCl	3 M	100 mM	67 ml
MgCl$_2$	1 M	1.5 mM	3 ml
EDTA (pH 8)	0.5 M	0.2 mM	0.8 ml

Autoclaved deionized water to 2000 ml

DTT	0.25 M	0.5 mM	4 ml
PMSF	0.1 M	0.5 mM	10 ml

Filter-sterilize the buffer before adding glycerol by vacuum filtration apparatus using bottle-top filter.

Always add DTT and PMSF just before using Roeder D buffer.

3.1.3 Step 3: RNA-Protein Complex Assembly and Purification

5 M NaCl:

58.44 g of NaCl in 200 ml of autoclaved deionized water

1 M HEPES (pH 7.9):

119.15 g of HEPES in 500 ml of autoclaved deionized water (adjust to pH 7.9 with KOH)

Gradient buffer without glycerol:

Component	Stock	Final concentration	Amount
NaCl	5 M	150 mM	15 ml
HEPES-KOH (pH 7.9)	1 M	20 mM	10 ml

| MgCl$_2$ | 1 M | 1.5 mM | 750 µl |

Autoclaved deionized water up to 500 ml

Filter-sterilize the buffer by vacuum filtration apparatus using bottle-top filter.

15 mM Maltose buffer:

0.11 g of maltose in 20 ml of gradient buffer without glycerol

3.1.4 Step 4: UV Cross-Linking

Buffers and solutions are not required for this step.

3.1.5 Step 5: Dissociation and Hydrolysis of Proteins

1% (w/v) SDS:

0.01 g of SDS in 1 ml of SE running buffer (pH 7.5)

3.1.6 Step 6: Isolation of Cross-Links

SE Running buffer:

Component	Final concentration	Amount
Tris base	20 mM	2.43 g
MgCl$_2$	1.5 mM	0.30 g
NaCl	150 mM	8.77 g

LiChrosolv water to 1000 ml

Adjust to pH 7.5 with HCl and filter-sterilize the buffer by vacuum filtration apparatus using bottle-top filter.

3.1.7 Step 7: Hydrolysis of RNA and Proteins

1 M Tris–Cl (pH 7.9):

2.43 g of Tris base in 20 ml of autoclaved deionized water (adjust to pH 7.9 with HCl)

50 mM Tris–Cl (pH 7.9):

Component	Stock	Final concentration	Amount
Tris–Cl (pH 7.9)	1 M	50 mM	75 µl

Autoclaved deionized water to 1.5 ml

4 M Urea:

0.24 g of Urea in 50 mM Tris–Cl (pH 7.9) to 1 ml

Filter-sterilize it by filtropur S syringe filter.

3.1.8 Step 8: Desalting and Removal of RNA Fragments
C18 reverse phase beads suspension:

Suspend C18 beads in 100% (v/v) MeOH by gently tapping the suspension in a microcentrifuge tube

95% (v/v) Acetonitrile, 0.1% (v/v) formic acid:

Component	Stock	Final concentration	Amount
Acetonitrile	100% (v/v)	90% (v/v)	47.5 ml
Formic acid	100% (v/v)	0.1% (v/v)	50 µl
LiChrosolv water to 50 ml			

80% (v/v) Acetonitrile, 0.1% (v/v) formic acid:

Component	Stock	Final concentration	Amount
Acetonitrile	100% (v/v)	80% (v/v)	40 ml
Formic acid	100% (v/v)	0.1% (v/v)	50 µl
LiChrosolv water to 50 ml			

50% (v/v) Acetonitrile, 0.1% (v/v) formic acid:

Component	Stock	Final concentration	Amount
Acetonitrile	100% (v/v)	50% (v/v)	25 ml
Formic acid	100% (v/v)	0.1% (v/v)	50 µl
LiChrosolv water to 50 ml			

20% (v/v) Acetonitrile, 0.1% (v/v) formic acid:

Component	Stock	Final concentration	Amount
Acetonitrile	100% (v/v)	20% (v/v)	10 ml
Formic acid	100% (v/v)	0.1% (v/v)	50 µl
LiChrosolv water to 50 ml			

0.1% (v/v) Formic acid:

Component	Stock	Final concentration	Amount
Formic acid	100% (v/v)	0.1% (v/v)	50 μl
LiChrosolv water to 50 ml			

10% (v/v) Formic acid:

Component	Stock	Final concentration	Amount
Formic acid	100% (v/v)	10% (v/v)	5 ml
LiChrosolv water to 50 ml			

3.1.9 Step 9: Enrichment of RNA–Protein Cross-Links

80% (v/v) Acetonitrile, 0.1% (v/v) trifluoroacetic acid:

Component	Stock	Final concentration	Amount
Acetonitrile	100% (v/v)	80% (v/v)	8 ml
Trifluoroacetic acid	100% (v/v)	0.1% (v/v)	10 μl
LiChrosolv water to 10 ml			

TiO_2 beads suspension:

Suspend TiO_2 beads in 80% (v/v) Acetonitrile, 0.1% (v/v) trifluoroacetic acid by gently tapping the suspension in a microcentrifuge tube

Buffer A:

200 mg of 2,5-dihydroxybenzoic acid in 1 ml buffer B

Buffer B:

Component	Stock	Final Concentration	Amount
Acetonitrile	100% (v/v)	80% (v/v)	40 ml
Trifluoroacetic acid	100% (v/v)	5% (v/v)	2.5 ml
LiChrosolv water to 50 ml			

Buffer C:

Component	Stock	Final Concentration	Amount
Ammonium hydroxide solution	28–30% (w/v)	0.3 N	4.2 ml
LiChrosolv water			95.8 ml

pH >10.5

3.1.10 Step 10: Analysis by LC-Coupled ESI MS
Buffer A:

Component	Stock	Final Concentration	Amount
Formic acid	100% (v/v)	0.1% (v/v)	1 ml
LiChrosolv water			999 ml

Buffer B:

Component	Stock	Final Concentration	Amount
Acetonitrile	100% (v/v)	95% (v/v)	950 ml
Formic acid	100% (v/v)		1 ml
LiChrosolv water			49 ml

3.1.11 Step 11: MS Data Analysis
Buffers and solutions are not required for this step.

4. PROTOCOL
4.1 Preparation

HeLa cells are grown in a fermenter (Hartmuth, van Santen, Rösel, Kastner, & Lührmann, 2012). We note that also harvesting *HeLa* cells in a spinner flask can be used. Harvest the cells freshly before the preparation of nuclear extract by centrifugation at 2000 rpm for 5 min in a Cryofuge 6000i swing bucket rotor. Prepare the template DNA (encoding the pre-mRNA transcript for assembly of the RNPs) by transformation of *Escherichia coli* (DH5α strain) cells with the plasmid containing the sequence for generation of MS2-tagged PM5 pre-mRNA (Bessonov et al., 2010) according to the protocol given by Sambrook, Fritsch, and Maniatis (1989). Linearize the plasmid DNA by using *Bam*HI-HF enzyme for *in vitro* transcription.

4.2 Duration

	Growing *HeLa* cells in a fermenter	2–4 weeks
Preparation	Transformation of *E. coli* (DH5α strain)	3 days
	Linearized plasmid DNA preparation	2 days
Step 1	*In vitro* transcription	2 days
Step 2	HeLa nuclear extract preparation	9–10 h
Step 3	RNA–protein complex assembly and purification	3–4 h
Step 4	UV cross-linking	3.5–4 h
Step 5	Dissociation and hydrolysis of proteins	16–17 h
Step 6	Isolation of cross-links	15–16 h
Step 7	Hydrolysis of RNA and proteins	18–19 h
Step 8	Desalting and removal of RNA fragments	2 h
Step 9	Enrichment of RNA–protein cross-links	2 h
Step 10	Analysis by LC-Coupled ESI MS	4–5 h
Step 11	MS data analysis	2–7 days

See Fig. 1 for the diagrammatic workflow of the complete protocol.

5. STEP 1—*IN VITRO* TRANSCRIPTION

5.1 Overview

Transcribe the MS2-tagged PM5 pre-mRNA by *in vitro* transcription using SP6 RNA polymerase and linearized plasmid DNA.

5.2 Duration

2 days

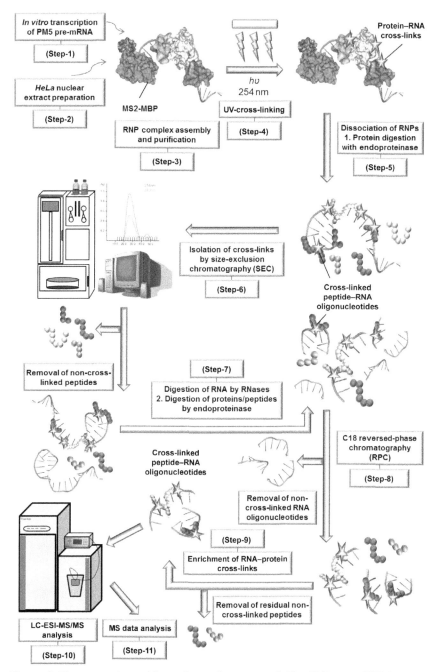

Figure 1 Diagrammatic workflow of complete protocol. The PM5 pre-mRNA is transcribed by *in vitro* transcription (Step-1). The *HeLa* nuclear extract is prepared and dialysed (Step-2). The PM5 pre-mRNA along with the *HeLa* nuclear extract is being used for the RNP complex assembly and purification (Step-3). The assembled complex is then UV-cross-linked (Step-4). The sample is digested with trypsin (Step-5) and administered to the SEC for the isolation of cross-links and removal of non-cross-linked peptides (Step-6). The RNA is digested by the RNases (Step-7) followed by the removal of non-cross-linked RNA oligonucleotides by RPC (Step-8). The RNA-protein cross-links are then enriched by the use of TiO_2 (Step-9). The sample is then analyzed by LC-ESI-MS/MS (Step-10). The MS data obtained are then analyzed by using RNP[xl] pipeline in an OpenMS environment (Step-11). (See the color plate.)

1.1 Set up the transcription reaction as follows:

Components	Final concentration	Amount
5 × Transcription buffer	1×	20 μl
0.1 M ATP	7.5 mM	7.5 μl
0.1 M UTP	7.5 mM	7.5 μl
0.1 M CTP	7.5 mM	7.5 μl
0.01 M GTP	1.3 mM	13 μl
152 mM m^{7}GpppG cap	5 mM	3.28 μl
1 M MgCl$_2$	20 mM	2 μl
1 M DTT	10 mM	1 μl
10 mg/ml BSA	0.1 mg/ml	1 μl
40 U/μl RNAsin	100 U	2.5 μl
SP6 RNA polymerase	200 U	10 μl
DNA template		10 μg
Make up to final volume with autoclaved deionized water	Total	100 μl

1.2 Mix gently and incubate the reaction mixture at 40 °C for 3–4 h in a thermomixer.
1.3 Add 10 μl of RQ1 DNase and incubate the reaction mixture for 20 min at 37 °C.
1.4 Mix 100 μl of RNA sample loading buffer with the reaction mixture.
1.5 Cast a 5% (w/v) polyacrylamide gel of 0.5 mm thickness and load the sample on the gel.
1.6 Run the gel in 1 × TBE buffer at 20 W for 45 min.
1.7 Visualize the transcribed RNA band by UV-shadowing (254 *nm*) and excise the band from the gel with the help of a scalpel.
1.8 To extract the RNA, crush the gel band in a microfuge tube containing 500 μl of RNA extraction buffer and leave it on a shaker overnight at 600 rpm.
1.9 Next day, purify the extracted RNA by PCI extraction. Add 1 μl of glycogen blue and 1 vol. of PCI and agitate it vigorously on a shaker for 15 min at room temperature.
1.10 Centrifuge for 5 min at 13,000 rpm at room temperature.

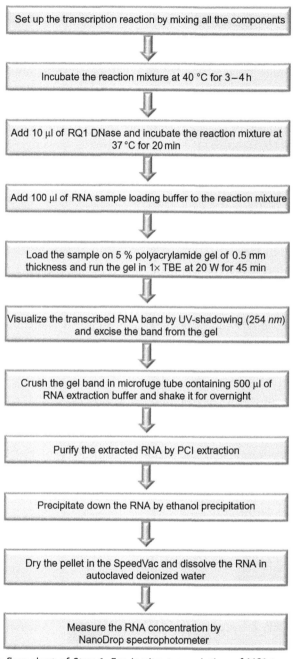

Figure 2 The flow chart of Step 1. For *in vitro* transcription of MS2-tagged PM5 pre-mRNA, the transcription reaction is incubated for 3–4 h at 40 °C. The DNA is digested by using RQ1 DNase. The sample is then run on 5% polyacrylamide gel. The transcribed RNA band is excised from the gel and crushed with the RNA extraction buffer. The extracted RNA is then purified by PCI extraction followed by ethanol precipitation. The RNA is dissolved in autoclaved deionized water and its concentration is determined by spectrophotometer.

1.11 Transfer the aqueous phase to a new microfuge tube. Add 1 vol. of chloroform to it and shake it vigorously at room temperature for 15 min.

1.12 Centrifuge again for 5 min at 13,000 rpm at room temperature.

1.13 Separate the aqueous phase into a new microfuge tube and precipitate down the RNA by ethanol precipitation.

1.14 Add 3 vol. of 100% ice-cold ethanol and 1/10 vol. of 3 M sodium acetate (pH 5.3) to the aqueous phase and leave the sample for at least 2 h at −25 °C.

1.15 Centrifuge at 13,000 rpm for 30 min at 4 °C.

1.16 Wash the pellet with 2 vol. of 80% (v/v) ice-cold ethanol.

1.17 Centrifuge at 13,000 rpm for 30 min at 4 °C.

1.18 Dry the pellet in the SpeedVac and dissolve the RNA in autoclaved deionized water.

1.19 Measure the RNA concentration with a NanoDrop spectrophotometer.

5.3 Tip

Use RNase-free tips and microfuge tubes to avoid contamination during *in vitro* transcription.

5.4 Tip

Separate the aqueous phase carefully while doing PCI extraction.

5.5 Tip

Sample can be left for overnight at −25 °C in 3 vol. of 100% ethanol and 1/10 vol. of 3 M sodium acetate (pH 5.3) for ethanol precipitation.

5.6 Tip

In order to get the required RNA concentration, several transcription reactions may be combined.

See Fig. 2 for the flowchart of Step 1.

6. STEP 2—*HeLa* NUCLEAR EXTRACT PREPARATION

6.1 Overview

Prepare the nuclear extract as described by Dignam, Lebovitz, and Roeder (1983).

6.2 Duration

9–10 h

- **2.1** Wash the cells twice with ice-cold 1 × PBS by gently pipetting the cells up and down with the help of a glass pipette and centrifuging at 2000 rpm for 10 min in 250 ml bottles in a Megafuge swing bucket rotor 2704.
- **2.2** Decant the PBS carefully and resuspend the cells in 1.25 × packed cell volume (PCV) of MC buffer.
- **2.3** Incubate the cells on ice for 5 min.
- **2.4** Transfer the cells into the glass douncer on ice and dounce the cells 18 times with a B-type pestle.
- **2.5** Determine the weight of empty Corex glass tubes. Fill them with dounced cells and centrifuge for 5 min at $18,000 \times g$ at 4 °C in SS-34 rotor.
- **2.6** Decant the supernatant (cytosolic fraction) or store at −80 °C and weigh the tubes again. Calculate the weight of the nuclei by the following formula

    Wt. of the tubes along with nuclei mass − *Wt. of the empty tubes* = *Wt. of Nuclei.*
- **2.7** Resuspend the pellet in 1.3-fold the weight of nuclei of Roeder C buffer and dounce 20 times in a glass douncer on ice.
- **2.8** Transfer the extract into a glass beaker and stir for 40 min at 4 °C.
- **2.9** Transfer the extract into Corex glass tubes again and centrifuge at 16,000 rpm for 30 min at 4 °C in SS-34 rotor.
- **2.10** Remove the nuclear membrane from the top by dipping the pipette tip into the supernatant.
- **2.11** Prepare the dialysis tube by washing it several times with autoclaved deionized water to remove any traces of the ethanol in which it was previously stored.
- **2.12** Fill the dialysis tube with nuclear extract while clipping one end of the dialysis tube.
- **2.13** After filling the nuclear extract, clip the other end of the membrane as well near to the nuclear extract surface.
- **2.14** Dialyze the nuclear extract three times for 2 h each, against 40 volumes of Roeder D buffer at 4 °C.
- **2.15** Transfer the dialyzed nuclear extract into the Corex glass tubes and centrifuge at $9000 \times g$ for 2 min at 4 °C in Hb-6 rotor.
- **2.16** Aliquot the supernatant in labeled Falcon tubes and freeze them in liquid nitrogen and store them at −80 °C.

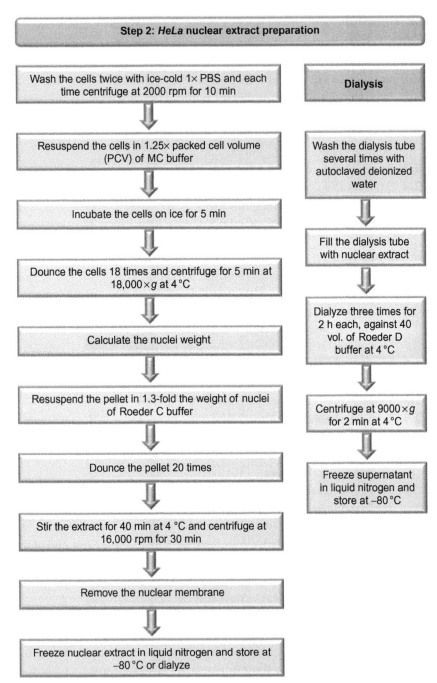

Figure 3 The flow chart of Step 2. *HeLa* cells are washed with PBS and resuspended in MC buffer. After incubation, the cells are dounced and centrifuged. The pellet is resuspended in Roeder C buffer, dounced, and stirred for 40 min. The extract is then centrifuged and dialyzed.

6.3 Tip

Always balance the Corex glass tubes before centrifugation.

6.4 Tip

Keep the glass douncer at 4 °C before preparing the nuclear extract, and carry out the whole nuclear extract preparation step on ice.

6.5 Tip

While douncing, the strokes should be in one direction, i.e., the pestle should not be twisted.

6.6 Tip

While dialyzing the nuclear extract, take care that the magnetic stir bar does not hit the clip of the dialysis tube.

See Fig. 3 for the flowchart of Step 2.

7. STEP 3—RNA–PROTEIN COMPLEX ASSEMBLY AND PURIFICATION

7.1 Overview

Incubate the PM5 pre-mRNA, prepared by transcription *in vitro*, with the HeLa nuclear extract to assemble the RNA–protein complex. Perform this step in the cold room.

7.2 Duration

3–4 h

3.1 Incubate 1 nmol of MS2-tagged PM5 pre-mRNA with a 15-fold excess of MS2-MPB fusion protein for 30 min on ice.

3.2 Incubate the PM5 pre-mRNA bound to MS2-MBP with 10 ml of HeLa nuclear extract in a Corex glass tube for 30 min on ice.

3.3 Take 1 ml of amylose beads in gravity flow disposable chromatographic column and wash them three times with 2 ml of gradient buffer without glycerol.

3.4 Load the sample onto the column and let it flow through under gravity.

3.5 Wash the column three times with 2 ml of gradient buffer without glycerol.

3.6 Elute the assembled complex with 2 ml of 15 mM maltose buffer by gravity flow.

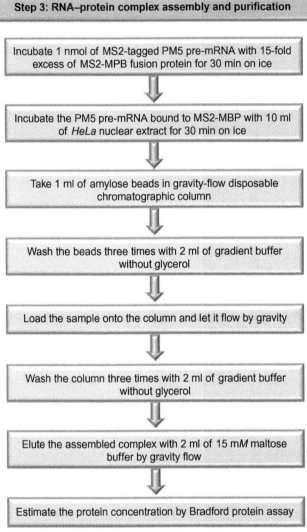

Figure 4 The flow chart of Step 3. The RNA–protein complex is assembled by incubating PM5 pre-mRNA bound to MS2-MBP with *HeLa* nuclear extract. The complex is then purified by elution with maltose buffer from the amylose beads. The protein concentration is determined by Bradford protein assay.

3.7 Estimate the protein concentration by Bradford protein assay (Bearden, 1978).

7.3 Tip
Never apply pressure to the column under gravity flow.

7.4 Tip
Do not let the column run dry.

7.5 Tip
Always use ice-cold buffers.

7.6 Tip
If required, repeat this step to prepare a control (non-UV-irradiated) sample.

7.7 Tip
Use Corex glass tubes made RNase-free by heating at 200 °C overnight. See Fig. 4 for the flowchart of Step 3.

8. STEP 4—UV CROSS-LINKING

8.1 Overview
UV cross-linking of the purified RNP complex in a cold room (4 °C). For the current protocol, UV cross-linking apparatus (constructed in-house) was used.

8.2 Duration
3.5–4 h

4.1 Switch on the UV lamps with the wavelength of 254 *nm*, 30 min before the start of cross-linking.

4.2 Pipette 1 ml of sample (protein concentration ~0.3 mg/ml) into precooled custom-made glass dishes, with a planar surface and an inner diameter of 3.5 cm, placed on an aluminum block in an ice-box, so that the depth of the sample solution is approximately 1 mm.

4.3 Place the sample under the UV lamp at a distance of 1 cm.

4.4 UV-irradiate the sample for 2 min (minimum) to 10 min (maximum).

4.5 Pool the sample in the Corex glass tube and ethanol-precipitate it as described above in Step 1 (*in vitro* transcription).

8.3 Tip
The sample can be left overnight during ethanol precipitation at −20 °C.

8.4 Tip

Usually, UV irradiation at 254 *nm* for 2 min is sufficient. Longer irradiation times can lead to a severe loss of sample material and to RNA damage, in particular when complex samples purified from cells are irradiated. Definitely, do not UV-irradiate the sample for more than 10 min. Ten minute irradiation should be used only for samples prepared by reconstitution *in vitro*.

8.5 Tip

Place the sample directly under the UV lamp.
   See Fig. 5 for the flowchart of Step 4.
   See Fig. 6 for a diagram of the apparatus for UV cross-linking.

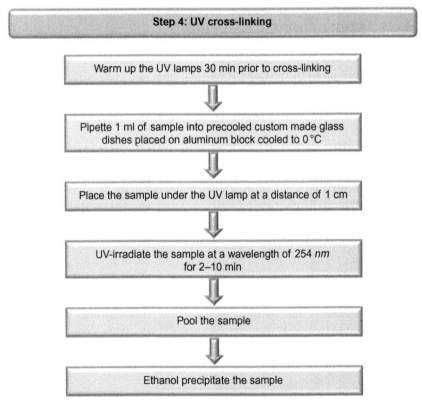

Figure 5 The flow chart of Step 4. The purified RNA–protein complex is cross-linked by UV irradiating at 254 *nm*. The sample is then ethanol-precipitated.

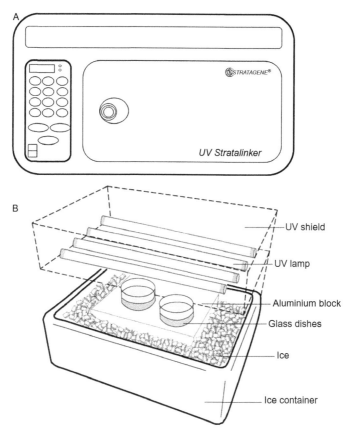

Figure 6 (A) Stratalinker UV cross-linker 2400. (B) UV-cross-linking apparatus (constructed in-house).

9. STEP 5—DISSOCIATION AND HYDROLYSIS OF PROTEINS

9.1 Overview

Dissociate the UV-irradiated RNP by using SDS. Digest the proteins with endoproteinase trypsin to generate an intact RNA that has cross-linked peptides attached.

9.2 Duration

16–17 h

5.1 Dry the pellet derived from ethanol precipitation in the air in the Corex glass tube and then dissolve the pellet in 100 µl of 1% (w/v) SDS by shaking.

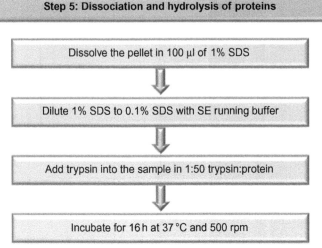

Figure 7 The flow chart of Step 5. The pellet is dissolved in 100 μl of 1% SDS and then diluted to 0.1% SDS with SE running buffer. The proteins are digested by using trypsin.

5.2 Dilute the 1% (w/v) SDS solution to an end concentration of 0.1% (w/v) SDS with SE running buffer.

5.3 Then add trypsin to the sample in a ratio 1:50 (w/w) trypsin:protein and incubate for 16 h at 37 °C while shaking at 500 rpm in a thermomixer.

9.3 Tip

Strictly avoid the production of bubbles while dissolving the pellet in 1% (w/v) SDS.

See Fig. 7 for the flowchart of Step 5.

10. STEP 6—ISOLATION OF CROSS-LINKS

10.1 Overview

Non-cross-linked RNA and RNA with cross-linked peptides are separated from the non-cross-linked peptides (generated by trypsin digestion) by SEC.

10.2 Duration

15–16 h

6.1 After digestion with trypsin, ethanol precipitate the sample as done before in Step 1 (*in vitro* transcription).

6.2 Then dissolve the pellet in 5 µl of 1% (w/v) SDS and dilute up to 0.1% (w/v) SDS with SE running buffer.

6.3 Inject the sample (50 µl) onto an S75 HR SE column mounted in the Smart System, running in SE running buffer with a flow rate of 40 µl/min at room temperature.

6.4 Monitor the absorbance at 260 and 280 *nm* and collect 100 µl fractions.

6.5 Take 10 µl from every second fraction and prepare the sample for NuPage® gel electrophoresis according to the manufacturer's instructions.

6.6 Run the gel by using MOPS running buffer with antioxidant for 50 min at 200 V.

6.7 Silver-stain the gel according to the protocol referred to Merril, Goldman, and Van Keuren (1982).

6.8 Pool the RNA-containing fractions (e.g., fractions 4–10, see Fig. 9) and ethanol precipitate the sample as mentioned in Step 1 (*in vitro* transcription).

Figure 9 shows the analysis of every second fraction derived from the SEC of control (i.e. non-UV-irradiated) and cross-linking (UV-irradiated) samples by SDS–PAGE and silver staining along with their respective SE chromatograms. The elution profiles of the SEC of both control (Fig. 9B) and cross-linking (Fig. 9D) samples do not reveal any noticeable difference. The silver-stained gel of control sample (Fig. 9A) shows in fractions 4–6, the sharp silver-stained band of the PM5 pre-mRNA and peptides (derived from the digestion of the sample with endoproteinase trypsin) eluting in fractions 10–16. The silver-stained gel of the cross-linking sample (Fig. 9C) shows a smear of the PM5 pre-mRNA in the fraction 4 that represents the UV-cross-linked PM5 pre-mRNA with cross-linked peptides attached to it. Fraction 8 shows a smear as well, presumably due to smaller RNA species cross-linked to the peptides.

10.3 Tip

Depending on sample concentration, several SEC runs must be performed.

10.4 Tip

Carefully monitor the column's back-pressure. Do not exceed the specified back-pressure limitation for the SE columns.

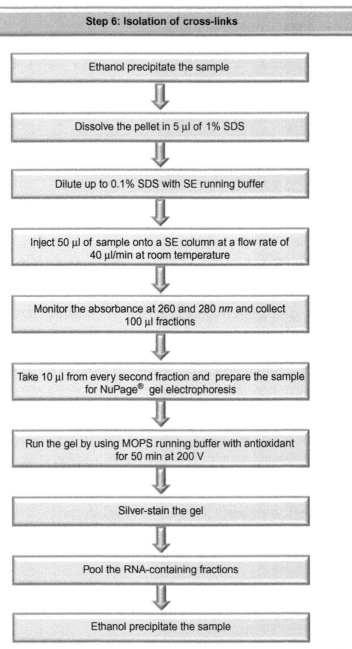

Figure 8 The flow chart of Step 6. The sample digested by trypsin is ethanol-precipitated. The pellet is then dissolved in 1% SDS and diluted to 0.1% SDS with SE running buffer. The sample is then administered to SEC. Every second fraction is run on the gel. The gel is stained by silver staining. The fractions containing RNA are pooled and ethanol-precipitated.

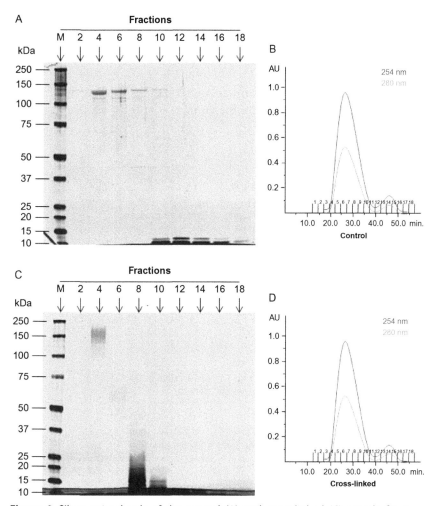

Figure 9 Silver-stained gels of the control (A) and cross-linked (C) sample fractions along with their respective size-exclusion chromatograms (B) and (D). Due to the low cross-linking yield and relatively low resolution of the column, there is no significant difference in the size-exclusion chromatograms however in the fourth fraction of the cross-linked sample, a smear can be seen on the silver-stained gel, resulting from cross-linking of the RNA, as compared to the fourth fraction of the control sample.

We note that the SMART chromatography systems are no longer produced and they are not supported by GE healthcare. The SMART systems have been replaced by ÄKTAmicro chromatography systems with the corresponding columns.

See Fig. 8 for the flowchart of Step 6.
See Fig. 9 for the silver-stained gels and SE chromatograms.

11. STEP 7—HYDROLYSIS OF RNA AND PROTEINS

11.1 Overview

Hydrolyze the RNA and perform an additional digest of potentially non-digested proteins and peptides in the sample.

11.2 Duration

18–19 h

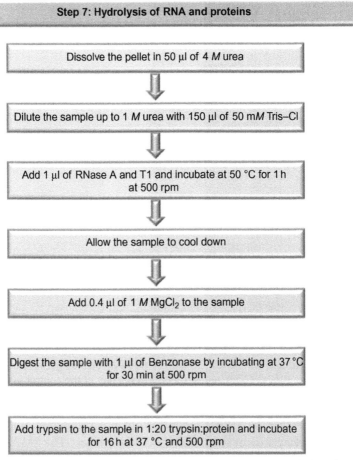

Figure 10 The flow chart of Step 7. The pellet is dissolved in 4 M urea and diluted up to 1 M urea with 50 mM Tris-Cl. The RNA is first digested with RNase A and T1 and then with benzonase. The sample is then redigested with trypsin.

7.1 After ethanol precipitation of the pooled fractions, dissolve the pellet in 50 μl of 4 M urea.
7.2 Dilute the sample with 150 μl of 50 mM Tris–Cl (pH 7.9) to an end concentration of 1 M urea making the final volume up to 200 μl.
7.3 Add 1 μl of RNase A and T1 and incubate at 52 °C for 1 h at 500 rpm.
7.4 Allow the sample to cool down.
7.5 Then add 0.4 μl of 1 M MgCl$_2$ to the sample.
7.6 Digest the sample with 1 μl of benzonase by incubating at 37 °C for 30 min at 500 rpm.
7.7 Add trypsin to the sample in 1:20 (w/w) trypsin:protein and incubate for 16 h at 37 °C and 500 rpm.

11.3 Tip

Allow the sample to cool down on ice before adding benzonase.
See Fig. 10 for the flowchart of Step 7.

12. STEP 8—DESALTING AND REMOVAL OF RNA FRAGMENTS

12.1 Overview

Remove the excess of non-cross-linked RNA from the cross-linked peptide–RNA heteroconjugates by C18 reversed-phase chromatography.

12.2 Duration

2 h

8.1 Start to prepare a column by fitting a 2 mm piece of coffee filter paper at the end of a 10-μl pipette tip as a frit material.
8.2 Fill the pipette tip with C18 suspension with the help of 1 ml combitip to give a bed height of about 3 mm.
8.3 Cut the lid of 2 ml microcentrifuge tube, make a hole in the lid, and fix the prepared C18 column into the hole. Place the lid with the column tightly onto the microcentrifuge tube.
8.4 Wash the column once with 60 μl of 95% (v/v) acetonitrile, 0.1% (v/v) formic acid, then with 80% (v/v) acetonitrile, 0.1% (v/v) formic acid followed by 50% (v/v) acetonitrile, 0.1% (v/v) formic acid, and finally with 0.1% (v/v) formic acid. Washing is always performed by applying the solution onto the column and centrifugation in a table centrifuge at 5000 rpm for 5 min.

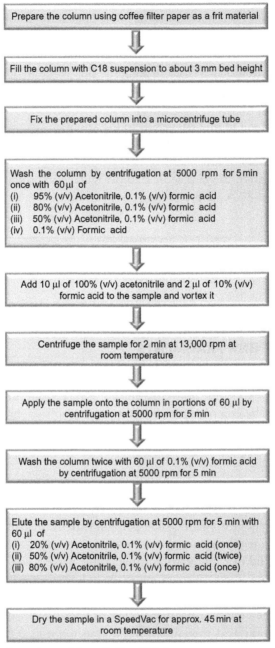

Figure 11 The flow chart of Step 8. The RNA fragments are removed by RPC. The C18 column is packed in-house. The column is washed with 95% (v/v) acetonitrile, 0.1% (v/v) formic acid, 80% (v/v) acetonitrile, 0.1% (v/v) formic acid, 50% (v/v) acetonitrile, 0.1% (v/v) formic acid, and 0.1% (v/v) formic acid. The sample is applied to the column and washed twice with 0.1% (v/v) formic acid. The sample is eluted with 20% (v/v) acetonitrile, 0.1% (v/v) formic acid, 50% (v/v) acetonitrile, 0.1% (v/v) formic acid, 80% (v/v) acetonitrile, and 0.1% (v/v) formic acid. The sample is then dried.

8.5 Meanwhile, add 10 μl of 100% (v/v) acetonitrile and 2 μl of 10% (v/v) formic acid to the sample volume of 200 μl to adjust the acetonitrile and formic acid end concentration to 5% (v/v) and 0.1% (v/v), respectively. Vortex the sample.
8.6 Centrifuge the sample for 2 min at 13,000 rpm at room temperature to remove precipitates.
8.7 Apply the sample onto the column in portions of 60 μl and after each sample application centrifuge at 5000 rpm for 5 min.
8.8 Wash the column twice with 60 μl of 0.1% (v/v) formic acid by centrifugation at 5000 rpm for 5 min.
8.9 Elute the sample once with 60 μl of 20% (v/v) acetonitrile, 0.1% (v/v) formic acid, twice with 50% (v/v) acetonitrile, 0.1% (v/v) formic and once with 80% (v/v) acetonitrile, 0.1% (v/v) formic acid by centrifugation at 5000 rpm for 5 min.
8.10 Dry the eluate in the SpeedVac for 45 min.

12.3 Tip

Do not increase the centrifugation speed to more than 5000 rpm.

12.4 Tip

Do not allow the column to run dry.

12.5 Tip

Do not let the sample dry for too long a time (e.g., hours or overnight) in the SpeedVac.

See Fig. 11 for the flowchart of Step 8.

13. STEP 9—ENRICHMENT OF RNA–PROTEIN CROSS-LINKS

13.1 Overview

This procedure makes use of the phosphate backbone of the peptide–RNA cross-links in order to enrich the cross-links from residual non-cross-linked peptides by the use of TiO_2.

13.2 Duration

2 h

9.1 Prepare the column using coffee filter paper as described for Step-8 (desalting and removal of RNA fragments).

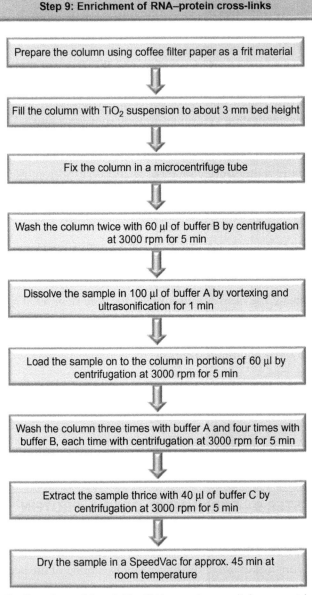

Figure 12 The flow chart of Step 9. The RNA–protein cross-links are enriched by using TiO_2-packed column (in-house). The column is washed with buffer B. The sample is loaded on to the column. The column is washed first with buffer A and then with buffer B. The sample is eluted with buffer C and is then dried.

9.2 Fill the column with TiO$_2$ suspension with the help of a 1 ml combitip to about 3 mm bed height.

9.3 Fix the column in a microcentrifuge tube as described in Step 8 (desalting and removal of RNA fragments).

9.4 Wash the column twice with 60 μl of buffer B by centrifugation at 3000 rpm for 5 min.

9.5 Dissolve the sample in 100 μl of buffer A by vortexing and ultrasonification for 1 min.

9.6 Load the sample onto the column in the portion of 60 μl by centrifugation at 3000 rpm for 5 min.

9.7 Wash the column three times with buffer A and four times with buffer B, each time with centrifugation at 3000 rpm for 5 min.

9.8 Extract the sample thrice with 40 μl of buffer C by centrifugation at 3000 rpm for 5 min.

9.9 Dry the sample in a SpeedVac for approximately 45 min at room temperature.

13.3 Tip

In order to remove DHB completely, the number of washes with buffer B can be increased after the sample has been loaded. Sometimes, the slightly brown color of the eluate, due to traces of DHB, can cause problems in subsequent MS analysis.

See Fig. 12 for the flowchart of Step 9.

14. STEP 10—ANALYSIS BY LC-COUPLED ESI MS

14.1 Overview

Dissolve the sample for the LC–MS/MS analysis and analyze the sample by MS. We note that any highly resolving LC-coupled ESI mass spectrometer (MS resolving power of >20,000 FWHM) can be used for this analysis. The trapping and analytical columns used in Agilent LC system are prepared in-house.

14.2 Duration

4–5 hr

10.1 Dissolve the dried sample in 2 μl of 50% (v/v) acetonitrile, 0.1% (v/v) formic acid by vortexing and ultrasonification for 1 min. Dilute the

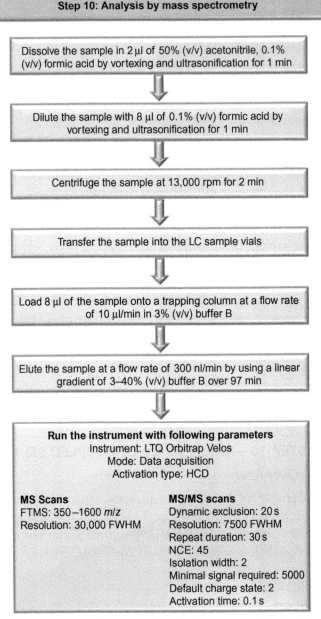

Figure 13 The flow chart of Step 10. The dried sample is dissolved in 50% (v/v) acetonitrile, 0.1% formic acid by vortexing and ultrasonification. The sample is diluted with 0.1% formic acid to an end concentration of 10% (v/v) acetonitrile, 0.1% formic acid. The sample is then analyzed by LTQ Orbitrap Velos.

sample with 8 μl of 0.1% (v/v) formic acid up to 10% (v/v) acetonitrile, 0.1% (v/v) formic acid by vortexing and ultrasonification for 1 min.

10.2 Centrifuge the sample at 13,000 rpm for 2 min and transfer the sample into the LC sample vials.

10.3 Place the vials in the autosampler (Agilent 1100 series; Agilent) and run the LC–MS/MS analysis on an LTQ Orbitrap Velos (Thermo) coupled to the nano-LC system (Agilent 1100 series; Agilent).

10.4 Load 8 μl of the sample onto a trapping column (constructed in-house; C18 5 μm material, 20 mm length, and 0.150 mm inner diameter) at a flow rate of 10 μl/min in 3% (v/v) buffer B.

10.5 Elute the sample on an analytical column (constructed in-house; C18 3 μm material, 150 mm length, and 0.075 mm inner diameter) at a flow rate of 300 nl/min by using a linear gradient of 3–36% (v/v) buffer B over 97 min.

10.6 Run the instrument in data-acquisition mode with top 10 higher energy collision dissociation (HCD) method.

10.7 Record the MS scans in the m/z range 350–1600 at a resolution setting of 30,000 FWHM.

10.8 Record the MS/MS scans with the dynamic exclusion of 20 s at a resolution setting of 7500 FWHM, repeat duration 30 s, normalized collision energy 45, isolation width 2, minimum signal required 5000, and default charge state 2, activation time 0.1 s.

14.3 Tip

Avoid bubble formation while transferring the sample into LC sample vials. See Fig. 13 for the flowchart of Step 10.

15. STEP 11—MS DATA ANALYSIS

15.1 Overview

Analyze the cross-linking data according to the details given by Kramer et al. (2014) by using RNP^{xl} pipeline in an OpenMS environment (http://www.openms.de/RNPxl). Inspect the MS/MS spectra manually to identify the cross-linked peptides. The presence of signals corresponding to RNA marker ions of the nucleic acid base and shift of b or y ion series by the mass of cross-linked RNA or its fragments indicate the cross-linked peptide or amino acid along with the cross-linked nucleotide.

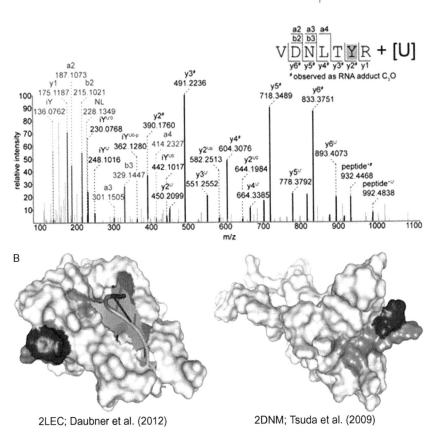

Figure 14 (A) MS/MS spectrum of peptide VDNLTYR (position V18-R24) cross-linked through Y23 to U. The mass shift of tyrosine immonium ion (iY) and the y-series starting with the y2 ion by RNA adducts [Uracil–H$_2$O (U$'^0$—94.0167 Da), Uracil (U$'$—112.0273 Da), Uridine–H$_2$O–HPO$_3$ (U^{0-P}—226.0590 Da), C$_3$O (#—51.9949 Da), Uridine–H$_2$O (U^0—306.0253 Da), Uridine–HPO$_3$ (U^P—244.0695 Da); in red] show that tyrosine (Y23) is the cross-linked amino acid. The peaks in blue show the ions without RNA. The cross-linked peptide lies in the RRM domain of the SRSF2 and SRSF8 proteins. (B) NMR structure of the proteins SRSF2 (2LEC) and SRSF8 (2DNM) showing the cross-linked peptide (dark gray), cross-linked amino acid (red) and RNA (multicolor). (See the color plate.)

Figure 14 shows an example for cross-linked peptide identification. The cross-linked peptide sequence VDNLTYR is located in the RRM domain of two proteins; serine/arginine-rich splicing factor 2 (SRSF2; amino acid position V18-R24) and serine/arginine-rich splicing factor 8 (SRSF8; amino acid position V18-R24). Both proteins have been described to be involved in pre-mRNA splicing by interacting with other splicing factors at the 5′ and 3′ end of pre-mRNA (Edmond et al., 2011). The MS/MS spectrum of the peptide (Fig. 14A) shows that the tyrosine (Y23) is the cross-linked amino acid. The three-dimensional structure of the SRSF2 and SRSF8 proteins (Fig. 14B) has been solved and depositioned in the pdb database (Berman et al., 2000) and described in PDB ID: 2LEC (Daubner, Cléry, Jayne, Stevenin, & Allain, 2012) and PDB ID: 2DNM (Tsuda et al., 2009), respectively.

15.2 Duration

2–7 days

See Fig. 14 for the example of annotated spectra of cross-linked peptide along with protein structures.

16. PROS AND CONS OF THE PRESENTED PROTOCOL

This protocol can be applied to any RNA–protein complex, i.e., *in vitro* reconstituted RNP consisting of single or multiple proteins bound to its cognate RNA or endogenous complexes, and also to *in vivo* assembled complexes (Kramer et al., 2014). The cross-linking of RNA–protein complex by UV irradiation at 254 *nm* produces a zero length cross-link that helps in the identification of direct RNA–protein interactions. Moreover, the MS analysis of the RNA–protein cross-links not only helps in determining the novel RNA-binding motifs but also in identifying of RNA–protein contact sites down to the peptide or even the amino acid level. This protocol is also suitable for the study of interactions between proteins and RNAs that contain modified nucleotides, like 4-thio-U or 6-thio-G or stable-isotope-labeled RNAs.

On the other hand, the UV irradiation of sample at 254 *nm* for long time can cause photodamage to the sample that can result in the reduced cross-linking yield. Due to low yield of cross-linking, certain amount of starting material is needed, e.g., for the current protocol 1 nmol of transcript is required to assemble enough complex for a confident identification of RNA–protein cross-links. In addition to this, extensive biochemical purification steps like SEC, etc., are indispensable. When working with endogenous RNPs, ribosomes should be separated because they are highly prone to UV cross-linking and will hamper the detection of other cross-links. The workflow is dedicated to identify cross-linked peptides. To unambiguously identify cross-linked peptides by MS, the respective cross-linked nucleotide moiety should be generated as small as possible. Of note, long stretches of cross-linked RNA fragments lead to intense fragment ions of the RNA moiety upon sequencing of the cross-links in the gas phase. These intense fragments suppress the detection of the fragmentation of the cross-linked peptide moiety and as a result the cross-inked peptide sequence cannot be determined. Therefore, we have chosen to hydrolyze the RNA with benzonase that generates very small stretches of RNA oligonucleotides (1–3 nucleotides). Furthermore, the TiO_2 enrichment step is not suitable for longer RNA oligonucleotides cross-linked to peptides due to the strong affinity of the RNA to the matrix. Moreover, taking RNA sequences of more than four nucleotides into account will drastically increase the search space in the MS data analysis. One needs a dedicated MS lab to perform analysis and data evaluation annotation.

Finally, it would be desirable to modify the existing workflow so that longer stretches of cross-linked RNA can be identified along with the corresponding peptide sequence, as the current cross-linked RNA length of 1–4 residues does often not allow confident identification of the RNA bound in a specific RNP. This would however require a complete reworked biochemical enrichment and LC–MS/MS analysis strategy suitable for the analysis of long RNA oligonucleotides. Moreover, the current database searching algorithms cannot cope with the extensive MS/MS fragmentation observed for cross-links with greater than four RNA residues.

ACKNOWLEDGMENTS

The authors wish to thank Thomas Conrad for the preparation of HeLa cells; Ulrich Steuerwald for the preparation of MS2–MBP protein; Monika Raabe and Uwe Pleßmann for their excellent technical assistance in the LC–MS/MS analysis. Katharina Kramer, Timo Sachsenberg, and Oliver Kohlbacher are acknowledged for generating the computational workflow that allows the database search of the cross-links. We thank Christof Lenz for critically reading the protocol.

REFERENCES

Antony, E., Weiland, E., Yuan, Q., Manhart, C. M., Nguyen, B., Kozlov, A. G., et al. (2013). Multiple C-terminal tails within a single E. coli SSB homotetramer coordinate DNA replication and repair. *Journal of Molecular Biology*, *425*(23), 4802–4819.

Bearden, J. C., Jr. (1978). Quantitation of submicrogram quantities of protein by an improved protein-dye binding assay. *Biochimica et Biophysica Acta*, *533*(2), 525–529.

Berman, H. M., Westbrook, J., Feng, Z., Gilliland, G., Bhat, T. N., Weissig, H., et al. (2000). The protein data bank. *Nucleic Acids Research, 28*(1), 235–242.

Bessonov, S., Anokhina, M., Krasauskas, A., Golas, M. M., Sander, B., Will, C. L., et al. (2010). Characterization of purified human Bact spliceosomal complexes reveals compositional and morphological changes during spliceosome activation and first step catalysis. *RNA, 16*(12), 2384–2403.

Blencowe, B. J. (2006). Alternative splicing: New insights from global analyses. *Cell, 126*(1), 37–47.

Castello, A., Fischer, B., Eichelbaum, K., Horos, R., Beckmann, B. M., Strein, C., et al. (2012). Insights into RNA biology from an atlas of mammalian mRNA-binding proteins. *Cell, 149*(6), 1393–1406.

Castello, A., Horos, R., Strein, C., Fischer, B., Eichelbaum, K., Steinmetz, L. M., et al. (2013). System-wide identification of RNA-binding proteins by interactome capture. *Nature Protocols, 8*(3), 491–500.

Castro-Roa, D., & Zenkin, N. (2011). Relations between replication and transcription. In J. Kusic-Tisma (Ed.), *Fundamental aspects of DNA replication* (pp. 289–306). Rijeka, Croatia: InTech.

Chen, Y. C., Sargsyan, K., Wright, J. D., Huang, Y.-S., & Lim, C. (2014). Identifying RNA-binding residues based on evolutionary conserved structural and energetic features. *Nucleic Acids Research, 42*(3), e15.

Daubner, G. M., Cléry, A., Jayne, S., Stevenin, J., & Allain, F. H. T. (2012). A syn-anti conformational difference allows SRSF2 to recognize guanines and cytosines equally well. *The EMBO Journal, 31*(1), 162–174.

Dignam, G. D., Lebovitz, R. M., & Roeder, R. G. (1983). Accurate transcription initiation by RNA polymerase II in a soluble extract from isolated mammalian nuclei. *Nucleic Acids Research, 11*(5), 1475–1489.

Edmond, V., Moysan, E., Khochbin, S., Matthias, P., Brambilla, C., Brambilla, E., et al. (2011). Acetylation and phosphorylation of SRSF2 control cell fate decision in response to cisplatin. *The EMBO Journal, 30*(3), 510–523.

Garneau, N. L., Wilusz, J., & Wilusz, C. J. (2007). The highways and byways of mRNA decay. *Nature Reviews. Molecular Cell Biology, 8*(2), 113–126.

Gleghorn, M. L., & Maquat, L. E. (2014). 'Black sheep' that don't leave the double-stranded RNA-binding domain fold. *Trends in Biochemical Sciences, 39*(7), 328–340.

Glisovic, T., Bachorik, J. L., Yong, J., & Dreyfuss, G. (2008). RNA-binding proteins and post-transcriptional gene regulation. *FEBS Letters, 582*(14), 1977–1986.

Hafner, M., Landthaler, M., Burger, L., Khorshid, M., Hausser, J., Berninger, P., et al. (2010). Transcriptome-wide identification of RNA-binding protein and microRNA target sites by PAR-CLIP. *Cell, 141*(1), 129–141.

Hartmuth, K., van Santen, M. A., Rösel, T., Kastner, B., & Lührmann, R. (2012). The preparation of HeLa cell nuclear extracts. In S. Stamm, C. W. J. Smith, & R. Lührmann (Eds.), *Alternative pre-mRNA splicing: Theory and protocols* (pp. 311–319). Weinheim, Germany: Wiley-VCH Verlag GmbH & Co. KGaA.

Hellman, L. M., & Fried, M. G. (2007). Electrophoretic mobility shift assay (EMSA) for detecting protein-nucleic acid interactions. *Nature Protocols, 2*(8), 1849–1861.

Henkin, T. M. (2014). The T box riboswitch: A novel regulatory RNA that utilizes tRNA as its ligand. *Biochimica et Biophysica Acta, 1839*(10), 959–963.

Hoell, J. I., Hafner, M., Landthaler, M., Ascano, M., Farazi, T. A., Wardle, G., et al. (2014). Transcriptome-wide identification of protein binding sites on RNA by PAR-CLIP (photoactivatable-ribonucleoside-enhanced crosslinking and immunoprecipitation). In R. K. Hartmann, A. Bindereif, A. Schön, & E. Westhof (Eds.), *Handbook of RNA biochemistry: Second, completely revised and enlarged edition* (pp. 877–898). Weinheim, Germany: Wiley-VCH Verlag GmbH & Co. KGaA.

Hogan, D. J., Riordan, D. P., Gerber, A. P., Herschlag, D., & Brown, P. O. (2008). Diverse RNA-binding proteins interact with functionally related sets of RNAs, suggesting an extensive regulatory system. *PLoS Biology, 6*(10), e255.

König, J., Zarnack, K., Luscombe, N. M., & Ule, J. (2012). Protein–RNA interactions: New genomic technologies and perspectives. *Nature Reviews Genetics, 13*(2), 77–83.

Landthaler, M., Gaidatzis, D., Rothballer, A., Chen, P. Y., Soll, S. J., Dinic, L., et al. (2008). Molecular characterization of human Argonaute-containing ribonucleoprotein complexes and their bound target mRNAs. *RNA, 14*(12), 2580–2596.

Lecuyer, E., Yoshida, H., & Krause, H. M. (2009). Global implications of mRNA localization pathways in cellular organization. *Current Opinion in Cell Biology, 21*(3), 409–415.

Merril, C. R., Goldman, D., & Van Keuren, M. L. (1982). Simplified silver protein detection and image enhancement methods in polyacrylamide gels. *Electrophoresis, 3*(1), 17–23.

Michaud, S., & Reed, R. (1991). An ATP-independent complex commits pre-mRNA to the mammalian spliceosome assembly pathway. *Genes & Development, 5*(12b), 2534–2546.

Sambrook, J., Fritsch, E. F., & Maniatis, T. (1989). *Molecular cloning: Vol. 2*. New York: Cold Spring Harbor Laboratory Press.

SenGupta, D. J., Zhang, B., Kraemer, B., Pochart, P., Fields, S., & Wickens, M. (1996). A three-hybrid system to detect RNA-protein interactions in vivo. *Proceedings of the National Academy of Sciences of the United States of America, 93*(16), 8496–8501.

Tadros, W., Goldman, A. L., Babak, T., Menzies, F., Vardy, L., Orr-Weaver, T., et al. (2007). SMAUG is a major regulator of maternal mRNA destabilization in Drosophila and its translation is activated by the PAN GU kinase. *Developmental Cell, 12*(1), 143–155.

Tenenbaum, S. A., Carson, C. C., Lager, P. J., & Keene, J. D. (2000). Identifying mRNA subsets in messenger ribonucleoprotein complexes by using cDNA arrays. *Proceedings of the National Academy of Sciences of the United States of America, 97*(26), 14085–14090.

Tsuda, K., Muto, Y., Inoue, M., Kigawa, T., Terada, T., Shirouzu, M., et al. (2009). Solution structure of RNA binding domain in SRp46 splicing factor. To be published. URL: http://www.rcsb.org/pdb/explore/pubmed.do?structureId=2DNM.

Ule, J., Jensen, K., Mele, A., & Darnell, R. B. (2005). CLIP: A method for identifying protein–RNA interaction sites in living cells. *Methods, 37*(4), 376–386.

Ule, J., Jensen, K. B., Ruggiu, M., Mele, A., Ule, A., & Darnell, R. B. (2003). CLIP identifies Nova-regulated RNA networks in the brain. *Science, 302*(5648), 1212–1215.

Wahl, M. C., Will, C. L., & Lührmann, R. (2009). The spliceosome: Design principles of a dynamic RNP machine. *Cell, 136*(4), 701–718.

Walser, C. B., & Lipshitz, H. D. (2011). Transcript clearance during the maternal-to-zygotic transition. *Current Opinion in Genetics & Development, 21*(4), 431–443.

Wang, I., Hennig, J., Jagtap, P. K. A., Sonntag, M., Valcárcel, J., & Sattler, M. (2014). Structure, dynamics and RNA binding of the multi-domain splicing factor TIA-1. *Nucleic Acids Research, 42*(9), 5949–5966.

Wilhelm, J. E., & Smibert, C. A. (2005). Mechanisms of translational regulation in Drosophila. *Biology of the Cell, 97*(4), 235–252.

Zhao, J., Ohsumi, T. K., Kung, J. T., Ogawa, Y., Grau, D. J., Sarma, K., et al. (2010). Genome-wide identification of polycomb-associated RNAs by RIP-seq. *Molecular Cell, 40*(6), 939–953.

SOURCE REFERENCE
Kramer, K., Sachsenberg, T., Beckmann, B. M., Qamar, S., Boon, K.-L., Hentze, M. W., et al. (2014). Photo-cross-linking and high-resolution mass spectrometry for assignment of RNA-binding sites in RNA-binding proteins. *Nature Methods, 11*(10), 1064–1070.

Referenced Protocol in Method Navigator
Plasmid transformation of *Escherichia coli* and other bacteria.
Quantification of protein concentration using UV absorbance and Coomassie dyes.
Silver staining of SDS-polyacrylamide gel.
In vitro transcription from plasmid or PCR-amplified DNA.
Purification of DNA oligos by denaturing polyacrylamide gel electrophoresis (PAGE).

CHAPTER FIFTEEN

RNA Bind-n-Seq: Measuring the Binding Affinity Landscape of RNA-Binding Proteins

Nicole J. Lambert*,[1], Alex D. Robertson*,[†,1,2], Christopher B. Burge*,[†,3]

*Department of Biology, MIT, Cambridge Massachusetts, USA
†Program in Computational and Systems Biology, MIT, Cambridge, Massachusetts, USA
[3]Corresponding author: e-mail address: cburge@mit.edu

Contents

1. Introduction	466
2. Experimental Method	468
3. Preparation of RBNS Reagents	469
3.1 *In Vitro*-Transcribed RNA Pool	469
3.2 Length of Library Random Region	470
3.3 Gel Purification of *In Vitro*-Transcribed RNA	471
3.4 Purified RBP QC	472
4. RBNS Assay	472
4.1 Prepare Buffers	472
4.2 Protein Equilibration	473
4.3 Temperature Considerations	473
4.4 Bead Preparation	473
4.5 RNA Binding	474
4.6 RBP Pull-Down	474
4.7 RBP Elution	474
4.8 Quantifying Concentration of RBP-Bound RNA (for Affinity Estimations)	475
4.9 Sequencing Library Preparation	475
4.10 Variations on Input RNA Pool Composition	476
5. Computational Analysis	477
6. Pipeline Inputs	482
7. *k*-mer Analysis	482
7.1 Naïve Method	482
7.2 Calculating B Values	483
8. Streaming *k*-mer Assignment	484

---

[1] These authors contributed equally.
[2] Current address: Counsyl, South San Francisco, CA 94080, USA.

9. Presence Fractions	487
9.1 RNA-Binding Equilibrium	487
10. Relative Binding Affinity Determination	487
11. RBNS Quality Control	489
12. Conclusions	491
References	492

Abstract

RNA-binding proteins (RBPs) coordinate post-transcriptional control of gene expression, often through sequence-specific recognition of primary transcripts or mature messenger RNAs. Hundreds of RBPs are encoded in the human genome, most with undefined or incompletely defined biological roles. Understanding the function of these factors will require the identification of each RBP's distinct RNA binding specificity. RNA Bind-n-Seq (RBNS) is a high-throughput, cost-effective *in vitro* method capable of resolving sequence and secondary structure preferences of RBPs. Dissociation constants can also be inferred from RBNS data when provided with additional experimental information. Here, we describe the experimental procedures to perform RBNS and discuss important parameters of the method and ways that the experiment can be tailored to the specific RBP under study. Additionally, we present the conceptual framework and execution of the freely available RBNS computational pipeline and describe the outputs of the pipeline. Different approaches to quantify binding specificity, quality control metrics, and estimation of binding constants are also covered.

1. INTRODUCTION

RNA-binding proteins (RBPs) guide RNA splicing, translation control, mRNA localization, and other forms of post-transcriptional gene regulation. Because of their key roles in proper gene expression and processing, genetic alterations to RBPs, the *cis*-elements they recognize, or changes in their cellular concentrations have important implications for development and disease.

Most RBPs are thought to bind to specific RNA sequences and/or secondary structures. RBPs of the developmentally regulated RBFOX, MBNL, and CELF splicing factor families each have high affinity for specific short RNA motifs, while other RBPs recognize more degenerate motifs, e.g., the pervasive splicing regulator SRSF2 binds diverse sequences containing SSNG (S=G or C, N=any base) with similar affinity(Daubner, Clery, Jayne, Stevenin, & Allain, 2012). Still other RBPs such as those of the adenosine deaminase acting on RNA (ADAR) family discriminate binding sites based on RNA secondary structure (Bass et al., 1997).

In vivo-binding data using UV cross-linking such as cross-linking immunoprecipitation (CLIP) identify directly bound RNA regions (Ule et al., 2003). In general, it is difficult to resolve comprehensive binding portraits of RBPs using these *in vivo* approaches because of the potential for interference from binding by nearby interacting or competing factors as well as inherent technical biases of crosslinking (Sugimoto et al., 2012). If the RBP has a known effect on mRNA stability, for example, then indirect information about RBP specificities can also be obtained by perturbing cellular RBP levels followed by global profiling of gene expression or decay rates.

Due to the difficulty in obtaining comprehensive binding spectra from *in vivo* derived data, *in vitro* approaches have been widely used to quantify the diversity of motifs recognized by RBPs and to quantify the binding strength of RBPs to individual binding sites. Standard techniques to quantify absolute affinities of RBPs for specific RNAs include isothermal calorimetry, surface plasmon resonance (SPR), and electrophoretic mobility shift assay (EMSA), all of which have low throughput. Recent efforts to capture binding specificity landscapes using higher throughput approaches have been employed, generally at the expense of quantitative measurement.

Several systematic high-throughput methods to identify RBP specificity have been developed, each with different strengths and weaknesses. Systematic evolution of ligands by exponential enrichment (SELEX) uses iterative cycles of binding, RNA isolation, and amplification followed by sequencing, allowing the identification of high-affinity binding sites (Tuerk & Gold, 1990). Multiround selection can be time consuming, and the nature of iterative binding/amplification cycles reduces the power to resolve moderate-affinity binding sites. However, in some cases, iterative binding steps may be necessary to detect very weak and/or degenerate binding signals. Recently described methods use an Illumina sequencer to directly measure fluorescent RBP binding to clusters of RNA molecules anchored to flowcell-attached DNA (Buenrostro et al., 2014; Tome et al., 2014). This methodology has a complicated experimental setup and is costly. However, affinity estimations may be more accurate due to the direct measurement of fluorescent RBP binding to RNA clusters. In a medium-throughput approach that uses a microfluidic platform, RNA-induced trapping of molecular interactions directly measures protein-bound sequences but requires specialized equipment and is not easily scalable (Martin et al., 2012). High-throughput identification of RBP-binding motifs can also be accomplished with RNAcompete, which uses an *in vitro* RNA-binding step followed by microarray hybridization of RBP-bound sequences (Ray et al., 2009).

RNA-compete has proved to be scalable and has been used to measure binding specificities of ~200 RBPs (Ray et al., 2013). Contextual and RNA secondary structural information important for RBP binding may be more difficult to obtain due to the low incubation temperature typically used and the limited set of structural contexts.

RNA Bind-n-Seq (RBNS) measures the sequence and structural preferences of RBPs (Lambert et al., 2014). Several binding reactions containing different quantities of a tagged RBP are incubated with random RNA, followed by high-throughput sequencing of RBP-bound RNA. RBNS is a relatively cost-effective, high-throughput assay suitable for simultaneously resolving weak and strong binding motifs as well as contextual information. Additionally, dissociation constants (K_d values) can be estimated from RBNS data, when supplemented by additional experiments for calibration. The direct sequencing of bound RNAs, the lack of iterative binding steps, and the relatively straightforward protocol distinguish RBNS from other high-throughput methods. Here, we will describe important parameters of an RBNS experiment and considerations to tailor RBNS to the experimenter's needs.

While RBNS can provide a wealth of information, careful analysis of the data is essential to quantify the specificity of the RBP. Furthermore, the subtle biases that must be accounted for differ greatly from genomic analyses. The analyses also differ from standard biochemical assays as the high-throughput nature necessitates modeling of many simultaneous equilibria. In this chapter, we walk through the computational analyses in detail and describe an open source implementation of these analyses. We discuss theory and practical considerations in running the computational analyses efficiently. Additionally, we describe analysis methods to calculate quality control metrics, motif enrichment values, and the estimation of binding affinities for RBP-bound motifs.

2. EXPERIMENTAL METHOD

In the RBNS assay, several individual binding reactions are performed in parallel, each reaction performed with a different RBP concentration (Fig. 1A). This is akin to an EMSA, where we anticipate RNA motifs to be bound in a RBP concentration-dependent manner. Unlike an EMSA assay, relative binding affinities can be determined for many oligonucleotides (k-mers) simultaneously. Additionally, two negative controls are typically performed: sequencing of the input library to assess the randomness of the input pool, and a no protein-binding experiment to detect potential

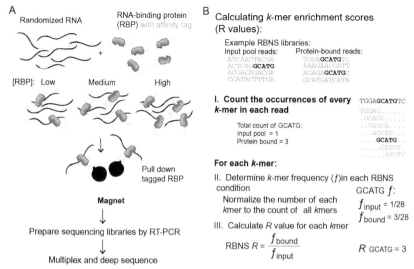

Figure 1 Overview of RBNS method and R value calculation. (A) RBNS method overview: several individual binding reactions (typically 5–10) are performed in parallel, each using a different RBP concentration. RNAs containing high- or low-affinity sites are shown. (B) Calculating enrichment (R) values from raw RBNS data.

apparatus-selected biases. RBNS can be used to measure an RBP's affinity among a pool of random sequences, a designed set of array-synthesized oligonucleotides or natural transcripts, as desired. The following details the experimental and computational methods of RBNS assuming the use of a randomized RNA pool.

3. PREPARATION OF RBNS REAGENTS

3.1 *In Vitro*-Transcribed RNA Pool

Input RNA is *in vitro* transcribed using the RBNS T7 template, a synthetic DNA oligo containing a random 40-mer flanked by Illumina primers and a T7 promoter sequence (primer sequences are listed in Table 1). If available, hand-mixed rather than machine-dispensed random nucleotides are recommended for the synthesis of the RBNS T7 template for more equal representation of the four nucleotides. To mimic a double-stranded T7 promoter, the T7 promoter oligo is annealed to an equimolar quantity of RBNS T7 template, by heating the two oligos at 65 °C for 5 min and then allowing the mixture to cool at room temperature for 2 min. This template is used as input to a T7 *in vitro* transcription reaction, using commercially available kits

Table 1 Primers Used in RBNS.

Primer Name	Sequence
T7 promoter oligo	TAATACGACTCACTATAGGG
RBNS T7 template	CCTTGACACCCGAGAATTCCA-(N)40-GATCGTCGGACTGTAGAACTCCCTATAGTGAGTCGTATTA
RT primer	GCCTTGGCACCCGAGAATTCCA
RNA PCR (RP1)	AATGATACGGCGACCACCGAGATCTACACGTTCAGAGTTCTACAGTCCGACGATC
Indexed PCR primer	CAAGCAGAAGACGGCATACGAGAT-barcode-GTGACTGGAGTTCCTTGGCACCCGAGAATTCCA

according to the manufacturer's instructions. Formamide is added to the *in vitro* transcription reaction (to a 25% final volume), which is then run on a 6% TBE-urea polyacrylamide gel. Dyes can be added for easier visualization during sample loading if the dyes are not expected to run near the expected product size.

3.2 Length of Library Random Region

The length of the input RNA is an important aspect of the experimental design. Increasing the length of the RNA pool has the effect of diluting the signal for simple linear motifs (discussed below) but increases power to detect the effect of RNA secondary structure on binding, since longer sequences have greater structure potential. For example, suppose that an RBP binds to a specific 5-mer within a 40-mer oligo. When analyzing enrichment of short motifs in the bound versus input pool, recovery of this 40-mer will increment the count of the specific 5-mer (the "signal") by one but will also increment the counts of the 35 ($=40-5$) other 5-mers present in the 40-mer (the "noise"), resulting in "dilution" of the signal by a factor of 1:36. However, if the oligo were of size 10, then the analogous dilution factor is only 1:6. We have compared RBNS results using different oligo sizes for RBFOX2. We observed that for the top motif, UGCAUG, the relative enrichment in the bound pool relative to the input pool—termed the "R value" (Fig. 1B)—was indeed several-fold higher when using the shorter library (Fig. 2A). However, importantly, the rank order of enriched motifs and the estimated binding affinities were very similar between the two oligo sizes. Therefore, when the signal above background is sufficient, using longer oligo sizes results in little loss of linear motif information, and 40-mers

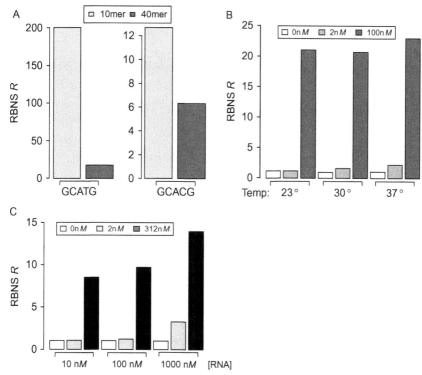

Figure 2 Effects of input oligonucleotide length, temperature, and input RNA concentration on RBNS R values. (A) R value of primary 5-mer motif, GCATG, and secondary motif, GCACG, for RBFOX2 using input oligonucleotide pools of length 10 and 40 nt (both at 1 μM total RNA concentration). (B) R value of top 6-mer motif, TGCATG, for RBFOX2 at different protein concentrations and temperatures (RT = room temperature). (C) R value of top 6-mer motif, TGCATG, at different protein concentrations (indicated by color (grayscale in the print version)) and different total input RNA concentrations (indicated below).

obviously contain a much greater diversity of secondary structures than 10-mers, so there is potential to learn more.

3.3 Gel Purification of *In Vitro*-Transcribed RNA

The gel is removed from the glass plates and placed on a fluor-coated TLC plate that is wrapped in plastic wrap. The band is visualized with a handheld UV light source and excised with a clean razor blade. In preparation for band isolation, use a clean 18- or 22-gauge needle to puncture a hole at the bottom of a 0.5-ml Eppendorf tube and then place into a larger 1.5-ml Eppendorf tube (gel-shredder). After isolating the band, place it into the gel-shredder and spin the gel slices at room temperature at maximum speed for 10 min. Shredded gel pieces should be pelleted at the bottom of the larger 1.5-ml tube. Remove and discard the smaller tube. Resuspend shredded gel pieces in 400 μl of gel extraction buffer. Extract the RNA by rotating at 4°C overnight. The following day, pipette gel and buffer solution into a microcentrifuge filtration tube and remove the gel pieces by spinning the mixture at 14,000 × g for 2 min. Place the flow-through into a clean 1.5-ml tube and add 400 μl of isopropanol. Place the mixture on dry ice for 20 min and spin for 25 min at 4°C at maximum speed on a tabletop centrifuge. Remove ethanol, and wash the pellet with cold 70% ethanol.

3.4 Purified RBP QC

A purified RBP with an affinity tag is a requirement for the assay. A tandem affinity glutathione S-transferase/streptavidin-binding peptide (GST–SBP) fusion tag is used in our system, though other tags can be used for this assay. We express the RBP in a bacterial system and purify it via the GST tag. Protein purification will not be covered due to the complex and variable nature of the process (see Burgess & Deutscher, 2009). The SBP tag is then used for protein capture in RBNS selection steps. Prior to performing RBNS, the RBP's affinity for RNA containing a known or putative binding site is determined by EMSA. This is an important step for confirming protein function and also for determining suitable protein concentrations to use for RBNS. For cases where no putative binding site is known for an RBP, we generally assume functionality if the purified protein is full length and soluble.

4. RBNS ASSAY

4.1 Prepare Buffers

Binding buffer[a]	25 mM Tris, pH 7.5, 150 mM KCl, 3 mM MgCl2, 0.01% Tween, 1 mg/ml BSA, 1 mM DTT
Wash buffer	25 mM Tris, pH 7.5, 150 mM KCl, 60 μg/ml BSA, 0.5 mM EDTA, 0.01% Tween
Elution buffer	10 mM Tris, pH 7.0, 1 mM EDTA, 1%SDS
Gel extraction buffer	10 mM Tris, pH 7.0, 300 mM NaCl, 2 mM EDTA
TE	10 mM Tris, pH 7.0, 1 mM EDTA

[a]30 μg/ml poly I/C can optionally be added to the binding buffer as a nonspecific competitor.

4.2 Protein Equilibration

Identical binding reactions are set up in parallel, each containing a different RBP concentration. Typical concentrations used in our lab are 0, 5, 20, 80, 320, and 1300 nM. These concentrations capture the range of binding affinities of most studied RBPs. If the RBP is expected to have a very high or low affinity for its binding substrates, the concentrations should be varied accordingly. As the protein concentration increases, R values will initially increase, reach a maximum at a moderate protein concentration, and then steadily decrease until reaching background levels, for reasons that we have outlined elsewhere (Lambert et al., 2014). Such a unimodal R value curve is observed in both simulated and actual RBNS data. It is important to select a wide range of RBP concentrations to capture the concentrations with maximum information, as each protein shows a different binding profile. Two negative control conditions are typically performed, a no-protein RBNS condition and sequencing of the input RNA pool. Sequencing the input pool controls for biases in nucleotide composition in the input RNA pool and the no-protein control can detect potential biases in binding to the apparatus. The protein concentrations are assembled in 250 μl of binding buffer and placed on a rotator for 30 min at a selected temperature.

4.3 Temperature Considerations

Protein equilibration, RNA binding, and RBP pull-down are performed at a single chosen temperature. To systematically explore the influence of temperature on the binding profile of RBFOX2, RBNS was performed at room temperature, 30 °C, and 37 °C (Fig. 2B). Temperature had modest effects on the enrichment of RBFOX's canonical and secondary motif, though the influence of temperature on binding may be somewhat RBP-specific. A slight increase in R values was observed at higher temperatures, likely because RNA secondary structure is reduced, increasing the accessibility of binding sites.

4.4 Bead Preparation

After protein equilibration reactions are assembled, streptavidin magnetic beads (Invitrogen) are washed and prepared for pull-down of the SBP-tagged RBP. For each binding reaction, 75 μl of streptavidin bead slurry is used. Sufficient bead slurry for all binding reactions is placed in a 1.5-ml Eppendorf tube. The tube is placed on a standing magnet until all visible magnetic beads are pulled out of solution, and the supernatant is removed and discarded.

The beads are washed three times with 1 ml of wash buffer, discarding the wash buffer after each wash step. The beads are then resuspended in binding buffer (the same volume initially removed from vial) along with Superasin (0.5 μl/reaction). The beads are then placed on a rotator until needed. As many beads are not guaranteed RNase-free, the equilibration of the beads with RNase inhibitors inactivates RNases that may be present. The inclusion of BSA will in theory coat nonspecific/"sticky" components of the binding apparatus.

4.5 RNA Binding

Following protein equilibration, the RNA pool is added to each of the equilibrated protein mixtures to a final concentration of 1 μM, along with 2 μl of Superasin. Each mixture is then placed back on the rotator at the temperature of choice and incubated for 1 h.

The final concentration of RNA used in each binding reaction can be varied based on the experimenter's wishes and reagent availability. To test the influence of input RNA concentration on the measurement of binding affinities, RBFOX2 RBNS experiments were performed with either 10, 100, or 1000 nM total RNA concentrations (Fig. 2C). The consensus motif was clearly detected in all tested conditions. However, higher input RNA concentrations yielded elevated R values for all RBNS conditions tested. For instance, at a low RBFOX concentration (2 nM), there was a threefold increase in R value for a top motif when the input RNA concentration was increased from 10 to 1000 nM (Fig. 2C). A possible explanation for this observation is that the top RNA motifs may be readily saturated when total RNA concentration is low, limiting the dynamic range of enrichment values.

4.6 RBP Pull-Down

After protein/RNA binding, the tagged protein is affinity-selected with the washed magnetic beads. In preparation for RBP selection, the beads are aliquoted into separate labeled tubes (75 μl/tube). The buffer is removed from the beads, and the RNA/protein mixture is directly added to the beads. The bead/RNA/RBP solution is placed back onto the rotator for 1 h.

4.7 RBP Elution

After the RBP mixture is incubated with streptavidin beads, the beads are pulled out of solution, washed, and the bound material eluted. First, the beads are placed on the magnetic strip and the supernatant is carefully removed and discarded. Then the beads are resuspended in 1 ml of wash buffer, placed back on the standing magnet, and the wash buffer discarded. The number and stringency of the wash steps have a considerable impact on the results. Highest R values for RBFOX were seen when the wash buffer contained low salts (data not shown). We typically do a single wash step in an attempt to capture moderate-affinity binding sites and maintain steady-state-like conditions.

After the beads are washed, the RBP and bound RNA are eluted. Immediately after washing the beads, 100 μl of elution buffer is added to the beads. The beads are then heated at 70 °C for 10 min and the supernatant collected and placed in a clean tube. Alternatively, bound RBP can be eluted with biotin. Bound RNA is then purified away from the RBP by phenol/chloroform isolation and ethanol precipitation (see above). The resulting pellet is then resuspended in 10 μl of TE.

4.8 Quantifying Concentration of RBP-Bound RNA (for Affinity Estimations)

If RBNS data will be used to estimate binding affinities, the amount of recovered RNA from each RBNS concentration must be quantified. Assuming that all tagged RBP was pulled out of solution and that all RBP-bound RNA was subsequently purified, the concentration of RBP bound to RNA can be estimated. Quantitation of bound RNA can be performed by running 20% of the purified RBP-bound RNA on a Bioanalyzer (Agilent), according to the manufacturer's instructions. If the peak corresponding to the expected library size is quantified, the concentration of RBP-bound RNA can be estimated by controlling for the fraction of the library loaded onto the Bioanalyzer and the volume of the intial binding reaction. This parameter can also be measured via quantitative RT-PCR.

4.9 Sequencing Library Preparation

Library preparation steps include reverse transcription and PCR. Half of the isolated RNA from each RBNS condition is input into a 20-μl Superscript III reverse transcription reaction (Invitrogen), with 2 pmol of RT primer. This reaction is incubated at 55 °C for 1 h. An additional RT reaction is conducted with 0.5 pmol of the input RNA pool as the RT template.

After reverse transcription, the resulting cDNA is amplified via PCR. Between 1 and 5 μl of cDNA is added into a 25 μl PCR reaction. The input quantity must be empirically determined because it is influenced by the amount of RNA bound by the protein and the amount of coeluting nonspecific RNA bound to the apparatus. The required volume of cDNA input into the PCR typically decreases as the RBP concentration used increases. As a starting point, use 1 μl of cDNA into each PCR reaction and perform 8–10 PCR cycles. A different barcoded PCR primer is used to amplify each RBNS condition for downstream sample multiplexing (Table 1).

After PCR, the sequencing library is size-selected to exclude primer dimers and PCR artifacts. Loading dye is added to each reaction and then loaded onto an 8% TBE gel (alternating with empty lanes) and run at 200 V for about 45 min. The gels are stained with SYBR gold for ~5 min. The sharp band at the expected size (at ~180 bp if using the 40-mer pool) is excised with a clean razor and gel-purified. Follow the same procedure outlined above except elute the DNA from the shredded gel pieces at 65 °C for 1 h. The pellet is resuspended in 10 μl of TE, samples are deep sequenced (Illumina single-end 40 nt, multiplexed to give ~20 million reads per concentration). PCR samples corresponding to higher RBP concentration conditions may need to be repeated due to over cycling. Over cycling is apparent from the appearance of PCR products with higher than expected molecular weights and, in extreme cases, the concomitant depletion of the correctly sized PCR product. This can be remedied by decreasing the number of PCR cycles and/or by decreasing the amount of cDNA used in the PCR reaction. Conditions with no or low RBP concentration may have very faint or even no PCR bands under initially tested conditions. In this case, the input cDNA quantity and/or the number of PCR cycles should be increased.

4.10 Variations on Input RNA Pool Composition

The RNA pool can be tailored to answer different questions about an RBP's binding affinity.
  i. *Randomized RNA* can be used for an unbiased motif inquiry. This approach tests the greatest diversity of RNA sequences, yielding unbiased analysis of RNA affinity.
  ii. *Predesigned sequences* can be used to resolve binding affinities to CLIP clusters, homologous sequences, mutated potential binding sites, etc. Thousands of DNA oligos with a T7 promoter can be synthesized on a custom microarray and the transcribed product used as input for RBNS.

iii. *Fragmented mRNA* could be size-selected and used as an RBNS input, limiting potential binding measurements to expressed RNA in a given cell type and to constrain binding to naturally occurring sequences.

The input RNA source should be tailored to the experimenter's interest, as the results will yield different types of information. An RBP's absolute biophysical preferences will be better assayed using randomized RNA, where all possible k-mers ≤ 10 nt are represented many times. The use of predesigned oligos or mRNA—which will typically have much lower diversity than random 40-mers—can yield an RBP's affinity for an extended oligo or RNA fragment rather than simply a motif. Depending on the complexity of the input pool, binding motifs may also be apparent, but more difficult to comprehensively survey using this design.

5. COMPUTATIONAL ANALYSIS

The computational analysis of RBNS data requires a specific set of software tools. We have organized the necessary computational tools for a basic analysis of an RBNS experiment into an open-sourced pipeline (https://github.com/alexrson/rbns_pipeline) available for general use. This section details the computational steps necessary for analyzing RBNS data, the usage of our computational pipeline implementing these steps, the sequence data format for input, and interpretation of the pipeline's output. For an RBNS experiment done on an RBP, the pipeline performs several analyses. The flow of the analyses of the data is illustrated in Fig. 3. Starting from a raw multiplexed fastQ file the pipeline can perform the following calculations, some of which are optional.

1. Demultiplex the reads by barcode
2. Count the k-mers for each barcode for each user-indicated value of k
3. Perform QC checks of the background and no input experiments
4. Calculate the enrichment (RBNS R) of each k-mer over background
5. Calculate the fraction of reads in which each k-mer is present
6. Run the streaming k-mer assignment algorithm (SKA) over the data
7. Analyze RNA structural preferences of the RBP
8. Calculate binding affinities (K_d values) for k-mers which show moderate to strong binding
9. Create various plots for the above analyses

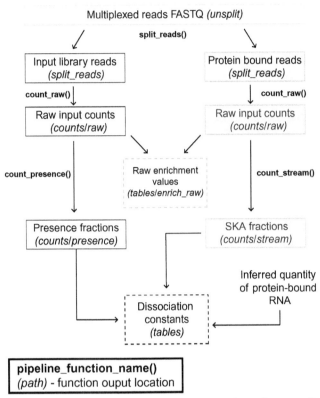

Figure 3 Detailed flowchart of the RBNS computational pipeline. Pipeline function names are followed by the paths to the associated files in parentheses.

As certain steps in the pipeline may prove time consuming for large data sets, the pipeline may be configured for parallelization across a compute cluster.

The pipeline can be downloaded and installed on a UNIX or Linux computer by following the instructions at https://github.com/alexrson/rbns_pipeline.

If you wish to perform structural calculations, you will need to install the Vienna RNAfold package, following the instructions on the providers' Web site: http://www.tbi.univie.ac.at/RNA/index.html#download.

Once the package and its dependencies have been installed, the RBNS pipeline can be run from source on a data set. The optional flags for the pipeline are listed in Table 2. Running the main script with the –h flag will display the help text for the RBNS pipeline along with the current options and settings ($ indicates the shell prompt): $ ./rbns_main.py –h.

Table 2 Description of Settings File Options for RBNS Pipeline Software.

Setting Name	Description
fastQ	The raw sequence data in fastQ format must be specified with a path to the file. It may optionally be compressed in gzip format. The multiplexed-sequencing barcodes are expected to be in the read identifiers as is usual with Illumina output
Mismatches allowed in barcode	The number of mismatches allowed in barcode (maximal Hamming distance from the known barcode) should be specified. This should be 1 in most cases, but possibly 0 if the barcodes are especially short (<5 nt) or 2 if they are especially long (>10 nt)
Barcodes	The barcodes relevant to an experiment must be specified. Since multiple may be run within a single HiSeq lane, only the barcodes for the protein of interest and the input RNA should be specified here. The order of the barcodes specifies the order of certain output plots. The barcode of the library must be input and included in this list
Read_len	The read length should be specified as the length of the randomized region of RNA used for the Bind-n-Seq experiment
3′ trim	If the actual read in the input fastQ is not the same length as the tested region, which end to trim should be specified with trim 3p and trim 5p. This may be the case, for example, when the size of the randomized oligo is smaller than the total read length and the sequencer reports the sequence of the oligo and primers concatenated
Experiment name	An experiment name used to label plots and the output directory of the whole run
Protein name	The name of the protein should be specified in the desired format for labeling of plots
Concentrations	The concentrations of protein in the Bind-n-Seq experiment should be specified in nanomolar units in the same order as the list of barcodes
Inhibitor concentration	The inhibitor (e.g., poly-IC) concentration (if any), input RNA concentration, number of washes, and temperature can also be set in this way. This setting only affects plot labels

Continued

Table 2 Description of Settings File Options for RBNS Pipeline Software.—cont'd

Setting Name	Description
Barcodes for sorting	In order to sort k-mers based on affinity one must select the concentration(s) of protein at which to do the sorting. This can be specified. Since it is unknown in advance where the greatest disparity between affinities is, one can refine after doing an initial run of the pipeline
RNA concentration	The concentration of RNA oligo in the binding reaction in nanomolar units
Relevant experimental variables	If the exact technical parameters of the experiment are still being worked out, this setting allows the user to specify which experimental variables (such as temperature or number of washes) should be labeled on the generated plots. Generally, these will be the ones that vary between library
Washes	Number of washes done for each library. We generally recommend one wash. This setting only affects plot labels
Temperature	The temperature of each binding reaction in the same order as the barcodes are specified
Library sequence barcode	The barcode sequence of the input library
k for concordance check	The RBNS pipeline does various concordance checks between libraries; the k-mer length for doing these checks should be specified
Number of k-mers for enrichment humps	The number of top k-mers to plot on various default plots can be specified
Naïve count	Doing the naïve count is optional and should be specified
Stream count	Doing the streamed counts aka SKA is optional and should be specified
Presence count	Doing the presence counts is optional and should be specified
k values for naïve/stream/presence counts	Which values of k to do for each counting method: naïve, SKA and presence. For each counting method the counting will take longer and the more values of k that are requested. It is not recommended to do values of $k > 10$ as the sampling is likely too low

Table 2 Description of Settings File Options for RBNS Pipeline Software.—cont'd

Setting Name	Description
Force naïve/stream/presence recount	Force the pipeline to recount rather than take the results already generated by a previous pipeline run. This should be done if, for example, the source code is changed. This can be specified individually for each counting method
Max reads to split	For testing purposes, one can do the whole pipeline on a subset of the reads. This can be done by specifying the number of reads one wishes to subsample
Known motif	If there is a known primary motif for the protein of interest it should be specified to display it differently in the output plots. If there are other motifs of interest, they may be specified as well
Motifs of intest	Additional motifs to include in output plots.
Free energy limits	To study the effects of RNA structure on protein binding, the RBNS pipeline optionally includes a simple analysis, binning of reads by free energy of folding. The edges of these free energy bins can be set here. k-mer Enrichment is analyzed separately in each of the free energy bins to give insight into the protein's structural preferences
Experiments to compare	The pipeline can also make automated comparisons to other RBNS experiments, such as a homolog of the protein. This can be set by specifying the path to the settings file for the other experiment(s)
Results Directory	The directory containing the output of the pipeline must be set
Work Directory	A work directory may also be set to store intermediate files during the process of running the pipeline. This directory is written to and read from frequently. Usually, this is the same as the results directory but can be changed to a local mount to tune performance
Error Directory	If the pipeline runs into an error the directory that it is logged in can be specified here

6. PIPELINE INPUTS

The pipeline requires two input files: the raw multiplexed sequence data for the experiment and a settings file detailing the settings for the experiment. In general, multiplexing all the concentrations needed for an RBNS experiment in a single Illumina HiSeq lane or NextSeq run provides adequate depth for analysis of k-mers up to about 10 bases in size, so the pipeline assumes that all the data for a factor are present in a single lane. For example, we typically run libraries for two proteins, with five nonzero protein concentrations each, plus a zero protein control (total of 11 libraries), in a single HiSeq 2000 lane, which typically yields over 200 million short reads. Given 11 multiplexed libraries, one can expect about 20 million reads per library, which is adequate for most analyses. Probing binding to k-mers of greater length requires exponentially greater read depth to obtain sufficient sampling, since the number of distinct k-mers grows as 4^k. For practical reasons, we choose to use short reads, typically of length 20–40 nt.

The settings are specified in a settings file that includes such information as the concentrations of the RBP in each experiment, the barcodes associated with each concentration, the desired values of k to be analyzed, and information to show in the outputted graphs. Several example settings files are included in the RBNS pipeline package, and the settings are summarized in Table 2.

7. *k*-mer ANALYSIS

The central question addressed by an RBNS experiment is usually: what RNA sequences does the protein bind to? Answering this question quantitatively requires a couple of algorithms, implemented in the pipeline.

7.1 Naïve Method

The naïve method of quantifying k-mer enrichment is intended to give a simple, straightforward estimate of the level of enrichment for all k-mers in the library over the input RNA pool. The naïve method simply counts the number of occurrences of each of the 4^k k-mers contained in all the reads in each RBNS condition. These counts are then normalized by dividing by the total count of all k-mers to give the k-mer frequency (fraction of total counts represented by the k-mer). If the input pool of oligonucleotides were perfectly uniformly distributed across k-mers, then these fractions would provide useful measures as is. However, since some biases are always present in the randomly synthesized input pool, these biases are corrected for by performing the same counting procedure on the input pool and then calculating the ratio of each k-mer's frequency in the selected pool to its frequency in the input pool (Fig. 1B). This gives the simple enrichment ratio, or RBNS R value, which is a good quantitative measure of a protein's relative affinity for different k-mers at a given protein concentration.

By default, the RBNS pipeline performs the naïve count and RBNS R calculations for all k-mers of specified sizes and creates tables of these values. It also optionally creates graphs of R values for k-mers of interest as a function of protein concentration, among other analyses.

7.2 Calculating B Values

One issue with R values is that there is an upper limit on the maximum possible R value that depends on the size of the k-mer relative to the oligonucleotide size used in the experiment (so R values underestimate affinity differences). This is because all of the k-mers present in the bound oligonucleotide are counted equally when measuring R values, though the protein might have bound to only one. RBNS R values of "offset" motifs (motifs that overlap a bound motif extensively, e.g., at $k-1$ positions) will also be enriched, even if they are not sufficient to confer binding. By making a few simple assumptions, a value called B can be calculated from the R value that provides a more accurate scale for relative binding affinity.

Consider the following simple model of an RBP interacting with a heterogeneous RNA population. Suppose that the RBP binds with high affinity to only one k-mer and with a low and constant affinity to all other k-mers. Suppose that in the binding equilibrium for a given concentration of RBP, the protein is B times more likely to bind its preferred k-mer than to any other. However, there are $4^k - 1$ times as many low-binding k-mers, assuming a uniform pool of oligonucleotides. Also, for every high-affinity k-mer pulled down, $\sim \lambda - k$ other k-mers will also be recovered, where λ is the length of the oligo. From these assumptions, one can calculate the value of R in terms of B as:

$$R = \sim 1 + \left(\frac{4^k B}{(B + 4^k)(\lambda - k + 1)} \right)$$

One can then solve for B in terms of R. The full derivation is provided in supplementary information of Lambert et al. (2014).

$$B = \frac{\sim (4^k - 1)(R - 1)(\lambda - k + 1)}{4^k + \lambda - k - R(\lambda - k + 1)}.$$

As R approaches its theoretical upper limit, the denominator becomes small and B becomes large. For example, RBFOX2 had an R value of 22 for the 6-mer UGCAUG in an experiment using a random 40-mer pool with 365 nM concentration of RBFOX2. Applying the equation above yields a B value of $\sim$900, implying $\sim$900-fold higher binding of RBFOX2 to UGCAUG than to nonspecific 6-mers under the tested conditions.

8. STREAMING k-mer ASSIGNMENT

There are several reasons why R values are not completely satisfactory from a theoretical perspective. First, they are highly dependent on the protein concentration: typically they increase with protein concentration, reach a maximum, and then decline. Second, they are related to but not directly proportional to the actual binding affinities of motifs. Third, a k-mer that highly overlaps with a strong binding k-mer will often have an elevated R value even if it does not bind to the protein of interest by itself. For these reasons and others, we seek to assign binding directly to different k-mers using a computational approach called streaming k-mer assignment or SKA (Fig. 4).

The gist of SKA is that every read in a selected library is fractionally assigned to the k-mers present in its sequence based on the continually updated knowledge of the weight of each k-mer in the library. This algorithm is based on the assumption that each read in the library is present because a protein molecule bound to one of the k-mers located within the bound oligonucleotide. Initially, we do not know which k-mer was specifically bound by the RBP so we equally credit each of the k-mers present in the read. However, as we process more and more k-mers, we achieve better and better knowledge of which k-mers are strongly binding and assign a larger fraction of each read's weight to the strongest k-mers present within the read's sequence. The set of weights for all k-mers converges to give the bound library fraction, the fraction of all protein-RNA binding sites located at each k-mer. This is a fairly important quantity as it gives insight into the equilibrium makeup of the protein and the set of all k-mer sequences. What follows is a much more in-depth description of the SKA algorithm and its implementation details in the RBNS pipeline.

A

Initalize weights (*w*) of all 5-mers

$w_{AAAAA} = 1$, $w_{AAAAC} = 1$, $w_{AAAAG} = 1$, ...

B

Process Read #1	Initial weight	Read #1 relative weight	Updated weight
A**UGCAUG**ANNNNNN			
AUGCA........	1	1/10	1.1
.UGCAU.......	1	1/10	1.1
..GCAUG......	1	1/10	1.1
...CAUGA.....	1	1/10	1.1
....AUGAN....	1	1/10	1.1
+ 5 more			

C

Process Read #2	Initial weight	Read #2 relative weight	Updated weight
NN**AGCAUG**GGNNNN			
NNAGC........	1	1/10.1	1.099
.NAGCA.......	1	1/10.1	1.099
..AGCAU......	1	1/10.1	1.099
...GCAUG.....	1	1.1/10.1	1.21
....CAUGG....	1	1/10.1	1.099
.....AUGGG...	1	1/10.1	1.099
+ 4 more			

Figure 4 Schematic of streaming *k*-mer assignment (SKA) algorithm. Illustrates how passes through the sequence reads are used to update the weights of *k*-mers in the SKA algorithm. An example of an enriched *k*-mer is shown in bold.

There are three steps in SKA:
1. Initialization,
2. First (rough) pass through reads, and
3. Subsequent (refining) passes through the reads.

Each step generates increasingly accurate estimates of the proportion of protein that is bound to each *k*-mer sequence. We call our current estimate of these proportions the *k*mers' weights. As the algorithm progresses, the weights converge. Initially, prior to looking at the data, all weights are initialized to a constant value. In our implementation, we choose to initialize with a starting value of 1, equivalent to saying that the *k*-mer is responsible for binding of one read (Fig. 4A).

The second step, the rough pass, steps through each of the bound reads and fractionally assigns its weight to the *k*-mers present in its sequence, in proportion to their current weights. So for each read, we identify all the *k*-mers present in its sequence. We then sum the current weights of each of the *k*-mers. The weights of these *k*-mers are then increased by their current weight normalized by the sum of the weights of all the *k*-mers present in the read. Thus, for example, if there are 10 *k*-mers present in single copy in the first read, then each of these *k*-mers' weights will be incremented by 1/10 to a value of 1.1. For the second read, however, if one of the *k*-mers present in the read's sequence was also present in the first, then it will be assigned a larger fraction of the second read's weight: 1.1/10.1, whereas each of the other *k*-mers will be incremented by only 1/10.1. Through these steps over millions of reads, the algorithm estimates the binding fraction for each *k*-mer. However, the fractional assignments are biased by the initialization of equal weight to all *k*-mers, especially for the early reads. For this reason, we do a subsequent pass or passes through the reads.

In subsequent passes, the weights are more finely honed. In each subsequent pass after the first, the reads are fractionally assigned to *k*-mers using the weights from the end of the previous pass. These fractional assignments are in turn used to fractionally assign reads in the next pass and so on until convergence. After convergence, the counts are normalized by the total number of reads for that library to give the fraction of bound RNA attributable to each *k*-mer. When analyzing RBFOX2 at a concentration of 370 n*M* binding to 6-mers, SKA estimated that 8.2% of RNA is attributable to binding to UGCAUG followed by other GCAUG- and GCACG-containing 6-mers. Note that this does not directly correspond to the expected fraction bound *in vivo*, since the pool of RNA differs in composition (e.g., GCACG is rarer than GCAUG, in part because it contains a hypermutable CpG dinucleotide).

The RBNS pipeline optionally runs SKA on all protein-selected libraries for specified values of *k*. We chose to analyze binding of 5-, 6-, and 7-mers of RBFOX2 using SKA. Running the algorithm is fairly computationally intensive and is ideally performed on a computer cluster, but the burden may be reduced by reducing the number of passes SKA makes through the reads. In general, SKA converges in fewer passes when the library is strongly biased (because the protein has high specificity at the given concentration) or when there are a large total number of reads in the library (>10 million). For RBFOX2 data with ~20 million reads per library, we found that the algorithm fully converged after only two passes. Larger numbers of passes may be required for proteins with lower sequence specificity or for data sets with lower coverage.

9. PRESENCE FRACTIONS

While the SKA algorithm estimates the fraction of bound k-mers that have each given k-mer, the presence fraction is the proportion of all reads, bound and unbound, that have each k-mer present somewhere within their sequence. The presence fraction is a measure directly proportional to the effective total concentration of each k-mer during binding equilibration and is required for properly controlling the sequence biases in the input library when calculating binding affinities.

The presence fractions can be easily determined by normalizing the naïve counts for each k-mer by the total number of reads in the input library. Note that the calculation of R values accounts for biases present in the input library, but values calculated by SKA do not control for these biases.

9.1 RNA-Binding Equilibrium

Throughout this chapter, it is worthwhile to bear in mind the following model of how the selection step of RBNS works, specifically with regard to the binding equilibrium.

In the experiment, an RBP is present and RNA oligos of length λ. The protein is assumed to bind RNA sequences of length k ($k < \lambda$) with a certain equilibrium constant specific to each k-mer. Binding and dissociation between the protein and the 4^k possible k-mers are assumed to be in simultaneous equilibrium, giving a system of 4^k dissociation equilibrium equations of the form:

$$K_{d(k\text{-mer})} = \frac{[\text{RBP}][\text{free } k\text{-mer}]}{[\text{RBP}:k\text{-mer}]}$$

During RBP capture, we magnetically select sequences containing a bound k-mer (corresponding to the denominators of the binding equilibria) and perform sequencing. The challenge then is to estimate from the full length read sequences of bound oligos what are the dissociation constants for the bound k-mers.

10. RELATIVE BINDING AFFINITY DETERMINATION

This section describes the calculations necessary to determine relative binding affinities for bound k-mers. Relative affinities can be estimated accurately over several orders of magnitude, but weakly bound motifs have higher signal to noise and cannot be estimated accurately. In practice, we do not consider motifs that bind with affinity more than $1000\times$ lower than the strongest binding motif.

We define the relative binding constant of a k-mer as the ratio of the k-mer's dissociation constant to the dissociation constant of the highest affinity k-mer. This can be thought of as how many fold weaker the binding is if all else (accessibility, copy number, and sequence context) is equal. The relative binding constant is expressible in terms of quantities that can be calculated from RBNS data:

$$K_{d,\text{relative}} = \frac{[L_{k\text{-mer, free}}]}{[L_{\text{best, free}}]} \times \frac{[RL_{\text{best}}]}{[RL_{k\text{-mer}}]}$$

In order to calculate dissociation constants for individual k-mers, one must first calculate the concentrations of bound k-mer and unbound k-mer for the strongest binding k-mer as well as other strong binders. This is done as follows. In the binding reaction, the total concentration of a k-mer is the concentration of oligos that contain that k-mer. We do not double count multiple occurrences in a single oligo, because we assume binding to one site is sufficient to ensure binding. Therefore, the total concentration of k-mer in solution is the concentration of the oligo times the fraction of oligos that contain each k-mer. This fraction is simply the k-mer presence fraction, which is calculated by the pipeline.

One also needs to know the total concentration of bound RNA oligo in the binding equilibrium step, which can be measured by taking a fluorescent trace after the selection step as described in Section 4.7. The concentration of bound k-mer is then simply the concentration of bound oligo multiplied by the SKA fraction for that k-mer.

The concentration of free RBP, which appears in the definition formula for absolute dissociation constants, could be determined by taking the difference between the concentration of bound oligo and the known concentration of protein. However, relative uncertainties in these values are increased in the subtraction operation, and it is usually not known what fraction of protein molecules is properly folded. Therefore, we recommend calculating *relative* dissociation constants, defined as the ratio of the K_d of a k-mer to the K_d of the highest affinity k-mer. In this case, the free RBP concentration term cancels out, and the effects of other systematic biases such as varying salt concentrations or proportions of unfolded protein are also reduced. This allows the relative K_d values to be compared to values from traditional low-throughput measurements such as gel shift (EMSA) and SPR. $K_{d,relative}$ values can be calculated from any selected library, in principle, but we find the values most accurate in the library where the enrichment of the strongest binding k-mer is maximal. This usually occurs at an intermediate concentration of RBP where the concentration of bound RNA is close to the K_d. Since this information is not known in advance, this is one of the advantages of performing RBNS with multiple concentrations in parallel.

An example calculation for RBFOX is shown in Table 3. The upper portion shows the input information about protein concentration and total RNA–protein complex concentration, which is needed to make the intermediate calculations for each k-mer. The middle portion of the table shows the calculation using the SKA data for UGCAUGU, the strongest binding 7-mer and UGCACGU, an additional 7-mer for which we generated SPR data. We show experiments here covering a wide range of RBP concentrations, but in practice a smaller number of experiments covering the middle of the RBP concentration range is sufficient. While the total concentration of RNA oligos containing UGCAUGU is assumed to be constant across experiments, the proportion of complexed UGCAUGU oligos increases to a maximum at 365 nM RBFOX. While most of the experiments we have shown have low signal and therefore underestimate $K_{d,relative}$, toward the middle of the concentration range (shaded row) is selected because the signal for this library is maximized. At this concentration, the $K_{d,relative}$ value of UGCACGU, 4.13, can be compared with the value calculated by SPR of 6.43. Overall, there is a close correlation between $K_{d,relative}$ measured by RBNS and SPR ($r=0.94$, $P<0.001$) (Lambert et al., 2014).

11. RBNS QUALITY CONTROL

There are several modes by which an RBNS experiment can fail. The following quality control checks should be performed:

1. Some k-mers should be significantly enriched above background (assuming there is reason to believe that the protein binds short RNA motifs)
2. RBNS R values of k-mers calculated for adjacent (similar) RBP concentrations should be well correlated, particularly for k-mers with R values that are above background

Table 3 Using SKA to Calculate Relative Binding Affinity

Bind-n-Seq Library	[RBFOX] (nM)	[L$_{total}$] (nM)	[RL$_{total}$] (nM)
1	1.5	1000	0.37
2	4.5	1000	0.40
3	14	1000	1.10
4	40.5	1000	0.73
5	121	1000	5.44
6	365	1000	5.92
7	1100	1000	12.92
8	3300	1000	16.97
9	9800	1000	28.92

UGCAUGU

Bind-n-Seq Library	Streaming Fraction	L$_{UGCAUGU}$ (nM)	RL$_{UGCAUGU}$ (nM)	L$_{UGCAUGU,free}$/RL$_{UGCAUGU}$
1	6.23×10^{-6}	1.4	2.27×10^{-6}	6.23×10^{5}
2	6.37×10^{-6}	1.4	2.56×10^{-6}	5.53×10^{5}
3	8.45×10^{-6}	1.4	9.25×10^{-6}	1.53×10^{5}
4	5.42×10^{-4}	1.4	3.96×10^{-4}	3.57×10^{3}
5	8.24×10^{-3}	1.4	4.48×10^{-2}	3.06×10^{1}
6	1.28×10^{-2}	1.4	7.57×10^{-2}	1.77×10^{1}
7	3.55×10^{-3}	1.4	4.59×10^{-2}	2.99×10^{1}
8	1.55×10^{-3}	1.4	2.63×10^{-2}	5.28×10^{1}
9	1.01×10^{-3}	1.4	2.93×10^{-2}	4.74×10^{1}

UGCACGU

Bind-n-Seq Library	Streaming Fraction	L$_{UGCACGU}$ (nM)	RL$_{UGCACGU}$ (nM)	L$_{UGCACGU,free}$/RL$_{UGCACGU}$	$K_{D,relative}$
1	6.9×10^{-6}	1.6	2.5×10^{-6}	6.21×10^{5}	1.0
2	7.0×10^{-6}	1.6	2.8×10^{-6}	5.53×10^{5}	1.0
3	8.5×10^{-6}	1.6	9.3×10^{-6}	1.67×10^{5}	1.1
4	1.2×10^{-5}	1.6	8.7×10^{-6}	1.79×10^{5}	49.9
5	7.82×10^{-4}	1.6	4.25×10^{-3}	3.65×10^{2}	11.9

Table 3 Using SKA to Calculate Relative Binding Affinity—cont'd
UGCACGU

Bind-n-Seq Library	Streaming Fraction	$L_{UGCACGU}$ (nM)	$RL_{UGCACGU}$ (nM)	$L_{UGCACGU,\ free}/RL_{UGCACGU}$	$K_{D,relative}$
6	3.5×10^{-3}	1.6	2.1×10^{-2}	7.32×10^{1}	4.13
7	2.1×10^{-3}	1.6	2.7×10^{-2}	5.62×10^{1}	1.9
8	1.1×10^{-3}	1.6	2.0×10^{-2}	7.88×10^{1}	1.5
9	9.8×10^{-4}	1.6	2.8×10^{-2}	5.37×10^{1}	1.1

Example $K_{d,relative}$ calculations for UGCACGU. In each section of the table, the shaded row indicates the RBNS experiment with maximal signal as measured by the maximum R value.
The upper portion of the table shows the experimental data that goes into calculating $K_{d,relative}$ for each RBNS experiment for our example of RBFOX. This includes the concentration of RBP, total concentration of RNA oligo, and the estimated total concentration of RNA–protein complex from Bioanalyzer trace.
The middle portion of the table shows calculations for how to use the output of SKA to calculate the ratio of bound to unbound oligo for the top 7-mer, UGCAUGU. The streaming fraction is taken from the output of SKA. The $total$ concentration of UGCAUGU containing oligo [$L_{UGCAUGU}$] is estimated from the input library experiment; it is the product of the presence fraction of the k-mer and the total concentration of oligo, [L_{total}]. The concentration of $bound$ UGCAUGU, [$RL_{UGCAUGU}$], containing oligo is estimated from the streaming fraction; it is the product of the streaming fraction and [RL_{total}] from the top section of the table. The final column shows the ratio of free to bound oligo.
The lower portion is similar to the middle portion but for the secondary k-mer UGCACGU, for which we have measured the K_d via SPR. The final column shows the $K_{d,relative}$ for UGCAUG. It is the ratio of free UGCACGU to bound UGCACGU divided by the ratio of free UGCAUGU to bound UGCAUGU (see formula in text).

3. If RBNS R values of top k-mers are plotted as a function of protein concentration across a broad range of concentrations, a unimodal curve should be evident

12. CONCLUSIONS

RBNS is a straightforward, fairly cost-effective experiment to quantitatively assess comprehensive sequence and structural specificities of RBPs. The experimental and computational methods outlined here enable identification of bound motifs and estimation of binding constants for an RBP of interest. The iterative binding and amplification used in SELEX approaches identify high-affinity motifs but have little power to measure moderate-affinity motifs and contextual binding preferences. We have previously shown that RBNS and CLIP data can provide complementary information. CLIP-seq surveys $in\ vivo$ binding sites, but is subject to cross-linking biases, has relatively high background and rarely if ever achieves saturation, limiting the identification of binding sites. RBNS more directly assesses binding specificity and can be used to distinguish likely true-positive and false-positive CLIP motifs.

Many variations can be introduced to tailor RBNS to the RBP or scientific question of interest. For example, potential synergistic or competitive binding can be assayed by performing RBNS in the presence of two recombinant RBPs, one tagged with the epitope used for pull down in RBNS and the other untagged (N. J. L. and C. B. B., unpublished data). RBNS performed in this manner has the potential to determine contexts where one RBP can displace another or enhance its binding. As discussed above, the input RNA in RBNS can also be varied. Instead of randomized RNA, fragmented mRNA or *in vitro*-transcribed RNA from custom array-synthesized oligonucleotides can be used in the binding reaction. This seemingly minor variation substantially alters the interpretation of the results; while the reduced sequence complexity relative to random sequences may decrease the precision of motif identification; it can also enable assessment of affinity relative to specific segments of the transcriptome, increasing the dynamic range of R values observed and potentially enabling assessment of subtle features such as cooperative binding (N. J. L. and C. B. B., unpublished data). The computational pipeline for analysis is under active development. Current efforts are aimed at automatic inference of the optimal k-mer size, of the number of distinct motif classes bound by the RBP, and of the influence of RNA secondary structure on binding. Updated versions of the pipeline will be posted at the github site referenced above.

REFERENCES

Bass, B. L., Nishikura, K., Keller, W., Seeburg, P. H., Emeson, R. B., O'Connell, M. A., et al. (1997). A standardized nomenclature for adenosine deaminases that act on RNA. *RNA*, *3*(9), 947–949.

Buenrostro, J. D., Araya, C. L., Chircus, L. M., Layton, C. J., Chang, H. Y., Snyder, M. P., et al. (2014). Quantitative analysis of RNA-protein interactions on a massively parallel array reveals biophysical and evolutionary landscapes. *Nature Biotechnology*, *32*(6), 562–568. http://dx.doi.org/10.1038/nbt.2880.

Burgess, R. R., & Deutscher, M. P. (Eds.), (2009). *Guide to protein purification* (2nd ed, Vol. 463, pp. 1–851). *Methods in Enzymology*. Academic Press.

Daubner, G. M., Clery, A., Jayne, S., Stevenin, J., & Allain, F. H. (2012). A syn-anti conformational difference allows SRSF2 to recognize guanines and cytosines equally well. *EMBO Journal*, *31*(1), 162–174. http://dx.doi.org/10.1038/emboj.2011.367.

Lambert, N., Robertson, A., Jangi, M., McGeary, S., Sharp, P. A., & Burge, C. B. (2014). RNA Bind-n-Seq: Quantitative assessment of the sequence and structural binding specificity of RNA binding proteins. *Molecular Cell*, *54*(5), 887–900. http://dx.doi.org/10.1016/j.molcel.2014.04.016.

Martin, L., Meier, M., Lyons, S. M., Sit, R. V., Marzluff, W. F., Quake, S. R., et al. (2012). Systematic reconstruction of RNA functional motifs with high-throughput microfluidics. *Nature Methods*, *9*(12), 1192–1194. http://dx.doi.org/10.1038/nmeth.2225.

Ray, D., Kazan, H., Chan, E. T., Pena Castillo, L., Chaudhry, S., Talukder, S., et al. (2009). Rapid and systematic analysis of the RNA recognition specificities of RNA-binding proteins. *Nature Biotechnology*, *27*(7), 667–670. http://dx.doi.org/10.1038/nbt.1550.

Ray, D., Kazan, H., Cook, K. B., Weirauch, M. T., Najafabadi, H. S., Li, X., et al. (2013). A compendium of RNA-binding motifs for decoding gene regulation. *Nature*, *499*(7457), 172–177. http://dx.doi.org/10.1038/nature12311.

Sugimoto, Y., Konig, J., Hussain, S., Zupan, B., Curk, T., Frye, M., et al. (2012). Analysis of CLIP and iCLIP methods for nucleotide-resolution studies of protein-RNA interactions. *Genome Biology*, *13*(8), R67. http://dx.doi.org/10.1186/gb-2012-13-8-r67.

Tome, J. M., Ozer, A., Pagano, J. M., Gheba, D., Schroth, G. P., & Lis, J. T. (2014). Comprehensive analysis of RNA-protein interactions by high-throughput sequencing-RNA affinity profiling. *Nature Methods*, *11*(6), 683–688. http://dx.doi.org/10.1038/nmeth.2970.

Tuerk, C., & Gold, L. (1990). Systematic evolution of ligands by exponential enrichment: RNA ligands to bacteriophage T4 DNA polymerase. *Science*, *249*(4968), 505–510.

Ule, J., Jensen, K. B., Ruggiu, M., Mele, A., Ule, A., & Darnell, R. B. (2003). CLIP identifies Nova-regulated RNA networks in the brain. *Science*, *302*(5648), 1212–1215. http://dx.doi.org/10.1126/science.1090095.

SECTION III

Large RNA-Protein Assemblies

CHAPTER SIXTEEN

Using Molecular Simulation to Model High-Resolution Cryo-EM Reconstructions

Serdal Kirmizialtin*,[†,‡], Justus Loerke[§], Elmar Behrmann[¶], Christian M.T. Spahn[§], Karissa Y. Sanbonmatsu[†,‡,1]

*Department of Chemistry, New York University, Abu Dhabi, United Arab Emirates
[†]New Mexico Consortium, Los Alamos, New Mexico, USA
[‡]Theoretical Biology and Biophysics, Theoretical Division, Los Alamos National Laboratory, Los Alamos, New Mexico, USA
[§]Institut für Medizinische Physik und Biophysik, Charité—Universitätsmedizin Berlin, Berlin, Germany
[¶]Structural Dynamics of Proteins, Center of Advanced European Studies and Research (CAESAR), Bonn, Germany
[1]Corresponding author: e-mail address: kys@lanl.gov

Contents

1. Introduction 498
2. Theory 501
   2.1 Molecular Model 501
   2.2 Scoring Function 502
   2.3 Masking the Cryo-EM Map 502
3. Methods 503
4. Results 505
5. Summary 508
Acknowledgments 510
References 510

Abstract

An explosion of new data from high-resolution cryo-electron microscopy (cryo-EM) studies has produced a large number of data sets for many species of ribosomes in various functional states over the past few years. While many methods exist to produce structural models for lower resolution cryo-EM reconstructions, high-resolution reconstructions are often modeled using crystallographic techniques and extensive manual intervention. Here, we present an automated fitting technique for high-resolution cryo-EM data sets that produces all-atom models highly consistent with the EM density. Using a molecular dynamics approach, atomic positions are optimized with a potential that includes the cross-correlation coefficient between the structural model and the cryo-EM electron density, as well as a biasing potential preserving the stereochemistry and secondary structure of the biomolecule. Specifically, we use a hybrid structure-based/*ab initio* molecular dynamics potential to extend molecular dynamics fitting. In addition, we find that simulated annealing integration, as opposed to straightforward molecular dynamics integration, significantly improves performance. We obtain atomistic models of the human ribosome consistent with high-resolution cryo-EM reconstructions of the human ribosome. Automated methods such as these have the potential to produce atomistic models for a large number of ribosome complexes simultaneously that can be subsequently refined manually.

1. INTRODUCTION

Mechanistic studies of the bacterial ribosome over the past decade have proceeded at a rapid rate, elucidating many of the steps of protein synthesis elongation. In addition to crystallographic (Dunkle et al., 2011; Tourigny, Fernandez, Kelley, & Ramakrishnan, 2013), cryo-electron microscopy (cryo-EM) (Dashti et al., 2014), single molecule (Blanchard, Kim, Gonzalez, Puglisi, & Chu, 2004; Munro, Wasserman, Altman, Wang, & Blanchard, 2010; Olivier et al., 2014; Wang et al., 2012) and rapid kinetics studies (Rodnina & Wintermeyer, 2011), computational studies by other groups have been performed on ratchet motion (Bock et al., 2013; Ishida & Matsumoto, 2014; Kurkcuoglu, Doruker, Sen, Kloczkowski, & Jernigan, 2008; Trylska, Konecny, Tama, Brooks, & McCammon, 2004), decoding (Adamczyk & Warshel, 2011; Zeng, Chugh, Casiano-Negroni, Al-Hashimi, & Brooks, 2014), protein translocation (Ishida & Hayward, 2008; Rychkova, Mukherjee, Bora, & Warshel, 2013), and other aspects of the ribosome (Baker, Sept, Joseph, Holst, & McCammon, 2001; Trabuco et al., 2010). Outstanding research has advanced molecular dynamics of RNA (Bergonzo et al., 2014; Cheatham & Case, 2013; Chen, Marucho, Baker, & Pappu, 2009; Henriksen, Davis, & Cheatham, 2012; Liu, Janowski, & Case, 2015). The publication of the first crystal structure of a eukaryotic ribosome in 2011 has opened the field to new possibilities, as variable regions of the eukaryotic ribosome are thought to be conduits to a large number of posttranscriptional regulatory pathways (Ben-Shem et al., 2011; Jenner et al., 2012; Melnikov et al., 2012). Cryo-EM has played a key role in elucidating the structure and function of ribosome complexes: large-scale conformational changes, movement of ligands through the ribosome, factor-binding interactions, along with many other aspects of ribosome function have been advanced by cryo-EM (Agrawal et al., 1996, 2000; Beckmann et al., 2001; Connell et al., 2008; Frank &

Agrawal, 2000; Frank, Radermacher, Wagenknecht, & Verschoor, 1988; Frank et al., 1995; Ratje et al., 2010; Schuette et al., 2009; Spahn et al., 2001, 2004; Valle et al., 2002, 2003; Villa et al., 2009; Wagenknecht, Carazo, Radermacher, & Frank, 1989). The recent advent of direct electron detectors has opened a new frontier for cryo-EM, producing 3D structures of the ribosome with comparable resolution to X-ray crystallography (Amunts et al., 2014; Fernandez, Bai, Murshudov, Scheres, & Ramakrishnan, 2014; Fernandez et al., 2013).

Over the past decade, a variety of techniques have been developed to produce 3D models of the ribosome consistent with cryo-EM reconstructions, including real-space refinement (Gao et al., 2003), normal mode fitting (Gorba, Miyashita, & Tama, 2008), and molecular dynamics simulation (Budkevich et al., 2014; Muhs et al., 2015; Orzechowski & Tama, 2008; Ratje et al., 2010; Trabuco, Villa, Mitra, Frank, & Schulten, 2008; Villa et al., 2009; Whitford et al., 2011), among others. Real-space refinement showed details of intersubunit rotation (Gao et al., 2003), while normal mode flexible fitting produced models of EF-G complexed with the ribosome in various functional states (Gorba et al., 2008). A landmark study used molecular dynamics simulation biased by cryo-EM electron density (MDFF) to produce more accurate models of the ternary complex bound to the ribosome (Trabuco et al., 2008; Villa et al., 2009). A similar method was developed simultaneously by Tama and coworkers (Orzechowski & Tama, 2008). The above method also produced new insights into the protein-conducting channel by a bound ribosome, and translational stalling in bacterial and eukaryotic systems (Armache et al., 2010; Becker et al., 2011; Seidelt et al., 2009). Structure-based molecular dynamics fitting (MDfit) methods allow molecular fitting that preserves stereochemistry present in initial starting structures, while offering the key advantage of running on a single desktop workstation (Muhs et al., 2015; Ratje et al., 2010; Whitford et al., 2011). This method revealed new translocational intermediates and a novel conformational change of the mammalian ribosome (Budkevich et al., 2014; Ratje et al., 2010).

More specifically, the MDfit method begins with a potential energy function based on the initial configuration. An energetic weight based on knowledge of the target is introduced, yielding a downhill energy profile, where the target configuration corresponds to the global minimum. For example, to obtain structural models of the bacterial ribosome in the rotated configuration, we defined a starting potential energy function based on a classical ribosome configuration. We defined a biasing term based on correlations between the cryo-EM reconstruction of the rotated state and the simulated density, determined from snapshots of the simulated structure throughout the simulation. This produced all-atom models highly consistent with cryo-EM reconstructions of the rotated states and also revealed the TI^{PRE} and TI^{POST} configurations (Ratje et al., 2010). The method was also used to construct one of the first all-atom models of the human ribosome, revealing a new conformational change specific to eukaryotic ribosomes (subunit rolling) (Budkevich et al., 2014).

The advent of direct-detector high-resolution cryo-EM studies and high-resolution eukaryotic ribosome structures is fueling a renaissance in ribosome mechanism (Amunts et al., 2014; Ben-Shem et al., 2011; Fernandez et al., 2013, 2014; Jenner et al., 2012). Higher resolution cryo-EM now allows many species of ribosomes to be captured in a wide range of functional states. With regard to eukaryotic ribosome research (80S ribosome), external regions of the ribosome have highly variable sequences and are thought to bind a variety of factors involved in posttranscriptional gene regulation (Jenner et al., 2012). In addition, comparison between human and bacterial ribosomes is essential to understand antibiotic action, as the ribosome is one of the main antibiotic targets. In light of the numerous new high-resolution cryo-EM structures of the ribosome (Amunts et al., 2014; Fernandez et al., 2013, 2014), there is high demand for methods capable of producing all-atom models consistent with these data. The methods described above are predominantly based on lower resolution data and leverage the lower resolution, either by using coarse-grained approaches, constraints, or native state biases (Ahmed, Whitford, Sanbonmatsu, & Tama, 2012; Ratje et al., 2010; Trabuco et al., 2008; Wang & Schroder, 2012). We note that one new successful and promising method uses self-guided Langevin dynamics (Wu, Subramaniam, Case, Wu, & Brooks, 2013). There are few fully automated methods tailored to produce all-atom models highly consistent with the new high-resolution cryo-EM data. Currently, some groups are using existing crystallography software (Amunts et al., 2014; Fernandez et al., 2013, 2014) and adjustments of domains and subregions of the macromolecular complex for the fitting process. Automating portions of this process has the potential to increase efficiency and accuracy. Here, we extend the MDfit method by combining an *ab initio* molecular dynamics potential with the all-atom structure-based potential, maintaining the capability to preserve the stereochemistry of initial models while simultaneously accessing alternative folds. Following the Korostelev et al. crystallographic refinement by simulating annealing and other treatments (Brunger, Kuriyan, & Karplus, 1987; Korostelev, Laurberg, & Noller, 2009), we employ a simulating annealing strategy, as opposed to a straightforward molecular dynamics sampling.

2. THEORY

The aim is to fit the atomic structure to high-resolution density derived from the cryo-EM reconstruction while maintaining correct stereochemistry. We merge two previous molecular simulation techniques in an effort to improve the fit: (i) an *ab initio* force field (Trabuco et al., 2008) and (ii) a structure-based potential that preserves the native stereochemistry and 3D fold of the initial structure (Ratje et al., 2010; Whitford et al., 2011). Specifically, we use a scoring function:

$$E = E(\mathbf{R}) + E^{NC}(\mathbf{R}) + w(1 - \text{CCC}) \quad (1)$$

here, $\mathbf{R}$ is the molecular coordinate vector, $E(\mathbf{R})$ is the molecular potential energy term, and $E^{NC}(\mathbf{R})$ is the native contact potential bias that restrains the atomic model to preserve its secondary structure and maintains tertiary native contacts. While this term is a localized potential, it effectively biases the system toward the overall 3D fold of the initial starting structure. The last term is the biasing function that creates a downhill potential, moving the atomic coordinates toward the cryo-EM map. Details of each term and their weighting are described below.

2.1 Molecular Model

The first term in Eq. (1) is the potential energy of the molecular model. This term includes the energy contribution from all bonding terms of the classical molecular mechanics force field, including bond, angle, torsion, and improper torsion, as well as a nonbonding term to account for volume exclusion. The nonbonding term is a pairwise summation of all-atom pairs, $\sum_{\text{pairs } i,j} C12_{ij}/r_{ij}^{12}$, where $C12_{ij}$ is a constant derived from van der Waals parameters of each atom pair i, j with $i - j > 4$. To maintain the native contacts and stereochemistry, an additional potential term was introduced. For a given reference structure coordinate set, $\mathbf{R_0}$ (for example, an atomic model that comes from crystallography experiments), we compute all pairwise contacts for pairs such that $i - j > 4$ and with a distance cutoff condition of $d_{ij0} = |r_{i0} - r_{j0}| \leq 4\text{Å}$. These contacts are protected by adding a term to the molecular potential. For a given configuration represented by coordinate

R, this term is a summation of all pairwise contacts $E^{NC}(\mathbf{R}) = \sum_{\text{pairs } i,j} w(d_{ij})$, where

$$w(d_{ij}) = \begin{cases} \frac{1}{2}\lambda(d_{ij} - d_{ij0})^2 & d_{ij} \leq 2d_{ij0} \\ \frac{1}{2}\lambda d_{ij0}(2d_{ij} - 3d_{ij0}) & d_{ij} > 2d_{ij0} \end{cases} \quad (2)$$

here, λ is the strength of native contact potential bias. We use $10 < \lambda < 50$ kJ/nm^2 in our study.

2.2 Scoring Function

To measure the quality of fit and bias the atomic model toward the cryo-EM map, we used a cross-correlation function, which is introduced in earlier publications (Gorba et al., 2008; Ratje et al., 2010). Cryo-EM reconstructions are represented by intensities on a cubic lattice stored in a vector $\rho^{EM}(k)$, where k is the index for grid space points in all directions. To measure the quality of fit between the map and the atomic model, a simulated map is computed from atomic coordinates using the same grid spacing as the experimental map and assuming a Gaussian distribution of electron density for each atom. The simulated electron density at grid site k due to atom j at $\mathbf{r}_j$ is

$$\rho(k, \mathbf{r}_j) = \exp\left[-\frac{1}{2}\left(\frac{\mathbf{r}_k - \mathbf{r}_j}{\sigma}\right)^2\right] \quad (3)$$

where 2σ is the resolution of cryo-EM map.

The similarity between the cryo-EM map and the simulated map is computed in a similar manner to MDfit with a cross-correlation function of the form:

$$\text{CCC} = \frac{\sum_k \rho^{EM}(k)\rho(k)}{\sum_k \rho^{EM}(k)^2 \sum_k \rho(k)^2} \quad (4)$$

2.3 Masking the Cryo-EM Map

In certain cases, a small region of cryo-EM electron density may correspond to an unresolved binding factor or a highly dynamic region of the ribosome that resides in multiple states that are difficult to separate during clustering.

Since a given atomic model represents only one conformational state in a fit, these regions cannot be fit by a single atomistic model. Elimination of these regions produces a cleaner map with less space to search for a better fit. To eliminate the extra occupancy in the electron density, we define a vector, $\mathbf{q} = (q_1, \ldots, q_N)$, where N is the total number of grid points in the simulated and experimental cryo-EM maps. Each element of the vector is filled as:

$$q(k) = \begin{cases} 0 & \rho(k) = 0 \\ 1 & \text{otherwise} \end{cases} \quad (5)$$

Our filtered map is given by $\rho'^{EM} = \rho^{EM} \odot \mathbf{q}$ where symbol $\odot$ denotes the element wise product.

3. METHODS

To effectively search the conformational space, we used the modified potential described above with a simulated annealing strategy. The constraints to the native topology are added by using the *genres* module of GROMACS. The attractive part of the Lennard–Jones interaction was set to zero in the program. The method used the MDfit code as a starting point, which itself is a modified version of gromacs 4.5.5. The calculations are performed with a desktop computer. No specialized hardware is required.

We integrate the equations of motion using stochastic dynamics where a friction term and a noise term are added to Newton's equation of motion as follows:

$$m_i \frac{d^2 \mathbf{r}_i}{dt^2} = -m_i \gamma_i \frac{d \mathbf{r}_i}{dt} + F(\mathbf{R}) + R_i \quad (6)$$

where γ_i is the friction coefficient for particle i and R_i is the noise term satisfying the fluctuation–dissipation theorem as $\langle R_i(t) R_j(t+s) \rangle = 2 m_i \gamma_i k_B T \delta_{ij}$.

For the *ab initio* portion of the potential, we use GROMOS force field with 53A6 parameter set (Oostenbrink, Villa, Mark, & Van Gunsteren, 2004). We emphasize that the method is not limited to a particular force field and that any *ab initio* molecular dynamics force field can be applied.

To effectively find the optimum fit, we implement simulated annealing. Here, the temperature of the simulation is reduced in n stages. At the beginning of each stage, we set the temperature to $T_n = T_0 / 2^{(n-1)}$, where T_0 is the initial temperature. Each simulated annealing simulation started with an initial temperature of $T_0 = 500\,\text{K}$. We cool the system from 500 K to a final temperature of 15 K and we repeat the procedure for three times in each fit.

Our protocol of fitting the high-resolution cryo-EM map is illustrated in Fig. 1. The method starts with rigid body fitting of the model to the cryo-EM map using Chimera (Goddard, Huang, & Ferrin, 2005). Next, we use simulated annealing molecular dynamics to sample the conformational space of the modified potential given in Eq. (1), slowly cooling the model from 500 to 15 K, repeating three times.

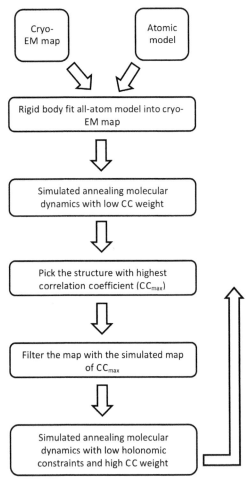

Figure 1 A schematic description of the multistep fitting of atomic models to cryo-EM map.

The rigid body fit may not produce a high correlation between the atomic model and the cryo-EM map. If there is a significant difference between the model and map, then a high bias toward the map may distort the structure. Therefore, the biasing strength toward the cryo-EM map is low in our first iteration. The bias to the native contacts is high at this stage. To achieve this, we use a smaller value of $w = N_{atoms}$ (see Eq. 1), where N_{atoms} is the number of atoms in the molecule. After the simulation is completed, we choose the structure with the highest cross-correlation coefficient (CCC) as the current model and move to the masking stage of our protocol. To mask the map, we created a vector of zeros and ones from the best-fitted model (see Eq. 5). This vector is multiplied pairwise with the original cryo-EM map as detailed above. The grid points that are zero in the simulated map become zero in the original map. The masked map is used to launch another simulated annealing molecular dynamics simulation. Here, we use a higher weight to bias our model toward the cryo-EM map. For that reason, we choose $w = mN_{atoms}$ where m starts at 2 and is incrementally increased in each iteration.

Importantly, our fits include two different biasing terms with adjustable weights. One corresponds to the bias toward native contacts, while the second bias is toward the cryo-EM reconstruct. To prevent overfitting, we chose these such that the two are on the same order of magnitude.

4. RESULTS

To test the protocol, we study two model systems. Model I is the h40–h44 region of the small subunit of the human ribosome. Figure 2 shows its structure before and after fitting to the cryo-EM reconstruction at near-atomic resolution. Model II is the entire small subunit of the protein–RNA complex of the human ribosome (Fig. 3). The initial atomic model for each structure used our previous all-atom model of the human ribosome as a starting point (Budkevich et al., 2014).

To effectively search the conformational space of the modified potential, we used stochastic dynamics. In obtaining the optimum value for the friction term in Eq. (6), we note that lower friction coefficients often improve sampling. However, this value also needs to be sufficiently high to remove the excess heat produced during minimization. Figure 4 shows the time evolution of cross-correlation coefficient (Eq. 4) for different friction coefficients. At the lowest value of friction coefficient $\gamma = 0.01(1/ps)$, the conformational search fails to sample the modified potential effectively. Higher

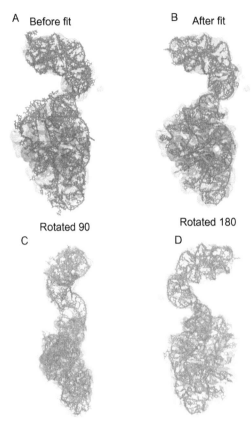

Figure 2 Atomic model of a small region from human ribosome before flexible fitting by the simulated annealing molecular dynamics (A), and after the fit (B–D).

friction coefficients such as $1 \leq \gamma \leq 100(1/ps)$, however, display an increase in the CCC.

We emphasize that our method uses a hybrid potential, generalizing the structure-based potential to also include *ab initio* molecular dynamics potential terms. To understand additional effects, such as (a) changing the integration to a simulated annealing protocol and (b) filtering the map to remove unresolved regions, we performed a comparison study with three cases: (i) straightforward molecular dynamics integration, (ii) simulated annealing, and (iii) simulated annealing plus filtering. In this comparison, we are interested in finding the method that gives the highest score in the correlation coefficient when sufficient sampling is performed (Fig. 5). Using an equal time of simulation in each protocol, we find that simulated annealing

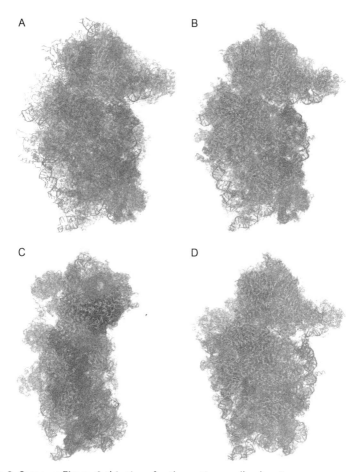

Figure 3 Same as Figure 2 this time for the entire small subunit.

molecular dynamics gives a higher correlation score than straightforward molecular dynamics integration. Interestingly, our protocol of filtering the map further improves the overall fit. A critical consideration is that for a given conformation, the correlation coefficient computed from iteration n will always be higher than when the map of the previous iteration $n-1$ is used. This follows from the fact that filtering reduces the number of nonzero grid points, thus reducing the denominator of the CCC in Eq. (4). This is not desirable since a configuration must have the same value of CCC between each iteration so that we can compare the progress over time. To account for this, we normalize the CCC of a given configuration, $\mathbf{R}$ in iteration n as $\overline{\text{CCC}}^n(\mathbf{R}) = (\text{CCC}^{n-1}(\mathbf{R}_0)/\text{CCC}^n(\mathbf{R}_0))\text{CCC}^n(\mathbf{R})$,

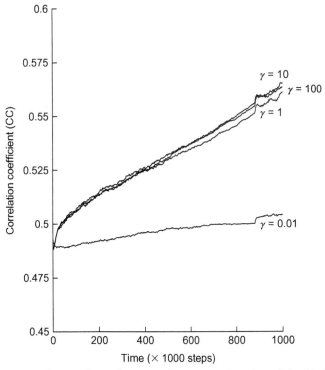

Figure 4 Time evolution of correlation coefficient as a function of the friction coefficient, γ used in stochastic molecular dynamics simulation method.

where $CCC^{n-1}(\mathbf{R}_0)$ is the correlation coefficient of a reference configuration $\mathbf{R}_0$ at iteration $n-1$, $CCC^n(\mathbf{R}_0)$ is its value in iteration n (due to the change in the map), and $CCC^n(\mathbf{R})$ is the correlation of $\mathbf{R}$ in iteration n before the normalization. Using this method, we compute the CCC between each iteration. Our results show an improvement in CCC scores for Model II. We obtain a similar performance in Model I. Our results show 0.812, 0.821, and 0.827 for the maximum CCC score for MD, simulated annealing, and simulated annealing with filtering, respectively. Figure 6 shows the fitted structure together with the change in correlation coefficient during the simulations.

5. SUMMARY

We have adapted the MDfit technique to high-resolution cryo-EM data by including an *ab initio* molecular dynamics potential, allowing

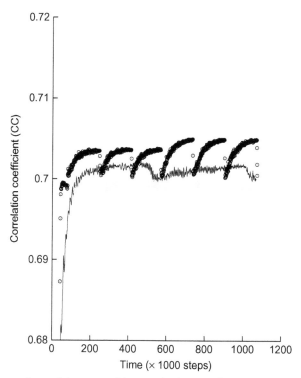

Figure 5 Comparison of three searching methods toward converging to their highest correlation coefficient value. Gray stars, molecular dynamics simulation at 300 K; solid lines, simulated annealing molecular dynamics; circles, simulated annealing with filtering.

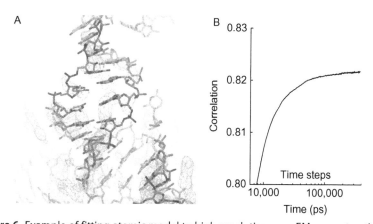

Figure 6 Example of fitting atomic model to high-resolution cryo-EM reconstruction of the human ribosome. (A) Cryo-EM map of h40–h44 segment. Fitted atomic model is represented by stick representation, while green mesh is the high-resolution cryo-EM map. The resulted fit is achieved by optimizing the correlation between the atomic model and cryo-EM map while preserving the stereochemistry. This is achieved by simulated annealing molecular dynamic simulations. (B) Time evolution of fit during simulations. (See the color plate.)

alternative folds from the native configuration. Employing a simulated annealing strategy, we improved performance, allowing us to automatically produce all-atom models highly consistent with high-resolution cryo-EM reconstructions, given an initial model of the complex. Such automated methods have the potential to enable rapid fitting of many ribosome complexes simultaneously, which can then be refined with manual manipulation.

The technique runs on a desktop computer with relatively modest compute requirements and can accommodate conformational changes of large molecular assemblies. A key advantage is the speed of sampling. We note that MDfit was previously used to fit a classical P/P tRNA configuration to a hybrid P/E-like configuration, requiring a tRNA movement of ~70 Å. Thus, one could either use the current method to attempt large conformational changes, or apply the standard MDfit to achieve large conformational changes and refine with the current technique, which allows *ab initio* fitting of local geometry based on classical mechanics force fields with multiple torsional minima, and thus enables a search of all rotamers. The fits show significant improvement on molprobity scores for the final fitted structures, reflecting improved agreement with local stereochemistry. We note that nonuniformity in map resolution and long-range electrostatic interactions are also important issues (Hayes et al., 2014).

ACKNOWLEDGMENTS

This work was supported by the National Science Foundation, the Human Frontiers Science Program, and the National Institutes of Health.

REFERENCES

Adamczyk, A. J., & Warshel, A. (2011). Converting structural information into an allosteric-energy-based picture for elongation factor Tu activation by the ribosome. *Proceedings of the National Academy of Sciences of the United States of America, 108*(24), 9827–9832.

Agrawal, R. K., Penczek, P., Grassucci, R. A., Li, Y. H., Leith, A., Nierhaus, K. H., et al. (1996). Direct visualization of A-; P-; and E-site transfer RNAs in the Escherichia coli ribosome. *Science, 271*(5251), 1000–1002.

Agrawal, R. K., Spahn, C. M., Penczek, P., Grassucci, R. A., Nierhaus, K. H., & Frank, J. (2000). Visualization of tRNA movements on the Escherichia coli 70S ribosome during the elongation cycle. *The Journal of Cell Biology, 150*(3), 447–460.

Ahmed, A., Whitford, P. C., Sanbonmatsu, K. Y., & Tama, F. (2012). Consensus among flexible fitting approaches improves the interpretation of cryo-EM data. *Journal of Structural Biology, 177*(2), 561–570.

Amunts, A., Brown, A., Bai, X. C., Llacer, J. L., Hussain, T., Emsley, P., et al. (2014). Structure of the yeast mitochondrial large ribosomal subunit. *Science, 343*(6178), 1485–1489.

Armache, J. P., Jarasch, A., Anger, A. M., Villa, E., Becker, T., Bhushan, S., et al. (2010). Cryo-EM structure and rRNA model of a translating eukaryotic 80S ribosome at 5.5-Å resolution. *Proceedings of the National Academy of Sciences of the United States of America*, 107(46), 19748–19753.

Baker, N. A., Sept, D., Joseph, S., Holst, M. J., & McCammon, J. A. (2001). Electrostatics of nanosystems: Application to microtubules and the ribosome. *Proceedings of the National Academy of Sciences of the United States of America*, 98(18), 10037–10041.

Becker, T., Armache, J. P., Jarasch, A., Anger, A. M., Villa, E., Sieber, H., et al. (2011). Structure of the no-go mRNA decay complex Dom34-Hbs1 bound to a stalled 80S ribosome. *Nature Structural & Molecular Biology*, 18(6), 715–720.

Beckmann, R., Spahn, C. M., Eswar, N., Helmers, J., Penczek, P. A., Sali, A., et al. (2001). Architecture of the protein-conducting channel associated with the translating 80S ribosome. *Cell*, 107(3), 361–372.

Ben-Shem, A., Garreau de Loubresse, N., Melnikov, S., Jenner, L., Yusupova, G., & Yusupov, M. (2011). The structure of the eukaryotic ribosome at 3.0 Å resolution. *Science*, 334(6062), 1524–1529.

Bergonzo, C., Henriksen, N. M., Roe, D. R., Swails, J. M., Roitberg, A. E., & Cheatham, T. E., 3rd. (2014). Multidimensional replica exchange molecular dynamics yields a converged ensemble of an RNA tetranucleotide. *Journal of Chemical Theory and Computation*, 10(1), 492–499.

Blanchard, S. C., Kim, H. D., Gonzalez, R. L., Jr., Puglisi, J. D., & Chu, S. (2004). tRNA dynamics on the ribosome during translation. *Proceedings of the National Academy of Sciences of the United States of America*, 101(35), 12893–12898.

Bock, L. V., Blau, C., Schroder, G. F., Davydov, I. I., Fischer, N., Stark, H., et al. (2013). Energy barriers and driving forces in tRNA translocation through the ribosome. *Nature Structural & Molecular Biology*, 20, 1390–1396.

Brunger, A. T., Kuriyan, J., & Karplus, M. (1987). Crystallographic R factor refinement by molecular dynamics. *Science*, 235(4787), 458–460.

Budkevich, T. V., Giesebrecht, J., Behrmann, E., Loerke, J., Ramrath, D. J., Mielke, T., et al. (2014). Regulation of the mammalian elongation cycle by subunit rolling: A eukaryotic-specific ribosome rearrangement. *Cell*, 158(1), 121–131.

Cheatham, T. E., 3rd., & Case, D. A. (2013). Twenty-five years of nucleic acid simulations. *Biopolymers*, 99(12), 969–977.

Chen, A. A., Marucho, M., Baker, N. A., & Pappu, R. V. (2009). Simulations of RNA interactions with monovalent ions. *Methods in Enzymology*, 469, 411–432.

Connell, S. R., Topf, M., Qin, Y., Wilson, D. N., Mielke, T., Fucini, P., et al. (2008). A new tRNA intermediate revealed on the ribosome during EF4-mediated back-translocation. *Nature Structural & Molecular Biology*, 15(9), 910–915.

Dashti, A., Schwander, P., Langlois, R., Fung, R., Li, W., Hosseinizadeh, A., et al. (2014). Trajectories of the ribosome as a Brownian nanomachine. *Proceedings of the National Academy of Sciences of the United States of America*, 111(49), 17492–17497.

Dunkle, J. A., Wang, L., Feldman, M. B., Pulk, A., Chen, V. B., Kapral, G. J., et al. (2011). Structures of the bacterial ribosome in classical and hybrid states of tRNA binding. *Science*, 332(6032), 981–984.

Fernandez, I. S., Bai, X. C., Hussain, T., Kelley, A. C., Lorsch, J. R., Ramakrishnan, V., et al. (2013). Molecular architecture of a eukaryotic translational initiation complex. *Science*, 342(6160), 1240585.

Fernandez, I. S., Bai, X. C., Murshudov, G., Scheres, S. H., & Ramakrishnan, V. (2014). Initiation of translation by cricket paralysis virus IRES requires its translocation in the ribosome. *Cell*, 157(4), 823–831.

Frank, J., & Agrawal, R. K. (2000). A ratchet-like inter-subunit reorganization of the ribosome during translocation. *Nature*, 406(6793), 318–322.

Frank, J., Radermacher, M., Wagenknecht, T., & Verschoor, A. (1988). Studying ribosome structure by electron microscopy and computer-image processing. *Methods in Enzymology*, *164*, 3–35.

Frank, J., Zhu, J., Penczek, P., Li, Y., Srivastava, S., Verschoor, A., et al. (1995). A model of protein synthesis based on cryo-electron microscopy of the *E. coli* ribosome. *Nature*, *376*(6539), 441–444.

Gao, H., Sengupta, J., Valle, M., Korostelev, A., Eswar, N., Stagg, S. M., et al. (2003). Study of the structural dynamics of the *E. coli* 70S ribosome using real-space refinement. *Cell*, *113*(6), 789–801.

Goddard, T. D., Huang, C. C., & Ferrin, T. E. (2005). Software extensions to UCSF chimera for interactive visualization of large molecular assemblies. *Structure*, *13*(3), 473–482.

Gorba, C., Miyashita, O., & Tama, F. (2008). Normal-mode flexible fitting of high-resolution structure of biological molecules toward one-dimensional low-resolution data. *Biophysical Journal*, *94*(5), 1589–1599.

Hayes, R. L., Noel, J. K., Whitford, P. C., Mohanty, U., Sanbonmatsu, K. Y., & Onuchic, J. N. (2014). Reduced model captures Mg(2+)-RNA interaction free energy of riboswitches. *Biophysical Journal*, *106*(7), 1508–1519.

Henriksen, N. M., Davis, D. R., & Cheatham, T. E., 3rd. (2012). Molecular dynamics re-refinement of two different small RNA loop structures using the original NMR data suggest a common structure. *Journal of Biomolecular NMR*, *53*(4), 321–339.

Ishida, H., & Hayward, S. (2008). Path of nascent polypeptide in exit tunnel revealed by molecular dynamics simulation of ribosome. *Biophysical Journal*, *95*(12), 5962–5973.

Ishida, H., & Matsumoto, A. (2014). Free-energy landscape of reverse tRNA translocation through the ribosome analyzed by electron microscopy density maps and molecular dynamics simulations. *PLoS One*, *9*(7), e101951.

Jenner, L., Melnikov, S., Garreau de Loubresse, N., Ben-Shem, A., Iskakova, M., Urzhumtsev, A., et al. (2012). Crystal structure of the 80S yeast ribosome. *Current Opinion in Structural Biology*, *22*(6), 759–767.

Korostelev, A., Laurberg, M., & Noller, H. F. (2009). Multistart simulated annealing refinement of the crystal structure of the 70S ribosome. *Proceedings of the National Academy of Sciences of the United States of America*, *106*(43), 18195–18200.

Kurkcuoglu, O., Doruker, P., Sen, T. Z., Kloczkowski, A., & Jernigan, R. L. (2008). The ribosome structure controls and directs mRNA entry, translocation and exit dynamics. *Physical Biology*, *5*(4), 046005.

Liu, C., Janowski, P. A., & Case, D. A. (2015). All-atom crystal simulations of DNA and RNA duplexes. *Biochimica et Biophysica Acta*, *1850*(5), 1059–1071.

Melnikov, S., Ben-Shem, A., Garreau de Loubresse, N., Jenner, L., Yusupova, G., & Yusupov, M. (2012). One core, two shells: Bacterial and eukaryotic ribosomes. *Nature Structural & Molecular Biology*, *19*(6), 560–567.

Muhs, M., Hilal, T., Mielke, T., Skabkin, M. A., Sanbonmatsu, K. Y., Pestova, T. V., et al. (2015). Cryo-EM of ribosomal 80S complexes with termination factors reveals the translocated cricket paralysis virus IRES. *Molecular Cell*, *57*(3), 422–432.

Munro, J. B., Wasserman, M. R., Altman, R. B., Wang, L., & Blanchard, S. C. (2010). Correlated conformational events in EF-G and the ribosome regulate translocation. *Nature Structural & Molecular Biology*, *17*(12), 1470–1477.

Olivier, N. B., Altman, R. B., Noeske, J., Basarab, G. S., Code, E., Ferguson, A. D., et al. (2014). Negamycin induces translational stalling and miscoding by binding to the small subunit head domain of the Escherichia coli ribosome. *Proceedings of the National Academy of Sciences of the United States of America*, *111*(46), 16274–16279.

Oostenbrink, C., Villa, A., Mark, A. E., & Van Gunsteren, W. F. (2004). A biomolecular force field based on the free enthalpy of hydration and solvation: The GROMOS force-field parameter sets 53A5 and 53A6. *Journal of Computational Chemistry*, *25*(13), 1656–1676.

Orzechowski, M., & Tama, F. (2008). Flexible fitting of high-resolution X-ray structures into cryoelectron microscopy maps using biased molecular dynamics simulations. *Biophysical Journal, 95*(12), 5692–5705.

Ratje, A. H., Loerke, J., Mikolajka, A., Brunner, M., Hildebrand, P. W., Starosta, A. L., et al. (2010). Head swivel on the ribosome facilitates translocation by means of intra-subunit tRNA hybrid sites. *Nature, 468*(7324), 713–716.

Rodnina, M. V., & Wintermeyer, W. (2011). The ribosome as a molecular machine: The mechanism of tRNA-mRNA movement in translocation. *Biochemical Society Transactions, 39*(2), 658–662.

Rychkova, A., Mukherjee, S., Bora, R. P., & Warshel, A. (2013). Simulating the pulling of stalled elongated peptide from the ribosome by the translocon. *Proceedings of the National Academy of Sciences of the United States of America, 110*(25), 10195–10200.

Schuette, J. C., Murphy, F. V., Kelley, A. C., Weir, J. R., Giesebrecht, J., Connell, S. R., et al. (2009). GTPase activation of elongation factor EF-Tu by the ribosome during decoding. *The EMBO Journal, 28*(6), 755–765.

Seidelt, B., Innis, C. A., Wilson, D. N., Gartmann, M., Armache, J. P., Villa, E., et al. (2009). Structural insight into nascent polypeptide chain-mediated translational stalling. *Science, 326*(5958), 1412–1415.

Spahn, C. M., Beckmann, R., Eswar, N., Penczek, P. A., Sali, A., Blobel, G., et al. (2001). Structure of the 80S ribosome from Saccharomyces cerevisiae-tRNA-ribosome and subunit-subunit interactions. *Cell, 107*(3), 373–386.

Spahn, C. M., Gomez-Lorenzo, M. G., Grassucci, R. A., Jorgensen, R., Andersen, G. R., Beckmann, R., et al. (2004). Domain movements of elongation factor eEF2 and the eukaryotic 80S ribosome facilitate tRNA translocation. *The EMBO Journal, 23*(5), 1008–1019.

Tourigny, D. S., Fernandez, I. S., Kelley, A. C., & Ramakrishnan, V. (2013). Elongation factor G bound to the ribosome in an intermediate state of translocation. *Science, 340*(6140), 1235490.

Trabuco, L. G., Schreiner, E., Eargle, J., Cornish, P., Ha, T., Luthey-Schulten, Z., et al. (2010). The role of L1 stalk-tRNA interaction in the ribosome elongation cycle. *Journal of Molecular Biology, 402*(4), 741–760.

Trabuco, L. G., Villa, E., Mitra, K., Frank, J., & Schulten, K. (2008). Flexible fitting of atomic structures into electron microscopy maps using molecular dynamics. *Structure, 16*(5), 673–683.

Trylska, J., Konecny, R., Tama, F., Brooks, C. L., 3rd., & McCammon, J. A. (2004). Ribosome motions modulate electrostatic properties. *Biopolymers, 74*(6), 423–431.

Valle, M., Sengupta, J., Swami, N. K., Grassucci, R. A., Burkhardt, N., Nierhaus, K. H., et al. (2002). Cryo-EM reveals an active role for aminoacyl-tRNA in the accommodation process. *The EMBO Journal, 21*(13), 3557–3567.

Valle, M., Zavialov, A., Li, W., Stagg, S. M., Sengupta, J., Nielsen, R. C., et al. (2003). Incorporation of aminoacyl-tRNA into the ribosome as seen by cryo-electron microscopy. *Nature Structural Biology, 10*(11), 899–906.

Villa, E., Sengupta, J., Trabuco, L. G., LeBarron, J., Baxter, W. T., Shaikh, T. R., et al. (2009). Ribosome-induced changes in elongation factor Tu conformation control GTP hydrolysis. *Proceedings of the National Academy of Sciences of the United States of America, 106*(4), 1063–1068.

Wagenknecht, T., Carazo, J. M., Radermacher, M., & Frank, J. (1989). Three-dimensional reconstruction of the ribosome from *Escherichia coli*. *Biophysical Journal, 55*(3), 455–464.

Wang, L., Pulk, A., Wasserman, M. R., Feldman, M. B., Altman, R. B., Cate, J. H., et al. (2012). Allosteric control of the ribosome by small-molecule antibiotics. *Nature Structural & Molecular Biology, 19*(9), 957–963.

Wang, Z., & Schroder, G. F. (2012). Real-space refinement with DireX: From global fitting to side-chain improvements. *Biopolymers*, *97*(9), 687–697.

Whitford, P. C., Ahmed, A., Yu, Y., Hennelly, S. P., Tama, F., Spahn, C. M., et al. (2011). Excited states of ribosome translocation revealed through integrative molecular modeling. *Proceedings of the National Academy of Sciences of the United States of America*, *108*(47), 18943–18948.

Wu, X., Subramaniam, S., Case, D. A., Wu, W. W., & Brooks, B. R. (2013). Targeted conformational search with map-restrained self-guided Langevin dynamics: Application to flexible fitting of electron microscopy images. *Journal of Structural Biology*, *183*(3), 429–440.

Zeng, X., Chugh, J., Casiano-Negroni, A., Al-Hashimi, H. M., & Brooks, C. L., 3rd. (2014). Flipping of the ribosomal A-site adenines provides a basis for tRNA selection. *Journal of Molecular Biology*, *426*(19), 3201–3213.

CHAPTER SEVENTEEN

In Vitro Reconstitution and Crystallization of Cas9 Endonuclease Bound to a Guide RNA and a DNA Target

Carolin Anders, Ole Niewoehner, Martin Jinek[1]
Department of Biochemistry, University of Zurich, Zurich, Switzerland
[1]Corresponding author: e-mail address: jinek@bioc.uzh.ch

Contents

1. Introduction 516
2. Electrophoretic Mobility Shift Assay 518
3. Fluorescence-Detection Size Exclusion Chromatography Assay 523
4. Crystallization of Cas9–RNA–DNA Complexes 530
5. Concluding Remarks 534
Acknowledgments 535
References 535

Abstract

The programmable RNA-guided DNA cleavage activity of the bacterial CRISPR-associated endonuclease Cas9 is the basis of genome editing applications in numerous model organisms and cell types. In a binary complex with a dual crRNA:tracrRNA guide or single-molecule guide RNA, Cas9 targets double-stranded DNAs harboring sequences complementary to a 20-nucleotide segment in the guide RNA. Recent structural studies of the enzyme have uncovered the molecular mechanism of RNA-guided DNA recognition. Here, we provide protocols for electrophoretic mobility shift and fluorescence-detection size exclusion chromatography assays used to probe DNA binding by Cas9 that allowed us to reconstitute and crystallize the enzyme in a ternary complex with a guide RNA and a *bona fide* target DNA. The procedures can be used for further mechanistic investigations of the Cas9 endonuclease family and are potentially applicable to other multicomponent protein–nucleic acid complexes.

1. INTRODUCTION

Cas9 is an RNA-guided endonuclease capable of cleaving double-stranded DNA. The substrate specificity of Cas9 is determined by a 20-nucleotide sequence within a bound guide RNA that directs the enzyme to complementary sequences in dsDNA through base-pairing interactions (Gasiunas, Barrangou, Horvath, & Siksnys, 2012; Jinek et al., 2012). The enzyme originates from bacterial type II CRISPR–Cas (clustered regularly interspaced palindromic repeats–CRISPR-associated) genome defense systems, in which it associates with CRISPR RNA (crRNA) guides and a trans-activating crRNA (tracrRNA) to target the DNA of invading viruses and other mobile genetic elements (Barrangou et al., 2007; Deltcheva et al., 2011; Garneau et al., 2010). Based on the dual crRNA:tracrRNA guide structure, single-molecule guide RNAs (sgRNAs) have been engineered to program the sequence specificity of Cas9 for DNA cleavage *in vitro* and *in vivo* (Cong et al., 2013; Jinek et al., 2012, 2013; Mali, Yang, et al., 2013). The resulting Cas9–sgRNA two-component system has been harnessed to form the core of emerging genome editing technologies, whereby double-strand breaks induced in a specific genomic locus by the Cas9–sgRNA complex are repaired by the nonhomologous end joining or homologous recombination pathways to engineer a desired modification in the target locus (Cong et al., 2013; Jinek et al., 2013; Mali, Yang, et al., 2013). Owing to its easy programmability and a high degree of specificity, CRISPR-Cas9 has brought about revolutionary advances in genetic engineering in numerous model organisms and cell types (reviewed in Doudna & Charpentier, 2014; Hsu, Lander, & Zhang, 2014; Mali, Esvelt, & Church, 2013). The Cas9–sgRNA system has also capabilities that go beyond genome modifications. Catalytically inactive variants of Cas9 (dCas9) that retain RNA-guided DNA-binding activity have been used for RNA-guided transcriptional control (Gilbert et al., 2014, 2013; Konermann et al., 2015; Mali, Aach, et al., 2013) and for marking genomic loci in imaging applications (Chen et al., 2013). Cas9 has also been adapted for RNA binding and cleavage (O'Connell et al., 2014). CRISPR-Cas9 thus represents a transformative technology with many promising applications not just in basic research but also in biotechnology and medicine.

At the molecular level, Cas9 associates with the guide RNA (either the naturally occurring crRNA:tracrRNA duplex structure or an engineered sgRNA) in a binary protein–RNA complex that uses a 20-nucleotide region at the 5′ end of the guide RNA to locate a matching sequence in a dsDNA target (Jinek et al., 2012; Karvelis et al., 2013). Watson–Crick base pairing between the guide sequence and the target DNA results in DNA strand separation and the formation of an RNA–DNA heteroduplex in an R-loop structure, positioning the two strands of the target DNA for site-specific cleavage. Cas9 contains two magnesium-dependent nuclease domains: an HNH domain that cleaves the complementary (target) strand of the DNA target, and a RuvC domain responsible for cleavage of the displaced non-complementary (non-target) DNA strand (Chen, Choi, & Bailey, 2014; Gasiunas et al., 2012; Jinek et al., 2012). DNA binding requires the presence of a short protospacer adjacent motif (PAM) in the vicinity of the target region in the DNA (Jinek et al., 2012; Sternberg, Redding, Jinek, Greene, & Doudna, 2014). PAM recognition is an initial step in target DNA binding that enables the Cas9–guide RNA complex to catalyze local strand separation in the target DNA and sequential guide RNA–target DNA heteroduplex formation (Sternberg et al., 2014). As a result, DNA sequences perfectly complementary to the guide RNA sequence but lacking the PAM are not recognized as cleavage substrates by Cas9 (Sternberg et al., 2014). Additionally, near-perfect complementarity between the guide RNA and the target DNA strand within a 8–12 base pair (bp) PAM-proximal "seed" region in the guide–target heteroduplex is a critical determinant of target DNA binding (Jinek et al., 2012). However, efficient DNA cleavage *in vivo* requires more extensive guide–target base pairing (Cencic et al., 2014; Kuscu, Arslan, Singh, Thorpe, & Adli, 2014; Wu et al., 2014). Mismatches within the guide–target heteroduplex are nevertheless tolerated in some positions, and this is the chief source of off-target activities in Cas9-based gene targeting applications (Fu et al., 2013; Hsu et al., 2013; Mali, Aach, et al., 2013; Pattanayak et al., 2013).

Structural studies have shed light on the molecular architecture and mechanism of Cas9. Crystal structures of apo-Cas9 defined the two structural lobes in the molecule, while crystal structures and electron microscopic reconstructions of nucleic acid-bound complexes revealed an extensive RNA-driven conformational rearrangement that primes the Cas9–guide RNA complex for target DNA binding (Jinek et al., 2014; Nishimasu et al., 2014). To obtain structural insights into the molecular mechanism of PAM-dependent target DNA recognition by Cas9, we recently determined the crystal structure of *Streptococcus pyogenes* Cas9 (SpyCas9) in complex with an sgRNA guide and a *bona fide* target DNA containing a canonical 5′-NGG-3′ PAM sequence (Anders, Niewoehner, Duerst, & Jinek, 2014).

The structure revealed that the PAM sequence is recognized by direct readout of the GG dinucleotide in the major groove of the PAM-containing DNA duplex. Furthermore, the structure suggested how PAM recognition by the Cas9–guide RNA complex might be coupled to strand separation in the target dsDNA and concomitant formation of the guide RNA–target DNA heteroduplex.

In this chapter, we present biochemical methods that we used to establish a robust protocol for the *in vitro* reconstitution of the Cas9–sgRNA–DNA complex and outline the procedure and considerations for its crystallization. To probe target DNA binding, we initially used electrophoretic mobility shift assays (EMSAs). We subsequently devised an analytical size exclusion chromatography (SEC) assay employing fluorescently labeled DNA oligonucleotides to identify minimal sgRNA and target DNA structures capable of supporting complex formation. These assays can be readily used to analyze target DNA binding by SpyCas9 as well as orthologous Cas9 proteins from other bacterial species. More generally, the methods may be applicable to other multicomponent protein–nucleic acid assemblies. The fluorescence-detection size exclusion chromatography (FSEC) assay permits high-throughput screening of multiple nucleic acid constructs and is particularly advantageous in situations where the analyzed macromolecular complex contains two or more nucleic acid molecules.

2. ELECTROPHORETIC MOBILITY SHIFT ASSAY

EMSAs are commonly used to characterize protein–nucleic acid interactions. These assays rely on the ability to separate bound and unbound fractions of a labeled nucleic acid probe based on their differential mobility through a nondenaturing polyacrylamide or agarose matrix, which is dependent on the molecular masses of the free nucleic acid and the bound protein, the shape of the complex, and overall electrical charge. Typically, nucleic acid probes used in EMSAs are radiolabeled at their 5′ ends. In this way, minute quantities of the nucleic acid can be used in the assay, and apparent dissociation constants can be determined by quantifying the ratio of bound and unbound molecules and fitting the data to a standard binding isotherm. The theory and the general considerations for practical implementation of EMSAs have been reviewed extensively elsewhere (Kerr, 1995; Mitchell & Lorsch, 2014; Ryder, Recht, & Williamson, 2008).

In our study, we sought to establish assays to probe target DNA binding by the Cas9–sgRNA complex. In an initial approach, we performed EMSAs using the catalytically inactive double mutant (D10A/H840A) of the S. pyogenes Cas9 protein (dCas9) (Fig. 1A and B). dCas9 was expressed and purified as described previously (Anders & Jinek, 2014; Jinek et al., 2012) from expression plasmid pMJ841 (available from Addgene). The sgRNA was prepared by *in vitro* transcription using T7 RNA polymerase and a fully double-stranded oligonucleotide DNA template, and was purified by denaturing polyacrylamide gel electrophoresis and ethanol precipitation (Anders & Jinek, 2014). The sgRNA consisted of a 5′-terminal 20-nucleotide guide segment, followed by a repeat:antirepeat duplex (capped with a GAAA tetraloop) and stem loops SL1 and SL2 in the 3′-terminal region. These features are known to be required for efficient DNA cleavage activity by Cas9 *in vitro* as well as *in vivo* (Briner et al., 2014; Hsu et al., 2013; Jinek et al., 2012). The target DNA was a 28-bp synthetic oligonucleotide duplex containing a sequence perfectly complementary to the guide segment of the sgRNA and a canonical PAM (5′-TGG-3′) in the non-target (i.e., noncomplementary) strand. The non-target strand in the duplex was labeled with the ATTO532 fluorophore covalently attached to its 3′ end. We decided to switch from radiolabeling to fluorescence detection for greater convenience, as custom fluorophore-labeled synthetic DNA oligonucleotides are readily available from commercial sources. The ATTO532 label was chosen for its photostability and superior quantum yield, allowing us to use the labeled duplex at a final concentration of 50 nM in our binding reactions. Although the working concentration is greater than the apparent dissociation constant of the DNA-bound Cas9–sgRNA complex, as measured previously using radiolabeled DNA probes (Jinek et al., 2012; Sternberg et al., 2014), and therefore does not permit accurate quantitation of binding affinities, the assay format nevertheless provides a reasonable indication of specific target DNA binding (Anders et al., 2014). In particular, the assay can be used to probe the PAM dependency of binding, since the absence of a cognate PAM interaction increases the apparent K_d by at least three orders of magnitude (Anders et al., 2014; Jinek et al., 2012; Sternberg et al., 2014).

In the EMSA, the target DNA duplex is titrated with increasing amounts of preassembled dCas9–sgRNA complex spanning a concentration range from 10 nM to 1 µM; a control reaction lacking the dCas9–sgRNA complex is also set up. The sequences of sgRNA and DNA oligonucleotides used in the assay are shown in Fig. 1B and listed in Table 1. The electrophoresis runs were performed in a cold room at 4 °C to minimize complex dissociation and resulting smearing of the gel bands. The following protocol describes the detailed procedure for the EMSA shown in Fig. 1C.

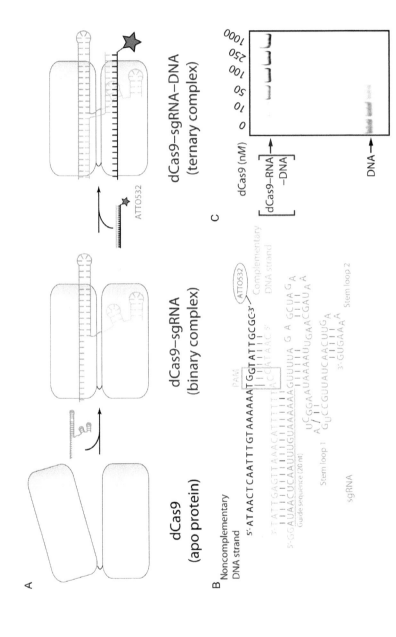

Figure 1 See legend on opposite page.

Table 1 DNA and RNA Sequences Used in EMSA

Description	Sequence[a]
sgRNA	5′-GGATAACTCAATTTGTAAAAAAGTTTTAGA GCTAGAAATAGCAAGTTAAAATAAGGCTAG TCCGTTATCAACTTGAAAAAGTG-3′
ssDNA template sgRNA	5′-CACTTTTTCAAGTTGATAACGGACTAGCCT TATTTTAACTTGCTATTTCTAGCTCTAAAA CTTTTTTACAAATTGAGTTATCCTATAGTG AGTCGTATTA-3′
Reverse primer for PCR amplification of ssDNA template for sgRNA	5′-CACTTTTTCAAGTTGA-3′
Complementary strand	5′-CAATA*CC*ATTTTTTACAAATTGAGTTAT-3′
Noncomplementary strand 1 3′-ATTO532	5′-<u>AAAA</u>**TGG**TATTGCGC-3′-ATTO532
Noncomplementary strand 2 3′-ATTO532	5′-<u>ATAACTCAATTTGTAAAAAA</u>**TGG**TATTGCGC-3′-ATTO532

[a]Nontarget strand PAM sites (5′-NGG-3′) are highlighted in bold. Target-strand complement of the PAM (5′-CCN-3′) is denoted in italic. Target sequence complementary to the guide RNA is underlined.

Figure 1—Cont'd (A) Schematic workflow of the electrophoretic mobility shift assay (EMSA) to probe for ternary complex formation. In a first step, apo-dCas9 and guide RNA are preincubated to form a binary complex. Addition of annealed duplex substrate leads to the assembly of ternary dCas9–guide RNA–target DNA complex. (B) Nucleic acid components used for the EMSA shown in panel (C). sgRNA is shown in orange (light gray in the print version) with the 5′-terminal GG dinucleotide originating from *in vitro* transcription using T7 RNA polymerase marked in gray. The complementary and noncomplementary DNA strands are colored in blue (light gray in the print version) and black, respectively. The noncomplementary strand is labeled at its 3′ end with an ATTO532 fluorophore. The PAM sequence and its complement are marked with a green (gray in the print version) box. (C) Target DNA duplex (50 nM) was titrated with increasing concentrations (0, 10, 50, 100, 250, and 1000 nM) of *in vitro* reconstituted dCas9–sgRNA complex. Binding reactions were analyzed using a native 8% polyacrylamide gel and visualized by detection of ATTO532 fluorescence using a laser scanner. The lower and upper bands represent the unbound and complex-bound DNA fractions, respectively.

1. Obtain synthetic oligonucleotides for the target and non-target strands of the DNA duplex. The non-target strand carries the ATTO532 label attached to the 3′-hydroxyl group. The labeled oligonucleotides should be PAGE purified prior to use. This can be done by the supplier or alternatively performed in the laboratory using a published protocol (Lopez-Gomollon & Nicolas, 2013). Determine oligonucleotide concentrations by measuring the absorbance at 260 nm. We calculate DNA concentrations using the OligoCalc server (Kibbe, 2007). Dilute both oligonucleotides to make 20 µM stock solutions (20 µl each).
2. Prepare the oligonucleotide DNA duplex probe by annealing target and nontarget strands in a 1.5:1 (target:nontarget) molar ratio to ensure complete hybridization of the labeled strand. In order to do this, mix 15 µl of the complementary strand stock solution (20 µM) with 10 µl of the non-complementary strand. Heat the mixture to 95 °C for 5 min and cool slowly to room temperature. Dilute the mixture with 975 µl binding buffer (20 mM HEPES, pH 7.5, 250 mM KCl, 2 mM MgCl$_2$, 0.01% Triton X-100, 0.1 mg ml^{-1} bovine serum albumin, 10% glycerol) to obtain 200 nM stock solution of the labeled DNA duplex.
3. Prepare a native 8% polyacrylamide gel containing 1 × TBE (89 mM tris(hydroxymethyl)aminomethane (Tris), 89 mM boric acid, 2 mM ethylenediaminetetraacetic acid, pH 8.0) supplemented with 2 mM MgCl$_2$. We typically pour gels with the dimensions of 300 mm × 180 mm × 1 mm (h × w × d) and use a well comb with 6 mm tooth width and 2 mm spacing.
4. Thaw a fresh aliquot of frozen dCas9 and dilute with storage buffer (20 mM HEPES, pH 7.5, 500 mM KCl) to prepare a 100 µM stock solution (50 µl). Prepare a stock solution of purified sgRNA by diluting to 400 µM with nuclease-free water. To reconstitute the dCas9–sgRNA complex in a 1:1.5 molar ratio, mix 8.0 µl dCas9 with 3.0 µl sgRNA and incubate for 10 min at room temperature. Dilute the mixture with 189 µl binding buffer to a final concentration of 4 µM and store on ice. Use the 4 µM complex solution to prepare additional dilutions at 1 µM (25 µl) and 200 nM (20 µl).
5. For each binding reaction, combine binding buffer and dCas9–sgRNA complex according to Table 2 and incubate for 10 min at room temperature. Add 5 µl oligonucleotide DNA duplex to each reaction and incubate for 10 min at 37 °C (Fig. 1A).
6. Load a 5 µl sample of each binding reaction onto the polyacrylamide gel. Load an additional lane with 5 µl binding buffer supplemented with

Table 2 Electrophoretic Mobility Shift Assay

Final dCas9–sgRNA Concentration	Binding Buffer (μl)	dCas9–sgRNA Mix[a]	Duplex Substrate (μl)[a]	Total Volume (μl)
0 nM	15	–	5	20
10 nM	14	1 μl 200 nM	5	20
50 nM	10	5 μl 200 nM	5	20
100 nM	13	2 μl 1 μM	5	20
250 nM	10	5 μl 1 μM	5	20
1 μM	10	5 μl 4 μM	5	20

[a]The target and nontarget DNA strands are annealed prior addition to the reaction. See Section 2, step 2 for the annealing procedure.

0.005% bromophenol blue to be able to monitor the progress of the electrophoresis run. Run the gel using 1 × TBE buffer supplemented with 2 mM MgCl$_2$ at 15 W at 4 °C for approximately 75 min, until the bromophenol blue dye front has reached halfway down the gel.

7. Scan the gel using a laser gel scanner (e.g., Typhoon FLA9500, GE Healthcare) with the appropriate excitation and emission wavelength settings for ATTO532 (532 nm laser for excitation and LPG emission filter for detection, respectively).

3. FLUORESCENCE-DETECTION SIZE EXCLUSION CHROMATOGRAPHY ASSAY

Successful crystallization of multicomponent macromolecular assemblies requires extensive coverage of the crystallization parameter space, varying not just the chemical composition of the crystallization buffer but also the precise forms of the components of the crystallized macromolecular complex. In the case of protein–nucleic acid complexes, variation of the sequence and length of the bound nucleic acids is an essential part of the crystallization strategy (Hollis, 2007; Obayashi, Oubridge, Pomeranz Krummel, & Nagai, 2007; Reyes, Garst, & Batey, 2009). This is because specific bases of the nucleic acid ligand often mediate crystal contact formation through base pairing, base stacking, and other tertiary interactions (Anderson, Ptashne, & Harrison, 1984, 1987). In some cases, this has been exploited to engineer crystal contacts for otherwise recalcitrant molecules (Berry, Waghray, Mortimer, Bai, & Doudna, 2011; Ferré-D'Amaré,

Zhou, & Doudna, 1998; Leung et al., 2010). However, this crystallization strategy relies on the ability to efficiently assess complex formation for a broad range of nucleic acid ligands. Although EMSAs are well suited to probe a specific protein–nucleic acid interaction and provide quantitative information about binding, this approach is very time consuming because the analysis of multiple nucleic acid ligands typically requires running many gels. Additionally, although EMSA can differentiate between bound monomers and multimers, the method otherwise yields few clues about the nature of the protein–nucleic acid complex. Finally, it is additionally difficult to determine whether binding is quantitative (i.e., whether all or only a fraction of the protein is active in binding to the ligand) and therefore whether stoichiometric amounts of complex can be reconstituted in large scale in order to prepare sufficient amounts of material for crystallization screening. Analytical SEC is an attractive alternative to EMSA because it provides additional information on the approximate molecular mass of the complex and the degree of its monodispersity (Winzor, 2003). Both protein and nucleic acid components can be monitored due to their UV absorption at 280 and 260 nm, respectively, and complex formation manifests in shifts in the elution volumes of the corresponding peaks and changes in their A_{260}/A_{280} ratios. By comparing the ratio of areas under the peaks in the chromatogram, the relative fractions of protein and/or nucleic acid ligand incorporated into the complex can be determined.

To obtain a crystal structure of Cas9 bound to a guide RNA and a PAM-containing DNA target, our efforts focused exclusively on SpyCas9 due to the availability of the crystal structure of the apo protein and EM reconstructions of its nucleic acid complexes. Aiming to generate dCas9–sgRNA–target DNA complexes suitable for crystallization, we sought to define the minimal features in the sgRNA and the target DNA that would support stable complex formation. To this end, we used analytical SEC to screen a panel of sgRNA structures and target DNAs. Due to the need to be able to monitor the simultaneous binding of two nucleic acid ligands to dCas9 (the sgRNA and the target DNA duplex), we devised a fluorescence-detection strategy in which the target DNA duplex carried a Cy3 fluorophore covalently attached to the 3′ end of the non-target strand. The sgRNA, being the larger of the two nucleic acid components in the ternary complex, was monitored by observing changes in the peak profile in the 260 nm channel. Our FSEC assay utilized an Agilent 1200 series high-performance liquid chromatography (HPLC) instrument equipped with UV–vis and single-channel fluorescence detectors connected in series. We analyzed samples on a

Table 3 Summary of Parameters for FSEC Assay Using a Superdex 200 5/150 Column

Column volume	3 ml
Flow rate	0.1 ml min^{-1}
Pressure limit	15 bar
Excitation wavelength	550 nm
Emission wavelength	570 nm
Maximum injection volume	50 μl
Minimum detectable amount of protein	125 pmol or 20 μg

Superdex 200 5/150 size exclusion column with a 3-ml bed volume and 15 bar pressure limit, eluting at a rate of 0.1 ml min^{-1} (see Table 3 for a summary of parameters). Using this method, as little as 20 μg of dCas9 can be analyzed in a single run within a span of 30 min. The availability of an autosampler allowed us to perform multiple runs in an automated, high-throughput manner. The eluate was fractionated and individual fractions were subsequently analyzed by native polyacrylamide gel electrophoresis.

Previous studies suggested that the Cas9–sgRNA complex recognizes the PAM on the non-target strand of the DNA target (Jinek et al., 2012; Sternberg et al., 2014). Furthermore, the presence of a 5′-truncated non-complementary strand that contained only the 5′-NGG-3′ PAM sequence in the partially duplexed target DNA was sufficient to stimulate cleavage of the complementary strand, suggesting that a truncated non-target strand is sufficient for stable complex formation (Sternberg et al., 2014). We sought to take advantage of these observations to prepare a minimal Cas9–sgRNA–target DNA ternary complex suitable for crystallization. To this end, we tested the binding of dCas9 complexed with a truncated 83-nucleotide sgRNA (containing stem loops SL1 and SL2 in the 3′-terminal region) to two target DNA duplexes that contained a 5′-TGG-3′ PAM sequence. The target DNA strand was identical in the two DNA ligands, whereas the non-target strand was either truncated at its 5′ end to mimic the Cas9 cleavage product (12 nt, Fig. 2A) or fully complementary to the target strand (31 nt, Fig. 2B). Control runs established that apo-dCas9 and unbound sgRNA eluted at 1.61 and 1.95 ml, respectively (Fig. 2C and D). Incubation of dCas9 with a slight excess of sgRNA led to efficient formation of the binary complex, as judged by a change in the A_{260}/A_{280} ratio of the 1.59-ml peak and proportional decrease in the size of the free sgRNA peak

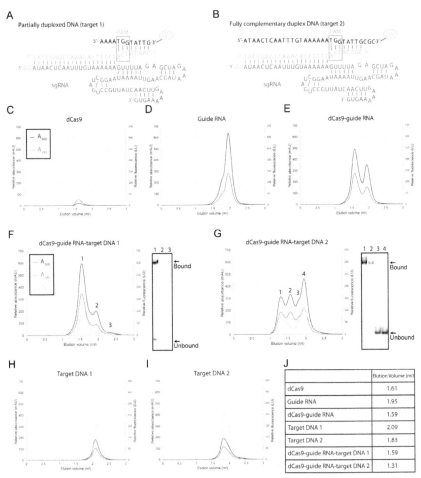

Figure 2 (A and B) Schematic representation of two target DNA variants used in fluorescence-detection size exclusion chromatography (FSEC). The complementary DNA strand (blue (light gray in the print version)) is identical in both DNA targets. The 3′-Cy3-labeled noncomplementary DNA strands (black) differ in length (panel A: target DNA 1, 15 nt; panel B: target DNA 2, 31 nt) to form either a partially or fully double-stranded target substrate. The PAM sequence and its complement are marked with a green (gray in the print version) box. The sgRNA is colored gray. (C–I) FSEC chromatograms for apo-dCas9 (C), sgRNA (D), binary dCas9–sgRNA complex (E), the ternary dCas9–sgRNA–target DNA 1 complex (F), ternary dCas9–sgRNA–target DNA 2 complex (G), target DNA 1 alone (H), and target DNA 2 alone (I). The measured absorbances at 260 and 280 nm are shown with red (black in the print version) and blue (light gray in the print version) lines, respectively. The yellow (light gray in the print version) line depicts the Cy3 fluorescence signal. For the ternary complexes, 20 μl of the labeled peak fractions were analyzed on a native polyacrylamide gel and detected by the Cy3 fluorescence (F and G). (J) Summary of elution volumes from FSEC experiments in (C–I).

(Fig. 2E). The observation that the elution volumes of apo-dCas9 and the dCas9–sgRNA complex were nearly identical was consistent with previous electron microscopic studies indicating that the overall molecular dimensions of the two species are quite similar, despite a major conformational rearrangement in the Cas9 protein upon sgRNA binding (Jinek et al., 2014). Addition of the target DNA duplex containing a truncated noncomplementary strand led to near-stoichiometric formation of the ternary dCas9–sgRNA–target DNA complex, as indicated by a shift of the Cy3 signal from the unbound peak at 2.09 ml to the complex peak at 1.59 ml (Fig. 2F and H). In contrast, addition of the fully complementary target DNA duplex also resulted in ternary complex formation, but only ~50% of the Cy3 label was incorporated in the ternary complex peak (Fig. 2G and I). Taken together, these experiments suggested that both target DNA duplexes were bound by the dCas9–sgRNA complex. However, only in the case of the DNA target containing a truncated noncomplementary strand was ternary complex formation stoichiometric, whereas a substantial fraction of the fully complementary target DNA duplex remained unbound. It is possible that in this case, strand separation by Cas9 was slow, despite the presence of a canonical PAM sequence, which precluded efficient formation of the target DNA-bound complex. Moreover, comparison of the elution volumes of the complexes containing the two target DNA duplexes (1.31 ml for the fully complementary target duplex and 1.59 ml for the partially duplexed target DNA, Fig. 2J) also suggested that the complex containing the partially duplexed target DNA likely adopted a more compact conformation. Together with the observed stoichiometric formation of the complex, these two factors were decisive in identifying the partially duplexed DNA target as the optimal choice for large-scale complex reconstitution and initial crystallization trials.

The following protocol describes the details of the FSEC assay. The sequences of the nucleic acids used in the assay are provided in Table 4. We recommend performing the procedure at 4 °C in a cold room, particularly if multiple complexes are to be analyzed.

1. Prepare 1 l of complex reconstitution buffer (20 mM HEPES, pH 7.5, 250 mM KCl, 2 mM MgCl$_2$). Set up the HPLC instrument and flush the pumps and flowpath with the complex reconstitution buffer.
2. Attach a Superdex 200 5/150 gel filtration column (GE Healthcare) to the HPLC instrument and equilibrate with complex reconstitution buffer at a flow rate of 0.1 ml min^{-1} (see Table 3 for a summary of parameters).

Table 4 DNA and RNA Sequences Used for FSEC Assays

Description	Sequence[a]
sgRNA	5'-GGATAACTCAATTTGTAAAAAAGTTTTAGA GCTAGAAATAGCAAGTTAAAATAAGGCTAG TCCGTTATCAACTTGAAAAAGTG-3'
ssDNA template sgRNA	5'-CACTTTTTCAAGTTGATAACGGACTAGCCT TATTTTAACTTGCTATTTCTAGCTCTAAAA CTTTTTTACAAATTGAGTTATCCTATAGTG AGTCGTATTA-3'
Reverse primer for PCR amplification of ssDNA template for sgRNA	5'-CACTTTTTCAAGTTGA-3'
Complementary strand	5'-CAATA*CC*ATTTTTTACAAATTGAGTTAT-3'
Noncomplementary strand 1 3'-Cy3	5'-<u>AAAA</u>**TGG**TATTG-3'-Cy3
Noncomplementary strand 2 3'-Cy3	5'-<u>ATAACTCAATTTGTAAAAAA</u>**TGG**TATTGCGC-3'-Cy3

[a]Nontarget strand PAM sites (5'-NGG-3') are highlighted in bold. Target-strand complement of the PAM (5'-CCN-3') is denoted in italic. Target DNA sequence complementary to the guide RNA is underlined.

3. Thaw a frozen stock solution of dCas9 and prepare 30 μl of a 100 μM (~16 mg ml^{-1}) solution of dCas9 by diluting with storage buffer (20 mM HEPES, pH 7.5, 500 mM KCl).

4. Prepare 30 μl of a 375 μM (~10 mg ml^{-1}) stock solution of sgRNA by diluting with water. RNA concentrations can be calculated using the OligoCalc server (Kibbe, 2007) from absorbance measured at 260 nm.

5. Mix 94 μl of complex reconstitution buffer with 2.5 μl 100 μM dCas9. Add 1.0 μl 375 μM of sgRNA (1.5-fold molar excess) to form the dCas9–sgRNA binary complex. Incubate for 10 min at room temperature. We recommend adding the sgRNA in 1.5-fold molar excess in order to ensure stoichiometric formation of the binary complex. The final concentrations of dCas9 and sgRNA in the reconstituted sample are 2.5 and 3.75 μM, respectively.

6. In the meantime, prepare double-stranded DNA substrate by annealing complementary and noncomplementary DNA strands at equimolar concentration. To this end, mix 10 μl of each oligonucleotide stock solution (200 μM), heat to 95 °C for 5 min, and cool slowly to room temperature.
7. Add 2.5 μl of 100 μM annealed duplex substrate to the dCas9–sgRNA complex to a final concentration of 2.5 μM. Incubate for 10 min at room temperature. The pipetting scheme of steps 5–7 is summarized in Table 5.
8. Centrifuge for 5 min at 14,000 rpm (16,900 × g) at 4 °C to remove any insoluble material from the sample prior to injection onto the Superdex 200 5/150 column.
9. Inject 50 μl of the sample to the equilibrated Superdex 200 5/150 column. Elute with 3 ml complex reconstitution buffer at a flow rate of 0.1 ml min^{-1}. Measure the UV absorbance of the eluate at 260 and 280 nm, and fluorescence by excitation at 550 nm and emission at 570 nm.
10. Fractionate by collecting 20 μl fractions in a 96-well microtiter plate. To analyze fractions by native polyacrylamide gel electrophoresis, add glycerol to each fraction to a final concentration of 5% glycerol and analyze 20 μl of the peak fractions on a native 8% polyacrylamide gel as described for EMSAs in Section 2. The gel can be scanned using a laser gel scanner (e.g., Typhoon FLA9500, GE Healthcare) with the appropriate excitation and emission wavelength settings for Cy3 (532 nm laser excitation and LPG emission filter) (Fig. 2).

Table 5 dCas9–sgRNA–Target DNA Complex Reconstitution for FSEC Assay

	Stock Concentration	Final Concentration	Volume (μl)[a]
Buffer	–	–	94
dCas9	100 μM	2.5 μM	2.5
sgRNA	375 μM	3.75 μM	1.0
Incubate for 10–15 min at room temperature. Then add annealed DNA duplex.			
Duplex substrate[a]	100 μM	2.5 μM	2.5
Total volume			100

[a]The target and nontarget DNA strands are annealed prior addition to the reaction. See Section 3, step 6 for the annealing procedure.

4. CRYSTALLIZATION OF CAS9–RNA–DNA COMPLEXES

Having identified partially duplexed target DNAs as the most promising candidates for initial crystallization trials, we proceeded to large-scale reconstitution of the ternary dCas9–sgRNA–DNA complex to generate milligram quantities of material for crystallization screening. We decided to follow a stepwise reconstitution procedure in which we first purified the binary dCas9–sgRNA complex, bound it to an excess of the target DNA duplex and repurified the resulting ternary complex by SEC. In initial reconstitution experiments, we attempted to reconstitute the dCas–sgRNA complex in 250 mM KCl. However, we found that the dCas9–sgRNA complex had limited solubility in low ionic strength solutions, and could not be concentrated to >1 mg ml^{-1} prior to SEC. For this reason, we repeated our FSEC assay using an elution buffer containing 500 mM KCl and found that the ternary complex could still be formed (data not shown). We therefore used 500 mM KCl in all subsequent purifications.

Our final purification procedure was as follows. dCas9 was recombinantly expressed and purified by nickel affinity and cation exchange chromatographic steps as described previously but the purification stopped short of the final SEC step (Anders & Jinek, 2014). To reconstitute the binary dCas9–sgRNA complex, cation exchange chromatography-purified Cas9 (~10 mg) was exchanged into SEC buffer (20 mM HEPES–KOH, pH 7.5, 500 mM KCl) by concentrating the protein to 5–10 mg ml^{-1} in a 50,000 molecular weight cut-off (MWCO) centrifugal filter at 3900 rpm (~3200 × g) and 4 °C, diluting with SEC buffer, and repeating the centrifugal concentration. The concentrated protein was mixed with *in vitro* transcribed sgRNA in a 1:1.5 molar ratio and incubated on ice for 10 min. The sample was centrifuged for 10 min at 14,000 rpm (16,900 × g) at 4 °C to remove any insoluble material and injected onto a Superdex 200 16/600 column (equilibrated with SEC buffer) using a 2-ml injection loop. The complex was eluted with 120 ml of SEC buffer at a flow rate of 1 ml min^{-1}. The binary complex typically eluted at a volume of ~69 ml. Peak fractions from the SEC step were pooled and the concentration of the binary complex was calculated from UV absorbance measured at 260 nm using the OlicoCalc server (assuming all of the absorbance was due to just the sgRNA component). MgCl$_2$ was added to a final concentration of 2 mM and the binary complex was combined with preannealed target DNA duplex added in 1.5-fold molar excess. The ternary complex sample was concentrated in a

50,000 MWCO centrifugal filter at 3900 rpm and 4 °C to <1.5 ml volume and subsequently centrifuged for 10 min at 14,000 rpm at 4 °C. The sample was again injected onto a Superdex 200 16/600 column equilibrated with SEC buffer 2 (20 mM HEPES–KOH, pH 7.5, 500 mM KCl, 2 mM MgCl$_2$) using a 2-ml injection loop and eluted with 120 ml SEC buffer 2 at a flow rate of 1 ml min^{-1} collecting 1 ml fractions. The ternary complex containing partially duplexed DNA target typically eluted at a volume of ~65 ml and had an A$_{260}$/A$_{280}$ ratio of ~1.8. Peak fractions from the SEC were pooled and concentrated in a 50,000 MWCO centrifugal filter at 3900 rpm and 4 °C to 10–15 mg ml^{-1} (assuming all of the absorbance at 260 nm was due to just the nucleic acid components). The concentrate was recovered, centrifuged for 10 min at 14,000 rpm at 4 °C, and flash frozen in 50 μl aliquots for storage at −80 °C. For crystallization experiments, the KCl concentration in the sample was reduced to 250 mM by diluting 1:1 with 20 mM HEPES pH 7.5, 2 mM MgCl$_2$ and then adjusting the protein concentration by further dilution with 20 mM HEPES pH 7.5, 250 mM KCl, 2 mM MgCl$_2$.

In our crystallization strategy, we settled on a set of eleven 96-well screens (Table 6) at 20 and 4 °C and explored a range of target DNAs obtained by annealing a small set of target and nontarget strand oligonucleotides. The screens were set up using a Phoenix nanoliter pipetting robot in the 100 + 100 nl format and imaged using a Formulatrix Rock Imager automated system. This screening approach was successful for a dCas9–sgRNA complex containing a partially duplexed target containing a blunt end at the 5′ end of the target strand (Fig. 3A). Initial hits were obtained with a ternary complex sample at a concentration of 1.8 mg ml^{-1} in a range of PEG-based crystallization solutions containing potassium thiocyanate or lithium sulfate (Fig. 3B and C). The outcome of the crystal screening strategy was in agreement with previous studies suggesting that many proteins and multisubunit complexes can be crystallized using a relatively limited set of crystallization screens and that extensive variation of protein constructs and/or nucleic acid ligands is a much more effective way of ensuring successful crystallization (Kimber et al., 2003; Page, Deacon, Lesley, & Stevens, 2005; Thomsen & Berger, 2012). Our future structural studies of Cas9 enzymes will take this into account. To improve crystal size and quality for diffraction experiments, the crystals were subsequently grown using the hanging drop method using Qiagen EasyXtal 15-well trays. In this format, a grid screen vapor diffusion varying the PEG3350 concentration from 12% to 16% (in 1% steps) and KSCN from 0.2 M to 0.3 M (in 50 mM steps) was typically

Table 6 Crystallization Screens Used

Clear Strategy™ Screen I, pH 5.5, pH 6.5, pH 7.5, and pH 8.5 (Molecular Dimensions MD1-14)
Clear Strategy™ Screen II, pH 5.5, pH 6.5, pH 7.5, and pH 8.5 (Molecular Dimensions MD1-15)
Structure Screen I (Molecular Dimensions MD1-01) and Structure Screen II (Molecular Dimensions MD1-02)
Crystallization Basic Kit for Membrane Proteins (Sigma-Fluka Art#73513) and Crystallization Low Ionic Kit for Proteins (Sigma-Fluka #86684)
NeXtal Tubes PACT Suite (Qiagen Art#130718)
NeXtal Tubes Anions Suite (Qiagen Art#130707)
NeXtal Tubes Cations Suite (Qiagen Art#130708)
Crystal Screen™ Kit (Hampton research Art#HR2-110) and Crystal Screen 2™ (Hampton research Art#HR2-112)
PEG4000, 0.2 M Li$_2$SO$_4$, PEG8000 5–30%, pH gradient:3–10 (custom screen)
PEG4000, 0.2 M Li$_2$SO$_4$, PEG6000 5–30%, pH gradient:3–10 (custom screen)
PEG400 15–45%//PEG4000 5–30%, pH gradient 4.5–9.4, K-phosphate//Li-Na-sulfate (custom screen)
PEG400 15–45%//PEG4000 5–30%, pH gradient 4.5–9.4, ammonium formate; zinc acetate, potassium iodide (custom screen)

performed with a range of complex concentrations. The resulting thin plate crystals grew best from complex solutions at 1.5 mg ml^{-1} and reached their final size in 1–2 weeks.

For diffraction experiments, the crystals were cryoprotected by a brief transfer into 0.1 M Tris–acetate, pH 8.5, 200 mM KSCN, 30% PEG3350, 10% ethylene glycol, and flash-cooled in liquid nitrogen. Crystal diffraction was tested at the Swiss Light Source (Paul Scherrer Institut, Villigen, Switzerland). Initially, the crystals diffracted mostly to 3.5 Å, but the diffraction limit could be improved to 2.6 Å by increasing crystal thickness. This was achieved (i) by switching from the catalytically inactive dCas9 protein to the single-mutant (H840A) Cas9 nickase and (ii) by iterative microseeding, whereby seeds generated from an initial batch of complex crystals were used to grow the subsequent batch of crystals for seed generation, until large single crystals of the complex could be obtained reproducibly (Fig. 3D and E). We also found empirically that reversing the traditional

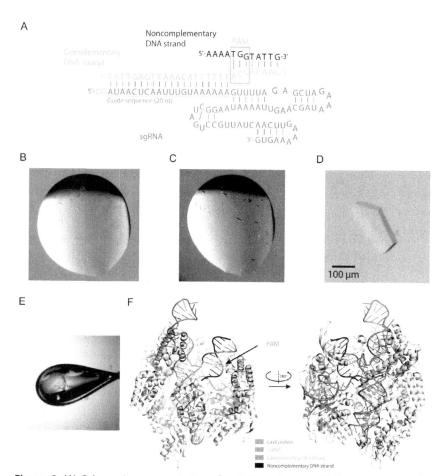

Figure 3 (A) Schematic representation of sgRNA and target DNA sequences in the dCas9–sgRNA–target DNA complex used for crystallization screening. The color coding is as in Fig. 2A. (B and C) Two examples of hits from initial crystallization screenings using 1.8 mg ml^{-1} ternary complex containing dCas9, sgRNA, and target DNA 1 (see panel A). The crystal screenings were set up at 20 °C and incubated either at 4 or 20 °C. Five out of 19 hits grew at 4 °C. Four out of 19 hits were found in KSCN, Tris–acetate, and polyethylene glycols of different chain lengths (PEG1500, PEG3350, and PEG4000). Thirteen out of 19 hits were found in Li$_2$SO$_4$, PEG4000, and different buffers spanning the pH range from pH 6.0 to 10.0. Crystals in (B) grew from 0.1 M Tris–acetate, pH 8.5, 0.15 M KSCN, 18% PEG3350. Crystals in (C) grew from 0.02 M CAPS, pH 10.0, 0.2 M Li$_2$SO$_4$, 15% PEG4000. (D) Optimized crystal after iterative rounds of seeding. The crystallization was performed at 20 °C with 1.5 mg ml^{-1} ternary complex in 20 mM HEPES, pH 7.5, 250 mM KCl, 5 mM MgCl$_2$. The crystal was grown from 0.1 M Tris–acetate, pH 8.5, 0.3 M KSCN, 16% PEG3350. (E) Ternary complex crystal harvested in a 200–300 μm nylon loop. (F) Front and rear views of dCas9–sgRNA–target DNA 1 complex shown in cartoon representation.

order of pipetting the protein and reservoir solutions (i.e., pipetting the reservoir solution first and adding protein complex solution second) led to consistently better crystals. Through these incremental improvements, we were able to obtain robust well-diffracting crystals of not just the native ternary complex but also crystals containing selenomethionine-substituted protein for phasing experiments that allowed us to solve the structure (Fig. 3F). We were subsequently able to crystallize dCas9 complexes containing mismatches between the sgRNA and the target strand. Together, the resulting structures have provided critical insights into the molecular mechanism of target DNA recognition by the Cas9–guide RNA complex and highlighted the key role of the PAM in this process (Anders et al., 2014).

5. CONCLUDING REMARKS

Within a short span of time, Cas9 has emerged as an extremely powerful molecular tool for genome editing and gene expression control that combines ease of use with a high degree of precision. Continued investigations of the molecular mechanism of Cas9 are critically dependent on structural studies of its complexes with guide RNAs and DNA substrates. These studies aim to provide detailed insights into the mechanism of target DNA recognition and cleavage and also to establish the structural framework needed for further development of Cas9-based applications. In particular, structure-based engineering of the Cas9–sgRNA platform could lead to improvements in targeting specificity, expand the spectrum of PAM sequences recognized by Cas9, and enhance the efficiency of the CRISPR-Cas9 system in transcriptional and epigenetic control (Konermann et al., 2015).

The assays outlined in this chapter have informed our work toward determining the crystal structure of Cas9 in complex with an sgRNA and a PAM-containing DNA target. Using both EMSA and FSEC approaches, we were able to analyze DNA binding by the Cas9–sgRNA binary complex and identify guide RNA and target DNA features necessary to support ternary complex formation. Using a focused set of target DNAs and crystallization screens, we crystallized the ternary complex and, through systematic improvement of the diffraction quality of the crystals, were able to solve its structure. Lessons learned from this study will influence our approach toward obtaining structures of additional functional states of Cas9 in future studies. Besides providing important tools for structural studies, our assays may prove invaluable for characterizing DNA binding by Cas9 in application development, particularly with respect to Cas9 orthologs from other bacterial species whose DNA targeting and cleavage activities are as yet uncharacterized. Likewise, we believe that these methods will be readily adaptable to other multicomponent protein–nucleic acid complexes.

ACKNOWLEDGMENTS

We thank A. Duerst for technical assistance. This work was supported by the University of Zurich and the European Research Council (ERC) Starting Grant ANTIVIRNA (337284).

REFERENCES

Anders, C., & Jinek, M. (2014). In vitro enzymology of Cas9. *Methods in Enzymology, 546*, 1–20.

Anders, C., Niewoehner, O., Duerst, A., & Jinek, M. (2014). Structural basis of PAM-dependent target DNA recognition by the Cas9 endonuclease. *Nature, 513*, 569–573.

Anderson, J., Ptashne, M., & Harrison, S. C. (1984). Cocrystals of the DNA-binding domain of phage 434 repressor and a synthetic phage 434 operator. *Proceedings of the National Academy of Sciences of the United States of America, 81*, 1307–1311.

Anderson, J. E., Ptashne, M., & Harrison, S. C. (1987). Structure of the repressor-operator complex of bacteriophage 434. *Nature, 326*, 846–852.

Barrangou, R., Fremaux, C., Deveau, H., Richards, M., Boyaval, P., Moineau, S., et al. (2007). CRISPR provides acquired resistance against viruses in prokaryotes. *Science, 315*, 1709–1712.

Berry, K. E., Waghray, S., Mortimer, S. A., Bai, Y., & Doudna, J. A. (2011). Crystal structure of the HCV IRES central domain reveals strategy for start-codon positioning. *Structure, 19*, 1456–1466.

Briner, A. E., Donohoue, P. D., Gomaa, A. A., Selle, K., Slorach, E. M., Nye, C. H., et al. (2014). Guide RNA functional modules direct Cas9 activity and orthogonality. *Molecular Cell, 56*, 333–339.

Cencic, R., Miura, H., Malina, A., Robert, F., Ethier, S., Schmeing, T. M., et al. (2014). Protospacer adjacent motif (PAM)-distal sequences engage CRISPR Cas9 DNA target cleavage. *PLoS One, 9*, e109213.

Chen, H., Choi, J., & Bailey, S. (2014). Cut site selection by the two nuclease domains of the Cas9 RNA-guided endonuclease. *Journal of Biological Chemistry, 289*, 13284–13294.

Chen, B., Gilbert, L. A., Cimini, B. A., Schnitzbauer, J., Zhang, W., Li, G.-W., et al. (2013). Dynamic imaging of genomic loci in living human cells by an optimized CRISPR/Cas system. *Cell, 155*, 1479–1491.

Cong, L., Ran, F. A., Cox, D., Lin, S., Barretto, R., Habib, N., et al. (2013). Multiplex genome engineering using CRISPR/Cas systems. *Science, 339*, 819–823.

Deltcheva, E., Chylinski, K., Sharma, C. M., Gonzales, K., Chao, Y., Pirzada, Z. A., et al. (2011). CRISPR RNA maturation by trans-encoded small RNA and host factor RNase III. *Nature, 471*, 602–607.

Doudna, J. A., & Charpentier, E. (2014). Genome editing. The new frontier of genome engineering with CRISPR-Cas9. *Science, 346*, 1258096.

Ferré-D'Amaré, A. R., Zhou, K., & Doudna, J. A. (1998). A general module for RNA crystallization. *Journal of Molecular Biology, 279*, 621–631.

Fu, Y., Foden, J. A., Khayter, C., Maeder, M. L., Reyon, D., Joung, J. K., et al. (2013). High-frequency off-target mutagenesis induced by CRISPR-Cas nucleases in human cells. *Nature Biotechnology, 31*, 822–826.

Garneau, J. E., Dupuis, M.-È., Villion, M., Romero, D. A., Barrangou, R., Boyaval, P., et al. (2010). The CRISPR/Cas bacterial immune system cleaves bacteriophage and plasmid DNA. *Nature, 468*, 67–71.

Gasiunas, G., Barrangou, R., Horvath, P., & Siksnys, V. (2012). Cas9-crRNA ribonucleoprotein complex mediates specific DNA cleavage for adaptive immunity in bacteria. *Proceedings of the National Academy of Sciences of the United States of America, 109*, E2579–E2586.

Gilbert, L. A., Horlbeck, M. A., Adamson, B., Villalta, J. E., Chen, Y., Whitehead, E. H., et al. (2014). Genome-scale CRISPR-mediated control of gene repression and activation. *Cell, 159*, 647–661.

Gilbert, L. A., Larson, M. H., Morsut, L., Liu, Z., Brar, G. A., Torres, S. E., et al. (2013). CRISPR-mediated modular RNA-guided regulation of transcription in eukaryotes. *Cell, 154*, 442–451.

Hollis, T. (2007). Crystallization of protein-DNA complexes. *Methods in Molecular Biology, 363*, 225–237.

Hsu, P. D., Lander, E. S., & Zhang, F. (2014). Development and applications of CRISPR-Cas9 for genome engineering. *Cell, 157*, 1262–1278.

Hsu, P. D., Scott, D. A., Weinstein, J. A., Ran, F. A., Konermann, S., Agarwala, V., et al. (2013). DNA targeting specificity of RNA-guided Cas9 nucleases. *Nature Biotechnology, 31*, 827–832.

Jinek, M., Chylinski, K., Fonfara, I., Hauer, M., Doudna, J. A., & Charpentier, E. (2012). A programmable dual-RNA-guided DNA endonuclease in adaptive bacterial immunity. *Science, 337*, 816–821.

Jinek, M., East, A., Cheng, A., Lin, S., Ma, E., & Doudna, J. (2013). RNA-programmed genome editing in human cells. *eLife, 2*, e00471.

Jinek, M., Jiang, F., Taylor, D. W., Sternberg, S. H., Kaya, E., Ma, E., et al. (2014). Structures of Cas9 endonucleases reveal RNA-mediated conformational activation. *Science, 343*, 1247997.

Karvelis, T., Gasiunas, G., Miksys, A., Barrangou, R., Horvath, P., & Siksnys, V. (2013). crRNA and tracrRNA guide Cas9-mediated DNA interference in Streptococcus thermophilus. *RNA Biology, 10*, 841–851.

Kerr, L. D. (1995). Electrophoretic mobility shift assay. *Methods in Enzymology, 254*, 619–632.

Kibbe, W. A. (2007). OligoCalc: An online oligonucleotide properties calculator. *Nucleic Acids Research, 35*, W43–W46.

Kimber, M. S., Vallee, F., Houston, S., Necakov, A., Skarina, T., Evdokimova, E., et al. (2003). Data mining crystallization databases: Knowledge-based approaches to optimize protein crystal screens. *Proteins, 51*, 562–568.

Konermann, S., Brigham, M. D., Trevino, A. E., Joung, J., Abudayyeh, O. O., Barcena, C., et al. (2015). Genome-scale transcriptional activation by an engineered CRISPR-Cas9 complex. *Nature, 517*, 583–588.

Kuscu, C., Arslan, S., Singh, R., Thorpe, J., & Adli, M. (2014). Genome-wide analysis reveals characteristics of off-target sites bound by the Cas9 endonuclease. *Nature Biotechnology, 32*, 677–683.

Leung, A. K. W., Kambach, C., Kondo, Y., Kampmann, M., Jinek, M., & Nagai, K. (2010). Use of RNA tertiary interaction modules for the crystallisation of the spliceosomal snRNP core domain. *Journal of Molecular Biology, 402*, 154–164.

Lopez-Gomollon, S., & Nicolas, F. E. (2013). Purification of DNA Oligos by denaturing polyacrylamide gel electrophoresis (PAGE). *Methods in Enzymology, 529*, 65–83.

Mali, P., Aach, J., Stranges, P. B., Esvelt, K. M., Moosburner, M., Kosuri, S., et al. (2013). Cas9 transcriptional activators for target specificity screening and paired nickases for cooperative genome engineering. *Nature Biotechnology, 31*, 833–838.

Mali, P., Esvelt, K. M., & Church, G. M. (2013). Cas9 as a versatile tool for engineering biology. *Nature Methods, 10*, 957–963.

Mali, P., Yang, L., Esvelt, K. M., Aach, J., Guell, M., DiCarlo, J. E., et al. (2013). RNA-guided human genome engineering via Cas9. *Science, 339*, 823–826.

Mitchell, S. F., & Lorsch, J. R. (2014). Standard in vitro assays for protein–nucleic acid interactions—Gel shift assays for RNA and DNA binding. *Methods in Enzymology, 541*, 179–196.

Nishimasu, H., Ran, F. A., Hsu, P. D., Konermann, S., Shehata, S. I., Dohmae, N., et al. (2014). Crystal structure of Cas9 in complex with guide RNA and target DNA. *Cell, 156*, 935–949.

Obayashi, E., Oubridge, C., Pomeranz Krummel, D., & Nagai, K. (2007). Crystallization of RNA-protein complexes. *Methods in Molecular Biology, 363*, 259–276.

O'Connell, M. R., Oakes, B. L., Sternberg, S. H., East-Seletsky, A., Kaplan, M., & Doudna, J. A. (2014). Programmable RNA recognition and cleavage by CRISPR/Cas9. *Nature, 516*, 263–266.

Page, R., Deacon, A. M., Lesley, S. A., & Stevens, R. C. (2005). Shotgun crystallization strategy for structural genomics II: Crystallization conditions that produce high resolution structures for T. maritima proteins. *Journal of Structural and Functional Genomics, 6*, 209–217.

Pattanayak, V., Lin, S., Guilinger, J. P., Ma, E., Doudna, J. A., & Liu, D. R. (2013). High-throughput profiling of off-target DNA cleavage reveals RNA-programmed Cas9 nuclease specificity. *Nature Biotechnology, 31*, 839–843.

Reyes, F. E., Garst, A. D., & Batey, R. T. (2009). Strategies in RNA crystallography. *Methods in Enzymology, 469*, 119–139.

Ryder, S. P., Recht, M. I., & Williamson, J. R. (2008). Quantitative analysis of protein-RNA interactions by gel mobility shift. *Methods in Molecular Biology, 488*, 99–115.

Sternberg, S. H., Redding, S., Jinek, M., Greene, E. C., & Doudna, J. A. (2014). DNA interrogation by the CRISPR RNA-guided endonuclease Cas9. *Nature, 507*, 62–67.

Thomsen, N. D., & Berger, J. M. (2012). Crystallization and X-ray structure determination of an RNA-dependent hexameric helicase. *Methods in Enzymology, 511*, 171–190.

Winzor, D. J. (2003). Analytical exclusion chromatography. *Journal of Biochemical and Biophysical Methods, 56*, 15–52.

Wu, X., Scott, D. A., Kriz, A. J., Chiu, A. C., Hsu, P. D., Dadon, D. B., et al. (2014). Genome-wide binding of the CRISPR endonuclease Cas9 in mammalian cells. *Nature Biotechnology, 32*, 670–676.

CHAPTER EIGHTEEN

Single-Molecule Pull-Down FRET to Dissect the Mechanisms of Biomolecular Machines

Matthew L. Kahlscheuer, Julia Widom, Nils G. Walter[1]

Single Molecule Analysis Group, Department of Chemistry, University of Michigan, Ann Arbor, Michigan, USA
[1]Corresponding author: e-mail address: nwalter@umich.edu

Contents

1. Introduction	540
2. Experimental Methods	543
2.1 Labeling and purification of pre-mRNA substrates for smFRET	543
2.2 Isolation of splicing complexes through pull-down	546
2.3 smFRET using prism-based TIRF microscopy	554
2.4 Experimental procedures for smFRET on the spliceosome	557
3. Data Analysis	559
3.1 FRET histograms	559
3.2 HMM and transition occupancy density plot analysis	559
3.3 Postsynchronized histograms	561
3.4 Clustering analysis	561
4. The Spliceosome as a Biased Brownian Ratchet Machine	562
5. Conclusions and Outlook	566
Acknowledgment	567
References	567

Abstract

Spliceosomes are multimegadalton RNA–protein complexes responsible for the faithful removal of noncoding segments (introns) from pre-messenger RNAs (pre-mRNAs), a process critical for the maturation of eukaryotic mRNAs for subsequent translation by the ribosome. Both the spliceosome and ribosome, as well as many other RNA and DNA processing machineries, contain central RNA components that endow biomolecular complexes with precise, sequence-specific nucleic acid recognition, and versatile structural dynamics. Single-molecule fluorescence (or Förster) resonance energy transfer (smFRET) microscopy is a powerful tool for the study of local and global conformational changes of both simple and complex biomolecular systems involving RNA. The integration of biochemical tools such as immunoprecipitation with advanced methods in smFRET microscopy and data analysis has opened up entirely new avenues toward studying the mechanisms of biomolecular machines isolated directly from complex biological specimens, such as cell extracts. Here, we detail the general steps for using prism-based total internal reflection fluorescence microscopy in exemplary single-molecule pull-down FRET studies of the yeast spliceosome and discuss the broad application potential of this technique.

1. INTRODUCTION

The spliceosome is the large protein–RNA complex responsible for the removal of introns from eukaryotic pre-messenger RNAs (pre-mRNAs) and the subsequent ligation of the remaining exons, generating continuous open reading frames that can be translated into protein (Wahl, Will, & Luhrmann, 2009). This process is a key step in gene expression, and defects in the splicing process are responsible for a significant fraction of known human genetic diseases (Wahl et al., 2009; Wang, Zhang, Li, Zhao, & Cui, 2012). Splicing consists of two chemical reactions, shown in Fig. 1A. In the first step of splicing, the $2'$ hydroxyl group of the branchpoint adenosine attacks the phosphodiester backbone at the $5'$ splice site, generating a free $5'$ exon and a lariat intermediate containing the intron and the $3'$ exon. In the second step, the liberated $3'$ hydroxyl group of the $5'$ splice site attacks a phosphate group at the $3'$ splice site, expelling the intron lariat and ligating together the two exons. The spliceosome is rather unique among macromolecular machines in that it lacks a preformed active site and becomes catalytically active through assembly and rearrangement steps on the template of the pre-mRNA. In addition to the pre-mRNA substrate, splicing requires the RNA and protein components of five small nuclear ribonucleoprotein particles (snRNPs) and in humans, over 100 additional non-snRNP proteins (Fabrizio et al., 2009). The splicing cycle (Fig. 1B) proceeds through a specific sequence of binding, dissociation, and rearrangement events involving many of these protein and RNA components.

Any technique used to study a large, dynamic macromolecular complex such as the spliceosome must be sufficiently sensitive to detect low concentrations of sample, sufficiently specific to address a particular location of interest amid a large background of other protein and RNA components, and sufficiently information-rich to allow rigorous testing of mechanistic hypotheses. One technique that meets these requirements is single-molecule fluorescence (or Förster) resonance energy transfer (smFRET) (Förster, 1948; Roy, Hohng, & Ha, 2008; Stryer, 1978; Walter, 2003). In a FRET

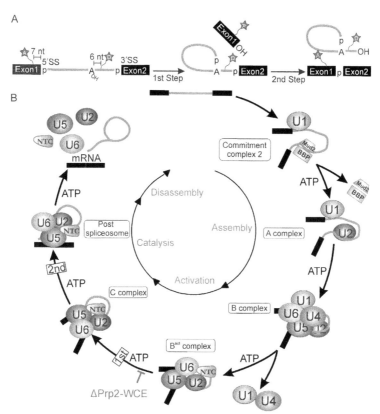

Figure 1 The canonical mechanism of pre-mRNA splicing catalyzed by the spliceosome. (A) The Ubc4 substrate contains donor (D, Cy3) and acceptor (A, Cy5) fluorophores near the BP and 5′SS, respectively, enabling observation of docking of the BP adenosine into the 5′SS during the first step of splicing. (B) Spliceosome assembly is thought to occur in a stepwise fashion with multiple ATP-dependent RNA–RNA and RNA–protein rearrangements between intermediary complexes, leading to the first and second chemical steps of splicing as indicated. In the Prp2-1,Cef1-TAP yeast strain, the B^{act} complex can be selectively formed through inactivation of the heat-sensitive ATPase Prp2 (red (gray in the print version) block).

experiment, a sample is labeled with a pair of fluorophores, chosen so that the emission spectrum of one (the "donor fluorophore") overlaps with the absorption spectrum of the other (the "acceptor fluorophore"). When the donor is excited, a dipole–dipole interaction between it and the acceptor permits the transfer of energy between the two, with the efficiency of this process depending on the distance and relative orientation between the two fluorophores, the fluorescence quantum yield of the donor, and the extent of spectral overlap between the donor's emission and the acceptor's absorption.

This distance dependence, which has a sensitivity range of ~10–100 Å, makes FRET a valuable technique for probing the conformations of biological macromolecules (Stryer, 1978; Walter, 2001, 2003). In smFRET, the molecule of interest is immobilized sparsely on a microscope slide so that the donor and acceptor fluorescence intensities, and thereby the FRET efficiency, can be measured for individual molecules (Roy et al., 2008). This is valuable because complex biological macromolecules often exist in multiple different conformations, and smFRET allows these conformations and their transitions to be observed, rather than reporting an ensemble average, which loses most of the information on transition kinetics, transient intermediates, and rare conformational states (Roy et al., 2008; Walter, Huang, Manzo, & Sobhy, 2008).

In recent years, smFRET has been applied to a number of protein–RNA complexes, including but not limited to the bacterial ribosome (Aitken, Petrov, & Puglisi, 2010; Chen, Tsai, O'Leary, Petrov, & Puglisi, 2012; Kim et al., 2014), the yeast spliceosome (Abelson, Blanco, et al., 2010; Crawford, Hoskins, Friedman, Gelles, & Moore, 2013; Krishnan et al., 2013), and human telomerase (Hwang et al., 2014; Parks & Stone, 2014). In most smFRET work, purified nucleic acid and protein components have been immobilized on slides through biotin–streptavidin linkages. Even the comparably small yeast spliceosome contains so many different components (five different snRNAs and ~40 different proteins, depending on the stage of splicing) (Fabrizio et al., 2009; Wahl et al., 2009) that it is not practical to purify every protein and RNA component in the spliceosome and reconstitute splicing "from scratch." Conversely, when working in cell extract, splicing can be stalled at many different stages of the splicing cycle using, for example, genetic manipulations that are readily available in yeast, leading to accumulation of certain intermediate complexes that can then be isolated and subjected to biochemical analysis. The gap between these two areas of inquiry (single-molecule observation of purified components and biochemical analysis of complexes isolated from cell extracts) was bridged by the technique of single-molecule pull-down (SiMPull), which was first demonstrated in 2011 (Jain, Liu, Xiang, & Ha, 2012; Jain et al., 2011). In this approach, a streptavidin-coated slide is incubated with a biotinylated antibody and extract is prepared from cells bearing a matching epitope, for example, a TAP or FLAG tag, on a protein of interest. This extract is incubated on the slide, allowing a particular complex to be "pulled down" from the extract onto the slide through the interaction between the antibody and the epitope. An extension of this approach termed SiMPull-FRET allows complexes to be studied via smFRET that otherwise may be difficult to purify, immobilize, and/or reconstitute, and offers the potential for them to be studied in cell extract, offering relatively *in vivo*-like conditions (Krishnan et al., 2013).

In this chapter, we present the method of SiMPull-FRET, applied to the yeast spliceosome. First, we discuss the experimental methods required, including pre-mRNA preparation, the isolation of spliceosomal complexes for smFRET and for biochemical analysis, slide preparation and microscopy, and collection of smFRET data. Second, we discuss SiMPull-FRET data analysis, including histograms, hidden Markov modeling (HMM), and single-molecule cluster analysis. Finally, we present an application of SiMPull-FRET to the spliceosome in which the conformational fluctuations of the pre-mRNA before and during the first step of splicing were investigated (Krishnan et al., 2013).

2. EXPERIMENTAL METHODS
2.1 Labeling and purification of pre-mRNA substrates for smFRET

In order to study the complex dynamic behavior of pre-mRNA substrates throughout splicing using smFRET, molecules have to be site-specifically labeled with donor and acceptor fluorophores at positions along the RNA the distance of which is of particular interest to monitor. The selection of pre-mRNA substrate, location and choice of dyes, and method of immobilization must all be taken into account in order to produce a molecular ruler sufficient for monitoring important changes in RNA structure. A short, efficiently spliced pre-mRNA substrate, such as the yeast intron Ubc4 (Abelson, Blanco, et al., 2010), is ideal. Pre-mRNA splicing *in vitro* using yeast whole cell extract (WCE) is quite inefficient and may become even more inefficient upon introduction of large, hydrophobic fluorophores into the RNA. In addition, attachment of a moiety for immobilization, such as biotin, may restrict movement of the pre-mRNA substrate and further decrease splicing efficiency. Fortunately, splicing-specific microarray analysis, which utilizes hundreds of transcript-specific DNA probes capable of distinguishing pre-mRNA from mRNA, has recently made available the splicing efficiency of nearly all known yeast pre-mRNA substrates (Clark, Sugnet, & Ares, 2002; Pleiss, Whitworth, Bergkessel, & Guthrie, 2007). Thus, the list of suitable pre-mRNA substrates is reduced dramatically. In addition to splicing efficiency, the pre-mRNA length and structure must also be taken into account. Total internal reflection fluorescence (TIRF) microscopy is limited to a depth of illumination of approximately 100–200 nm on the slide surface. More importantly, the synthesis and labeling of pre-mRNA substrates becomes increasingly difficult as the length of the RNA increases. As such, it is important to choose a pre-mRNA substrate short enough (<~400 nucleotides, nt) to be efficiently synthesized and labeled. Choosing a pre-mRNA substrate with significant secondary structure is also desired so that, depending upon the labeling sites, the assembly effects of the spliceosome (i.e., unfolding) can be observed through monitoring large changes in FRET efficiency. SHAPE-directed structure probing (McGinnis, Duncan, & Weeks, 2009) and structure prediction software provide a reasonable starting point in selecting the proper pre-mRNA substrate.

Once a pre-mRNA substrate has been chosen, optimal sites of labeling must next be chosen. Depending upon the desired experiment, FRET probes can be placed in such a way as to allow for the observation of particular assembly or catalytic steps in the splicing cycle. For example, fluorophores positioned near the 5'SS and BP allow for observation of docking of the BP adenosine near the 5'SS during the first step of splicing (Krishnan et al., 2013), while labeling near the 5'SS and 3'SS will allow for observation of docking of the 5' exon near the intron–exon junction during the second step of splicing (Abelson, Blanco, et al., 2010). In either case, several sites should be tested to ensure the substrate splices with high efficiency upon incorporation of the donor and acceptor dyes. Many nucleotides and RNA sequences within a pre-mRNA (5'SS, BP, 3'SS, polypyrimidine tract, etc.) are evolutionarily conserved and participate in essential hydrogen bonding interactions with the spliceosome. Direct labeling of these sites should thus be avoided. In addition, particular structural motifs are often required for efficient splicing, further limiting the location of fluorescent probes. Taking all of these factors into account will ensure that addition of large, somewhat bulky fluorophores will have minimal effects on spliceosome assembly or catalysis.

There are several methods available for the site-specific, internal labeling of RNA with fluorophores for smFRET (Rinaldi, Suddala, & Walter, 2015; Solomatin & Herschlag, 2009; Walter, 2003; Walter & Burke, 2000). Chemical synthesis allows for incorporation of site-specific modifications and fluorophores directly during synthesis. Unfortunately, most pre-mRNA substrates are larger than 100 nt in length and thus exceed the typical length limitations of chemical synthesis of ~80 nt. Perhaps the most common method to overcome this limitation is to use splint-mediated RNA ligation

(Abelson, Blanco, et al., 2010; Abelson, Hadjivassiliou, & Guthrie, 2010; Crawford, Hoskins, Friedman, Gelles, & Moore, 2008; Krishnan et al., 2013; Moore & Query, 2000). In this approach, an RNA substrate is chemically synthesized in several segments, two of which contain aminoallyl uridine in the locations where the FRET probes are to be attached. Fluorophores are conjugated to the RNA by incubation with the N-hydroxysuccinimidyl ester fluorophore under slightly basic conditions. Unconjugated free dye is removed by ethanol precipitation of the RNA. Further purification of labeled from unlabeled RNA can be achieved by taking advantage of the hydrophobic nature of the attached fluorophore. Benzoylated naphthoylated DEAE-cellulose is a medium that has an affinity for hydrophobic material and thus will bind more tightly to fluorophore-containing RNA (Abelson, Blanco, et al., 2010; Krishnan et al., 2013). Fully conjugated RNA segments are then ligated together using DNA splints and RNA ligase 1, followed by polyacrylamide gel electrophoresis (PAGE) purification, yielding nearly 100% labeled RNA. Alternative methods of internal labeling include using the 10DM24 deoxyribozymes to site-specifically attach fluorophore-modified guanosine triphosphate analogs to specific adenosine residues on the *in vitro*-transcribed RNA backbone (Buttner, Javadi-Zarnaghi, & Hobartner, 2014). RNA length, structure, and location of adenosines can limit this method, and it is not yet clear whether the lariat-debranching enzyme found in yeast WCE may remove the label; however, it should provide a cheaper approach to labeling than chemical synthesis. If a longer pre-mRNA substrate is desired, labeling can be achieved through annealing of fluorescently labeled oligonucleotides to *in vitro*-transcribed pre-mRNA, as long as the resulting hybrid is sufficiently stable for the intended application (Fiegland, Garst, Batey, & Nesbitt, 2012).

A number of approaches also exist to modify and label the $5'$ and $3'$ ends of RNA (Ohrt et al., 2012; Qin & Pyle, 1999; Rinaldi et al., 2015). Incubation of RNA with sodium periodate results in the formation of $3'$ aldehydes that can be conjugated with hydrazide derivatives of fluorophores or biotin (Newby Lambert et al., 2006). Alternatively, the free phosphate on the $5'$ end of the RNA can be activated upon incubation with EDC (1-ethyl-3-[3-dimethylaminopropyl]carbodiimide hydrochloride). Incubation with imidazole and ethylenediamine produces the required primary amine for labeling with an NHS-ester fluorophore or biotin (Qin & Pyle, 1999; Rinaldi et al., 2015). Lastly, pre-mRNA transcripts can be $5'$ end labeled through incorporation of $5'$-GMPS (guanosine-$5'$-O-monophosphorothioate) during *in vitro* transcription followed by labeling with maleimide-derivative fluorophores or biotin (Ohrt et al., 2012; Rueda, Hsieh, Day-Storms, Fierke, & Walter, 2005). While these methods of end-terminal labeling are efficient and relatively cost effective, labeling the $5'$ or $3'$ end of RNA does not offer the flexibility of internal labeling, potentially precluding the observation of important conformational dynamics required for assembly or catalysis.

A large number of fluorophores are available for smFRET studies, containing a variety of chemical properties (Roy et al., 2008). Perhaps, the most widely used FRET pair for the study of RNA dynamics is Cy3 and Cy5 (Abelson, Blanco, et al., 2010; Krishnan et al., 2013), although a number of improved derivatives are becoming increasingly available that exhibit increased photostability and higher quantum yield (Zheng et al., 2014). In addition, fluorophore lifetimes can be greatly increased by the use of an oxygen scavenging system (OSS), such as glucose, glucose oxidase, and catalase, or protocatechuic acid (PCA) and protocatechuate-3,4-dehydrogenase (PCD) (Aitken, Marshall, & Puglisi, 2008). In these systems, an enzyme and its substrate are added to the sample, chosen such that oxygen is consumed through the enzyme's reaction cycle. In addition, trolox is added as a triplet state quencher, which significantly reduces the occurrence of photoblinking. Care must be exercised when choosing the OSS for a particular smFRET experiment—for example, the addition of glucose to yeast WCE results in ATP depletion due to the action of endogenous hexokinase (Tatei, Kimura, & Ohshima, 1989), making PCA/PCD the preferred OSS for experiments performed in extract. Because fluorophore lifetimes are greatly reduced in the presence of the protein and RNA components within the splicing complex of interest as well as in the extract, choosing the proper FRET pair and OSS is crucial to an effective smFRET study. One interesting recent approach covalently attaches a triplet quencher directly to the fluorophore to affect "self-healing" through intramolecular triplet state quenching (Zheng et al., 2014); however, the added bulkiness of the so appended fluorophore has to be taken into consideration.

2.2 Isolation of splicing complexes through pull-down

Successful isolation and pull-down of a specific protein–protein or protein–RNA complex is central to the SiMPull-FRET methodology. SiMPull requires many of the same reagents that would be needed for coimmunoprecipitation experiments followed by Western blot analysis but is typically much cheaper, more sensitive, and less time consuming.

Selective pull-down is achieved through capture of a bait protein using a high-affinity antibody either specific for the bait protein itself or to a purification tag appended to the protein (TAP, Flag, etc.). The immobilized bait is then typically used to capture one or more, fluorescently labeled prey protein (as well as any interacting partners that might form a complex) in order to study stoichiometry of the bait–prey complex (Bharill, Fu, Palty, & Isacoff, 2014; Jain et al., 2011; Panter, Jain, Leonhardt, Ha, & Cresswell, 2012; Peterson et al., 2014; Shen et al., 2012) or conformational dynamics of a protein or protein complex (Zhou, Kunzelmann, Webb, & Ha, 2011; Zhou, Zhang, Bochman, Zakian, & Ha, 2014).

The bait-functionalized surface must specifically bind to the protein or complex of interest while rejecting all other biomolecules. In the case of SiMPull-FRET, the bait can be, for example, a TAP-tagged derivative of one of the components of the NineTeen Complex (NTC), Cef1, known to be present in the spliceosome during formation of the B^{act} complex (Lardelli, Thompson, Yates, & Stevens, 2010; Warkocki et al., 2009). The prey are the remaining protein, snRNA, and fluorescent substrate components known to be associated with the bait during B^{act} formation (Fig. 2A). As Cef1 does not interact with the spliceosome or the fluorescent pre-mRNA substrate prior to the B^{act} stage, any free Cef1 protein or Cef1-containing NTC that becomes captured by the antibody will not be visible upon Cy3 excitation.

The bait protein should be one that is stably associated within the larger complex (slow dissociation constant) to ensure that once immobilized, the larger complex will remain on the slide surface long enough for smFRET experimentation. Similarly, the antibody needs to feature high affinity together with slow epitope dissociation not to artificially shorten the observation of the immobilized complex, as well as exquisite specificity with little cross-reactivity with other WCE components. These requirements imply that not every Western blot validated antibody may work and that instead a more stringent selection criterion has to be applied such as suitability for immunofluorescence applications. In essence, pull-down of a complex onto a nonporous slide is a separation using only a single equilibrium binding stage (or "theoretical plate") that does not afford the benefits of, for example, an affinity purification column with its many theoretical plates that increase the separation efficiency between specifically and nonspecifically bound proteins. In addition, the bait should be known to join the spliceosome at a defined point in the assembly pathway and remains associated throughout all splicing steps of interest. Further enrichment for a specific splicing

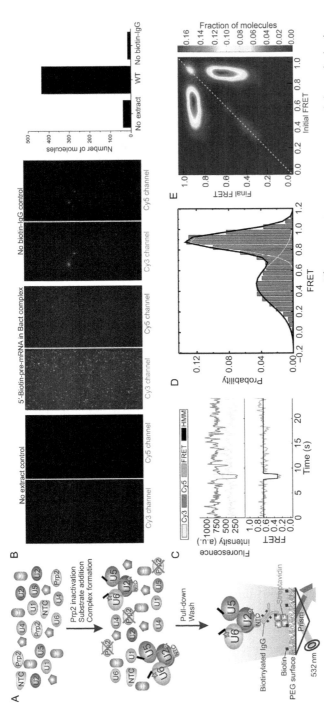

Figure 2 Schematic and analysis methodology of the SiMPull-FRET technique. (A) B^{act} complex formation is promoted through the heat inactivation of Prp2 prior to incubation of the whole cell extract with the fluorescent pre-mRNA substrate. Complexes are then pulled down for smFRET analysis on a biotin-PEG slide surface coated with streptavidin and biotinylated IgG antibody utilizing the TAP-tagged NTC component Cef1. (B) Representative fields of view showing the selective binding of the fluorescent substrate to the slide surface only when contained in the B^{act} complex and when slide surfaces have been saturated with IgG–biotin. Quantifications of the number of molecules binding under each slide condition are shown on the far right. (C) Representative Cy3 donor, Cy5 acceptor, FRET, and idealized hidden Markov model (HMM) traces from an smFRET experiment. (D) FRET probability distribution analysis used to determine the dominant FRET states of a population of single molecules. (E) Transition occupancy density plots (TODPs) are scaled by the fraction of all molecules that exhibit transitions from a particular initial FRET state (plotted along the x-axis) to a particular final FRET state (plotted along the y-axis). *Panel B was modified from Krishnan et al. (2013).* (See the color plate.)

intermediate is achieved through introduction of a biochemical or genetic stall known to arrest splicing at the assembly stage of interest. Fortunately, *Saccharomyces cerevisiae* is conducive to a host of genetic and biochemical modifications that allow for the introduction of point mutations and purification tags directly into the genome. Large yeast libraries are commercially available containing TAP and GFP tags on nearly every known protein. Of course, modern gene modification tools promise to make such tags much more readily available (Gaj, Gersbach, & Barbas, 2013; Tanenbaum, Gilbert, Qi, Weissman, & Vale, 2014). The IgG antibody–TAP interaction is one of the most specific protein–protein interactions, eliminating the need to raise antibodies to a protein of interest if one is not already available. In addition to the large availability and specificity, the IgG–TAP interaction provides a lengthy "spacer" between the splicing complex and the slide surface. It may be desired to isolate a particular complex and observe changes in FRET as spliceosomes progress through assembly and catalytic steps upon addition of the required proteins (Krishnan et al., 2013). Providing the spliceosome sufficient freedom to move about may improve the spliceosome's ability to function and progress along its assembly pathway on the slide surface.

Furthermore, years of experimentation have revealed a number of heat-sensitive and dominant-negative mutations in several essential splicing protein factors (Edwalds-Gilbert et al., 2000; Kim & Rossi, 1999; Plumpton, McGarvey, & Beggs, 1994; Schneider, Hotz, & Schwer, 2002; Vijayraghavan, Company, & Abelson, 1989). Heat-sensitive mutations allow for the inactivation of a protein component upon heating to the restricted temperature. Typically, the protein carrying this mutation is one required for progression beyond a specific assembly step in the splicing cycle. Yeast WCE from strains carrying the temperature-sensitive mutation can be made, allowing for the inactivation of the target protein prior to the *in vitro* splicing assay in much the same way that the strain itself is raised to a nonpermissive temperature. Once pre-mRNA substrate is introduced to this protein-inactivated extract, spliceosomes will accumulate at the assembly stage of interest. For example, B^{act} complex enrichment can be achieved through utilization of the Prp2-1,Cef1-TAP strain of yeast (Fig. 2A). Extract from this strain can be heated to the nonpermissive temperature, destroying the ATPase Prp2 required to progress beyond the B^{act} stage and preventing formation of complexes beyond B^{act}. Cef1 is known to only join the spliceosome upon B^{act} formation, thus preventing purification of pre-B^{act} complexes during TAP purification and thus leading to further enrichment of the correct target on the slide (Fig. 1B; Lardelli et al., 2010). Alternatively, dominant-negative mutations are most typically introduced into a protein for recombinant overexpression and purification. Upon incubation of WCE with the recombinant mutant protein, splicing becomes stalled at the point where the wild-type protein is known to function. Both stalling techniques are known to be very efficient and allow for significant enrichment of a complex of interest. Finding the right combination of stalling and complex isolation will ensure a highly specific and efficient pull-down for smFRET.

2.2.1 Isolation of splicing complexes for biochemical control experiments

Regardless of the complex to be studied, biochemical verification of the specificity of complex isolation and function is required. Such biochemical validation may be as simple as isolating the RNA from the *in vitro*-assembled complex and looking for pre-mRNA, first-step, or second-step products via denaturing PAGE. For earlier assembly complexes (CC2, A, and B complexes), other verification approaches such as native gel analysis are required (Konarska, 1989; Legrain, Seraphin, & Rosbash, 1988). Alternatively, Northern and Western blot analyses can be used to identify which snRNAs and proteins, respectively, are assembled in the purified complex. For example, upon A complex formation, the presence of U1 and U2 snRNA would be expected, while U4/U6·U5 recruitment would not be expected until formation of the B complex or later.

A number of parameters must be optimized during this biochemical analysis phase to ensure efficient purification of a specific complex that is absent of all nonspecific binding partners. Because the pull-down itself will be performed on a slide surface, it is best to use conditions as similar as possible. One method is to use magnetic beads coated with streptavidin or neutravidin. Using a magnetic tube strip, the beads can easily be isolated from solution and washed of all unbound material. For example, in our work (Krishnan et al., 2013), streptavidin-coated beads were first incubated with biotinylated IgG, and subsequently, washed with a mild salt-containing buffer to remove excess antibody prior to incubation with yeast WCE containing the complex of interest. The immobilized complexes were then isolated by pulling the beads down using the magnetic strip and removing excess lysate with the supernatant. It is important to optimize the quantities of RNA, extract, and magnetic beads used to ensure efficient complex formation and isolation. In addition, complex formation may be performed prior to the addition of the beads to improve the efficiency of formation.

Removal of unbound and nonspecifically bound WCE components is perhaps the most crucial step to SiMPull-FRET. A suitable buffer should allow the complex to remain intact while removing protein and RNA loosely associated with the complex. Increasing salt concentrations and addition of mild detergents such as NP-40 will better remove nonspecifically bound components. However, it is important to verify the activity of the remaining complex upon washing. For situations in which a complex will be transitioning through assembly or catalytic steps upon addition of recombinant protein, wash steps must be stringent enough to prevent progression in the absence of recombinant proteins but mild enough to allow efficient transition (Krishnan et al., 2013).

Lastly, pre-mRNA and splicing products can be isolated from the magnetic beads using proteinase K digestion in order to remove all protein components of the spliceosome. Efficient degradation by proteinase K will result in release of the RNA into solution for retrieval by phenol–chloroform extraction and ethanol precipitation. Recovered material can then be dissolved and visualized via denaturing PAGE to identify the appropriate splicing intermediates or snRNA factors.

A general protocol for the purification of the B^{act} complex for biochemical validation purposes as performed in Krishnan et al. (2013) is as follows:

1. Inactivate Prp2-1,Cef1-TAP extract by heating at 37 °C for 45 min. Immediately place on ice
2. During extract inactivation, prepare streptavidin-coated magnetic beads (Dynabeads MyOne Streptavidin C1, Invitrogen)
    a. Equilibrate 200 μL of beads per 135 μL splicing reaction in T50 buffer (10 mM Tris–HCl, pH 8.0, 50 mM NaCl)
    b. Add an equal volume of 0.5 mg/mL biotin–IgG (ZyMAX rabbit anti-mouse IgG (H+L)—BT (ZyMAX Grade)) in T50 and incubate at room temperature (RT) for 30 min
    c. Pull-down the beads using a magnet and discard the supernatant
    d. (Optional) If using biotinylated pre-mRNA, incubate the beads with excess free biotin at 1.5 mg/mL in T50 buffer for 20 min at RT
    e. Pull-down beads, wash with T50, and equilibrate in splicing buffer
3. In a 135-μL reaction volume, incubate 40% (v/v) inactivated extract with 0.7–1.0 nM fluorescent pre-mRNA substrate and 2 mM ATP in splicing buffer (60 mM $K_i(PO_4)$, pH 7.0, 2 mM $MgCl_2$, 3% (w/v) poly(ethylene glycol) (PEG)) for 15 min at RT to allow accumulation of the B^{act} complex

4. Add complex formation reaction to prepared magnetic beads and continue to incubate at RT for 30 min
5. Pull-down beads and remove unbound supernatant
6. Thoroughly wash beads three times with wash buffer A (20 mM HEPES-KOH, pH 7.9, 120 mM KCl, 0.01% NP40, 1.5 mM MgCl$_2$, 5% (v/v) glycerol), and once with splicing buffer
7. For reconstitution with purified proteins, incubate beads with the proteins of interest (Prp2, Spp2, and Cwc25) at each 100 nM final concentration in splicing buffer in the presence of 2 mM ATP for 30 min at RT
8. Isolate pre-mRNA and splicing products by incubating each 200 μL splicing reaction with 30 mM EDTA, 0.5% SDS, and 20 μg Proteinase K (Life Technologies) at 42 °C for 20 min
9. Phenol–chloroform extract protein and ethanol precipitate RNA for analysis on a denaturing, 7 M urea, 15% polyacrylamide gel

2.2.2 Isolation of splicing complexes for smFRET

Isolation of the complex of interest for smFRET analysis is very similar to that for biochemical validation. Slide surfaces functionalized with a small amount of biotinylated PEG are first incubated with streptavidin, producing the functionalized surface that serves to take the place of the magnetic beads used for biochemical isolation. As with the biochemical purification, the slide surface is then coated with biotinylated IgG that will serve as the binding partner for the TAP-tagged splicing protein contained in the complex of interest. It is important to thoroughly wash away any unbound antibody to ensure all bait–prey complexes only bind to antibody coupled to the slide surface. A mild T50 buffer (10 mM Tris, pH 8.0, 50 mM NaCl) is usually sufficient for this step. Again, complex formation is typically performed in a test tube away from the slide to allow for unimpeded assembly on the fluorescent substrate. If formation reactions are flowed onto slides too early, the TAP-tagged protein will immediately start to become immobilized on the surface, restricting the space in which the complex can properly assemble. Conveniently, when using Cef1-TAP as the bait to isolate B^{act} any immobilized protein that has not been assembled into the proper splicing complex will lack fluorescence and thus be dark during the smFRET experiments. After each slide functionalization step, slide surfaces should be buffer exchanged into the buffer used for the subsequent immobilization in order to provide optimal conditions to support efficient binding of the complex of interest to the slide surface. Ideally, the surface density should be 300–400 molecules per field of view. A 100-μL reaction of 0.5–1 nM fluorescent Ubc4 incubated with 40% (v/v) Prp2-inactivated

WCE for 30 min was found to provide sufficient density upon introduction of the complex formation reaction to the slide surface (Krishnan et al., 2013). Depending on the efficiency of complex formation and the volume of the splicing reaction, it may also be necessary to test dilutions of the formation reaction to achieve proper single-molecule density on the slide surface. Complex formation reactions can also be allowed to incubate on the slide surface for a longer period of time until the proper surface density is achieved. Once the desired surface density is achieved, slide surfaces are washed with 300–400 μL of the previously optimized wash buffer to ensure removal of all nonspecifically bound complexes as well as loosely associated protein and RNA.

Several control experiments are required to verify the specificity of the pull-down. This is most easily and quickly performed on the slide surface because one simply has to detect enrichment in the number of fluorescent molecules bound to the slide surface in the presence of all binding partners. The required controls will primarily involve exclusion of the secondary antibody or binding partner responsible for identifying and pulling down the complex, a condition that should significantly reduce the number of binding events on the slide surface (Fig. 2B). Alternatively, inclusion of an alternative antibody lacking an affinity to the bait protein can serve as a negative control. In addition, when using biotinylated IgG as the bait binding partner, slides can be preincubated with Protein A, the target protein for the antibody. If slides are cleaned and functionalized properly, complex pull-down should be largely inhibited in the presence of excess Protein A.

A general protocol for the purification of the B^{act} complex for smFRET analysis as performed in Krishnan et al. (2013) is as follows:

1. Inactivate Prp2-1,Cef1-TAP extract by heating at 37 °C for 45 min. Immediately place on ice
2. In a 100-μL reaction volume, incubate 40% (v/v) inactivated extract with 0.7–1.0 nM fluorescent pre-mRNA substrate and 2 mM ATP in splicing buffer for 15 min at RT
3. Prepare functionalized slides (see Section 2.3.2)
    a. Hydrate PEGylated slides with 100 μL T50 buffer
    b. React slides with 0.2 mg/mL streptavidin in T50 buffer for 15 min. Wash slides with 100 μL T50 buffer
    c. Incubate slides with 100 μL of 0.5 mg/mL biotin–IgG in T50 buffer for 20 min
    d. (Optional) If using biotinylated pre-mRNA, incubate slide with free biotin at 1.5 mg/mL in T50 buffer for 15 min
    e. Wash slide with T50 and equilibrate in splicing buffer

4. Add complex formation reaction to prepared slide and continue to incubate until optimal density is achieved
5. Flow out splicing reaction with splicing buffer and wash extensively with 400 μL of wash buffer followed by equilibration in splicing buffer
6. If slide is to be imaged, include one further wash with splicing buffer containing the proper OSS
7. For reconstitution, incubate slide with the proteins of interest (Prp2, Spp2, and Cwc25) at a final concentration of 100 nM protein in splicing buffer, 2 mM ATP, and OSS

2.3 smFRET using prism-based TIRF microscopy
2.3.1 Summary of smFRET microscopy

Obtaining single-molecule sensitivity in a microscopy setup requires the separation of the desired signal from background. This is accomplished spectrally, by selecting optical filters that transmit only the red-shifted fluorescence emission of the fluorophore of interest, and spatially, by confining the excitation volume to the location where the fluorophore is immobilized. The latter is typically accomplished through TIRF microscopy (Axelrod, Burghardt, & Thompson, 1984), in which a laser beam is directed at the sample slide at an angle that generates total internal reflection at the glass–liquid interface. Under this condition, the laser generates an evanescent field within the liquid that penetrates only 50–150 nm from the interface (Axelrod et al., 1984; Walter et al., 2008), thus limiting the detected signal-to-fluorescence resulting from this region. This approach greatly enhances the signal detected from fluorophores immobilized on the slide surface relative to any that are in solution, significantly improving signal-to-background.

The basic optical setup for smFRET requires a laser to excite the donor fluorophore, an optional laser to excite the acceptor fluorophore, a microscope, optics to separate the donor and acceptor fluorescence signals, and a low-background camera such as an EMCCD (electron multiplying charge-coupled device). In many smFRET setups, total internal reflection is accomplished by directing the laser into a prism that sits on the microscope slide. Due to index matching oil placed between the prism and the slide, the laser is transmitted directly into the slide before undergoing total internal reflection at the slide–sample interface. Fluorescence is collected through the coverslip by a microscope objective, and a high-efficiency dichroic beamsplitter is used to separate donor and acceptor emission. Through the use of mirrors and a second dichroic, the donor and acceptor signals are redirected to be parallel but displaced from each other, and are imaged onto adjacent regions of the active area of the EMCCD camera. Raw smFRET data consist of movies in which time-dependent images of the donor and acceptor regions of the camera are recorded while the donor fluorophore is being excited, and the intensity traces of individual points in each channel are then extracted (an example trace is shown in Fig. 2C). The details of further data analysis vary depending on the application and are detailed below.

2.3.2 Slide preparation and surface attachment

Because of the high sensitivity of single-molecule fluorescence microscopy, there are stringent requirements for the microscope slides and sample chambers used for imaging. Slides used for single-molecule fluorescence microscopy must be rigorously cleaned to remove any fluorescence contaminants that might otherwise contribute spurious signals. The surface of the slide must be passivated to prevent components of the sample, particularly proteins, from nonspecifically adhering to it. Finally, a means of immobilizing the sample on the slide surface must be incorporated into the slide preparation procedure (Roy et al., 2008). A variant of the latter procedure, briefly described here, was used for imaging B^{act} complex. The slide is boiled in water for 10 min, allowing tape, glue, and coverslips from previous experiments to be removed. It is then cleaned through sonication in Alconox detergent for 30 min, methanol for 10 min, and 1 M KOH for 20 min, and finally by boiling in "basic piranha solution," which consists of 14% (v/v) ammonium hydroxide and 4% (w/w) hydrogen peroxide in water. Aminosilanization of the slide surface is accomplished by immersing the slides in a 2% solution of (3-aminopropyl)triethoxysilane in acetone for 20 min with 1 min of sonication. The slides are then reacted for 2 h with a solution of 176 mg/mL NHS-ester functionalized PEG in 0.1 M sodium bicarbonate, including a 1:10 ratio of biotinylated to nonbiotinylated PEG. This step serves the dual purposes of passivating the slide surface against nonspecific binding and providing biotins that can later be used for sample immobilization. In a final step, the surface is reacted for 30 min with a solution of 20 mg/mL disulfosuccinimidyl tartrate (sulfo-DST) in 1 M sodium bicarbonate to passivate any unreacted amino groups. Sample chambers are constructed by attaching a coverslip to the slide using double-sided tape and sealing the ends with epoxy glue, and assembling tubing to allow sample to be injected into holes predrilled into the slide (Michelotti, de Silva, Johnson-Buck, Manzo, & Walter, 2010). As described above, the sample chamber is incubated with a solution of streptavidin immediately before use, allowing a biotinylated sample to be immobilized.

A general protocol for the preparation of PEG-modified slides for use in smFRET experiments follows. This procedure can be used to prepare new slides or regenerate used slides for repeated use. The volumes listed assume that one is preparing five slides at a time, and the assembly instructions assume that one is using a slide that already has two holes drilled into it, so that the sample chamber will run between the two holes. All steps are carried out at RT except where indicated.

1. Clean quartz slides
   a. Boil in water for 10 min or until glue from previous use turns yellow. Use a razor blade to scrape off glue and coverslips
   b. Make a thick paste with Alconox powder and a small amount of water. Use fingers to scrub each slide with paste for 30 s
   c. Place slides in coplin jar with Alconox on slide surface. Fill with water, add 1 mL of concentrated cuvette cleaner, and sonicate for 30 min
   d. Rinse with water and methanol, then sonicate for 10 min in methanol
   e. Rinse with water, then sonicate for 20 min in 1 M KOH
   f. Rinse with water, then boil for at least 20 min in basic Piranha solution (14% (w/v) NH_3 and 4% (w/v) H_2O_2 in water)
   g. Rinse slides and coplin jar with water, then completely dry slides with N_2
2. Aminosilanate the slide surface
   a. Rinse slides and coplin jar with acetone
   b. Combine 70 mL acetone with 2 mL (3-aminopropyl)triethoxysilane (APTES) in coplin jar with slides
   c. Incubate for 10 min, then sonicate for 1 min, and then incubate for 10 min
   d. Rinse slides thoroughly with water, then dry slides completely with N_2
3. PEGylate the slide surface
   a. Prepare empty pipette tip boxes to hold slides during reaction by cleaning and placing a small amount of water in the bottom to maintain a humid atmosphere
   b. Prepare PEG reaction solution immediately before use by combining 8 mg biotin–PEG–succinimidyl valerate (MW 5000), 80 mg mPEG–succinimidyl valerate (MW 5000), and 500 μL PEGylation buffer (0.1 M $NaHCO_3$). Vortex to dissolve and centrifuge for 1 min at 10,000 rpm to pellet any undissolved material. Sterile filter using 0.2 μm syringe filter

c. Place slides in prepared pipette tip boxes, pipette 70 μL of PEG solution onto the region of each slide that will form the sample chamber, and carefully place a coverslip over liquid, avoiding the formation of bubbles
d. Incubate for at least 2 h in a dark place. Then rinse thoroughly with water and dry with N_2
4. React any remaining $-NH_2$ groups on the slide surface with disuccinimidyl tartarate (DST)
   a. Prepare DST reaction solution immediately before use by combining 10 mg DST with 500 μL DST buffer (1 M NaHCO$_3$). Vortex to dissolve and centrifuge for 1 min at 10,000 rpm to pellet any undissolved material. Sterile filter using 0.2 μm syringe filter
   b. Place slides in prepared pipette tip boxes, pipette 70 μL of DST solution onto each slide, and carefully place a coverslip over liquid, avoiding the formation of bubbles. Make sure to place the solution on the surface of the slide that the DST reaction solution was placed on
   c. Incubate for at least 30 min in a dark place. Then rinse thoroughly with water and dry with N_2
5. Assemble slide according to application. For example:
   a. Form a sample chamber between two strips of double-sided tape and place a coverslip over the tape. Make sure that the surface of the slide that was reacted with PEG and DST faces inward. Use epoxy to seal any edges that are not sealed by the double-sided tape
   b. Working on the opposite side of the slide, create inlet and outlet tubes by cutting pipette tips as needed to fit into the drilled holes and connecting the tips with rubber tubing. Use epoxy to seal the regions where the tips contact the slide and where the tubing contacts the tips
   c. Store slides in a dry, dark place

2.4 Experimental procedures for smFRET on the spliceosome

Once selective isolation and immobilization of the complex of interest have been achieved, SiMPull-FRET experiments are performed in a very similar manner to that of classical smFRET experiments. FRET is monitored through excitation of the donor fluorophore and detection of the subsequent fluorescence from the donor and acceptor fluorophores. As many complexes may contain only one of the two fluorophores (in its fluorescent form), it will be important to include a direct excitation of the higher wavelength fluorophore near the end of the movie for a significant amount of time (at least 100 imaging frames on the camera). This allows low-FRET states to be distinguished from molecules that completely lack the acceptor fluorophore. In addition, donor excitation should be allowed to proceed until most of the field of view has bleached by the end of the observation to ensure identification of single complexes through single-step photobleaching. The detection of multiple photobleaching steps is usually attributable to the binding of two or more complexes very close to one another and FRET from these complexes should be ignored.

Due to the presence of extract or high concentrations of proteins, photobleaching will play a significant role in the longevity of the fluorophores. Observation times prior to photobleaching will last anywhere from seconds to several minutes depending upon the condition and concentration of protein, even in the presence of an efficient oxygen scavenging system. Laser power settings can be adjusted to extend the lifetime of the fluorophores. However, if the laser power is too low, the signal-to-noise will be very low making the confident identification of dynamics difficult. Several excitation laser powers should be tested in order to maximize fluorophore lifetime (>10 s) while still yielding a high enough signal-to-noise to detect FRET dynamics. Because of this short lifetime, it will also be important to record smFRET from at least five fields of view in order to gain good confidence in the data. For an equilibrium experiment, this can be done on the same slide, for a nonequilibrium (i.e., time-lapse) experiment, separate slides may have to be used.

The best smFRET experiments on the spliceosome will be those that allow for observation of progression through further assembly and catalytic steps upon addition of the required proteins and/or ATP (Krishnan et al., 2013). Having the ability to add a cofactor to release a specific block and thus "chase" splicing complexes through subsequent splicing steps opens up the door to experiments investigating proofreading, protein function, kinetics, etc. This ability can be confirmed during the biochemical validation experiments. However, dynamics and changes in dynamic behavior during smFRET experiments alone can often already be attributed to successful "chase" of protein–RNA complexes. The complex at equilibrium can be studied by incubation with the required proteins and ATP for 15 min prior to visualization. Alternatively, a progression in FRET states and the associated dynamics can be observed overtime as the protein acts on the isolated complexes by recording smFRET from several fields of view immediately after protein addition. Once dynamic behaviors have been assigned to particular proteins, this activity can be confirmed through exclusion of ATP from the "chase" solution (in ATP-dependent processes) or addition of a mutant form of the added protein. Common yeast mutant proteins often are deficient in ATP binding, ATP hydrolysis, or protein/complex binding and as a result, become completely inactive.

3. DATA ANALYSIS
3.1 FRET histograms

One of the simplest ways to visualize a large number of smFRET traces in aggregate is to create a histogram (see Fig. 2D for an example). To create a histogram, the first step is to truncate each of the traces from a given experimental condition to its first 100 frames. This ensures that each trace contributes equally to the histogram, regardless of the total length of the trace (due to photobleaching, most molecules do not persist for the entire length of the movie, but they should persist for longer than 100 frames). All of the traces are combined into a single file in which the 100 frames contributed by each molecule are listed one after the other, with columns for time and FRET efficiency. This file is imported into a program such as OriginLab, in which the FRET efficiency values are grouped into discreet bins and the fraction of time spent in each FRET efficiency bin (collectively, between all of the traces considered) is determined. These data are displayed as a bar graph, indicating the relative frequency with which each bin of FRET states is observed. The histogram will typically include one or more peaks, whose centers and relative intensities can be estimated by fitting with the appropriate number of Gaussians. In Fig. 2D, for example, the FRET histogram indicates two populations, one with a broad range of FRET efficiencies centered near 0.5 and other with a narrow range of FRET efficiencies centered near 0.9.

3.2 HMM and transition occupancy density plot analysis

The histogram allows one to quickly determine the distribution of FRET states that are visited by the entire ensemble of molecules. The histogram says nothing, however, about transitions between the different populations observed. This information exists in the raw smFRET traces (as seen in Fig. 2C), but extracting it in a systematic way requires one to differentiate legitimate transitions from the noise that is inherent in single-molecule measurements. One approach to this problem is the technique of HMM. When applied to smFRET, HMM requires the construction of a model that is defined by three probability matrices (Blanco & Walter, 2010). The "transition probability matrix" defines the probabilities of a given FRET state transitioning to another state in the next time step. The "emission probability matrix" defines the probability of a given FRET signal resulting from a given discreet FRET state. The "initiation probability matrix" defines the probability of a given trace starting in each of the possible FRET states. This model is "trained" on a dataset consisting of many individual molecules observed under a particular experimental condition, and the resulting optimization provides the model that best describes the data, as well as the most likely path between discreet FRET states in each single-molecule trace. This path is an "idealization" of the data, as shown in the lower panel of Fig. 2C, in which the fluctuating signal has been reduced to a series of discreet FRET states. In doing this, HMM clearly identifies transitions between different FRET states that may otherwise be hidden in noisy data.

HMM particularly shines when applied to complex systems like the spliceosome, in which many different FRET states are present, and it is difficult to identify transitions in a consistent and unbiased manner. Something as simple as the number of FRET states present can often be unclear in data such as these. In the work presented in the next section, the number of FRET states was determined by fitting the data using HMM models with different numbers of FRET states and using the Bayesian information criterion (BIC) to determine the appropriate number of states. BIC is based on both the likelihood score of the fit, which inevitably increases with an increasing number of states (and therefore fitting parameters), and a penalty term that increases with increasing number of parameters. Choosing the model with the number of FRET states that minimizes the BIC obtains a balance between model parsimony and likelihood score (Blanco & Walter, 2010).

One of the greatest benefits of HMM is that it clearly reveals transitions between different FRET states. The results of HMM can be presented in transition occupancy density plots (TODPs), which are two-dimensional plots indicating the fraction of molecules exhibiting at least one transition between a particular initial FRET efficiency (plotted along the x-axis) and a particular final FRET efficiency (plotted along the y-axis). An example is shown in Fig. 2E. In this plot, which was obtained from the same dataset as the histogram in Fig. 2D, a majority of molecules undergo transitions between FRET states centered at approximately 0.9 and 0.6. In addition, a small population remains stably in the 0.9 FRET state and an even smaller population remains stably in the 0.6 FRET state. For each off-diagonal peak in the TODP, HMM yields a set of dwell times in the initial FRET state. These dwell times can be used to obtain the rate of each transition represented in the smFRET data. While TODPs offer a convenient visual representation of the dynamics of a population of molecules, rate constants allow quantitative comparison of the dynamics between different experimental conditions.

3.3 Postsynchronized histograms

Another analysis technique often employed for characterizing single-molecule behavior is postsynchronized histogram (PSH) analysis (Blanchard, Gonzalez, Kim, Chu, & Puglisi, 2004; Lee, Blanchard, Kim, Puglisi, & Chu, 2007; Senavirathne et al., 2012). PSHs are constructed by "postsynchronizing" smFRET time traces or HMM-fitted time trajectories to start at the first observation of a particular FRET state for every molecule in a certain condition. Trajectories are then binned within particular FRET bins and time windows to determine the number of trajectories located at a given FRET and time value. This approach is often used to remove the blurring effect resulting from asynchronous binding of a ligand to a target. By synchronizing all smFRET traces to the moment of the initial binding event, the subsequent changes in FRET can more easily be compared and the change in behavior of the entire population can more easily be inferred (Blanchard et al., 2004; Lee et al., 2007; Senavirathne et al., 2012). The initial starting FRET state can simply be a threshold of which FRET must be above or below, or a particular FRET efficiency. The constructed histograms can then be compiled to show how quickly a population progresses out of a particular FRET state. PSHs can also be applied to the data output of SiMPull-FRET (Krishnan et al., 2013). In this case, PSHs were constructed by synchronizing individual FRET events to the first occurrence of one of the macrostates. Such an approach allowed for a visually appealing method to determine whether a dataset was more likely to remain in the starting FRET state or transition to an alternative FRET state and, if so, how quickly this transition took place.

3.4 Clustering analysis

While histogram and TODP analyses provide the average dominant FRET conformations of a population of molecules as well as the most common two-state transitions, these methods are not sufficient to provide an in-depth dissection of the complex dynamics often observed in smFRET studies. More complex systems often contain multiple interconverting FRET states with varying kinetics of transition between them. In addition, an ordered progression of states is often required for the proper function of many protein and RNA biomolecular machines. Unfortunately, techniques like TODP analysis assume that transitions are independent from one another, losing valuable information on molecules containing a characteristic series of transitions between multiple FRET states. Single-Molecule Cluster Analysis (SiMCAn) is a recently developed analysis tool capable of dissecting this complex behavior (Blanco et al., under review). SiMCAn utilizes hierarchical clustering from bioinformatics to group and sort complex smFRET traces based upon the FRET states present in a population as well as the kinetics between those FRET states. Interestingly, application of SiMCAn to a prior experimental dataset (Krishnan et al., 2013) produced similar conclusions to the previously only manually identified subpopulations of pre-mRNA molecules found to be within each splicing complex (Blanco et al., under review). SiMCAn thus promises to become a powerful analysis tool capable of unbiased extraction of the FRET states and multistep kinetics from single-molecule trajectories acquired using SiMPull-FRET.

4. THE SPLICEOSOME AS A BIASED BROWNIAN RATCHET MACHINE

SiMPull-FRET is a combination of classical smFRET and SiMPull. In an illustration of the type of mechanistic insight it can yield, it was used for a detailed analysis of the pre-mRNA conformational changes associated with the activation of the spliceosome during the first step of splicing (Krishnan et al., 2013). In this case, donor and acceptor fluorophores placed near the 5'SS and BP allowed for the observation of pre-mRNA conformational changes associated with the activity of specific proteins during progression of the B^{act} to the C complex. Utilizing a modified yeast strain, the B^{act} complex was selectively isolated from yeast WCE through inactivation of the ATPase Prp2 and affinity purification using a TAP-tagged Cef1 protein. Upon addition of the required recombinant proteins Prp2, Spp2, Cwc25, and ATP, protein and ATP-dependent changes in pre-mRNA conformation were observed, resulting from progression to the catalytically active B* intermediate and eventually the post-first-step C complex.

Histogram and TODP analysis of SiMPull-FRET data reveal that the isolated B^{act} complex remains locked in a static low-FRET state in which the 5'SS and BP are held stably apart from one another in order to prevent premature splicing activity (Fig. 3A–D, left column). Addition of Prp2, Spp2, and ATP (B* condition) resulted in the dynamic association of the BP with the 5'SS as indicated by rapid transitions from the low-FRET state of B^{act} into and out of a high-FRET state (Fig. 3A–D, center column). The ATPase activity of Prp2 is thought to weaken the binding of several BP-associated proteins (such as SF3a/b), proteins that presumably prevent the premature attack of the BP on the 5'SS. Accordingly, removal of such a complex would allow for the pre-mRNA to transiently and reversibly visit the higher FRET states indicative of a more proximal 5'SS and BP. Low levels of splicing are detected under these B* conditions, but only upon addition of Cwc25 do the 5'SS and BP become stably associated with one another (C condition), resulting in the significant enhancement in splicing efficiency observed in biochemical assays in the presence of Cwc25 (Fig. 3A–D, right column). Using smFRET between the pre-mRNA and protein, we found that Cwc25 binds near the BP, which then slows particularly the rate constant of the high- to mid-FRET transition, leading to longer dwell times in the precatalytic, stabilized high-FRET conformation required for first-step chemistry (Fig. 3E). PSH analysis showed that molecules under C complex conditions rapidly transition out of the M state and become stably locked in the H state (Fig. 3E). As a result, Cwc25 binding enhances the progression to the static high-FRET state associated with the postcatalytic C complex.

This behavior, observed using SiMPull-FRET, exemplifies how the spliceosome as a biomolecular machine couples chemical energy through ATP hydrolysis by the DExD/H-box helicase Prp2 into the release of a tightly bound road block ("pawl," in this case SF3a/b) stalling the B^{act} complex in a distal, low-FRET conformation. This release then allows for large-scale, intrinsic, random, and entirely thermally driven fluctuations of the pre-mRNA substrate in the B* complex (Fig. 4A). Such stochastic (Brownian) "ratcheting" is then rectified to produce directional ("biased") motion by introduction of a new, differently binding pawl (Cwc25) that brings the splice sites into close proximity and likely requires another helicase for its removal down the road. That is, chemical energy is primarily needed to set free the intrinsic conformational fluctuations of the substrate–spliceosome complex, followed by tight binding of a pawl to a newly accessible conformation, restricting motions again, and providing

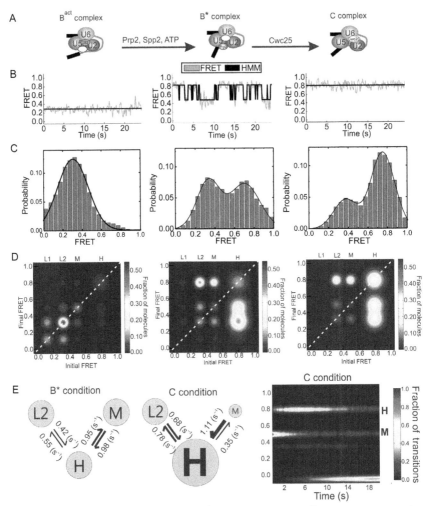

Figure 3 SiMPull-FRET analysis of the B^{act} complex. (A) The B^{act} complex is allowed to progress to the B* complex upon addition of Prp2, Spp2, and ATP, and then completes the first step of splicing into the C complex upon addition of Cwc25. (B–D) Representative FRET and idealized FRET (HMM) traces (B), FRET probability distributions (C), and TODPs (D) for the B^{act}, B*, and C complexes. (E) Kinetic analysis shows enrichment of the M-to-H transition under C complex conditions (middle) as compared to B* conditions (left). Postsynchronized histogram (PSH) showing rapid transition to, and stabilization of, the H state under C complex conditions. The accumulating zero FRET state represents the fraction of molecules that has photobleached overtime. *Modified from Krishnan et al. (2013).*

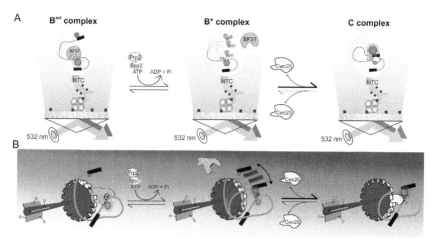

Figure 4 Biochemical (A) and mechanical (B) representations of the biased Brownian ratchet mechanism utilized by the spliceosome to promote first-step splicing. Binding of the SF3a/b complex (cyan) (gray in the print version) acts as a pawl to prevent docking of the BP and 5'SS in the B^{act} complex. Addition of Prp2, Spp2, and ATP results in the ATP-dependent release of SF3a/b from the spliceosome, allowing for dynamic docking and undocking of the 5'SS and BP and low levels of first-step splicing in the B* complex. Last, Cwc25 (yellow) (light gray in the print version) acts as a new pawl, stabilizing proximal 5'SS and BP in the C complex and allowing for more efficient first-step splicing.

directionality to the reaction pathway. Such a biased Brownian ratchet machine (Fig. 4B) was first envisioned by famed American physicist and Nobel laureate Richard Feynman. In this model, a gear is free to rotate in either direction due to random thermal fluctuations, but is held in place by a pawl (SF3a/b). An energy source releases this first pawl, allowing for the free, random rotation of the gear before once again becoming stalled at a new position through specific capture by a second pawl (Cwc25). Such a mechanism stands in stark contrast to the way most macroscopic machines function, such as a car engine, where burning of fossil fuel directly generates a force that propels the car forward by exploiting the inertia of its mechanical parts. A better macroscopic analogy for a biased Brownian ratchet machine may be a trapeze artist. The acrobat uses jump energy to propel off one platform, allowing intrinsic gravitational forces to freely carry them back and forth between the two platforms. Directional motion is achieved once the trapeze artist comes in contact with a second acrobat who catches them on the opposite platform. According to Feynman, a biased Brownian ratchet machine is the most likely process to affect directed motion at the nanoscale where neither inertia nor friction plays the dominant role they do in our macroscopic world (Feynman, 1963).

5. CONCLUSIONS AND OUTLOOK

Characterizing complex RNA and protein conformational changes is crucial to fully understand the function of large biomolecular machines. Here, we have provided details of how SiMPull-FRET integrates biochemical and biophysical approaches for the study of RNA-based machines like the spliceosome. Rather than purify every component of a complex system to reconstitute activity *in vitro*, SiMPull-FRET allows for the selective purification, immobilization, and characterization of macromolecular machines assembled from native components. As an example, application of SiMPull-FRET to selectively isolated, activated spliceosomes (B^{act} complex) allowed for the characterization of the pre-mRNA dynamics associated with the first step of splicing, revealing a biased Brownian ratcheting mechanism through which the spliceosome achieves efficient and accurate first-step splicing.

Conceivably, SiMPull-FRET can be used to study the pre-mRNA dynamics associated with most other splicing complexes using Ubc4 as the model substrate, although stages with helicase-driven conformational rearrangements, such as the Prp16- and Prp22-dependent rearrangements during the second catalytic step, will contain the most dynamic information and thus be most insightful. Most snRNA components are relatively short and are thus also suitable for fluorescent labeling in order to study snRNA–snRNA and snRNA–pre-mRNA rearrangements that are often critical for proper assembly and proofreading (Staley & Guthrie, 1998). In addition, such an approach could be used to more intimately study the mechanisms of alternative splicing in yeast and higher eukaryotic systems. In addition, as long as an efficient antibody and FRET fluorophore pair can be utilized for the system, SiMPull-FRET is suitable to study the mechanism of other molecular motor-driven processes from a variety of organisms, including—but not limited to—DNA replication, transcription, translation, DNA repair, chromatin dynamics, and RNA metabolism and export (Enguita & Leitao, 2014; von Hippel & Delagoutte, 2001).

As further RNA and protein labeling strategies become available that allow for the site-specific incorporation of ever improved donor and acceptor fluorophores and antibody–antigen interactions, we anticipate that SiMPull-FRET will become an increasingly valuable tool for identifying and quantifying the conformational dynamics associated with the folding and function of the individual proteins, RNAs, and large RNA–protein complexes that dominate the plethora of RNA-mediated processes in the eukaryotic cell (Pitchiaya, Heinicke, Custer, & Walter, 2014).

ACKNOWLEDGMENT

This work was supported by NIH Grant GM098023 to N.G.W.

REFERENCES

Abelson, J., Blanco, M., Ditzler, M. A., Fuller, F., Aravamudhan, P., Wood, M., et al. (2010). Conformational dynamics of single pre-mRNA molecules during in vitro splicing. *Nature Structural & Molecular Biology*, 17(4), 504–512.

Abelson, J., Hadjivassiliou, H., & Guthrie, C. (2010). Preparation of fluorescent pre-mRNA substrates for an smFRET study of pre-mRNA splicing in yeast. *Methods in Enzymology*, 472, 31–40.

Aitken, C. E., Marshall, R. A., & Puglisi, J. D. (2008). An oxygen scavenging system for improvement of dye stability in single-molecule fluorescence experiments. *Biophysical Journal*, 94(5), 1826–1835.

Aitken, C. E., Petrov, A., & Puglisi, J. D. (2010). Single ribosome dynamics and the mechanism of translation. *Annual Review of Biophysics*, 39, 491–513.

Axelrod, D., Burghardt, T. P., & Thompson, N. L. (1984). Total internal reflection fluorescence. *Annual Review of Biophysics and Bioengineering*, 13, 247–268.

Bharill, S., Fu, Z., Palty, R., & Isacoff, E. Y. (2014). Stoichiometry and specific assembly of best ion channels. *Proceedings of the National Academy of Sciences of the United States of America*, 111(17), 6491–6496.

Blanchard, S. C., Gonzalez, R. L., Kim, H. D., Chu, S., & Puglisi, J. D. (2004). tRNA selection and kinetic proofreading in translation. *Nature Structural & Molecular Biology*, 11(10), 1008–1014.

Blanco, M. R., Martin, J., Kahlscheuer, M. L., Krishnan, R., Abelson, J., Laederach, A., et al. (2015). Single molecule cluster analysis identifies signature dynamic conformations along the splicing pathway. *Nature Methods*, under revision.

Blanco, M., & Walter, N. G. (2010). Analysis of complex single-molecule FRET time trajectories. *Methods in Enzymology*, 472, 153–178.

Buttner, L., Javadi-Zarnaghi, F., & Hobartner, C. (2014). Site-specific labeling of RNA at internal ribose hydroxyl groups: Terbium-assisted deoxyribozymes at work. *Journal of the American Chemical Society*, 136(22), 8131–8137.

Chen, J., Tsai, A., O'Leary, S. E., Petrov, A., & Puglisi, J. D. (2012). Unraveling the dynamics of ribosome translocation. *Current Opinion in Structural Biology*, 22(6), 804–814.

Clark, T. A., Sugnet, C. W., & Ares, M., Jr. (2002). Genomewide analysis of mRNA processing in yeast using splicing-specific microarrays. *Science*, 296(5569), 907–920.

Crawford, D. J., Hoskins, A. A., Friedman, L. J., Gelles, J., & Moore, M. J. (2008). Visualizing the splicing of single pre-mRNA molecules in whole cell extract. *RNA*, 14(1), 170–179.

Crawford, D. J., Hoskins, A. A., Friedman, L. J., Gelles, J., & Moore, M. J. (2013). Single-molecule colocalization FRET evidence that spliceosome activation precedes stable approach of 5′ splice site and branch site. *Proceedings of the National Academy of Sciences of the United States of America*, 110(17), 6783–6788.

Edwalds-Gilbert, G., Kim, D. H., Kim, S. H., Tseng, Y. H., Yu, Y., & Lin, R. J. (2000). Dominant negative mutants of the yeast splicing factor Prp2 map to a putative cleft region in the helicase domain of DExD/H-box proteins. *RNA*, 6(8), 1106–1119.

Enguita, F. J., & Leitao, A. L. (2014). The art of unwinding: RNA helicases at the crossroads of cell biology and human disease. *Journal of Biochemical and Pharmacological Research*, 2(3), 144–158.

Fabrizio, P., Dannenberg, J., Dube, P., Kastner, B., Stark, H., Urlaub, H., et al. (2009). The evolutionarily conserved core design of the catalytic activation step of the yeast spliceosome. *Molecular Cell*, 36(4), 593–608.

Feynman, R. P. (1963). *The Feynman lectures on physics*. Massachusetts, USA: Addison-Wesley.
Fiegland, L. R., Garst, A. D., Batey, R. T., & Nesbitt, D. J. (2012). Single-molecule studies of the lysine riboswitch reveal effector-dependent conformational dynamics of the aptamer domain. *Biochemistry, 51*(45), 9223–9233.
Förster, T. (1948). Zwischenmolekulare Energiewanderung Und Fluoreszenz. *Annalen der Physik, 2*(1–2), 55–75.
Gaj, T., Gersbach, C. A., & Barbas, C. F., 3rd. (2013). ZFN, TALEN, and CRISPR/Cas-based methods for genome engineering. *Trends in Biotechnology, 31*(7), 397–405.
Hwang, H., Kreig, A., Calvert, J., Lormand, J., Kwon, Y., Daley, J. M., et al. (2014). Telomeric overhang length determines structural dynamics and accessibility to telomerase and ALT-associated proteins. *Structure, 22*(6), 842–853.
Jain, A., Liu, R., Ramani, B., Arauz, E., Ishitsuka, Y., Ragunathan, K., et al. (2011). Probing cellular protein complexes using single-molecule pull-down. *Nature, 473*(7348), 484–488.
Jain, A., Liu, R., Xiang, Y. K., & Ha, T. (2012). Single-molecule pull-down for studying protein interactions. *Nature Protocols, 7*(3), 445–452.
Kim, H., Abeysirigunawarden, S. C., Chen, K., Mayerle, M., Ragunathan, K., Luthey-Schulten, Z., et al. (2014). Protein-guided RNA dynamics during early ribosome assembly. *Nature, 506*(7488), 334–338.
Kim, D. H., & Rossi, J. J. (1999). The first ATPase domain of the yeast 246-kDa protein is required for in vivo unwinding of the U4/U6 duplex. *RNA, 5*(7), 959–971.
Konarska, M. M. (1989). Analysis of splicing complexes and small nuclear ribonucleoprotein particles by native gel electrophoresis. *Methods in Enzymology, 180*, 442–453.
Krishnan, R., Blanco, M. R., Kahlscheuer, M. L., Abelson, J., Guthrie, C., & Walter, N. G. (2013). Biased Brownian ratcheting leads to pre-mRNA remodeling and capture prior to first-step splicing. *Nature Structural & Molecular Biology, 20*(12), 1450–1457.
Lardelli, R. M., Thompson, J. X., Yates, J. R., 3rd., & Stevens, S. W. (2010). Release of SF3 from the intron branchpoint activates the first step of pre-mRNA splicing. *RNA, 16*(3), 516–528.
Lee, T. H., Blanchard, S. C., Kim, H. D., Puglisi, J. D., & Chu, S. (2007). The role of fluctuations in tRNA selection by the ribosome. *Proceedings of the National Academy of Sciences of the United States of America, 104*(34), 13661–13665.
Legrain, P., Seraphin, B., & Rosbash, M. (1988). Early commitment of yeast pre-mRNA to the spliceosome pathway. *Molecular and Cellular Biology, 8*(9), 3755–3760.
McGinnis, J. L., Duncan, C. D., & Weeks, K. M. (2009). High-throughput SHAPE and hydroxyl radical analysis of RNA structure and ribonucleoprotein assembly. *Methods in Enzymology, 468*, 67–89.
Michelotti, N., de Silva, C., Johnson-Buck, A. E., Manzo, A. J., & Walter, N. G. (2010). A bird's eye view tracking slow nanometer-scale movements of single molecular nano-assemblies. *Methods in Enzymology, 475*, 121–148.
Moore, M. J., & Query, C. C. (2000). Joining of RNAs by splinted ligation. *Methods in Enzymology, 317*, 109–123.
Newby Lambert, M., Vocker, E., Blumberg, S., Redemann, S., Gajraj, A., Meiners, J. C., et al. (2006). Mg^{2+}-induced compaction of single RNA molecules monitored by tethered particle microscopy. *Biophysical Journal, 90*(10), 3672–3685.
Ohrt, T., Prior, M., Dannenberg, J., Odenwalder, P., Dybkov, O., Rasche, N., et al. (2012). Prp2-mediated protein rearrangements at the catalytic core of the spliceosome as revealed by dcFCCS. *RNA, 18*(6), 1244–1256.
Panter, M. S., Jain, A., Leonhardt, R. M., Ha, T., & Cresswell, P. (2012). Dynamics of major histocompatibility complex class I association with the human peptide-loading complex. *The Journal of Biological Chemistry, 287*(37), 31172–31184.

Parks, J. W., & Stone, M. D. (2014). Coordinated DNA dynamics during the human telomerase catalytic cycle. *Nature Communications, 5*, 4146.

Peterson, J. R., Labhsetwar, P., Ellermeier, J. R., Kohler, P. R., Jain, A., Ha, T., et al. (2014). Towards a computational model of a methane producing archaeum. *Archaea, 2014*, 898453.

Pitchiaya, S., Heinicke, L. A., Custer, T. C., & Walter, N. G. (2014). Single molecule fluorescence approaches shed light on intracellular RNAs. *Chemical Reviews, 114*(6), 3224–3265.

Pleiss, J. A., Whitworth, G. B., Bergkessel, M., & Guthrie, C. (2007). Transcript specificity in yeast pre-mRNA splicing revealed by mutations in core spliceosomal components. *PLoS Biology, 5*(4), e90.

Plumpton, M., McGarvey, M., & Beggs, J. D. (1994). A dominant negative mutation in the conserved RNA helicase motif "SAT" causes splicing factor PRP2 to stall in spliceosomes. *The EMBO Journal, 13*(4), 879–887.

Qin, P. Z., & Pyle, A. M. (1999). Site-specific labeling of RNA with fluorophores and other structural probes. *Methods, 18*(1), 60–70.

Rinaldi, A. J., Suddala, K. C., & Walter, N. G. (2015). Native purification and labeling of RNA for single molecule fluorescence studies. *Methods in Molecular Biology, 1240*, 63–95.

Roy, R., Hohng, S., & Ha, T. (2008). A practical guide to single-molecule FRET. *Nature Methods, 5*(6), 507–516.

Rueda, D., Hsieh, J., Day-Storms, J. J., Fierke, C. A., & Walter, N. G. (2005). The 5' leader of precursor tRNAAsp bound to the *Bacillus subtilis* RNase P holoenzyme has an extended conformation. *Biochemistry, 44*(49), 16130–16139.

Schneider, S., Hotz, H. R., & Schwer, B. (2002). Characterization of dominant-negative mutants of the DEAH-box splicing factors Prp22 and Prp16. *The Journal of Biological Chemistry, 277*(18), 15452–15458.

Senavirathne, G., Jaszczur, M., Auerbach, P. A., Upton, T. G., Chelico, L., Goodman, M. F., et al. (2012). Single-stranded DNA scanning and deamination by APOBEC3G cytidine deaminase at single molecule resolution. *The Journal of Biological Chemistry, 287*(19), 15826–15835.

Shen, Z., Chakraborty, A., Jain, A., Giri, S., Ha, T., Prasanth, K. V., et al. (2012). Dynamic association of ORCA with prereplicative complex components regulates DNA replication initiation. *Molecular and Cellular Biology, 32*(15), 3107–3120.

Solomatin, S., & Herschlag, D. (2009). Methods of site-specific labeling of RNA with fluorescent dyes. *Methods in Enzymology, 469*, 47–68.

Staley, J. P., & Guthrie, C. (1998). Mechanical devices of the spliceosome: Motors, clocks, springs, and things. *Cell, 92*(3), 315–326.

Stryer, L. (1978). Fluorescence energy transfer as a spectroscopic ruler. *Annual Review of Biochemistry, 47*, 819–846.

Tanenbaum, M. E., Gilbert, L. A., Qi, L. S., Weissman, J. S., & Vale, R. D. (2014). *A protein-tagging system for signal amplification in gene expression and fluorescence imaging*. Cell.

Tatei, K., Kimura, K., & Ohshima, Y. (1989). New methods to investigate ATP requirement for pre-mRNA splicing: Inhibition by hexokinase/glucose or an ATP-binding site blocker. *Journal of Biochemistry, 106*(3), 372–375.

Vijayraghavan, U., Company, M., & Abelson, J. (1989). Isolation and characterization of pre-mRNA splicing mutants of *Saccharomyces cerevisiae*. *Genes & Development, 3*(8), 1206–1216.

von Hippel, P. H., & Delagoutte, E. (2001). A general model for nucleic acid helicases and their "coupling" within macromolecular machines. *Cell, 104*(2), 177–190.

Wahl, M. C., Will, C. L., & Luhrmann, R. (2009). The spliceosome: Design principles of a dynamic RNP machine. *Cell, 136*(4), 701–718.

Walter, N. G. (2001). Structural dynamics of catalytic RNA highlighted by fluorescence resonance energy transfer. *Methods, 25*(1), 19–30.

Walter, N. G. (2003). Probing RNA structural dynamics and function by fluorescence resonance energy transfer (FRET). *Current Protocols in Nucleic Acid Chemistry*, Chapter 11, Unit 11.10.

Walter, N. G., & Burke, J. M. (2000). Fluorescence assays to study structure, dynamics, and function of RNA and RNA-ligand complexes. *Methods in Enzymology, 317,* 409–440.

Walter, N. G., Huang, C. Y., Manzo, A. J., & Sobhy, M. A. (2008). Do-it-yourself guide: How to use the modern single-molecule toolkit. *Nature Methods, 5*(6), 475–489.

Wang, J., Zhang, J., Li, K., Zhao, W., & Cui, Q. (2012). SpliceDisease database: Linking RNA splicing and disease. *Nucleic Acids Research, 40*(Database issue), D1055–D1059.

Warkocki, Z., Odenwalder, P., Schmitzova, J., Platzmann, F., Stark, H., Urlaub, H., et al. (2009). Reconstitution of both steps of *Saccharomyces cerevisiae* splicing with purified spliceosomal components. *Nature Structural & Molecular Biology, 16*(12), 1237–1243.

Zheng, Q., Juette, M. F., Jockusch, S., Wasserman, M. R., Zhou, Z., Altman, R. B., et al. (2014). Ultra-stable organic fluorophores for single-molecule research. *Chemical Society Reviews, 43*(4), 1044–1056.

Zhou, R., Kunzelmann, S., Webb, M. R., & Ha, T. (2011). Detecting intramolecular conformational dynamics of single molecules in short distance range with subnanometer sensitivity. *Nano Letters, 11*(12), 5482–5488.

Zhou, R., Zhang, J., Bochman, M. L., Zakian, V. A., & Ha, T. (2014). Periodic DNA patrolling underlies diverse functions of Pif1 on R-loops and G-rich DNA. *Elife, 3,* e02190.

CHAPTER NINETEEN

Single-Molecule Imaging of RNA Splicing in Live Cells

José Rino, Robert M. Martin, Célia Carvalho, Ana C. de Jesus, Maria Carmo-Fonseca[1]

Instituto de Medicina Molecular, Faculdade de Medicina, Universidade de Lisboa, Lisboa, Portugal
[1]Corresponding author: e-mail address: carmo.fonseca@medicina.ulisboa.pt

Contents

1. Introduction	572
2. Overview of the Method	574
3. Detailed Protocol	575
3.1 Construction of reporter genes	575
3.2 Genomic integration of the reporter genes	576
3.3 Generation and expression of fluorescent fusion proteins	578
3.4 Preparation of cells for imaging	578
3.5 Choosing a suitable microscopic device	579
3.6 Cell imaging	579
3.7 Image processing and data analysis	580
3.8 Direct measurement of intron lifetime	582
4. Concluding Remarks	583
Acknowledgments	584
References	584

Abstract

Expression of genetic information in eukaryotes involves a series of interconnected processes that ultimately determine the quality and amount of proteins in the cell. Many individual steps in gene expression are kinetically coupled, but tools are lacking to determine how temporal relationships between chemical reactions contribute to the output of the final gene product. Here, we describe a strategy that permits direct measurements of intron dynamics in single pre-mRNA molecules in live cells. This approach reveals that splicing can occur much faster than previously proposed and opens new avenues for studying how kinetic mechanisms impact on RNA biogenesis.

1. INTRODUCTION

The vast majority of human protein-coding genes contain up to 90% of noncoding sequence in the form of introns that must be removed from the primary transcripts or pre-mRNAs (Braunschweig, Gueroussov, Plocik, Graveley, & Blencowe, 2013). There are over 200,000 different introns in the human genome, ranging in size from approximately 100 to 700,000 nucleotides. The intron–exon structure of genes is thought to have played an important role in the generation of new genes during evolution (Keren, Lev-Maor, & Ast, 2010). Moreover, diverse forms of mRNAs encoded from a single gene are created by the differential use of splice sites and alternative splicing seems to be rapidly evolving, particularly among physiologically equivalent organs from vertebrate species (Barbosa-Morais et al., 2012). Excision of introns is carried out by the spliceosome, a macromolecular machine composed of uridine-rich small nuclear RNAs that recognize, through base pairing, short consensus sequence motifs located at splice sites. However, RNA–RNA interactions formed within the spliceosome are weak and require the assistance of auxiliary proteins that bind to specific sequences in exons and introns. The combination of multiple weak RNA–RNA, protein–RNA, and protein–protein interactions is crucial for the flexibility of the spliceosome, in particular during regulated splicing decisions (Chen & Moore, 2014).

Although spliceosomes accurately recognize the correct splice sites among many other similar sequences, splicing errors can arise stochastically at any time in a normal cell. In that case, aberrantly processed RNAs are recognized and degraded. Several studies done mainly in yeasts suggest that RNA quality control mechanisms act cotranscriptionally and operate by kinetic competition: a delay in splicing would suffice to promote degradation (Porrua & Libri, 2013). Additional models to explain splicing fidelity posit that ATP-dependent RNA helicases are implicated in a kinetic proofreading mechanism that prevents the use of suboptimal splicing substrates (Semlow & Staley, 2012). According to a recent view, the energy of ATP hydrolysis may drive conformational changes that promote rejection of the first splicing chemical step at aberrant splice sites but are much slower acting at correct splice sites, allowing the first chemical step to be completed (Koodathingal, Novak, Piccirilli, & Staley, 2010). Reversibility of the splicing reactions (Hoskins et al., 2011; Tseng & Cheng, 2008) could be combined with rates of forward and reverse/reject pathways to ensure a kinetic proofreading mechanism for splicing: increasing the rate of the reverse reactions when spliceosomes form at incorrect splice sites would decrease the probability of producing aberrantly spliced RNAs. Testing these models requires making direct experimental measurements of the kinetic rates of the splicing process. However, our ability to measure RNA splicing kinetics is still limited.

Using electron microscopy to visualize dispersed chromatin with attached nascent transcripts, Beyer and Osheim showed that the spliceosome formed shortly after synthesis of the 3' splice site and that splicing of pre-mRNA often occurred on the nascent transcript (Beyer & Osheim, 1988). Assuming an elongation rate of 1500 nucleotides per minute, they inferred that splicing of introns removed cotranscriptionally occurred within 3 min after synthesis of the 3' splice site (Beyer & Osheim, 1988). Similar conclusions were obtained when studying actively transcribed genes in *Chironomus tentans* salivary gland polytene chromosomes by electron microscopy (Baurén & Wieslander, 1994). *In vivo* splicing rates in the range of 5 min or less were also measured biochemically for adenovirus and β-globin transcripts (Audibert, Weil, & Dautry, 2002; Curtis, Mantei, & Weissmann, 1978; Gattoni, Keohavong, & Stévenin, 1986). More recently, the advent of genetically encoded fluorescent protein tags combined with microscopes with increasingly higher spatial and temporal resolution made it possible to visualize protein dynamics in living cells. The development of methods such as FRAP (fluorescence recovery after photobleaching) unraveled the kinetic properties of splicing proteins in the nucleus of live cells (Phair & Misteli, 2000). FRAP analysis revealed that core spliceosomal snRNP proteins have a residence time of 15–30 s in the nucleoplasm, where spliceosomal snRNPs are thought to interact predominantly with pre-mRNA (Huranová et al., 2010). Based on these results, it was suggested that splicing can be accomplished within 30 s (Huranová et al., 2010). A method was further developed to image RNAs in living cells by genetically inserting the binding sites for the MS2 bacteriophage coat protein in the RNA of interest (Bertrand et al., 1998). The resulting reporter gene is integrated in the genome of cells that express the MS2 coat protein fused to GFP. Insertion of the MS2-binding sites in the terminal exon of reporter genes revealed kinetic properties of the entire mRNA life cycle, from transcription to transport in the nucleus and export to the cytoplasm (Janicki et al., 2004; Shav-Tal et al., 2004), while insertion of binding sites for phage coat proteins in introns has been used to visualize splicing in real time (Coulon et al., 2014; Schmidt et al., 2011). These studies analyzed an ensemble population of pre-mRNAs synthesized from a gene cluster comprising multiple copies of the reporter gene. Thus, multiple nascent RNAs were simultaneously detected, necessitating a modeling approach to infer kinetic information. To circumvent these significant limitations and potential problems in data interpretation, we developed a strategy that permits direct tracking of single pre-mRNA molecules in live cells (Martin, Rino, Carvalho, Kirchhausen, & Carmo-Fonseca, 2013). Here, we describe in detail the method designed to measure intron dynamics at single-molecule level.

2. OVERVIEW OF THE METHOD

The protocol starts with construction of a reporter gene (Fig. 1, step 1) that is integrated as a single copy in the genome of human cells (Fig. 1, step 2).

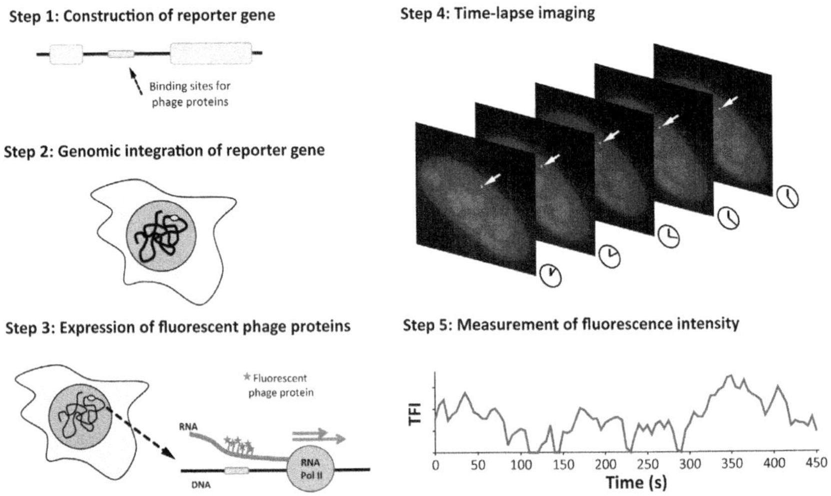

Figure 1 Protocol overview. To visualize pre-mRNA molecules in living cells, binding sites for a bacteriophage protein are genetically inserted in an intron of the gene of interest (step 1) and a single copy of the resulting transgene is integrated at a specific site in the genome of a host human cell (step 2). The host cells are additionally transfected with plasmids encoding the phage protein fused to a fluorescent protein (step 3). As soon as the binding sites are transcribed by RNA polymerase II (Pol II), the fluorescent fusion proteins bind to the RNA hairpins, rendering the nascent pre-mRNA visible. Time-lapse imaging of cells in 3D (step 4) reveals a diffraction-limited spot (arrows) that corresponds to nascent pre-mRNAs synthesized at the transgene locus. Images shown for each time point result from maximum intensity projections of eight optical planes. The fluorescence observed at the transcription site (arrows) fluctuates temporally. The total fluorescence intensity (TFI) at the transcription site is quantitated by Gaussian fitting and plotted over time (step 5).

Cells are then transiently transfected to express fluorescent fusion proteins (Fig. 1, step 3) and maintained live in a spinning-disk confocal microscope stage for time-lapse imaging (Fig. 1, step 4). Nascent transcripts emanating from the reporter gene are imaged in 4D and the fluorescence intensity at the transcription site is measured as a function of time (Fig. 1, step 5).

3. DETAILED PROTOCOL
3.1 Construction of reporter genes

In order to make single pre-mRNA molecules visible in living cells, binding sites for a bacteriophage protein are genetically inserted in the gene of interest. The most commonly used binding site is a hairpin from the genomic RNA of bacteriophage MS2, which binds with strong affinity to the phage coat protein (*in vitro* dissociation constant in the range of 5 nM; Johansson et al., 1998). For dual-color labeling, an additional sequence is inserted encoding an RNA hairpin that binds either the coat protein of bacteriophage PP7 ($K_d \sim 1$ nM; Lim, Downey, & Peabody, 2001) or the antiterminator protein N of bacteriophage λ ($K_d \sim 200$ pM; Austin, Xia, Ren, Takahashi, & Roberts, 2002). Table 1 provides a list of available

Table 1 Reagents available for RNA imaging

Construct name	References	www link
MS2-coat protein fused to GFP	Fusco et al. (2003)	https://www.addgene.org/27121/
MS2-coat protein fused to mCherry	Hocine, Raymond, Zenklusen, Chao, and Singer (2013)	http://www.addgene.org/45930/
LambdaN protein fused to GFP	Daigle and Ellenberg (2007)	http://web.uni-frankfurt.de/fb15/mikro/euroscarf/data/ellenberg.html
PP7-coat protein fused to GFP	Larson, Zenklusen, Wu, Chao, and Singer (2011)	http://www.addgene.org/35194/
25 × BoxB hairpin cassette	Martin et al. (2013)	https://www.addgene.org/60817/
24 × MS2 hairpin cassette	Bertrand et al. (1998)	http://www.addgene.org/31865/
24 × PP7 hairpin cassette	Larson et al. (2011)	http://www.addgene.org/31864/

reagents. This approach allows simultaneous imaging of two distinct introns or an intron and an exon of the same pre-mRNA (Fig. 2A). Binding sites are inserted as 24 tandem repeats to ensure sufficient fluorescence for imaging a single molecule of pre-mRNA in live cells. A problem with this approach is that the presence of multiple foreign RNA domains and the artificial protein tethering may interfere with the dynamics of gene expression and RNA processing. To assess the functional impact of the RNA labeling system on gene expression, reporter genes can be engineered to encode a cyan fluorescent protein fused to a peroxisomal targeting signal at the carboxyl terminus (Janicki et al., 2004); this sequence is inserted in frame with spliced exons, so that the visualization of cyan fluorescence in peroxisomes confirms that labeled pre-mRNAs were correctly spliced and exported to the cytoplasm (Fig. 2A and B). RT-PCR should also be performed to assess splicing efficiency. As shown in Fig. 2C, the distance between the phage RNA hairpins and the pre-mRNA sequence elements required for splicing influences the efficiency of tagged intron excision.

3.2 Genomic integration of the reporter genes

Random genomic integration of exogenous genes in mammalian cells tends to lead to the formation of gene arrays with multiple copies of the transcriptional unit integrated in tandem (Janicki et al., 2004; Shav-Tal et al., 2004). Moreover, the site of integration is not controlled, making it difficult to compare cell lines expressing different reporter genes. Lack of control over the copy number and position of the integrated transgenes is avoided by using a heterologous site-specific recombinase strategy (Yunger, Rosenfeld, Garini, & Shav-Tal, 2010). The method involves the site-specific recombinase FLP (flippase; named for its ability to invert or "flip" a DNA segment in *Saccharomyces cerevisiae*), which recognizes a short DNA target site named FRT (for *F*lp *R*ecombination *T*arget). As parental cell line to receive and express the reporter gene, we use Flp-In™ T-REx™-293 cells (Invitrogen Life Technologies) that have integrated in their genomes a unique FRT site. These cells are cotransfected with a plasmid that codes for the FLP recombinase and a vector with a FRT site, into which the reporter gene has been cloned according to the manufacturer's protocol. The expressed FLP recombinase mediates a DNA recombination event between the FRT sites such that a single exogenous gene is inserted into the genome at the unique integrated FRT site (Sauer, 1994). Thus, all cell lines generated have the reporter gene integrated at the

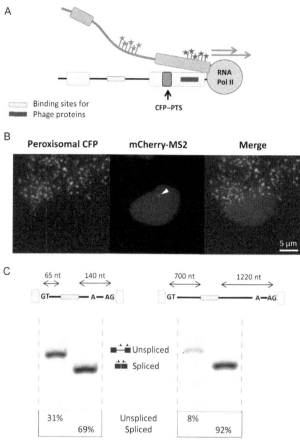

Figure 2 Splicing of labeled introns. (A) Schematic of a typical reporter gene construct harboring two cassettes that encode distinct RNA hairpins, one located in an intron and the other in an exon. Immediately after transcription of each cassette, the corresponding phage proteins fused to either GFP or mCherry bind to the pre-mRNA, resulting in dual-color labeling of two different parts of the nascent transcript. This reporter gene was additionally engineered to encode a cyan fluorescent protein (CFP) fused to a peroxisomal targeting signal (PTS) at the carboxyl terminus, and the CFP–PTS sequence was inserted in frame with spliced exons. (B) Maximum intensity projections of 3D images showing nascent pre-mRNA (arrow) detected by mCherry-MS2 fluorescence and cytoplasmic peroxisomes labeled with cyan fluorescence. (C) RT-PCR analysis shows how splicing efficiency is affected by positioning of the hairpin cassette relative to splice sites. In this experiment, RNA was isolated from cells expressing MS2–GFP and reverse transcription was carried out using an oligonucleotide that is complementary to the sequence downstream of the poly(A) site in order to detect predominantly uncleaved nascent RNAs. (See the color plate.)

same site in the genome and hence have the same genetic background. Transcription of the transgene can be under endogenous or viral promoter control. We use the human cytomegalovirus (CMV) immediate early promoter with conditional expression regulated by a system derived from the tetracycline-resistance operon (TetO) of the bacterial transposon Tn *10* (Gossen & Bujard, 1992). The parental Flp-In™ T-REx™-293 cells express the tetracycline repressor, which prevents transcription by binding to operator sequences positioned downstream the CMV promoter. Tetracycline (or doxycycline) dissociates this interaction, enabling transgene expression.

3.3 Generation and expression of fluorescent fusion proteins

Among all the fluorescent protein variants, the monomeric version of the enhanced green fluorescent protein (mEGFP) is recommended for experiments in mammalian cells. For dual-color labeling experiments, mEGFP is combined with the red fluorescent protein mCherry. Cells are transfected with plasmids encoding the fluorescent protein fused in frame to the 5′ or 3′ ends of either MS2 or PP7 coat protein, or λN protein (Table 1). The fusion proteins contain a nuclear localization signal that confines the chimera to the nucleus. For the detection of single transcripts, it is important to avoid overexpression of the fluorescent fusion proteins (Fusco et al., 2003; Hocine et al., 2013; Larson et al., 2011).

3.4 Preparation of cells for imaging

Lines derived from the parental Flp-In™ T-REx™-293 cells are grown as monolayer in Dulbecco's modified Eagle medium (D-MEM) supplemented with 10% fetal bovine serum and 2 mM L-glutamine, and maintained under selective pressure in the presence of 200 μg/ml hygromycin B and 15 μg/ml blasticidin. For live-cell imaging, cells are plated on 25 mm diameter glass coverslips coated with 0.01% poly-L-lysine and grown until they reach a confluence of approximately 60–70%. Cells are then transfected with plasmids encoding fluorescent fusion proteins using Lipofectamine™ 2000 (Invitrogen) according to the manufacturer's protocol. After overnight incubation, transgene expression is induced by addition of either 1 μg/ml tetracycline or 0.5–1 μg/ml doxycycline for at least 3 h. Before imaging, the medium is changed to α-MEM without phenol red (Invitrogen) supplemented with 20 mM HEPES, pH 7.4, and 5% FBS. Each coverslip is mounted into a perfusion chamber and placed in a heated sample holder (20/20 Technology, Inc., Wilmington, NC) mounted on the microscope stage. The microscope stage and objective lenses are maintained inside an environmental chamber set at 37 °C with 5% CO_2 and 100% humidity. The average temperature inside the cell chamber is controlled using a thermocouple.

3.5 Choosing a suitable microscopic device

Commonly used devices for live-cell imaging of gene expression are custom build wide-field fluorescence microscopes equipped with cooled CCD cameras. An alternative system that is widely used to analyze intracellular dynamics of single molecules with high spatial and temporal resolution is spinning-disk confocal microscopy, which combines high sensitivity with high speed optical sectioning and minimal photobleaching. We use the spinning-disk confocal imaging system commercially provided by 3i (Intelligent Imaging Innovations, Inc.). The 3i Marianas SDC system includes an Axio Observer Z1 inverted microscope (Carl Zeiss Micro-Imaging, Inc., Germany) equipped with a Yokogawa CSU-X1 spinning-disk confocal head (Yokogawa Electric, Tokyo, Japan) and 100 mW solid state lasers (Coherent, Inc., Santa Clara, CA) coupled to an acoustic-optical tunable filter. Excitation sources are 488 nm for GFP and 561 nm for mCherry. Images are acquired using either $63\times$ or $100\times$ (Plan-Apo, 1.4 NA) oil immersion objectives (Carl Zeiss, Inc.) under control of Slidebook 6 software (Intelligent Imaging Innovations, Denver, CO). Digital images (16-bit) are obtained using a back-thinned air-cooled EMCCD camera (Evolve 512, Photometrics, Tucson, AZ) with acquisition times between 15 and 30 ms, and Slidebook parameters Intensification and Gain set to 1000 and 1, respectively. The axial position of the sample is controlled with a piezo-driven stage (Applied Scientific Instrumentation, Eugene, OR).

3.6 Cell imaging

Imaging of cells that express the fluorescent fusion protein but have not been treated with doxycycline to induce transgene transcription reveals diffuse nuclear fluorescence, with a tendency to accumulate in nucleoli. Following transcriptional induction, a diffraction-limited spot is detected in the nucleoplasm (Fig. 1, arrows). Because fluorescent MS2 proteins are diffusing throughout the nucleus, as soon as the MS2-binding RNA hairpins are transcribed, the proteins bind making the nascent transcripts visible. Thus, the spot corresponds to nascent pre-mRNAs synthesized by RNA polymerase II at the transgene locus. The majority of cells expressing pre-mRNAs

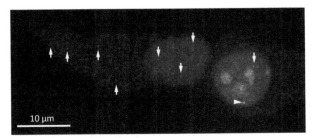

Figure 3 Imaging β-globin transcripts. HEK-293 cells containing a β-globin reporter gene tagged with a cassette of 24 × MS2-binding sites in the 3′ UTR were transiently transfected with a plasmid encoding MS2-GFP. An arrowhead indicates the transcription site and arrows indicate mature mRNAs diffusing in the nucleus and cytoplasm.

tagged with intronic hairpins have a unique spot in the nucleus, consistent with the single integration site. However, some cells present two closely located spots, probably resulting from replication of the transgene locus. Less frequently, more than two spots are present per nucleus, most likely reflecting cellular aneuploidy. In contrast, when the MS2-binding hairpins are inserted in the 3′ UTR, the mature mRNAs carry the bound fluorescent proteins after they are released from the transcription site and are visible as multiple diffusing spots throughout the nucleus and the cytoplasm (Fig. 3).

3.7 Image processing and data analysis

To monitor intron dynamics at the site of transcription, stacks of optical sections centered on the transcription site are recorded at 5 s intervals (Fig. 4A). The following sequential steps are then performed to analyze each transcription site in a time-lapse sequence. (1) The XY position of the transcription site in the first frame of the sequence is determined as a local intensity maximum, and a volume of interest is defined based on the acquired Z-stack. The Z plane corresponding to the highest fluorescence intensity value at the XY coordinates of the transcription site is recorded as its Z position. (2) The XYZ coordinates of the transcription site for subsequent time points are automatically determined by recentering the volume of interest at the site for each time point. This is done by searching for a local intensity maximum, provided the signal-to-noise ratio of fluorescence in the volume of interest is higher than a given threshold (typically 6). (3) The total fluorescence intensity (TFI) of the transcription site is calculated for each time point by performing a 2D Gaussian fit on the volume of interest at the Z plane corresponding to the highest intensity value (Fig. 4A), with a modified implementation of the Gaussian fitting function developed by David Kolin

Single-Molecule Imaging of RNA Splicing in Live Cells 581

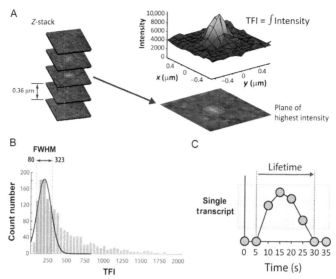

Figure 4 Quantification of fluorescence intensity. (A) A Z-stack of optical sections centered on the diffraction-limited spot corresponding to nascent pre-mRNAs at the transcription site is recorded every 5 s. At each time point, the plane where the spot appears with highest intensity is selected. In this plane, the spot is fitted by a 2D Gaussian function, the integral of which yields the total fluorescence intensity (TFI) value. (B) Histogram depicting TFI values of individual mRNA molecules diffusing in the nucleoplasm. The distribution was fitted to a Gaussian curve and the full width at half-maximum was calculated to estimate the range of TFI values corresponding to a single transcript. This range is then applied to time traces to select cycles of fluorescence gain and loss corresponding to transcription and splicing of a single transcript (C). Intron lifetime is directly estimated as the time from the moment fluorescence increases above background until it returns to background level.

(http://www.cellmigration-gateway.com/resource/imaging/icsmatlab/ICSTutorial.html) using the formula:

$$I(x,y) = A + I_0 e^{-\frac{\left(\frac{x-x_0}{a}\right)^2 + \left(\frac{y-y_0}{b}\right)^2}{2}}$$

where A is the background nuclear fluorescence intensity, I_0 is the peak intensity, x_0 and y_0 are the coordinates of the peak, a is the Gaussian width in x, and b is the Gaussian width in y. All these parameters are estimated by the fitting function. The TFI of the transcription site is calculated as the integral of the Gaussian curve: TFI $= 2\pi ab I_0$ (Rust, Bates, & Zhuang, 2006). For transcription sites in which the difference between the maximum intensity value I_0 and the background A is smaller than the standard deviation of the background σ_A estimated at the edges of the volume of interest (i.e., $|I_0 - A| < \sigma_A$), the TFI is set to zero. Transcription sites for which the Z plane of highest fluorescence intensity corresponded either to the first or last planes of the Z-stack are discarded from the analysis. (4) Finally, the estimated TFI values for each transcription site are plotted over time (Fig. 1).

3.8 Direct measurement of intron lifetime

Time traces show fluctuations in fluorescence intensity, with periods of increment in fluorescence followed by periods of fluorescence loss. Increments in fluorescence intensity are due to *de novo* synthesis of MS2-binding sites immediately followed by binding of fluorescent MS2 fusion proteins, whereas fluorescence loss can result from either splicing followed by rapid degradation of the excised intron, or release of unspliced transcripts from the site of transcription. To determine whether introns tagged with the MS2 system are spliced cotranscriptionally (i.e., before release of the mRNA from the transcription site), we use a variety of experimental approaches. First, RNA isolated from cells coexpressing the MS2-tagged reporter gene and fluorescent MS2 protein is analyzed by RT-PCR. The RNA is reversed transcribed using an oligonucleotide that is complementary to a sequence downstream of the transgene poly(A) site; the resulting cDNA is then PCR amplified using primers that specifically detect spliced and unspliced transcripts. This RT-PCR experiment detects transcripts that have not yet been cleaved at the poly(A) site and are therefore still associated with the gene template. We observed that uncleaved transcripts are predominantly spliced, arguing for cotranscriptional splicing (Martin et al., 2013). Second, we reason that if mRNAs were released unspliced, spots should be detected emanating from the transcription site. Consistent with this view, live-cell imaging of cells treated with spliceostatin A, a potent splicing inhibitor, revealed a multitude of diffraction-limited objects diffusing in the nucleus (Martin et al., 2013). We therefore conclude that if MS2-tagged introns are exclusively detected at the transcription site, it is most likely that they are cotranscriptionally spliced. Third, the dynamics of two introns in the same pre-mRNA was simultaneously visualized by double-labeling experiments. If transcripts were released unspliced, then fluorescence associated with the first intron should increase before fluorescence associated with the second intron and both fluorescent signals should decrease simultaneously. Rather, we observed that the first intron was excised, while the second intron was still present in the nascent transcript, arguing against release of unspliced transcripts (Martin et al., 2013). Determining whether splicing of the reporter gene occurs predominantly co- or post-transcriptionally is critical for the interpretation of fluorescence fluctuations. In the case of cotranscriptional splicing, loss of fluorescence at the transcription site reflects intron excision. When splicing occurs cotranscriptionally, cycles of fluorescence gain and loss (starting and ending at fluorescence background levels) reflect the intron lifetime, from transcription of the MS2-binding hairpins to intron excision and degradation (Fig. 4). However, multiple pre-mRNAs can be simultaneous present at the transcription site. To identify fluctuations in fluorescence intensity corresponding to single pre-mRNAs, the TFI of individual transcripts is estimated. This is achieved by measuring the TFI of either pre-mRNAs that are seen diffusing in the nucleoplasm after treatment of cells with a splicing inhibitor (Martin et al., 2013) or mature mRNAs labeled in the 3' UTR that are released from the transcription site and diffuse in the nucleus and in the cytoplasm (Fig. 3). A single-peaked distribution of fluorescence intensity values (Fig. 4B) indicates a unique population of diffusing diffraction-limited objects. From this distribution, the most frequent TFI values are used to identify single-transcript fluctuations in the time traces (Fig. 4C).

4. CONCLUDING REMARKS

The combination of genomic integration of a single reporter gene in human cells, intron labeling with the MS2 detection technique, and spinning-disk confocal microscopy enables the characterization of splicing kinetics at a new level of detail. Using this methodology, it is possible to visualize individual pre-mRNA molecules from transcription of MS2-binding sites located within an intron to completion of splicing and disappearance of the excised intron. By measuring the intron lifetime in single pre-mRNA reporter molecules, we have shown that different introns have distinct splicing kinetics, depending on their relative position in the gene, their length and the strength of their splice sites (Martin et al., 2013). This methodology further revealed that intron removal can occur within a few seconds after transcription, contrasting with quantitative RT-PCR measurements indicating that splicing takes 5–10 min for completion (Singh & Padgett, 2009). To understand this discrepancy, it is important to note first that time-lapse images are recorded every 5 s while in the Singh and Padgett method RNA samples are collected with 5 min intervals and second, that analysis of individual pre-mRNA molecules provides a dynamic level of information that is not possible to obtain in ensemble measurements, which typically reflect steady-state conditions. The discrepant results obtained by these two methods highlight how time and single-molecule resolution can be rate limiting when addressing questions related to the kinetics of gene expression.

ACKNOWLEDGMENTS

We gratefully acknowledge Tomas Kirchhausen and members of the Kirchhausen lab for advice and support during the development of this protocol. This work was supported by Fundação para a Ciência e Tecnologia, Portugal (PTDC/SAU-GMG/118180/2010; SFRH/BPD/66611/2009), and the Harvard Medical School-Portugal Program in Translational Research and Information.

REFERENCES

Audibert, A., Weil, D., & Dautry, F. (2002). In vivo kinetics of mRNA splicing and transport in mammalian cells. *Molecular and Cellular Biology*, 22, 6706–6718.
Austin, R. J., Xia, T., Ren, J., Takahashi, T. T., & Roberts, R. W. (2002). Designed arginine-rich RNA-binding peptides with picomolar affinity. *Journal of the American Chemical Society*, 124, 10966–10967.
Barbosa-Morais, N. L., Irimia, M., Pan, Q., Xiong, H. Y., Gueroussov, S., Lee, L. J., et al. (2012). The evolutionary landscape of alternative splicing in vertebrate species. *Science*, 338, 1587–1593.
Baurén, G., & Wieslander, L. (1994). Splicing of balbiani ring 1 gene pre-mRNA occurs simultaneously with transcription. *Cell*, 76, 183–192.
Bertrand, E., Chartrand, P., Schaefer, M., Shenoy, S. M., Singer, R. H., & Long, R. M. (1998). Localization of ASH1 mRNA particles in living yeast. *Molecular Cell*, 2, 437–445.
Beyer, A. L., & Osheim, Y. N. (1988). Splice site selection, rate of splicing, and alternative splicing on nascent transcripts. *Genes & Development*, 2, 754–765.
Braunschweig, U., Gueroussov, S., Plocik, A. M., Graveley, B. R., & Blencowe, B. J. (2013). Dynamic integration of splicing within gene regulatory pathways. *Cell*, 152, 1252–1269.
Chen, W., & Moore, M. J. (2014). The spliceosome: Disorder and dynamics defined. *Current Opinion in Structural Biology*, 24, 141–149.
Coulon, A., Ferguson, M. L., de Turris, V., Palangat, M., Chow, C. C., & Larson, D. R. (2014). Kinetic competition during the transcription cycle results in stochastic RNA processing. *eLife*, 3, e03939.
Curtis, P. J., Mantei, N., & Weissmann, C. (1978). Characterization and kinetics of synthesis of 15S β-globin RNA, a putative precursor of β-globin mRNA. *Cold Spring Harbor Symposia on Quantitative Biology*, 42, 971–984.
Daigle, N., & Ellenberg, J. (2007). λ_N-GFP: An RNA reporter system for live-cell imaging. *Nature Methods*, 4, 633–636.
Fusco, D., Accornero, N., Lavoie, B., Shenoy, S. M., Blanchard, J.-M., Singer, R. H., et al. (2003). Single mRNA molecules demonstrate probabilistic movement in living mammalian cells. *Current Biology*, 13, 161–167.
Gattoni, R., Keohavong, P., & Stévenin, J. (1986). Splicing of the E2A premessenger RNA of adenovirus serotype 2. Multiple pathways in spite of excision of the entire large intron. *Journal of Molecular Biology*, 187, 379–397.
Gossen, M., & Bujard, H. (1992). Tight control of gene expression in mammalian cells by tetracycline-responsive promoters. *Proceedings of the National Academy of Sciences of the United States of America*, 89, 5547–5551.

Hocine, S., Raymond, P., Zenklusen, D., Chao, J. A., & Singer, R. H. (2013). Single-molecule analysis of gene expression using two-color RNA labeling in live yeast. *Nature Methods, 10*, 119–121.
Hoskins, A. A., Friedman, L. J., Gallagher, S. S., Crawford, D. J., Anderson, E. G., Wombacher, R., et al. (2011). Ordered and dynamic assembly of single spliceosomes. *Science, 331*, 1289–1295.
Huranová, M., Ivani, I., Benda, A., Poser, I., Brody, Y., Hof, M., et al. (2010). The differential interaction of snRNPs with pre-mRNA reveals splicing kinetics in living cells. *The Journal of Cell Biology, 191*, 75–86.
Janicki, S. M., Tsukamoto, T., Salghetti, S. E., Tansey, W. P., Sachidanandam, R., Prasanth, K. V., et al. (2004). From silencing to gene expression: Real-time analysis in single cells. *Cell, 116*, 683–698.
Johansson, H. E., Dertinger, D., LeCuyer, K. A., Behlen, L. S., Greef, C. H., & Uhlenbeck, O. C. (1998). A thermodynamic analysis of the sequence-specific binding of RNA by bacteriophage MS2 coat protein. *Proceedings of the National Academy of Sciences of the United States of America, 95*, 9244–9249.
Keren, H., Lev-Maor, G., & Ast, G. (2010). Alternative splicing and evolution: Diversification, exon definition and function. *Nature Reviews. Genetics, 11*, 345–355.
Koodathingal, P., Novak, T., Piccirilli, J. A., & Staley, J. P. (2010). The DEAH box ATPases Prp16 and Prp43 cooperate to proofread 5′ splice site cleavage during pre-mRNA splicing. *Molecular Cell, 39*, 385–395.
Larson, D. R., Zenklusen, D., Wu, B., Chao, J. A., & Singer, R. H. (2011). Real-time observation of transcription initiation and elongation on an endogenous yeast gene. *Science, 332*, 475–478.
Lim, F., Downey, T. P., & Peabody, D. S. (2001). Translational repression and specific RNA binding by the coat protein of the Pseudomonas phage PP7. *The Journal of Biological Chemistry, 276*, 22507–22513.
Martin, R. M., Rino, J., Carvalho, C., Kirchhausen, T., & Carmo-Fonseca, M. (2013). Live-cell visualization of pre-mRNA splicing with single-molecule sensitivity. *Cell Reports, 4*, 1144–1155.
Phair, R. D., & Misteli, T. (2000). High mobility of proteins in the mammalian cell nucleus. *Nature, 404*, 604–609.
Porrua, O., & Libri, D. (2013). RNA quality control in the nucleus: The Angels' share of RNA. *Biochimica et Biophysica Acta, 1829*, 604–611.
Rust, M. J., Bates, M., & Zhuang, X. (2006). Sub-diffraction-limit imaging by stochastic optical reconstruction microscopy (STORM). *Nature Methods, 3*, 793–795.
Sauer, B. (1994). Site-specific recombination: Developments and applications. *Current Opinion in Biotechnology, 5*, 521–527.
Schmidt, U., Basyuk, E., Robert, M. C., Yoshida, M., Villemin, J. P., Auboeuf, D., et al. (2011). Real-time imaging of cotranscriptional splicing reveals a kinetic model that reduces noise: Implications for alternative splicing regulation. *The Journal of Cell Biology, 193*, 819–829.
Semlow, D. R., & Staley, J. P. (2012). Staying on message: Ensuring fidelity in pre-mRNA splicing. *Trends in Biochemical Sciences, 37*, 263–273.
Shav-Tal, Y., Darzacq, X., Shenoy, S. M., Fusco, D., Janicki, S. M., Spector, D. L., et al. (2004). Dynamics of single mRNPs in nuclei of living cells. *Science, 304*, 1797–1800.
Singh, J., & Padgett, R. A. (2009). Rates of in situ transcription and splicing in large human genes. *Nature Structural & Molecular Biology, 16*, 1128–1133.
Tseng, C. K., & Cheng, S. C. (2008). Both catalytic steps of nuclear pre-mRNA splicing are reversible. *Science, 320*, 1782–1784.
Yunger, S., Rosenfeld, L., Garini, Y., & Shav-Tal, Y. (2010). Single-allele analysis of transcription kinetics in living mammalian cells. *Nature Methods, 7*, 631–633.

AUTHOR INDEX

Note: Page numbers followed by "*f*" indicate figures and "*t*" indicate tables.

A

Aach, J., 516–517
Abagyan, R., 280–281
Abe, R., 262*t*
Abelson, J., 542–553, 558–559, 561–562
Abeysirigunawarden, S.C., 542–543
Aboul-ela, F., 42–43
Aboyoun, P., 176
Abudayyeh, O.O., 516, 534
Accornero, N., 578
Adamczyk, A.J., 498–499
Adams, J., 284
Adams, M.D., 236
Adams, P.D., 185–186, 188, 204–205, 224–225, 345–346, 407–408
Adamson, B., 516
Adli, M., 516–517
Aeschbacher, T., 295
Afonine, P.V., 185, 201, 204, 208, 224–225
Afroz, T., 236–268, 241*t*, 243*f*, 247*f*, 257*f*
Agami, R., 40–41
Agarwala, V., 516–519
Agrawal, A., 264–265, 265*f*
Agrawal, R.K., 76, 498–499
Ahmed, A., 499–501
Aigner, M., 100–101
Aime, S., 300
Aitken, C.E., 542–543, 546
Akaike, H., 59
Akke, M., 46
Alber, T., 252–254
Alemán, E.A., 100–101
Al-Hashimi, H.M., 40–68, 49*f*, 76–78, 281, 409–410, 498–499
Alivisatos, A.P., 81–84
Allain, F.H.T., 223, 236–268, 241*t*, 243*f*, 244*t*, 247*f*, 250*f*, 253*f*, 257*f*, 280–324, 334–335, 337, 346, 410, 459, 466–467
Allen, F.H., 188, 191–192
Allen, K.N., 336
Allen, M., 337
Altenbach, C., 300

Altman, R.B., 226, 498–499, 546
Altuvia, S., 280
Alvarado, L.J., 51
Amata, I., 62–64
Ames, J.M., 261–264
Ames, T.D., 223
Amode, M.R., 163
Ampe, C., 42–43
Amrane, S., 241*t*, 246–248
Amunts, A., 498–501
Anczuków, O., 250–251, 250*f*, 260–264
Anders, C., 516–535
Andersen, G.R., 498–499
Anderson, D.E., 343
Anderson, D.L., 224
Anderson, E.G., 572–573
Anderson, J.E., 523–524
Andersson, T., 21
Andreu, J.M., 368, 402
Andrews, K.L., 237–239, 241*t*, 257*f*, 258–259
Andricioaei, I., 61, 68
Androphy, E.J., 246–248
Anger, A.M., 499
Anglister, J., 344
Anko, M.-L., 262*t*
Anokhina, M., 421, 433–434
Antony, E., 419
Antson, A.A., 189
Appolaire, A., 410–411
Arauz, E., 542–543, 546–547
Aravamudhan, P., 542–546
Araya, C.L., 467–468
Arendall, B.W., 204–205
Arendall, W.B., 182–183, 185–186, 193–194
Ares, M., 134, 543–544
Armache, J.P., 499
Arnold, E., 202
Arnold, J.D., 261–264
Arslan, S., 516–517
Artsimovitch, I., 221

587

Asai, K., 266
Ascano, M., 264–265, 420–421
Assmann, S.M., 127–128
Ast, G., 572
Astashyn, A., 163
Athavale, S.S., 5–6
Auboeuf, D., 573–574
Audibert, A., 573–574
Audic, Y., 262t
Auerbach, P.A., 561
Austin, K.S., 256, 260
Austin, R.J., 575–576
Auweter, S.D., 236–248, 238f, 241t, 244t, 247f, 250f, 251–254, 256, 257f, 260–264
Au-Yeung, S.C.F., 62–64
Aviran, S., 127, 169
Avis, J.M., 237–239
Axelrod, D., 554

B

Babak, T., 419
Babcock, H.P., 4
Babon, J., 237–239, 238f
Bachorik, J.L., 419
Bacikova, V., 261–264
Bacolla, A., 76–77
Bae, E., 252–254
Baejen, C., 262t
Bagby, S.C., 223
Bai, X.C., 498–501
Bai, Y., 523–524
Bailey, S., 516–517
Baird, N.J., 223
Baker, D., 185, 204–206
Baker, M., 185, 205–206, 280
Baker, N.A., 498–499
Baldwin, A.J., 281
Baldwin, J.P., 387–388, 392
Ban, N., 214
Bandwar, R.P., 5, 7
Banerjee, H., 262t, 347
Banham, J., 284–285, 291, 305–306
Baralle, F.E., 262t
Barbas, C.F., 547–549
Barberato, C., 392–393, 402, 404f, 406–408
Barbosa-Morais, N.L., 572
Barcena, C., 516, 534
Barrangou, R., 516–517

Barraud, P., 252–254
Barrell, D., 163
Barretto, R., 516
Barrick, J.E., 221
Bartel, D.P., 223
Barthelmes, K., 336
Bartoli, K.M., 170–171, 177
Bartunik, H.D., 334–335
Basarab, G.S., 498–499
Bass, B.L., 40–41, 466–467
Bastet, L., 214–216
Basyuk, E., 573–574
Bateman, A., 216–218
Bates, D.M., 176
Bates, M., 580–582
Batey, R.T., 4–5, 42, 100–101, 183–185, 215–216, 218–219, 222, 224, 226, 523–524, 544–545
Battiste, J.L., 335–336, 343–344
Baudin, F., 154–159
Baudy, P., 392
Bauer, C., 301–302
Bauer, W.J., 375, 377
Baum, J., 48
Baumeister, W., 334–335
Baumgartner, M., 334–335
Baurén, G., 573–574
Bax, A., 56, 65–67, 282, 316, 344, 346
Baxter, W.T., 498–499
Bayer, P., 336–337
Beal, K., 163
Bearden, J.C., 442
Beauchamp, K.A., 79, 82–84, 94–96
Becker, S., 77–78
Becker, T., 499
Beckmann, B.M., 419–421, 459
Beckmann, J., 410–411
Beckmann, R., 498–499
Beggs, J.D., 549–550
Behlen, L.S., 575–576
Behlke, M.A., 22
Behrmann, E., 498–510
Behrouzi, R., 5–6
Bellaousov, S., 150
Bellapadrona, G., 300
Belloc, E., 236–246, 238f, 243f, 247f, 252–254, 256–258, 257f, 260
Benda, A., 573–574

Bender, A., 289–291, 311
Benderska, N., 246–248, 247f, 260–264
Beneken, J., 189
Ben-Shem, A., 223–224, 498–501
Berger, J.M., 531–532
Berger, S., 62
Bergkessel, M., 543–544
Bergonzo, C., 498–499
Berkhout, B., 5, 40–41
Berliner, L.J., 337–338
Berman, H.M., 185, 459
Bernado, P., 410
Bernhart, S.H., 216–218, 221–222
Bernier, S., 192f
Berninger, P., 264–265, 419–420
Bernstein, D.A., 177
Berry, K.E., 523–524
Berry, S.M., 336
Berthault, P., 48
Bertini, I., 335–336, 343, 345–346
Bertrand, E., 573–574
Bess, J.W., 5–6, 30–32
Bessi, I., 337
Bessonov, S., 334–335, 421, 433–434
Best, A., 261–264
Bevilacqua, P.C., 127–128
Beyer, A.L., 573–574
Bharill, S., 546–547
Bhat, T.N., 185, 459
Bhatnagar, J., 296, 310–311
Bhushan, S., 499
Bicout, D., 401
Bida, J.P., 169
Billeter, M., 247f
Billis, K., 163
Bindereif, A., 420–421
Bingman, C.A., 252–254
Birney, E., 250–251
Black, D.L., 244t, 250f, 251–252, 261–264
Blackledge, M., 346, 410
Blaha, G., 183f, 202f, 206–207
Blanchard, J.-M., 578
Blanchard, L., 410
Blanchard, S.C., 100–101, 226, 498–499, 561
Blanco, M.R., 542–553, 558–562
Blanquet, S., 392
Blatter, M., 236, 256, 295

Blau, C., 498–499
Blaum, B.S., 335
Blencowe, B.J., 419, 572
Blobel, G., 498–499
Block, S.M., 4–5, 226
Blomberg, F., 62
Blouin, S., 100–101, 215–216
Blumberg, S., 545–546
Bobst, A.M., 337
Bochman, M.L., 546–547
Bochtler, M., 334–335
Bock, L.V., 498–499
Bodenhausen, G., 49–50
Boelens, R., 46, 48
Bogh, L., 22
Boisbouvier, J., 65–67
Bolstad, B., 176
Bolton, D.R., 301–302
Bonilla, S., 76–95
Bonanno, J.B., 241t, 246–248, 253f, 254–255
Bonnal, S., 236–237, 252–258, 257f, 281–283, 334–335, 337, 347, 349, 351–353
Bonneau, E., 337
Bonvin, A.M.J.J., 338, 343
Boon, K.-L., 419–421, 459
Bora, R.P., 498–499
Borbas, J., 300, 336
Borbat, P.P., 286, 289–291, 296, 310–311, 322–323
Bordignon, E., 286, 297–298, 301–306, 309
Boseley, P.G., 392
Bothe, J.R., 43, 46–47, 49–50, 59–60
Boudet, J., 334–335
Bougault, C.M., 346
Bourbigot, S., 337
Bourgeois, C.F., 241t, 246–248, 247f, 262t
Bouvet, P., 241t, 244t, 252–255, 253f
Bouvignies, G., 47–48
Bowman, A., 303
Bowman, J.C., 5–6
Bowman, M.K., 304–305
Boyaval, P., 516
Boyd, M., 164–165
Bozdag, D., 162–163
Bradbury, E.M., 387–388, 392
Bram, S., 392

Brambilla, C., 459
Brambilla, E., 459
Brand, T., 62
Brandl, H., 262t
Brar, G.A., 516
Braunschweig, U., 572
Breaker, R.R., 40–41, 214–216, 221, 223
Brenner, S., 167, 173
Brenowitz, M., 4–5, 20, 119
Brent, S., 163
Briber, R.M., 5–6, 410–411
Briese, M., 262t
Brigham, M.D., 516, 534
Briner, A.E., 518–519
Brody, Y., 573–574
Brookes, E., 383
Brooks, B.R., 500–501
Brooks, C.L., 409–410, 498–499
Brow, D.A., 237–239, 238f, 244t
Brown, A., 498–501
Brown, D.T., 405
Brown, G.R., 163
Brown, K.L., 214
Brown, M.K., 266
Brown, P.O., 262t, 419
Brunel, C., 134
Brunger, A.T., 185–187, 198–199, 204–205, 345–346, 407–408, 500–501
Brunner, M., 498–502
Brustad, E.M., 300
Brutscher, B., 65–67
Büchner, P., 62
Buck, J., 40–41
Buckroyd, A.N., 256
Budkevich, T.V., 499–500, 505
Buenrostro, J.D., 467–468
Bujard, H., 576–578
Bulkley, D., 334–335
Bullock, S.L., 295
Bulyk, M.L., 265–266
Bunkoczi, G., 185, 201, 204, 224–225
Buratti, E., 241t, 253f, 255–256, 261–264, 262t
Burd, C.G., 262t
Burge, C.B., 466–492
Burger, L., 264–265, 419–420
Burghardt, T.P., 554
Burke, J.E., 282

Burke, J.M., 544–545
Burkhardt, N., 498–499
Burley, S.K., 237–239, 238f, 241t, 246–248, 253f, 254–255
Busan, S., 127
Bushnell, D.A., 334–335
Bushweller, J.H., 343
Bussotti, G., 5
Bustin, M., 343
Butcher, S.E., 42–43, 237–239, 238f, 241t, 244t, 257f, 258–259, 282–283, 295–296
Butler, E.B., 223
Butler-Browne, G., 392
Buttner, L., 299, 337, 544–545
Byrne, R.T., 189

C

Cai, Q., 309, 311
Calderone, V., 346
Calero, G., 239–240
Caliskan, G., 116–117
Calnan, B.J., 42–43
Calvert, J., 542–543
Canny, M.D., 282
Capel, M.S., 81, 387–388
Carazo, J.M., 498–499
Cardinali, B., 383
Carey, V.J., 176
Carlile, T.M., 170–171, 177
Carlomagno, T., 62–64, 282–283, 335, 343, 346, 392–393, 405, 407–408
Carlson, M., 176
Carmo-Fonseca, M., 573–574, 582–584
Caron, M.P., 214–215
Carr, H.Y., 46
Carroll, K.L., 261–264
Carson, C.C., 419–420
Cartegni, L., 262t
Carvalho, C., 572–584
Cary, P.D., 237–239, 238f
Casbon, J.A., 167, 173
Cascio, D., 249, 250f, 300
Case, D.A., 62–64, 282–283, 296, 498–501
Casey, J.L., 40–41
Casiano-Negroni, A., 42–43, 48–50, 49f, 52–53, 55–56, 64–65, 67, 77–78, 498–499

Castello, A., 419–421
Castillo, L.P., 262t
Castro-Roa, D., 419
Catalyurek, U.V., 162–163
Cate, J.H.D., 164–165, 171–172, 205f, 206, 498–499
Cavaloc, Y., 262t
Cech, T.R., 40, 154–159
Cekan, P., 264–265, 298
Cellitti, S.E., 336
Cencic, R., 516–517
Cenik, B., 262t
Cenik, C., 262t
Cereda, M., 262t
Cerione, R.A., 239–240
Chabot, B., 249–250, 250f
Chacon, P., 365, 368, 402
Chakraborty, A., 546–547
Chan, E.T., 262t, 467–468
Chan, Y.L., 189
Chance, M.R., 4–5, 119
Chang, A.T., 223–224
Chang, H.Y., 168, 467–468
Chao, J.A., 578
Chao, Y., 516
Charpentier, E., 516–519, 525–527
Chartrand, P., 573–574
Chasin, L.A., 265–266
Chateigner, D., 188
Chaudhry, S., 262t, 467–468
Chauvin, C., 392
Cheah, M.T., 40–41, 214–216
Cheatham, T.E., 296, 498–499
Chechik, V., 284–285, 291, 305–306
Chelico, L., 561
Chen, A.A., 498–499
Chen, B., 51, 516
Chen, C., 280–281
Chen, D., 42
Chen, H., 410–411, 516–517
Chen, J., 542–543
Chen, K., 542–543
Chen, M.C., 183–185
Chen, P.Y., 419–421
Chen, V.B., 183f, 185, 189, 201–202, 204, 224–225, 498–499
Chen, W., 262t, 572
Chen, Y., 215, 218–222, 224–226, 516

Chen, Y.C., 215, 218–222, 224–226, 419–420, 516
Chenevert, R., 192f
Cheng, A., 516
Cheng, S.C., 572–573
Chew, S.L., 262t
Chiang, C.-H., 262t
Chiang, Y., 289–291
Chiarparin, E., 49–50
Chillón, I., 4–35
Chircus, L.M., 467–468
Chiu, A.C., 516–517
Chiu, W., 185, 205–206
Choi, C.P., 249
Choi, J., 516–517
Chou, F.-C., 185, 204–205, 224–225
Chou, J.J., 280–281
Chow, C.C., 573–574
Chowdhury, A., 215–219, 221, 225–226
Chu, S., 4, 498–499, 561
Chugh, J., 40, 42–43, 46, 58, 64–65, 67–68, 77–78, 281, 498–499
Church, G.M., 516
Chylinski, K., 516–519, 525–527
Cienikova, Z., 236–268, 243f, 262t
Cimini, B.A., 516
Clardy, J.C., 239–240
Claridge, S.A., 81–84
Clark, T.A., 543–544
Clarkson, B.K., 282–283
Cléry, A., 236–268, 241t, 247f, 250f, 334–335, 459, 466–467
Clingman, C.C., 236
Cloonan, N., 266
Clore, G.M., 47–48, 281–282, 284, 335, 342–346, 364, 407–409
Clos, L., 42–43
Clos, L.J., 237–239, 238f
Code, E., 498–499
Cohen-Chalamish, S., 214
Cole, J.L., 5–6
Cole, T.D., 100–101
Collins, J.A., 214–215
Collins, K., 249, 250f
Colombo, M., 410–411
Colombrita, C., 262t
Comellas-Aragones, M., 405
Company, M., 549–550

Compton, E.L., 410
Cong, L., 516
Conn, G.L., 100–103, 105–106
Connell, S.R., 498–499
Conti, E., 334–335
Cook, K.B., 262t, 334–335, 467–468
Cooper, D.N., 262t
Corcoran, David L., 262t
Corden, J.L., 261–264
Cordero, P., 43, 150
Cornilescu, G., 316
Cornish, P., 498–499
Correll, C.C., 189
Corrionero, A., 250–251, 250f, 260–264
Cort, J.R., 310–311
Cortez, D., 364–365, 379–380
Cosson, B., 262t
Coufal, N.G., 262t
Coughlan, J.L., 154–159
Coulon, A., 573–574
Coutinho-Mansfield, G., 261–264
Cowan, S.W., 187–188
Cowtan, K., 203–204, 224–225
Cox, D., 516
Cramer, P., 262t, 334–335
Craven, C.J., 250–251
Crawford, D.J., 542–545, 572–573
Creamer, T.J., 262t
Cresswell, P., 546–547
Crothers, D.M., 42–43, 215–216
Crowder, S.M., 252–254
Cruickshank, P.A.S., 301–302
Cruz, J.A., 40
Cui, Q., 540
Cukier, C.D., 252–254
Curk, T., 167, 177, 262t, 264–265, 265f, 466–467
Curran, E.C., 237–239, 241t, 257f, 258–259
Curry, S., 237–239, 238f, 256
Curtis, J.E., 393–396
Curtis, P.J., 573–574
Cusack, S., 239–240, 246–248
Custer, T.C., 566

D

Dadon, D.B., 516–517
Dahlquist, F.W., 46–48
Dai, H., 21

Daish, T., 5
Daley, J.M., 542–543
Dalgliesh, C., 261–264
Dalluge, J.J., 201
Damaschun, G., 385
Damberger, F., 240–248, 243f, 247f, 261–266
Dammer, E.B., 262t
Dang, K.K., 5–6, 30–32
Dannenberg, J., 540, 542–543, 545–546
Danos, O., 262t
Daragan, V.A., 344
Darby, M.M., 262t
Darden, T., 296
Darnell, J.C., 280–281
Darnell, R.B., 260–261, 264–265, 419–420, 466–467
Darty, K., 35, 176
Darzacq, X., 573–574, 576–578
Das, R., 43, 76, 79, 81–84, 93–96, 150, 169, 185, 204–206, 224–225, 409–410
Dashti, A., 498–499
Daubner, G.M., 236–237, 241t, 246–248, 250–251, 250f, 256, 260–264, 334–335, 459, 466–467
Daughters, R.S., 261–264
Daujotyte, D., 241t, 253f, 255–256, 261–264
Dautry, F., 573–574
Davis, D.R., 201, 498–499
Davis, I.W., 185, 202, 204–205, 224–225
Davis, J.H., 283, 295–296
Davis, W., 347
Davydov, I.I., 498–499
Dayie, T.K., 51
Day-Storms, J.J., 545–546
de Graaf, R.A., 46, 48
De Maria Antolinos, A., 397
de Jesus, A.C., 572–584
de Moor, C.H., 284
de Silva, C., 100–101, 555
de Turris, V., 573–574
de Vries, S.J., 338
Deacon, A.M., 531–532
Deckert, J., 5
Deigan, K.E., 154–159
Deis, L.N., 188–189
Dejaegere, A.P., 62–64

Del Campo, M., 364–365
Delaglio, F., 56, 316
Delagoutte, E., 566
Delaittre, G., 405
DeLano, W.L., 345–346, 407–408
Deltcheva, E., 516
Denise, A., 35, 176
DeNunzio, N.J., 336
Denysenkov, V.P., 298
Deo, R.C., 241t, 246–248, 253f, 254–255
Deo, S., 364–365
Derrien, T., 5
Dertinger, D., 575–576
Desnoyers, G., 215–216
Dessen, P., 392
Desvaux, H., 48
Dethoff, E.A., 40, 42–43, 50, 58, 64–68, 77–78, 281
Dettling, M., 176
Deveau, H., 516
Deveau, L.M., 236
Deverell, C., 46
Devor, E.J., 22
Dey, A., 283, 295–296
Di Gregorio, E., 300
Diarra dit Konte, N., 281–283
Diaz, J.F., 368, 402
Diaz-Moreno, I., 252–254
Dibrov, S.M., 185, 204–205, 224–225
DiCarlo, J.E., 516
DiDonato, C.J., 246–248
Dieckmann, T., 254–255, 337–338
Diederichs, S., 5
Diedrich, G., 410–411
DiMaio, F., 185, 204–206
Ding, F., 154–159, 205f, 206, 215–219, 221–222, 224–226
Ding, J., 241t, 243f, 261–264
Ding, Y., 127–128
Dinic, L., 419–421
Ditzel, L., 334–335
Ditzler, M.A., 542–546
Djebali, S., 5
Dohmae, N., 517–518
Dokholyan, N.V., 154–159
Doll, A., 323
Dominguez, C., 246–250, 250f, 260–264, 262t, 282–283, 295–296, 316, 323

Doniach, S., 4, 81
Donohoue, P.D., 518–519
Doruker, P., 498–499
Dotsch, V., 343, 345–346
Doudeva, L.G., 244t, 262t
Doudna, J.A., 187f, 205f, 206, 218–219, 280, 516–519, 523–527
Downey, T.P., 575–576
Downs, R.T., 188
Doye, J.P.K., 59
Draper, D.E., 100–103, 105–106, 109–112, 116–117
Drescher, M., 300, 336
Dreyfuss, G., 262t, 419
D'Souza, V., 40–41, 283, 295–296
Duarte, C.M., 203–204
Dube, H., 303–306
Dube, P., 540, 542–543
Dubin, P., 21
Dubois, D.Y., 192f
Duchardt-Ferner, E., 223–224
Dudoit, S., 176
Duerst, A., 517–519, 532–534
Dumas, P., 184f, 200f
Duncan, C.D., 543–544
Dunker, A.K., 396–397
Dunkle, J.A., 164–165, 171–172, 183f, 189, 498–499
Dupuis, M.-È., 516
Dupuy, D., 241t, 246–248
Dura, M.A., 410–411
Durney, M.A., 40–41
Duss, O., 236–237, 280–324, 335, 337, 346, 410
Dussault, A.L., 100–101
Dybkov, O., 545–546
Dyson, H.J., 282–283
Dzananovic, E., 364–365

E

Eargle, J., 498–499
East, A., 516
Easton, L.E., 51
East-Seletsky, A., 516
Ebel, J.P., 154–159
Ebert, M.S., 214
Ebner, T.J., 261–264
Ebrahimi, M., 62–64

Echols, N., 185, 201, 204, 208, 224–225
Eddy, S.R., 216–218
Edmond, V., 459
Edwalds-Gilbert, G., 549–550
Edwards, J.M., 284
Edwards, T.A., 236–237, 240–248, 241t, 243f
Edwards, T.E., 184f, 223
Ehresmann, B., 154–159
Ehresmann, C., 154–159
Eichelbaum, K., 419–421
Eichhorn, C.D., 43, 46, 223–224
Elkon, R., 40–41
Ellermeier, J.R., 546–547
Ellman, G.L., 338–342
El-Mkami, H., 303
Emeson, R.B., 466–467
Emsley, J., 284
Emsley, P., 186, 203–204, 224–225, 498–501
Endeward, B., 323
Endo, H., 262t
Engelman, D.M., 81, 364, 385, 387–388, 392
Engels, J.W., 337
Enguita, F.J., 566
Ennifar, E., 100–101, 184f, 200f
Enokizono, Y., 241t
Epel, B., 301–302
Erat, M., 237–239, 238f, 246–248, 247f, 252–254, 257f, 260–264
Erdmann, V.A., 392
Erlacher, M., 183–185
Ermolaeva, O., 163
Essig, K., 262t
Esteban-Martin, S., 76
Esteve, V., 256
Esvelt, K.M., 516–517
Eswar, N., 498–499
Ethier, S., 516–517
Eustermann, S., 264–265, 265f
Evans, M.E., 223
Evans, P.R., 223, 236–237, 239–240, 241t
Evdokimova, E., 531–532
Eykyn, T.R., 49–50

F

Fabrizio, P., 540, 542–543
Farazi, T.A., 420–421
Fares, C., 62–64, 77–78
Farjon, J., 65–67
Fasan, R., 244t, 250f, 251–252, 261–264
Faulkner, G.J., 266
Favorov, O.V., 22, 33, 169
Fawzi, N.L., 47–48
Fedorova, O., 4–6
Feigin, L.A., 81, 393, 401–402
Feigon, J., 43, 223–224, 241t, 244t, 249, 252–255, 253f, 337
Feintuch, A., 300
Feldkamp, U., 76
Feldman, M.B., 183f, 189, 498–499
Feldmann, E.A., 310–311
Felli, I.C., 335, 343, 345–346
Feng, H., 343
Feng, Z., 185, 459
Fennell, T., 148
Fenwick, R.B., 76
Férec, C., 280
Ferguson, A.D., 498–499
Ferguson, M.L., 573–574
Fernandez, I.S., 498–501
Ferner, J., 40
Ferrage, F., 49–50
Ferré-D'Amaré, A.R., 4–5, 183–185, 184f, 223–224, 384–385, 523–524
Ferrero, C., 397
Ferrin, T.E., 504
Fersht, A.R., 61
Feynman, R.P., 563–565
Fichou, Y., 280
Fiegland, L.R., 226, 544–545
Field, Y., 76–77
Fields, S., 419–420
Fierke, C.A., 545–546
Findeisen, M., 62
Findlay, H.E., 410
Finger, L.D., 241t, 282–283
Fischer, B., 419–421
Fischer, N., 76, 498–499
Fisette, J.F., 249–250, 250f
Fisher, C.K., 77–78, 281
Flamm, C., 216–218, 221–222
Fleissner, M.R., 300
Flicek, P., 163
Flynn, R.A., 154–159, 168
Foden, J.A., 516–517

Fodor, S.P., 167, 173
Fondufe-Mittendorf, Y., 76–77
Fonfara, I., 516–519, 525–527
Forrer, J., 303–306
Forrest, A.R.R., 266
Förster, T., 540–542
Forthmann, S., 410–411
Foster, D.A., 226
Fournet, G., 393, 398
Fox, G.E., 101–103
Fragai, M., 346
Frangakis, A.S., 334–335
Frank, J., 76, 498–501
Franke, D., 365, 373–374, 402
Frankel, A.D., 42–43
Franz, K.J., 336
Frato, K.E., 240–246, 241t, 243f, 261–264, 280–281, 349
Freed, J.H., 286, 289–291, 296, 310–311, 322–323
Freisz, S., 184f, 200f
Fremaux, C., 516
Fresco, J.R., 4–5
Freund, S., 284
Fried, M.G., 419–420
Frieda, K.L., 4–5, 226
Friedersdorf, M.B., 260–261
Friedman, L.J., 542–545, 572–573
Frye, M., 466–467
Fu, G.K., 167, 173
Fu, J., 334–335
Fu, X.-D., 262t
Fu, Y., 516–517
Fu, Z., 546–547
Fuchs, R.T., 215–218, 221–222, 225–226
Fucini, P., 498–499
Fukunaga, T., 266
Fuller, F., 542–546
Fung, R., 498–499
Furey, T.S., 159
Furtig, B., 40–41, 51
Fusco, D., 573–574, 576–578

G

Gabel, F., 282–284, 335, 343, 346, 351–353, 364, 392–411
Gacho, G.P.C., 309, 311
Gaffney, B.L., 62
Gage, F.H., 262t
Gagnon, M.G., 334–335
Gaidatzis, D., 419–421
Gaj, T., 547–549
Gajda, M., 365
Gajraj, A., 545–546
Gallagher, S.S., 572–573
Gandra, S., 286, 301–306
Ganser, L.R., 65–67
Gao, H., 499
Gao, W., 261–264
Gardiner, B.B.A., 266
Gardner, K.H., 335
Garini, Y., 576–578
Garneau, J.E., 516
Garneau, N.L., 419
Garreau de Loubresse, N., 498–501
Garst, A.D., 42, 100–101, 215–216, 222, 226, 523–524, 544–545
Gartmann, M., 499
Gasch, A., 236–237, 252–258, 257f, 281–283, 334–335, 337, 347, 349, 351–353
Gasiunas, G., 516–517
Gattoni, R., 573–574
Gautheret, D., 219–221
Gaw, H.Y., 240–246, 241t, 243f, 261–264
Gee, S.L., 262t
Geerlof, A., 259, 264–265, 334–355, 405–407
Geissler, P.L., 81–84
Gelis, I., 343
Gelles, J., 542–545
Genga, R.M., 236
Gentleman, R.C., 176
Georgiev, S., 262t
Gerber, A.P., 262t, 419
Gerisch, G., 334–335
Gersbach, C.A., 547–549
Gershernzon, N.A., 323
Gerstberger, S., 264–265
Gerstein, M., 266, 393–396, 402
Gheba, D., 467–468
Ghirlando, R., 47–48, 261–264, 282, 335, 346
Giddings, M.C., 22, 33
Giegé, R., 392
Giesebrecht, J., 498–500, 505

Gilbert, D.E., 241t, 244t, 252–254, 253f
Gilbert, L.A., 516, 547–549
Gilbert, S.D., 183–185
Gilbert, W.V., 154–159, 170–171, 177
Gill, D., 46
Gillespie, D., 335
Gillilan, R.E., 410–411
Gilliland, G., 185, 459
Girard, E., 410–411
Giri, S., 546–547
Gittis, A.G., 100–103
Glaser, S.J., 323
Glatter, O., 368, 393, 401–402, 409–410
Gleghorn, M.L., 419–420
Glisovic, T., 419
Glover, J.R., 335
Gluick, T.C., 100
Gnatt, A.L., 334–335
Gobl, C., 335–336, 343–346
Goddard, T.D., 504
Godt, A., 284–291, 300–306, 311, 323
Goecks, J., 159
Goff, S.P., 40–41
Gohlke, H., 296
Golas, M.M., 334–335, 421, 433–434
Gold, L., 467–468
Goldfarb, D., 300–302, 323
Goldman, A.L., 419
Goldstone, D.C., 262t
Golovanov, A.P., 241t, 261–264
Gomaa, A.A., 518–519
Gomez-Lorenzo, M.G., 498–499
Gonzales, K., 516
Gonzalez, R.L., 498–499, 561
Gonzalez-Bonet, G., 296, 310–311
Goodman, M.F., 561
Gorba, C., 499, 502
Gore, S., 185, 187–188
Gorelick, R.J., 5–6, 30–32
Gorin, A.A., 76
Görlach, M., 62–64, 262t
Gorlich, D., 282
Gorodetsky, Y., 301–302
Gorup, C., 262t
Gossen, M., 576–578
Gossett, A.J., 76–77
Gossett, J.J., 5–6
Goswami, B., 62

Gottardo, F.L., 48–50, 49f, 68
Gottstein, D., 343, 345–346
Gouridis, G., 343
Grace, C.R.R., 295
Granneman, S., 260–264, 262t
Grant, G.P., 309
Grant, R.A., 223
Grassucci, R.A., 498–499
Grau, D.J., 260–261, 419–420
Graveley, B.R., 572
Grazulis, S., 188
Greef, C.H., 575–576
Green, M.R., 237–246, 241t, 243f, 261–265, 262t, 265f, 280–281, 347, 349
Green, N.J., 219–221
Green, R., 183–185
Greenbaum, J.A., 154–159
Greene, E.C., 516–519, 525–527
Greenleaf, W.J., 226
Greer, B., 364–365, 379–380
Gregory, R.I., 280–281
Grellscheid, S.N., 246–248, 247f, 261–264, 262t
Gressel, S., 262t
Griffiths-Jones, S., 216–218
Grigg, J.C., 214–227
Grilley, D., 116–117
Grillo, I., 410–411
Grishaev, A., 65–67, 282, 335, 346, 374
Groebe, D.R., 103
Groll, M., 334–335
Gros, P., 198–199, 345–346, 407–408
Gross, A., 300
Grosse-Kunstleve, R.W., 186, 188, 201, 345–346, 407–408
Grote, M., 334–335
Grundy, F.J., 40–41, 215–222, 224–226
Grunert, S., 284
Grunwald, J., 337–338
Grzesiek, S., 56, 344
Guariento, M., 335
Guell, M., 516
Guenneugues, M., 48
Gueroussov, S., 572
Guerry, P., 351–353
Guest, C.R., 100–101
Guex, N., 22, 33
Guilinger, J.P., 516–517

Guillen-Boixet, J., 236–246, 238f, 243f, 247f, 252–254, 256–258, 257f, 260
Guinier, A., 393, 398–399
Gunderson, S.I., 282–283
Güntert, P., 296, 309–310, 313, 343, 345–346
Guo, L., 364–365, 410–411
Gupta, A., 264–265, 265f, 375, 377
Gural, G., 183f, 202f, 206–207
Gusarov, I., 214
Gutell, R.R., 100
Guthrie, C., 542–553, 558–559, 561–562, 566
Gutschner, T., 5
Guttler, T., 282
Guttman, M., 280
Guymon, R., 100–103
Gvozdev, R.I., 81

H

Ha, T., 5, 7, 498–499, 540–543, 546–547, 555
Haas, J., 392
Habeck, M., 345–346
Habib, D., 283, 295–296
Habib, N., 516
Hadjivassiliou, H., 544–545
Haertlein, M., 396, 410–411
Hafner, M., 262t, 264–265, 419–421
Hagelueken, G., 309
Hainard, A., 171–172
Hajdin, C.E., 150
Hall, I.M., 148
Hall, J., 240–248, 243f, 247f, 261–266
Hall, K.B., 100–122
Haller, A., 40–41, 100–101, 226
Hamma, T., 223
Hammel, M., 364, 374, 396–397, 410–411
Han, Y., 262t
Handa, N., 243f, 253f, 254–255, 353
Handsaker, B., 148
Hankovszky, H.O., 337–338
Hankovszky, O.H., 337–338
Hansen, A.L., 43, 46–50, 49f, 52–53, 55–56
Hansen, J., 214
Hao, H., 262t
Hao, X., 336
Harbison, G.S., 62–64
Harbury, P.A.B., 76–96

Harbury, P.B., 79, 82–84, 94–96
Hardin, J.W., 244t, 250–251
Hardman, S.J., 100–101, 110
Hargous, Y., 237–239, 241t, 246–248, 247f, 252–254, 257f, 260–264
Harris, M.R., 187–188, 196–197
Harrison, S.C., 523–524
Hartmann, R.K., 420–421
Hartmuth, K., 433–434
Harvey, S.C., 5–6
Hashimoto, K., 262t
Hashizume, T., 201
Hass, M.A., 282
Hastings, C., 170–171
Hatem, A., 162–163
Hatmal, M.M., 309
Hauer, M., 516–519, 525–527
Haumann, S., 62–64
Hausser, J., 264–265, 419–420
Hautbergue, G.M., 241t, 250–251, 261–264
Haworth, I.S., 309, 311
Hay, S.A., 236
Hayashi, M.K., 238f, 241t, 243f, 261–264
Hayes, R.L., 510
Hayward, S., 498–499
He, F., 241t, 246–248, 257f, 259, 264–265
He, L., 405
He, Q., 4–5
Headd, J.J., 185, 188–189, 193–195, 201, 208
Hector, R., 261–264, 262t
Hegde, B.G., 309–311
Hegde, P.B., 309
Heilman-Miller, S.L., 4–5
Heinicke, L.A., 566
Heller, W.T., 405
Hellman, L.M., 419–420
Helmers, J., 498–499
Helmling, C., 337
Hemminga, M.A., 284–285, 322–323
Henderson, S.J., 410
Henkin, T.M., 40–41, 215–222, 224–226, 419–420
Hennelly, S.P., 5–6, 499, 501
Hennig, J., 259, 264–265, 334–355, 392–393, 405–407, 419–420
Henning, A., 237–239, 238f, 246–248, 247f, 252–254, 257f, 260–264
Henrick, K., 185

Henriksen, N.M., 498–499
Henry, I., 262t
Hentze, M.W., 419–421, 459
Heppell, B., 100–101, 215–216
Herbert, A.P., 335
Herbst, R.H., 177
Hermann, T., 185, 204–205, 224–225
Hernandez, R., 405
Heroux, A., 215–216
Herschlag, D., 42, 76–96, 262t, 409–410, 419, 544–545
Hewitt Klenø, K., 410–411
Hiatt, S.M., 163
Hideg, K., 309, 311, 337–338
Hilal, T., 499
Hildebrand, P.W., 498–502
Hilger, D., 288–289, 296, 300, 311–313
Hinderberger, D., 288–289
Hiruma, Y., 336
Hjelm, R.P., 387–388
Hobart, D.E., 410
Hobartner, C., 299, 337, 544–545
Hobor, F., 261–264
Hochstrasser, R.A., 100–101
Hocine, S., 578
Hock, L.M., 76
Hodgson, K.O., 81
Hodil, S.E., 219–221
Hoell, J.I., 420–421
Hof, M., 573–574
Hofmann, Y., 246–248, 261–264
Hogan, D.J., 262t, 419
Hohng, S., 540–542, 546, 555
Hollingworth, D., 252–254
Hollis, T., 523–524
Holmes, W.M., 215, 221, 225–226
Holst, M.J., 498–499
Holtzmann, E., 342
Homer, N., 148
Hon, B., 46
Honer Zu Siederdissen, C., 216–218, 221–222
Honig, B., 76–77
Hood, J.L., 282–283
Hoogstraten, C.G., 42–43, 47–48
Hopkins, J.B., 410–411
Hoppe, W., 392, 410
Horlbeck, M.A., 516

Horner, M., 241t, 244t, 262t
Horos, R., 419–421
Horvath, P., 516–517
Hoskins, A.A., 542–545, 572–573
Hosseinizadeh, A., 498–499
Hotz, H.R., 549–550
Hou, V.C., 262t
Houck-Loomis, B., 40–41
Hounslow, A.M., 250–251
Houston, S., 531–532
Howe, P.W., 223, 237–239
Hrossova, D., 261–264
Hsieh, J., 545–546
Hsu, P.D., 516–519
Hu, D., 246–248
Hu, J., 167, 173
Hu, Y.X., 244t, 250–251
Huang, C.C., 504
Huang, C.Y., 540–542, 554
Huang, H., 223
Huang, J.R., 346, 351–353
Huang, L., 22
Huang, Y.-S., 419–420
Huang, Z., 224
Hubbell, W.L., 284–285, 300, 309, 311, 322–323, 337
Huber, R.G., 337
Huber, W., 176
Hud, N.V., 5–6
Hudson, B.P., 282–283
Huelga, S.C., 261–264, 262t
Huggins, W., 150
Hughes, J.M., 134
Hung, M.L., 250–251
Hunter, R.I., 301–302
Hura, G.L., 81–84, 364, 374, 396–397, 410–411
Huranová, M., 573–574
Hurwitz, R., 5
Hussain, S., 466–467
Hussain, T., 498–501
Huthoff, H., 5, 40–41
Hutt, K.R., 262t
Hwang, H., 542–543

I

Ibel, K., 387–388, 392, 399, 401–402
Igel, A.H., 134

Igumenova, T.I., 48, 51
Ikeda, Y., 261–264
Imai, T., 246–248
Immormino, R.M., 185, 193–195, 208
Imperiali, B., 336
Ince-Dunn, G., 261–264
Innis, C.A., 499
Inoue, M., 241t, 244t, 246–248
Ionita, P., 284–285, 291, 305–306
Irimia, M., 572
Irobalieva, R.N., 282
Irving, T., 410–411
Isacoff, E.Y., 546–547
Ish-Horowicz, D., 295
Ishida, H., 498–499
Ishikawa, F., 241t
Ishitsuka, Y., 542–543, 546–547
Ishiura, S., 244t
Iskakova, M., 498–501
Ito, N., 223, 241t
Ivani, I., 573–574
Iwahara, J., 281–282, 284, 335, 342–345
Iwasaki, W., 266
Iyalla, K., 282
Izaurralde, E., 244t

J

Jacks, A., 237–239, 238f
Jacques, D.A., 374, 396–397
Jacques, V., 284–285
Jacrot, B., 392–398, 402
Jaeger, L., 283, 295–296
Jaffery, S.R., 183–185
Jäger, H., 284–285, 322–323
Jagtap, P.K.A., 419–420
Jahnz, M., 336
Jain, A., 542–543, 546–547
Jain, S., 182–208, 187f, 205f
Jaiswal, R., 346
Jakubik, M., 261–264
Jamonnak, N., 262t
Jangi, M., 468, 473, 489
Janicki, S.M., 573–578
Janowski, P.A., 498–499
Janssen, B.J.C., 198–199
Jao, C.C., 309–311
Jarasch, A., 499
Jardine, P.J., 224
Jares-Erijman, E.A., 78–79
Jarman, S.N., 146
Jarmoskaite, I., 364–365
Jasnin, M., 396
Jaszczur, M., 561
Javadi-Zarnaghi, F., 299, 337, 544–545
Jayne, S., 246–248, 247f, 260–264, 459, 466–467
Jean, J.M., 100–101, 117
Jenkins, H.T., 236–237, 240–248, 241t, 243f
Jenkins, J.L., 264–265, 265f, 351
Jenner, L., 223–224, 498–501
Jens, M., 262t
Jensen, K.B., 261–264, 419–420, 466–467
Jensen, M.R., 342, 351–353
Jernigan, R.L., 498–499
Jeschke, G., 236–237, 280–324, 335, 337, 346
Jessen, T.H., 236–237
Jiang, F., 517–518, 525–527
Jin, X., 76–77
Jin, Y., 262t
Jinek, M., 516–535
Jockusch, S., 546
Johansson, C., 241t, 282–283
Johansson, H.E., 575–576
John, M., 335–336
Johnson, E., 48–50
Johnson, J.E., 47–48, 224
Johnson, R., 5
Johnson-Buck, A.E., 555
Jonas, U., 284–285, 289–291, 306
Jones, D.H., 336
Jones, R.A., 62
Jones, T.A., 187–188, 196–197
Jonikas, M.A., 409–410
Jonker, H.R.A., 336–337
Joosten, K., 197–198
Joosten, R.P., 186, 197–198
Jorgensen, R., 498–499
Joseph, S., 498–499
Joshi, A., 256
Joshi, R., 76–77
Jossinet, F., 219–221
Joung, J.K., 516–517, 534
Jourdan, M., 282–283
Jovanovic, M., 177

Jovin, T.M., 78–79
Juette, M.F., 546
Jung, H., 288–289, 296, 300, 311–313

K

Kaback, H.R., 284–285
Kahlscheuer, M.L., 540–565
Kalai, T., 300
Kalina, J., 185
Kambach, C., 523–524
Kampmann, M., 523–524
Kanaar, R., 244t, 252–254
Kanelis, V., 335
Kang, M., 43, 223–224
Kaplan, M., 516
Kaplan, N., 76–77
Kapral, G.J., 183f, 185, 187f, 189, 193–195, 203, 205f, 206, 208, 498–499
Kaptein, R., 46, 48
Karabiber, F., 22, 33, 169
Karamanou, S., 343
Karim, C.B., 344–345, 349–351
Karn, J., 42–43
Karplus, M., 500–501
Karvelis, T., 516–517
Kastner, B., 334–335, 433–434, 540, 542–543
Kasumaj, B., 303–306
Kato, H., 343
Kattnig, D.R., 288–289
Katzman, S., 168
Kay, L.E., 42–43, 46–50, 53, 61, 77–78, 281–283, 295, 335, 343, 346
Kaya, E., 517–518, 525–527
Kayikci, M., 167, 177, 262t, 264–265, 265f
Kazan, H., 262t, 334–335, 467–468
Ke, A., 205f, 206, 214–227
Keating, K.S., 4–6, 203–204, 224–225
Kedde, M., 40–41
Keedy, D.A., 185
Keene, J.D., 260–261, 419–420
Keizers, P.H., 336
Keller, W., 466–467
Kelley, A.C., 498–501
Kellis, M., 127, 154–159, 163–165, 170–171, 175f
Kellogg, D., 40–68
Kelly, G., 237–239, 238f, 252–254

Kent, W.J., 159
Keohavong, P., 573–574
Keramisanou, D., 343
Keren, H., 572
Kerr, L.D., 518
Kertesz, M., 168
Keyes, R.S., 337
Khanna, A., 216–218
Khayter, C., 516–517
Khin, N.W., 244t
Khochbin, S., 459
Khorshid, M., 264–265, 419–420
Khvorova, A., 337
Kibbe, W.A., 522, 528
Kieffer, B., 337
Kieft, J.S., 5
Kielkopf, C.L., 237–246, 238f, 241t, 243f, 261–265, 265f, 280–281, 349, 351
Kielpinski, L.J., 154–177
Kikhney, A.G., 365
Kilburn, D., 5–6
Kilburn, J.D., 410–411
Kim, D.H., 549–550
Kim, E., 61, 68
Kim, H.D., 498–499, 542–543, 561
Kim, H.S., 396, 405, 408–409
Kim, I., 11, 243f, 253f, 254–255, 282–283, 353
Kim, S., 48, 241t, 282–283
Kim, S.H., 549–550
Kimber, M.S., 531–532
Kimsey, I.J., 40–68
Kimura, K., 546
King, S.M., 405
Kino, Y., 244t
Kirby, T.L., 344–345, 349–351
Kirchhausen, T., 573–574, 582–584
Kirmizialtin, S., 498–510
Kirste, R.G., 392, 399
Kiryu, H., 266
Kishore, A.I., 346
Kisseleva, N., 337
Kister, L., 241t, 246–248, 247f, 262t
Kjeldgaard, M., 81, 187–188, 387–388
Kladwang, W., 43, 150, 169
Klee, C.B., 344
Klein, D.J., 183–185
Kleywegt, G.J., 185, 187–188, 196–197

Author Index

Kloczkowski, A., 498–499
Kloiber, K., 337
Klostermeier, D., 250–251, 284–285
Kneale, G.G., 387–388
Kobayashi, N., 246–248
Koch, A., 284–285, 289–291, 306
Koch, M.H.J., 364, 392–393, 396–397, 402, 404f, 406–408
Kohlbrecher, J., 410–411
Kohler, P.R., 546–547
Koldobskaya, Y., 223
Kolle, G., 266
Komisarow, J.M., 260–261
Konarska, M.M., 550
Kondo, Y., 523–524
Konecny, R., 498–499
Konermann, S., 516–519, 534
Konevega, A.L., 76, 189
König, J., 167, 177, 264–265, 265f, 420–421, 466–467
Konishi, Y., 241t
Koo, B.K., 249, 250f
Koodathingal, P., 572–573
Koradi, R., 247f
Korostelev, A., 499–501
Korzhnev, D.M., 43, 46–50, 53, 61
Kosen, P.A., 344–345
Kosuri, S., 516–517
Kothary, R., 246–248
Koukaki, M., 343
Kozlov, A.G., 419
Kraemer, B., 419–420
Krainer, A.R., 238f, 241t, 243f, 250–251, 261–264, 262t
Kramer, K., 419–460
Krasauskas, A., 421, 433–434
Kratky, O., 368, 393, 401–402, 409–410
Krause, H.M., 419
Kreig, A., 542–543
Kreneva, R.A., 214
Kreutz, C., 51, 100–101, 183f, 337
Krishnan, R., 542–553, 558–559, 561–562
Kriz, A.J., 516–517
Kroenke, C.D., 43, 46–48, 51
Krueger, S., 393–396
Krumkacheva, O.A., 322–323
Kubicek, K., 261–264
Kucukural, A., 262t

Kudla, G., 260–264, 262t
Kulshina, N., 223
Kumar, S., 250–251
Kung, J.T., 260–261, 419–420
Kunjir, N.C., 322–323
Kunzelmann, S., 546–547
Kuo, P.-H., 244t, 262t
Kuprin, S., 392–393, 396, 402, 406–408
Kurimoto, K., 243f, 253f, 254–255, 353
Kuriyan, J., 500–501
Kurkcuoglu, O., 498–499
Kurshev, V.V., 304–308
Kuscu, C., 516–517
Kusic-Tisma, J., 419
Kusnetzow, A.K., 309, 311
Kuwasako, K., 241t, 246–248, 257f, 259, 264–265
Kuzhelev, A.A., 322–323
Kwan, S.S., 252–254
Kwok, C.K., 127–128
Kwon, Y., 542–543
Kynde, S., 410–411

L

LaBean, T.H., 185, 188–189
Labhsetwar, P., 546–547
Laederbach, A., 409–410
Lafontaine, D.A., 214–215, 226
Lager, P.J., 419–420
Lagier-Tourenne, C., 262t
Lai, D., 4–5, 42
Laing, L.G., 100
Laird, K.M., 351
Laird-Offringa, I.A., 241t, 244t, 282–283
Lakomek, N.-A., 77–78
Lakowicz, J.R., 110
Lambert, N., 466–492
Lamichhane, R., 256, 260
Lamond, A.I., 260–261
LaMontaine, D.A., 100–101
Lander, E.S., 516
Landthaler, M., 262t, 264–265, 419–421
Lang, K., 100–101, 183–185, 184f, 200f
Lange, O.F., 77–78
Langen, R., 309–311
Langer, J.A., 387–388
Langlois, R., 498–499
Langmead, B., 148, 162–163

Lapinaite, A., 282–283, 335, 343, 346, 405, 407–408
Lapointe, J., 192f
Lardelli, R.M., 547, 549–550
Larson, D.R., 573–574, 578
Larson, M.H., 516
Lary, J.W., 5–6
Latham, J.A., 154–159
Lattman, E.E., 100–103
Laue, T.M., 5–6
Laughrea, M., 387–388
Laurberg, M., 500–501
Lavender, C.A., 154–159
Lavoie, B., 578
Law, M.J., 244t
Lawrence, M., 176
Layton, C.J., 467–468
Le bec, C., 262t
LeBarron, J., 498–499
Lebars, I., 337
Lebedeva, S., 262t
LeBlanc, R.M., 51
Lebowitz, J., 17–20
Lecuyer, E., 419
LeCuyer, K.A., 575–576
Lee, A.L., 244t
Lee, B.M., 282–283
Lee, D., 252–254
Lee, J., 42–43, 50, 64–67
Lee, J.E., 262t
Lee, J.Y., 262t
Lee, K.S., 5, 7
Lee, L.J., 572
Lee, T.H., 561
Leeper, T.C., 241t, 261–264, 282–284
Lefmann, K., 410–411
Legault, P., 337
Leger, P., 303–306
Legiewicz, M., 4–35
Legrain, P., 550
Lehmann, J., 219–221
Lehnert, U., 401
Leipply, D., 103, 109–112
Leitao, A.L., 566
Leith, A., 498–499
Lemay, J.F., 215–216, 226
Leonard, C.W., 5–6, 25–27, 30–32, 150
Leonhardt, R.M., 546–547

Leon-Ricardo, B.X., 177
Leontis, N.B., 189
Lersch, R.A., 262t
Lesley, S.A., 531–532
Leslie, C.S., 265–266
Leung, A.K.W., 523–524
Lev-Maor, G., 572
Lewis, M.S., 17–20
Lex, L., 337–338
Li, F., 183f
Li, G.-W., 516
Li, H., 148, 261–264
Li, J., 236–237
Li, K., 540
Li, N.S., 223
Li, P.W., 236
Li, S., 214–215
Li, T.W., 154–159
Li, W., 498–499
Li, X., 262t, 266, 334–335, 467–468
Li, Y.H., 498–499
Li, Y.Y., 309
Lian, L.Y., 250–251
Liang, B., 343
Liang, T.Y., 261–264, 262t
Liao, H.H., 237–239, 241t, 257f, 258–259
Libri, D., 572–573
Lichtenstein, C.P., 167, 173
Liechti, A., 5
Liedl, K.R., 337
Likhtenshtein, G.I., 81
Lilley, D.M., 226
Lim, C., 419–420
Lim, F., 575–576
Lin, C.H., 262t
Lin, J., 334–335
Lin, P., 244t
Lin, R.J., 549–550
Lin, S., 516–517
Lin, Y., 127–151, 164–165
Linden, M.H., 250–251
Linge, J.P., 345–346
Linn, S., 100–101
Lipchock, S.V., 223
Lipkin, Y., 301–302
Lipshitz, H.D., 266, 419
Lipsitz, R.S., 282
Lis, J.T., 467–468

Lisacek, F., 171–172
Littler, S.J., 244t
Liu, C., 498–499
Liu, D.R., 516–517
Liu, F., 4–35
Liu, H., 240–246, 241t, 243f, 252–255, 253f, 261–264
Liu, H.-X., 262t
Liu, P.P., 261–264
Liu, Q., 240–246, 241t, 243f, 252–255, 253f, 261–264
Liu, R., 542–543, 546–547
Liu, X., 237–239, 238f
Liu, Y., 237–239, 252–254, 256, 261–264, 322–323
Liu, Z., 240–246, 241t, 243f, 252–255, 253f, 261–264, 516
Livingston, A.L., 223
Llacer, J.L., 498–501
Loeliger, K., 282
Loerke, J., 498–510
Lohkamp, B., 203–204, 224–225
Lomant, A.J., 4–5
Lomzov, A.A., 322–323
Long, D., 47–48
Long, J., 284
Long, R.M., 573–574
Longhini, A.P., 51
Lopez, C.J., 300
Lopez, L.E., 214
Lopez-Gomollon, S., 522
Lorenz, R., 216–218, 221–222
Lorenz, S., 392
Loria, J.P., 43, 46–48, 51
Lormand, J., 542–543
Lorsch, J.R., 498–501, 518
Lorson, C.L., 246–248
Lovell, S.C., 185, 188–189
Lowe, J., 334–335
Lowenhaupt, K., 59
Lu, C., 215–219, 221–222, 224–226, 241t, 261–264, 282–284
Lu, K., 283, 295–296
Lu, M., 100, 105–106, 109, 280–281
Lu, X.J., 76, 295
Lu, Y., 336
Lubbers, M., 189
Luchinat, C., 336, 346

Lucks, J.B., 127, 169
Lueders, P., 284–285, 322–323
Lührmann, R., 347, 421, 433–434, 540, 542–543
Lukavsky, P.J., 51, 241t, 253f, 255–256, 261–264, 282–283, 295–296, 410
Lunde, B.M., 241t, 244t, 262t, 334–335
Lundstrom, P., 61
Luo, R., 296
Luo, S., 127, 169
Luscombe, N.M., 420–421
Lussier, A., 214–215
Luthey-Schulten, Z., 498–499, 542–543
Lutterotti, L., 188
Lv, H., 237–239, 252–254, 256
Lvov, Y.M., 81
Ly, T., 239–240
Lyons, S.M., 467–468

M

Ma, E., 516–518, 525–527
Mackereth, C.D., 236–237, 241t, 246–248, 252–258, 257f, 280–283, 334–355
MacMillan, F., 335
Madl, T., 236–237, 252–258, 257f, 280–284, 334–337, 343–347, 349–353, 392–393, 405
Maeder, C., 105–106
Maeder, M.L., 516–517
Maegawa, S., 262t
Maier, M.A., 183f
Maier-Davis, B., 334–335
Mainzer, J.E., 168
Major, F., 64–65, 66f, 189
Makino, D.L., 334–335
Makunin, I.V., 280
Mali, P., 516–517
Malina, A., 516–517
Malinina, L., 280–281
Malkova, B., 236–237, 240–248, 241t, 243f
Mallam, A.L., 364–365
Manatschal, C., 256, 260
Manceau, V., 375, 377
Manche, L., 238f, 241t, 243f, 261–264
Mandal, M., 40
Manhart, C.M., 419
Maniatis, T., 262t
Mankin, A.S., 164–165, 171–172

Manley, J.L., 262t
Mann, M., 260–261
Mann, R.S., 76–77
Manolaridis, I., 237–239, 238f
Manor, O., 154–159, 168
Mantei, N., 573–574
Manzo, A.J., 540–542, 554–555
Maquat, L.E., 419–420
Marakushev, S.A., 81
Marcia, M., 4–35
Margraf, D., 298
Mariani, P., 397
Marion, D., 410
Maris, C., 240–248, 243f, 247f, 261–266, 283, 295–296, 299, 337
Marius Clore, G., 343
Mark, A.E., 503
Marko, A., 298
Markwick, P., 351–353
Marquis, J., 262t
Marshall, M., 216–218
Marshall, R.A., 546
Martin, L., 467–468
Martin, M., 148, 159–161
Martin, R.M., 572–584
Martin, S.R., 252–254
Martincorena, I., 264–265, 265f
Martinez-Yamout, M.A., 282–283
Martin-Tumasz, S., 237–239, 238f, 244t
Martorana, A., 300
Marucho, M., 498–499
Maryasov, A.G., 304–305
Marzluff, W.F., 467–468
Mason, A.C., 364–365, 379–380
Masse, E., 214–215
Massi, F., 43, 46–50, 57–58, 236
Mastroianni, A.J., 81–84
Mathew-Fenn, R.S., 76, 79, 81–84, 93–96
Mathews, D.H., 35, 150, 154–159, 168
Mathieu, S.L., 183f
Matsumoto, A., 498–499
Mattaj, I.W., 239–240, 246–248, 282–283
Matthias, P., 459
Mattick, J.S., 280
Maucuer, A., 375, 377
Maus, V., 284–285
May, G.E., 127–151, 164–165
May, R.P., 392–393, 397, 410–411

Mayer, O., 42
Mayerle, M., 542–543
Mayo, K.H., 344
Mazor, E., 168
Mazza, C., 239–240, 246–248
McAfee, J.G., 262t
McBrairty, M., 40–68
McCammon, J.A., 498–499
McCarrick, R.M., 310–311
McCloskey, J.A., 201
McCluckey, K., 100–101
McCluskey, K., 226
McConnell, H.M., 59
McCoy, A.J., 208
McCutcheon, J.P., 100–103
McDaniel, B.A., 221
McEleney, K., 364–365
McGarvey, M., 549–550
McGeary, S., 468, 473, 489
McGinnis, J.L., 22, 33, 169, 543–544
McKay, D.B., 244t, 250–251
McKenna, S.A., 11, 364–365
McManus, C.J., 127–151, 164–165
Meares, C.F., 284–285
Medalia, O., 334–335
Megiorni, F., 262t
Meiboom, S., 46
Meier, M., 467–468
Meilleur, F., 405
Meiners, J.C., 545–546
Meinhart, A., 241t, 244t, 262t
Meisburger, S.P., 410–411
Mele, A., 419–420, 466–467
Melikian, M., 346
Melnikov, S., 498–501
Mendez, R., 236–246, 238f, 243f, 247f, 252–254, 256–258, 257f, 260
Mendillo, M.L., 81–84
Menon, A.L., 410–411
Menzies, F., 419
Merino, E.J., 154–159, 218–219
Merz, K.M., 296
Meyer, I.M., 4–5, 42
Miakelye, R.C., 81
Michaud, S., 421
Michel, E., 236–237, 281–283, 292, 295–296, 306–308, 313–315, 318–319, 322, 335, 337, 346, 410

Michelotti, N., 555
Micklem, D.R., 284
Micura, R., 40–41, 100–101, 183–185, 184f, 200f, 226
Mielke, T., 498–500, 505
Mierendorf, R.C., 6
Mikolajka, A., 498–502
Miksys, A., 516–517
Militti, C., 259, 264–265, 334–335, 346, 353, 405–407
Millar, D.P., 100–101, 284–285
Miller, A., 84–85
Miller, M.D., 208
Millet, O., 46
Millett, I.S., 4
Milligan, J.F., 103
Miloushev, V.Z., 57–58
Milov, A.D., 285–286, 288–289
Mironov, A.S., 214
Misra, V.K., 100–103, 116–117
Misteli, T., 573–574
Mitchell, D., 4
Mitchell, S.F., 518
Mitra, K., 499–501
Mitra, S., 5–6, 17–20
Mittermaier, A., 77–78
Miura, H., 516–517
Miyashita, O., 499, 502
Miyazaki, Y., 282–283, 295–296
Mohanty, U., 510
Moineau, S., 516
Mollica, L., 351–353
Moltke, I., 175f
Montange, R.K., 183–185, 222
Montemayor, E.J., 237–239, 241t, 257f, 258–259
Moody, T.P., 5–6
Mooney, S.D., 262t
Moore, B.P., 392
Moore, C., 241t, 261–264, 282–284, 334–335
Moore, I.K., 76–77
Moore, M.J., 542–545, 572
Moore, P.B., 81, 183–185, 183f, 202f, 206–207, 214, 364, 371–372, 385, 387–388, 392
Moore, S.A., 262t
Moosburner, M., 516–517

Moradi, S., 335
Moran, F., 368, 402
Moran, J., 262t
Moras, D., 392
Morgan, H.P., 335
Morgan, R.E., 46
Mori, D., 244t
Moriarty, N.W., 188–189
Morris, K.V., 280
Morris, Q., 266
Morris, S., 4–5
Morshed, N., 208
Morsut, L., 516
Mort, M., 262t
Mortensen, K., 410–411
Mortimer, S.A., 25–27, 127, 169, 523–524
Moseley, M.L., 261–264
Mosteller, F., 170–171
Mougel, M., 154–159
Moulin, M., 396, 410–411
Moursy, A., 241t, 246–248, 250–251, 250f, 260–264
Moxon, S., 216–218
Moysan, E., 459
Mu, Y., 337
Mueller, M.W., 164–165
Mueller, T.D., 241t, 282–283
Muhandiram, R., 47–48
Muhlbacher, J., 100–101
Muhs, M., 499
Mukherjee, N., 262t
Mukherjee, S., 498–499
Mulder, F.A.A., 46, 48
Mulhbacher, J., 226
Muller, J.J., 385
Muller-McNicoll, M., 262t
Mumbach, M.R., 177
Munari, F., 237–239, 238f
Mund, M., 335
Munishkin, A., 189
Munro, J.B., 498–499
Mural, R.J., 236
Murphy, E.C., 343
Murphy, F.V., 498–499
Murray, L.W., 182–183, 185, 193–195, 203, 208
Murshudov, G., 498–501
Murshudov, G.N., 197–198

Musier-Forsyth, K., 100–101
Mustoe, A.M., 40, 58, 68, 281, 409–410
Muth, G.W., 214
Muto, Y., 241t, 244t
Myers, E.W., 236
Myers, J.C., 262t
Mylonas, E., 397–398, 410
Myszka, D.G., 244t

N

Nagai, K., 223, 236–240, 241t, 523–524
Nagalakshmi, U., 266
Nagaswamy, U., 101–103
Nagata, K., 241t
Nagata, T., 246–248
Nagle, J.M., 40–41
Nagy, G., 410–411
Nahvi, A., 40–41, 214–215, 221
Naismith, J.H., 309
Najafabadi, H.S., 262t, 334–335, 467–468
Nakamura, H., 185
Nalepa, A.I., 323
Nam, Y., 280–281
Nandakumar, D., 5, 7
Necakov, A., 531–532
Necsulea, A., 5
Nekrutenko, A., 159
Nelson, J.A.E., 127
Nesbitt, D.J., 226, 544–545
Neuhaus, D., 223, 237–239, 282–283
Newby Lambert, M., 545–546
Nguyen, B., 419
Nguyen, P., 337
Ni, S., 310–311
Nicastro, D., 334–335
Niccolai, N., 42–43
Nicolas, F.E., 522
Nielsen, R.C., 498–499
Niemeyer, C.M., 76
Nierhaus, K.H., 364–365, 392–393, 402, 406–407, 498–499
Niewoehner, O., 516–535
Nikolova, E.N., 43, 46, 48–50, 49f, 52–53, 55–56, 61, 68
Nikonowicz, E.P., 223–224
Nilges, M., 280–282, 343–347, 349–351, 364
Nishikura, K., 466–467

Nishimasu, H., 517–518
Nissen, P., 214
Nitz, M., 336
Niu, C., 237–239, 238f
Niu, L., 237–246, 241t, 243f, 252–256, 253f, 261–264
Nodet, G., 342, 351–353
Noel, J.K., 510
Noeske, J., 40–41, 187f, 205f, 206, 498–499
Noller, H.F., 154–159, 500–501
Norman, D.G., 303
Novak, T., 572–573
Novikova, I.V., 5–6
Nudler, E., 215
Nureki, O., 192f, 243f, 253f, 254–255, 353
Nusbaum, Jeffrey D., 262t
Nutter, R.C., 168
Nye, C.H., 518–519

O

Oakes, B.L., 516
Obayashi, E., 523–524
Oberstrass, F.C., 236–248, 241t, 244t, 252–254, 257f, 260–264, 262t, 282–283
O'Brien, P.J., 61, 68
Ochmann, A., 299, 337
O'Connell, M.A., 466–467
O'Connell, M.R., 516
Odenwalder, P., 545–547
Ogawa, S., 262t
Ogawa, Y., 260–261, 419–420
Ohlenschlager, O., 62–64, 223–224
Ohrt, T., 545–546
Ohshima, Y., 546
Ohsumi, T.K., 260–261, 419–420
Ohyama, T., 246–248
Okano, H., 246–248
Okano, H.J., 261–264
O'Leary, S.E., 542–543
Oliver, P.L., 280
Olivier, N.B., 498–499
Olson, W.K., 76, 295
Omelka, M., 185
Onesto, E., 262t
Onodera, C.S., 168
Onuchic, J.N., 510
Oostenbrink, C., 503
Orekhov, V.Y., 48–50, 53

Orr-Weaver, T., 419
Ortega Roldan, J.L., 351–353
Ortoleva-Donnelly, L., 214
Ortore, M.G., 397
Orzechowski, M., 499
Osborne, R.J., 167, 173
Osheim, Y.N., 573–574
Osman, R., 100–101
Otting, G., 335–337
Otto, G.A., 282–283
Oubridge, C., 223, 236–237, 241t, 523–524
Oude Vrielink, J.A., 40–41
Ouhashi, K., 241t
Ouyang, Z., 154–159, 168
Overbeck, J.H., 335
Overhand, M., 336
Owen-Hughes, T., 303
Ozaki, H., 266
Ozenne, V., 342, 351–353
Ozer, A., 467–468

P

Padan, E., 288–289, 296, 311–313
Padgett, R.A., 583–584
Pagano, J.M., 467–468
Page, K., 410
Page, R., 531–532
Pages, H., 176
Paillard, L., 262t
Palangat, M., 573–574
Palecek, E., 76–77
Pallan, P.S., 183f
Palmer, A.G., 43, 46–51, 54, 57–58, 77–78
Palmer, P.D., 410
Palty, R., 546–547
Pan, Q., 572
Pan, T., 5–6
Pancevac, C., 262t
Pandit, S., 261–264
Panek, G., 289–291, 311
Pannier, M., 285–286
Panter, M.S., 546–547
Pantos, E., 368, 402
Pappu, R.V., 498–499
Paranawithana, S.R., 240–246, 241t, 243f, 261–264, 280–281, 349
Pardi, A., 42–43, 65–67, 282
Pardon, J.F., 392

Parfait, R., 392
Parigi, G., 336
Parisien, M., 64–65, 66f, 189
Park, S., 244t
Park, S.Y., 296, 310–311
Park, W.-Y., 261–264
Parker, B.J., 175f
Parks, J.W., 542–543
Patel, A., 249, 250f
Patel, D.J., 40–41, 222, 240–246, 241t, 243f, 244t, 261–264
Patel, T.R., 364–365
Patrick, E.M., 100–101
Pattanayak, V., 516–517
Patton, J.G., 262t, 347
Paulsen, H., 289–291, 311
Peabody, D.S., 575–576
Pearlman, S., 409–410
Peattie, D.A., 154–159
Peisach, E., 336
Pelikan, M., 364
Pelupessy, P., 49–50
Pena Castillo, L., 467–468
Penczek, P.A., 498–499
Penedo, J.C., 100–101, 226
Peng, G.E., 262t
Perez, C.F., 226
Pérez, I., 262t
Perez, J., 383
Perez-Canadillas, J.M., 237–239, 241t, 252–255, 253f, 262t
Pergoli, R., 261–264
Perona, J.J., 402
Perrakis, A., 197–198
Pervushin, K., 335
Pestova, T.V., 499
Peters, F.B., 300
Peterson, J.R., 546–547
Peterson, R.D., 43, 223–224
Petfalski, E., 260–261
Petoukhov, M.V., 364–365, 392–393, 406–407, 410
Petrov, A., 51, 542–543
Petzold, K., 42–43, 64–65, 67, 77–78
Pfeifer, J., 56
Pfister, T.D., 336
Phair, R.D., 573–574
Phillips, G.N., 252–254

Piccirilli, J.A., 223, 572–573
Pierattelli, R., 336
Pintacuda, G., 335–336
Piotto, M., 50
Piper, A., 405
Pirzada, Z.A., 516
Pitchiaya, S., 566
Piton, N., 337
Pitsch, S., 244t, 250f, 251–252, 261–264
Pizzuti, A., 262t
Plantinga, M.J., 189
Platzmann, F., 547
Pleiss, J.A., 543–544
Plocik, A.M., 572
Plumpton, M., 549–550
Pochart, P., 419–420
Polacek, N., 183–185
Polaski, J.T., 224
Pollack, L., 215, 218–222, 224–226, 396, 410–411
Pollard, A.M., 296, 310–311
Polson, A.G., 40–41
Polyhach, Y., 286, 288–289, 296–298, 300–306, 309, 311–313
Polymenidou, M., 262t
Pomeranz Krummel, D., 523–524
Ponthier, J.L., 262t
Ponting, C.P., 280
Ponty, Y., 35, 176
Poole, F.L.N., 410–411
Pop, M., 148
Popowicz, G.M., 259, 264–265, 334–335, 346, 353, 405–407
Porrua, O., 572–573
Porter, E.B., 100–101, 222
Poser, I., 573–574
Potapov, A., 301–302
Powers, T., 154–159
Pradhan, V., 215, 221, 225–226
Prasanth, K.V., 546–547, 573–578
Presley, B.K., 185, 188–189
Prestegard, J.H., 346
Pribitzer, S., 323
Price, E.A., 309, 311
Price, I.R., 215, 226
Price, S.R., 239–240, 241t
Pringle, T.H., 159
Prior, M., 545–546
Prisant, M.G., 188–189
Prisner, T.F., 284–285, 296, 298, 323, 335, 337
Pritchett, M., 364–365, 379–380
Proctor, J.R., 4–5, 42
Proctor, M.R., 284
Profumo, A., 383
Pruitt, K.D., 163
Ptashne, M., 523–524
Puglisi, E.V., 51
Puglisi, J.D., 11, 42–43, 51, 105–106, 282–283, 498–499, 542–543, 546, 561
Pulk, A., 183f, 189, 498–499
Pullman, B., 62
Purcell, E.M., 46
Purschel, H.V., 385
Putnam, C.D., 364, 374, 396–397
Pyle, A.M., 4–35, 203–204, 224–225, 545–546

Q

Qamar, S., 419–460
Qi, L.S., 547–549
Qi, M., 300, 323
Qin, P.Z., 298–299, 309, 311, 337–338, 545–546
Qin, Y., 498–499
Qiu, J., 261–264
Qu, K., 154–159, 168
Qu, X., 241t, 261–264, 282–284
Quake, S.R., 467–468
Query, C.C., 544–545
Quinlan, A.R., 148
Quiros, M., 188
Quon, G., 266

R

Rabenstein, M.D., 284–285
Rachofsky, E.I., 100–101
Radermacher, M., 498–499
Radmer, R.J., 409–410
Rafikov, R., 214
Ragunathan, K., 542–543, 546–547
Raha, D., 266
Rahn, A., 347
Rai, P., 100–101
Raitsimring, A.M., 301–302, 304–308

Rajashankar, K.R., 215–218, 221–222, 224–226
Rajeev, K.G., 183f
Rajkowitsch, L., 42
Rakwalska-Bange, M., 282–283, 335, 343, 346, 405, 407–408
Ramachandran, R., 62–64
Ramakrishnan, V., 100–103, 498–501
Ramani, B., 542–543, 546–547
Rambo, R.P., 5–6, 215–216, 364–388, 410–411
Ramelot, T.A., 310–311
Ramos, A., 252–254, 262t, 284, 298, 336–337
Ramrath, D.J., 499–500, 505
Ran, F.A., 516–519
Rance, M., 46, 48–51
Rao, J.N., 310–311
Rappsilber, J., 260–261
Rasche, N., 545–546
Ratje, A.H., 498–502
Ravindranathan, S., 282–283, 295–296, 316, 323
Rau, M.J., 100–122
Ray, D., 262t, 334–335, 467–468
Raymond, P., 578
Razzaghi, S., 284–285, 323
Read, R.J., 185–188, 198–199, 204–205, 208
Rebora, K., 241t, 246–248
Recht, M.I., 518
Reckel, S., 343, 345–346
Redding, S., 516–519, 525–527
Redemann, S., 545–546
Reed, R., 421
Reginsson, G.W., 322–323
Reich, E., 100–101
Reichenwallner, J., 288–289
Reid, David W., 262t
Reider, U., 100–101
Reik, W., 280
Reiter, N.J., 244t, 252–254
Religa, T.L., 61, 283, 295, 335
Ren, A., 222
Ren, H., 344
Ren, J., 237–239, 238f, 575–576
Ren, Z., 189
Reuter, J.S., 35, 150

Reyes, F.E., 224, 523–524
Reyes, R.E., 374
Reymond, L., 244t, 250f, 251–252, 261–264
Reynolds, A.M., 336
Reyon, D., 516–517
Rice, A.J., 244t
Rice, G.M., 25–27, 127
Richard, S., 392–393, 396, 402, 406–408
Richards, B.M., 392
Richards, M., 516
Richardson, D.C., 182–208, 187f, 205f
Richardson, J.S., 182–208, 187f, 205f
Richie, A.C., 237–239, 238f
Richter, C., 51, 337
Rieder, R., 5, 100–101
Rieder, U., 100–101
Riek, R., 295, 335
Rieping, W., 345–346
Rinaldi, A.J., 544–546
Ringel, I., 295
Rinn, J.L., 168, 280
Rinnenthal, J., 40
Rino, J, 572–584
Rio, D.C., 244t, 252–254
Riordan, D.P., 262t, 419
Rios-Steiner, J.L., 239–240
Robert, F., 516–517
Robert, M.C., 573–574
Roberts, R.W., 575–576
Robertson, A., 468, 473, 489
Robertson, D.A., 301–302
Robertson, J.D., 466–492
Robin, X., 171–172
Rocco, M., 383
Rodionova, N.A., 237–239, 238f
Rodnina, M.V., 76, 189, 498–499
Roe, D.R., 498–499
Rogelj, B., 262t
Rogers, C., 62–64
Rogozhnikova, O.Y., 322–323
Roh, J.H., 5–6, 410–411
Rohrer, M., 335
Rohs, R., 76–77
Roitberg, A.E., 498–499
Rojas-Duran, M.F., 170–171, 177
Rokop, S.E., 410
Rollins, S.M., 219–221

Romby, P., 134, 154–159
Romero, D.A., 516
Rosbash, M., 550
Rose, S., 22
Rösel, T., 433–434
Rosen, M.K., 335
Rosenfeld, L., 576–578
Rosenzweig, R., 335
Roskin, K.M., 159
Ross, F.B., 100–101
Rossi, J.J., 549–550
Rossi, P., 62–64
Rossmann, M.G., 202
Rot, G., 167, 177, 264–265, 265f
Roth, A., 175f, 214–215
Rothballer, A., 419–421
Rould, M.A., 402
Rouskin, S., 127, 154–159, 163–165, 170–171
Roy, R., 5, 7, 540–542, 546, 555
Ruan, J., 148
Rudner, D.Z., 244t
Rudolph, M.G., 250–251
Rueda, D., 100–101, 545–546
Ruggiu, M., 419–420, 466–467
Ruigrok, R.W., 410
Russell, R., 4
Rust, M.J., 580–582
Rüterjans, H., 62
Rychkova, A., 498–499
Ryder, S.P., 518
Ryder, U., 260–261

S

Sachidanandam, R., 573–578
Sachsenberg, T., 419–421, 459
Saenger, W., 190
Sagar, M.B., 218–219
Said, N., 5
Sakamoto, H., 241t, 243f, 244t, 253f, 254–255, 262t, 353
Sakashita, E., 262t
Salghetti, S.E., 573–578
Salguero, C., 40–41
Sali, A., 498–499
Salikhov, K.M., 285–286
Salmon, L., 76, 342, 351–353
Salvatella, X., 76

Salzberg, S.L., 148, 162–163
Sanbonmatsu, K.Y., 5–6, 498–510
Sanchez, C., 346, 351–353
Sanchez, J.C., 171–172
Sandelin, A., 164–165
Sander, B., 334–335, 421, 433–434
Sanford, J.R., 262t
Santangelo, M.G., 303–306
Sarachan, K.L., 393–396
Saragliadis, A., 336
Sargsyan, K., 419–420
Sarma, K., 260–261, 419–420
Sasagawa, N., 244t
Sashital, D.G., 282
Sathyamoorthy, B., 40–68
Sattler, M., 236–237, 280–282, 284, 334–355, 364, 392–393, 405, 419–420
Sauau, P., 387–388
Saudek, V., 50
Sauer, B., 576–578
Savinov, A., 226
Savitsky, A., 323
Sayers, Z., 392–393, 396, 402, 406–408
Schaal, T.D., 262t
Schaefer, M., 573–574
Schanda, P., 65–67
Schaughency, P., 262t
Schenborn, E.T., 6
Scheres, S.H., 498–501
Schiemann, O., 284–285, 296, 298, 309, 322–323, 337
Schluepen, C., 262t
Schmeing, T.M., 183–185, 516–517
Schmidt, C.Q., 335
Schmidt, E., 295
Schmidt, M.J., 300, 336
Schmidt, U., 573–574
Schmidt, W.M., 164–165
Schmidtke, S.R., 223–224
Schmitzova, J., 547
Schneider, B., 185, 193–195, 208
Schneider, C., 237–239, 238f
Schneider, S., 549–550
Schnitzbauer, J., 516
Schnorr, K.A., 337
Schoenborn, B.P., 385, 392
Schön, A., 420–421
Schreiner, E., 498–499

Schroder, G.F., 498–501
Schroeder, G.F., 77–78
Schroeder, S.J., 183*f*, 202*f*, 206–207
Schroth, G.P., 127, 169, 467–468
Schubert, M., 236–237, 281–283, 292, 295–296, 306–308, 313–316, 318–319, 322–323, 335, 337, 346
Schuck, P., 17–20
Schuette, J.C., 498–499
Schulte, M., 287–288
Schulten, K., 499–501
Schultz, P.G., 336
Schultz, S.C., 42–43
Schwalbe, H., 40–41, 51, 336
Schwander, P., 498–499
Schwanhäusser, B., 262*t*
Schwartz, C.R., 374
Schwartz, S., 177
Schweiger, A., 304
Schwer, B., 549–550
Schwieters, C.D., 282, 284, 335, 342, 345–346, 364, 408–409
Sclavi, B., 119
Scott, D.A., 516–519
Scott, L.G., 283, 295–296
Scott, W.G., 183–185, 203–204, 224–225
Searle, M.S., 284
Seeburg, P.H., 466–467
Seeger, P.A., 410
Seetin, M.G., 169
Segref, A., 239–240, 246–248
Seidelt, B., 499
Seifert, S., 364–365
Seikowski, J., 299, 337
Sekhar, A., 42–43
Sekine, S., 192*f*
Selbach, M., 262*t*
Selle, K., 518–519
Semlow, D.R., 572–573
Semrad, K., 42
Sen, T.Z., 498–499
Senavirathne, G., 561
SenGupta, D.J., 419–420
Sengupta, J., 498–499
Sephton, C.F., 262*t*
Sept, D., 498–499
Seraphin, B., 550
Serganov, A., 40–41, 215

Shaikh, T.R., 498–499
Shamoo, Y., 262*t*
Shanahan, C.A., 223
Shandilya, S.M.D., 236
Shankar, N., 40–41
Sharma, C.M., 516
Sharp, P.A., 468, 473, 489
Shatalin, K., 214
Shav-Tal, Y., 573–574, 576–578
Shaw, E., 100–101, 226
Shaw, J.J., 183–185
Shchirov, M.D., 285–286
Shechner, D.M., 223
Shehata, S.I., 517–518
Shelke, S.A., 223, 337
Shen, C.-K.J., 244*t*
Shen, H., 240–246, 241*t*, 243*f*, 261–264, 280–281, 349
Shen, Y., 316
Shen, Z., 546–547
Sheng, J., 224
Shenoy, S.M., 573–574, 576–578
Shevelev, G.Y., 322–323
Shi, X.S., 76–96
Shibata, Y., 51
Shimura, Y., 243*f*, 253*f*, 254–255, 353
Shin, H., 170–171, 177
Shin, Y.K., 284–285
Shiue, L., 261–264
Shkumatov, A.V., 365
Shou, C., 266
Sickmier, E.A., 240–246, 241*t*, 243*f*, 261–264, 280–281, 349
Sidiropoulos, N., 154–177
Sieber, H., 499
Siegfried, N.A., 127, 218–219
Sigurdsson, S.T., 298, 322–323, 337
Sikkema, F.D., 405
Siksnys, V., 516–517
Silver, P.A., 262*t*
Silverman, J.A., 79, 81, 93–96
Simon, B., 236–237, 252–258, 257*f*, 280–283, 334–355, 364, 405, 407–408
Simon, I., 77–78
Simoneau-Roy, M., 214–215
Simorre, J.P., 65–67
Singer, R.H., 573–574, 578
Singh, J., 583–584

Singh, M., 249, 250f
Singh, R., 262t, 347, 516–517
Sinha, R., 250–251, 250f, 260–264
Sit, R.V., 467–468
Sivak, D.A., 81–84
Skabkin, M.A., 499
Skarina, T., 531–532
Skinner, T.E., 323
Skjaerven, L., 282–283, 335, 343, 346, 405, 407–408
Sklenar, V., 50
Skrisovska, L., 236–248, 238f, 241t, 243f, 247f, 252–254, 256–258, 257f, 260–264, 282–283, 295, 410
Skrynnikov, N.R., 46–48
Sliz, P., 280–281
Slorach, E.M., 518–519
Smalls-Mantey, A., 282
Smibert, C.A., 419
Smith, A.M., 215–219, 221–222, 225–226
Smith, C.W.J., 433–434
Smith, G.M., 301–302
Smith, J.M., 81–84
Smith, K.D., 223
Snoeyink, J., 203
Snyder, M.P., 467–468
Sobhy, M.A., 540–542, 554
Söding, J., 262t
Soll, S.J., 419–421
Solomatin, S., 544–545
Somarowthu, S., 4–35
Someya, T., 241t, 246–248
Sonenberg, N., 241t, 246–248, 253f, 254–255
Song, J., 240–246, 241t, 243f, 261–264, 280–281
Song, W., 183–185
Sonntag, M., 259, 264–265, 334–335, 346, 353, 405–407, 419–420
Sopchik, A.E., 201
Sosnick, T.R., 5–6
Souliere, M.F., 40–41, 100–101, 226
Soumillon, M., 5
Sowa, G.Z., 299, 309
Sowers, L.C., 100–101
Sozudogru, E., 303
Spahn, C.M.E., 498–510

Spector, D.L., 573–574, 576–578
Spiess, H.W., 284–286
Spindler, P.E., 323
Spinozzi, F., 397
Spitzer, R., 337
Sprangers, R., 283, 295, 335
Sripakdeevong, P., 185, 204–205, 224–225
Srivastava, S., 498–499
Stadler, P.F., 216–218, 221–222
Stagg, S.M., 498–499
Staley, J.P., 566, 572–573
Stamm, S., 246–248, 247f, 260–264, 433–434
Stampfl, S., 42
Stark, H., 76, 498–499, 540, 542–543, 547
Starosta, A.L., 498–502
Steen, K.A., 218–219
Stefl, R., 241t, 246–248, 247f, 282–283
Steimer, L., 250–251
Stein, Z.W, 40–68
Steinmetz, L.M., 419–421
Steitz, J.A., 40
Steitz, T.A., 42–43, 183–185, 183f, 189, 202f, 206–207, 214, 334–335, 402
Stelzer, A.C., 77–78, 281
Sternberg, S.H., 280, 516–519, 525–527
Stetefeld, J., 364–365
Stévant, I., 264–265, 265f
Stévenin, J., 241t, 246–248, 260–264, 262t, 459, 466–467, 573–574
Stevens, R.C., 531–532
Stevens, S.W., 547, 549–550
Stivers, J.T., 100–101
Stock, D., 334–335
Stock, G., 337
Stoeckel, P., 392
Stombaugh, J., 189
Stone, M.D., 542–543
Storbeck, M., 261–264
Storz, G., 40, 280
St-Pierre, P., 100–101, 215–216, 226
Straek, R.L., 183–185
Strange, J.H., 46
Stranges, P.B., 516–517
Strein, C., 419–421
Strobel, S.A., 183–185, 214, 223
Stryer, L., 100–101, 540–542
Stuani, C., 241t, 255–256, 261–264

Stuhrmann, H.B., 84–85, 364, 392, 399, 401–402, 410–411
Su, L.J., 6, 20
Su, X.C., 335–337
Subramaniam, S., 500–501
Sudarsan, N., 40–41, 214–215, 221
Suddala, K.C., 544–546
Suess, B., 223–224
Sugano, S., 262t
Sugimoto, Y., 466–467
Sugnet, C.W., 159, 543–544
Suh, J.Y., 282, 346
Sullivan, M., 119
Summerer, D., 336
Summers, M.F., 40–41, 283, 295–296
Suslov, N.B., 223
Sutch, B.T., 309
Sutton, G.G., 236
Suzuki, H., 262t
Svergun, D.I., 364–365, 368, 373–374, 388, 392–393, 396–398, 401–402, 404f, 406–408, 410
Svozil, D., 185
Swails, J.M., 498–499
Swami, N.K., 498–499
Swanstrom, R., 5–6, 30–32
Szathmáry, E., 280

T

Tacke, R., 262t
Tadros, W., 419
Tafer, H., 216–218, 221–222
Tainer, J.A., 5–6, 364–365, 371–372, 374, 379–380, 384–385, 388, 396–397
Tajnik, M., 264–265, 265f
Takagaki, Y., 262t
Takahashi, M., 241t, 246–248, 257f, 259, 264–265
Takahashi, T.T., 575–576
Takamoto, K., 4–5
Takayama, Y., 282, 335, 346
Talkish, J., 127, 143, 148, 164–165
Talukder, S., 262t, 467–468
Tama, F., 498–502
Tamm, L.K., 343
Tan, R., 42–43
Tanaka Hall, T.M., 240–246, 241t, 243f, 254–255, 261–264, 262t

Tanenbaum, M.E., 547–549
Tang, C., 281, 284, 342–343
Tang, G.Q., 5, 7
Tang, Y., 127–128
Tansey, W.P., 573–578
Tanzer, A., 5
Tatchell, K., 392
Tatei, K., 546
Taylor, D.W., 517–518, 525–527
Taylor, H.C., 185, 188–189
Taylor, I.A., 262t
Taylor, J., 159
Taylor, S.D., 4–6
Taylor, T.C., 187–188, 196–197
Tehei, M., 396, 401
Tempel, W., 204–205
Tenenbaum, S.A., 419–420
Teo, C.H., 223, 241t
Teplov, A., 240–246, 241t, 243f, 261–264
Teplova, M., 240–246, 241t, 243f, 244t, 261–264, 280–281
Terada, T., 241t, 246–248
Terai, G., 266
Terry, A.E., 405
Theil, K., 262t
Thibaud-Nissen, F., 163
Thickman, K.R., 375, 377
Thierry, J.-C., 392
Thomas, D.D., 344–345, 349–351
Thomas-Crusells, J., 256, 260
Thompson, E., 100–101
Thompson, J.X., 547, 549–550
Thompson, K.C., 100–101, 110
Thompson, N.E., 334–335
Thompson, N.L., 554
Thomsen, N.D., 531–532
Thorn, A., 183–185
Thorpe, J., 516–517
Tian, B., 262t
Tian, S., 43
Tiberti, N., 171–172
Tijerina, P., 364–365
Tilgner, H., 5
Tillo, D., 76–77
Timmins, P.A., 393, 397, 410
Tinoco, I., 105–106
Tintaru, A.M., 241t, 250–251, 261–264
Tjandra, N., 282, 346

Tohyama, M., 262t
Toland, A.E., 162–163
Tolbert, B.S., 282
Tollervey, D., 260–261, 262t
Tollervey, J.R., 241t, 253f, 255–256, 261–264, 262t
Tome, J.M., 467–468
Tomsic, J., 215, 221, 225–226
Tonelli, M., 283, 295–296
Toor, N., 4–6
Topf, M., 498–499
Torchia, D.A., 46–48
Torkler, P., 262t
Torres, S.E., 516
Tourigny, D.S., 498–499
Trabuco, L.G., 498–501
Trantirek, L., 241t, 282–283
Trapnell, C., 127, 148, 169
Treba, C.N., 237–239, 241t, 257f, 258–259
Tremblay, R., 226
Trevino, A.E., 516, 534
Trewhella, J., 282, 346, 374, 396–397, 410
Tria, G., 365
Trott, O., 47–48, 54, 58
Trowitzsch, S., 237–239, 238f
Trukhin, D.V., 322–323
Trylska, J., 498–499
Tsai, A., 542–543
Tsai, C.L., 81–84
Tschaggelar, R., 284–286, 301–306, 323
Tseng, C.K., 572–573
Tseng, Y.H., 549–550
Tsuda, K., 241t, 246–248, 257f, 259, 264–265
Tsukamoto, T., 573–578
Tsutakawa, S.E., 410–411
Tsvetkov, Y.D., 288–289, 304–308
Tuchsen, E., 77–78
Tuck, Alex C., 262t
Tucker, B.J., 40–41
Tuerk, C., 467–468
Tugarinov, V., 335, 346
Tukey, J.W., 170–171
Tullius, T.D., 154–159
Tung, C.S., 309
Turck, N., 171–172
Turek, P., 337
Tuschl, T., 264–265

Tuttle, D.L., 261–264
Tyka, M., 185, 205–206
Tzakos, A.G., 295

U

Ubbink, M., 282, 336, 342
Uesugi, S., 241t
Uhlenbeck, O.C., 4–5, 103, 575–576
Ule, A., 419–420, 466–467
Ule, J., 241t, 253f, 255–256, 261–264, 419–421, 466–467
Ullmann, J., 335
Ulmer, T.S., 310–311
Underwood, J.G., 168, 244t, 250f, 251–252, 261–264
Unzai, S., 241t, 246–248, 257f, 259, 264–265
Upton, T.G., 561
Urlaub, H., 5, 419–460, 540, 542–543, 547
Urzhumtsev, A., 498–501
Uversky, N., 396–397
Uzilov, A.V., 168

V

Vachette, P., 383
Vainshtein, B.K., 81
Valcárcel, J., 262t, 419–420
Vale, R.D., 547–549
Valentini, S.R., 262t
Valle, M., 498–499
Vallee, F., 531–532
Vallurupalli, P., 47–48
van Dijk, M., 338
Van Eps, N., 309, 311
Van Gunsteren, W.F., 503
Van Holde, K.E., 392
van Ingen, H., 343
van Kouwenhove, M., 40–41
van Nuland, N.A., 351–353
van Santen, M.A., 433–434
VanDuyn, N., 262t
VanLang, C.C., 150
Varani, G., 42–43, 223, 237–239, 241t, 261–264, 282–284, 298, 334–337
Varani, L., 282–283, 337
Vardy, L., 419
Vasa, S.M., 22, 33
Veit, S., 285–286

Velankar, S., 185, 187–188
Velyvis, A., 283, 295, 335
Venditti, V., 42–43
Venter, J.C., 236
Verma, V., 188–189
Verschoor, A., 498–499
Vestheim, H., 146
Viani Puglisi, E., 11
Videau, L.L., 188–189
Vijayraghavan, U., 549–550
Vileno, B., 337
Villa, A., 503
Villa, E., 498–501
Villalta, J.E., 516
Villemin, J.P., 573–574
Villion, M., 516
Vinther, J., 154–177
Visser, D., 405
Vitali, F., 252–254
Vocker, E., 545–546
Vogel, J., 5
Volkov, A.N., 342
von Hippel, P.H., 566
von Schroetter, C., 283, 295–296, 299, 337
Voss, J., 284–285
Voss, N.R., 393–396, 402
Vu, Anthony Q., 261–264
Vuister, G.W., 56

W

Wachter, A., 40–41, 214–215
Wacker, A., 40, 337
Waern, K., 266
Wagenknecht, T., 498–499
Waghray, S., 523–524
Wagner, G., 335–336, 343–344
Wahl, M.C., 237–239, 238f, 347, 421, 540, 542–543
Wahlby, A., 187–188, 196–197
Waldsich, C., 4–6, 42
Wales, D.J., 59
Walker, P.A., 79, 81, 93–96
Walser, C.B., 419
Walstrum, S.A., 4–5
Walter, K.F.A., 77–78
Walter, N.G., 540–566
Wan, Y., 4, 154–159, 168
Wang, C., 48–50

Wang, H., 240–246, 241t, 243f, 252–255, 253f, 261–264
Wang, I., 259, 264–265, 334–335, 346, 353, 405–407, 419–420
Wang, J., 65–67, 223, 540
Wang, L., 183f, 189, 336, 498–499
Wang, P.H., 167, 173
Wang, S., 204–205
Wang, W., 375, 377
Wang, X., 203, 240–246, 241t, 243f, 254–255, 261–264, 262t
Wang, Y.-T., 244t, 262t
Wang, Y.X., 65–67, 282
Wang, Z., 249, 250f, 266, 500–501
Wank, J.R., 42–43
Ward, D.C., 100–101
Ward, R., 303, 309
Wardle, G., 420–421
Warkentin, M., 410–411
Warkocki, Z., 547
Warnefors, M., 5
Warner, K.D., 183–185
Warner, L.R., 334–355
Warshel, A., 498–499
Washietl, S., 127, 154–159, 163–165, 170–171, 175f
Wassarman, K.M., 280
Wasserman, M.R., 498–499, 546
Watts, J.M., 5–6, 30–32
Wawrzyniak, K., 299, 337
Webb, M.R., 546–547
Webb, S., 261–264, 262t
Weber, I., 334–335
Weeks, K.M., 22, 25–27, 33, 42–43, 127, 150, 154–159, 169, 218–219, 543–544
Weigand, J.E., 223–224
Weik, M., 401
Weil, D., 573–574
Weiland, E., 419
Weinstein, J.A., 516–519
Weir, J.R., 498–499
Weirauch, M.T., 262t, 334–335, 467–468
Weiss, V.H., 262t
Weissig, H., 185, 459
Weissman, J.S., 127, 154–159, 163–165, 170–171, 547–549
Weissmann, C., 573–574
Wells, R.D., 76–77

Wells, S.E., 134
Wemmer, D.E., 244*t*
Wen, J., 175*f*
Wenter, P., 237–239, 238*f*, 246–248, 247*f*, 257*f*, 260–264, 262*t*
West, S.M., 76–77
Westbrook, J., 185, 459
Westhof, E., 40, 189, 337, 420–421
Wheelan, S.J., 262*t*
White, S.W., 100–103
Whitehead, E.H., 516
Whitford, P.C., 499–501, 510
Whitworth, G.B., 543–544
Wickens, M., 419–420
Wickiser, J.K., 215–216
Wider, G., 335
Wiedenheft, B., 280
Wieslander, L., 573–574
Wilhelm, J.E., 419
Wilkinson, K.A., 22, 33, 154–159, 218–219
Will, C.L., 347, 421, 433–434, 540, 542–543
Williams, D.J., 100–101
Williams, G.J., 81–84
Williams, L.D., 5–6
Williamson, J.R., 42–43, 183–185, 283, 295–296, 518
Willumeit, R., 410–411
Wilson, D.N., 183–185, 498–499
Wilson, K.F., 239–240
Wilusz, C.J., 262*t*, 419
Wilusz, J., 262*t*, 419
Wimberley, B.T., 100–103
Winkler, W.C., 40–41, 214–216, 221
Winsor, C.P., 170–171
Winter, R., 410–411
Wintermeyer, W., 76, 498–499
Winzor, D.J., 523–524
Wirth, B., 246–248, 261–264
Wise, A.A., 61, 68
Witherell, G.W., 103
Wittmann, H.G., 392
Witz, J., 392
Woese, C.C., 100
Wohlbold, L., 244*t*
Wohnert, J., 40–41, 51, 223–224, 250–251, 336
Wolf, B., 392
Wolf, E., 334–335
Wolff, P., 337
Wombacher, R., 572–573
Wood, M, 540–565
Woodside, M.T., 226
Woodson, S.A., 4–6, 119, 410–411
Woodward, C., 77–78
Wool, I.G., 189
Wooley, J.C., 392
Woolford, J.L., 127, 143, 148, 164–165
Worcester, D.L., 392
Word, J.M., 185, 188–189
Worrall, J.A., 342
Worthmann, W., 410
Wriggers, W., 365
Wright, J.D., 419–420
Wright, P.E., 282–283
Wu, B., 578
Wu, J., 282, 284–285, 346
Wu, T., 51
Wu, W.W., 500–501
Wu, X., 500–501, 516–517
Wunderlich, C.H., 337
Wurm, J.P., 250–251
Wüthrich, K., 247*f*, 335, 410
Wylde, R.J., 301–302
Wysoczanski, P., 237–239, 238*f*
Wysoker, A., 148

X

Xia, T., 575–576
Xiang, S., 237–239, 238*f*
Xiang, Y.K., 542–543
Xiao, H., 223
Xie, J., 336
Xie, X., 237–239, 238*f*
Xiong, H.Y., 572
Xiong, L., 164–165, 171–172
Xiong, Y., 223
Xu, J., 282–283
Xu, R.-M., 238*f*, 241*t*, 243*f*, 261–264
Xu, X.P., 62–64
Xue, Y., 40–68

Y

Yamamoto, K., 262*t*
Yamazaki, T., 47–48
Yan, Bernice Y., 261–264
Yang, L., 516

Yang, S., 76
Yang, Y., 310–311
Yang, Z.Y., 300, 322–323
Yates, J.R., 547, 549–550
Yeates, T.O., 186
Yeo, G.W., 262t
Yin, G., 342, 351–353
Ying, J., 47–48, 65–67, 282
Yokochi, A.T., 188
Yokoyama, S., 241t, 244t
Yong, J., 419
Yoshida, H., 419
Yoshida, M., 573–574
Yoshikawa, S., 241t, 246–248, 257f, 259, 264–265
Young, P.J., 246–248
Yu, M., 244t
Yu, P., 65–67
Yu, Y., 499, 501, 549–550
Yuan, H.S., 244t, 262t
Yuan, Q., 419
Yulikov, M., 236–237, 280–324, 335, 337, 346
Yunger, S., 576–578
Yusupov, M., 223–224, 498–501
Yusupova, G., 223–224, 498–501

Z

Zaccai, G., 392–393, 396–398, 401–402, 406–408
Zachariae, U., 410
Zaher, H.S., 183–185
Zahler, A.M., 159
Zakian, V.A., 546–547
Zalis, M.E., 185, 188–189
Zamore, P.D., 347
Zanier, K., 236–237, 252–258, 257f, 281–283, 334–335, 337, 347, 349, 351–353
Zarnack, K., 167, 177, 264–265, 265f, 420–421
Zarrine-Afsar, A., 335
Zavialov, A., 498–499
Zeller, U., 5
Zeng, F., 237–246, 241t, 243f, 252–256, 253f, 261–264
Zeng, X., 498–499
Zenkin, N., 419
Zenklusen, D., 578
Zhang, B., 419–420

Zhang, C., 264–265
Zhang, F., 516
Zhang, J., 4–5, 223–224, 237–239, 238f, 384–385, 540, 546–547
Zhang, M.Q., 262t
Zhang, Q., 43, 47–48, 50, 77–78, 281, 336
Zhang, Q.C., 154–159, 168
Zhang, W., 237–239, 252–254, 256, 516
Zhang, X., 298, 309
Zhang, X.H.F., 265–266
Zhang, Y., 127–128, 238f, 241t, 243f, 261–264, 323
Zhao, B., 43, 47–48, 50
Zhao, J., 260–261, 419–420
Zhao, W., 224, 540
Zhao, Y., 237–239, 252–254, 256
Zheng, Q., 546
Zhong, R., 261–264
Zhou, B.R., 343
Zhou, K., 205f, 206, 523–524
Zhou, R., 546–547
Zhou, W., 204–205
Zhou, Y., 261–264
Zhou, Z., 546
Zhu, G., 56
Zhu, H., 237–239, 238f
Zhu, J., 498–499
Zhuang, X., 4, 580–582
Zhurkin, V.B., 76
Zimmermann, H., 284–285, 287–288, 291, 305–306
Zimmermann, M., 261–264, 282–283
Zinn-Justin, S., 48
Zinshteyn, B., 170–171, 177
Zipper, P., 401–402, 409–410
Zirbel, C.L., 189
Zniber, I., 241t, 246–248
Zou, J.Y., 187–188, 196–197
Zou, X., 214
Zubradt, M., 127, 154–159, 163–165, 170–171
Zuker, M., 64–65, 216–218, 221–222
Zuo, X., 282
Zupan, B., 167, 177, 264–265, 265f, 466–467
Zwart, P.H., 186
Zwart, W., 40–41
Zweckstetter, M., 342, 351–353
Zweier, J.L., 322–323

SUBJECT INDEX

Note: Page numbers followed by "*f*" indicate figures and "*t*" indicate tables.

A

A1089, 103, 109, 113–114
A1089AP, 114–115, 121
 fluorescence intensity of, 120–121
A1089AP fluorescence, temperature dependence of, 112*f*
Agilent Tapestation image, of Mod-seq libraries, 147, 148*f*
Amicon ultra centrifugal filters, 103–105
Amide proton PREs, 343
2-Aminopurine fluorescence
 circular dichroism spectropolarimetry, 108
 fluorescence probe sites selection, 108–109
 GAC folding, 121–122
 nucleic acid structure and structural perturbations, 100–101
 quantum yield, 100–101
 RNA riboswitch properties, 100–101
 sample preparation, 103–105
 steady-state fluorescence, 109–113
 stopped-flow fluorescence, 117–121
 TCSPC/TRA, 113–117
 thermal denaturation, 105–108
Analytical size-exclusion chromatography, 20–22
Anisotropy decays, 117
2AP fluorescence
 decays, 114–115
 temperature dependence of, 111–112
2AP GAC RNAs, 109
2AP, in solution, 110
Applied Photophysics PMT R6095, 119–120
Applied Photophysics SX-20 Stopped-Flow spectrometer, 119–120
2AP-RNA, 109–110
 TCSPC trace of, 114*f*
2AP triphosphate fluorescence, 110

Atomic scattering, spatial distribution of, 371–372
Au-labeled oligonucleotides, 90, 91*f*

B

Backbone conformers, RNA, 208
 advantages, 195
 rotameric, 193–194
 suite division, 194–195, 194*f*
 Suitename, 195
 suitestrings, 195
Biased Brownian ratchet machine, 562–565
Binary protein–RNA complex, 372

C

Calibrated distance, 79
χ angle, 190–191
Capillary electrophoresis, 32–33
Cas9
 Cas9–sgRNA platform, 534
 CRISPR-Cas9, 516
 crystallization, 530–534
 crystal structures, 517–518
 electrophoretic mobility shift assay
  oligonucleotides, 519–523, 520*f*
  protein–nucleic acid interactions, 518
  sgRNA, 518–519
 electrophoretic mobility shift assays, 518
 fluorescence-detection size exclusion chromatography assay
  crystallization parameter space, 523–524
  dCas9–sgRNA–target DNA complex, 525–527
  fluorescence-detection strategy, 524–525
  native polyacrylamide gel electrophoresis, 524–525
  nucleic acid ligands, 523–524
  protocol, 527–530
 magnesium-dependent nuclease domains, 516–517
 20-nucleotide sequence, 516
 size exclusion chromatography, 518

619

CD spectropolarimetry. *See* Circular dichroism (CD) spectropolarimetry
C2'-*endo* pucker, 191–192
C3'-*endo* pucker, 191–192
Chemical probing, LncRNAs
  capillary electrophoresis, 32–33
  data analysis, 33–35
  DMS reaction, 27–28
  mobility shift correction, 30
  primer extension reaction, 28–29
  primers, 22–23
  sequencing ladder reactions, 23–25
  SHAPE reaction, 25–26
  spectral calibration, 30–31
Chemical shift perturbations (CSPs), 335
Circular dichroism (CD) spectropolarimetry, 108
Clustering analysis, 561–562
Cochran–Mantel–Haenszel test, 150
Contrast variation, principle of, 399–400
Cross-links, isolation of
  RNA–protein interactions, of PM5 pre-mRNA complexes
    cautions, 447–449
    duration, 446–447
    flow chart of, 448f
    silver-stained gel, of control sample, 447, 449f
    size-exclusion chromatograms, 449, 449f
    solutions and buffers, 430
CUUGUAA nucleotides, 216–218
CYANA structure protocol
  additional restraints, 318
  EPR constraints, 317
  GLY-dummy residues, 317–318
  NMR hydrogen-bond restraints, 316
  NMR NOE restraints, 316
  NMR torsion angles, 316
  sequence file, 315–316
  simple extension, 318–319
  structure calculation protocol, 313–315

D
Deuterium-labeled tRNA (d-tRNA), 403–405
Dimethyl sulfate (DMS), 27–28, 154–159
Dithiothreitol (DTT), 337–338

DMS-Seq and Pseudo-Seq, RNA structure investigation, 170–171
DNA plasmid linearization, 6–7
DNase digestion, 9
DNA structure validation, 185
Double electron–electron resonance (DEER), 284–285
Δ termination-coverage ratio (Δ TCR) method, 169–170
Dulbecco's modified Eagle medium (D-MEM), 578–579

E
E. coli U1061A GAC, secondary structure of, 101–103, 102f
Electron Density Server (EDS), 196–197
Electron paramagnetic resonance (EPR) theory
  dipolar interaction and analysis, 287–292
  measurement settings, 285–286
Electrophoretic mobility shift assay (EMSA)
  oligonucleotides, 519–523, 520f
  protein–nucleic acid interactions, 518
  sgRNA, 518–519
Ellman's test, for free thiols, 338–340
Enrichment, of RNA–protein cross-links
  cautions, 455
  duration, 453–455
  flow chart of, 454f
  solutions and buffers, 432–433
Ensembles, 76. *See also* X-ray scattering interferometry (XSI)
  experimental challenge, 77
  fluorescence resonance energy transfer, 78–79
  molecular rulers, 78–79
  NMR relaxation measurements, 77–78
  nucleic acid conformational ensemble, 76–77
  residual dipolar couplings, 77–78
  three adenosine DNA bulge, 82–84, 83f
Enterococcus faecalis S_{MK} box riboswitch, 221
EPR-aided approach
  CYANA structure protocol
    additional restraints, 318
    EPR constraints, 317
    GLY-dummy residues, 317–318

Subject Index

NMR hydrogen-bond restraints, 316
NMR NOE restraints, 316
NMR torsion angles, 316
sequence file, 315–316
simple extension, 318–319
structure calculation protocol, 313–315
DEER EPR distance distributions
  detection frequencies, 301–302
  maximum measurable length, 303
  nitroxide spin labels, 300–301
  optimization parameters, 301
  protocol, 303–308
  quality factor resonators, 302
distance constraints
  experimental mean distance, 311–312
  interelectron, 310–311
  interspin distances, 310
  long-range distance distribution constraints, 312–313
global structure determination, 296–297
homodimeric RsmE protein, 292, 293f
70 kDa protein–RNA complex, 292
large doubly spin-labeled RNAs, 298–299
long-range distance distribution constraints, 293–294
protein resonances, 295
RNA resonances, 295–296
spin-label conformations, 308–310
spin-labeled proteins production, 300
spin-labeling sites, 297–298
structures of isolated domains, 295
structure validation, 319–321
ERRASER, 224–225
  correction efficiency of, 204–205, 205f
  kink-turn suite, single-residue correction of, 206–207, 207f
  rebuilding process, 205–206
  ribose-pucker correction, 206, 206f
  for RNA/protein complexes, 207–208
Exogenous genes, genomic integration of, 576–578

F

Flippase (FLP), 576–578
Flp Recombination Target (FRT) sites, 576–578
Fluctuation-dissipation theorem, 503

Fluorescence-detection size exclusion chromatography assay
  crystallization parameter space, 523–524
  dCas9–sgRNA–target DNA complex, 525–527
  fluorescence-detection strategy, 524–525
  native polyacrylamide gel electrophoresis, 524–525
  nucleic acid ligands, 523–524
  protocol, 527–530
Fluorescence lifetime decays, 117
Fluorescence recovery after photobleaching (FRAP) methods, 573–574
Folded U1061A GAC, crystal structure of, 103
FRET histograms, 559

G

GAC
  A1089AP, stopped-flow fluorescence traces, 118f
  denaturation transitions, 105–106
  folding, 121–122
  RNA, 108, 115–116, 122
  secondary structure, 113–114
  tertiary structure of, 104f
  ultraviolet spectra and circular dichroism spectra of, 106f
Gilford thermoprogrammer 2527, 108
Guanosines, 105–106

H

HeLa nuclear extract preparation, RNA–protein interactions, 433–434
  cautions, 441
  duration, 434, 439–440
  flow chart of, 440f
  solutions and buffers, 427–429
Heterologous site-specific recombinase strategy, 576–578
Higher order ribonucleoprotein complexes
  multi-RRM proteins, 258–259
  RRM–RNA interactions, 259–260
High-molecular-weight protein complexes, structural analysis of, 335

High-resolution cryo-EM reconstructions
  ab initio molecular dynamics potential, 508–510
  direct-detector high-resolution cryo-EM studies, 500–501
  eukaryotic ribosome, 498–499
  MDfit method, 499–500
  methods, 503–505
  molecular dynamics simulation, 499
  results, 505–508
  theory
    masking, 502–503
    molecular model, 501–502
    scoring function, 502
High-throughput sequencing-based methods, 154–159, 157t
Human protein-coding genes, 572
Hydroxyl radical probing, 154–159

I

Illumina MiSeq, 127–128
3i Marianas SDC system, 579
Introns, 572
In vitro-transcribed RNA pool, 469–470
In vitro transcription, RNA–protein interactions
  aqueous phase separation, 438
  duration, 434–438
  flow chart of, 437f
  microfuge tube usage, 438
  RNase-free tips usage, 438
  schematic illustration, 435f
  solutions and buffers, 424–427
  transcription reactions, 438
IPSL spin label, 337–338, 339f
IRF, 117

K

390 kDa box C/D ribonucleoprotein enzyme structure, 407–408
k-mer counting methods
  B values, 483–484
  naïve method, 482–483
K-turn binding proteins, 223

L

Lanthanide-binding tags (LBTs), 336
Large doubly spin-labeled RNAs, 298–299

Large RNAs and protein–RNA complexes
  challenges and implications, 322–324
  EPR-aided approach
    CYANA structure protocol, 313–319
    DEER EPR distance distributions, 300–308
    distance constraints, 310–313
    global structure determination, 296–297
    homodimeric RsmE protein, 292, 293f
    70 kDa protein–RNA complex, 292
    large doubly spin-labeled RNAs, 298–299
    long-range distance distribution constraints, 293–294
    protein resonances, 295
    RNA resonances, 295–296
    spin-label conformations, 308–310
    spin-labeled proteins production, 300
    spin-labeling sites, 297–298
    structures of isolated domains, 295
    structure validation, 319–321
  EPR theory
    dipolar interaction and analysis, 287–292
    measurement settings, 285–286
  limitations of current techniques
    dipole–dipole coupling, 284–285
    double electron–electron resonance, 284–285
    intermediate-size RNAs, 282–283
    isotopically labeled protein, 284
    low-resolution techniques, 284
    nuclear overhauser effects, 282
    protein resonances, 284
    RNA-labeling schemes, 283
    unfavorable relaxation properties, 282
  macromolecules, 281
  protein-producing machines, 280
  transient macromolecular interactions, 281
  X-ray crystallography, 280–281
Library preparation, Mod-seq, 127–128, 134
  caution, 135
  circularization, of cDNA products, 143–144
  duration, 133
  ethanol precipitation, 143
  flowchart of, 133, 134f
  3' linker ligation, 136–137
  5' linker ligation, 137–138

5' linker selection, 138–139
Mod-seeker analysis, 148–151, 149f
PAGE purification, 141–143
PCR amplification, 146–147
phosphatase and kinase, RNA, 135–136
purification and analysis, of PCR
    products, 147–148
reverse transcription, 140, 141f
RNA fragmentation, 135
subtractive hybridization, 144–146
Long noncoding RNAs (LncRNAs)
    analytical size-exclusion chromatography,
        20–22
    biophysical analysis, 5
    catalytic and noncatalytic, 4
    chemical probing
        capillary electrophoresis, 32–33
        data analysis, 33–35
        DMS reaction, 27–28
        mobility shift correction, 30
        primer extension reaction, 28–29
        primers, 22–23
        sequencing ladder reactions, 23–25
        SHAPE reaction, 25–26
        spectral calibration, 30–31
    native purification
        buffer exchange and purification,
            10–11
        construct design, 6
        DNA plasmid linearization, 6–7
        DNase digestion, 9
        EDTA chelation of divalent ions, 9–10
        proteinase K treatment, 9
        size-exclusion chromatography, 11–12
        in vitro transcription, 7–9
    sedimentation velocity analytical
        ultracentrifugation
        data analysis, 17–20
        experiment setup, 15–17
        optical cells, sample loading, and
            instrument setup, 14–15
        sample preparation, 13–14
    T7 polymerase construct, 5

M

Macromolecule, SAXS profile of, 368
Manual RNA backbone correction,
    202–203, 202f

Mass spectrometry analysis, RNA–protein
    interactions, 457–459
    cautions, 457
    duration, 455–457
    flow chart of, 456f
    solutions and buffers, 433
Methanethiosulfonate spin label (MTSL),
    300
Mg^{2+} titrations, 110–111, 111f
Mod-seeker, 128, 148–151, 149f
Mod-seeker-map.py, 148–149
Mod-seeker-stats.py, 150
Mod-seq
    custom oligonucleotides and adapters,
        132–133
    description, 127–128
    equipments, 128–129
    library preparation, 127–128, 134
        caution, 135
        circularization, of cDNA products,
            143–144
        duration, 133
        ethanol precipitation, 143
        flowchart of, 133, 134f
        3' linker ligation, 136–137
        5' linker ligation, 137–138
        5' linker selection, 138–139
        Mod-seeker analysis, 148–151,
            149f
        PAGE purification, 141–143
        PCR amplification, 146–147
        phosphatase and kinase, RNA,
            135–136
        purification and analysis, of PCR
            products, 147–148
        reverse transcription, 140, 141f
        RNA fragmentation, 135
        subtractive hybridization, 144–146
    materials required, 129–133
    solutions and buffers, 130–132
Molecular mass, 397–398
MolProbity clashscore, 188–189
Monomeric version of enhanced
    green fluorescent protein (mEGFP),
    578
MONSA, 365, 373, 402, 403t
    script for, 367t
    using pseudocomponents, 386f
MTSL spin label, 337–338, 339f

Multiphase volumetric modeling, 364–365
  SAXS data analysis workflow for, 371f
  of SF1-U2AF 65 splicing complex, 381f
  of SMARCAL1 protein–DNA complex, 382f
  software for, 366t
Multi-RRM proteins, 258–259

N

Native cysteine residues, 338–340
Native purification, LncRNAs
  buffer exchange and purification, 10–11
  construct design, 6
  DNA plasmid linearization, 6–7
  DNase digestion, 9
  EDTA chelation of divalent ions, 9–10
  proteinase K treatment, 9
  size-exclusion chromatography, 11–12
  in vitro transcription, 7–9
Neomycin-sensing riboswitch, 223–224
NMR relaxation dispersion
  chemical exchange, 43–46
  data analysis
    algebraic equations, 57–59
    Bloch–McConnell equations, 59
    chemical shifts, 59–60
    kinetic-thermodynamic analysis, 61–62
    monoexponentials, 56–57
    RD profiles, 60
    uncertainties, 60–61
  inferring structures of RNA ESs
    ^{13}C and ^{15}N chemical shift–structure relationships, 62–64
    secondary structure prediction, 64–65
  low-to-high SL fields, 48–50
  protein chaperones, 42
  RD experiments, 46–48
  riboswitches, 40
  RNA-targeting therapeutics, 43
  RNA transitions, 42–43
  splicing, 40–41
  testing model RNA ESs
    mutations, 65–67
    pH, 68
    single-atom substitutions, 68
  uniformly labeled nucleic acid samples
    construct design, 50–51
  ^{13}C $R_1\rho$ data measurement, 55–56
  ^{15}N $R_1\rho$ data measurement, 53–55
  sample preparation and purification, 51
  SL power calibration, 51–52
  trouble shooting $R_1\rho$, 56
Non-A-form backbone conformations, 183–185, 184f
Norm2bedgraph() function, 176
Nuclear Overhauser effects (NOEs), 282
Nucleic acid conformational ensembles, 76–77
Nucleic acid constructs, XSI
  Au nanocrystals
    labeled constructs, generation of, 88–92
    synthesis and purification of, 85–87
  choice of, 84–85
  data analysis, 93–94
  data collection, 92–93
  design, 84–85
Nucleic acid macromolecule, SAXS profile of, 368

O

Open source bioconductor repository, 177
Oxygen scavenging system (OSS), 546

P

Paramagnetic relaxation enhancement (PRE) data
  for amide and methyl protons, 343
  ARIA/CNS structure calculation software, 345–346
  distance restraints, 345
  quality factor, 346
  RRM domains, in apo U2AF65-RRM1,2, 351–353, 352f
  semiquantitative analysis of, 343
  structural analysis, 342
  Sxl–Unr mRNA regulatory complex, 353–355, 354f
  U2AF65 RRM1,2-U9 RNA complex structure, 347–351, 348f, 350f
PARS and FragSeq methods, 170–171
P(r)-distribution, 372, 373f
Photon Technology International (PTI), 113
plotRNA() function, 176

Polyacrylamide gel electrophoresis (PAGE), 218–219, 544–545
Postsynchronized histograms, 561
Primer extension reaction, 28–29
Prism-based TIRF microscopy, 554–557
Proteinase K treatment, 9
Protein equilibration, 473
Protein macromolecule, 368
Protein resonances, 295
Protein–RNA complexes
  case studies
    SF1–U2AF65 splicing complex, 375–379
    SMARCAL1–DNA complex, 379–383
    ternary T-box•tRNA•YbxF complex, 383–387
  PRE data
    for amide and methyl protons, 343
    ARIA/CNS structure calculation software, 345–346
    distance restraints, 345
    quality factor, 346
    RRM domains, in apo U2AF65-RRM1,2, 351–353, 352f
    semiquantitative analysis of, 343
    structural analysis, 342
    Sxl–Unr mRNA regulatory complex, 353–355, 354f
    U2AF65 RRM1,2-U9 RNA complex structure, 347–351, 348f, 350f
  SANS, 396–397
    experiments and modeling of, 401
    spatial distributions of contrast, 400, 406f
  scripts, 365, 367t
  softwares, 365, 366t
  spin labeling
    attachment sites, 337–340
    covalent attachment to reactive thiol group, 336
    IPSL spin label, 337–338, 339f
    lanthanide-binding tags, 336
    MTSL spin label, 337–338, 339f
    native cysteines, 338–340
    nitroxide spin labels, 336
    NMR experiments, of paramagnetic sample, 341–342
    paramagnetic metal-binding tags, 336
    pH values, 340–341
    PRE effects, 338
    protocol and sample handling, 340–342
    site-specific placement, 336
    spin label design, 340
    Tris buffer, 340–341
  technical requirements, 365
  theoretical considerations, 365–373
  wwPDB summary slider validation plot, 186–187, 187f
Proteins, dissociation and hydrolysis of
  cautions, 446
  duration, 445–446
  flow chart of, 446f
  solutions and buffers, 430
Proton PREs, 343
  distance restraints from, 343–345
Pseudo-RRM, 250–251
PTI. *See* Photon Technology International (PTI)
Purified RBP QC
  bead preparation, 473–474
  buffer preparation, 472
  concentration, 475
  input RNA pool composition, 476–477
  library preparation, 475–476
  protein equilibration, 473
  RBP elution, 474–475
  RBP pull-down, 474
  RNA binding, 474
  temperature considerations, 473

Q

Quasi-RRM, 249–250

R

Radius of gyration, 398–399
RBNS. *See* RNA Bind-n-Seq (RBNS)
RCrane, 224–225
  Coot software, 204f
Reagents, for RNA imaging, 575–576, 575t
Relative binding affinity determination
  bioanalyzer trace, 490t
  dissociation constants, 488
  fluorescent trace, 488
  RBFOX, 489, 490t
  relative binding constant, 488

Relative binding affinity determination
(Continued)
  relative dissociation constants, 488–489
  UGCACGU, 490t
R-factor, 186–187
Riboswitches
  aptamer and effector domains, 214–215, 215f
  crystallization and structure determination, 223–225
  description, 214
  rational construct design
    biochemistry, 218–221
    crystallization, 221–222
    rational construct engineering, 222–223
    RNA-binding proteins, 223
    sequence gazing, 216–218
  structure–function validation
    functional states, 226
    single-molecule measurement and molecular simulations, 226
    in vivo and *in vitro* experiments, 225–226
RNABC tool, 203
RNA-binding domains (RBDs), 419. *See also* RNA-recognition motif (RRM)
RNA-binding equilibrium, 487
RNA-binding motifs (RBMs), 419–420
RNA-binding proteins (RBPs), 223, 334–335, 419
RNA Bind-n-Seq (RBNS)
  computational analysis, 477–481
  cross-linking immunoprecipitation, 466–467
  dissociation constants, 468
  experimental method, 468–469
  k-mer counting methods
    B values, 483–484
    naïve method, 482–483
    pipeline inputs, 482
  purified RBP QC
    bead preparation, 473–474
    buffer preparation, 472
    concentration, 475
    input RNA pool composition, 476–477
    library preparation, 475–476
    protein equilibration, 473
    RBP elution, 474–475
    RBP pull-down, 474
    RNA binding, 474
    temperature considerations, 473
  quality control, 489–491
  reagents
    gel purification, 471–472
    length of library random region, 470–471
    in vitro-transcribed RNA pool, 469–470
  relative binding affinity determination
    bioanalyzer trace, 490t
    dissociation constants, 488
    fluorescent trace, 488
    RBFOX, 489, 490t
    relative binding constant, 488
    relative dissociation constants, 488–489
    UGCACGU, 490t
  RNA-binding equilibrium, 487
  streaming k-mer assignment, 484–486
  systematic evolution of ligands by exponential enrichment, 467–468
RNA crystallography, 182–183
RNA fragments, desalting and removal of
  cautions, 453
  duration, 451–453
  flow chart of, 452f
  solutions and buffers, 431–432
RNA oligonucleotides, paramagnetic labeling of, 337
RNA–protein complex assembly and purification
  cautions, 442–443
  duration, 441–442
  flow chart of, 442f
  solutions and buffers, 429–430
RNA–protein interactions, of PM5 pre-mRNA complexes
  cross-linked peptide identification, 458f, 459
  cross-links, isolation of
    cautions, 447–449
    duration, 446–447
    flow chart of, 448f
    silver-stained gel, of control sample, 447, 449f

Subject Index

size-exclusion chromatograms, 449, 449f
solutions and buffers, 430
enrichment, of RNA–protein cross-links
  cautions, 455
  duration, 453–455
  flow chart of, 454f
  solutions and buffers, 432–433
equipment, 421–423
HeLa nuclear extract preparation, 433–434
  cautions, 441
  duration, 434, 439–440
  flow chart of, 440f
  solutions and buffers, 427–429
hydrolysis, of RNA and proteins
  cautions, 450f
  duration, 450–451
  flow chart of, 450f
  solutions and buffers, 430–431
mass spectrometry analysis, 457–459
  cautions, 457
  duration, 455–457
  flow chart of, 456f
  solutions and buffers, 433
materials required, 423–433
pros and cons of, 459–460
proteins, dissociation and hydrolysis of
  cautions, 446
  duration, 445–446
  flow chart of, 446f
  solutions and buffers, 430
RNA fragments, desalting and removal of
  cautions, 453
  duration, 451–453
  flow chart of, 452f
  solutions and buffers, 431–432
RNA–protein complex assembly and purification
  cautions, 442–443
  duration, 441–442
  flow chart of, 442f
  solutions and buffers, 429–430
UV cross-linking
  apparatus, 443–444, 445f
  cautions, 443–444
  duration, 443
  flow chart of, 444f
  solutions and buffers, 430
in vitro transcription
  aqueous phase separation, 438
  duration, 434–438
  flow chart of, 437f
  microfuge tube usage, 438
  RNase-free tips usage, 438
  schematic illustration, 435f
  solutions and buffers, 424–427
  transcription reactions, 438
RNA-recognition motif (RRM)
  canonical β—sheet interface, 267
  higher order ribonucleoprotein complexes
    multi-RRM proteins, 258–259
    RRM–RNA interactions, 259–260
  high-molecular-weight complexes, 267–268
  maskstabilize longer RNA sequences, 267
  numerous variations, 237–239
  RNA sequences
    β-sheet surface, 239–240
    dissociation constant, 239–240, 244t
    domain-containing proteins, 239–240, 241t
    5′-GA-3′, 249
    5′-GGA-3′, 250–251
    5′-GGG-3′, 249–250
    sequence-specific contacts, 240–246, 243f
    sequence-specificity, 246–248
    uracils, 240–246
  specificity and affinity, 251–254
  structural and genome-wide approaches
    adenines, 261–264
    cross-linked nucleotide, 264–265, 265f
    eukaryotic posttranscriptional control, 264–265
    hnRNP C transcriptome-wide binding behavior, 266
    isolated protein/mRNA systems, 260–261
    mutational analyses, 260
    promiscuous proteins, 261–264, 262t
    RNP immunoprecipitation, 260–261
    X-ray crystallography, 264–265
  structural investigations, 236–237
  tandem, 254–258

RNA resonances, 295–296
RNAs, 103–105
  thermal denaturation, 105–108
RNA sequences
  β-sheet surface, 239–240
  dissociation constant, 239–240, 244t
  domain-containing proteins, 239–240, 241t
  5′-GA-3′, 249
  5′-GGA-3′, 250–251
  5′-GGG-3′, 249–250
  sequence-specific contacts, 240–246, 243f
  sequence-specificity, 246–248
  uracils, 240–246
RNA splicing, in live cells
  cell preparation, 578–579
  Dulbecco's modified Eagle medium, 578–579
  fluorescence intensity, 574–575, 574f, 581f
  fluorescent fusion protein expression, 574–575, 574f, 578
  genomic integration, of reporter gene, 574–578, 574f
  image processing and data analysis, 580–582
  intron lifetime measurement, 582–583
  live-cell imaging, 578–579
  MS2-binding RNA hairpins, 579–580
  reporter gene construction
    binding sites, 575–576
    dual-color labeling, 575–576
    labeled introns, splicing of, 575–576, 577f
    schematic illustration, 574–575, 574f
  RT-PCR, 575–576, 582–583
  spinning-disk confocal microscopy, 579
  time-lapse imaging, 574–575, 574f
  transcription site, in time-lapse sequence, 580–582
RNA structure. *See also* 2-Aminopurine fluorescence
  backbone conformers, 208
    advantages, 195
    rotameric, 193–194
    suite division, 194–195, 194f
    Suitename, 195
    suitestrings, 195

better ion identification, 208
χ angle, 190–191
C2′-*endo* pucker, 191–192
C3′-*endo* pucker, 191–192
data validation, 186
Electron Density Server, 196–197
ERRASER
  correction efficiency of, 204–205, 205f
  kink-turn suite, single-residue correction of, 206–207, 207f
  rebuilding process, 205–206
  ribose-pucker correction, 206, 206f
  for RNA/protein complexes, 207–208
glycosidic bond, 190
KiNG software, 196–197
LLDF score, 199
local validation criteria, 196
manual RNA backbone correction, 202–203, 202f
model-to-data match, 186–188
model validation, 188–189
MolProbity validation, 200–201
PHENIX refinement, with pucker-specific targets, 204
pucker outliers, 192, 192f
PyMol, 196–197
RCrane tool, in Coot software, 203–204, 204f
real-space correlation coefficient, 187–188
real-space residual, 187–188
ribose pucker, 191–193
RNA bases, 189–191
RNABC tool, 203
RSR-Z scores, 199
secondary structure (*see also* Mod-seq)
  chemical modification method, 127
  SHAPE-seq, 127
steric interactions, 188–189
twinning, 186
validation results
  end user interpretation, 195–198
  journal review, 198–200
  structural biologist interpretation, 200–201
van der Waals interactions, 188–189
RNA-targeting therapeutics, 43
R package, 159

RRM. *See* RNA-recognition motif (RRM)
RRM–RNA interactions, 259–260

S

S-adenosyl-L-methionine (SAM), 216–218
SAM-I riboswitch, 223
SANS. *See* Small-angle neutron scattering (SANS)
SAXS. *See* Small-angle X-ray scattering (SAXS)
SAXS data analysis workflow, multiphase volumetric modeling, 371*f*
SAXS Debye equations, 372*f*
SEC. *See* Size-exclusion chromatography (SEC)
Sedimentation velocity analytical ultracentrifugation (SV-AUC)
    data analysis, 17–20
    experiment setup, 15–17
    optical cells, sample loading, and instrument setup, 14–15
    sample preparation, 13–14
Selective 2′-hydroxyl acylation analyzed by primer extension (SHAPE), 25–26, 154–159
Sequencing-based RNA probing data
    data export and visualization
        Galaxy environment, 175
        RNAprobR, 176
        UCSC Genome Browser, 174, 175*f*
        VARNA applet, 176
    mapping
        alignment, 162–163
        Bowtie 2 algorithm, 162–164
        FASTA file, 163
        stringency, 162–163
    normalization
        BedGraph file, 173
        comparison of, 171–172, 172*f*
        count of cDNA molecules, 169
        Δ termination-coverage ratio method, 169–170
        Estimated Unique Counts calculation, 172–173
        FASTA file, 173
        logarithm of the ratio approach, 168
        Normalize tool, 172–173
        pooled two-sample Z-test, 168
        RNAprobR package, 173–174
        Smooth Log_2 Ratio procedure, 168
    PCR amplification, 167
    preprocessing
        Cutadapt tool, 159–162
        paired-end sequencing case, 161
        Preprocessing tool, 161–162
        procedure, 159–161
    schematic illustration, 154–159, 156*f*
    Summarize Unique Barcodes tool
        command line, 166–167
        function, 164–165
        output, 166
        running of, 166
        trimming untemplated nucleotide, 164–165, 165*f*
    workflow, 159, 160*f*
SF1–U2AF65 splicing complex, 375–379
Single-molecule fluorescence resonance energy transfer (smFRET), 226
Single-molecule guide RNAs (sgRNAs), 516
Single-molecule pull-down FRET
    biased Brownian ratchet machine, 562–565
    data analysis
        clustering analysis, 561–562
        FRET histograms, 559
        HMM and transition occupancy density plot analysis, 559–561
        postsynchronized histograms, 561
    dipole–dipole interaction, 540–542
    dynamic macromolecular complex, 540–542
    experimental methods
        isolation of splicing complexes, 546–554
        labeling and purification of pre-mRNA substrates, 543–546
    gene expression, 540
    prism-based TIRF microscopy
        experimental procedures, 557–559
        slide preparation and surface attachment, 555–557
        summary, 554–555
    spliceosome, 542–543
    splicing cycle, 540, 541*f*
Single-stranded RNA, 108

Size-exclusion chromatography (SEC), 11–12, 380–381
Slide-A-Lyzer dialysis, 103–105
Small-angle neutron scattering (SANS), 392–393
  ab initio approach, 402–405
  advantages, 396
  with atomic-resolution techniques, 408–409
  concentration series, 396
  contrast variation, principle of, 399–400
  at cryotemperatures, 410–411
  deuterium-labeled tRNA, 403–405
  experimental setup, 393, 394f
  innovative labeling schemes, 410
  low-resolution nucleosome models, 392
  modulus of momentum transfer, 393–396
  molecular mass, 397–398
  MONSA, 402, 403t
  neutron scattering length density, 393–396, 395f, 399–402
  NMR technique, 405
  principles of, 393–396
  protein–RNA complex, 396–397
    experiments and modeling of, 401
    spatial distributions of, 400
  radius of gyration, 398–399
  signal/noise limitations, 410–411
  Stuhrmann plot, 397, 399
  tRNA-synthetase complex, 3D crystal structure of, 403, 404f
Small-angle X-ray scattering (SAXS), 364
  of protein–RNA complexes (see Protein–RNA complexes)
  sample quality assessment, 374
SMARCAL1–DNA complex, 379–383
Smooth Winsorization procedure, 170–171
Spin-labeled proteins, 300
Spin labeling, protein–RNA complexes
  attachment sites, 337–340
  covalent attachment to reactive thiol group, 336
  IPSL spin label, 337–338, 339f
  lanthanide-binding tags, 336
  MTSL spin label, 337–338, 339f
  native cysteines, 338–340
  nitroxide spin labels, 336

NMR experiments, of paramagnetic sample, 341–342
  paramagnetic metal-binding tags, 336
  pH values, 340–341
  PRE effects, 338
  protocol and sample handling, 340–342
  site-specific placement, 336
  spin label design, 340
  Tris buffer, 340–341
Spinning-disk confocal microscopy, 579
Spliceosomes, 572–573
Steady-state fluorescence, 109–113
Stopped-flow fluorescence, 117–121
Stopped-flow kinetic parameters, 119t
Stopped-flow kinetics data, 117–119
Stratalinker UV cross-linker 2400, 445f
Streaming k-mer assignment, 484–486
Streptococcus pyogenes Cas9 (SpyCas9), 517–518
Stuhrmann plot, 397, 399
Suite division, 194–195, 194f
Suitename, 195
Suitestrings, 195
Summarize Unique Barcodes tool
  command line, 166–167
  function, 164–165
  output, 166
  running of, 166
  trimming untemplated nucleotide, 164–165, 165f
SUPCOMB transformation matrices, 369t
Systematic evolution of ligands by exponential enrichment (SELEX), 467–468

T

Tandem RRMs, 254–258
TCSPC. *See* Time-correlated single photon counting (TCSPC)
Temperature-dependent fluorescence experiments, 113
Ternary 34 kDa Sxl–Unr–*msl2* ribonucleoprotein complex structure, 406–407, 406f
Ternary T-box•tRNA•YbxF complex, 383–387
Theoretical rotational correlation time, 115–116

Thermal denaturation, 2-aminopurine fluorescence, 105–108
Time-correlated single photon counting (TCSPC), 113–117
Time-resolved anisotropy (TRA), 113–117
Ti:Sapphire laser, 117
Tme-resolved fluorescence parameters, 114t
Total fluorescence intensity (TFI), of transcription site, 580–582, 581f
Total internal reflection fluorescence (TIRF), 543–544
TRA. *See* Time-resolved anisotropy (TRA)
Transition occupancy density plot analysis, 559–561
Transverse $R_{2,PRE}$ relaxation rates, 343–344
Tris(2-carboxyethyl)phosphine (TCEP), 337–338
tRNA-synthetase complex, 3D crystal structure of, 403, 404f

U

U2AF65 RRM1,2-U9 RNA complex structure, 347–351, 348f, 350f
Uniformly labeled nucleic acid samples
  construct design, 50–51
  ^{13}C $R_1\rho$ data measurement, 55–56
  ^{15}N $R_1\rho$ data measurement, 53–55
  sample preparation and purification, 51
  SL power calibration, 51–52
  trouble shooting $R_1\rho$, 56
U1SnRNP U1A protein, 223
UV cross-linking, of PM5 pre-mRNA complexes
  apparatus, 443–444, 445f
  cautions, 443–444
  duration, 443
  flow chart of, 444f
  solutions and buffers, 430
"UV melting", 105

V

VARNA applet, 176

X

X-ray scattering interferometry (XSI)
  calibrated distance, 79
  complication, 81
  distance distribution function, 80–81
  DNA bulge experiments, 82–84, 83f
  experimental factors, 81
  nucleic acid constructs
    Au nanocrystal labeled constructs, generation of, 88–92
    Au nanocrystals, synthesis and purification of, 85–87
    choice of, 84–85
    data analysis, 93–94
    data collection, 92–93
    design, 84–85
  principle of, 79
  probe-probe distance distribution, 80, 80f
  scattering form factor, 80–81
  spherical gold nanocrystals, 81–82, 82f
  strength of, 79
X-ray scattering signal, 364
xRRM, 249
XSI. *See* X-ray scattering interferometry (XSI)

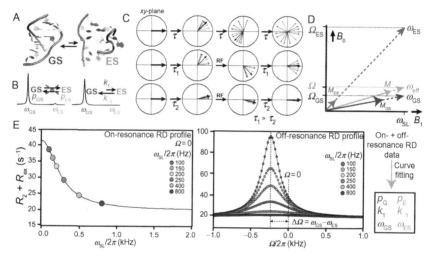

Yi Xue et al., Figure 3 Characterizing chemical exchange using NMR relaxation dispersion. (A) Example of an equilibrium between a GS and ES. (B) Chemical exchange between GS and ES leads to broadening of resonances and disappearance of minor ES signal. (C) Bulk magnetization aligned along the Y-axis of the lab frame, followed by magnetization dephasing due to transverse relaxation. The middle and lower panel show the dephasing of bulk magnetization due to chemical exchange suppressed by application of RF fields for long (τ_1) and short (τ_2) delays in the CPMG experiment. (D) Resonance offset (Ω), effective field (ω), and magnetization (M) vectors for the GS and ES in $R_{1\rho}$ experiment (denoted by the subscripts GS and ES, respectively). ω_{eff}, Ω, and M represent the effective field, average offset and magnetization, respectively. (E) Simulated examples of on-resonance (left) and off-resonance (right) RD profiles showing the dependence $R_{1\rho}$ on spin-lock power ω_{SL} and offset Ω.

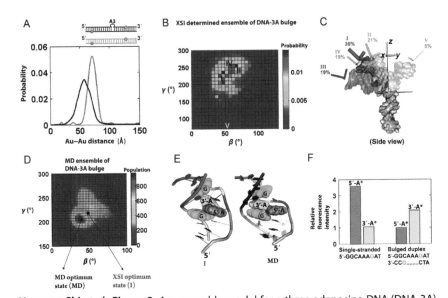

Xuesong Shi et al., **Figure 3** An ensemble model for a three adenosine DNA (DNA-3A) bulge based on XSI data. (A) Comparison of the Au–Au center-to-center distance distributions for a 26-base pair DNA helix (gray) and for DNA-3A, a 26-base pair DNA helix with a three adenosine bulge (black). (B) A heat map showing the conformational ensemble of the DNA-3A bulge determined by XSI. β and γ represent, respectively, the bending angle and the bending direction of the top helix relative to the bottom helix (see panel C). The model is based on six XSI distance distributions with different labeling positions for the Au nanocrystal probes. The distribution in panel A is one of the six data sets. (C) Five representative conformations of the DNA-3A ensemble (I–V) are shown as three-dimensional models. (D) A heat map showing the conformational ensemble of the DNA-3A bulge generated by an molecular dynamics (MD) ensemble simulation. The MD ensemble differs significantly from the XSI ensemble. For example, the most highly populated conformer in the XSI ensemble (state I, black star) is sparsely populated in the MD ensemble. (E) Atomic-level models of the most populated conformer from the XSI ensemble (state I) and from the MD. The XSI model predicts that the 5′ adenosine of the bulge is stacked on a flanking guanine, whereas the MD model predicts that the 3′ adenosine of the bulge is stacked on the flanking guanine. (F) Test of atomic-level models. The 5′ and 3′ adenosines of the bulge were replaced with fluorescent 2-aminopurine analogs. Stacking of 2-aminopurine on a guanine base quenches its fluorescence. Thus, the XSI model predicts that the 5′ 2-aminopurine fluorophore (A*) should be quenched when the bulge is formed from single-stranded oligonucleotide precursors, whereas the MD model predicts that the 3′ 2-aminopurine fluorophore (A*) should be quenched. Experimentally, quenching is observed at the 5′ position, consistent with the XSI model. *Panels (A–F) are reproduced from figures 1D, S6A, 4C, S12, and 5 of Shi et al. (2014).*

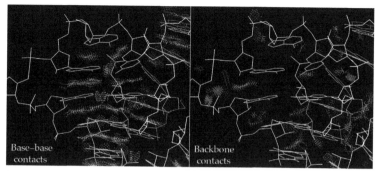

Swati Jain et al., Figure 2 Base versus backbone accuracy as shown by all-atom steric contacts. (A) Well-fit base–base contacts, shown by favorable H-bond and van der Waals contact dots (green and blue). (B) Fairly frequent backbone problems shown by spike clusters of steric clash overlap ≥ 0.4 Å (hotpink). *From PDB 2HOJ at 2.5 Å (Edwards & Ferré-D'Amaré, 2006).*

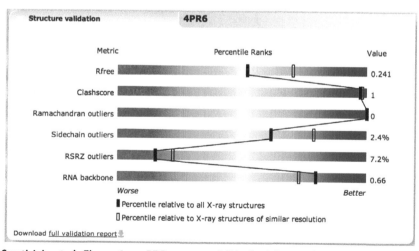

Swati Jain et al., Figure 4 wwPDB summary "slider" validation plot for an RNA/protein complex (4PR6 at 2.3 Å; Kapral et al., 2014). Percentile scores on six validation criteria are plotted versus all PDB X-ray structures (filled bars) and versus the cohort at similar resolution (open bars).

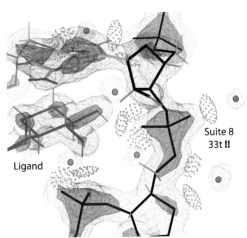

Swati Jain et al., Figure 10 A valid !! "outlier" suite conformer, with excellent confirming electron density and favorable interactions. The 2F$_o$-F$_c$ density is shown at 1.2σ as pale gray mesh and at 3.2σ as purple. The water molecules are shown as separate balls and hydrogen bonds as pillows of dots. *From PDB 3C3Z, HIV-1 RNA/antibiotic complex at 1.5 Å (Freisz et al., 2008), near residue B 8.*

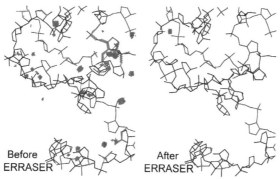

Swati Jain et al., Figure 13 Overall correction efficiency of ERRASER, supplemented with other tools. The active-site region of the uncleaved HDV ribozyme structure, with the RNA backbone in black. At left, the original structure (PDB ID: 1VC7; Ke, Zhou, Ding, Cate, & Doudna, 2004), with many clashes (hotpink spikes), bond-length outliers (red and blue spirals), bond-angle outliers (red and blue fans), and ribose-pucker outliers (magenta crosses). At right, the rebuilt structure (PDB ID: 4PRF; Kapral et al., 2014), with essentially all validation outliers corrected.

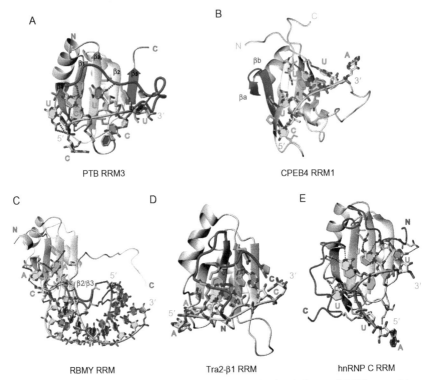

Tariq Afroz et al., Figure 3 Use of elements outside the β-sheet of RRMs to achieve sequence-specificity. (A) Structure of PTB RRM3 in complex with the 5′-CUCUCU-3′ RNA (Oberstrass et al., 2005). The additional β5-strand, which is involved in the RRM–RNA interaction, is shown in red. (B) Solution structure of CPEB4 RRMs bound to 5′-CUUUA-3′ (Afroz et al., 2014). For clarity, only RRM1 domain is shown. The insertion of β-strands is shown in red. (C) Structure of RBMY RRM in complex with a stem-loop RNA capped by a 5′-CACAA-3′ pentaloop (Skrisovska et al., 2007). The β2/β3 loop that binds the RNA stem is in red. (D) Solution structure of Tra2-β1 RRM bound to 5′-AAGAAC-3′ RNA (Cléry et al., 2011). The N- and C-terminal extremities of the RRM are in red. (E) Solution structure of hnRNP C RRM bound to 5′-AUUUUU-3′ RNA (Cienikova et al., 2014) (PDB ID: 2MXY). The N- and C-terminal extremities of the RRM are in red. In all the figures, the ribbon of the RRM is shown in gray, the RNA nucleotides are in yellow, and the protein side-chains in green. The N, O, and P atoms are in blue, red, and orange, respectively. The N- and C-terminal extensions of the RRM and 5′- and 3′-end of RNA are indicated. Hydrogen bonds are represented by purple dashed lines. All the figures were generated by the program MOLMOL (Koradi, Billeter, & Wuthrich, 1996).

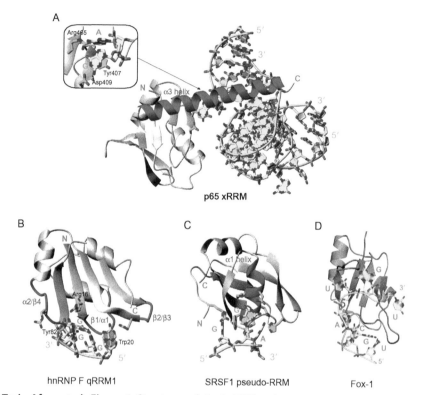

Tariq Afroz et al., Figure 4 Structures of single RRMs using an unusual mode of interaction with RNA. (A) Crystal structure of p65 xRRM bound to the telomerase RNA stem IV (Singh et al., 2012). The α3-helix is shown in red. (B) Solution structure of the hnRNP F qRRM1 bound to 5′-AGGGAU-3′ RNA (Dominguez et al., 2010). Loops involved in RNA binding are in red. (C) Solution structure of SRSF1 pseudo-RRM bound to RNA (Cléry et al., 2013). The α1-helix, which is primarily involved in RNA binding, is in red. (D) Structure of Fox-1 RRM in complex with RNA (Auweter, Fasan, et al., 2006). For all figures, the color code is the same as in Fig. 3.

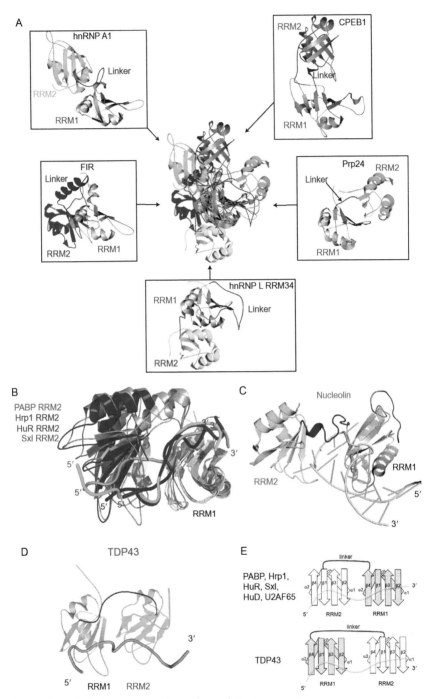

Tariq Afroz et al., Figure 5 See figure legend on next page.

Tariq Afroz et al., Figure 5 (A) Overlay of CPEB1 RRM12 with four previously characterized tandem RRMs in the free state—hnRNP L (3S01), FIR (2KXF), hnRNP A1 (2LYV), and Prp24 (2GO9). All the structures were overlaid on the RRM1 domain (shown in gray for all structures). RRM2 domain is shown in yellow, blue, cyan, orange, and green for hnRNP L, FIR, hnRNP A1, CPEB1, and Prp24, respectively. For clarity, the loops in the structures have been smoothened. (B) Overlay of PABP (Deo et al., 1999), Hrp1 (Perez-Canadillas, 2006), HuR (Wang et al., 2013), and Sxl (Handa et al., 1999) tandem RRMs in complex with RNA. Structures have been overlaid on the first RRM of each structure (shown in gray). RRM2 of PABP, Hrp1, HuR, and Sxl are shown in orange, magenta, blue, and green, respectively. RNA is shown in tube representation in the same color as RRM2 for individual structures. (C) Structure of Nucleolin RRMs bound to RNA (Allain, Gilbert, Bouvet, & Feigon, 2000). (D) Structure of TDP-43 tandem RRMs in complex with RNA (Lukavsky et al., 2013). (E) Schematic representation to illustrate the two modes of RNA binding based on the directionality of RNA binding by tandem RRMs.

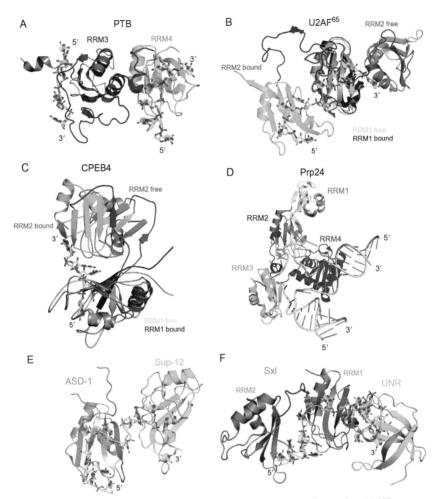

Tariq Afroz et al., Figure 6 (A) Structure of PTB RRM34 in complex with RNA (Oberstrass et al., 2005). RRM3 (dark gray), RRM4 (green) with RNA (in yellow) binding on individual RRMs. (B and C) Overlay of structures of U2AF[65] (B) (Mackereth et al., 2011) and CPEB4 (C) (Afroz et al., 2014) tandem RRMs in their free form and in complex with RNA. The structures have been overlaid on RRM1 (shown in gray). RRM2 in the free state is depicted in orange, while in green in complex with RNA. RNA is shown in stick representation in yellow. (D) Crystal structure of yeast U6-Prp24 complex (Montemayor et al., 2014). Prp24 RRM1–4 are shown in ribbon representation in yellow, gray, green, and blue, respectively. RNA is shown in tube representation in yellow with base pairs as ladders. (E) Structure of ternary complex of ASD-1 and SUP-12 RRMs in complex with RNA (Kuwasako et al., 2014). (F) Crystal structure of ternary complex of Sxl tandem RRMs (in gray), UNR CSD1 (in cyan) and RNA (shown as yellow sticks) (Hennig et al., 2014).

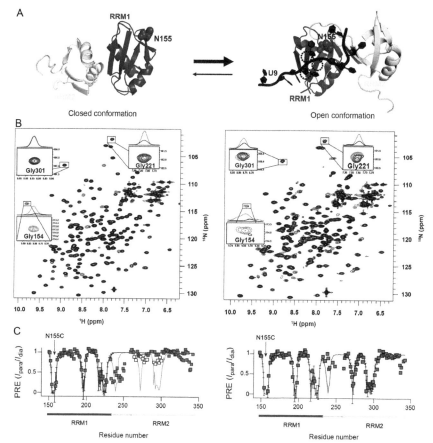

Janosch Hennig et al., **Figure 2** Paramagnetic relaxation enhancement data for structural analysis of U2AF65 RRM1,2 free and bound to U9 RNA. (A) NMR solution structures of RRM1,2 free (left, RRM1: purple; RRM2: yellow, PDB ID: 2YH0) and in complex with U9 RNA (right, RRM1: purple; RRM2: yellow; RNA, black; PDB ID: 2YH1) are shown with the RNP1 and RNP2 of RRM1 indicated with dotted circles and the N155 side chain carbons as a sphere. N155 was chosen as a spin label site because it is attached to a rigid backbone and solvent exposed. N155 is located far enough from the RNA-binding β-sheet surface of RRM1 and thus does not interfere with RNA binding. (B) ^{1}H,^{15}N HSQC spectra of the oxidized (black) and reduced (red) N155C IPSL labeled RRM1,2, left: free and right: bound to U9 RNA. NMR signals of three residues that show a maximal (G154), intermediate (G221), and minimal (G301) PRE effect are shown as zoomed insets together with the ^{1}H 1D traces (paramagnetic: black line, diamagnetic: red dashed line) at the ^{15}N peak maximum. (C) Experimentally measured amide proton PRE effects plotted as I_{para}/I_{dia} for RRM1,2 with a proxyl spin label attached to residue 155 (SL155), left: free; right: bound to U9 RNA. Filled blue and red boxes are experimental data in rigid and flexible regions of the protein backbone, respectively. Open white boxes are data indicate the presence of the open conformation. Only blue filled boxes were used for structure calculations.

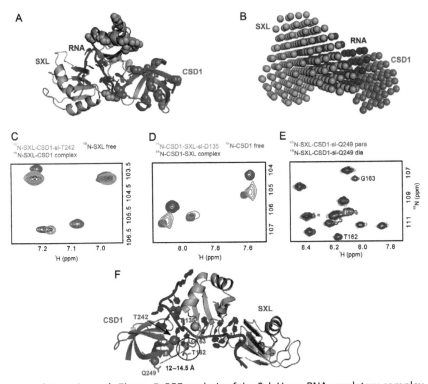

Janosch Hennig et al., Figure 5 PRE analysis of the Sxl–Unr mRNA regulatory complex. NMR chemical shift perturbations (CSPs), PREs, and small-angle scattering data for the SXL–CSD1 RNA complex. (A) HADDOCK model based on CSPs. The SXL–CSD1 interface residues, determined by CSPs, are shown as spheres (SXL: green, CSD1: blue). The RNA binds along a continuous surface provided by both proteins, which is stabilized by protein–protein contacts. (B) Small-angle neutron scattering of the complex reveals that the RNA is sandwiched by the two proteins. (C and D) Spin labels CSD1-T242 (C) and SXL-D135 (D) could not form the ternary complex as can be seen by the lack of CSP of CSD1 residues in sample with RNA and SXL-D135-spin label (yellow), compared to CSD1 chemical shifts in complex with RNA and SXL (blue). (E) Spin label CSD1-Q249 yields significant PRE effect for SXL residues T162 and G163. (F) The side chains of the three spin labels are indicated as a sphere in the crystal structure of the ternary complex (PDB ID: 4QQB). PREs observed for the CSD1-Q249 spin label to Sxl T162 and G163 (black circle) are indicated by dotted lines and the corresponding distances are annotated.

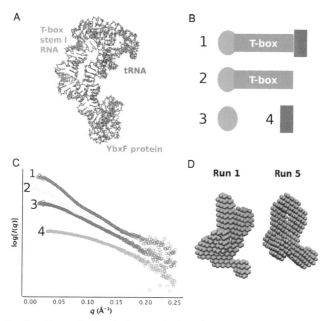

Robert P. Rambo, Figure 8 Simulations with the T-box stem I•tRNA•YbxF ternary complex. (A) Crystal structure (PDB: 4TZP) shows the tRNA (orange) is parallel along one arm of the bent T-box stem I RNA (green). YbxF protein (cyan) binds distal to the tRNA arm away from the particle's center-of-mass. (B) Cartoon representation of the ternary complex (1), binary T-box stem I•YbxF complex (2), YbxF protein (3), and tRNA (4). (C) SAXS profiles were simulated of the four components in B using coordinates derived from PDB code: 4TZP. Simulations were performed with CRYSOL using default parameters. Noise was added using a transformation method from an empirical datasets. (D) Two example MONSA models. Fifteen independent runs were performed and each failed to reconstruct the true particle. In each case, the small protein was found near the particle's center-of-mass with the tRNA flipping between two sides of the T-box arm.

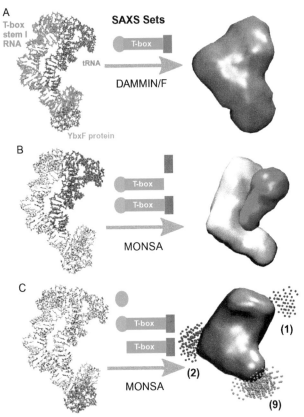

Robert P. Rambo, Figure 9 MONSA modeling runs using pseudocomponents. Cartoon above each arrow indicates the simulated SAXS data of the complex or components used in the multiphase volumetric modeling. (A) Averaged DAMMIN/F model calculated from 13 independent runs. (B) MONSA modeling treating the binary T-box stem I•YbxF complex as a single component of mixed phase with a volume-weighted contrast of 1.8. The pseudocomponent is the reference phase during averaging and alignment. This arrangement of SAXS data determines the orientation of the tRNA within the complex. (C) MONSA modeling to determine the position of the YbxF protein. T-box•tRNA binary complex is treated as a single RNA particle (reference phase). Therefore, the MONSA modeling should position the protein. Alignment of the 12 protein phases after averaging the reference phase shows three possible placements of the protein. Most likely position (cyan is consistent with the true position and can be inferred from the median position (region with a frequency >50%). False positives (red) are artifacts of the modeling algorithm.

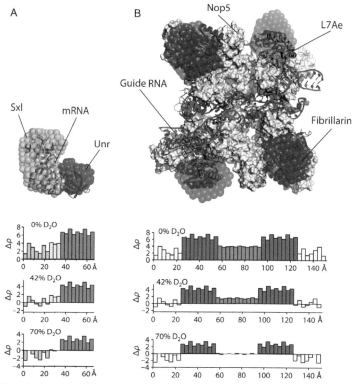

Frank Gabel, Figure 4 (A) Sxl–Unr–*msl2* complex. The two protein partners and the mRNA are shown in cartoon representation, superposed with their respective MONSA envelopes. The neutron contrasts $\Delta\rho$ of hydrogenated Sxl, hydrogenated mRNA, and deuterated Unr are shown schematically along an imaginary cross section of the complex for several $H_2O:D_2O$ ratios. (B) Box C/D apo-complex (in the absence of substrate rRNA). Fibrillarin, L7Ae, and the guide RNA are depicted as cartoons, Nop5 in space-filled mode. The MONSA envelopes of fibrillarin are superposed with their respective atomic-resolution copies. The neutron contrasts $\Delta\rho$ of hydrogenated Nop5, hydrogenated guide RNA, and deuterated fibrillarin are shown schematically along an imaginary cross section for several $H_2O:D_2O$ ratios. For simplicity, L7Ae is not shown and overlapping regions with contributions from several molecules are equally simplified.

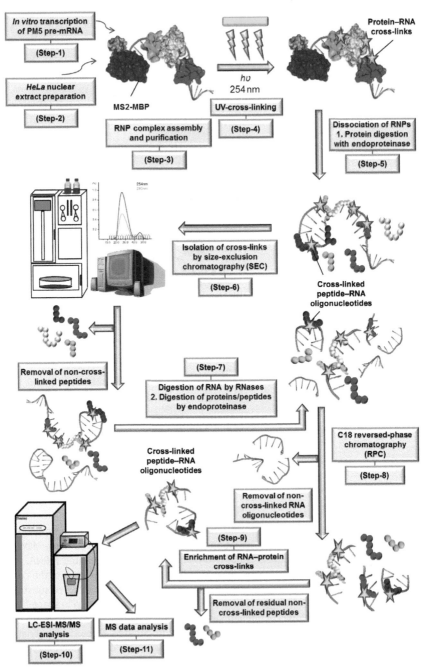

Saadia Qamar et al., Figure 1 Diagrammatic workflow of complete protocol. The PM5 pre-mRNA is transcribed by *in vitro* transcription (Step-1). The *HeLa* nuclear extract is prepared and dialysed (Step-2). The PM5 pre-mRNA along with the *HeLa* nuclear extract is being used for the RNP complex assembly and purification (Step-3). The assembled complex is then UV-cross-linked (Step-4). The sample is digested with trypsin (Step-5) and administered to the SEC for the isolation of cross-links and removal of non-cross-linked peptides (Step-6). The RNA is digested by the RNases (Step-7) followed by the removal of non-cross-linked RNA oligonucleotides by RPC (Step-8). The RNA-protein cross-links are then enriched by the use of TiO_2 (Step-9). The sample is then analyzed by LC-ESI-MS/MS (Step-10). The MS data obtained are then analyzed by using RNP[xl] pipeline in an OpenMS environment (Step-11).

A

VDNLTYR + [U]
SRSF2 (Q01130)/ SRSF8 (Q9BRL6)
(V18-R24)

Spectrum details

m(calc) [Da]	peptide	RNA	cross-link	[M+2H]²⁺	m/z (exp)	Δm
	879.4450	324.0359	1203.4809	602.7483	602.7473	1.58 ppm

B

2LEC; Daubner et al. (2012)   2DNM; Tsuda et al. (2009)

Saadia Qamar et al., Figure 14 (A) MS/MS spectrum of peptide VDNLTYR (position V18-R24) cross-linked through Y23 to U. The mass shift of tyrosine immonium ion (iY) and the y-series starting with the y2 ion by RNA adducts [Uracil–H_2O (U'^0—94.0167 Da), Uracil (U'—112.0273 Da), Uridine–H_2O–HPO_3 (U^{0-P}—226.0590 Da), C_3O (#—51.9949 Da), Uridine–H_2O (U^0—306.0253 Da), Uridine–HPO_3 (U^P—244.0695 Da); in red] show that tyrosine (Y23) is the cross-linked amino acid. The peaks in blue show the ions without RNA. The cross-linked peptide lies in the RRM domain of the SRSF2 and SRSF8 proteins. (B) NMR structure of the proteins SRSF2 (2LEC) and SRSF8 (2DNM) showing the cross-linked peptide (dark gray), cross-linked amino acid (red) and RNA (multicolor).

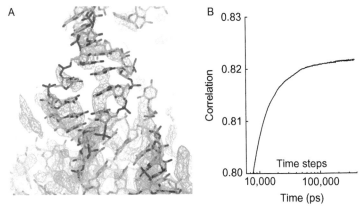

Serdal Kirmizialtin et al., Figure 6 Example of fitting atomic model to high-resolution cryo-EM reconstruction of the human ribosome. (A) Cryo-EM map of H40–H44 segment. Fitted atomic model is represented by stick representation, while green mesh is the high-resolution cryo-EM map. The resulted fit is achieved by optimizing the correlation between the atomic model and cryo-EM map while preserving the stereochemistry. This is achieved by simulated annealing molecular dynamic simulations. (B) Time evolution of fit during simulations.

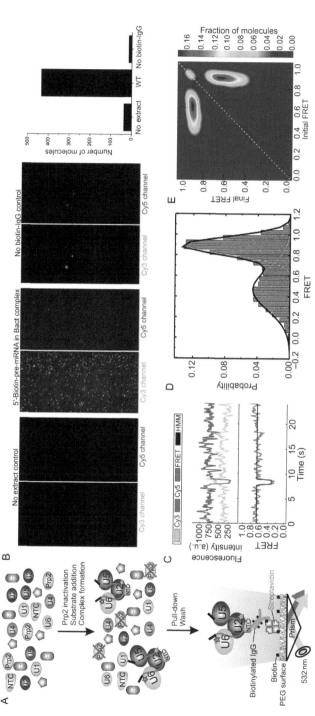

Matthew L. Kahlscheuer et al., Figure 2 Schematic and analysis methodology of the SiMPull-FRET technique. (A) B^{act} complex formation is promoted through the heat inactivation of Prp2 prior to incubation of the whole cell extract with the fluorescent pre-mRNA substrate. Complexes are then pulled down for smFRET analysis on a biotin-PEG slide surface coated with streptavidin and biotinylated IgG antibody utilizing the TAP-tagged NTC component Cef1. (B) Representative fields of view showing the selective binding of the fluorescent substrate to the slide surface only when contained in the B^{act} complex and when slide surfaces have been saturated with IgG–biotin. Quantifications of the number of molecules binding under each slide condition are shown on the far right. (C) Representative Cy3 donor, Cy5 acceptor, FRET, and idealized hidden Markov model (HMM) traces from an smFRET experiment. (D) FRET probability distribution analysis used to determine the dominant FRET states of a population of single molecules. (E) Transition occupancy density plots (TODPs) are scaled by the fraction of all molecules that exhibit transitions from a particular initial FRET state (plotted along the x-axis) to a particular final FRET state (plotted along the y-axis). *Panel B was modified from Krishnan et al. (2013)*.

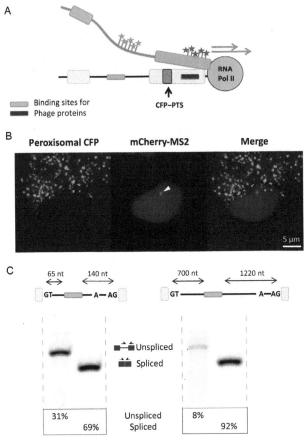

José Rino et al., Figure 2 Splicing of labeled introns. (A) Schematic of a typical reporter gene construct harboring two cassettes that encode distinct RNA hairpins, one located in an intron and the other in an exon. Immediately after transcription of each cassette, the corresponding phage proteins fused to either GFP or mCherry bind to the pre-mRNA, resulting in dual-color labeling of two different parts of the nascent transcript. This reporter gene was additionally engineered to encode a cyan fluorescent protein (CFP) fused to a peroxisomal targeting signal (PTS) at the carboxyl terminus, and the CFP–PTS sequence was inserted in frame with spliced exons. (B) Maximum intensity projections of 3D images showing nascent pre-mRNA (arrow) detected by mCherry-MS2 fluorescence and cytoplasmic peroxisomes labeled with cyan fluorescence. (C) RT-PCR analysis shows how splicing efficiency is affected by positioning of the hairpin cassette relative to splice sites. In this experiment, RNA was isolated from cells expressing MS2–GFP and reverse transcription was carried out using an oligonucleotide that is complementary to the sequence downstream of the poly(A) site in order to detect predominantly uncleaved nascent RNAs.

Edwards Brothers Malloy
Ann Arbor MI. USA
May 26, 2015